Surface Chemistry
of Froth Flotation

Surface Chemistry of Froth Flotation

Jan Leja

University of British Columbia
Vancouver, British Columbia, Canada

PLENUM PRESS • NEW YORK AND LONDON

058/5391

Library of Congress Cataloging in Publication Data

Leja, Jan.
  Surface chemistry of froth flotation.

  Bibliography: p.
  Includes index.
  1. Flotation. 2. Surface chemistry. I. Title.
TN523.L37   622'.752   81-21152
ISBN 0-306-40588-1   AACR2

© 1982 Plenum Press, New York
A Division of Plenum Publishing Corporation
233 Spring Street, New York, N.Y. 10013

Printed in the United States of America

Preface

The process of froth flotation is an outstanding example of applied surface chemistry. It is extensively used in the mining, mineral, metallurgical, and chemical industries for separation and selective concentration of individual minerals and other solids. Substances so concentrated serve as raw materials for producing appropriate metals and chemicals. The importance of flotation in technology is chiefly due to the ease with which it can be made selective and versatile and to the economy of the process.

The objective of this book is to review the fundamentals of surface chemistry together with the relevant aspects of organic and inorganic chemistry that—in the opinion of the author—are important to the control of the froth flotation process. The review updates the information that had been available in books by Sutherland and Wark (1955), Gaudin (1957), Klassen and Mokrousov (1963), and Glembotsky *et al.* (1963). It emphasizes mainly the surface chemical aspects of the process, leaving other relevant topics such as hydrodynamics, mechanical and electrical technology, circuit design and engineering, operations research, instrumentation technology, modeling, etc., to appropriate specialized treatments.

In Chapter 1 a brief review of the flotation field with its main areas of applications is given. Chapters 2, 3, and 4 deal, respectively, with basic concepts of physical and chemical bonding, an outline of the crystallography of solid minerals, and an outline of the structure of water and aqueous solutions. Chapter 5 is devoted to a discussion of organic reagents known as surfactants. Some of these constitute the most important flotation additives, called collectors, which control the conversion of hydrophilic into hydrophobic solids. Other surfactants affect the kinetics of attachment of particles to air bubbles and are known as flotation frothers. The basic aspects of the heterogeneous nature of solid surfaces are presented in

v

Chapter 6, while some of the electrical characteristics of solid/liquid interfaces are discussed in Chapter 7. Different qualities involved in the adsorption of surfactants at the various solid/liquid interfaces determine the extent of separation selectivity that can be achieved in froth flotation. These adsorption mechanisms are considered in Chapter 8. The formation of foams and froths and the mechanisms of thinning of liquid films interposed between two approaching particulates (such as those existing between a solid particle colliding with an air bubble) are covered in Chapter 9. Inorganic additives such as salts, acids, and bases are extensively employed as the so-called modifying agents, or activators and depressants. They are indispensable adjuncts to the achievement of selective separations in complex mixtures. Their role and mode of action are reviewed in Chapter 10.

This book is intended, primarily, for research workers in the flotation field and, in part, for honor students in mineral processing courses or operators in industrial flotation plants. Its purpose is to complement the books dealing with practical aspects of flotation. The book may also contribute to the compilations of papers presented at various international or national congresses and symposia. The reader will find, however, neither ready answers nor formulated recipes for carrying out specific separations by flotation. The latter objective is better served by some of the books listed at the end of Chapter 1. Again, the references included at the end of this book are not intended to be all-embracing. That purpose is served by the IMM Abstracts, Dow's Flotation Index, and ACS Chemical Abstract Selects.

It is a great pleasure to acknowledge the most helpful comments and suggestions received from my friends and colleagues during various stages of the manuscript preparation. In particular I wish to express my thanks to Dr. N. P. Finkelstein (formerly at the NIM, Johannesburg, R.S.A.; presently at the IMC Institute for Research, Haifa, Israel), Professor J. S. Forsyth (University of British Columbia), Professor R. N. O'Brien (University of Victoria, B.C.), Dr. A. Pomianowski (PAN, Krakow, Poland), Dr. R. S. Rao (University of Saskatchewan, Saskatoon, Canada), Dr. J. Szczypa (UMCS, Lublin, Poland), and my colleagues in the Department of Mining and Mineral Process Engineering at the University of British Columbia: Professors G. W. Poling, A. Mular, and J. B. Evans (presently in the Department of Mining & Metallurgy, University of Melbourne, Australia).

Technical help was provided over periods of time by Sally Finora, Dave Hornsby, June Wrobel-Randle, Connie Velestuk, and Melba Weber. To all of them I extend my sincere appreciation.

I feel greatly indebted to the late Dr. J. H. Schulman, who had kindled my interest in surface chemistry when he was Reader in the Colloid Science Department, Cambridge University, England. It was his suggestion (made in 1967) that initiated the tardy process of this book preparation.

To my former graduate students and postdoctorate associates, who shared with me their interest and ideas in research, I express my sincere thanks for their invaluable cooperation.

The assistance received from the editorial staff of Plenum Publishing Corporation is greatly appreciated.

Department of Mining   Jan Leja
and Mineral Process Engineering
University of British Columbia
Vancouver, Canada

Contents

Class III: Nonionic Surfactants

Notation

α Phase, for example, a solvent or a solid

α Capillary constant

α Term due to lateral interactions between adsorbates in de Boer equation (6.18)

α Charge transfer coefficient

α_M Activity in the bulk metal phase, taken as unity

$\alpha_0, \alpha_1, \alpha_2$ Polarizability

$\alpha_1{}^*, \alpha_2{}^*$ Effective polarizability

β Phase; species, for example, a solute

β Constant, characteristic of a given homologous series of surfactant

β Coarea of adsorbate molecule (species) $= \frac{1}{2}\pi d^2$

β Charge transfer coefficient

β_{exp} Ratio of molecular volume of a new adsorbate to that of benzene

γ Surface tension

$\Gamma_i, \Gamma_1, \Gamma_2$ Surface excess of species i or components 1 and 2

γ_c Critical surface tension of the solid

γ^d Constituent contribution due to dispersion forces

γ^h Constituent contribution to surface tension due to hydrogen bonding

$\gamma_{\mathrm{s/v}}, \gamma_{\mathrm{a/s}}, \gamma_{\mathrm{s/l}}, \gamma_{\mathrm{l/a}}$ Interfacial tension of solid/vapor, air/solid, solid/liquid, and liquid/air interfaces

γ_0 Work to create a unit area of the close-packed low-index face

ε Work to create a ledge in a solid surface

$\varepsilon, \varepsilon_1, \varepsilon_2$ Dielectric permeability of a substance

ε_i Polanyi's adsorption potential

ε_s Static dielectric constant

ε_{00} Dielectric permittivity independent of frequency

ζ Zeta potential

η Overpotential

η Viscosity (Newton-second per square meter)

θ Contact angle, or an angle between two crystal faces

θ Coverage, fractional or percentage

θ Fraction of ions in ion-pair formation

θ_c Critical coverage

θ_r, θ_a Receding and advancing contact angle

θ_1, θ_2 Angles between dipole projections

$\varkappa$ Reciprocal of the electrical double-layer thickness

$\varkappa$ Reciprocal of the Debye length of ionic atmosphere

λ Wavelength

λ Constant related to Young's modulus of elasticity

λ Electrical conductivity of solution in streaming potential

λ_i Characteristic absorption wavelength of species i

μ Dipole moment

μ_h Chemical potential of the thin film of thickness h

μ_i, μ_A, μ_B Chemical potential of species i, A, B

μ_{liq} Chemical potential of the bulk liquid

μ_i° Standard chemical potential of i at a concentration of 1 mol/liter

$\mu_i^{\dagger}$ Standard chemical potential of i when its partial pressure is 1 atm

ν Kinematic viscosity, $\nu = \eta/\varrho$ (square meters per second)

ν_1, ν_2 Frequencies of harmonic oscillations

Π Disjoining pressure

Π_{el} Electrical contribution in thin films due to an overlap of two electrical double layers

Π_h Disjoining pressure at thickness h corresponding to equilibrium with external pressure P

Π_{SH} Steric hindrance contribution in thin films

Π_W van der Waals contribution to disjoining pressure in thin films

$\varrho, \varrho_1, \varrho_2$ Density

ϱ_r Charge density

τ Sticking coefficient (condensation coefficient)

τ Dielectric relaxation time

τ Induction time for bubble–particle attachment

τ_{cr} Induction time for thinning from h_0 to h_{cr}

τ_0 Induction time for thinning of a thin film from macro-thickness to h_0 thickness, determined by the hydro-dynamic or capillary forces

ϕ Constant in Good and Girifalco's equation

ϕ Surface or film pressure ($\phi = \gamma - \gamma_v$)

ϕ^α, ϕ_M, ϕ^β Inner potential of phase α, M, or β

ϕ_1, ϕ_2 Angle of inclination

χ^α, χ^β Surface potential of phase α, β

χ_1, χ_2 Magnetic susceptibility of an atom

ψ_d Surface potential (potential at OHP)

ψ^α, ψ^β Outer potential of the phase α, β

$\Delta\psi$ Volta potential difference

ω Cohesive energy per methylene group frequency

A Constant in the differential capacitance equation

A Constant in the mean ionic activity coefficient

A, A^s Area of surface, of film

A_c Critical surface area per molecule

A_h Hamaker constant

A_p Polarization constant

a Parameter representing the extent of lateral interactions between adsorbates

a Constant in the van der Waals equation

a Work done in transporting a unit charge

a_A, a_i Activity of species A or i

a_s Activity of a surfactant

a_1, a_2 Constants

B_h Hamaker constant for retarded forces

B_p Polarization constant

b Constant in the van der Waals equation

b Constant in the Langmuir adsorption isotherm

b_{ij} Constant for the given ionic compound

b_1, b_2 Constants

C Constant in the BET equation

C London dispersion forces constant

C.N.   Coordination number

C_{b1} Concentration of surfactant for initial formation of (first) black film

C_H, C_{G-C} Capacitance of the Helmholtz (C_H) or Gouy–Chapman layer

c Velocity of light (in vacuum $= 2.99795 \times 10^8 \, \mathrm{ms^{-1}}$)

c_h Concentration in the thin film

c_i Concentration of (ionic) species i

c_s Concentration of a surfactant

c_0 Initial concentration of (ions) species in the bulk solution

D Diffusion coefficient (square centimeter per second)

d Diameter of particle, molecule

d Separation distance of capacitor's plates

d Distance between centers of ions

d_M Spacing between atoms in a metallic lattice

d_{sc} Spacing between atoms in a semiconductor

E Elastic constant

E Applied potential or measured potential

E Edge effect function for sessile drops

E_L Heat of liquefaction

Eo Eotvos number

E_1 Heat of adsorption in the first layer

e Electronic charge $= 4.8 \times 10^{-10} \, \mathrm{esu} = 1.602 \times 10^{-19} C$

F Helmholtz free energy

F Faraday (a gram mole of ions)

F Factor converting electron volts into kilocalories per gram mole or gram equivalent

f_A Activity coefficient

f^s Specific free interfacial energy associated with a thin film existing between two interfaces

G Gibbs free energy

$G^\dagger$ Standard Gibbs free energy at the standard pressure $p^\dagger$

g Acceleration due to gravity

h Planck's constant ($6.6256 \times 10^{-34} \, \mathrm{J \, s}$)

h Height of the single ledge in a low-index face

h Thickness of residual stable layer remaining on a solid after rupture of thin film of liquid

h_{cr} Critical thickness of thin film at the moment of rupture

I_1, I_2 Ionization energies of the atoms

i Any species or adsorbate

i_0 Exchange current density

K Equilibrium constant

K_A Association constant

K_D Dissociation constant

K_g Experimental constant

K_T Constant in Temkin's equation

K_0 Constant

K_1 London force constant for retarded forces

$K_{11}, K_{12}, K_{12}^{\alpha}$ van der Waals–Hamaker constants for thin films between phases 1 and 2 with an adsorbed layer α

k Boltzmann's constant $= 1.3805 \times 10^{-16}$ erg deg^{-1} $= 1.3805 \times 10^{-23}$ JK^{-1}

k Permeability coefficient (in mass transfer)

L Characteristic dimension (e.g., of the orifice, of the bubble)

l Distance between centers of negative and positive charges in a dipole

l Distance between electrodes

l_{sc} Thickness of space charge layer

M Molecular weight

M   Mole per liter (mol/dm^3)

m Mass of a molecule

m "Salt" parameter in Gibbs adsorption equation (2.83)

m Capillary rate (of dropping mercury electrode) (grams per second)

N Avogadro's number $= 6.02252 \times 10^{23}$ mol^{-1}

n Exponent of the electronic configuration

n Concentration of 1:1 electrolyte in molecules per cubic meter (Chapter 9)

n_a Number of molecules striking a unit surface every second

n_d Number of desorbing molecules per second

n_r Refractive index

P, p Pressure

p_A Partial pressure, $p_A = x_A P$

P_M Molar polarization

Q Heat, energy

Q^s Partition function related to Helmholtz free energy F^s

q Moles of material transferred in time t

q_c Energy (heat) of chemisorption

q_{el} Electrolyte countercharge

q_p Energy (heat) of physical adsorption

q_{sc} Space charge

q_{ss} Surface state charge

R Gas constant $= 1.9872$ cal deg^{-1} mol$^{-1} = 8.3143$ J K$^{-1} = 0.082057$ liter atm deg^{-1} mol^{-1}

R_a Rate of adsorption

R_d Rate of desorption

Re   Reynolds number

R_1, R_2 Radii of spheres, bubbles, the curvature of the surface, etc.

r Distance from the central ion or from the surface

r_0 Distance at $U(r) = 0$

S Entropy

S Spreading coefficient

S Surface area of a drop

S_0, S_1, S_2, ...   Fractions of surfaces with no adsorbate, one layer, two layers, etc.

T Absolute temperature (°K)

T_{2c}, T_{3c} Critical temperature for two-dimensional and three-dimensional phases

t Thickness of film

t Lifetime of adsorbate on the surface

t_c Time of contact (in collision, attachment, etc.)

t_0 Time period related to atomic vibrations in the solid

U Internal energy

U_μ Keesom effect (energy of dipole interaction)

U_0 Characteristic velocity of the fluid motion

V Volume

V_m Molar volume of a liquid

V_m Amount of gas for a monolayer coverage

V_0 Specific volume of a 1-cm^2 monolayer

v Velocity of a particle or bubble; electrophoretic velocity of a colloid

v Volume of gas adsorbed on a surface

W_A Work of adhesion

W_C Work of cohesion

X Proportionality factor

X Displacement depth (centimeters)

X_A, X_B Mole fraction of A and B

X_s Mole fraction of a surfactant

x Integer

Y Modulus of elasticity

y Integer

z Valence

z Number of electrons

$z_a e$ Charge carried by a

z^+, z^- Number of positive, negative charges on ions

1

Introduction

1.1. Scope of Froth Flotation

Heterogeneous mixtures of finely subdivided solid phases frequently require an efficient technique for separation of components. Separations of solids can be achieved, for appropriate particle sizes denoted by diameter d, by techniques exploiting the following physical and physicochemical characteristics:

Color	In *sorting* (of lump coal or other distinct minerals, when $d > \sim 50$ mm)
Specific gravity	In wet or dry *gravity concentration* techniques ($d > 0.1$ mm)
Shape and size	In *screening* and *classification* ($d > 0.04$ mm)
Electrical charge	In *electrostatic separation* of well-dried materials ($0.05 < d < 5$ mm)
Magnetic susceptibility	In wet or dry *magnetic concentration* devices ($0.1 < d < 5$ mm)
Radioactivity	In *radiometric sorting* (5 mm $< d < 1000$ mm)
Surface properties	In *froth flotation, spherical agglomeration, selective flocculation*

Of the preceding, it is the surface properties exploited in separations by froth flotation and spherical agglomeration that we shall deal with in this text. Spherical agglomeration and selective flocculation processes have not yet assumed the same dominance in the industrial field as froth flotation. The underlying surface properties required for all three techniques are, however, somewhat similar.

Froth flotation is frequently employed for separations of solids encountered particularly in the primary mineral and chemical industries. Materials mined from any deposit within the earth's crust usually represent a highly heterogeneous mixture of solidified phases; these are mostly crystalline and represent various minerals. Occasionally, they are noncrystalline (amorphous) such as, for example, coal, glasses, resins, and opal. Crushing and grinding operations are employed to free the individual phases from their neighbors, that is, *to liberate the mineral species* and, occasionally, to reduce the size of solids to the range suitable for the intended separation technique. Once mineral species are liberated, an economic recovery of the valuable component(s) contained in the original mixture depends greatly on the application of the most appropriate separation or concentration process.

Of the surface properties, flotation and spherical agglomeration exploit two parameters, *surface energy* and *surface excess charge* (or surface potential). At constant temperature, any differences in these two parameters exhibited by various minerals can be effectively exploited if the solids are contacted with either two immiscible liquids, such as oil and water, or with two fluid phases, that is, a liquid and a gas. The differences in interfacial tensions [defined as derivatives of interfacial free energies with respect to surface area; see equation (2.46), Section 2.9] lead to a preferential nonwetting of some minerals by one fluid, namely water. This differentiation enables the nonwetted (hydrophobic) solids to be separated in a froth if the nonwetted solids are allowed to contact air bubbles as in froth flotation, or, alternatively, if an oil is employed as in oil flotation or in spherical agglomeration, the nonwetted solid is collected in the oil phase.

Separation processes exploiting the surface characteristics of solids differ from all other processes in the relative ease of adjusting and controlling the respective interfacial tensions and surface charges. This possibility of varying the relative properties of different interfaces contained in the system makes the froth flotation and spherical agglomeration processes more versatile and universal than any other industrial separation process. The physical parameters being exploited in the latter separation techniques, e.g., specific gravity, magnetic susceptibility, etc., cannot be changed.

For convenience, froth flotation of solids will be referred to henceforth as the separation process. Most of the desired surface properties under discussion will relate, however, to all flotation systems regardless of whether separating crystalline solids or amorphous phases. Similarly, nearly identical surface characteristics will relate to spherical agglomeration which exploits them in an alternative way.

The surface properties of different solids are controlled by numerous regulating agents, ionic or dipolar in character, and by surfactants. The latter are used primarily to convert water-wetted solids to nonwetted ones. In addition to chemical reagent control, the overall process of selective separation by froth flotation depends on many other factors, such as, for example, the physical and the mechanical features of thinning of liquid films, the hydrodynamics of solid-in-water slurries, the kinetics of chemical reactions and of physical processes such as wetting, and the attachment of solid and fluid phases.

The flotation process is applied on an ever-increasing scale in the concentration of nonferrous metallic ores of Cu, Pb, Zn, Co, Ni, Mo, Sb, etc., whether sulfides, oxides, or carbonates, as a preliminary step to metal production. It is also used in concentrating the "nonmetallic" minerals which are utilized as raw materials in the chemical industry, for example, CaF_2, $BaSO_4$, NaCl, KCl, S, alumina, silica, clay, etc. In the treatment of solid fuels, flotation is employed to remove from coal fines the ash-producing shale and rock and the associated metallic sulfides, which cause air pollution by SO_2 during combustion.

Different mixtures of solid particles encountered at various stages of production in the chemical industries can be, and frequently are, separated using flotation techniques. For example, ink can be removed from recycled paper in the manufacture of new paper; the Kodak Company employs froth flotation to recover metallic silver from photographic residues; "cement copper," which is precipitated on scrap iron from acid mine effluents or from solutions derived in leaching oxide copper ores, is frequently recovered by flotation. Flotation of heavy crude oil from the tar sands in the Athabasca River deposits of northern Alberta (Canada) represents a unique process in that liquid oil is separated from solids (mostly silica sand) in water. Recent applications of the flotation technique are the various antipollution measures in water, sewage, and effluent treatment. Colloidal suspensions, bacteria, and traces of harmful metallic ions are removed by a modified flotation procedure, referred to as ion flotation or precipitate flotation [Gaudin (1957), Sebba (1962), Lemlich (1972)].[†]

[†] There is a tendency to introduce new terms for each modification of the same technique. When the levitation of particles heavier than water was achieved in the process of concentration by means of large quantities of oil, the process was called *bulk oil flotation*. When small quantities of oil were employed, it was called *skin flotation*. When the oil was replaced by small quantities of surfactants and the particles became levitated by bubbles forming a layer of froth, the process became *froth flotation*. The term *flotation* superseded froth flotation. While *ion flotation* is distinct from flotation of particulates and is thus justified, the new terms such as colloid, micro-, and precipitate flotation appear to the author to be unjustified.

1.2. *An Outline of a Mineral Flotation System, Definitions*

The technique of selective separation by froth flotation can be efficiently applied only to those mixtures in which the particulates to be separated are present as liberated (i.e., individually free) grains. To become levitated by the buoyancy of the bubble, these particles must be smaller than a maximum size. This maximum size is generally below 300 μm (a micrometer, 1 μm $= 10^4$ Å), and the limiting parameter is usually the available adhesion between the particle and the bubble. As regards the minimum size, there appears to be one in the flotation of sulfide minerals, where flotation becomes less effective when the particle size is reduced *by grinding* to below $\sim$10 μm. (Some of the causes of this behavior are discussed in Section 1.5.) But, on the other hand, metallic ions and colloidal-size precipitates are readily floated, so their behavior in effect signifies that no lower limit of particle size exists.

If the solid phase to be floated is not fully liberated from other solid phases in the mixture, the resultant composite particles can still be floated, but the separation is then not as completely *selective* as it should be, since the resulting float product is unavoidably contaminated with the adjoining solid components.

The mixture of appropriately sized and liberated solids from which selected particles are to be separated is agitated in water, and this suspension constitutes the flotation pulp or flotation feed. The pulp is then placed in a suitable container called a flotation cell. There are about 40 designs of flotation cells used all over the world. Most of these represent modifications of three main types, which are exemplified in Figures 1.1 to 1.3:

1. A *mechanical cell* (Figure 1.1) equipped with a stator and a rotor to keep the mineral particles in suspension and to disperse air supplied (partly by suction and partly by compression) through a central pipe around the shaft for the rotor. The stator may be attached to the air pipe or to the cell walls. In some cell designs, the stator is omitted.

2. A *pneumatic cell*, in which suspension of solid particles in water is achieved by the compressed air being suitably dispersed throughout the volume of the cell. The latest modification of pneumatic cells, used in the *froth separation* technique [Malinowskii *et al.* (1974)],[†] employs a perforated grid (or pipes) arranged in an appropriate

---

[†] See footnote on p. 36 with regard to spelling.

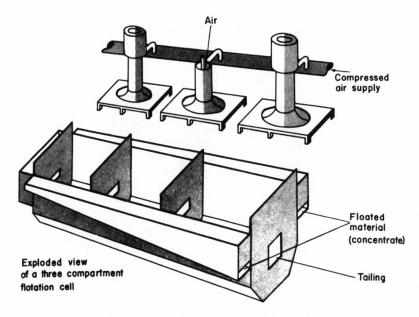

Figure 1.1. An exploded view of a three-compartment, *mechanical-type cell*. Flotation feed enters through a feed box (left end) and passes from one compartment to the next through appropriate openings. Several cells (containing one, two, three, or four compartments) may be joined together to form a bank of cells in which a particular flotation stage is accomplished, such as roughing, scavenging, cleaning, etc. The floating particulates overflow to launders positioned along the sides of the cells. The tailing from a given bank exits through the last opening into a tailing box (not shown) and hence to a next bank of cells or to a tailing disposal unit. Stators within each compartment may be fixed to the walls of the compartment or to the air pipes surrounding the shafts of rotors. Nowadays most of the mechanical cells are supplied with compressed air.

position near the top of the cell. This arrangement allows a thick bed of froth to be formed. The flotation pulp, appropriately prepared and ready for separation, is fed with a minimum of agitation on top of this bed of froth (Figure 1.2).

3. A *cyclone cell*, into which the feed is delivered (together with air) through a cyclone feeder, under pressure (Figure 1.3).

Independently of the type, the size of each of these cells is varied from a laboratory model of 2–8 liter volume (capable of treating 500–2000 g of solids in a batch test) to large commercial cells of 15–20 m^3 volume.

Details of cell design are given in appropriate pamphlets provided by each manufacturer [such as Denver, Wemco, Agitair, Sala (Boliden), Outokumpu, Mekhanobr, Zinc Corporation (Davcra) cell, H & P cyclocell,

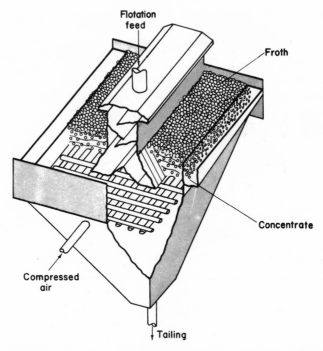

Figure 1.2. A cutaway view of a *pneumatic cell* developed for *froth separation* flotation [after Malinowskii *et al.* (1974)]. Flotation feed is gently deposited (through a central feed well) onto a thick layer of froth, created by a grid of perforated pipes supplying air. Floating solids remain on top of the froth, while nonfloating tailing solids pass through the grid to the conical discharge in the bottom of each cell. Lack of agitation within the cell favors flotability of coarser-than-normal particles and those particles which adhere weakly to air bubbles.

etc.] and, in addition, in texts such as Taggart (1945), Pryor (1965), and Gaudin (1957). Some papers, e.g., Flint (1973), Maxwell (1972), Malinow-skii *et al.* (1974), Cusack (1968), Harris and Lepetic (1966), and Arbiter and Harris (1962), deal with specific aspects of flotation cell design or kinetics.

For the solid particles to float, their surfaces must be made *hydrophobic*, that is, wetted only partially by water. An indication of hydrophobic character is given by a restricted area of contact made when a drop of water is deposited on a flat solid surface exposed to air, as in Figure 1.5. The water drop silhouette may be highly rounded, as it is when the solid is highly hydrophobic, or it may spread partly to form a flattish cap, when the solid is less hydrophobic in character. The *contact angle* θ, measured through the liquid phase, between the solid surface and the tangent to the liquid surface at the three-phase contact, is often referred to as a measure of hydrophobicity. (In the opinion of the author, the contact angle is an indicator but is not a measure of the hydrophobic character; for a more detailed discussion, see Chapter 2.) When the solid is wetted completely by water, drops of water spread over the surface of the solid in a thin film, as shown in Figure 1.4. Such solids are referred to as *hydrophilic* (water-wet). When small particles are submerged in water and then contacted with a captive bubble (held at the end of a tube), the particles which are hydrophobic

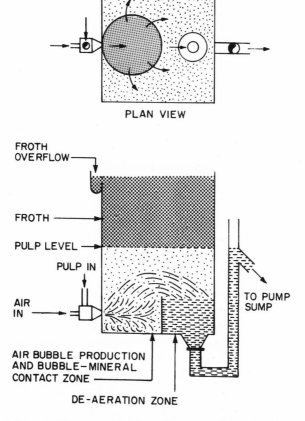

Figure. 1.3. A Davcra cell, an example of a *cyclone-type cell*, has no rotor and no special grids or mats for air dispersal. Highly pressurized flotation feed is delivered tangentially into a cyclone feeder, together with axially supplied air. A thorough mixing of air and pulp ensures ample particle-bubble contact. [After Cusack (1971). Copyright McGraw-Hill Inc.]

(A) SESSILE LIQUID DROP ON PLANE SOLID IN
        AIR   SPREADS COMPLETELY,
     SHOWING NO CONTACT ANGLE
                    θ = 0°

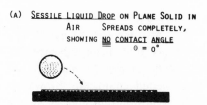

(B) AIR BUBBLE CLINGING TO SUBMERGED PLANAR
              SOLID SURFACE

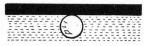

NO ADHESION TO
HYDROPHILIC SURFACES

(c) CAPTIVE BUBBLE PICK-UP TEST OF SOLID
                          PARTICLES

NO ATTACHMENT OF
HYDROPHILIC PARTICLES

Figure 1.4. *Hydrophilic solids.* Spreading and no bubble attachment. Water wets the hydrophilic solid completely, spreading readily to cover large areas of flat solids. Air bubbles neither attach themselves to flat hydrophilic solids nor pick up particles which are hydrophilic.

stick to the bubble (Figure 1.5), while those which are hydrophilic do not (Figure 1.4). If the surface of the solid to be floated does not possess the requisite hydrophobic characteristic, it must be made to acquire such hydrophobicity by the adsorption of suitable surfactants, which are called *collectors.* (Surfactants are defined and discussed in detail in Chapter 5.) For any selective flotation to occur, there must be a difference in the degree of wetting and nonwetting of the solid components in the mixture. Flotation is, simply, one of the techniques available for separating the hydrophobic constituent particulates (in a mixture of solid phases) from the hydrophilic ones.

In separations from highly complex multiphase mixtures additions of various "modifying" agents may be required, some of which help to keep selected components hydrophilic in character, while others are added to reinforce the action of the given collector. The former modifying agents are called *depressants* and the latter, *activators.* (Later in this chapter, these reagents are referred to again and are more extensively discussed in Chapter 10.)

In froth flotation, an addition of another surfactant, acting as a *frother*, is usually needed. When the pulp within the cell becomes adequately aerated, the hydrophobic solid particles attach to air bubbles and are buoyed by these to the surface of the suspension. Aggregated air bubbles with the attached solid particles constitute a *mineralized froth*, which builds up atop the pulp and overflows the lip of the flotation cell. This material represents the separated particles and is called the *concentrate*. The hydrophilic solids remain in the cell. In batch flotation, when all floatable particles are removed with the overflowing froth, the suspension of the hydrophilic solids remaining in the cell constitutes the final *tailing* of the separation process. In a continuous flotation process, several individual cells are joined to form a multicompartment unit (Figure 1.1), which is referred to as a bank of cells. From each cell in such a bank only a portion of floating solids is removed to a launder, and the intermediate tailing is discharged by an opening to the next cell.

Aggregates of particles attached to air bubbles in the process of being buoyed to the surface of the aqueous suspension in the cell are

WEAKLY HYDROPHOBIC STRONGLY HYDROPHOBIC

(A) SESSILE LIQUID DROP - on Plane Solid
in Air → Partial Spreading with:

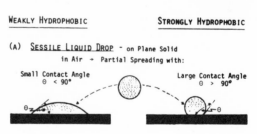

Figure 1.5. *Hydrophobic solids.* Incomplete wetting, no spreading, attachment of bubbles. When the degree of hydrophobicity is small, the angles formed by drops of water on the surface of flat solids (or by bubbles attached to submerged solids) are also small, $0 < \theta \ll 90°$. When the solid is highly hydrophobic, the contact angles are high, $\theta \sim 90°$ or even $\theta > 90°$. Hydrophobic particles immersed in aqueous solutions are more readily picked up by static bubbles contacted with them if their hydrophobicity is higher.

(B) AIR BUBBLE CLINGING TO SUBMERGED PLANAR SOLID SURFACE:

(C) CAPTIVE BUBBLE PICK-UP TEST OF SOLID PARTICLES:

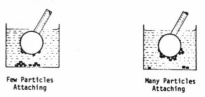

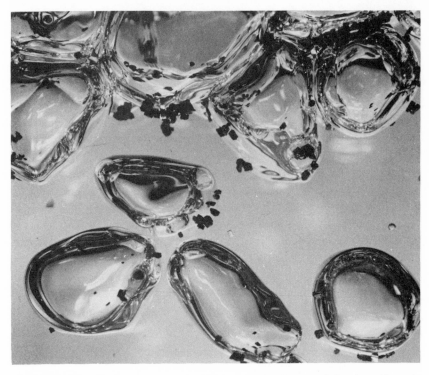

Figure 1.6. Hydrophobic particles of xanthate-coated galena are being buoyed to the top of the cell.

shown in Figure 1.6. Thus, the components of a *flotation system* are the following:

Cell	A container with an impeller or an aeration device, capable of keeping the solids in suspension and providing aeration for frequent air bubble–particle collisions.
Feed	A mixture of solids to be separated, suspended in water at, usually, about 1:3 solids to water, by weight; referred to as flotation pulp.
Regulating or modifying agents	Ions such as H^+ and OH^- (to control pH), dissolved oxygen or oxidizing species, HS^-, HCO_3^-, CN^-, and metal ions (derived from partial dissolution of some solids or added purposely to act as depressants and activators); also, specifically added organic compounds for depressing or activating action.
Surfactants	Minimum of two are usually required—collector and frother.

These reagents are added in quantities of 0.05–100 g/tonne of solids (0.02–35 ppm of the pulp or, assuming 200 MW, concentrations of $\sim 10^{-7} \rightarrow 10^{-5}$ M). In some flotation systems one surfactant is apparently used; however, as will be seen later, the pH of the pulp is such that two forms of the surfactant, ionized and nonionized, are present (one of these two forms acts as a collector, the other as a frother). Examples of the latter situation are the flotation of CaF_2 using fatty acids and the flotation of quartz using alkyl amines. Only when the solid particles to be separated are naturally hydrophobic is but one surfactant, acting as frother, required.

Air   Drawn in by the suction of the impeller and/or injected under pressure into the pulp (in amounts equivalent to $\frac{1}{3}$–$\frac{2}{3}$ cell volume per minute).

The purpose of regulating or modifying agents is to prepare the surfaces of the various solids for the subsequent selective adsorption of the surfactant(s) in such a way that only the desired particles are made hydrophobic. This preparation of surfaces is achieved through pH control, through establishment of an appropriate oxidation–reduction level within the pulp, and through control of the concentration of ions. In the absence of suitable modifying reagents, the adsorption of surfactant(s) may be insufficient in producing hydrophobic surfaces or may not be restricted to the surfaces of the desired particles. (For details, see Chapter 8.)

A strict control of the pH of the pulp is usually the first important requirement for a successful selective separation by flotation. All surfactants that can act as collectors do so only within a certain narrow range of the pH scale. When such surfactant species adsorb as collectors within this pH range on several solid phases, specific inorganic or organic ions (or nonionized multipolar species) known as depressants have to be added to suppress flotation of those phases whose separation with the chosen solid is not required. When the surfactant adsorption on a selected solid phase requires a reinforcing action of a suitable ion, called an activator, the latter usually acts as a "bridge" between the surface of the chosen solid and the adsorbing surfactant. For example, in the flotation of sphalerite (ZnS) by ethyl xanthate, copper sulfate is added ahead of the collector species, and the cupric ions act as an activator, preadsorbing on sphalerite. Without such preadsorbed copper species ethyl xanthate is not capable of acting as a collector for sphalerite.

The action of a surfactant performing the role of a collector is to change the surface character of the solid phase to be separated from the initially hydrophilic state (water-wet) to a partly hydrophobic state (incompletely water-wet). This change in the surface characteristics is achieved through an adsorption of collector species in such a manner that air bubbles can attach to the surface of this solid particle under the dynamic conditions existing in the flotation cell. A special degree and extent of collector adsorption, and of hydrophobicity, are required for these dynamic conditions, and the cooperation of a second surfactant, the frother, helps to achieve particle–bubble attachment. Under one set of conditions, a purely physical adsorption of an appropriate surfactant may be sufficient to obtain the requisite hydrophobic character and its collector action. Other types of surfactants may need to be chemisorbed before they act as collectors. Yet, chemisorption, which usually involves a transfer of electrons between the adsorbate (surfactant) and the substrate (solid surface), may *sometimes* cause such an *excessive disturbance* among the bonds surrounding the site of adsorption (i.e., *within the surface* of the solid) that the floatability of these particles is damaged. The perturbed bonds surrounding the site to which the surfactant is adsorbed may become too weak to support the particle during flotation, particularly if flotation is carried out in a mechanically agitated cell. The highly hydrophobic patch represented by surfactants chemisorbed to the weakly held underlying atoms of the adsorption site may be torn out of the solid by the strong centrifugal forces generated in a mechanical or a cyclone-type cell. The rupture occurs across the weakened bonds within the solid lattice itself and not across the adsorption bonds. Such gross perturbations are likely to occur during chemisorption of surfactants on most ionically bonded solids. Under quiescent conditions of gentle aeration, as in some pneumatic cells, particles with so chemisorbed surfactants would remain floatable.

A second surfactant added as a frother facilitates the attachment between an air bubble and a hydrophobic solid during the extremely short contact time available in collision in an agitated pulp. Many modes of effecting an attachment between an air bubble and a hydrophobic particle are presumed to be involved in flotation. Gas may be precipitated on hydrophobic particles by an *in situ* electrolysis as in electroflotation[†] or under vacuum conditions generated in the wake of an impeller blade. Particles and bubbles may flow next to each other or slip past each other, or particles

---

[†] A modification of the flotation process in which electrodes inserted into flotation pulp generate gas bubbles by decomposition of water.

may be drawn into the wake of a receding bubble surface during its oscillations or its motion through the pulp. But a direct collision during agitation is probably responsible for the major portion of successful attachments.

The collision contact time is of the order of milliseconds. This has been established by high-speed photography. Studies of the induction time (the time that is needed for a rupture of a thin liquid film existing between the air bubble and the approaching solid) confirmed this order of magnitude of attachment time. It appears that the frother does affect the kinetics of attachment by two actions. One of these is the development of multiple van der Waals bonding by the frother molecules with the partly hydrophobic solid (or molecular interactions between the frother molecules and the collector species adsorbed at the solid surface). The other action is a possible replacement of any electrostatic repulsion by attraction, brought about by relaxation of dipoles of frother species adsorbed at the air bubbles. (See Section 9.3.)

The buoyancy of the bubbles lifts the attached solid particles to the top of the flotation cell, where a mineralized froth is formed. This is continuously separated from the rest of the pulp by allowing it to overflow into a launder adjoining the cell (Figure 1.1). The mineralized froth should possess a restricted degree of stability; it should be stable enough to overflow the cell without losing the attached solid particles, but it should break down after entering the launder. The frother–collector interactions strongly affect the degree of froth stabilization that is desired.

As can be inferred from the preceding outline of a flotation system, the success of the separation process depends heavily on the proper control of surfaces and interfaces involved in the highly complex mixture of several solid phases suspended in aqueous solution through which air bubbles are being dispersed. A judicious use of specific modifying agents, collectors, and frothers should enable effective separations of different component solid phases to be carried out selectively and in succession. For example, complex synthetic mixtures containing Cu_2S, $CuFeS_2$, PbS, ZnS, FeS_2, CaF_2, and SiO_2 can be separated into relatively clean, individual components, in turn, if each component is fully liberated. Complex ores containing the same mineral components would not be as readily or as cleanly separated. This behavior is caused by the interlocking of mineral phases and the contaminating consequences of grinding, which may make efficient separation very difficult.

The efficient operation of a complex flotation process is also greatly dependent on other parameters which are not physicochemical or surface chemical in nature. The hydrodynamics of the flotation pulp in different cells, the technology of flotation circuits, and an optimum integration of procedures comprising the particular industrial operation greatly influence the extent of the overall separation performance. Ultimately, the efficiency and the economy of the process are determined by a compromise among the various parameters. Yet the role of surface chemistry is, frequently, the decisive factor in the success or the failure of a given separation.

1.3. *Typical Flotation Procedures and Flowsheets*

The application of the flotation technique to the separation of individual minerals has developed entirely through empirical testing. Theoretical understanding lagged behind practice and, despite a great deal of progress made since the beginning, many aspects are still not clear. As yet, it is impossible to predict unequivocally the exact conditions that are necessary for a successful separation of a particular solid phase from complex mixtures of solids represented by new ores. This situation is due to the fact that the presence of one single additional constituent, whether as a solid phase or a dissolved ion, even in an extremely small proportion with respect to the rest of the constituents, may radically change the behavior of the system.

The scope of applications in the mineral industry may best be considered in terms of minerals which are separated by different types of reagents (these reagents are discussed in Chapter 5). Ores containing sulfide minerals are treated most efficiently by thiocollectors such as xanthates. Nonsulfide ores of heavy metals (oxides, carbonates, and sulfates) are recovered either by xanthates after a preliminary sulfidization or by carboxylates and sulfonates. The so-called nonmetallic minerals (silicates and insoluble salts such as $BaSO_4$, calcium phosphates, etc.) are floated using alkyl amines, alkyl sulfates, or carboxylates. The readily soluble sodium and potassium salts are being separated from each other by flotation in highly concentrated brine solutions using *n*-alkyl amines or fatty acids. Flotation processes separating sulfur, coal, or bitumen from tar sands form yet another distinct group of systems employing mostly nonionized surfactants.

The discovery by C. H. Keller in 1923 that xanthates are very effective and efficient collectors for sulfide minerals led to the development of highly selective separations of individual mineral sulfides. This selectivity in separations was made possible by a strict control of pH and of different levels of oxidizing conditions and through adjustment of various modifying additives such as CN^-, HS^-, Cu^{2+}, silicate ion, etc. The real significance of the role played by each of the preceding parameters is becoming recognized only gradually.

It can readily be shown that, when tested separately as pure particles, each of the three different minerals A, B, and C may require a different threshold collector concentration ($C_A < C_C < C_B$) for, say, 95% recovery (Figure 1.7). Theoretically, it would thus appear that starting with the lowest level of threshold collector concentration, such as C_A, a selective separation of mineral A could be attained first without any appreciable quantity of the minerals C and B being floated. An increase of the collector concentration to C_C (Figure 1.7) should then allow mineral C to be floated next, leaving mineral B to be floated last, after the concentration of collector is raised above C_B.

However, as soon as the three minerals are mixed together, they often begin to interact, mutually altering their individual surface characteristics. In consequence, the theoretically possible selectivity based on the use of an appropriate threshold collector concentration alone is not commonly attained.

The effects of mutual alteration of surface characteristics may be ameliorated by the control of pH and by additions of modifying agents. With a proper control of these parameters a sequential selective flotation has been found possible for a large number of ores, comprising solidified mixtures of sulfides liberated at relatively coarse sizes from nonsulfide gangue minerals. An example of a generalized flowsheet, that is, a sequence

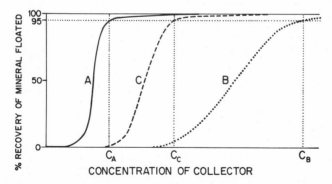

Figure 1.7. *Threshold concentration of collector.* Particles of each solid phase A, B, and C, tested individually for floatability in a series of collector solutions of increasing concentrations, show a distinct minimum concentration specific to each solid, viz., C_A, C_C, and C_B, for a 95% recovery. Such differentiation is the basis for selective separations by flotation. However, in practice the selectivity is achieved not so much through different concentrations of collector alone but through a combined effect of activating and depressing agents together with the collector species.

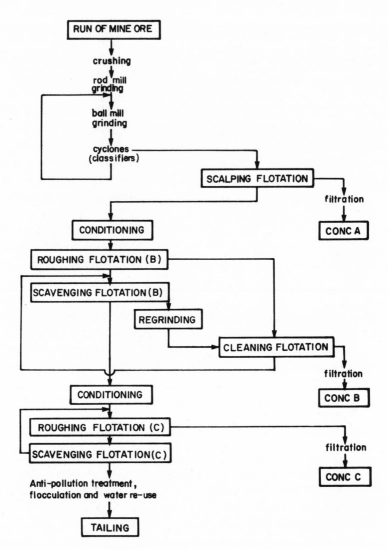

Figure 1.8. Sequential selective flotation of valuable minerals B and C, preceded by a scalping removal by flotation of a naturally hydrophobic contaminant mineral A. Applicable to most coarsely liberated sulfide and nonsulfide ores.

of integrated operational procedures representing this type of sequential selective flotation, is shown in Figure 1.8. Numerous variants of such flow sheets are adopted in industrial practice.

The ore supplied from the mine to the flotation plant (or mill) is referred to as the *run-of-mine ore*. It represents a mixture of valuable minerals (let these be minerals B and C) distributed within the host rock. The latter represents the valueless components, the so-called *gangue* minerals. Some additional valueless rock material (waste) may also be picked up during mining operations from the boundary regions delineating the ore deposit itself.

Occasionally, one of the minor constituent minerals is hydrophobic. It may be represented by a valuable component such as MoS_2 in Cu–Mo ores, or it may be a valueless contaminant solid, such as talc, graphite, or bitumen. In Figure 1.8, let mineral A denote such a hydrophobic component. Its presence may necessitate a preconcentration treatment, a so-called "scalping" flotation, ahead of the separation of the valuable minerals B and C. Otherwise it would be floated together with the first mineral to be concentrated.

Invariably, with all naturally associated multiphase solids, the first operation preceding flotation is the grinding operation, carried out primarily to obtain the liberation of minerals. Occasionally, attrition grinding is carried out simply to clean the surfaces of particles. Liberation is usually carried out in two stages, in a rod mill followed by a ball mill; the latter is in a closed circuit with a size separation device, such as a cyclone or a classifier. The cost of grinding may account for a large portion (40–60%) of the total treatment costs involved in recovering the minerals from the ore. With the very low grades[†] of the ores presently available (for example, 0.4% Cu ore or 0.025% MoS_2 ore) it becomes imperative to reduce the amount of grinding to a minimum.

Reduction in grinding cost can be achieved by adopting various approaches. One of these may consist of rejecting from the grinding circuit the coarse particles of gangue minerals as soon as they are liberated, that is, freed from valuable mineral. With some ores, an appreciable proportion of liberated gangue can be rejected by screening the discharge of the rod mill; occasionally, even larger-size gangue can be rejected by gravity separation using heavy media on the discharge of the crushing unit. In other approaches, the following procedure is adopted in treating low-grade ores

---

[†] The *grade* of a material, whether it is an ore, a concentrate, or a tailing, with respect to a given metallic element is the percentage content of this metal in the material. Thus, a grade of, say, 65.3% Pb means 65.3% lead content (or lead *assay*). The *recovery obtained* in a particular separation process denotes that proportion of the valuable component which is separated as a concentrate. It is expressed as a percentage of the total metal content in the ore feed. Thus, for example, of 100% metal (Pb, Cu, Zn, etc.) in the ore feed, the concentrate may represent, say, 87% *recovered* metal, and the rest, 13%, is lost in the tailing or is distributed among other products of separation.

of suitable liberation characteristics. Instead of carrying out the grinding to a stage when nearly all valuable minerals are fully liberated, flotation is conducted on relatively coarse and incompletely liberated valuable mineral particles. The prevailing coarse-size gangue minerals, constituting the bulk of the ore, may then be sufficiently free from the valuable minerals to be rejected as the final flotation tailing without an undue loss of valuable minerals. The savings in grinding cost so effected may be considerable. It is obvious that this procedure cannot be applied to ores with a fine uniform dissemination of valuable minerals throughout the gangue.

When grinding provides sufficient amounts of liberated valuable minerals, these are concentrated, first, in the so-called *rougher* stage flotation as the *rougher concentrate*. The remaining unfloated minerals, which are incompletely liberated, are then concentrated in a subsequent flotation section, called the *scavenger* stage. These unliberated minerals usually require a longer time of treatment and higher reagent additions for their flotation.

Note: If the grinding is insufficient and provides mostly unliberated valuable minerals but free gangue particles, the rougher and scavenger concentrates will require regrinding and reflotation.

For fully liberated flotation systems, the so-called cleaning and re-cleaning floats (or stages) carried out on rougher concentrates accomplish an upgrading action through the removal of mechanically admixed (entrained) gangue particles. This entrainment of nonfloatable gangue particles occurs during the froth-building stage atop the cell. The bubbles arriving at the top of the cell tend to imprison some of the intervening pulp containing nonfloating particles (see Chapter 9).

Since flotation is a statistical process, dilution obtained in cleaning and recleaning circuits produces the desired effect of upgrading only if the valuable mineral particles are sufficiently fully liberated.

When two or more valuable minerals are too intimately associated with each other, a sequential selective flotation procedure, outlined in Figure 1.8, may not be either feasible or the most economical. A modification of the flowsheet is then adopted, whereby the grinding is carried out to the extent that would enable a separation of the two (or more) valuable minerals together as a bulk float away from the gangue minerals. This flowsheet is outlined in Figure 1.9.

Depending on the extent to which the two valuable minerals are liberated from each other by the time they are in the bulk concentrate, the latter may have to be subjected to a controlled regrinding necessary for liberation of the two. A chemical treatment with suitable additives is then applied to the bulk concentrate to destroy the collector coating on either one or both minerals. If after such treatment one of the valuable minerals in the bulk concentrate is still hydrophobic, a very simple reflotation step with a frother enables its separation from the other mineral, producing two valuable products, one as a concentrate and the other as a "tailing." If both minerals become hydrophilic after the destruction of the collector coatings, reflotation will require an addition of a suitable collector for one of the minerals, at appropriate conditions for the suppression of the other mineral, and a frother.

Ores containing complex sulfides of Cu–Ni, Cu–Mo, Cu–Zn, and Pb–Sb and even some finely disseminated Pb–Zn ores may be more easily separated by the procedure outlined in Figure 1.9 than by the sequential selective flotation procedure of Figure 1.8.

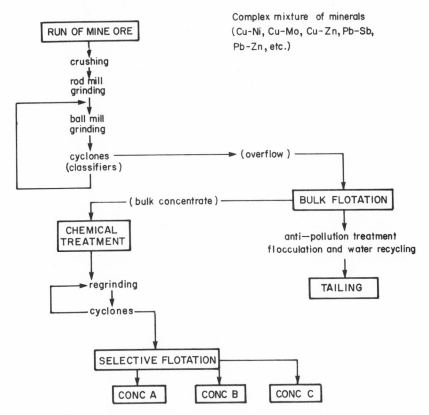

Figure 1.9. Bulk flotation of sulfides followed by *chemical treatment, regrinding,* and, finally, *selective separations* by *flotation.* Adopted for finely disseminated complex sulfide ores, for example, Cu–Mo, Cu–Ni, Ni–Co–Cu, etc.

Flowsheets of the type shown in Figure 1.9 are not too common in nonsulfide flotation systems. The reason for this situation is the difficulty encountered in attempts to destroy the non-thio collector coatings as quickly and effectively as can be done in systems containing thio compounds. The non-thio collector may be stripped off a solid surface, frequently by a change in pH, but it is not readily decomposed and remains in the pulp available for readsorption. Lengthy biodegradation reactions utilizing enzymes and bacteria, or drastic chemical treatments, are required to destroy the non-thio collector molecules and their adsorption products. Thio collectors, on the other hand, are fairly readily decomposed (forming derivatives such as metallic sulfides and alcohols) even after a relatively mild treatment with small additions of NaHS, NaOH, or NaCN [Sheikh (1972)]. A modification of a bulk flotation procedure outlined in Figure 1.9 is sometimes possible and is practiced in nonsulfide flotation systems. It consists of carrying out a bulk float of two or more minerals (which cannot be readily differentiated under a given set of flotation conditions) and such a bulk concentrate is subsequently treated under a different pH with a high dosage of depressants in order to float selectively one of the constituents. Such a modification does not involve chemical destruction of the collector, only its desorption. An example of such treatment is the procedure of cleaning calcium phosphates from the cofloated calcium and magnesium carbonates and gypsum using high concentrations of soluble pyrophosphates as depressants [Johnston and Leja (1978)].

1.4. *Mineral Liberation Size*

When an ore is mined, the crystalline grains of various minerals, representing the valuable mineral species to be concentrated and the valueless gangue components to be rejected, are all cemented together. Their intergrowth, that is, the manner in which the valuable minerals are distributed among the gangue minerals (which act as a matrix or host), and the size representing the majority of these valuable grains are of paramount importance for the success of their separation and their subsequent concentration. Before any concentration technique can be applied to the ore, the minerals must be freed from the adjoining foreign phases.

The manner in which the adjoining mineral phases become free, or liberated, from each other is of great significance. The separation may occur along the "true" interface, with no atoms of the opposing phase left on either surface produced after separation. Such separations rarely occur and are possible only when the adhesion between the two phases is smaller than the cohesion within either of the two phases (see Section 2.10). More common is a separation that occurs along a shear plane residing predominantly within the interfacial region of one phase, thus incorporating some atoms of the adjoining phase (see Chapter 7 for a discussion of shear plane position). These residual layers of the adjoining phases have to be removed (before flotation) either by abrasion or by dissolution.

The friability of the valuable mineral and/or the extreme softness of a minor hydrophobic component such as molybdenite or bitumen (and its relevant ability to become indiscriminately smeared on gangue minerals) may become extremely important parameters under some conditions of the separation technique employed.

Crushing and grinding are employed to accomplish the liberation of the valuable minerals. If the size of valuable mineral grains is large, as, for example, in Figure 1.10(a), the grinding operation will readily free the valuable minerals, and the subsequent selective separation from the gangue components should not be difficult. However, if the valuable minerals are highly interlocked and very finely disseminated, their separation by flotation may present major difficulties. For example, very finely crystalline Pb and Zn sulfides disseminated in massive quantities of pyrite and pyrrhotite, as present in some deposits in New Brunswick, Canada or in Queensland, Australia, still await a really satisfactory treatment by flotation or by other techniques.

The various types of association between mineral phases (mineral intergrowth) have been classified by Amstutz (1962) into a total of nine groups of intergrowth patterns. The most characteristic ones are shown in Figure 1.11. These patterns take into account the relative size ratio of disseminated mineral to the overall particle size. What they fail to emphasize is the fact that, in addition to that ratio, the particle size itself may vary over two orders of magnitude. Thus, the intergrowth pattern exemplified in Figure 1.11(a) may create no major problems in liberation or in subsequent separation by flotation if the size of the particle represented is 300 μm. However, if the same drawing represents a particle of 3 μm or less, the liberation of so disseminated valuable mineral is not feasible by grinding, and its subsequent concentration by flotation is likely to be unsuccessful. In any ore deposit, several of the interlocking patterns shown in Figure 1.11 may be represented, as exemplified by Figure 1.10.

It is clear that a systematic optical microscopic examination of the ore itself and of the products obtained at different stages of the treatment in grinding and flotation circuits is indispensable in a rational testing and

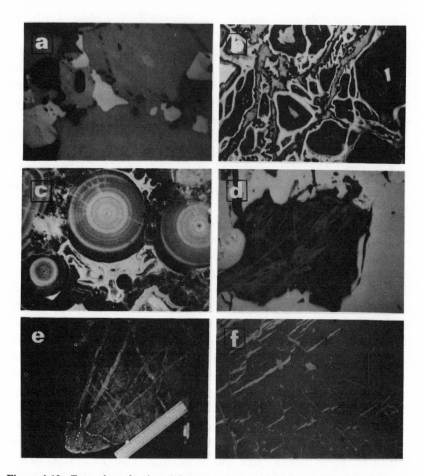

Figure 1.10. Examples of mineral intergrowths. (a) *Stannite*, Cu_2FeSnS_4 [olive gray with finely disseminated blebs of chalcopyrite (yellow) and sphalerite (dark gray)]; *pyrargyrite*, $3Ag_2S \cdot Sb_2S_3$ (bluish gray); *sphalerite*, ZnS (dark gray); *chalcopyrite*, $CuFeS_2$ (yellow); magnification 160×; locality, Cornwall, England. (b) *Chalcopyrite*, $CuFeS_2$ (yellow) oxidized to goethite, FeO(OH) (gray), and *malachite*, $CuCO_3 \cdot Cu(OH)_2$ (greenish); magnification 40×; locality unknown. (c) Botryoidal *sphalerite*, ZnS (light to dark brown), and galena, PbS (black); magnification 5×; locality, Pine Point, North West Territories, Canada. (d) Acicular *graphite*, C (light to dark gray), in a cavity in pyrrhotite FeS_{1+x} (white); magnification 160×; locality, Saskatchewan Province, Canada. (e) Molybdenum ore from Climax, Colorado; molybdenite, MoS_2 (dark veins in quartz, SiO_2), and *pyrite*, FeS_2 (yellow veins). (f) *Chalcopyrite*, $CuFeS_2$ (yellow), and chalcocite, Cu_2S (bluish gray), in *bornite*, Cu_5FeS_4 (purplish pink); magnification 8×; locality, Red Devil, Alaska. (All photographs courtesy of Dr. A. Sinclair, University of British Columbia. Reproduced at 60%.)

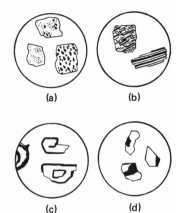

Figure 1.11. The most characteristic intergrowth patterns. (a) Finely disseminated intergrowth; (b) layer intergrowth, single or complex fissures; (c) rim or shell intergrowth; (d) simple, coarse intergrowth.

control of any flotation separation process. Such microscopic examinations should be made routinely but especially during periods of an unusual decrease in grades of concentrates or in recoveries, periods which have the tendency to occur with undesired and unexpected frequency in all operations. A suitable text (or several of them) on ore microscopy should be available in the mill laboratory to help the recognition and the interpretation of observations. Examples of such texts are Short (1940), Schouten (1962), Jones and Fleming (1965), Heinrich (1965), Freund (1966), Uytenbogaardt (1968), and Ramdohr (1969).

In any preliminary studies on samples representing possible ores (the so-called feasibility studies) or in research and development of a concentrating treatment procedure, the new techniques of electron microprobe and scanning electron microscope prove to be invaluable adjuncts supplementing or replacing the optical microscopic examination technique. By using these techniques, the identification of various components in the sample is often much simpler and faster than that possible under the optical microscope. Frequently, the elements dissolved in the mineral, or in isomorphic substitution, can be revealed and more readily identified, as exemplified in Figure 1.12. Also, if a coating of a different composition surrounds a particle and has a thickness in excess of the limit for detection of a phase by the microprobe (this is, at present approximately 0.5 μm), it can be readily detected and identified, provided that a cross section of such a coated particle is viewed.

The extent of liberation of valuable minerals achieved in any industrial grinding–flotation plant is generally the result of trial and error experimentation. The ultimate criteria are economic. These include a salable grade and an optimum recovery of the valuable mineral, obtainable with the

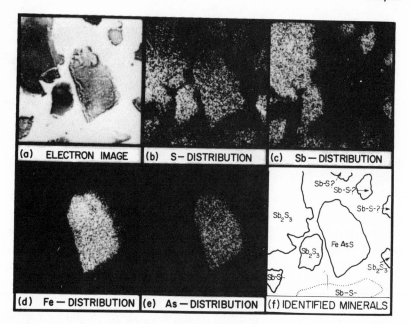

Figure 1.12. An example of microprobe scans (of a bulk flotation concentrate) and an interpretation of mineral grains based on elemental scans.

available equipment. With the enormous increase in tonnages being treated in modern flotation plants, an appropriate instrumentation of the grinding circuits, together with their modeling and computer control, is becoming indispensable. Information on these topics can be found in references such as King (1976), Mular (1976), Smith (1976), Whiteside (1974), and Mular and Bhappu (1978). Whatever the size of the treatment plant and the state of instrumentation and modeling, the control of grinding circuits and liberation is greatly facilitated by the application of visual techniques, such as optical microscopic examination and/or microprobe and scanning electron microscopy. Unfortunately, these are at present too infrequently employed in industrial plants to the detriment of some operational results.

1.5. *Some Problems Encountered in Separations of Minerals*

A number of difficulties have been, and are being, encountered in the treatment of some of the available ore deposits, both sulfide and nonsulfide in character. These difficulties can be grouped under the following headings:

1. Those relevant to sulfide deposits:
   a. An extensive surface oxidation of individual sulfide grains, as exemplified in Figure 1.10(b)
   b. The coexistence of sulfide and oxidized minerals (such as oxides, carbonates, sulfates)
   c. A high degree of interlocking and an extremely fine dissemination of the valuable sulfides within the body of massive pyrrhotite and pyrite
2. Those relevant to both sulfide and nonsulfide deposits:
   a. Fine dissemination within the matrix of kindred minerals
   b. The coexistence of highly hydrophobic minerals (such as talc, graphite, molybdenite, carbonaceous and tarry residues, elemental sulfur, stibnite, etc.) with the valuable sulfide or nonsulfide minerals
   c. The coexistence of clayey minerals, especially in nonsulfide ores, producing large quantities of interfering slimes

To provide an effective solution for a separation appropriate to the particular ore under treatment, each of the preceding situations has to be analyzed and dealt with separately. With respect to:

1a. An extensive, superficial oxidation of valuable sulfide minerals is generally counteracted by sulfidization using judicious amounts of HS^- to convert the surfaces at the oxidized layers to appropriate sulfides. This conversion has to be achieved without leaving an excessive concentration of HS^- (because its excess is usually detrimental to flotation when using xanthates) or without creating unnecessary losses by an insufficient action.

1b. The coexistence of sulfides with homogeneous grains of fully oxidized valuable minerals calls for a sequence of flotation steps in which the sulfides may be floated off first, either selectively or as a bulk sulfide float; then a separation of the oxidized minerals using suitable activators and/or special collectors is carried out. When the oxidized minerals are represented by carbonates or simple oxides, it may be possible to float them using sulfidization and xanthate-type collectors. If other oxidized minerals are present, each type of mineral may require an appropriate activator–collector combination for a satisfactory separation.

1c and 2a. A high degree of interlocking of the valuable minerals within the gangue minerals represents a major obstacle to any flotation separation. This is particularly so if the intergrowth shown in Figures

1.10(b), (c), and (f) or Figures 1.11(a), (b), and (c) is presented in very fine-size particles. The difficulty is compounded, however, if the host gangue minerals are similar in chemical character to the valuable minerals. Such difficulties are encountered in the recovery of Pb, Zn, or Cu sulfides disseminated in pyrite or pyrrhotite. Even with a coarse-size liberation, whenever pyrite and pyrrhotite exceed approximately 15% by weight, the separation of Pb, Zn, or Cu sulfides by flotation, using xanthates, becomes less selective or entirely nonselective. If, in addition to a high content of iron sulfides, the nonferrous sulfides are finely disseminated and can only be liberated at very fine sizes, the separation by flotation becomes unsuccessful. Similar examples of separation difficulties caused by fine dissemination in matrix of kindred minerals occur with silicates and oxides.

Numerous explanations have been proffered for these difficulties experienced with separations of sulfide minerals. The most common is that fine particles, of sizes less than about 5–10 μm, are too small to float. Theoretical treatments of the problem by Sutherland (1948), Klassen (1952), and Derjaguin and Dukhin (1961) led to the conclusion that small particles acquire insufficient energy of impact to disrupt the water layer existing between the two colliding particulates. (For more details, refer to the thinning of films between particles and air bubbles in Section 9.2.2.) Laboratory and in-plant investigations attempting to correlate particle size with floatability established that, after grinding, the very coarse- and the very fine-size sulfide particles do not float efficiently. Gaudin *et al.* (1931) established separation curves in laboratory tests that were very similar to the curves obtained by Cameron *et al.* (1971) some 40 years later for plant flotation circuits. (See Figure 1.13.)

A pronounced falloff in the recovery of metallic minerals for sizes below ∼10 μm has been recorded in all investigations, the two already listed and those of Clement *et al.* (1970), Suwanasing and Salman (1970), Trahar (1976), Goodman and Trahar (1977), and Woodburn *et al.* (1971).

The poor recovery in flotation of coarse particles is readily explained by the inability of bubbles to buoy large particles to the top of the flotation cell. Under conditions of high agitation in a mechanical cell, the high centrifugal forces prevailing in agitated pulp tend to separate large floating particles from their bubbles. However, the falloff in the recovery of fine-size particles cannot be readily accounted for by the argument of insufficient impact energy that we have mentioned. Were it really so, then the techniques of ion flotation and precipitate flotation [Sebba (1962), Mahne (1971), Lemlich (1972)], in which coursing bubbles pick up and float ionic species a few angstroms in size and particulates of sizes very much smaller than

10 μm, could not possibly function. Also, when metallic sulfides or metals, such as "cement" copper, are precipitated as extremely fine-size particles, these precipitates can be and are efficiently recovered by flotation. An extensive review of fine particles' response to flotation, of various theories proposed to explain the low flotation rates of slimes, and of methods which claim to improve the flotation of ultrafine particles has been published by Trahar and Warren (1976).

There is no doubt that very fine-size particles show decreased rates of flotation: These particles possess much higher specific surface areas and may require specific concentrations of reagents, longer times of treatment, or a different mode of particle–bubble encounter (such as, for example, gas precipitation in vacuum flotation). However, even if all these factors are suitably modified, the apparent depressing effect caused by prolonged grinding of sulfide minerals in steel ball mills still persists.[†]

Lin and Somasundaran (1972), Lin *et al.* (1975), and Gammage and Glasson (1976) have shown that a prolonged grinding causes major physical and chemical changes in the solid surface; it becomes progressively more amorphous, the crystalline lattice of layers underlying the surface becomes distorted, and polymorphic transformations may occur which may result in chemical reactions between solid and liquid constituents. At the same time, a continued wear of the grinding media takes place: rods, balls, and liners in the mills are progressively worn out. This fact is known to the operators of the industrial mineral concentration plants only too well, as the consumption of steel (in the form of grinding media and liners) varies directly with the fineness of grind aimed for. For a relatively coarse grind (say, 40% passing 150 mesh size), the steel consumption by a given ore may be ~100–150 g/tonne. When a very fine grind is called for (say, 80% passing 325 mesh), the same ore may consume 2–5 kg of steel per tonne of ore ground. Such steel consumption is an order or even two orders of magnitude higher than the consumption of flotation reagents.

The loss of metal is accounted for by abrasion and, to some extent, corrosion [Lui and Hoey (1975)]. The abrasion (wear) of grinding media results in deposition of tiny flecks of metallic iron (steel) on all particles undergoing grinding. Whenever this metallic phase is deposited on the surfaces of semimetallic sulfides, well known for their semiconducting properties, minute galvanic couples are set up. These galvanic couples become

---

[†] Wells (1973) found that synthetic mixtures of fully liberated galena, PbS, and quartz, SiO_2, ground for prolonged periods of time in steel mills, provide recoveries steadily decreasing with the increase in grinding time.

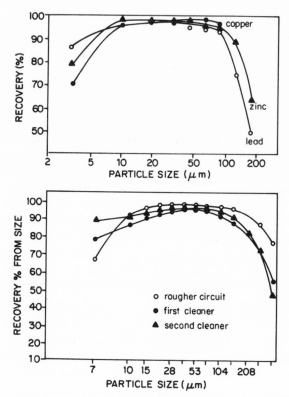

Figure 1.13. Dependence of flotation recovery on particle sizes, produced by grinding: (top) results obtained by Gaudin *et al.* (1931); in *AIME Techn. Publ.* **414**, reproduced with the permission of the American Institute of Mining and Metallurgical Engineers, Littleton, Colorado; (bottom) results of Cameron *et al.* (1971). *Proc. Australas. IMM* **240**, reproduced with the permission of the Australasian Institute of Mining and Metallurgy, Parkville, Victoria, Australia.

highly active in the presence of an electrolyte such as the liquid phase of the ore pulp containing inorganic ions from dissolved mineral species, collector ions, and modifying agents. With a coarse grind, the metallic flecks are only sparsely distributed over surfaces of sulfide particles, and their galvanic coupling effect is insignificant. When a very fine grinding is carried out, however, the proportion of metallic flecks abraded onto and covering sulfide surfaces becomes comparable in extent to, or even larger than, the area of the uncovered sulfide mineral. Then the galvanic mixed potential of the sulfide becomes progressively lowered and may reach that of the metallic iron; that is, it may become highly reducing in character [Rao *et al.* (1976)]. As will be seen later (Chapter 8), the reducing reactions thus fostered are not conducive to the adsorption of xanthate in the form that is needed to convert the sulfide surface to hydrophobic character. Thus, it would appear that the very finely ground sulfide particles, obtained in grinding highly disseminated ores with steel rods and balls, may have surfaces too contaminated with abraded iron, which would make them incapable of becoming sufficiently hydrophobic in xanthate solutions and thus unfloatable.

2b. The presence of even minor quantities of naturally hydrophobic constituents in an ore makes the selective separation process more difficult. Either an effective and highly selective depressant is required or, preferably, a scalping separation of the hydrophobic component with the use of a frother only before any collector additions are made. Such a separation of the hydrophobic mineral is usually carried out as soon as possible, preferably on the discharge from the coarse grinding mill, as, for example, reported by Lyon and Fewings (1971). However, there are always some complications likely to arise, caused by the softness of all inherently hydrophobic minerals. During grinding, the soft mineral becomes indiscriminately smeared on other solid particles, making them locally hydrophobic and prone to float under favorable conditions. When such smearing occurs, neither the use of a depressant nor a scalping flotation with a frother alone is capable of providing a sufficient degree of selectivity, especially whenever extensive grinding is needed for liberation of minerals. Some amelioration of such smearing effects may be realized by making the additions of the depressant (or of the frother) to the grinding circuit, with a concomitant reduction in the pulp density in the mills. In special cases, a bulk flotation of the sulfide minerals together with the inherently hydrophobic mineral may be carried out and the collector coating on the sulfide minerals destroyed by a suitable chemical treatment. This treatment is then followed by flotation of the inherently hydrophobic mineral, leaving the recovered sulfide as the "tailing" (similar to the flowsheet in Figure 1.9).

2c. Clayey minerals are easily hydrated by water penetrating their layer structures. In addition, they also exhibit cation exchange characteristics with many inorganic and organic ions in solution (see Section 3.5). On being agitated with water, clays produce extremely fine particles, slimes, which not only increase the viscosity of pulps but, due to their very high specific areas, readily abstract most additives used to control the flotation systems. A de-sliming of the pulp before flotation is the most common countermeasure that is adopted in practice. It is effective when the valuable minerals are liberated at a coarse size. With finely disseminated valuable minerals, or in potash flotation, such de-sliming causes high losses of valuable minerals and, therefore, is uneconomic and impractical. The adsorption of collector on clay and the ingress of water between the clay layers could be prevented in some systems by blocking off the edges of clay particles with a suitable depressant added as early as possible with the water. A successful blocking treatment of this type, resulting in clay depression, was applied in potash flotation by Arsentiev and Gorlovskii (1974).

Most of the problems we have considered are created by the type of intergrowth and the nature of associated minerals within the ore lumps. They stem from the inability of existing grinding procedures to liberate the interlocked phases cleanly along the interfaces. When some novel chemical comminution techniques to be introduced in the future enable the cleavage of associated minerals to be carried out along their interfaces, the difficulties encountered at present in liberation and in flotation of finely disseminated minerals may disappear. The phenomenon of stress–corrosion–cracking of metallic alloys in a specific environment and that of intergranular disintegration of some solid metals when wetted by other liquid metals indicate clearly that comminution by a specific chemical environment is feasible. In fact, ash-producing minerals can be liberated from coals by a suitable treatment in solutions of NaOH in methanol (U.S. Patent 3815826, June 11, 1974, to R. G. Aldrich, D. V. Keller, and R. G. Sawyer). Preliminary studies on chemical grinding aids and chemical comminution of ores have been reported by Revnivtsev *et al.* (1977) and Klimpel and Manfroy (1977).

1.6. *Examples of Industrial Separations of Sulfide Minerals*

The most important raw materials for the production of the nonferrous metals such as copper, lead, zinc, nickel, cobalt, molybdenum, mercury, and antimony are those ores which contain these metals as sulfides. The approximate figures for the 1976–1977 world production of the four major nonferrous metals are ~ 8 million tonnes of metallic copper, ~ 6 million tonnes of zinc, ~ 4 million tonnes of lead, and 0.7 million tonnes of nickel. All these metals are derived from ores that are subjected to flotation treatment. The grades of ores are generally low; hence the quantities of metallic ores treated by flotation are more than two orders of magnitude greater than the metals produced. In addition, ~ 22 million tonnes of K_2O equivalent potash fertilizer are produced, and a fairly large proportion of the $\sim 2.5 \times 10^9$ tonnes of coal is being treated to recover fines (data from

Canadian Mining Journal, Feb. 1978). Thus, it is estimated that over 2.0×10^9 tonnes of ores are being treated annually by flotation processes.

The treatment of each ore is determined largely by the degree of interlocking and the liberation achieved in grinding. Occasionally, the type of associated gangue minerals may impose severe restrictions on the treatment. For example, dolomite gangue excludes treatment under acidic conditions, while large proportions (more than 15%) of pyrrhotite and pyrite react with the oxygen in the pulp so avidly that in effect they create reducing conditions during grinding and flotation.

The initial industrial applications of froth flotation to relatively simple sulfide ores achieved separations of all sulfide minerals from the nonsulfide gangue minerals. These were, then, the bulk sulfide flotation separations, and they were frequently accomplished in acid circuits [Diamond (1967)]. The first attempt at selective separation of galena from sphalerite was made in 1912 at the Zinc Corporation mill in Broken Hill, Australia [Diamond (1967)] and that of a selective separation of chalcopyrite from pyrite by the Granby Company at Anyox, British Columbia, Canada, in 1918 [Petersen (1967)]. The reagents employed were distillation products of coal or wood: coal tar creosote and wood tar or wood oils.[†]

Once xanthates were introduced in 1923, the selective flotation separations of individual sulfides proved unusually successful from ores containing dolomitic gangue minerals, especially when the valuable sulfides were liberating at relatively coarse sizes. Difficulties were being encountered with more complex and finely disseminated ores. However, selective separation of relatively coarsely liberated galena and sphalerite, using cyanide as a depressant and copper sulfate as an activator, really signaled the era of sequential selective flotation.

1.6.1. *Lead–Zinc Ores*

Generally, both Pb and Zn minerals occur together in sufficient quantities to justify their selective concentration.

The main lead minerals are: *galena*, PbS (86.6% Pb); *cerussite*, $PbCO_3$ (83.5% Pb), and anglesite, $PbSO_4$ (73.6% Pb); less common are the complex sulfides of Pb–Sb, Pb–As, and Pb–Bi; also crocoite, $PbCrO_4$, and wulfenite,

---

[†] Historical reviews on the development of flotation as a separation process can be found in Gaudin (1957) and in three chapters in Fuerstenau (1962), viz., Hines (1962), Hines and Vincent (1962), and Crabtree and Vincent (1962). Also see Pryor (1961) and Anon (1961).

$PbMoO_4$. (The minerals which are concentrated on an industrial scale are in italics.)

The main zinc minerals consist of *sphalerite* (or zinc blende), ZnS (67.0% Zn), and its alternative crystalline form *wurtzite*; *marmatite*, (Zn, Fe)S, with a varying Fe content from less than 1% to 20% Fe; willemite, Zn_2SiO_4 (58.0% Zn); *smithsonite*, $ZnCO_3$ (52.1%); calamine or hemimorphite $Zn_4Si_2O_7(OH)_2 \cdot H_2O$ or $H_2Zn_2SiO_5$ (54.2% Zn); zincite, ZnO; and franklinite, $(Fe,Mn,Zn)O \cdot (Fe,Mn)_2O_3$.

The crystalline structures of minerals are briefly reviewed in Chapter 3. Some Cu may be dissolved (in solid solution) in marmatite, or a finely disseminated chalcopyrite ($CuFeS_2$) may be associated with the Pb and Zn sulfides, together with pyrrhotite ($Fe_{1-x}S$). In deposits containing galena in the form of very finely crystalline mineral, minor quantities of dissolved Au or Ag may be present (acting as nuclei for crystallization). Large single crystals of galena usually indicate lack of dissolved precious metals. The precious metals (Au, Ag) may still be dissolved in other sulfides such as pyrite or may be present as separate phases at the grain boundaries. Pyrite (FeS_2) and pyrrhotite ($Fe_{1-x}S$) are the two most frequently encountered sulfide–gangue minerals. Calcite ($CaCO_3$) and dolomite ($(Ca,Mg)CO_3$, together with siderite ($FeCO_3$), form the basic-in-character gangue, while quartz, silicates, $BaSO_4$, etc., form the acid-in-character gangue.

The grade of commercial Pb–Zn ores treated nowadays varies from about 3% to 15–20% combined Pb and Zn. The objective of the selective flotation treatment is to produce

1. A Pb concentrate of more than $\sim 55\%$ Pb with, preferably, less than about 3% Zn and/or less than 0.5–1% Cu, and
2. A separate Zn concentrate with more than about $\sim 50\%$ Zn content and a minimum Pb content.

The quality of products which are attempted depends on the details of a particular smelter contract that the mill is able to secure. A smelter contract is an agreement establishing conditions for sale of concentrates to a smelter plant where the metals are being recovered. It establishes the price and specifies the minimum grade of the main metallic element and the maximum content of all contaminating elements whose presence may have a critical effect on the performance of the given smelting operation.

In coarse-grained ores, galena is floated first under alkaline pH of between pH 8 to pH 10 together with any copper sulfide mineral that may be present. A minimum addition of ethyl or propyl xanthate is used as a collector and one of the short-chain alcohols or a low-MW polyoxypropylene (e.g., Dowfroth 250) as a frother. An addition of sodium sulfite, zinc sulfate, and/or sodium cyanide is occasionally made to depress zinc minerals. After the lead sulfide is floated off, the pH is raised to 11–12 in order to depress iron sulfides, and $CuSO_4 \cdot 5H_2O$ is added to activate sphalerite and marmatite. Invariably, an additional quantity of xanthate is required before zinc sulfide can be floated. Following the flotation of zinc sulfides, iron sulfides may be separated out on reducing the pH to neutral or acid and using dixanthogen (or an oxidized thiocompound) as a collector.

The higher-grade deposits are usually liberated at coarser sizes and are easier to concentrate by flotation, but such deposits become progressively exhausted. As the intergrowth increases and the liberated particles of galena and sphalerite become finer, a modified selective separation must be practiced more and more frequently. It is of the type outlined in Figure 1.9, and its objective is to reduce the total amount of grinding required. A bulk Pb–Zn float, as a single rougher–scavenger concentrate, is made after a preliminary grind aimed at liberating the gangue minerals only, without completing the liberation of the two valuable minerals from each other. A regrinding of this concentrate is then followed by a steam treatment at temperatures of 40–80°C in a highly alkaline medium containing some suitable additives such as NaHS, NaCN, or Na sulfite. Depending on the intensity, this treatment destroys partly or completely the collector coating on one or both sulfides. Flotation of the sulfide whose hydrophobic character can be more readily regenerated is carried out subsequently. When such a treatment is applied to an impure Pb concentrate which is too high in its zinc content, it is known as a "de-zincing" treatment.

1.6.2. *Copper Ores*

The sulfide minerals of copper fall into three distinct groups:

1. Pure sulfides: Cu_2S *chalcocite* (79.8% Cu) and CuS *covellite* (66.4% Cu)
2. The Fe–Cu sulfides: $CuFeS_2$ *chalcopyrite* (34.5% Cu and 30.5% Fe) and Cu_5FeS_4 *bornite* (63.3% Cu and 11.1% Fe)
3. Complex sulfides: Cu_3AsS_4 enargite, Cu_2FeSnS_4 stannite, and Cu_3SbS_3 tetrahedrite.

Occasionally, traces of native copper scattered through an ore deposit are also encountered.

The oxidized minerals comprise Cu_2O cuprite, CuO tenorite, $CuCO_3 \cdot Cu(OH)_2$ *malachite*, $2CuCO_3 \cdot Cu(OH)_2$ *azurite*, and $CuO \cdot n\text{-}SiO_2 \cdot m\text{-}H_2O$ *chrysocolla*. In some deposits in arid countries, minerals possessing a fair or even high solubility in water may also be present; for example, in Peru, minerals such as brochantite $[CuSO_4 \cdot 3Cu(OH)_2]$, antlerite $[CuSO_4 \cdot Cu(OH)_2]$, or chalcanthite $(CuSO_4 \cdot 5H_2O)$ are occasionally encountered.

The associated minerals are FeS_2 pyrite, $Fe_{1-x}S$ pyrrhotite, $(Fe,Ni)S$ pentlandite, MoS_2 molybdenite, $(Co_3S_4 + CuS)$ carrolite, PbS galena, and $(Zn,Fe)S$ marmatite. Two major types of copper ore deposits are recognized: porphyry deposits (defined as of hydrothermal origin, in acidic intrusive rocks, representing low grades, 0.4–0.8% Cu, but very large tonnages—tens and hundreds of million tonnes) and vein-type sedimentary deposits. Porphyry copper deposits contain mainly chalcopyrite $(CuFeS_2)$ and pyrite (FeS_2) with at least traces of molybdenite (MoS_2); sometimes chalcocite (Cu_2S) and bornite (Cu_5FeS_4) with minute traces of Ag and Au are also present. Sutulov (1974) gives detailed accounts of the world's porphyry deposits, the associated processing technology, and metallurgical production data.

At present, porphyry copper deposits are responsible for approximately half of the total copper production. The need to supply the market requirements from such low-grade ores resulted in a spectacular development of the open pit mining technology and, at the same time, in an equally spectacular increase in throughput of associated flotation plants. Some 50 years ago, flotation plants had 100–300 tonne/day capacities and $\sim$25 years ago, 1000–3000 tonne/day capacities, whereas most of the recently constructed plants range from 20,000 to 60,000 tonnes/day, with a few which are in the 100,000–175,000 tonne/day range.

The details of treatment are determined by the mineralogical composition of the ore. Pure copper sulfides and Fe–Cu sulfides, particularly in porphyry ores, are readily and efficiently floated with small additions of alkyl xanthates, or Z-200 (isopropyl ethylthionocarbamate; see Chapter 5) if the proportion of the associated iron sulfides (pyrite and pyrrhotite) is not too high.

With higher proportions of pyrite and pyrrhotite in the ore, a general guideline (based on past experience) is to extend the conditioning with air and to raise the pH to 10–12 (with lime!). The air oxidation treatment is designed to bring about the depression of iron sulfides (creating high $CaSO_4$ concentrations which appear necessary for iron sulfide depression) and is often reinforced with small additions of sodium or calcium cyanide. Larger quantities of cyanides tend to depress copper minerals as well.

When pentlandite, molybdenite, or carrolite are the associated minerals, a procedure similar to that outlined in Figure 1.9 is usually followed. The details of the treatment procedure may be quite complicated, particularly if there is a very fine dissemination of some minerals, a high pyrrhotite or pyrite content, and/or an appreciable quantity (0.5–2% by weight) of a naturally hydrophobic component (bitumen, graphite, or talc) in the run-of-mine ore.

The most difficult to separate are the finely disseminated Cu–Pb–Zn and Cu–Pb–Zn–Fe ores. Surface oxidation, taking place during mining of the ore and its grinding, contributes copper ions which activate the zinc and iron sulfides. The use of cyanide, sulfite, and SO_2 in judicious quantities is relied upon to separate selectively the various sulfides, with a greater or lesser degree of success. Bushell *et al.* (1971) described a successful method for the separation of Cu and Zn from the ore of the Prieska Mine. This consists of a bulk float combining Cu and Zn minerals followed by re-grinding, deactivation, and subsequent separation of Cu from Zn.[†]

The grades of copper concentrates produced depend very much on the mineralogical composition and the details of the smelting contract; usually 25% Cu content in the concentrate is aimed at when recovering chalcopyrite from ores.

1.7. Examples of Industrial Separations of Oxidized Ores of Copper, Lead, Zinc, Iron, and "Nonmetallic" Minerals

1.7.1. Separations of the Superficially Oxidized and Oxide-Type Minerals

There are, generally, two procedures available for the separation of superficially oxidized sulfide minerals and of certain oxide-type minerals:

1. Flotation with fatty acids as collectors (without intentional activation)

[†] Despite the fact that flotation was initiated using fatty acids as collectors in acid circuits, not many—if any—selective flotation separations of sulfides are accomplished under acid conditions using the oxidized forms of thio compounds (like dixanthogens and their analogues of dithiophosphates, dithiocarbamates, etc.). With complex ores not conducive to H_2S evolution and containing acidic gangue minerals, separations that are possible under acidic conditions may be much more selective than those achieved under alkaline conditions.

2. Activation using sodium sulfide to convert the oxide surface to sulfide (a treatment referred to as sulfidization) followed by flotation with thio collectors, such as xanthates

The first procedure, flotation with fatty acids, is not very selective and can be applied to only a few ores that have relatively insoluble valuable minerals; otherwise the metallic ions released into solution create an undesirable activation of gangue minerals. Also, for fatty acid flotation the gangue minerals should not comprise pyrite or pyrrhotite (which float readily with fatty acids) and should be acidic in character. The second procedure, that of controlled sulfidization followed by xanthate flotation, is by far the more selective and efficient technique of separation. Even then, difficulties are experienced whenever complex oxidized minerals are associated with clays and/or altered pyrite and pyrrhotite.

With some oxidized minerals, very high additions of xanthates (or other thio compound collectors) of the order of 1–2 kg/tonne are necessary for an effective flotation without sulfidization, for example, in flotation of vanadinite [Fleming (1952)].

Cerussite ($PbCO_3$), malachite [$CuCO_3 \cdot Cu(OH)_2$], and azurite [$2CuCO_3 \cdot Cu(OH)_2$] can be sulfidized fairly readily with sodium sulfide and floated with xanthates in ores containing dolomitic gangue and no iron oxides. Mitrofanov[†] *et al.* (1955) appear to be the only ones who have studied the rates of sulfidization in great detail.

Many of the siliceous copper ores of low copper content do not contain distinct copper silicate minerals but are in reality solid solutions of Cu ions impregnating silicates to a variable extent. Sulfidization attempts on such solid solutions of copper in silicates result in "leaching" the copper ions

---

[†] The spelling of names translated from languages using the Cyrillic alphabet (Russian, Bulgarian, Serbian) varies considerably. According to the English or the German system of transliteration the same names will be spelled, for example, Sheludko or Scheludko; Exerova or Ekserowa; Volkova or Wolkowa; Yanis or Janis; Deryaguin or Derjaguin; Lifshits, Livshitz, or Lifshitz; Malinowskii, Malinovskij, or Malinowsky; etc. Often, when English names have been translated phonetically into Russian and then retransliterated into English, they become unrecognizable. For example, Wark, I. W. and Sutherland, K. L. appears as Uork I. V. and Sazerlend, K. L. Since a number of Russian papers have been published in English in a variety of journals, the names of the same authors frequently appear spelled differently. In this book a uniform spelling is adopted regardless of transliteration in the journal quoted; thus, Lifshits, Glembotsky, Gorodetskaya, Derjaguin, Gorlovskii, etc. are used throughout. Similarly a strict transliteration of the Soviet Union (CCCP in Cyrillic) is SSSR, whereas USSR is customarily used. To differentiate the above from the transliteration for the Ukrainien Republic, Ukr. SSR is adopted.

out of the silicate matrix. The copper ions which diffuse to the surface of solid particles are complexed by sulfide ions and/or collector ions added later on, and these complexes are then mostly separated out in the subsequent flotation procedures [Scott and Poling (1973)]. Raghavan and Fuerstenau (1977) emphasize the porosity and the fibrous structure of chrysocolla. Only when the siliceous copper mineral has a higher copper content than $\sim 13\%$ Cu is a truly crystalline variety of chrysocolla, $CuO \cdot n\text{-}SiO_2 \cdot m\text{-}H_2O$, believed to be represented.

Anglesite, $PbSO_4$, can be sulfidized with difficulty and floated with xanthates. Fatty acids or alkyl sulfonates are also used as collectors for anglesite, in ores with acidic gangue, in the absence of excessive iron oxides or other insoluble sulfates.

Oxidized zinc ores [containing smithsonite, $ZnCO_3$; willemite, Zn_2SiO_4; and hemimorphite (calamine), $2ZnO \cdot SiO_2 \cdot H_2O$] have been treated by flotation since about 1950 using numerous regulating agents in order to disperse and to depress the deleterious "slimes." Na_2CO_3, Na silicates of varying $Na_2O : SiO_2$ ratio, sodium sulfide, and NaCN are used for these purposes. The collectors used are alkyl amine salts with pine oil as a frother [Rey *et al.* (1954)] and/or 8-hydroxyquinoline (oxine). A number of industrial concentration procedures for oxidized lead and zinc ores are described in AIME (1970). A review of different treatment procedures available for oxidized copper–lead–zinc ores was presented by Abramov *et al.* (1969). Rinelli and Marabini (1973) investigated the use of several chelating agents for oxide–sulfide ores of lead and zinc.

The oxides of Cu (cuprite, Cu_2O, and tenorite, CuO) are not readily sulfidized and floated with xanthates in alkaline solutions. However, no attempts appear to have been made to use acidic solutions after sulfidization in ores with siliceous gangue minerals. Qualitative laboratory tests [Leja (1956)] indicated that such conditions may be effective for flotation of these oxides using dixanthogen. De Cuyper (1977) has reviewed flotation of copper oxide ores.

Metals such as aluminum, tin, manganese, and iron are extracted from their oxide minerals. These minerals are at present concentrated mostly by various nonflotation techniques (gravity, magnetic, electrostatic). When ores containing coarsely liberated oxides of these metals become scarce, the industry begins gradually to adopt flotation as a separation technique for much finer materials.

Gravity methods have been used since the 1930s to recover chromite, $FeCr_2O_4$, from ores in which it is associated with olivine, $(Fe,Mg)_2SiO_4$, and serpentine, $Mg_3Si_2O_5(OH)_4$. A comprehensive review of chromite flotation was published by Sobieraj and Laskowski (1973). Dogan (1973) dealt with the concentration of Turkish chromium ore, and Lukkarinen and Heikkila (1974) with that of a Finnish ore.

Cassiterite, SnO_2, has been the object of very intensive flotation studies for the last two to three decades. Adsorption studies of various surfactants on cassiterite and stannite and on the effects of these surfactants on floatability were carried out in a number of laboratories in the United States, the USSR, Australia, the United Kingdom, and Germany. Varying degrees of success were achieved for different types of collectors depending on the grade of the ores, the size of the cassiterite grains, and their mineral environment. The most numerous were tests with alkyl carboxylates [Taggart and Arbiter (1943), Gaudin *et al.* (1946), Schuhmann and Prakash (1950)], alkyl sulfates [Hergt *et al.* (1946), Edwards and Ewers (1951), Jaycock *et al.* (1964), Mitrofanov and Rozin (1955), Evans *et al.* (1962), Bruce and Yacksic (1967), *p*-tolyl arsonic acid [Gründer (1955, 1960), Töpfer (1960, 1967)], alkyl phosphonic acids [Wottgen (1969), Collins *et al.* (1968)], salicylaldehyde [Rinelli *et al.* (1976)] and sulfidization and lead activation followed by xanthate flotation [Shorsher (1946), Pryor and Wrobel (1951), Wrobel (1971)]. Trahar (1965, 1970) and Blazy *et al.* (1969) reviewed the status of the laboratory studies, while Bogdanov *et al.* (1974), Moncrieff *et al.* (1974), and Pol'kin *et al.* (1974) described also the subsequent pilot plant testing and the industrial scale operations in the USSR and the United Kingdom developed for the treatment of cassiterite slimes by flotation.

Magnetic oxides of iron are concentrated most economically by magnetic techniques. Nonmagnetic ores and some magnetic fines are upgraded by flotation, either through the removal of associated quartz and silicates or by flotation of specular hematite and taconites. (For a review of recent status in the U.S. iron ore concentration industry, see *Mining Engineering New York*, November 1975 issue.) Carboxylic acids and petroleum sulfonates are the main collectors employed, with an occasional use of amine salts as frother-acting surfactants in acid circuits. Cationic reagents are employed in alkaline circuits for the removal of silicates.

Following their concentration, the iron oxide fines are subjected to pelletizing and sintering in order to produce a high-strength but highly porous sintered product which will be reduced at high rates in blast furnace smelting.

With the opening of enormous deposits containing high-grade iron ores in western Australia and in Brazil, interest in flotation of the Mesabi Range taconites (Minnesota) has abated somewhat. This may be an explanation for the lack of a monograph on iron ore flotation in English;

there is one on this subject available in Russian, by Glembotsky and Bechtle (1964).

1.7.2. *Separations of the "Nonmetallic" Industrial Minerals*

Fluorspar (CaF_2), scheelite ($CaWO_4$), barytes (or barite) ($BaSO_4$), and phosphates [$Ca_5(PO_4)_3(F,CO_3)_1$] are the main "nonmetallic" minerals successfully concentrated on an industrial scale by flotation. The name "nonmetallic" is applied to these insoluble mineral species because they are concentrated for use as chemicals rather than for extraction of their metallic components in smelting operations. Thus, barytes are used as an additive to oil-drilling mud, phosphates as fertilizer and for production of phosphoric acid, and fluorspar as a flux in steel making or for production of hydrofluoric acid.

The principles of nonmetallic minerals flotation have been reviewed by Aplan and Fuerstenau (1962) and, more recently, by Hanna and Somasundaran (1976). Clement *et al.* (1973) and Carta *et al.* (1973) discussed flotation of fluorite and barite. Flotation of oxides and silicates has been reviewed by Smith and Akhtar (1976) and Fuerstenau and Palmer (1976).

Fluorspar (or fluorite) and barite are floated with fatty-acid-type collectors (oleic or linoleic acids) after a careful dosage of depressants: Na_2CO_3, sodium silicate, tannic acid, or quebracho (used as a depressant for calcite). Usually, elevated temperatures, 75°–85°C (employing steam injection to obtain such temperatures in the pulp), are used in fluorspar flotation in order to bring about the desired selectivity of separation from silica and calcite and to achieve a satisfactory recovery. The grade of the fluorspar concentrate to be produced in the separation process depends on the use for which the concentration product is destined. If the concentrate is to be used for the production of hydrofluoric acid, the requirement is that it contain at least 95% CaF_2 and no more than 1.1% SiO_2 and 1.25% $CaCO_3$. Such a concentrate represents the highest-quality product, the so-called acid-grade fluorspar. If the subsequent use of fluorspar is that of a flux in the manufacture of steel, the requirements are far less stringent. The role of fluorspar in smelting and refining of metals is discussed by See (1976) and its industrial significance by Gössling and McCulloch (1974). When the flotation pulp is maintained at temperatures near the boiling point of water, the selective separation of CaF_2 from silica and calcite is greatly improved. Hukki (1973) discussed improvements in selectivity using hot flotation processes. Martinez *et al.* (1975) applied new techniques to develop a flotation process for barite.

Scheelite is most often floated using fatty acids as collectors with numerous additives: Na_2CO_3 for controlling pH and various depressants or dispersing agents to prevent deleterious action of slimes. Depending on the composition of ores and the liberation size of the valuable minerals, gravity separation techniques may be employed ahead of flotation. Frequently, a suitable chemical treatment (e.g., acid leaching) is necessary for the removal of some minor quantities of contaminants such as phosphates and carbonates from the final scheelite flotation concentrate in order to meet the exacting specifications for the product (e.g., minimum 65% WO_3 with 0.05% P and 0.3% S maximum). Bogdanov *et al.* (1974) described the results of laboratory and plant tests using pelargonic hydroxamic acid reagent, IM-50, in the flotation of wolframite, cassiterite, and pyrochlore. Vazquez *et al.* (1976) dealt with a selective flotation process developed for a $\sim 0.5\%$ WO_3 scheelite ore in which lime additions were found necessary for selectivity.

The main problem to be overcome in the flotation of phosphates is due to the necessity of an effective removal of acid-consuming components such as calcite, $CaCO_3$, and dolomite, $(Ca,Mg)CO_3$, which are also readily floated by the fatty acid collectors used. Besides lowering the grade of the concentrate, these carbonate minerals create difficulties during acid digestion in the manufacture of phosphoric acid. If gypsum, $CaSO_4 \cdot 2H_2O$, is present as well in the ore, its relatively high solubility causes excessive consumption of fatty acids due to calcium soap precipitation. Rougher concentrates produced are cleaned using orthophosphate or fluorosilicate ions to depress phosphate minerals while removing carbonates by flotation. For the production of phosphoric acid, a minimum grade of $\sim 35\%$ P_2O_5 is required. A comprehensive study of different aspects encountered in the treatment of Moroccan phosphate ores has been presented by Smani *et al.* (1975). Golovanov *et al.* (1968) described the treatment of apatite ores in the USSR, while Lamont *et al.* (1972), Davenport *et al.* (1969), and Sun *et al.* (1957) dealt with flotation characteristics of U.S. apatite ores. Difficulties in producing satisfactory froths during flotation of apatite from Transvaal deposits, South Africa, are discussed by Lovell (1976).

1.8. *Separation of "Soluble Salts" from Saturated Brines*

The components of potash deposits usually comprise a complex mixture of halides—halite, NaCl; sylvite, KCl; carnallite, $KCl \cdot MgCl_2 \cdot 6H_2O$—

and/or sulfates—langbeinite, $2MgSO_4 \cdot K_2SO_4$; kainite, $MgSO_4 \cdot KCl \cdot 3H_2O$; etc. Previously these salts were recovered from the mixtures comprising natural deposits by a lengthy fractional crystallization (Germany, New Mexico, etc.).

The most important use of the potassium salts is as fertilizer (potash). The fertilizer quality of the various potassium salts is expressed in terms of an equivalent K_2O percentage: Sylvite (KCl) contains 63.2% K_2O equivalent; langbeineite ($K_2SO_4 \cdot 2MgSO_4$), 22.7% K_2O equivalent; kainite ($MgSO_4 \cdot KCl \cdot 3H_2O$), 18.9% K_2O equivalent; carnallite ($KCl \cdot MgCl_2 \cdot 6H_2O$), 16.9% K_2O equivalent. In addition to potassium salts, the potash deposits contain sodium salts, mainly halite (NaCl) (with which KCl forms a physical mixture known as sylvinite); hydrated sodium sulfate, known as Glauber salt, $Na_2SO_4 \cdot 10H_2O$; complex alums [i.e., potash alum, $KAl(SO_4)_2 \cdot 12H_2O$; soda alum, $NaAl(SO_4)_2 \cdot 12H_2O$; and ammonia alum, $(NH_4)Al(SO_4)_2 \cdot 12H_2O$] and $MgSO_4 \cdot 7H_2O$, known as epsomite or Epsom salts. The grades of the various deposits run $\sim18\%$ K_2O equiv. in New Mexico, 10–20% K_2O in Germany, and 25–35% K_2O in Saskatchewan (Canada), and in the USSR.

The concentration of the potash minerals by selective flotation in a saturated brine solution is preceded by the removal of clay, using hydrocyclones, followed by thickeners and/or inclined hydroseparators. Clays cause unduly high consumptions of all flotation reagents, increase the viscosity of saturated solutions, and—if desliming is attempted to remove them—create substantial losses of KCl. With ores containing sylvinite and carnallite, the separation of clay is followed by selective flotation of either sylvite (using n-alkyl amines or sulfonates as collectors) or halite (using n-alkyl carboxylates of heavy metal ions and a nonpolar oil addition). The concentrate grade aimed for is 60–63% K_2O equiv. With ores containing other salts as well as sylvite and halite, the selective flotation is much more complicated, and the grade of concentrates produced is lower. The sulfate salts of Mg and K can also be separated from sodium sulfates, though less efficiently than chlorides.

The technological details as regards (1) the extent of grinding necessary for a suitable degree of liberation, (2) the screening into various size fractions (fines are used for making the saturated brine solution, while coarser fractions undergo pretreatment with reagents before flotation), and (3) the final selection of activating and collector-acting reagents depend on the nature of salts and gangue minerals comprising the particular deposit and vary from one operation to another.

Saturated brine solutions represent ~7–$10\ M$ concentrations of electrolytes (K^+, Na^+, Cl^-, SO_4^{2-}, etc.). Flotation in saturated brine solutions is thus carried out in an environment of ionic strength an order or so higher than that when flotation in seawater is conducted ($\sim 0.7\ M$ electrolyte concentrations). Most other flotation systems represent concentrations of 10^{-3}–$10^{-1}\ M$ electrolytes.

1.9. Flotation of Silicate Minerals

Many of the so-called "space age" metals, such as berylium, lithium, titanium, zirconium, and niobium, exist as (or are associated with) complex silicates in relatively low-grade deposits. Both anionic and cationic collectors are employed in the flotation of these silicate minerals, with fluoride ion additions used as the most active modifying agent. The role of fluoride in the flotation of feldspar was investigated by Warren and Kitchener (1972), and the effect of alkyl chain length and its structure on the flotation characteristics of quartz was discussed by Smith (1973) and Bleier *et al.* (1976). The details of the fluoride ionic equilibria, or of the electrical characteristics such as streaming potential and points of zero charge in relation to the flotation of individual silicate minerals, are given in two publications from the Warren Spring Laboratory, England: Read and Manser (1975) and Manser (1975). These two publications represent the most comprehensive compilation of information published to date on silicate flotation, together with a detailed list of references used in evaluating the data. A list giving the composition of flotation reagents used in the USSR is also included. The minerals dealt with in the book by Manser comprise orthosilicates (andalusite, axinite, beryl, kyanite, olivine, topaz, tourmaline, zircon), pyroxenes (augite, diopside, spodumene, wollastonite), amphiboles (actinolite, hornblende, tremolite), sheet silicates (chlorite, micas—biotite, lepidolite, muscovite—talc, serpentine), and framework silicates (felspars, nepheline, quartz). Anionic flotation of silicates (and oxides) is also extensively discussed by Fuerstenau and Palmer (1976), while cationic flotation is dealt with by Smith and Akhtar (1976). Zeta potential studies on many silicate minerals are presented in the monograph by Ney (1973).

Silicates are separated by selective flotation, first, for the purpose of concentrating a given component, e.g., quartz for glass manufacture, felspar for ceramic use, and spodumene for production of Li. An alternative objective of selective silicate flotation may be to remove the undesirable contaminant, such as, for example, sodium and potassium felspars from quartz intended for Portland cement, or to remove mica, etc. Usually, a flowsheet of the type shown in Figure 1.8 is used. The success of selective separation depends entirely on the use of modifying agents, control of pH, and the choice of collectors of suitable structure.

1.10. *Separation of Naturally Hydrophobic Minerals*

A number of solids exhibit varying degrees of hydrophobic character-istics when their surfaces are freshly formed. The solids may be either organic, such as, for example, hydrocarbons, waxes, coals, graphite, tars, bitumen, and various synthetic plastics, or inorganic, such as sulfur, talc, and molybdenite. As long as the surfaces of these solids remain uncon-taminated (by oxidation or nitration reactions or by embedded hydrophilic colloidal matter), the fine particles of such solids can be easily separated from the associated hydrophilic particles. For such separations only the use of frother-acting surfactant is required. However, the hydrophobic character of these solids invariably decreases on lengthy exposure to oxi-dizing environment, whether liquid or gaseous. Hence, additions of nonpolar oils (which invariably contain traces of surfactants!) or of a collector-acting surfactant, supplementary to frother additions, may be required to achieve separation of such solids with reduced hydrophobicity. Heavily oxidized coals, bitumen, molybdenite, and sulfur are no longer naturally hydrophobic and may require elaborate additives to effect their flotation. Some of these additives may be performing a cleansing action by dissolving the hydrophilic coating and removing it from the surface. Ammonium hydrosulfide may be acting in this manner with the oxidized molybdenite. Others may be needed to re-establish a new hydrophobic surface or to improve the highly reduced hydrophobicity by adsorption on the hydrophilic sites.

As already mentioned in Section 1.5, the presence of naturally hydro-phobic minerals within an ore body may create difficulties in selectivity of separations caused by their softness. In grinding, the soft mineral will tend to smear on other gangue minerals, making them partly hydrophobic. In turn, fine particles of gangue minerals will tend to become embedded in hydrophobic particles, making them less hydrophobic.

Extensive investigations on the action and mechanism of coflotation of hydrophilic slimes with native sulfur have been carried out by Wak-smundzki *et al.* (1971, 1972). The effects of slime coatings in galena flotation were discussed by Gaudin *et al.* (1960). The characteristics of iron oxide slime coatings in flotation of quartz and corundum were evaluated by Fuerstenau *et al.* (1958) and by Iwasaki *et al.* (1962a, b). The influence of inorganic electrolytes on the floatability of naturally hydrophobic minerals (different quality coals, sulfur, talc, and graphite) was examined theoretically and experimentally by Laskowski (1966). He explained the improved flotation (observed in higher electrolyte concentration when slight additions of nonpolar oils are made) by the precipitation of microbubbles on hydrophobic surfaces. Such precipitation is facilitated by the decreased thickness of the electrical double layer, resulting from increased electrolyte concentrations.

Coal is one of the naturally hydrophobic minerals that is now concentrated on a very large scale. It is estimated that from 20 to 40% of the total coal output (which was 2.5×10^9 tonnes in 1973) is undergoing some concentration process, usually consisting of washing in heavy media or jigging in order to remove shale; only a relatively small portion of the total coal, namely fines smaller than ~ 3 mm, is subjected to concentration by flotation. The main difficulties associated with the treatment of coal fines are the following:

1. About one-third of the recovered coal may have to be expended in driving off the moisture from the flotation concentrate. It thus becomes uneconomical to carry out an operation that yields, effectively, slightly more than 50% of the coal content in the feed (even if it is a high-rank, superficially altered but still hydrophobic coal).

2. For heavily oxidized coal fines the recoveries by flotation drop off drastically, and new, more effective collector-type reagents are needed to get satisfactory results; such reagents are yet to be developed and tested.

General characteristics of coal are dealt with in Gould (1966); *Coal Preparation*, Leonard and Mitchell (1968), contains a few pages devoted to coal flotation [Zimmerman (1968)]. The floatability of coarse coal and the effects of coal oxidation were dealt with by Sun and Zimmerman (1950), Sun (1952, 1954a, b), and Whelan (1953). General reviews of coal flotation were presented by Brown (1962), Zimmerman (1968), and, most recently and very comprehensively, Aplan (1976). In Russian, there is a textbook devoted specifically to coal flotation, viz., Klassen (1963). Spherical agglomeration promises to be of great industrial importance in the recovery of coal fines. Spherical agglomeration has been developed by Puddington and his coworkers at the National Research Council of Canada [Smith and Puddington (1960), Farnand *et al.* (1961), Meadus *et al.* (1966), Mular and Puddington (1968), and Capes *et al.* (1974)]. It exploits the characteristics of the solid/liquid interfaces in a manner similar to that used in flotation, differentiating between the hydrophobic and the hydrophilic solids, but utilizes oil, instead of air, as the water-displacing phase. The greater hydrophobic character of the oil phase (than that of the air phase) allows sharper and more delicate separations to be achieved by spherical agglomeration than those possible in flotation. The need to use 10–15% of oil (by weight) in the formation of agglomerates represents, to a degree, a drawback of this technique. The high oil content of agglomerates necessitates a subsequent retreatment for oil recovery. Only when the oil represents a needed additive that improves the quality of the agglomerated coal product (as in separating low-rank coal fines away from the ash-producing inorganic sulfides, oxides, and silicates) may the retreatment for oil recovery be superfluous.

1.11. *Recent Developments in Industrial Flotation Operations and Research Techniques*

Taking the last two to three decades as the span of "recent" developments, the following remarks could be made about changes that occurred during that period:

1. As a result of a general decrease in grade of ores treated and a higher market demand, the tonnages treated by individual flotation plants have increased from the range of 500–2000 tonnes daily to 20,000–100,000 tonnes daily (in tar sands flotation, even 350,000 tonnes daily).

2. The separation difficulties (due to intricate interlocking of minerals and their variable association and dissemination characteristics) have increased greatly, yet they are being gradually resolved.

3. The additions of reagents into flotation circuits have been lowered by approximately an order of magnitude in comparison with the pre-World War II additions. A large number of new reagents have been brought into use.

4. The sizes of all equipment used in flotation plants (and in mining) have increased severalfold in order to cope with the demands of throughput. Thus, for example, autogenous grinding mills are often 10–11 m in diameter (32–36 ft), as compared to the 3–5 m (10–16 ft) of the prewar grinding units. Flotation cells have become very large indeed, increasing from the $\sim 1\frac{1}{2}$ m^3 (~ 50 ft^3) cell volume (which 30 years ago was considered to be the maximum) to the 15–21 m^3 (500–700 ft^3) cells used nowadays.

5. Some pieces of equipment, such as Dorr rake classifiers and spiral classifiers, used in size separation, have been nearly completely replaced by cyclones. Similarly, inclined lamella settling tanks are gradually replacing the voluminous rake thickeners used in all treatment plants hitherto.

6. Numerous instruments and devices for quick, on-stream determinations of pulp flow, size analysis, chemical analysis, mineralogical evaluations, etc., have been introduced. Many of these instruments are interphased with computers, making possible a rigid control of individual operations.

7. Modeling, optimization, and automation of specific unit operations are being developed and introduced into industrial operations on an ever-increasing scale.

Parallel with the preceding changes, there has been a continued adjustment of treatment approaches to suit the requirements of individual systems. To reduce grinding costs and to facilitate difficult separations, mixtures of sulfides in complex ores are treated more frequently by procedures interposing flotation and chemical treatments in alternating sequences.

In an attempt to improve floatability of coarse particles, a so-called "froth separation" technique was introduced by Malinowskii *et al.* (1974) (Figure 1.2). Considerable improvements in separations of coarse sizes in KCl and coal flotation and a decrease in the cost of treatment are claimed for this technique.

The role of flotation is gradually changing; in most cases it is still the unit process employed as the main technique to concentrate components of the naturally occurring mixtures of minerals. However, more and more frequently, flotation is called upon to play a role as an auxiliary step in a highly integrated sequence of chemical and metallurgical operations to separate synthetic solids produced in one of the preceding steps of operations. Examples of such applications are Anaconda's modified process for Cu, Ag, and Au ores, known as Arbiter's process, or Inco's matte flotation.

Anaconda's Arbiter process [Kuhn *et al.* (1974)] offers a number of options depending on the mineral composition of the feed, the economics of by-product recovery, and the cost of energy. Mixed concentrates of copper, zinc, nickel, and cobalt sulfides, containing traces of Au and Ag, are treated by low-pressure ammonia–oxygen leach, at 65°–80°C, to dissolve most of the copper sulfides as $CuNH_4SO_4$. The leach liquor is then subjected to a solvent ion exchange and stripped to provide copper sulfate for electrowinning of copper or for precipitation by SO_2 gas. The ammonia leach residue is washed and treated by flotation to recover the rest of the (undissolved) copper, precious metals, and other sulfides. The by-product metals, sulfur, and ammonia are then recovered by a variety of alternative treatments.

Inco's process of matte flotation is incorporated in a sequence of pyrometallurgical processes [Sproule *et al.* (1961), Tipman *et al.* (1976)]. A Cu–Ni matte produced during smelting of Cu–Ni–Fe concentrates is subjected to a very slow cooling process during which separate crystals of copper sulfide and nickel sulfide are produced. The cold, recrystallized matte is crushed and ground and subjected to flotation in order to separate the two sulfide products.

A number of new reagents have been introduced, such as polyoxyethylene and polyoxypropylene alcohols, which are used as frothers (e.g., Dowfroth 200 and 250). Also, alkyl thionocarbamate esters (Reagent Z-200) are used for flotation of Cu sulfides, and alkyl thiocyclanes and thiobicyclanes (recovered from high-sulfur crude oils) are beginning to be used in flotation of Cu–Ni ores. Hydroxamic acids, alkyl arsonic acids, and alkyl phosphonic acids were introduced for flotation of cassiterite, wolframite, and pyrochlore. 8-Hydroxyquinoline (oxine), together with nonpolar oils, is employed in flotation of Italian calamines. Numerous multipolar surfactants, in particular "amphoteric" ones, with anionic and cationic polar groups such as aminocarboxylates or aminosulfonates, are being tested as prospective flotation reagents. High-molecular-weight polymers, used as depressants and flocculating agents, comprise the major portion of surfactant production. Flocculating agents in particular are widely used in all antipollution treatments of effluents and in water reclamation.

The role of surfaces in accomplishing the act of flotation was recognized from the very beginning of flotation practice [Hildebrand (1916), Smith (1916), Coghill (1916), Taggart and Beach (1916)]. However, the realization that flotation is but one of numerous applications of surface chemistry has occurred only within the last three decades. With this realization, new surface chemical research tools and techniques have been gradually brought into flotation investigations. The more successful of such techniques are the electrokinetic studies on zeta potentials in different environments and the surface balance (Langmuir trough) studies on solidification of surfactant monolayers under the influence of organic and inorganic additives. The interactions between two types of surfactants used in flotation and the role played by oxygen in the flotation of sulfide minerals have been recognized and are being documented. The electrochemical nature of reactions between some surfactants and solid surfaces of semiconductors have been identified. The most relevant have become the (initially) abstract studies on the unique character of thin liquid layers adhering to solids or existing as free liquid films. The rupturing of such films occurs in the process of establishing a particle–bubble attachment during flotation, in aggregate formation during coagulation, or in breaking emulsions and froths. It has become recognized as a common phenomenon in numerous applications of surface chemistry.

Electron diffraction and electron microscopy were applied to study the oxidation of mineral surfaces and, subsequently, to study the nature of the adsorbed layers of collectors. UV spectroscopy was utilized to follow the reactions of some surfactants in solutions, and infrared spectroscopy was used to study the structure of collector species before and after adsorption onto mineral surfaces. Other spectroscopic techniques such as low-energy electron diffraction (LEED), Auger electron spectroscopy (AES), electron paramagnetic resonance (EPR), Mössbauer spectroscopy, electron spectroscopy for chemical analysis (ESCA), etc., are being applied to flotation systems more and more frequently.[†]

Parallel with the studies of the chemistry of flotation, the kinetics of the process has been investigated. A concise account of the results published up to the early 1950s is given in Gaudin (1957, Chapter 12, pp. 369–392). Since then, papers by Kelsall (1961), Bushell (1962), Arbiter and Harris (1962), Imaizumi and Inoue (1965), Loveday (1966), and Haynman (1975) have dealt with the various aspects of flotation kinetics.

The availability of computers stimulated not only an extension of such kinetic studies but also a development of mathematical models and automatic process control of integrated grinding and flotation circuits. Empirical modeling and optimization methods suitable for mineral processing were reviewed by Mular (1972). A comprehensive set of course notes on this

---

[†] The conditions under which the individual physical and chemical techniques are applied to surface studies are restricted and specific to each technique. These conditions do not coincide with the conditions under which flotation occurs. Therefore, an uncritical transfer of results obtained with individual techniques to flotation systems may lead to misinterpretation. Occasionally, the limitations imposed by parameters operative in a dynamic flotation system nullify the inferences drawn from the use of static techniques. The resulting misinterpretations may lead to unwarranted generalizations.

Similar remarks apply to unrestricted extrapolations of results from simplified systems (such as Hallimond tube flotation experiments, or zeta potential determinations, on individual pure minerals) to highly complex mixtures represented by ores and industrial intermediate products. A consequent failure to fulfill the tacit expectations in replication of simplified systems creates confusion and distrust in all fundamental studies.

subject has been published by Mular and Bull (1969, 1970, 1971). Davis (1964) presented a model for a lead flotation circuit, Woodburn (1970) discussed modeling of any flotation process, and a two-phase distributed-parameter model of the flotation process was presented by Ball and Fuerstenau (1970).

A symposium organized in 1972 in Johannesburg, South Africa, dealt specifically with the applications of computer methods in the mineral industry [proceedings edited by Salamon and Lancaster (1973)]. Papers on modeling of flotation circuits were presented by King (1973), Loveday and Marchant (1973), and Amsden *et al.* (1973).

Instrumentation used in the processing of minerals is described in publication edited by Whiteside (1974). The impact of the monitoring instrumentation on the operation of the flotation process has been far more spectacular than the understanding of the chemistry involved. The emphasis has shifted toward modeling and computer control in a modern flotation plant. The grinding circuits are using computer-based control systems and effect a 5–10% increase in grinding capacity. Flotation plants are feeding reagents using a computer control system, for example, at Ecstall Mining Limited, Timmins, Ontario [Amsden (1973)] and at Outokumpu, Finland [Tanila *et al.* (1973)]. The result is an increase in metal recovery by up to 2% and a reduction in reagent costs.

Plant design aspects of mineral processing industry are dealt with in the publication edited by Mular and Bhappu (1978).

1.12. *Flotation Literature*

The developments in applications of flotation to industrial separations and in modern research techniques have led to a spectacular increase in flotation literature. It is impossible to do justice to all the numerous publications that have appeared in the last two decades or so concerning the variety of research topics and practical developments in flotation. These have to be traced individually from year to year. Some journals publish yearly reviews of progress in fundamentals of flotation or of mineral processing. For example, *World Mining*, in its Catalogue and Directory number (issued every June), presents a thorough review of developments in mineral processing for the preceding year (authored during the last few years by M. Doleźil and J. Reznicék). This review deals with topics from comminution, gravity and magnetic separation, flotation, automation, and chemical treatment to smelting and direct reduction. Similar reviews are presented in the February issues of *Mining Engineering (AIME)*, and *Canadian Mining Journal*, each topic reviewed by different authors.

Dow Chemical Company publishes a yearly issue of *Flotation Index* (the 46th edition appeared for 1975), which lists titles of publications on research and industrial accomplishments for all applications of flotation [in alphabetical order for each metal (from Al to Zn) whose ores are treated by flotation]. The index is derived from the Chemical Abstracts Service as well as the following publications:

AIME Transactions, Society of Mining Engineers of AIME, New York
The Canadian Mining and Metallurgical Bulletin, Montreal
Canadian Mining Journal: *Canadian Mining Manual*, Quebec
Deco Trefoil, Denver Equipment Division, Joy Mfg. Co., Denver
Engineering and Mining Journal, New York
The Institution of Mining and Metallurgy Transactions, London
Mining Congress Journal, Washington, D.C.
Mining Engineering, New York
Mining and Minerals Engineering, London
Pit and Quarry, Chicago
U.S. Bureau of Mines Reports of Investigations, Pittsburgh
World Mining, San Francisco
Erzmetall, Gesellschaft Deutscher Metallhuetten-Und Bergleute, Clausthal-
    Zellerfeld, Germany

A bimonthly survey of world literature, covering subjects from geology, mining, and mineral processing to extraction metallurgy, is published by the Institution of Mining and Metallurgy, London, as the *IMM Abstracts* (Vol. 31 for 1981).

A monthly abstracting service is available in Russian literature under *Referativnyj Zhurnal* (Mining, Mineral Processing). The topics covered consist of descriptions of plants, gravity separations, flotation, electrostatic and magnetic separations, control of processing, and agglomerations.

All these publications provide invaluable help in tracing available information relevant to a particular theoretical and practical development in the field of flotation. The large numbers of items listed each year, some 800 to 1300, testify to the difficulty of keeping abreast with all aspects of the field.

Other sources of information are the proceedings of international mineral processing congresses (IMPCs) and those on surface activity. There are also numerous national conferences and special symposia organized to discuss selected topics. As of March 1981, thirteen proceedings of international mining processing congresses have been published since 1952, and ten commonwealth mining and metallurgical congresses were held (resulting in valuable publications concerned with technological developments in individual countries wherever the meetings were held). Also, seven proceedings of international congresses of surface activity appeared, invariably containing publications related to applications of surfactants to flotation. Special semi-international conferences and symposia are held nearly every year in the United States or the Soviet Union, in one of the Mid-European countries, or in South America, Australia, India, or Japan.

The most recent publications resulting from such symposia and dealing with both theory and practice of flotation are the two volumes of *Flotation, A. M. Gaudin Memorial Volume*, edited by M. C. Fuerstenau (1976), *Advances in Interfacial Phenomena of Particulate/Solution/Gas Systems, Application to Flotation Research*, edited by Somasundaran and Grieves (1975) and *Fine Particles Processing*, vols. I and II, edited by Somasundaran (1980). These three publications were preceded by *Froth-Flotation, 50th Anniversary Volume*, edited by D. W. Fuerstenau (1962), and *Fiftieth Anniversary of Froth Flotation in the USA*, published in the *Colorado School of Mines Quarterly* **56**, (1961), pp. 1–239.

Publications resulting from symposia on more restricted topics are, for example, *Mining and Concentrating of Lead and Zinc*, edited by D. O. Rausch and B. C. Mariacher and published by AIME (1970), and *Broken Hill Mines*, published by the Australasian Institute of Mining and Metallurgy, Melbourne (1968). Less widely publicized national symposia frequently produce publications of more restricted circulation, such as the proceedings of annual meetings of the Canadian Mineral Processors (the thirteenth was held in 1981) or special issues of *Journal of Mines, Metals and Fuels* (Calcutta) (for example, the 1960 issue No. 7 devoted to mineral beneficiation) or of *Indian Mining Journal* (1957 Special Issue devoted to mineral beneficiation and extractive metallurgical techniques).

In addition to journals dealing specifically with the topics of mineral processing and flotation, such as, for example, *International Journal of Mineral Processing, Transactions of the AIME, Transactions of the IMM*, and *Minerals Science and Engineering*, there are numerous publications of surface chemistry which contain, from time to time, papers on topics relevant to flotation. Thus, *Journal of Colloid and Interface Science, Separation Science, Kolloidnyj Zhurnal, Colloid and Polymer Science* (formerly *Kolloid Zeitschrift*), *Tenside*, etc., have to be referred to by those concerned with the theoretical developments and research in flotation.

1.13. Selected Readings

1.13.1. Flotation Theory

Aleksandrovich, K. M. (1973), *Principles of the Use of Reagents During Flotation of Potassium Ores* (in Russian), Nauka i Tekhnika, Minsk, Beloruss. SSR.

Aleksandrovich, K. M. (1975), *Flotation of Water-Soluble Salts* (in Russian), Nauka, Moscow.

Aplan, F. F. (1966), Flotation, in *Kirk–Othmer Encyclopedia of Chemical Technology*, Vol. 9, 2nd ed. Wiley-Interscience, New York, pp. 380-398.

Beloglazov, K. F. (1956), *Basic Rules of the Flotation Process* (in Russian), Rep. Leningr. Min. Institute, Metallurgizdat, Moscow.

Belov, V. N., and Sokolov, A. F. (1971), *Extraction and Processing of Potassium Salts* (in Russian), Khimiya, Leningrad.

Blazy, P. (1970), *La Valorisation des Minerais*, Presses Universitaires de France, Paris, Chaps. V–VIII, pp. 233-333.

Bogdanov, O. S. (1959), *Problems in the Theory and Technology of Flotation* (in Russian), Issue No. 124, Mekhanobr, Leningrad.

Bogdanov, O. S. (1965) *Studies of flotation reagents and their action* (in Russian) Issue N. 135, Mekhanobr, Leningrad.

Castro, S., and Alvarez, J., eds. (1977), *Avances en Flotacion* (Advances in Flotation): Vol. 3, Universidad de Concepcion, Chile.

Cooke, S. R. B. (1950), Flotation, in *Advances in Colloid Science*, Vol. 3, H. Mark and G. J. W. Verwey, eds., Wiley-Interscience, New York, pp. 321-374.

Eigeles, M. A. (1950), *Theoretical Basis for the Flotation of Non-Sulfide Minerals* (in Russian), Metallurgizdat, Moscow.

Eigeles, M. A. (1952), *Concentration of Non-Metallic Minerals* (in Russian), Promstroizdat, Moscow.

Felstead, J. E. (1956), Flotation, in *Chemical Engineering Practice*, H. W. Cremer and T. Davies, eds., Academic Press, New York, pp. 209-247.

Fuerstenau, D. W. ed. (1962), *Froth Flotation, 50th Anniversary Volume*, AIME, New York.

Fuerstenau, D. W., and Healy, T. W. (1972), Principles of mineral flotation, in *Adsorptive Bubble Separation Techniques*, R. Lemlich, ed., Academic Press, New York, pp. 91-131.

Fuerstenau, M. C., ed. (1976), *Flotation—A. M. Gaudin Memorial Volume*, Vols. 1 and 2, AIME, New York.

Gaudin, A. M. (1957), *Flotation*, McGraw-Hill, New York.

Glembotsky, V. A. (1972), *Physical Chemistry of Flotation Processes* (in Russian), Nauka, Moscow.

Glembotsky, V. A., and Bechtle, G. A. (1964), *Flotation of Iron Ores* (in Russian), Nedra, Moscow.

Glembotsky, V. A., and Klassen, V. I. (1973), *Flotatsya* (in Russian), Nedra, Moscow.

Glembotsky, V. A., Klassen, V. I. and Plaksin, I. N. (1961), *Flotatsya* [trans. by R. E. Hammond, *Flotation* (1963)], Primary Sources, New York.

Glembotsky, V. A., Dmitrieva, G. N., Sorokin, M. M. (1970), *Flotation Agents and Effects* (trans. from the 1968 Russian ed., Nauka, Moscow), Israel Program for Scientific Translations, Jerusalem.

Glembotsky, V. A., Popov, E. L., and Solozhenkin, P. M. (1972), *Flotation of Alkaline Earths Metal Sulfates and Carbonates* (in Russian), Dushanbe, Tadzh. SSR.

Glembotsky, V. A., Sokolov, E. S., and Solozhenkin, P. M. (1972), *Beneficiation of Bismuth-Containing Ores* (in Russian), Dushanbe, Tadzh. SSR.

Havre, H. (1952), *Preparation Mecanique et Concentration des Minerais par Flotation et Sur Liquefeurs Denses*, C. Béranger, Paris.

Joy, A. S., and Robinson, A. J. (1964), Flotation, in *Recent Progress in Surface Science*, Vol. 2, J. F. Danielli, K. G. A. Pankhurst, and A. C. Riddiford, eds., Academic Press, New York, pp. 169–260.

Karaseva, T. P., Gershenkop, A. S., and Sychuk, V. F. (1972), *Beneficiation of Vermiculite Ores* (in Russian), Nauka, Leningrad.

Klassen, V. I., ed. (1956), *Flotatsionnye Reagenty i ikh Svoistva* (Flotation Reagents and Their Properties), Izdatel. Akad. Nauk, USSR, Moscow.

Klassen, V. I. (1963), *Flotatsya Ugley* (Flotation of Coals), Gosgortechizdat, Moscow.

Klassen, V. I., and Mokrousov, V. A. (1963), *Introduction to the Theory of Flotation* (trans. from the Russian 1959 ed.), Butterworths, London.

Laskowski, J. (1969), *Physical Chemistry in Mineral Processing Technology* (in Polish), Slask, Katowice, Poland.

Laskowski, J., and Lekki, J., eds. (1976), *Physicochemical Problems of Mineral Processing No. 10* (No. 1 in 1967) (in Polish), Wroclaw, Poland.

Lemlich, R., ed. (1972), *Adsorptive Bubble Separation Techniques*, Academic Press, New York.

Manser, R. M. (1975), *Handbook of Silicate Flotation*, Warren Spring Laboratory, Stevenage, Hertfordshire, England.

McGlashan, D. W. (1961), 50th Anniversary of froth flotation in the USA, *Colo. Sch. Mines Q.* **56** (3), p. 1–12.

Mitrofanov, S. I. (1967), *Selectivnaya Flotatsya* (Selective Flotation), Nedra, Moscow.

Mitrofanov, S. I., ed. (1970), *Combination Treatment Methods for Oxidized and Mixed Copper Ores* (in Russian), Nedra, Moscow.

Plaksin, I. N., and Myasnikova, G. A. (1974), *Investigations of Flotation Properties of Tungsten Minerals* (in Russian), Nauka, Moscow.

Plaksin, I. N., and Solnyshkin, V. I. (1966), *Infrared Spectroscopy of Surface Layers of Reagents on Minerals* (in Russian), Nauka, Moscow.

Pryor, E. J. (1965), *Mineral Processing*, Elsevier, Amsterdam, pp. 457–570.

Rao, S. R. (1971), *Xanthates and Related Compounds*, Marcel Dekker, New York.

Rao, S. R., ed. (1972), *Surface Phenomena*, Hutchinson, London.

Razumov, K. A. (1975), *Flotatsjonnyj Metod Obogashchenia* (Flotation Concentration Method), Publ. L.G.I., Leningrad.

Read, A. D., and Manser, R. M. (1975), *The Action of Fluoride as a Modifying Agent in Silicate Flotation*, Warren Spring Laboratory, Stevenage, Hertfordshire, England.

Somasundaran, P., and Grieves, R. B., eds. (1975), *Advances in Interfacial Phenomena of Particulate/Solution/Gas Systems, Application to Flotation Research*, AIChE Symposium Series No. 150, Vol. 71, American Institute of Chemical Engineers, New York.

Somasundaran, P., ed. (1980), *Fine Particles Processing*, AIME, New York.

Sutherland, K. L., and Wark, I. W. (1955), *Principles of Flotation*, Australasian Institute of Mining and Metallurgy, Melbourne.

Taggart, A. F. (1951), *Elements of Ore Dressing*, John Wiley, New York, Chap. 13–18, pp. 234–333.

Troitsky, A. V. (1956), *Die Flotation*: *Aufbereitung der Erze im Nassverfahren* (trans. from the Russian), M. Nijhoff, The Hague, The Netherlands.

Whelan, P. F. (1956), Froth flotation—A half-century review, *Ind. Chem.* **32**, 315–318, 409–411, 489–491.

Zhelnin, A. A. (1973), *Theoretical Principles and Practice of Flotation of Potassium Salts* (in Russian), Khimiya, Leningrad.

1.13.2. Flotation Practice

Arbiter, N., ed. (1964), *Milling Methods in the Americas*, 20 papers, Gordon & Breach, New York. Centennial Celebration of the School of Mines at Columbia University and 7th International Mineral Processing Congress, New York.

CIM—Canadian Institute of Mining and Metallurgy (1957), *The Milling of Canadian Ores* (6th Commonwealth Mining and Metallurgical Congress), Montreal.

CIM (1974), *Mineral Industries in Western Canada* (10th Commonwealth Mining and Metallurgical Congress, Sept. 1974), Montreal.

Denver Equipment Co., *Flowsheet Studies*, Bulletins Nos. M7-F1 to M7-F124, Denver, Colorado.

Fuerstenau, M. C., ed. (1976), *Flotation—A. M. Gaudin Memorial Volume*, Vol. 2, Plant Practice, AIME, New York.

Gründer, W., ed. (1955), *Erzaufbereitungsanlagen Westdeutschland* (III International Mineral Processing Congress, Goslar), Springer, Berlin.

IMM—Institution of Mining and Metallurgy (1960), Session IV (5 papers), in *Proceedings V International Mineral Processing Congress*, London.

Michell, F. B. (1950), *The Practice of Mineral Dressing*, Electrical Press, London.

Mular, A. L., and Bhappu, R. B. (1978), *Mineral Processing, Plant Design and Practice*, AIME, New York.

Mular, A. L., and Bull, W. R. (1969, 1970, 1971), *Mineral Processes, Their Analysis, Optimisation and Control*, Canadian Institute of Mining and Metallurgy, Montreal.

OEEC—Organisation for European Economic Co-operation (1953), *Non-Ferrous Ore Dressing in the USA*, Report No. 54, Paris.

OEEC (1955), *The Mining and Dressing of Low-Grade Ores in Europe*, Report No. 127, Paris.

Pickett, D. E. (1978), *Milling Practice in Canada*, Canadian Institute of Mining and Metallurgy, Montreal.

Rabone, P. (1957), *Flotation Plant Practice*, 4th ed., Mining Publications, London.

Radmanovich, M., and Woodcock, J. T. (1968), *Broken Hill Mines—1968* (9 papers on concentration), Australasian Institute of Mining and Metallurgy, Melbourne.

Rausch, D. O., and Mariacher, B. C. (1970), *AIME World Symposium on Mining and Metallurgy of Lead and Zinc*, Vol. 1 (20 papers on flotation), AIME, New York.

Rothelius, E., ed. (1957), *Swedish Mineral Dressing Mills*, 15 papers on iron ores, 9 papers on other ores (IV International Mineral Dressing Congress, Stockholm, 1957).

Salamon, M. D. G., and Lancaster, G. R. (1973), *Applications of Computer Methods in the Mineral Industry*, South African Institute of Mining and Metallurgy, Johannesburg.

Thomas, R., ed. (1977), *E/MJ Operating Handbook of Mineral Processing*, McGraw-Hill, New York.

1.13.3. *International Conferences*

The International Mineral Processing Congresses (IMPCs)

1. England, London, 1952
   Recent Developments in Mineral Dressing
   The Institution of Mining and Metallurgy (1953)—10 papers on the theory and practice of flotation

2. France, Paris, 1953
   Section 12 Mines et Minerais—Premier Congres Mondial de la Detergence
   Chambre Syndicale Tramagras—6 papers on flotation reagent application

3. Germany, Goslar, 1955
   "Internationaler Kongress für Erzaufbereitung," *Z. Erzbergbau Metallhuettenwes.*
   VIII (1955)
   Dr. Riederer Verlag, GmbH, Stuttgart—6 papers on flotation theory and practice

4. Sweden, Stockholm, 1957
   Progress in Mineral Dressing
   Edited by "Svenska Gruvföreningen and Jernkontoret"
   Almquist & Wiksell, Uppsala (1958)—12 papers on flotation: theory and applications to sulfide and nonsulfide ores

5. England, London, 1960
   International Mineral Processing Congress
   The Institution of Mining and Metallurgy, London—8 papers on flotation research, 5 on flotation practice

6. France, Cannes, 1963
   Mineral Processing
   Edited by A. Roberts
   Pergamon, London—14 papers on flotation topics

7. United States, New York, 1964
   Seventh International Mineral Processing Congress
   Edited by N. Arbiter
   Gordon & Breach, New York (1965)—5 papers on flotation kinetics, 4 on flotation practice, 7 on chemistry of flotation, 7 on surface chemistry in mineral processing

8. USSR, Leningrad, 1968
   VIII Mezhdunarodnyj Kongress po Obogashcheniu Poleznykh Iskopajemykh
   Mekhanobr, USSR (1969)—17 papers on flotation theory, 27 papers on flotation technology and flotation machines

9. Czechoslovakia, Prague, 1970
   Proceedings, Ninth International Mineral Processing Congress
   Ustav Pro Vyzkum Rud, Praha—13 papers on flotation theory and applications

10. England, London, 1973
    Tenth International Mineral Processing Congress
    Edited by M. J. Jones
    The Institution of Mining and Metallurgy, London (1974)—12 papers on flotation theory and technology, 3 papers on plant design, 3 papers on fine particle processing

11. Italy, Cagliari, 1975
    Eleventh International Mineral Processing Congress
    Edited by M. Carta
    Istituto di Arte Mineraria, Università di Cagliari, Cagliari, Italy (1976)—21 papers
    on flotation topics, additional 8 papers in a special volume
12. Brazil, Sao Paulo, 1977
    Proceedings of the Twelfth International Mineral Processing Congress
    NACIONAL Publicacöes e Publicidade—General session: 15 papers on applied
    surface chemistry, theory, and practice of industrial processes; special Vols. 1
    and 2: 7 and 8 papers on copper; Vols. 1 and 2: 9 and 9 papers on iron; also
    round table discussion: 5 papers on phosphates.
13. Poland, Warszawa, 1979
    Mineral Processing, Thirteenth International Mineral Processing Congress
    Edited by J. Laskowski
    Polish Scientific Publishers, Warszawa and Elsevier, Amsterdam (1981)—2 Volumes
    (Parts A and B)

The International Congresses on Surface Activity

1. France, Paris, 1953
   Premier Congrès Mondial de la Detergence et des Produits Tensio-actifs, Tome 3,
   Section 12 Mines et Minerais
   Chambre Syndicate Tramagras—6 papers on flotation, 14 papers on use of sur-
   factants in mining, etc.
2. England, London, 1957
   Proceedings of the Second International Congress of Surface Activity, Vols. 1–4
   Edited by J. H. Schulman
   Butterworths, London—Vol 3: 17 papers on flotation, 9 papers on adsorption, 9
   papers on contact angle, spreading, etc.
3. West Germany, Köln, 1960
   *Vorträge in Originalfassung der III Internationalen Kongresses für Grenzflächenaktive
   Stoffe*, Vols. 1 and 2
   Verlag der Universitätsdruckerei Mainz, GmbH—papers on synthesis of surfactants,
   their solutions and adsorption, films, electrokinetic and interface potentials,
   wetting, etc.
4. Belgium, Brussels, 1964
   Chemistry, Physics and Application of Surface Active Substances, Vols. 1–3 (Fourth
   Congress)
   Edited by F. Asinger, J. Th. G. Overbeek, and C. Paquot
   Gordon & Breach, New York (1967)—papers on electrical properties of interfaces,
   cohesion–adhesion, solutions of surfactants, adsorption, foams, films, etc.
5. Spain, Barcelona, 1968
   Proceedings Fifth International Congress on Surface Active Substancès, Vols. 1–3
   Ediciones Unidas, S.A., Barcelona—26 papers on surfactants, a number of which
   are related to flotation
6. Switzerland, Zürich, 1972
   Chemie, Physikalische Chemie und Anwendungstechnik der Grenzflächenaktiven Stoffe,
   Vols. 1–3 (Sixth International Congress)

Carl Hanser Verlag, München (1973)—11 papers on adsorption and monomolecular layers, 15 papers on properties of relevance to flotation, 5 papers on application to flotation

7. USSR, Moscow, 1976—no publication has appeared.
[A new organization. IACIS (International Association of Colloid and Interface Scientists) plans a meeting in Potsdam, N.Y. in 1985.]

Proceedings of Special Meetings and Symposia on Wetting, Colloids, Interfaces, and Surfactants

1964   *Contact Angle, Wettability and Adhesion*, ACS Advances in Chemistry Series No. 43, ed. R. F. Gould, American Chemical Society, Washington, D. C. (1964)

1966   *Internationale Vorträgung über Grenzflächenaktive Stoffe*, Abhandlungen der Deutschen Akademie der Wissenschaften zu Berlin, Akademie Verlag, Berlin (1967)

1967   *Wetting*, symposium by the Society of Chemical Industry at the University of Bristol, Sept. 1966, Society of Chemical Industry and Gordon & Breach, New York

1970   *Thin Liquid Films and Boundary Layers*, Special Discussions of the Faraday Society No. 1, Cambridge, England, Academic Press, New York

1975   *Proceedings of the International Conference on Colloid and Surface Science*, Parts 1–4, ed. E. Wolfram, Elsevier, Amsterdam (1975)

1975   *Advances in Interfacial Phenomena of Particulate/Solution/Gas Systems, Applications to Flotation Research*, AIChE Symposium Series Vol. 71, eds. P. Somasundaran and R. B. Grieves, American Institute of Chemical Engineers, New York (1975)

1976   *Colloid and Interface Science*, Vols. 1–5, ed. Milton Kerker, Academic Press, New York. (*Proceedings of the International Conference on Colloids and Surfaces—Fiftieth Colloid and Surface Science Symposium in Puerto Rico*, June 1976) 10 plenary and 34 invited lectures on surface thermodynamics, rheology of disperse systems, catalysis, water at interfaces, solid surfaces, surfactants, forces at interfaces—these appear in Vol. 1; additional 221 papers appear in Vols. 2–5; emulsions and surfactants in Vol. 2; adsorption, catalysis, wetting, and surface tension in Vol. 3; hydrosols and rheology in Vol. 4, and monolayers, polymers, etc., in Vol. 5

1978   *Wetting, Spreading and Adhesion*, symposium by the Society of Chemical Industry at Loughborough University, Sept. 1976, ed. J. F. Padday, Academic Press, New York (1978).

1.13.4. *List of Periodicals Regularly Publishing Papers on Flotation Theory and Practice*

Canadian Mining and Metallurgical Bulletin, Montreal, Quebec
Canadian Mining Journal, Montreal, Quebec
Colorado School of Mines, Mineral Industries Bulletin, Golden, Colorado
Deco Trefoil, Denver Equipment Co., Denver

Engineering and Mining Journal, New York
Erzmetall, Stuttgart
Fizykochemiczne Problemy Przerobki Kopalin, Gliwice, Poland
Freiberger Forschungshefte, Leipzig
International Journal of Mineral Processing, Amsterdam
Izvestiya Akademii Nauk USSR, Moscow
Journal of the South African Institute of Mining & Metallurgy, Johannesburg
Mine and Quarry Engineering, London
Minerals Science and Engineering, Johannesburg
Mines et Metallurgie, Paris
Mines Magazine, Denver
Mining and Minerals Engineering, London
Mining Congress Journal, Washington, D.C.
Mining Engineer, London
Mining Engineering, Littleton, Colorado
Mining Magazine, London
Mining World, San Francisco
Nippon Kogyo Kaishi, Tokyo
Obogashchenie Rud, Leningrad
Prace Ustavu pro Vyzkum a Vyuziti Paliv, Prague
Proceedings of the Australasian Institute of Mining and Metallurgy, Melbourne
Revue Roumaine de Metallurgie Academie de la Republique Populaire, Bucharest
Rudy, Prague
Sbornik Nauchnykh Rabot Moskovskogo Gornogo Instituta, Moscow
Sbornik Nauchnykh Trudov, Gosudarstvennyi Nauchno-Issledovatel'skii *Institut Tsvetnykh Metallov*, Moscow
Tohoku Daigaku Senko Seinren, Sendai, Japan
Transactions of the Canadian Institute of Mining and Metallurgy, Montreal, Quebec
Transactions of the Indian Institute of Metals, New Delhi
Transactions of the Institution of Mining and Metallurgy, London
Tsvetnaya Metallurgya, Moscow
Tsvetnye Metally, Moscow
Zeszyty Naukowe Akademii Gorniczo-Hutniczej, Krakow
United States Bureau of Mines, Reports of Investigations, Washington, D.C.

1.13.5. *List of Periodicals Publishing Occasional Papers on Flotation Chemistry*

Abhandlungen der Deutschen Akademie der Wissenschaften zu Berlin, Klasse fuer Chemie, Geologie und Biologie, Berlin
Acta Chemica Scandinavica, Munksgaard, Copenhagen
Acta Metallurgica, Toronto
Acta Polytechnica, Royal Institute of Technology, Stockholm
AIChE Journal, New York
Analytical Chemistry, Easton, Pennsylvania
Angewandte Chemie, Weinheim, ERG
Annalen der Chemie, Weinheim, ERG

Annales de Chimie, Paris

Atti della Accademia Nazionale dei Lincei, Memorie, Classe di Scienze Fisiche, Matematiche e Naturali Sezione, Rome

Canadian Journal of Chemistry, Montreal

Canadian Metallurgical Quarterly, Montreal, Quebec

Chemical Engineering Progress, New York

Chemical Engineering Science, London

Chemicke Listy, Prague

Chemisch Weekblad, Hilversum, Netherlands

Chemistry and Industry, London

Chimie et Industrie, Paris

Collection of Czechoslovak Chemical Communications, Prague

Colloid Journal of the USSR [transl. Kolloidnyi Zh.], New York

Colloid and Polymer Science, Darmstadt

Communications de la Faculte des Sciences de l'Universite d'Ankara, Serie B, Ankara

Discussions of the Faraday Society, London

Doklady Akademii Nauk USSR, Moscow

Endeavour, London

Fuel, London

Gazzetta Chimica Italiana, Rome

Gornyi Zhurnal, Sverdlovsk, USSR

Helvetica Chimica Acta, Basel

Indian Journal of Applied Chemistry, Calcutta

Indian Journal of Chemistry, New Dehli

Indian Journal of Technology, New Delhi

Industrial and Engineering Chemistry, Washington, D.C.

Izvestiya Akademii Nauk USSR, Moscow

Journal of Colloid and Interface Science, New York

Roczniki Chemii, Warszawa

Separation Science, New York

Surface Science, Amsterdam

Svensk Kemisk Tidskrift, Stockholm

Tenside Detergents, München

Transactions of the Faraday Society, London

Zeitschrift fuer Elektrochemie, Leipzig

Zeitschrift fuer Physikalische Chemie, Frankfurt am Main

Zeitschrift fuer Physikalische Chemie, Leipzig

Zhurnal Fizicheskoi Khimii, Moscow

2

Chemical and Molecular Bonding. Interfacial Energetics

Flotation pulps consist of mixtures of solid particles in an aqueous medium, with small additions of surface active agents (inorganic and organic), and air bubbles. Thus, all three types of phases—solid, liquid, and gaseous—are present. The properties of the bulk phases influence and determine the characteristics of the interfaces formed between the adjoining phases. The bonds across interfaces are a direct consequence of bonds within each phase and of the electronic structures in the participating atoms. Any account of interfacial properties must take cognizance of different types of bonding, their relative strengths, and their cooperative effects.

Since every standard book on structural chemistry contains a treatment of chemical and molecular bonds, only those aspects of bonding relevant to the interpretation of surface phenomena will be mentioned here.

2.1. Ionic Bonding

The simplest, the *ionic bond*, is formed between a positively and a negatively charged ion. It acts over large distances (i.e., constitutes a long-range force) and is nondirectional; an ion always attracts counterions from every direction, regardless of their relative positions. Also, in an ionically bonded neutral aggregate composed of equal numbers of oppositely charged ions, the bonds do not become saturated, but further ionic bonds can be developed by each of the terminal ions.

Depending on the electronic structure of an atom, it can either give up its valence electrons to become a positive ion, or it can accept electrons from any available source to become a negatively charged ion.

For an isolated pair of oppositely charged ions the energy of the bond developed between them is the sum of an attractive and a repulsive contribution to the overall potential energy:

$$U_{\text{ionic}} = -\frac{z^+z^-e^2}{d} + \frac{b_{ij}e^2}{d^n} \qquad (2.1)$$

$$\underset{\substack{\text{(attractive} \\ \text{energy)}}}{} \qquad \underset{\substack{\text{(repulsive} \\ \text{energy)}}}{}$$

where z^+ and z^- represent the numbers of positive and negative charges on the two ions; e is the electronic charge, 4.8×10^{-10} esu $= 1.602 \times 10^{-19}$ C (coulomb); d is the distance between the centers of ions; b_{ij} is a constant for the given ionic compound; and n is an exponent depending on the electronic configuration of ions such that for the configuration of electrons in: He, $n = 5$; Ne, $n = 7$; Ar, $n = 9$; Kr, $n = 11$; Xe, $n = 12$; Rn, $n = 14$.

When an ionic crystal is placed in a liquid of a high dielectric permeability[†] ε (that is, a polar liquid), the attractive forces between the ions are diminished in proportion to $1/\varepsilon$, causing breakdown of the crystal lattice to produce solute species. In water these solute species are usually in the form of solvated ions (see Chapter 4).

The lattice energy calculated for different compounds differs in many cases from the value obtained experimentally. This difference indicates departures from a purely ionic character of the bonding in the given compounds. The departures are due to a transition toward covalent bonding and to polarization effects. Even when the differences between the calculated and the experimental lattice energy values are small, as in Table 2.1, there are other than ionic contributions involved which may nearly cancel each other.

[†] The dielectric permeability (permittivity) ε of a substance expresses the response of that substance to an electric field applied across it. For static electric fields (dc) the dielectric permeability is constant. For fields of low frequency it is nearly constant and reflects the ability of permanent dipoles within the substance to follow an ac applied field. The dielectric relaxation time τ (that is, the time required by the given permanent dipoles within the substance to reorient themselves with the ac field) depends on the structure developed by the dipoles within the substance. For example, the dielectric relaxation time for water dipoles in liquid phase is of the order of 10^{-11} sec; for the same water dipoles in their ice I phase $\tau = \sim 10^{-4}$ sec, and for ice III, V, and VI phases $\tau = \sim 10^{-6}$ sec [Franks (1972, Vol. I, Chapter 4, Figure 12, p. 140)]. The contributions to dielectric properties of the substance arising from induced dipoles (as opposed to permanent ones) make the dielectric permeability frequency-dependent whenever electric fields of frequencies higher than those of the relaxation processes are employed. (For more details on this aspect, see Section 4.1.) The dielectric constant is dependent on temperature [as indicated by equations (2.6) and (2.7)].

Table 2.1. Estimated Contributions to Lattice Energies (kcal mol^{-1})^{a,b}

Type of energy	AgF	AgCl	AgBr
Coulomb attraction	−235.0	−209.0	−202.0
London attraction	−23.7	−29.2	−27.4
Born repulsion	+38.2	+34.5	+31.9
Zero-point energy	+1.4	+1.3	+0.9
Total (calculated) lattice energy	−219.0	−202.0	−197.0
Experimental lattice energy	−231.0	−219.0	−217.0

^{a} From: Phillips and Williams (1965).
^{b} 1 cal ≡ 4.184 J.

Whenever ionic solids fracture, the new surfaces thus created are always highly hydrophilic, polar, attracting ions or dipoles from the surroundings.

2.2. Covalent Bonding

Covalent bonds form at much shorter distances than ionic bonds whenever the two approaching atoms can share one pair (or more) of electrons, each atom contributing one electron of each pair.

Changes in the energy levels of the constituent atoms occurring when a covalent bond is formed as a result of sharing of electrons can be represented using either the *valence bond theory* and the concomitant concept of *resonance energy* or the *molecular theory* and the concept of overlapping of orbitals. *Hybridization* (mixing) of orbitals leads to the formation of orbitals with the characteristic directions in space and a change in the interbond angles. For example, from one $1s$ and one $2p$ orbital, two sp hybrid orbitals can be formed, which point in diametrically opposite directions; one lobe of the hybrid orbital is much larger and the other lobe much smaller than the lobes of the original orbital [Figure 2.1(a)].

When a $1s$ orbital is combined with $2p_x$ and $2p_y$ orbitals, three sp^2 hybrid orbitals are formed. These are directed at 120° to each other in one plane, resulting in the trigonal shape of the molecule formed by the four atoms; see Figure 2.1(b).

Hybridization of $2s$ and three $2p$ orbitals produces four sp^3 hybrid orbitals which are directed toward the corners of a regular tetrahedron (with interbond angle 109°28′, e.g., CH_4, CCl_4). If the electrons in the atomic orbitals are unpaired, the sp^3 hybridization of the s and p electronic

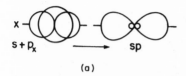

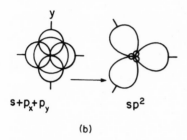

Figure 2.1. Hybridization of orbitals in covalent bonding: (a) the formation of *sp* hybrid orbitals from $1s$ and $2p_x$ orbitals; (b) the formation of sp^2 hybrid orbitals from $1s$, $2p_x$, and $2p_y$ orbitals.

orbitals results in a decreased bond angle because of an extra repulsion of the lone pairs. Thus, in the H_2O molecule, the oxygen atom has the electronic structure $1s^2 2s^2 2p^4$, resulting in one p orbital with paired electrons and two unpaired p orbitals containing one electron each with parallel spins; there is a bond angle of 104°30′ between the two hydrogens instead of a character- istic tetrahedral bond angle of 109°28′. Similarly, in ammonia, NH_3, one unpaired electron of nitrogen causes a decrease of the interbond angle to 106°45′. When d orbitals become available for interatomic bonding, the hybridization involves s, p, and d orbitals and leads to the spatial arrange- ment of hybridized orbitals indicated in Table 2.2 (see also Table 3.3).

When a double bond is formed between two carbon atoms, as in ethylene, C_2H_4, the σ-type carbon–carbon bond is formed by the overlap of two sp^2 hybrid orbitals, one from each atom, and the four carbon– hydrogen bonds are formed by the overlap of the remaining sp^2 hybrid

Table 2.2. Hybrid Orbitals

Hybridization	Arrangement of hybrid orbitals	Example
sp	Linear	$HgCl_2$
sp^2	Trigonal planar	BCl_3
sp^3	Tetrahedral	CH_4
dsp^2	Square planar	$K_2Ni(CN)_4$
d^2sp^3	Octahedral	$K_4Fe(CN)_6$
dsp^3	Trigonal–bipyramidal	PCl_5
d^4sp^3	Dodecahedral	$K_4Mo(CN)_8$

orbitals with hydrogen s orbitals. The two remaining unhybridized p-type orbitals (p_z) overlap, giving a π bond. This is weaker than the σ bond (giving rise to chemical reactivity of carbon compounds containing similar double bonds), and its presence prevents a free rotation of methylene groups around the C=C bond. The π bond is not cylindrically symmetrical along the bond axis but is represented as a "bent" or "banana-like" bond disposed above and below the plane of the C—C nuclei. If a molecule contains one or more carbon–carbon single bonds, rotation about the bond axis can usually occur freely, and this can give rise to different conformations. With a double carbon–carbon bond, rotation is prevented, and if each carbon is bonded to different groups, two types of configuration of the molecule (geometrical isomers) may exist: either *cis* form or *trans* form.

A triple bond between two carbon atoms, as in acetylene, C_2H_2, consists of sp hybrid orbitals forming σ bonds and two π bonds (formed by the unused $2p_y$ and $2p_z$ orbitals of the atoms). This arrangement makes the electron charge cloud of acetylene symmetrical about the C—C axis.

A molecule containing two atomic nuclei is discussed in terms of "localized molecular orbitals"; when several atomic nuclei are present, as, for example, in benzene, C_6H_6, the π bonds are considered to result from mutual overlapping of all the $2p_z$ orbitals, endowing the molecule with an extra chemical stability due to "delocalization" of molecular orbitals. The delocalization energy corresponds to the resonance energy of the valence bond theory.

The methods of treating covalent bonding by the valence-bond or the molecular orbital theory represent equivalent approximations: Neither can be called correct or incorrect. Their utility depends on the degree of clarity of bond representation and their ability to predict the behavior of molecules.

2.3. *Partial Ionic–Covalent Bonds*

As a result of different *electronegativities* of atoms (Table 2.3), that is, different tendencies to attract electrons by atoms in a molecule, the majority of atomic aggregates possess bonds which are partially ionic and covalent in character. Mulliken (1932) attempted to relate the electronegativity to the ionization potential, but this differentiation was of limited use. Pauling has based his representation of electronegativities on the following considerations:

A diatomic molecule A—B is purely covalent only if A and B have the same electronegativity. If A and B have different electronegativities,

Table 2.3. Pauling Electronegativities (H = 2.1)[a,b]

Li 1.0	Be 1.5											B 2.0	C 2.5	N 3.0	O 3.5	F 4.0
Na 0.9	Mg 1.2											Al 1.5	Si 1.8	P 2.1	S 2.5	Cl 3.0
K 0.8	Ca 1.0	Sc 1.3	Ti 1.5	V 1.6	Cr 1.6	Mn 1.5	Fe 1.8	Co 1.8	Ni 1.8	Cu 1.9	Zn 1.6	Ga 1.6	Ge 1.8	As 2.0	Se 2.4	Br 2.8
Rb 0.8	Sr 1.0	Y 1.2	Zr 1.4	Nb 1.6	Mo 1.8	Tc 1.9	Ru 2.2	Rh 2.2	Pd 2.2	Ag 1.9	Cd 1.7	In 1.7	Sn 1.8	Sb 1.9	Te 2.1	I 2.5
Cs 0.7	Ba 0.9	La–Lu 1.1–1.2	Hf 1.3	Ta 1.5	W 1.7	Re 1.9	Os 2.2	Ir 2.2	Pt 2.2	Au 2.4	Hg 1.9	Tl 1.8	Pb 1.8	Bi 1.9	Po 2.0	At 2.2
Fr 0.7	Ra 0.9	Ac 1.1	Th 1.3	Pa 1.5	U 1.7	Np 1.3										

[a] Source: Phillips and Williams (1965).

[b] The values given refer to the common oxidation states of the elements. For some elements variation of the electronegativity with oxidation number is observed, for example, Fe(III), 1.8; Fe(II), 1.8; Cu(I), 1.9; Cu(II), 2.0; Sn(II), 1.8; Sn(IV), 1.9 [Pauling (1948)]. There have been other methods proposed for estimating electronegativity values. A comparison of three sets of values [Alfred and Rochov, Mulliken, and recalculated Pauling-type values] is given in Cotton and Wilkinson (1972, p. 115).

the bond A—B will no longer be purely covalent, and the energy of the bond, D_{AB} (experimental), is greater than the calculated D_{AB}(calc.)—which is usually taken as the arithmetic mean of the bond energies of the covalent molecules A—A and B—B:

$$D_{AB}(\text{calc.}) = \tfrac{1}{2}(D_{AA} + D_{BB}) \tag{2.2}$$

The difference between the experimental and the calculated values of bond energy has been related by Pauling to the difference in electronegativities by the following empirical expression:

$$\varDelta_{AB} = D_{AB}(\text{exp}) - \tfrac{1}{2}(D_{AA} + D_{BB}) = 23.06(x_A - x_B)^2 \tag{2.3}$$

where x_A and x_B are the electronegativities of A and B and 23.06 is the conversion factor from electron volts to kilocalories. To obtain comparative values, an arbitrary value of $x_H = 2.1$ was assigned by Pauling to hydrogen, giving relative values for other elements, as shown in Table 2.3.

Ionic bonding is favored when the difference in electronegativities is high, and covalent bonding is favored by small differences in electronegativity values. The percentage of ionic character of a bond in a diatomic molecule can be calculated by an empirical formula [due to Hannay and Smyth (1946)],

$$\% \text{ ionic character} = 16(x_A - x_B) + 3.5(x_A - x_B)^2 \tag{2.4}$$

and the values obtained by equation (2.4) for hydrogen halides agree well with the ionic character determined for these compounds from dipole moments.

The fracture of a covalent bond in a solid leads to ionization of the atoms which had been sharing electrons. Thus, any new surface created by fracture of covalent bonds, originally present within a covalently bonded solid, should become ionized, polar in character, and hydrophilic (unless the ionization is immediately neutralized, for example, by adsorption from the environment).

2.4. *Dipole Moments and Dipole–Dipole Interactions*

Every polyatomic molecule in which the bonds are not symmetrical about the central atom possesses a dipole moment and is known as a polar molecule. In symmetrical molecules such as CCl_4, CH_4, CO_2, etc., the center of positive charges coincides with the center of negative charges, no dipole moment exists, and the molecule is known as nonpolar. When the centers of positive and negative charges do not coincide but are separated by a distance l, the charge displacement is measured by the dipole moment μ, where

$$\mu = zel \tag{2.5}$$

in debye units [1 debye unit $= 10^{-18}$ esu cm $= 3.335 \times 10^{-30}$ C m (coulomb meter)]. A dipole is a vector quantity, and the overall dipole moment of a molecule is the resultant of the component vectors.

Dipole moments are valuable parameters in determining the shapes of molecules and complexes. They contribute to intermolecular bonding forces through dipole–dipole interactions. At interfaces, dipole–dipole interactions play a very significant role, particularly in the adsorption of surfactants and in the mutual interactions between molecules of different surfactant species.

For covalently bonded compounds which can be studied in vapor phase, the dipole moments are evaluated from properties, such as their static dielectric constant ε. If the capacitance of a condenser (capacitor) in vacuum is C_0 and it changes to C when the vapor of the compound is placed between the plates of the condenser, the dielectric constant of the compound is $\varepsilon = C/C_0$. *Clausius and Mossotti* related the molecular (or molar) polarization of a compound to the dielectric constant by the following expression:

$$P_M = \frac{\varepsilon - 1}{\varepsilon + 2} \frac{M}{\varrho} \tag{2.6}$$

where M is the molecular weight and ϱ is the density. Molar polarization is a quantity representing the dipole moment per unit volume of a continuous substance. When a polar compound (with a permanent dipole) is placed in a uniform electric field, the permanent dipoles tend to become oriented with the field; the field also *induces* a dipole which lasts only as long as the field is present and whose magnitude depends on properties, such as the molecular *polarizability* α_0 of the substance. In fact, molecular polarizability can be defined as the magnitude of the induced dipole created by a unit field. Thermal vibrations of molecules tend to counteract the orientation of both permanent and induced dipoles. Debye (1929) showed that the molecular polarization P_M is a function of temperature T,

$$P_M = A_p + \frac{B_p}{T} \tag{2.7}$$

where

$$A_p = \frac{4\pi}{3} N\alpha_0 \tag{2.8}$$

(N is Avogadro's number) is the contribution due to the induced dipole and

$$B_p = \frac{4\pi}{9} N \frac{\mu^2}{k} \tag{2.9}$$

(k = Boltzmann's constant = 1.3805×10^{-16} erg deg^{-1} $\equiv 1.3805 \times 10^{-23}$ JK^{-1}) is the orientation contribution due to the permanent dipole μ. Thus, measurements of the static dielectric constant ε as a function of temperature T enable [using equations (2.6)-(2.9)] both the polarizability α_0 and the permanent dipole moment μ to be evaluated. If the constant B_p in equation (2.7) is zero, there is no permanent dipole moment, $\mu = 0$.

For liquids, solutions, and solids the application of equation (2.7) is restricted because the assumption of pairwise additivity does not apply to simultaneous interactions in many-body systems. For these condensed phases the evaluation of the permanent dipole μ used to be carried out by an indirect method: The molecular polarization P_M was evaluated by equation (2.6) from the dielectric constant measurements carried out using a static (dc) electric field and an ac field of appropriate frequency. For fields of low frequency (for example, less than $\sim 10^8$ Hz for water) the permanent dipoles can orient themselves with the changing electric field. Above such frequency, the permanent dipoles can no longer follow the field. The *Lorentz–Lorenz* equation, based on an assumption that molar polarization is due to the induced dipoles only, was then utilized to evaluate α_0:

$$\frac{4\pi}{3} N\alpha_0 = \frac{n_r^2 - 1}{n_r^2 - 2} \frac{M}{\varrho} \dots \tag{2.10}$$

where n_r is the refractive index, defined as the ratio of phase velocity of light in vacuum to that in the given medium (liquid or solid).

A second assumption that, in the absence of high magnetic polarizability, $n_r^2 = \varepsilon$ is also made. [These assumptions were found (later on) to hold for substances having cubic or isotropic lattice symmetry but not for most substances, including water.] Thus, from measurements of ε at a single temperature, the evaluation of B_p (i.e., the orientation contribution due to the permanent dipole) and the evaluation of μ could be carried out.

Le Fèvre (1953) gave a comprehensive treatment of dipole moments and their measurement. Some values of ε, μ, and α_0 are given in Table 2.4.

Table 2.4. Selected Values of Dipole Moments and Polarizabilities

Substance	Dielectric constant ε (25°C)	Dipole moment μ [debye units, 3.335×10^{-30} C m]
Water	79.5	1.85
Benzene	2.27	0
Ethanol	24.3	1.7
Hexane	1.90	0
CCl_4	2.238	0
CH_3NO (formamide)	109	3.25
HCN (cyanic acid)	158.1	2.9–3.0

The polarizabilities α_0 of some ions,[a] molecules, and groups (10^{-24} cm^3)

Li^+	0.03	Be^{2+}	0.01	Ag^+	1.9	F^-	0.81
Na^+	0.24	Mg^{2+}	0.10	Tl^+	3.9	Cl^-	2.98
K^+	1.00	Ca^{2+}	0.60	Zn^{2+}	0.5	Br^-	4.24
Rb^+	1.50	Sr^{2+}	0.90	Cd^{2+}	1.15	I^-	6.45
Cs^+	2.40	Ba^{2+}	1.69	Hg^{2+}	2.45	OH	1.89
NH_4^+	1.65	La^{3+}	1.3	Pb^{2+}	3.6	O^{2-}	3

HCl	2.63	CH_2 (in polyethylene)	1.76
CH_4	2.58	C_2H_3Cl (in polymer)	3.31
CH_3Cl	4.56	CCl_4	10.5
C_6H_6	10.32	H_2O	1.48

[a] Source: Phillips and Williams (1965).

[Since ∼1950 a new approach has been developed for interaction between dielectric materials. It is based on electrodynamics of continuous media, neglecting the atomic "graininess" and utilizing charge fluctuations (absorption frequencies) of interacting substances (Section 2.6).]

Reliable data on permanent dipole moments of amphipatic compounds (compounds possessing polar and nonpolar portions, more commonly known as surfactants; see Chapter 5) are scarce. And what information is available in the literature may, in some instances, be suspect because of two effects:

1. The possibility of aggregates existing in certain solvents used in dipole determinations

2. The belief that the length and the structure of the hydrocarbon portion of the molecule have no influence on the dipole moment

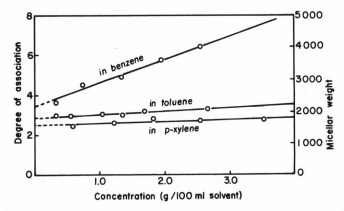

Figure 2.2. Micellar weight and degree of association of zinc–oleate in organic solvents: benzene, toluene, and *p*-xylene. [After Gilmour *et al.* (1953).]

The use of benzene as a solvent for amphipatic compounds is common in evaluating dipole moments from the measurements of the molar polarization of solute. However, as clearly established by Gilmour *et al.* (1953), aggregates are evident in benzene solutions of many metal–surfactant compounds (see Figure 2.2). Thus, the measured polarizations, and hence the dipole moment values reported in the literature should be considered approximate and may represent average values for the different sizes of aggregates present in the solutions.

The effect of the hydrocarbon chain length and its structure on the dipole moment, as quoted by Glembotsky *et al.* (1968) after Chichibabin (1953), is shown in Table 2.5. It appears that the most noticeable increase in the value of the dipole moment occurs in the first four or so homologues.

Regardless of the chemical character of the species representing dipoles, whenever two or more dipoles occur in the vicinity of each other, there is an interaction taking place between them. The dipoles will always orient themselves in such a way as to attract each other. This behavior is known as the *orientation or Keesom effect*, after Keesom, who showed in 1921 that the energy of interaction is given by

$$U_\mu = \frac{-2\mu_1{}^2\mu_2{}^2}{3kTd^6} \tag{2.11}$$

for high temperatures, when

$$kT \gg \frac{\mu_1\mu_2}{d^3}$$

and by

$$U_\mu = \frac{-2\mu_1\mu_2}{d^3} \tag{2.12}$$

Table 2.5. Dipole Moments (in Debye Units) of Some Organic Compounds with Various Substituents

After Chichibabin (1953) as quoted by Glembotsky et al. (1968)

Alkyl Group, R	Substituents						
	—F	—Cl	—Br	—I	—NO$_3$	—C≡N	—OH
CH$_3$	1.81	1.87	1.78	1.59	3.50	3.94	1.64
CH$_3$—CH$_2$—	1.92	1.05	2.02	1.90	3.70	4.04	1.63
CH$_3$—CH$_2$—CH$_2$—	—	2.10	2.15	2.01	3.72	4.05	1.66
(CH$_3$)$_2$—CH—	—	2.15	2.19	—	3.73	—	—
CH$_3$—CH$_2$—CH$_2$—CH$_2$—	—	2.09	2.14	2.08	—	4.09	1.65
(CH$_3$)$_2$—CH—CH$_2$	—	2.04	—	—	—	—	—
CH$_3$—CH$_2$—CH—(CH$_3$)$_2$	—	2.12	2.20	—	—	—	1.72
(CH$_3$)$_3$—C—	—	2.13	—	—	3.71	—	—
CH$_3$—CH$_2$—CH$_2$—CH$_2$—CH$_2$—	—	2.12	—	—	—	—	—
(CH$_3$)$_2$—CH—CH$_2$—CH$_2$—	—	—	—	—	—	—	1.85

From McClellan (1963, 1974)

Surfactant	Compound					
Alcohols	Cyclohexanol 1.9		o-Cresol 1.45	m-Cresol 1.61	p-Cresol 1.65	
Fatty acids		Caproic 1.13	Palmitic 0.73	Oleic 1.69	Linoleic 1.52	
Metal oleates	Cu^{2+} 1.21		Fe^{2+} 2.88	Cr^{2+} 4.39	Pb^{2+} 4.32	Zn^{2+} 0.29
Metal xanthates	Fe^{3+} triethyl 2.8		Zn^{2+} dibutyl 2.5	As^{3+} triethyl 1.5		
Metal dithiocarbamates	Fe^{2+} di-isopropyl 1.62		Cu^{2+} di-isopropyl 1.68	Zn^{2+} dibutyl 2.20		

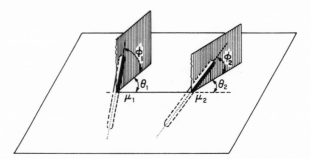

Figure 2.3. Parameters in the Debye formula for the interaction of two oblique dipoles. ϕ_1 and ϕ_2: angles of inclination with respect to a plane bisecting the dipoles along their midpoints; θ_1 and θ_2: angles between dipole projections and the axis joining the midpoints of each dipole; d: distance of dipole separation.

for low temperatures, when

$$kT \ll \frac{\mu_1\mu_2}{d^3}$$

Debye (1929) showed that the energy of association (interaction) between two permanent dipoles μ_1 and μ_2 is given, in a general case of nonplanar dipoles, by (see Figure 2.3 for notation of angles)

$$U_\mu = \frac{\mu_1\mu_2}{d^3} \left[2 \cos \theta_1 \cos \theta_2 - \sin \theta_1 \sin \theta_2 \cos(\phi_1 - \phi_2)\right] \qquad (2.13)$$

Debye also showed in 1920 that an atom or molecule with a permanent dipole moment will polarize a nearby neutral (nonpolar) atom or molecule, resulting in an induced dipole,

$$\mu_i = \frac{\mu\alpha_0}{d^3} \qquad (2.14)$$

being created in the nonpolar molecule. An interaction between the polar and the nonpolar molecules is then due to an attraction between the permanent and the induced dipoles and is expressed by

$$U_i = \frac{-\mu^2\alpha_0}{d^6} \qquad (2.15)$$

This is known as an *induction force interaction or Debye effect*.

Examples of dipole interaction energies at room temperature are given in Table 2.6. It is seen that the interaction energy of the two dipoles may

Table 2.6. Examples of Dipole–Dipole Interaction Energies

Interacting dipoles	μ_1 (debye)	μ_2 (debye)	Assumed intermolecular distance (Å)	Energy[a] [erg $(10^{-7}\,\mathrm{J})$]	
				$-2\mu_1{}^2\mu_2{}^2/3kTd^6$	$-2\mu_1\mu_2/d^3$
H_2O—H_2O	1.85	1.85	3.0	2.6×10^{-13}	2.5×10^{-13}
CH_3Cl—H_2O	1.94	1.85	3.5	7.8×10^{-14}	1.7×10^{-13}
$CHCl_3$—$CHCl_3$	1.1	1.1	4.5	6×10^{-15}	3.8×10^{-14}
CHI_3—CHI_3	0.9	0.9	5.0	7×10^{-16}	1.3×10^{-14}

$(kT = 4.113 \times 10^{-14}\text{ erg at }25^\circ\text{C})$

[a] The negative sign in equations (2.11) and (2.12) indicates that the interaction between dipoles is always an attraction.

be an order of magnitude smaller or greater than kT, thermal agitation energy.

The dipole moments of selected surfactants are listed in Table 2.5. During adsorption at solid/liquid or liquid/gas interfaces, there are Keesom and Debye types of interactions occurring between dipoles of water and dipoles (permanent and induced) of surfactants on one hand and those of surfactant with surfactant on the other. Their relative strengths determine the changes from hydrophilic to hydrophobic surfaces.

Since dipoles are vector quantities, the energy of interaction of several dipoles is not always additive; it is additive only for end-to-end or parallel orientation. For other orientations, Debye's expression, given by equation (2.13), must be used.

2.5. *Hydrogen Bonding*

When hydrogen is bonded to the very electronegative elements (F, O, N, Cl) or the highly electronegative groups (such as —CCl_3 or —CN), a strong tendency for association of the molecules exists. This is evident in gases: For example, hydrogen fluoride vapor phase contains short zigzag chains of polymeric aggregates up to $(HF)_5$ at temperatures below about 60°C. Association of molecules is also evident in liquids such as water, ammonia, alcohols, and amides, which represent the so-called polar solvents and in which either clusters of molecules, as, for example, clusters in water, or dimers, as in alcohols and carboxylic acids, exist:

R denotes a hydrocarbon chain, C_nH_{2n+1}.

The hydrogen-bonded clusters are temporary, since the thermal energies of molecules are usually sufficient to cause these bonds to break very rapidly (in less than milliseconds).

In many biological systems, in proteins, and in solid phases (ice, crystals of $NaHCO_3$, $CuSO_4 \cdot 5H_2O$) hydrogen bonding is an important factor in the overall crystalline structure. The presence of hydrogen bonding is responsible for the high dielectric constants of these substances, the high boiling points of their liquid phases, and the low vapor pressures whenever these materials are compared with compounds of other less electronegative elements in the respective groups.

The structure of some crystalline substances is determined by hydrogen bonding, for example:

1. Discrete hydrogen-bonded HF_2^- anions (and K^+) are found to constitute the ionic crystal of KHF_2.
2. Infinite chains of HCO_3^- are linked by hydrogen bonds in $NaHCO_3$.
3. Infinite layers of planar $(BO_3)^{3-}$ groups are obtained by hydrogen bonding in boric acid, H_3BO_3.
4. Water molecules in ice I (stable at low temperature, $0° \rightarrow -20°C$) adopt a three-dimensional "open" structure (similar to that of wurtzite).

Hydrogen bonding plays an important role in many interfacial phenomena in systems involving oxides and oxidized solids, carbonates, silicates, and water. Although the strength of a hydrogen bond (3–10 kcal/mol) is much lower than that of a covalent bond, it is frequently sufficient to "tip the scale" whenever energetically comparable interactions (per unit area) are involved at interfaces.

Extensive reviews and treatments of hydrogen bonding are given by Pimentel and McClellan (1960), Hamilton and Ibers (1968), Rich and Davidson (1968), and Zundel (1970). Hydrogen bonding has been much studied by infrared and Raman spectroscopy. When an atom A enters into $A \cdots H$ hydrogen bonding, changes are observed in the infrared spectra in comparison with the spectrum given by the free molecule containing atom

Table 2.7. Some Data on Hydrogen Bonds

Covalent bond length	Substance	Lowering of stretching frequency (cm^{-1})	Bond energy (kcal/mol)	Hydrogen bond lengths (between atoms A and B)
F—H, 1.13 Å	H_6F_6	~2700	6.7	F···H—F in HF_2^-, 2.26 Å
O—H, 0.99 Å	H_2CO_3	—	—	O···H—O in HCO_3^-, 2.61 Å
	H_2O	~430	4.5	O···H—O in H_2O ice, 2.76 Å
	ROH^a	—	6.2	O···H—O in ROH (alcohol), 2.70 Å
N—H, 1.05 Å				N···H—O (proteins), 2.67-3.07 Å
	$(HCN)_3$	~180	4.4	N···H—C, $(HCN)_n$, 3.2 Å
	NH_3	—	1.3	N···H···N, 3.38 Å
	NH_4F	—	5	N—H···F, 2.63 Å

^{a} $R \equiv C_nH_{2n+1}$.

A. These changes consist of lowering the stretching frequency, a substantial broadening of the stretching frequency band (up to several hundred reciprocal centimeters), and raising of the bending frequency. The interpretation of these changes enables the energies of the hydrogen bonding in different atomic groupings to be evaluated. Bellamy (1968) and Zundel (1970), give detailed accounts of the infrared spectral changes and the consequent evaluations of various hydrogen bond energies.

Both the molecular orbital theory and the electrostatic approach have been used to explain hydrogen bond formation. The electrostatic approach is by far the simpler. A covalent bond between hydrogen and a strongly electronegative atom *A* is highly polarized; i.e., a large dipole is created which tends to interact with any other strongly electronegative atom *B* by aligning the positive end,

$$A^{\ominus}\!-\!H^{\oplus}\!\cdots\!B^{\ominus}\!-\!$$

giving in effect a dipole–dipole interaction. Lone pair electrons play an important part in determining the strength and the direction of hydrogen bonds; also, in many cases, hydrogen bonds may have some covalent character.

In comparison with the lengths of covalent bonds, hydrogen bond lengths are much larger (Table 2.7).

2.6. *London Dispersion Forces. van der Waals Interactions*

Depending on temperature and pressure, some degree of aggregation of all covalently bonded nonpolar molecules occurs even in the gaseous state, and a complete aggregation takes place on condensation into liquids or in solidification. Such aggregation results from the action of *dispersion forces*. Initial information about these forces can be obtained from the constants *a* and *b* in the van der Waals thermal equation of state for the nonideal behavior of gases:

$$\left(P + \frac{a}{V^2}\right)(V - b) = RT \tag{2.16}$$

This equation gives a qualitative description of the behavior of gases (and also of liquids) by a suitable choice of *a* and *b* for low pressures and high temperatures. It cannot be made to fit the experimental data over any wide range of *P* and *T*. The term a/V^2 in equation (2.16) is a measure of attractive intermolecular forces causing departures from the ideal behavior of gases.

All intermolecular forces have their origin in the electromagnetic dispersion forces since they are closely related to optical dispersion. Wang (1927) showed that two nonpolar atoms will attract each other. London (1930) used the quantum mechanical approach to show that the continuous motion of electrons in molecules creates rapidly fluctuating dipoles which interact with each other. He obtained an expression for the attractive interaction energy between two atoms or molecules that do not have permanent dipole moments:

$$U_{\text{dispersion}} = -\frac{C_1}{d^6} - \frac{C_2}{d^8} - \frac{C_3}{d^{10}} \cdots \tag{2.17}$$

where the first term, $-(C_1/d^6)$, represents the interaction between the two induced dipoles; the second term, $-(C_2/d^8)$, the dipole–quadrupole interaction; the third term, $-(C_3/d^{10})$, the quadrupole–quadrupole interaction; etc. Dipole–quadrupole interactions become negligible when $d > 8$ Å, while the third and higher terms in equation (2.17) are usually very small and are neglected.

For two nonpolar atoms the constant C_1 was evaluated by London:

$$C_1 = \tfrac{3}{2}\alpha_1\alpha_2 \frac{\nu_1\nu_2}{\nu_1 + \nu_2} h \tag{2.18}$$

where α_1 and α_2 are the polarizabilities and ν_1 and ν_2 are the frequencies of harmonic oscillations of the two atoms; $h =$ Planck's constant $= 6.6256 \times 10^{-34}$ J s.

Since $h\nu_1$, $h\nu_2$ are often approximated by the ionization energies of the atoms, I_1 and I_2, the constant C_1 may be expressed by

$$C_1 = \tfrac{3}{2}\alpha_1\alpha_2 \frac{I_1 I_2}{I_1 + I_2} \tag{2.19}$$

For the hydrogen atom the constant $C_1 \cong 6 \times 10^{-60}$ erg cm^6, and for the molecule CH_4 the constant $C_1 \cong 144 \times 10^{-60}$ erg cm^6.

Kirkwood (1932) and Müller (1936) derived another expression for C_1 also involving contributions from magnetic dipole fluctuations as well as electronic polarizabilities:

$$C_{\text{KM}} = 6mc^2 \frac{\alpha_1\alpha_2}{(\alpha_1/\chi_1) + (\alpha_2/\chi_2)} \tag{2.20}$$

where m is the mass of the electron (9.1091×10^{-31} kg), c the velocity of light $(2.99795 \times 10^8 \text{ m s}^{-1})$, and χ_1, χ_2 the magnetic susceptibilities of the atoms. Constants C_2 and C_3 may be evaluated by similar methods.

When the distance separating the two interacting nonpolar atoms increases to several hundred angstroms, the time taken for the electrostatic field of the first atom to reach the second atom and return may be comparable with the fluctuation period. The induced dipoles will no longer be in phase, and the resultant interaction (known as the *retarded van der Waals interaction*) was shown by Casimir and Polder (1948) to be proportional to the d^{-7} instead of the d^{-6} power law obtained for the *nonretarded van der Waals interaction*. Thus, when $d \gg \lambda_i/2\pi$ (where λ_i are the characteristic absorption wavelengths of the atoms), the energy of dispersive interactions is

$$U_{12} = -\frac{K_{12}}{d^7} \tag{2.21}$$

where the subscript *12* denotes the corresponding parameter for interaction of species *1* and *2*.

For separations smaller than ~ 4 Å, in addition to attractive forces, the two interacting nonpolar molecules are subject to mutual repulsion as the electronic clouds begin to interpenetrate. The empirical equation

$$U_{\text{repulsion}} = \frac{C_4}{d^{12}} \tag{2.22}$$

is most frequently employed to fit the experimental data, the exponent 12 being commonly used, though without any theoretical basis. An equation representing the combined attraction and repulsion terms of interaction energy is known as the Lennard-Jones potential when it comprises two terms: the $-(C_1/d^6)$ term for the nonretarded dispersion forces attractive interaction and the $+(C_4/d^{12})$ term for the repulsive interaction. Generally, the total interaction potential between inert gas molecules and those between various nonpolar molecules with nonionic surfaces is represented by several terms:

$$U_{\text{interaction}} = -\frac{C_1}{d^6} - \frac{C_2}{d^8} + \frac{C_4}{d^{12}} \tag{2.23}$$

However, the assumptions made in theoretical calculations are often not fulfilled except for the simplest inert gases (for example, the repulsion term C^4/d^{12} frequently cancels the dipole–quadrupole dispersion term C_2/d^8), so in many cases the simplest form,

$$U = -\frac{C_1}{d^6} \tag{2.24}$$

is the best approximation used for the nonretarded dispersion forces interaction.

Except in the case of strongly polar molecules, the dispersion forces dominate over the Keesom orientation and the Debye induction forces. The total van der Waals interactions between two atoms or molecules are given by the sum of those due to orientation, induction, and dispersion forces. For the highly polar H_2O molecules (dipole moment, 1.85 debye; polarizability, $\alpha = 1.48$ Å^3) the ratios of dispersion, orientation, and induction forces are 4:20:1; for the HCl molecule (dipole moment, 1.03 debye; polarizability, $\alpha = 2.63$ Å^3) the corresponding ratios are $10:2:\frac{1}{2}$; and for the nonpolar CO_2 molecule the ratios are 7:0.003:0.006—all on the same scale [data from Israelachvili and Tabor (1973); another comprehensive review of recent studies concerning van der Waals forces is that of Parsegian (1975)].

The expressions for interactions that have been given have been derived for atoms and molecules in vacuum. When a third medium is present between the interacting atoms, molecules, or species, a general theory of electronic dispersion forces presented by McLachlan (1963, 1965), called the susceptibility theory, provides expressions for retarded and nonretarded interactions in terms of d^{-7} and d^{-6} and the so-called effective polarizabilities, α_1^* and α_2^* (for species 1 and 2). These expressions can be simplified to

$$U = -\frac{C_{132}}{d^6}\frac{\alpha_1^*\alpha_2^*}{\varepsilon^2} \tag{2.25}$$

for the nonretarded dispersion forces and to

$$U = -\frac{K_{132}}{d^7}\frac{\alpha_1^*\alpha_2^*}{\varepsilon^{5/2}} \tag{2.26}$$

for the retarded forces.

The effective polarizabilities are defined for a macroscopic-in-size species of radius a_1 or a_2 (which must be small in comparison with the separation distance d) using the macroscopic property of the corresponding dielectric permeabilities ε_1 and ε_2 in the following manner:

$$\alpha_1^* = \varepsilon\,\frac{\varepsilon_1 - \varepsilon}{\varepsilon_1 + 2\varepsilon}\,a_1^3 \tag{2.27}$$

$$\alpha_2^* = \varepsilon\,\frac{\varepsilon_2 - \varepsilon}{\varepsilon_2 + 2\varepsilon}\,a_2^3 \tag{2.28}$$

where ε is the dielectric permeability of the medium present between the interacting species. The effective polarizabilities α_1^* and α_2^* and ε of the medium are all functions of absorption frequencies exhibited by the phases 1, 2, and 3.

When the species 1 or 2 has the same dielectric constant as the medium $\varepsilon_1 = \varepsilon$ or $\varepsilon_2 = \varepsilon$, its effective polarizability α^* is zero, and the species is invisible in the surrounding medium and does not experience a dispersion force.

The expressions obtained using the susceptibility theory show the following:

1. *Two similar particles* $(\varepsilon_1 = \varepsilon_2)$ *always attract each other regardless of the nature of the surrounding medium as long as they are not charged.* Thus, two uncharged air bubbles or identical colloidal particles invariably attract each other.

2. *When* $\varepsilon_1 \neq \varepsilon_2 \neq \varepsilon$, *the two species 1 and 2 attract each other when* ε_1 *and* ε_2 *are both either greater or smaller than* ε *and repel each other when* ε *is intermediate between* ε_1 *and* ε_2. Finally, the retarded and nonretarded interactions need not be of the same sign; the non-retarded forces may be repulsive, while the retarded ones are attractive, or vice versa.

Table 2.8. Comparison of Forces Involved in Intermolecular Interactions[a]

Substance	μ (debye)	α_0 (Å^3)	I (eV)	Interactions in 10^{-60} erg cm^6		
				Dipole–dipole force, $2\mu^4/3kT^b$	Dipole-induced dipole force, $2\alpha\mu^2$	London dispersion force, $\frac{3}{4}\alpha^2 I$
Ar	0	1.63	15.8	—	—	50
CH_4	0	2.58	13.1	—	—	97
CO_2	0	2.86	13.85	—	—	136
Na	0	29.7	5.15	—	—	5340
HI	0.78	3.85	10.3	6.2	4.0	191
HCl	1.03	2.63	12.78	18.6	5.4	106
CH_3Cl	1.6	4.56	11.35	109	23	284
H_2O	1.85	1.48	12.67	190	10	33
CH_3NO_2	3.1	3.01	11.34	1530	58	124

[a] 1 eV $= 1.602 \times 10^{-19}$ J; 1 erg $= 10^{-7}$ J.
[b] Quadrupole forces are important for CO_2. For highly polar molecules, the energy expression for oriented dipoles should be used, $2\mu^2/d^3$.

A comparison of energies in intermolecular interactions is given in Table 2.8.

2.6.1. *Dispersion Forces Between Macroscopic-in-Size Aggregates of Atoms*

When an atom (or a molecule or a colloidal particle) approaches an extended aggregate of atoms (a cylinder, a sphere, or a flat surface in the form of a slab representing a solid phase or a liquid), the total interaction between these bodies is evaluated on an assumption of additivity[†] by a summation of interactions carried over in three directions, or by a triple integration. Interactions caused by an overlap of the electrical double layers are considered in the theory developed independently by Derjaguin and Landau (1941) and Verwey and Overbeek (1948). The theory is often referred to as the DLVO theory (see Chapter 7). It is presented in the text by Verwey and Overbeek (1948) and, in addition, briefly reviewed by Haydon (1964) and more extensively by Dukhin and Derjaguin (1974). Interactions due to nonpolar London dispersion forces have been evaluated by de Boer (1936) and Hamaker (1937) for an atom–slab and a slab–slab system using the summation approach. Since then, Casimir and Polder (1948) and others have shown that electromagnetic fluctuations in two different bodies may be described using black body radiation theory (a continuum approach). Lifshits (1955) introduced a general theory of dispersion forces for macroscopic bodies in vacuum using the concept of continuum for large assemblies of atoms, and Dzyaloshinskii *et al.* (1961) extended it to systems comprising a third medium between the interacting bodies. Numerous expressions have been derived to evaluate forces acting between bodies of various geometries. Adamson (1967, p. 331), Parsegian (1975), and Israelachvili and Tabor (1973) list several of these expressions, some of which are reproduced in Table 2.9.

For n atoms in a flat slab interacting with an atom at a distance d in vacuum, de Boer (1936) derived the following formula for the dispersion force,

$$F = -\frac{\pi n C_1}{6} \frac{1}{d^3} \qquad (2.29)$$

[†] The assumption of pairwise additivity of forces or interaction energies is reasonably satisfactory for rarefied media such as gases, but it breaks down for condensed phases. In the latter cases, when electron clouds begin to overlap, the polarizabilities of atoms are affected.

Table 2.9. Nonretarded and Retarded Forces Between Macroscopic
Bodies in Vacuum, Calculated on the Basis of Pairwise Additivity[a]

System	Nonretarded	Retarded
Atom–atom at distance d	energy $U = -(C/d^6)$ force $F = 6C/d^7$	$U = -(K/d^7)$ $F = 7K/d^8$
Atom–flat surface containing n atoms at distance d	$F = \dfrac{n\pi C_1}{2d^4}$	$F = \dfrac{4n\pi K_1}{10d^5}$
Sphere R_1–sphere R_2 at $d \ll R_1, R_2$	$F = \dfrac{A_h R_1 R_2}{6d^2(R_1 + R_2)}$	$F = \dfrac{2}{3}\dfrac{\pi B_h R_1 R_2}{d^3(R_1 + R_2)}$
Sphere R_1–flat surface at distance $d \ll R_1$	$F = \dfrac{A_h R_1}{6d^2}$	$F = \dfrac{2\pi B_h R_1}{3d^3}$
Cylinder R–cylinder R at distance $d \ll R$	$F = \dfrac{A_h R}{6d^2}$	$F = \dfrac{2\pi B_h R}{3d^3}$
Two flat surfaces at distance d, specific force per unit area	$f = \dfrac{A_h}{6\pi d^3}$	$f = \dfrac{B_h}{d^4}$

[a] Constants C, C_1 are the London dispersion force constants for nonretarded forces; constant A_h is the Hamaker constant for nonretarded forces $= n_1 n_2 \pi^2 C_1$, where n_1 and n_2 are the numbers of atoms in the two flat surfaces; constant B_h is the Hamaker constant for retarded forces $= n_1 n_2 \pi K_1/10$; constants K, K_1 are the London force constant for retarded forces; R, R_1, R_2 are radii of spheres.

and for a slab with n_1 atoms interacting with a slab containing n_2 atoms, Hamaker (1937) obtained the following formula for the slab–slab interaction energy:

$$U = \frac{\pi n_1 n_2 C_1}{12} \frac{1}{d^2} = \frac{1}{12\pi} \frac{A_h}{d^2} \qquad (2.30)$$

where $A_h = \pi^2 n_1 n_2 C_1$ is known as the Hamaker constant, and it is of the order of 10^{-12}–10^{-13} erg.

The expressions obtained subsequently by Dzyaloshinskii *et al.* (1961), utilizing the continuum approach, are very complicated derivations of quantum field theory. They can be simplified for the two limiting cases of nonretarded and retarded dispersion forces when the distance d between the two reacting bodies is much smaller than the characteristic absorption wavelengths of the three media. A contribution to the dispersion forces interaction, due to temperature dependence, is usually a small fraction when the interaction takes place in vacuum. However, in a third medium contribution of the temperature-dependent dispersion forces may assume major proportions. For example, for two water phases separated by a hydrocarbon film [as in water-in-oil (w/o) emulsions], the temperature-dependent contribution to the Hamaker constant A_h [in an equation analogous to (2.30)] may be about half of its total value.

For systems of macroscopic-in-size bodies (relative to molecular or atomic scale), comprising either spherical or cylindrical particles, large flat slabs, or blocks, the expressions for interaction forces involve the corresponding dielectric permeabilities ε_1 and ε_2 of the two interacting phases and that of the medium, ε (if other than vacuum). All such interactions due to dispersion forces depend on the differences in effective polarizabilities [equations (2.27) and (2.28)], which in reality means the differences in respective dielectric permeabilities: $\varepsilon_1 - \varepsilon$ and $\varepsilon_2 - \varepsilon$. When the ratios (involving these differences) $(\varepsilon_i - \varepsilon)/(\varepsilon_i + \varepsilon)$, where $i = 1$ or 2, are small compared with 1, $(\varepsilon_i - \varepsilon)/(\varepsilon_i + \varepsilon) \ll 1$, the interaction force derived from the expressions obtained by Dzyaloshinskii *et al.* (1961) or McLachlan (1963, 1965) has the same dependence on the separation distance d as the force indicated in Table 2.9 for Hamaker summation for an appropriate configuration of the interaction bodies. When these ratios are not negligibly small, the expressions derived by the continuum approach have to be employed in evaluations of interaction forces, instead of the Hamaker summation.

Depending in the relative values of the dielectric permeabilities ε_1, ε_2, and ε, an uncharged spherical particle approaching an interphase (interfacial region) between two immiscible liquids (or phases) may behave in one of four ways:

1. The particle may be repelled from the boundary.
2. The particle may be attracted from either side toward the boundary and remains within the boundary upon reaching it.
3. The particle may be attracted from one side to the boundary and repelled from the boundary on the other side.
4. At a certain distance from the boundary, the attraction may change to repulsion, keeping the particle at this distance from the boundary.

These patterns of particle behavior are of particular importance to mineralized froths in flotation systems and to solid-containing emulsion systems which are to be stabilized or destabilized (solids of colloidal size are used in paints, creams, some lubricant formulations, etc.).

The dielectric characteristics of the interacting bodies and of the intervening medium are, therefore, of paramount importance in dispersion force interactions. If the dielectric behavior is such that only a one-frequency dependence is indicated, for example, by a peak somewhere in the UV region[†] around 1000 Å, the dielectric constant can be replaced by its optical value, which is the square of the refractive index, $n_r^2 = \varepsilon$. However, most dielectrics have infrared and microwave absorptions as well, and these contribute to the nonretarded dispersion force interactions. This contribution may be small for interactions in vacuum, about 20% maximum. For a third medium such as water, which has a strong absorption in the low-frequency infrared and microwave region, these low-frequency absorptions, rather than the UV, determine much of the van der Waals force interactions. For more details, see the reviews of Israelachvili and Tabor (1973) or Parsegian (1975) and references therein to original calculations.

When the interacting bodies are covered by surface layers (either physically adsorbed or chemisorbed, at least one molecular layer thick) whose dielectric properties are different from those of the bulk, the concept of a single Hamaker constant breaks down, because it (the Hamaker constant) changes with the distance d between the interacting bodies. For thick layers, the interaction occurs as if the layers themselves were the interacting bodies. For very thin layers, the interaction is proportional to the so-called *effective* Hamaker constant (which comprises a dependence on t/d where t = thickness of the film and d = distance between the reacting film-covered bodies) and inversely proportional to the square of distance d. Details and references are provided by Israelachvili and Tabor (1973).[‡]

[†] The different regions of electromagnetic radiation can be expressed in frequency, wavelength or wave number. The following relationships obtain: The frequency ν (Hz, s^{-1}) $= C/\lambda$, where $C = 2.99795 \times 10^8$ (m s^{-1}) is the velocity of light and λ (m) is the wavelength; $1/\lambda$ (m^{-1} or cm^{-1}) is the wave number. Since electromagnetic radiation transmits energy, each frequency is associated with energy $\Delta E = h\nu$, where h is Planck's constant $= 6.6256 \times 10^{-27}$ erg sec $= 6.6256 \times 10^{-34}$ J s; 1 eV (electron volt) $= 1.602 \times 10^{-19}$ J $= 8066$ cm^{-1}. The region of 10^8–10^{11} Hz is known as the microwave frequency region, 10^{11}–10^{14} Hz as the infrared region, and that of frequencies greater than 10^{14} Hz as the UV region.

[‡] What appears to have been excluded from the theoretical considerations so far is the *nonuniform distribution of the film* covering the interacting bodies. Islands of adsorbed molecules in a monolayer, scattered over the surface of the macroscopic body, or islands of multilayers (patches) represent a most likely state of the film covering a highly heterogeneous surface of real solids. Evidence of this type of nonuniformity of adsorbed film is steadily mounting. Preliminary calculations, carried out by Baratin (1976), indicate that such nonuniformly distributed patches of adsorbed surfactants may exert a much greater attraction at large distances than a uniformly adsorbed surfactant.

2.6.2. *Practical Significance of van der Waals Interaction*

In all multiphase systems van der Waals forces are always present and exert a major influence in both long- and short-range interactions. These interactions include adsorption and adhesion of polyatomic aggregates (nonpolar and polar molecules or aggregates such as colloidal particles) to surfaces of solids and to liquid/gas or liquid/liquid interfaces. Also, the thinning of liquid films, which occurs in the drainage of froth systems and in the processes of aggregation of colloidal particles during flocculation, involves interactions which are due mostly to van der Waals forces. The stability of colloidal dispersion is determined by the balance between the attractive van der Waals forces and the repulsive forces between the charge electrical double layers surrounding each colloidal particulate (liquid droplets in aerosols and emulsions, solid particles, or gas dispersions). Similarly, the stability of mineralized (mineral-laden) froths (which are found on surfaces of containers storing floated materials) depends on the balance of the same two types of forces: van der Waals and electrical. The main difference between the dispersions and the mineralized froths is the far greater heterogeneity of the latter, since three types of interfaces (solid/gas, liquid/gas, and solid/liquid) are participating in their structure instead of just two as in dispersions.

van der Waals forces may be repulsive as well as attractive—the relative values of the dielectric constants of the two interacting phases and of the intervening third medium determine which type of interaction occurs. Small particles may be induced to migrate toward or away from different types of interfaces as a consequence of the action of van der Waals forces. Previously, the electrostatic charge–charge interactions (with the interaction energy being proportional to the inverse of distance $[U \propto (1/d)]$ were considered to be the only ones responsible for the long-range interactions. In systems containing surfactants, the charge–dipole interactions $[U \propto (1/d^2)]$, the dipole–dipole interactions $[U \propto (1/d^3)]$, and the induction interactions between the various groups of surfactant and colloidal particulates $[U \propto (1/d^2) - (1/d^6)]$ are responsible for aggregation. (The bonds between the hydrocarbon chains in micelles, in isolated islands of condensed surfactants at interfaces, and in the close-packed films are solely dispersion force interactions.)

The measurements of the force of attraction between two bodies as a function of the distance between them were first carried out in the 1950s and 1960s. Initial results were obscured by difficulties due to vibrations, surface asperities, and extraneous electrostatic charges on the surfaces. The first successful results obtained around 1970 showed a good agreement between the theoretically predicted Hamaker constants and the quantities evaluated from the measurements [see Israelachvili and Tabor (1973), Buscall and Ottewill (1975), Richmond (1975), and Parsegian (1975)]. For a cylindrical mica sheet approaching another cylindrical mica sheet the transition from the nonretarded van der Waals force interaction with the power of d^{-2} to the retarded one with the power law d^{-3} occurred within 120–500 Å distance between two cylinders.

The Hamaker constants of mica and of stearic acid monolayers were found to be similar and close to 1×10^{-12} erg. Yet the surface energies of hydrocarbons and mica, calculated (using this value of Hamaker constant) to be about 30 erg/cm^2, are in good agreement with the experimentally determined surface energy for hydrocarbons but are approximately an order of magnitude too low in comparison with the experimental values of surface energy for mica. These comparisons indicate that with hydrocarbons the adhesion and cohesion forces involve only van der Waals interactions, whereas with mica there must be strong forces operative at less than 4-Å distance. The latter, nondispersive type of forces dominate the cohesion of mica and, presumably, of all polar solids whose surface energies are in the 100–2000 erg/cm^2 range.

2.7. Metallic Bonding. Band Theory of Solids

Occasionally, flotation systems involve separations of metallic phases, such as, for example, flotation of fine free gold, native copper, or "cement" copper. Detailed treatments of metallic bonding can be found in standard metallurgical texts, such as Hume-Rothery (1936), Barrett (1952), or Seitz (1940).

Metals have many physical properties quite different from those of other solids. The most characteristic ones are high electrical conductance (decreasing with increasing temperature), high thermal conductance, strength, and ductility. These properties are due to the existence of a special type of bonding called metallic bonding, which is neither ionic nor covalent nor any combination thereof.

An explanation of differences in electrical conductivity between metals, insulators, and semiconductors has emerged during the period 1930–1950 from gradual developments of the so-called band theory of solids. When the Schroedinger equation for the motion of an electron began to be applied by Bloch, Brillouin, and others to electrons in a crystalline lattice, the calculations indicated splitting of energy levels and existence of forbidden zones.

When two atoms approach each other, each energy state of these atoms splits into two states. When n atoms are coming together to form a lattice of a solid, each energy state of individual atoms splits into n states (levels), some of which may be degenerate (i.e., have the same energy). With some 10^{22}–10^{23} atoms/cm^3 of the solid the energy levels are spaced so closely together that instead of discrete levels a continuous band of allowed energy levels is visualized. The valence electrons (electrons in the incomplete outer shell of an isolated atom, which are responsible for bonding between atoms) provide two bands. The lower of these two levels of energy results from splitting of the filled valence orbitals and is known as the valence band. The upper (higher) energy band is known as the conduction band and incorporates partially filled or empty levels (resulting from splitting of valence electronic levels of the atoms involved).

Since the splitting of energy states occurs as a result of an overlap between the electronic clouds of participating atoms, the electrons in the inner (core) shells, which are completely filled, do not overlap and show no band structure. Further, the distances between the atoms, the so-called lattice constant d for metallic solids, plays a critical role in determining the diffraction of electron waves and the division of energy space into permitted and forbidden zones.

When the interatomic spacing in the lattice is sufficiently small, the conduction and the valence bands may overlap to a considerable extent (at the spacing d_M), endowing the solid with high electronic conductivity and metallic characteristics (Figure 2.4).

For much larger interatomic spacing d_I, the energy gap separating the valence from the conduction band may be so much greater than the thermal energy of electrons kT that no electrons can be excited thermally from the filled valence band to the conduction band level. The material behaves as an insulator.

For intermediate spacings d_{sc}, the energy gap is comparable with the thermal energy kT, electrons can be excited into the conduction band from the valence band by photoillumination or thermally, and the material behaves as a semiconductor. For more details on semiconductors, see Section 7.6.

The other characteristics of metals such as, for example, strength (or cohesive energy, nearly equal to the enthalpy of atomization) increases from alkali metals to a maximum for transition metals with partially filled d bands (Hf, Ta, W, Re, Os) and then decreases again for Zn, Hg, and Cd.

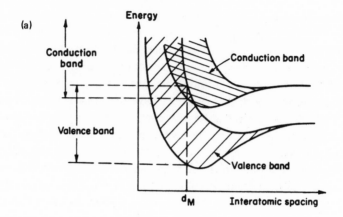

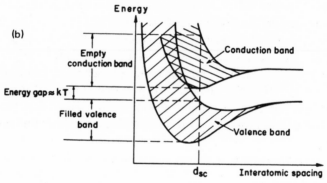

Figure 2.4. Energy bands in (a) a metallic conductor (with overlapping conduction and valence bands at a spacing d_M) and (b) a semiconductor (at a spacing d_{sc}). [From Bockris and Reddy (1970), *Modern Electrochemistry*, Plenum Press, New York.]

The electronic structure and the extent of overlapping that occurs on close packing of atomic aggregates determine their cohesive (and other) properties.

2.8. *The Importance of Steric (Size) Parameters in Interfacial Bonding*

The strength of bonding (or the energy of interaction) between atomic species is usually expressed in appropriate units measured or evaluated for 1 mol (6.023×10^{23}) of bonds. In all applications of surface chemistry, the total area of surface or of interface in a given system is a major factor determining the behavior of this system. Hence the important aspect of bonding within interfacial regions is not so much its type or its relative strength per mole of bonds but the relative specific magnitude of bonding *per unit area of the interface.* The *density of bonds* developed per unit area may change severalfold, occasionally even within two orders of magnitude, with only a minor change in the principal variable, such as partial pressure or concentration. Thus, in addition to *the strength* (energy) *of bonding, its distribution on the surface,* which is ultimately dependent on the *size* of the species developing a particular bond, becomes an important parameter. This feature (the steric parameter) in interfacial bonding makes it possible for the weaker but more densely distributed bonding to compete with and, occasionally, to overcome the effects of a much stronger but less densely distributed bonding.

All the formulas that have been derived for the forces or energies in the ionic, dipole, and dispersion interactions reflect the size parameter to some extent insofar as the distance of interaction d is always involved. However, this parameter relates to the direction perpendicular to the interface, but it *does not reflect the distribution* of the interacting species parallel to the interface. The mean area occupied by the species developing a particular bond, or the number of bonds developed per unit area, or the energy of interaction per unit area (the specific interaction energy) is the parameter which is primarily involved in all comparisons of interfacial reactions. Stereochemistry of interfacial regions and/or the steric parameters of all components of interfaces have to be considered in all surface chemical phenomena whenever relative values decide the behavior of the system as a whole. Numerous instances of this effect of steric parameters are encountered in various applications of surface chemistry. Only when the steric effects are taken into account do some of the surface characteristics become readily understood.

2.9. *Interfacial Thermodynamics, Definitions and Concepts*

The thermodynamic relationships that are normally developed for bulk systems omit variables which are specifically associated with gravitational, magnetic, or electrical fields and with all phase boundaries. For all systems with finely divided phases and a large extent of interfaces, the surface parameters and the associated effects of electrical charges and potentials can no longer be neglected.

The most frequently employed thermodynamic potentials (that is, functional relationships) for bulk systems are defined in terms of the following variables:

Internal energy: $U = F + TS$ (2.31)

Helmholtz free energy
(Helmholtz function): $F = U - TS$ (2.32)

Enthalpy: $H = U + PV$ (2.33)

Gibbs free energy
(Gibbs function): $G = F + PV = U - TS + PV$ (2.34)

where T is the absolute temperature, P is the pressure, S is the entropy, and V is the volume. Parameters T and P are intensive variables (independent of the quantities in the system), whereas S, V, and U and F, H, and G are all extensive variables, dependent on the amount of matter in the system.

For systems involving phase boundaries, two additional parameters are needed to define the state of the system: the surface area A^s, an extensive variable, and the interfacial (surface) tension γ, an intensive variable. Therefore, to each of the functions (2.31)–(2.34) the product $-\gamma A^s$ is added[†].

In the case of enthalpy and Gibbs (free-energy) functions, two pairs of conjugate mechanical variables P, V and γ, A^s are being used; it is possible then to derive three sets of H and G functions depending on the choice of these mechanical variables, e.g.,

$$\bar{H} = U + PV \qquad \text{and} \qquad \bar{G} = F + PV$$
$$\hat{H} = U - \gamma A^s \qquad \text{and} \qquad \hat{G} = F - \gamma A^s \qquad (2.35)$$
$$H = U + PV - \gamma A^s \qquad \text{and} \qquad G = F + PV - \gamma A^s$$

[†] Two conventions are being used in stating the first law of thermodynamics:

$$\Delta U = q + W \qquad \text{and} \qquad \Delta U = q - W$$

where q is the heat absorbed by the system and W is the work done *on* the system or work done *by* the system—hence it is used with a negative sign to state the same law of the conservation of energy.

Similar duality of signs is used with respect to other parameters, such as surface work γA^s, changes in chemical or electrochemical work, etc. As long as there is an internal consistency in the use of signs, the subsequent interrelationships derived between parameters are not affected by these conventions. However, it can be highly confusing when intermediate thermodynamic relationships taken from books using different conventions are being compared.

Of these sets, the functions H and G (but not $\hat{H}$ or $\bar{H}$ and $\hat{G}$ or $\bar{G}$) are the ones most frequently employed.

For a concise treatment of thermodynamic principles applied to chemistry, refer to Everett (1959). The classic treatments of thermodynamics of surfaces and interfaces are those by Gibbs (1878, 1928) and Hill (1960, 1963, 1968). A manual of symbols and terminology in colloid and surface chemistry adopted by IUPAC (International Union of Pure and Applied Chemistry) has also been published by Everett (1972). Schay (1969, 1975) and Goodrich (1969) provide excellent accounts of the application of thermodynamics to adsorption. Thermodynamic aspects of flotation are discussed by Fuerstenau and Raghavan (1976).

A system containing a phase boundary (interface) can be treated either by

1. The so-called Gibbs approach, in which the thermodynamic properties of the interface are defined in terms of *excess* quantities relative to a suitably chosen reference, *a dividing surface*, separating the two phases, or

2. The Hill approach, in which the boundary (distinguished by the term *interphase* in contrast to the geometrical interface) is assigned a definite thickness τ and a volume $V^s = \tau A^s$, and the thermodynamic functions are assigned to this volume as a portion of the total system, interacting with the adjoining bulk phases.

Thus, for a system consisting of phase α, interphase s, and phase β, all thermodynamic functions and variables are expressed as sums (if extensive variables) of composite portions α, β, and s; then

Interfacial energy (surface energy):

$$U^s = U - U^\alpha - U^\beta = F^s + TS^s \tag{2.36}$$

Interfacial Helmholtz free energy:

$$F^s = F - F^\alpha - F^\beta = U^s - TS^s \tag{2.37}$$

Interfacial enthalpy (usually H^s):

$$H^s = H - H^\alpha - H^\beta = U^s + PV^s - \gamma A^s \tag{2.38}$$

Interfacial Gibbs free energy (usually G^s):

$$G^s = G - G^\alpha - G^\beta = F^s + PV^s - \gamma A^s \tag{2.39}$$

where *no superscripts* denote the function for the total system and a super-script α, β, or s denotes the function for the phase α or β or the interphase s. Whatever approach is employed, the final relationships derived for the phase boundaries establish the precise functional dependence among the various thermodynamic variables under clearly specified sets of conditions.

All the preceding interfacial functions can be expressed as the corresponding quantities per unit area of the interface and are usually denoted by lowercase letters:

$$u^s = \frac{U^s}{A^s}$$

$$s^s = \frac{S^s}{A^s}$$

(2.40)

$$\vdots$$

These quantities are then known as the *specific interfacial energy u^s* or the *specific interfacial Gibbs free energy g^s*, etc.

For solid/gas, solid/liquid, and solid/solid interfaces, all the preceding interfacial thermodynamic quantities (with the superscript s, whether specific or total for the interface) *depend on the crystallographic orientation of the solid phase*. In Section 3.2 we shall deal with this aspect in greater detail.

In chemical systems, when matter is exchanged between the constituent portions of the system, the chemical work is expressed in terms of the chemical potential μ_i per mole of species i and the number of moles n_i. The most convenient form of expressing a reversible change in the state of the system is through the differentials, which, for interfacial systems, include the surface work contributions:

$$dU = T\,dS - P\,dV + \gamma\,dA^s + \sum \mu_i\,dn_i \qquad (2.41)$$

$$dF = -S\,dT - P\,dV + \gamma\,dA^s + \sum \mu_i\,dn_i \qquad (2.42)$$

$$dH = T\,dS + V\,dP - A^s\,d\gamma - \sum \Gamma_i\,d\mu_i \qquad (2.43)$$

$$dG = -S\,dT + V\,dP - A^s\,d\gamma - \sum \Gamma_i\,d\mu_i \qquad (2.44)$$

(for μ_i *chemical potential* and Γ_i *surface excess*, see below). For different portions of the system, α, β, or s, the expressions (2.41)–(2.44) contain variables with appropriate superscripts.

The definitions of quantities characteristic of interfacial region(s) follow from the differential relationship (2.41)–(2.44) employing appropriate superscripts:

Surface chemical potential:

$$\mu_i^s = \left(\frac{\partial F^s}{\partial n_i^s}\right)_{T,V^s,A^s,n_j^s} = \left(\frac{\partial G^s}{\partial n_i^s}\right)_{T,P,\gamma,n_j^s} \tag{2.45}$$

Surface tension:

$$\gamma = \left(\frac{\partial F^s}{\partial A^s}\right)_{T,V^s,n_j^s} = \left(\frac{\partial F}{\partial A^s}\right)_{T,V,n_j} \tag{2.46}$$

Surface tension denotes the work necessary to *form* (to create) 1 cm² of a new surface under *conditions of adsorption* if the system is a multi-component one. If the system is a one-component system, then

$$\gamma = f^s = \frac{F^s}{A^s}$$

but whenever there is another component present, e.g., a solute in solution, two gases in the gaseous phase, etc., $\gamma \neq f^s$. At constant T and constant interfacial volume V^s the system will attain equilibrium if the integral $\int \gamma \, dA^s$ attains a minimum. Thus,

$$\gamma A^s = F^s - \sum \mu_i^s \, dn_i^s \quad \text{or} \quad \gamma = f^s - \sum \Gamma_i \mu_i^s \tag{2.47}$$

where

$$\Gamma_i = \frac{n_i^s}{A^s}$$

represents the *surface excess concentration* of species i in the interfacial layers over and above that which would have existed without adsorption (or desorption) at the interface.

From (2.47) it is clear that only when $\sum \Gamma_i \mu_i^s = 0$ does $\gamma = f^s$.

Surface pressure π^s is defined as the difference between the surface tension of a pure substance, γ_0, and the same substance under conditions of adsorption, i.e., γ; thus,

$$\pi^s = \gamma_0 - \gamma = \sum \Gamma_i \mu_i^s \tag{2.48}$$

Interfacial systems containing a solid phase require a definition of an additional quantity, viz. *surface stress*, which denotes the *energy necessary to stretch the surface by 1 cm²*. Surface stress was defined by Shuttleworth in 1950 [see Mullins (1963)] as a *two-dimensional tensor, representing an excess over bulk stresses*:

$$f_{xy} = n_{xy}\gamma + \frac{d\gamma}{d\varepsilon_{xy}} \tag{2.49}$$

where x, y are the axes of a Cartesian reference frame in a plane of the surface. The directions in between axes x and y are denoted by xy; ε_{xy} is the strain in the direction xy; the parameter $n_{xy} = 1$ for the directions xx and yy (i.e., directions parallel to x and y axes), but $n_{xy} = 0$ for $x \neq y$ (i.e., directions not parallel to x or y).

For liquids the term $\partial\gamma/\partial\varepsilon_{xy} = 0$; hence surface tension is identical with surface stress. In solids, either *compressive* or *tensile surface stresses* are possible in the surface region, and these are compensated by the interior elastic stresses.

Most of the interfacial systems involve mobile charges and, thus, electrical double layers at interfaces. Therefore, the expressions for the thermodynamic potentials (2.41)–(2.44) have to be modified by replacing the contributions due to chemical work, $\mu_i n_i$, by the corresponding electrochemical work:

$$(\mu_i + z_i e\phi_p)n_i$$

where ϕ_p is the inner potential (see equation 7.1, p. 435).

2.9.1. *Thermodynamic Relationships in Chemical and Electrochemical Systems*

The conditions leading to establishing an equilibrium in a system, and the free-energy changes in any process occurring reversibly within the system, are of great importance to an understanding of the system.

For a perfect or an ideal gas (defined as one that obeys the $pV = nRT$ gas equation) the free energy G of n moles of this gas at pressure p and temperature T is given by

$$G = G^\dagger + nRT \ln \frac{p}{p^\dagger} \tag{2.50}$$

where $G^\dagger$ is the standard free energy at $p^\dagger$, the standard pressure, which for convenience is usually chosen as unit pressure. $G^\dagger$ is a function of temperature T.

The *chemical potential* of pure substance is defined as the free energy per one mole; for an ideal gas

$$\mu = \frac{G}{n} = \frac{G^\dagger}{n} + RT \ln p \tag{2.51}$$

or

$$\mu = \mu^\dagger + RT \ln p \tag{2.52}$$

For a mixture of gases n_A moles of A and n_B moles of B, the free energy of the system is defined as

$$G = \mu_A n_A + \mu_B n_B \tag{2.53}$$

and the free energy per mole of mixture as

$$\frac{G}{n_A + n_B} = \mu_A X_A + \mu_B X_B \tag{2.54}$$

where

$$X_A = \frac{n_A}{n_A + n_B} \quad \text{and} \quad X_B = \frac{n_B}{n_A + n_B} \tag{2.55}$$

are the *mole fractions* of the components in the mixture.

Hence, for an ideal gas mixture, the chemical potential of substance A is

$$\mu_A = \mu_A{}^\dagger + RT \ln p_A \tag{2.56}$$

where p_A is the *partial pressure* $p_A = X_A p$ and $\mu_A{}^\dagger$ is the *standard chemical potential* of A when its partial pressure is 1 atm. For *ideal solutions*, analogous expressions for the chemical potential (of solute A) are obtained:

$$\mu_A = \mu_A{}^\circ + RT \ln X_A \tag{2.57}$$

where $\mu_A{}^\circ$ is the standard chemical potential of A in solution at 1-mol/liter concentration.

The ideal solution is *perfect* if relationship (2.57) holds for all $0 < X_A < 1$. If equation (2.57) holds only for a limited range of $0 < X_A \ll 1$, then it is referred to as an *ideal dilute solution* of solute A. For the X_A range when the ideal dilute solution relationship (2.57) does not hold, a correction factor f_A is applied known as the *activity coefficient* which is a measure of departure from ideality such that

$$\mu_A = \mu_A{}^\circ + RT \ln X_A f_A \tag{2.58}$$

and the quantity

$$a_A = X_A f_A \tag{2.59}$$

is then known as the *activity* of A in solution.

The criterion of establishing an equilibrium for a reversible reaction

$$a_A A + b_B B \rightleftarrows c_C C + d_D D \tag{2.60}$$

in any system is that the change in the free energy of the whole system be zero with respect to the *dn* moles undergoing reaction:

$$\frac{dG}{dn} = \mu_C + \mu_D - \mu_A - \mu_B = 0 \tag{2.61}$$

which converts to the equilibrium condition

$$\Delta G° = -RT \ln \frac{c_C d_D}{a_A d_B} \tag{2.62}$$

or

$$\Delta G° = -RT \ln K \tag{2.63}$$

where

$$K = \frac{c_C d_D}{a_A b_B} = equilibrium \ \ constant \tag{2.64}$$

and $\Delta G°$ is *the standard free-energy change* realized when 1 mole of each pure substance A and B are reacted to give pure C and D in the mixture.

The equilibrium constant K in (2.63) and (2.64) becomes the usual mass–action constant (Guldberg–Ostwald), or Henry constant for vapor pressure relationship, or solubility product, dissociation constant, etc.

By convention, the activities of the products are written as the numerator in the expression for K, and the standard free-energy change $\Delta G°$ of reaction (2.60) is defined as the difference between the standard free-energy change of the products less that of the reactants.

The standard free-energy change varies with temperature in a way similar to the variation of the equilibrium constant K with temperature. Since $G = H - TS$ [equations (2.33) and (2.34)], the way in which the equilibrium constant varies with temperature has been derived as the so-called *Gibbs–Helmholtz equation*:

$$\Delta G° = \Delta H° + T \frac{d(\Delta G°)}{dT} \tag{2.65}$$

or

$$\frac{d(\ln K)}{dT} = \frac{\Delta H°}{RT^2} \tag{2.66}$$

(known as the Van't Hoff isochore).

If the heat of the reaction is independent of temperature, a simplified integrated form of the preceding relationship

$$\ln K = -\frac{\Delta H}{RT} + constant \tag{2.67}$$

is obtained, serving as a basis for all graphical representations of equilibrium constants (for example, Ellingham graphs of metal oxide or sulfide formation relating $\Delta G°$ vs. T). When a half reaction involves a release of electrons during the formation of a product, it is known as an *oxidizing reaction*, for example,

$$H_2 \rightarrow 2H^+ + 2e \qquad \text{(homogeneous system)}$$

$$Zn \rightarrow Zn^{2+} + 2e \qquad \text{(heterogeneous system)}$$

When electrons are consumed by the reactant forming a product, the reaction is a *reducing* one, for example,

$$Zn^{2+} + 2e \rightarrow Zn$$

Reactions and systems which involve ions and electrons are known as *electrochemical*, while those which do not involve ionized species are referred to as chemical. The electrochemical reactions may be either homogeneous, such as reduction–oxidation between dissolved species and/or gases, or heterogeneous, at solid/gas, solid/solid, or solid/solution interfaces. The solids may be highly conducting metals or nonmetals (e.g., graphite) acting as *electrodes*, or they may be semiconductors like metallic sulfides and some oxides. The solution must be conducting, and for that it must contain ionized species usually derived from easily dissociating compounds like ionic-bonded salts; it is known as an *electrolyte*.

Electrons participating in an electrochemical reaction are considered a separate chemical component of the system. Of the two possible conventions of presenting electrochemical data in terms of potentials, IUPAC adopted the convention of writing any reaction involving electrons in the form

$$\text{oxidized species} + ne \rightleftarrows \text{reduced species} \qquad (2.68)$$

for example,

$$Zn^{2+} + 2e \rightarrow Zn$$

$$Cu^{2+} + e \rightarrow Cu^+$$

The associated potential, representing the additional term in the free-energy change of such an electrochemical reaction (2.68), is thus a reduction product potential. It represents the additional free energy involved in the transfer of electronic charge across the solid/liquid interface or from one ionic species to another. There is a difference in two outer potentials (see equation 7.2, p. 435 for details on outer potentials) associated with the interface or with variation of species concentrations. Thus, for a reaction (2.68),

$$\Delta G = \Delta G^\circ - nFE \qquad \text{(IUPAC reduction scale)} \qquad (2.69)$$

$$= \Delta G^\circ + nFE \qquad \text{(oxidation scale used by Pourbaix)} \qquad (2.69')$$

where n is the number of electrons involved and $F =$ Faraday constant $=$ 96 487 C mol^{-1}. When an equilibrium is established,

$$\Delta G = 0 \qquad \text{and} \qquad \Delta G^\circ = nFE^\circ \qquad (2.70)$$

$$\text{or} \quad \Delta G^\circ = -nFE^\circ \qquad (2.70')$$

Using Pourbaix oxidation scale,

$$\Delta G^\circ = \ln K = \ln \frac{a_{\text{ox}}}{a_{\text{red}}} \qquad (2.71)$$

For temperature 25°C $(T = 298.1°K)$ and $R = 1.987$ cal deg^{-1}mol^{-1}, denoting by μ° the standard chemical potential at 25°C and changing from natural to

$$\log_{10} a = \frac{\ln a}{2.3026}$$

relationship (2.69) becomes

$$\Delta G = \log K - \frac{nE}{0.0591} \qquad (2.72)$$

and

$$\log K = -\frac{\sum \mu^\circ}{1363} \qquad (2.73)$$

$$E^\circ = \frac{\sum \mu^\circ}{23,060n} \qquad (2.74)$$

as derived by Pourbaix (1949) and used subsequently by Garrels and Christ (1965) and others.

Since there is no reliable method of measuring the absolute potential of a single electrode (solid/solution interface), another electrode is incorporated in the measurement circuit, serving as a reference electrode, and a potential difference between these two electrodes is measured. The standard hydrogen electrode (consisting of Pt in solution of unit activity of H$^+$, at 1 atm H$_2$ gas pressure) is arbitrarily chosen as the reference electrode representing zero potential on the hydrogen scale of potentials. For convenience, substitute reference electrodes are often employed, such as one of the three calomel electrodes. These consist of either

1. 0.1 N KCl, $Hg_2Cl_2(s)|Hg$, with a $+0.3338$-V potential on the hydrogen scale, or

2. 1.0 N KCl, $Hg_2Cl_2|Hg$, with a $+0.2800$-V potential on the hydrogen scale, or

3. Saturated KCl, $Hg_2Cl_2|Hg$, with a $+0.2415$-V potential on the hydrogen scale, all potentials at 25°C.

Other substitute reference electrodes are $Ag|AgCl(s)$ in (a) 1 N Cl^- ion concentration with a $+0.222$-V potential and (b) saturated NaCl solution with a $+0.225$-V potential, all in the reduction scale.

Application of equation (2.71) to homogeneous oxidation–reduction (or redox) potentials of ideal solutions (in which activities are equal to concentration) provides data represented graphically in Figure 2.5(a). Heterogeneous redox systems (representing reversible oxidation–reduction of metals and their appropriate ions in the solution) give, for an ideal solution behavior, the linear relationship represented in Figure 2.5(b).

2.9.2. *Thermodynamics of Adsorption at an Air/Liquid Interface*

For an equilibrium in a two-component system, the differential form of equation (2.47) is

$$dy + \Gamma_1^n d\mu_1 + \Gamma_2^n d\mu_2 = 0 \qquad (2.75)$$

where the superscript n in the surface excess quantities Γ of components 1 and 2 refers to a constant *total number of moles in the system*. The surface excess thus defined is independent of the amount of bulk phases present— *provided these extend far away from the interfacial region*. Thus, any change in the number of moles of component 1, n_1, is accompanied by, and at the expense of, a counterchange in n_2, i.e.,

$$\Delta n_1 + \Delta n_2 = 0 \qquad \text{or} \qquad \Gamma_1^n + \Gamma_2^n = 0 \qquad (2.76)$$

signifying that any positive excess (adsorption) of one constituent at the air/liquid interface is accompanied by an equivalent negative excess of the other component, its expulsion from the interface.

Another way of defining a surface excess is by referring to a system containing *the same number of moles of one component*, for example, the component β (and not the total number of moles in the system); if x_α is

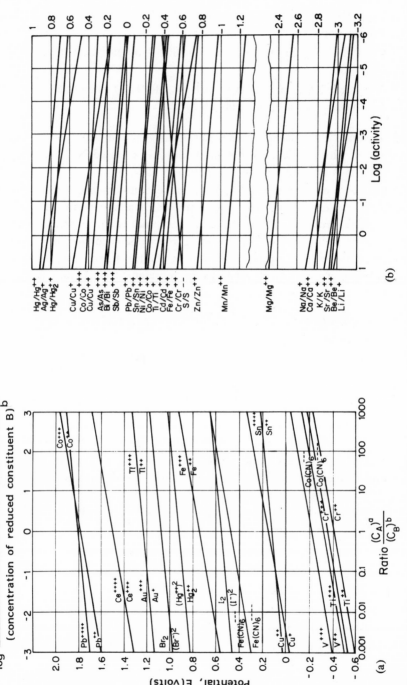

Figure 2.5. (a) Influence of potential on equilibria in homogeneous systems (oxidation–reduction potentials). [From Pourbaix (1949), with the permission of the author.] (b) metal–solution potentials (oxidation–reduction potentials in heterogeneous systems.) Oxidation scale potentials are not IUPAC. [From Pourbaix (1949), with the permission of the author.]

the mole fraction of component α in the binary liquid mixture and x_β is the mole fraction of component β, then

$$\left(\frac{\partial \gamma}{d\mu_\alpha}\right)_T = \Gamma_\alpha^\beta \cong \frac{\Gamma_\beta^n}{x_\beta} \quad \text{and} \quad \left(\frac{\partial \gamma}{\partial \mu_\beta}\right)_T = \Gamma_\beta^\alpha \cong \frac{\Gamma_\beta^n}{x_\alpha} \qquad (2.77)$$

where Γ_α^β with the superscript β denotes the surface excess of component α with respect to the total number of moles of component β. It follows that

$$x_\beta \Gamma_\alpha^\beta + x_\alpha \Gamma_\beta^\alpha = 0 \qquad (2.78)$$

For dilute aqueous solutions the distinction between these two types of surface excesses is not of much importance, since the mole fraction of water is then close to unity. However, for other solvents, both mole fractions may be varied from 0 to 1, and the surface excess quantities will differ from Γ_1^n and Γ_2^n.

Figures 2.6(a), (b), and (c) show examples of surface tension–concentration relationships for some inorganic salts and alcohols, methyl and ethyl.

Preferential adsorption of surfactants at the air/solution interface causes a lowering of surface tension in dilute solutions of surfactants. For a *nonionized* surfactant *the Gibbs equation* relates the change in surface tension to the logarithm of activity and the surface excess of adsorbed surfactant:

$$-d\gamma = \sum RT(\Gamma_s - x_s \Gamma_{H_2O}) \, d\ln a_s \qquad (2.79)$$

This simplifies for very dilute solutions to

$$-d\gamma = RT\Gamma_s \, d\ln a_s \cong RT\Gamma_s \, d\ln c_s \qquad (2.80)$$

where γ is the surface tension, Γ_s is the surface excess of surfactant, Γ_{H_2O} is the surface excess of water, a_s is the activity of surfactant, c_s is the concentration of surfactant, and x_s is the mole fraction of surfactant.

The magnitude of surface excess Γ_s at any c_s is obtained as the slope $d\gamma/d\ln c_s$ multiplied by RT. For an *ionized* surfactant, such as a sodium salt of a surfactant, Na^+-s^-, the Gibbs adsorption equation takes the form

$$d\gamma = -RT\Gamma_{Na^+} \, d\ln a_{Na^+} - RT\Gamma_{s^-} \, d\ln a_{s^-}$$

where a_{s^-} denotes the activity of charged surfactant species s^- (anion); the condition of electrical neutrality at the interface requires that

$$\Gamma_{Na^+} = \Gamma_{s^-}$$

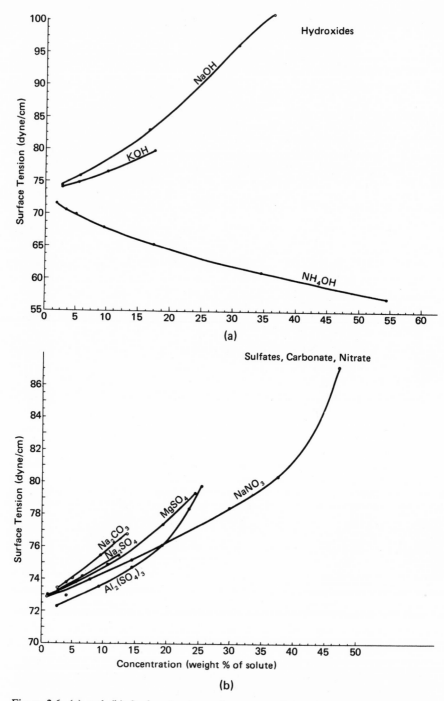

Figure 2.6. (a) and (b) Surface tensions of aqueous solutions of inorganic salts; (c) surface tensions of methyl and ethyl alcohols.

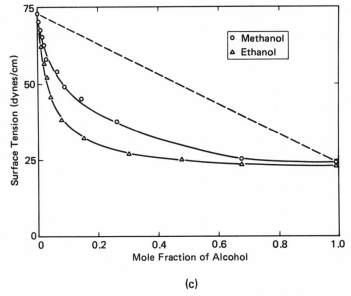

(c)

Figure 2.6 (*continued*).

and further that

$$d \ln a_{\mathrm{Na}^+} = d \ln a_{s^-}$$

Hence,

$$d\gamma = -2RT\Gamma_{\mathrm{Na}-s} \, d \ln a_{\mathrm{Na}-s} \tag{2.81}$$

If an excess of a neutral electrolyte containing an ion in common with the surfactant, such as NaCl with *sodium* alkyl sulfates or carboxylates, is added to the solution, then

$$d \ln a_{\mathrm{Na}^+} \neq d \ln a_{s^-}$$

and the concentration of Na^+ counterions remains practically constant, being several orders of magnitude greater than the surfactant concentration. Thus the adsorption equation becomes

$$d\gamma = -RT\Gamma_s \, d \ln a_{s^-} \tag{2.82}$$

In the absence of indifferent electrolyte, a factor 2 appears in the Gibbs equation (2.81); in the presence of an excess of salt electrolyte the factor is unity [equation (2.82)]. For small additions of electrolyte, comparable in order to the concentration of surfactant, the Gibbs equation becomes

$$d\gamma = -mRT\Gamma_s \, d \ln a_s \tag{2.83}$$

where the "salt" parameter m was shown by Matijevic and Pethica (1958) to become

$$m = 1 + \frac{C_s}{C_s + C_{\text{NaCl}}} \tag{2.84}$$

where C_s is the concentration of surfactant and C_{NaCl} is the concentration of neutral salt, NaCl.

2.9.3. *Thermodynamics of Adsorption at a Solid/Liquid Interface*

The basic assumption is that the solid is insoluble in the (binary) liquid mixture, that is, does not contribute any species to the liquid. An exact interpretation of the interfacial tension at the solid/gas and the solid/liquid interfaces is still lacking (because of unknown contributions due to surface stress, an extra entropy contribution due to structural and orientation effects on the liquid side of the interface, etc.) Neither $\gamma_{a/s}$ nor $\gamma_{s/1}$ can be measured. Calorimetrically measured heats of immersion are proportional to the extent of interfacial area provided the whole solid surface is uniform and is equally accessible to the components of the solution (not porous or with restricted access as in molecular sieves).

For a two-component liquid (binary mixture) $\alpha + \beta$, the adsorption on the solid is a displacement of one component by the other; a positive surface excess of one must be accompanied by an equivalent negative excess of the other. Since $x_\alpha + x_\beta = 1$, then $\Delta n_\alpha + \Delta n_\beta = 0$, and the surface excesses which can be determined experimentally are

$$\Gamma_\alpha{}^\beta = \frac{\Delta n_\alpha}{A x_\beta} \quad \text{and} \quad \Gamma_\beta{}^\alpha = \frac{\Delta n_\beta}{A x_\alpha} \tag{2.85}$$

where A is the surface area of the solid and

$$\left(\frac{\partial \gamma}{\partial \mu_\alpha} \right)_{T,P} = -\Gamma_\alpha{}^\beta \quad \text{or} \quad \left(\frac{\partial \gamma}{\partial \mu_\beta} \right)_{T,P} = -\Gamma_\beta{}^\alpha \tag{2.86}$$

The final changes in $\gamma_{s/1}$ can be computed by graphical integration of either of the preceding relationships if the dependence of chemical potential on mole fraction and the activity coefficients are known. The change in γ is negative for a positive surface excess of one adsorbing component, say solute β, but this negative change is achieved not by adsorption alone but as a result of *the desorption of an* equivalent amount of solvent α *and the adsorption* of solute β.

2.9.4. *Thermodynamics of Thin Liquid Layers*

Thin layers of liquids (formed between the bubbles in a froth or underneath a bubble attaching to a hydrophobic solid or between flocculated solids or between stabilized emulsion droplets) possess special thermodynamic properties in that *their chemical potential depends on the thickness h of the thin layer* and *differs* from the chemical potential of the bulk liquid, μ_{liq}.

The differential of the Helmholtz free energy for a thick layer of a one-component system is

$$dF = -S\, dT - P\, dV + \gamma\, dA^s + \mu\, dn \qquad (2.87)$$

which for an incompressible liquid ($P\, dV = 0$) gives

$$\mu = \left(\frac{\partial F}{\partial n}\right)_{T,A^s} \qquad (2.88)$$

When such a layer thins into a film of thickness h, a change in the chemical potential with thickness h occurs, and this change has been defined in terms of an extra pressure Π acting in the film over and above that existing within the bulk of the liquid.

The various forms of the definition are assembled as follows:

$$\Delta\mu_h = \mu_h - \mu_{\text{liq}} = v_m\Pi = v_m\left(\frac{\partial F}{\partial h}\right)_{T,A^s} = \frac{RT \ln P}{P_s} \qquad (2.89)$$

where μ_h is the chemical potential of the thin film of thickness h, μ_{liq} is the chemical potential of the bulk liquid, v_m is the molar volume of liquid, P is the vapor pressure in equilibrium with the thin film, and P_s is the saturated vapor pressure at T. This extra pressure Π has been called the *disjoining pressure* by Derjaguin (1935). It tends to zero, $\Pi \rightarrow 0$, when the layer becomes thick, $h \rightarrow \infty$, and then $\mu_h \rightarrow \mu_{\text{liq}}$.

For an ideal solution of bulk liquid concentration C_{liq} the change in the chemical potential of the film is

$$\Delta\mu_h = RT \ln \frac{C_h}{C_{\text{liq}}} \qquad (2.90)$$

where C_h is the concentration in the thin film.

An indication of the magnitudes of the disjoining pressure operative in thin films is given by Padday (1970) in the form of a comparative set of figures which show that, for example, films 1500–4000 Å thick possess a disjoining pressure $\Pi \cong 10^3$ dyn/cm² that would raise a column of liquid on a Wilhelmy plate by approximately 1 cm; films 400–100 Å thick have $\Pi \cong 10^4$ dyn/cm² and would raise the liquid 10 cm; and films 60–20 Å thick have $\Pi = 10^8$ dyn/cm² that would raise the liquid 10^5 cm.

The disjoining pressure is composed of several contributions (electrical, van der Waals, steric hindrance); hence the thickness of the films varies with the type of liquid for the same value of disjoining pressure. The characteristics of thin layers are discussed in greater detail in Section 9.2. It will be seen there that the disjoining pressure may acquire negative values for some thin films under specific conditions, giving nonequilibrium films undergoing sudden rupture.

2.10. *Interfacial Wetting and Contact Angle Relationships*

The term *hydrophilic solid* denotes a solid which is completely wetted by the bulk water phase (or an aqueous solution), while *hydrophobic* denotes a partial or incomplete wettability by the water phase. Wetting signifies an adhesion of a liquid phase to another phase (a solid or a liquid) such that the cohesion of the spreading (wetting) phase is exceeded.

Theoretically, when a column of a pure liquid, 1 cm² in cross section, is separated by a direct pull to form two surfaces, each 1 cm² in area, the energy of cohesion of this liquid, W_C, is converted to the surface energy,

$$W_C = 2\gamma_0 \tag{2.91}$$

where $\gamma_0 = f^s = $ the specific surface free energy of the liquid.

If the liquid contains a surface active component, the work of creating the two new surfaces is equal to the change in the surface free energy:

$$W_C = 2(\gamma_0 + \mu_i \Gamma_i) = 2\gamma \tag{2.92}$$

where γ denotes the *surface tension* measured at constant μ_1.

When a column of two different liquids, *a* and *b*, insoluble in each other and in contact across 1 cm², is separated along the interface, the *work of adhesion* W_A to be overcome in this separation has been derived by Dupré (1867):

$$W_A = \gamma_a + \gamma_b - \gamma_{ab} \tag{2.93}$$

indicating that the two new surface tensions γ_a and γ_b are created, with a loss of the initially existing interfacial tension γ_{ab}. See Figure 2.7.

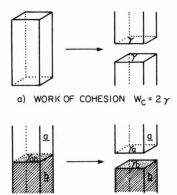

Figure 2.7. Work of cohesion and work of adhesion—definitions.

Harkins (1952) extended this approach to the spreading of one insoluble liquid over another liquid or solid phase, introducing a *spreading coefficient*, defined as

$$S = W_A - W_C > 0 \tag{2.94}$$

which has to be positive in value for a spontaneous spreading to occur. If γ_b denotes the surface tension of the spreading liquid and γ_a that of the substrate, then

$$S = \gamma_a - \gamma_b - \gamma_{ab} > 0 \tag{2.95}$$

For systems consisting of two insoluble liquids, all interfacial tensions can be determined, and the predictions of the spreading coefficient can be verified experimentally. Harkins (1952) has confirmed experimentally that whenever $S > 0$, the spreading does take place.

Quite frequently, however, for some liquids the initial value of S is found to be positive, but when each phase becomes saturated with the vapor of the other phase, the final S' becomes negative, $S' < 0$, and the initially spread phase retracts; some examples can be seen in the spreading of oils on water at 20°C in Table 2.10 [from Harkins (1952)].

When $S < 0$ for a liquid b with respect to liquid a, liquid b does not spread but forms a lens floating on a (if the density of b is lower than that of a), establishing the so-called Neumann's triangle of surface forces (Figure 2.8). For an equilibrium in this system, the following relationships, involving two sets of equations, must be satisfied:

$$\frac{\gamma_a}{\sin \alpha} = \frac{\gamma_b}{\sin \beta} = \frac{\gamma_{ab}}{\sin[360 - (\alpha + \beta)]} \tag{2.96}$$

Table 2.10. Initial and Final Spreading Coefficients of Oils on Water
(dyn/cm)

Oil	Initial S	Final S'
n-Heptyl alcohol	37.8 (spreading)	−5.9 (retraction)
Benzene	9.8 (spreading)	−1.5 (retraction)
Isoamyl alcohol	44.0 (spreading)	−2.6 (retraction)
Carbon disulfide	−4.2 (no spreading)	−10.0 (no spreading)
Methylene iodide	−26.5 (no spreading)	−24.2 (no spreading)

Buff and Saltburg (1957) discussed the thermodynamics of such a system of nonspreading liquids and came to the conclusion that for a complete evaluation of the system an additional piece of information, a length parameter of the lens, L, is also required.

When substrate a in a nonspreading system is a flat solid, the conditions depicted in Figure 2.9 obtain. The existence of a negative spreading coefficient,

$$S = \gamma_{a/s} - \gamma_{a/l} - \gamma_{s/l} < 0 \qquad (2.95')$$

establishes a definite contact angle θ across the liquid phase which is undergoing deformation (subscripts a/s, a/l, and s/l refer to the air/solid, air/liquid, and solid/liquid interfaces). The value of cos θ is frequently used (together with $\gamma_{a/l}$) in evaluating the difference in the two unknown surface quantities ($\gamma_{a/s}$ and $\gamma_{s/l}$) of the solid substrate, that is,

$$\gamma_{a/s} - \gamma_{s/l} = \gamma_{a/l} \cos \theta \qquad (2.97)$$

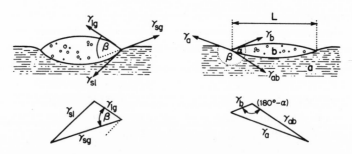

Figure 2.8. Contact angles in a system of a nonspreading liquid on another immiscible liquid. Equilibrium of surface tension vectors for a lens of nonspreading oil on water.

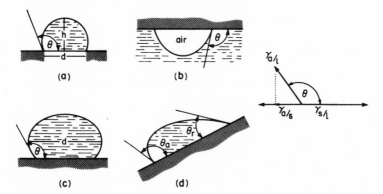

Figure 2.9. Contact angles for a system of a nonspreading liquid on flat surfaces. Hysteresis of contact angle on an inclined plane. Traditional misrepresentation of vectorial relationship between respective surface tensions.

The preceding relationship is frequently referred to as the Young equation. Its use in expressions (2.95') and (2.93) results in the relationships

$$S = \gamma_{a/l}(\cos \theta - 1) \leqq 0 \tag{2.98}$$

and

$$W_A = \gamma_{a/l}(\cos \theta + 1) \tag{2.99}$$

for the presumed negative spreading coefficient and the work of adhesion, respectively, of a nonwetting liquid on a solid.[†]

[†] The concept of spreading is of direct consequence to the process of liberation of solids in grinding and that of particle-to-bubble attachment in flotation. *As regards liberation*: So that two adjoining solid phases 1 and 2 will *separate along their interface*, the work of adhesion

$$W_A = \gamma_1 + \gamma_2 - \gamma_{12}$$

must be smaller than the cohesion of either of the two phases; that is,

$$W_A < 2\gamma_1 \quad \text{and} \quad W_A < 2\gamma_2$$

Therefore, two inequalities must hold:

$$\gamma_1 - \gamma_2 - \gamma_{12} < 0$$

and

$$\gamma_2 - \gamma_1 - \gamma_{12} < 0$$

Hence,

$$\gamma_{12} > \gamma_1 - \gamma_2 \quad \text{and} \quad \gamma_{12} > \gamma_2 - \gamma_1$$

Whichever of the two solids has a higher γ, say $\gamma_1 > \gamma_2$, then $\gamma_{12} > \gamma_1 - \gamma_2$ becomes

2.10.1. *Interpretations of Contact Angle*

Whenever a particle attaches to an air bubble, there is always in the profile of the three-phase contact a definite "angle" across the liquid phase. This angle is assumed to be directly related to the contact angle established by the same liquid on a flat surface of the same solid. In fact, in the development of the theory of flotation (since 1916) the conditions leading to the formation of contact angle on a flat solid were often considered to be synonymous with collector adsorption and with flotation conditions. The first book by Wark (1938) on the principles of flotation provides numerous evidence regarding the significance of contact angle studies in flotation research.

Gradually, however, some inconsistencies have been observed in the approach identifying floatability with the contact angle development. In the last two decades a considerable amount of controversy has arisen on the subject of the contact angle relationship itself. The effect of gravity on equation (2.97) was discussed at the 2nd International Congress on Surface Activity in 1957, and since then relationships (2.98) and (2.99) were examined by different authors: Adam (1957), Adam and Livingston (1958), Johnson (1959), Collins and Cooke (1959), McNutt and Andes (1959), Leja and Poling (1960), Kitchener (1960), Johnson and Dettre (1964, 1969), and Johnson *et al.* (1977). An extensive treatment of contact angles is given by Adamson (1967). In 1963, a special symposium devoted to various aspects of contact angle and adhesion was organized by the American Chemical Society and its proceedings published as ACS Chemistry Series No. 43 [Gould (1964)].

A number of papers on contact angles, presented at the symposium on wetting, held in September 1966 at Bristol University, are published as *Wetting*, Society of Chemical Industry Monograph No. 25, Gordon & Breach, New York (1967).

---

a necessary and sufficient condition for separation during grinding along the interface.

If, as is more often the case, $W_A \geq 2\gamma_1$ and $W_A \geq 2\gamma_2$, liberation occurs within the weaker of the adjoining phases. The residual layer of the phase which has yielded during liberation must be subsequently removed by dissolution and/or abrasion.

As regards *particle–bubble attachment*: So that an air bubble will displace water from the surface of the particle, the spreading coefficient of water over the solid/gas must be less than zero; that is,

$$S = \gamma_{a/s} - \gamma_{a/l} - \gamma_{s/l} < 0 \quad \text{or} \quad \frac{\gamma_{a/s} - \gamma_{s/l}}{\gamma_{a/l}} < 1$$

Sheludko *et al.* (1970b) studied the expansion of the contact area in relation to flotation. Schulze (1977) examined, theoretically and experimentally, the stability of particle–bubble aggregates in relation to particle size floatability. Oliver and Mason (1977) used an electron scanning microscope to view microspreading of liquids on rough surfaces; Pethica (1977) suggested that the problem of contact angle equilibrium be thoroughly reexamined experimentally; Neumann and Good (1972) discussed the thermodynamics and the effect of surface heterogeneity on contact angles; Eick *et al.* (1975) and Huh and Mason (1977) discussed the effects of roughness; while Good (1973) compared the different interpretations of surface tension parameters involved in the expression for contact angle. Neumann (1974) published an exhaustive review of temperature dependence, interpretation, and applications of contact angles. The most recently published papers on various aspects of contact angle relationship can be found in the proceedings of a symposium organized at Loughborough University, Leicestershire (England) [Padday (1978)]. The arguments and experimental evidence presented by different authors reaffirm that the topic of contact angle and its equilibrium and hysteresis is an ever-recurring subject of discussion.

There are many indications that equilibrium in a system involving contact angle is frequently not established. Hysteresis of contact angle readily occurs, and two different values of contact angle are usually measured, depending on the manner in which the angle is allowed to become established, that is, a receding contact angle θ_r, formed when the liquid is withdrawn from the solid that has been partly wetted, and an advancing contact angle θ_a, formed when a partly wetting liquid is made to flow over the solid/gas interface. The most common example of hysteresis and of the effect of gravity on the magnitude of θ is a drop of liquid on an inclined planar solid. It is always found that $\theta_a > \theta_r$, and the so-called "equilibrium" contact angle is an intermediate value: $\theta_r < \theta_e < \theta_a$. (There is no unambiguous rule to establish θ_e. Note also that along the circumference of the drop on an inclined surface [Figure 2.9(d)] there must occur all angles intermediate between the values θ_r and θ_a.) Further, the so-called "rugosity" of the solid surface, or surface roughness, calls for corrections to be applied to the Young formula (2.97). The usual correction is in terms of a coefficient r_W, giving the ratio of the actual to the geometrical surface area; thus the Young equation (2.97) becomes

$$\gamma_{a/s} - \gamma_{s/l} = r_W \gamma_{a/l} \cos \theta \qquad (2.100)$$

(often referred to as the Wenzel formula). The well-documented heterogeneity of solid surfaces (which makes $\gamma_{a/s}$ and $\gamma_{s/l}$ differ from their average value assumed for a uniform surface) must contribute to the contact angle hysteresis; it must also be responsible, in part, for a discontinuous change in the contour of the three-phase contact, often observed in real systems [see, for example, Oliver and Mason (1977)].

For flotation systems, the heterogeneity of the solid surfaces and the development of scattered patches of hydrophobic areas (produced by adsorption of collector-acting species on such heterogeneous surfaces) make it necessary to use a more realistic model of surface than the polished flat surface employed hitherto. Dettre and Johnson (1967) discussed the development of contact angles on porous surfaces with several configurations of regularly spaced arrays of solid asperities (or pores). Their theoretical predictions have been subsequently compared with measurements of advancing and receding contact angles and have indicated a reasonable degree of agreement. However, the true surfaces are not composed of uniformly arrayed hydrophobic–hydrophilic areas. The degree of agreement with the behavior predicted on the basis of theoretical models may not be as high as indicated.

When a comparison is made of the relationship (2.96) describing an equilibrium in a nonspreading liquid–liquid–air system and the relationship (2.97) for a solid–liquid–air system, it becomes apparent that the arbitrary choice of a flat undeformable solid surface in deriving equation (2.97) is equivalent to stating that

$$\gamma_{a/l} : \sin 180° = \gamma_{a/s} : \sin \theta$$

Since $\gamma_{a/l} : \sin 180°$ is indefinable (because of $\sin 180° = 0$), the single expression (2.97) cannot unequivocally describe the equilibrium of the solid–liquid–gas system. [Note that *two equations* are necessary to express the equilibrium (2.96) in a deformable liquid–liquid–gas system.]

Analyzing the process of bubble attachment to a flat solid surface, Leja and Poling (1960) came to the conclusion that the contact angle is simply an indication of the extent to which the given solid–liquid–air system is utilizing the available free energy of interfaces in the bubble deformation process [Figure 2.10(a)].

If there are external agencies helping or hindering such a deformation, the magnitude of the contact angle changes accordingly, resulting in hysteresis. The experimental values of contact angles reflect, therefore, not only the extent of the free energy of interfaces available for some spontaneous transformation in the system but also the nonequilibrium and the non-

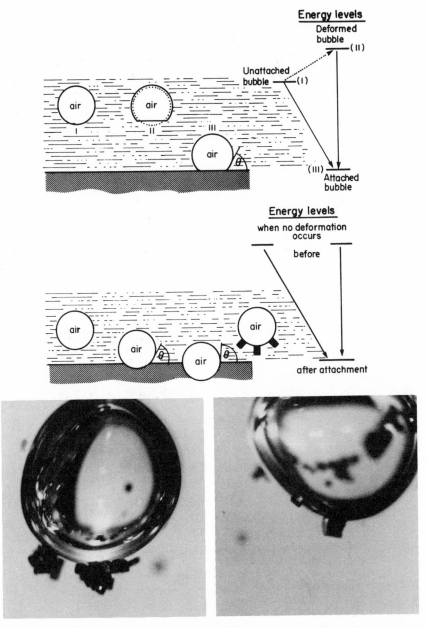

Figure 2.10. (a) Deformation of a bubble on attachment to a rigid flat surface. Relationship between respective energy levels before and after deformation and attachment. (b) Attachment of bubbles to solids without deformation of the air/liquid interface. (c) Photographs of particle–bubble attachment without deformation of bubble shape at the three-phase contact.

interfacial (kinetic and mechanical) characteristics of the deformable phase, that is, the liquid/gas or liquid/liquid interfaces. Hence, the viscosity of the liquid itself and that of the liquid/gas interface, the elasticity of the interface, etc., are also reflected. The theoretical contact "angle" (representing only the thermodynamic parameters as the ratio

$$\frac{\gamma_{a/s} - \gamma_{s/l}}{\gamma_{a/l}}$$

of the system) is obtained in all mathematical derivations as the *limiting* value, or the maximum proportion of the available free energy of interfaces, to be utilized in the combined process of deformation and adhesion. In attachment without deformation [Figure 2.10(b)], no meaningful contact angle is involved. When the concept of nonspreading is applied to the solid–liquid–gas systems, the expression

$$S = \gamma_{a/s} - \gamma_{s/l} - \gamma_{a/l} < 0 \tag{2.95'}$$

becomes

$$\frac{\gamma_{a/s} - \gamma_{s/l}}{\gamma_{a/l}} < 1 \tag{2.101}$$

This inequality is obviously different from expression (2.97), which, by identifying the ratio of interfacial tensions with $\cos \theta$,

$$\frac{\gamma_{a/s} - \gamma_{s/l}}{\gamma_{a/l}} = \cos \theta \tag{2.102}$$

automatically restricts the preceding ratio to the limits -1 and $+1$, that is,

$$-1 \leqq \frac{\gamma_{a/s} - \gamma_{s/l}}{\gamma_{a/l}} \leqq 1 \tag{2.103}$$

with no justification whatever.

Still more significant is the fact that the electrical parameters pertaining to the respective interfaces are omitted in all relationships developed so far for the work of wetting, adhesion, spreading, etc. This omission effectively restricts all these formulas to the conditions in the system when all three interfaces involved are, simultaneously, at the point of zero charge, pzc (see Chapter 7). The fundamental equation for polarizable interfaces (see Section 7.3),

$$d\gamma = -q_M \, dE - \frac{q_M}{z_j F} \, d\mu_j - \sum \Gamma_i \, d\mu_i \tag{7.10}$$

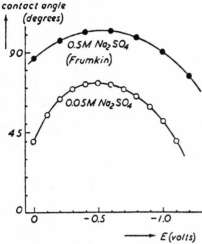

Figure 2.11. Contact angles on mercury as a function of potential vs. SCE (saturated calomel electrode), no surface active agent; $\bigcirc$, calculated points ($0.05M$ Na_2SO_4); $\bullet$, experimental points found by Frumkin ($0.5M$ Na_2SO_4). [From Smolders (1961), with the permission of the author.]

establishes clearly the contribution of charge q_M to the surface tension. However, a glance at any electrocapillary curve (Figure 7.6) shows that, except for γ_{max}, for which $q_M = 0$, every value of γ is associated with a definite negative or positive charge q_M, so both γ and q_M (or E) are necessary to describe the state of the interface unequivocally. Already in 1908, Möller (1908) obtained for the noble metal/solution interface some experimental evidence confirming the effect of electrical potential on contact angle. Frumkin and coworkers [Frumkin *et al.* (1932), Kabanov and Frumkin (1933), Gorodetskaya and Kabanov (1934), Kabanov and Ivanishenko (1937), Frumkin (1938), Frumkin and Gorodetskaya (1938), Tverdovskiy and Frumkin (1947)] have established the parabolic character of the contact angle–potential curves, similar to the electrocapillary curves; however, the subsequent work by Smolders (1961) showed large quantitative differences (Figure 2.11) between the experimental values of contact angle measured by Frumkin *et al.* (1932) and the values calculated by Smolders from the measured interfacial tension values using an expression derived from the equilibrium of forces in a mercury–liquid–gas system:

$$\cos \theta = \frac{\gamma_{a/m}^2 - \gamma_{m/l}^2 - \gamma_{a/l}^2}{2\gamma_{m/l}\gamma_{a/l}} \tag{2.104}$$

where $\gamma_{a/m}$ is the air/mercury interfacial tension, $\gamma_{m/l}$ is the mercury/electrolyte liquid interfacial tension, and $\gamma_{a/l}$ is the air/electrolyte interfacial tension.

Smaller-in-magnitude but still significant discrepancies have been observed by Smolders (1961) between the measured and the calculated contact angles in the systems gas–mercury–aqueous solutions containing

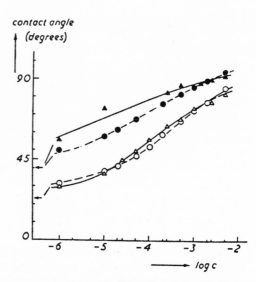

Figure 2.12. Contact angles on mercury as a function of sodium dodecyl sulfate concentration. No potential applied. Calculated values: $\bigcirc$, Na_2SO_4 added; $\bullet$, no Na_2SO_4 present: experimental values: $\triangle$, Na_2SO_4 added; $\blacktriangle$, no Na_2SO_4 present. [From Smolders (1961), with the permission of the author.]

sodium dodecyl sulfonate with and without the application of the electrical potential (Figures 2.12 and 2.13). These discrepancies are observed in systems devoid of problems associated with a rigid solid surface and its heterogeneity. Neither can they be ascribed to experimental errors in view of the precautions taken by Smolders in his work.

For conditions other than those at the pzc (where $\gamma = \gamma_{max}$ and $q_M = 0$) there exist, in general, two possibilities for each value of $\gamma_{s/l}$ and

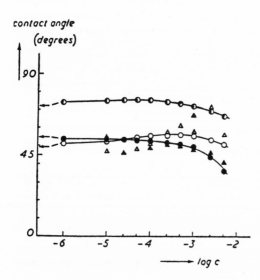

Figure 2.13. Contact angles on mercury as a function of sodium dodecyl sulfate concentration. $\bigcirc$ and $\triangle$, calculated and experimental values for -0.060 V vs. SCE; $\pmb{\mathbb{O}}$, calculated values for -0.460 V (no experimental values given); $\bullet$ and $\blacktriangle$, calculated and experimental values for -0.960 V. [From Smolders (1961), with the permission of the author.]

$\gamma_{a/l}$ regarding their associated charge densities $q_{s/l}$ and $q_{a/l}$ during an approach of a particle to an air bubble: Either

1. These two interfaces possess charge densities $q_{s/l}$ and $q_{a/l}$ of the same sign and then a repulsion term opposes their approach and inhibits the spontaneous process of establishing a particle–bubble contact, or

2. The charges corresponding to the two approaching interfaces are opposite in sign and then an additional electrostatic attraction facilitates the particle–bubble attachment.

These two possibilities have been considered by Leja and Bowman (1968) in discussing some puzzling aspects observed in oil droplets flotation and nonflotation. Electrical contributions may also play a role in static contact angle development between bubbles and solid surfaces: Sometimes a prolonged application of pressure on the bubble is required before attachment takes place; under other conditions, the approaching bubble appears to "jump" to the surface from a considerable distance away.

The problem of electrical effects in interfacial relationships is treated by Derjaguin in terms of electrical contribution to the disjoining pressure created in the thinning film underneath the air bubble (see Section 9.3).

In conclusion, the contact angle is a very useful parameter approximating the relative magnitudes of the three interfacial energies involved, but it does not reproduce the exact thermodynamic relationship it is purported to represent. Under some (but indeterminate) conditions there may be a fairly close agreement between a given theoretical interfacial relationship and its measure expressed with the use of contact angle; under other conditions large differences between the two exist. This situation, however, does not detract from the general utility of the contact angle as an extremely useful *indicator* (but not a *measure*) of the hydrophobic character of solids.

2.10.2. *Dynamic Aspects of Wetting and De-wetting*

The primary act in flotation is the selective attachment of hydrophobic solid particles to air bubbles *under dynamic conditions*. It is useful, therefore, to consider the extent to which dynamic parameters play a role in the process.

In fluid mechanics, dimensionless parameters are employed to facilitate evaluation and comparisons of behavior in various systems. Such parameters of direct relevance to flotation systems are those which relate capillary forces to viscosity, gravity, and flow (inertia) forces. The following dimensionless parameters may be mentioned as examples:

$$\text{capillary number} = \text{cap} = \frac{\mu v}{\gamma} = \frac{\text{We}}{\text{Re}} \qquad (2.105)$$

(representing ratio of viscous to surface tension forces),

$$\text{Weber number} = \text{We} = \frac{x\varrho v^2}{\gamma} \qquad (2.106)$$

(ratio of inertia to surface tension forces),

$$\text{Reynolds number} = \text{Re} = \frac{x\varrho v}{\mu} \qquad (2.107)$$

(ratio of viscous to inertia forces)

$$\text{Eotvos (or Bond) number} = \text{Eo or Bo} = \frac{x^2\varrho}{\gamma g} \qquad (2.108)$$

(ratio of acceleration or gravity to surface tension forces), where x is the displacement depth (cm), μ is the fluid viscosity (g cm^{-1} s^{-1}), v is the velocity of the particle or bubble (cm s^{-1}), γ is the surface tension (g s^{-2}), ϱ is the fluid density (g cm^{-3}), and g is the acceleration due to gravity (cm s^{-2}).

For systems consisting of two fluid phases (air/liquid or liquid/liquid), a relationship between three dimensionless parameters involving interfacial tension is represented graphically in Figure 2.14. It indicates clearly that capillary forces in such systems may be easily surpassed by inertia or gravity forces. The resultant effects are visible in departures of the fluid/fluid in-

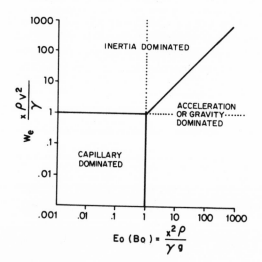

Figure 2.14. Hydrodynamic domains. [After Otto (1966).]

terfaces from their equilibrium shapes, which are described by the Laplace equation

$$\Delta P = \gamma\left(\frac{1}{R_1} + \frac{1}{R_2}\right) \tag{2.109}$$

where ΔP is the difference in pressure across the interface and R_1 and R_2 are the principal radii of curvature. For air bubbles in liquids and drops of liquids in air the equilibrium shape is spherical. Any departures from sphericity are due to deformations incurred by forces other than surface tension of the fluid/fluid interface. Whenever Eotvos (Bond) and Weber or Reynolds numbers exceed 1, ellipsoidal bubbles or drops are obtained (compare Figure 9.5); when these numbers exceed 10^2 or 10^3, spherical-cap bubbles (Figure 9.36) are formed [for details, see Clift *et al.* (1978, Figure 2.4, p. 25)].

When systems comprising three phases, either three fluids or two fluids and one solid phase, are considered, the departures from equilibrium shapes, under static or dynamic conditions, occur within the most easily deformable (the lowest in energy) interface. Thus, with a mercury–water–air system, the air/water interface is likely to be deformed, whereas with a solid–water–air or a solid–oil–water system, it will be the fluid/fluid interface. The deformation is likely to arise regardless of the origin of the excess force— whether due to surface tensions of the solid phase, dynamic parameters (see Figure 2.15), or interactions of electrical charges at respective interfaces.

Since the ratio of Weber and Reynolds numbers gives the capillary number [equation (2.105)], it is evident that the viscosity of the flotation pulp may play an important role in selective separations. Indeed, such an effect is amply documented in practice by the empirical finding that there is an optimum pulp density (which, among others, controls viscosity as well) for each flotation system. Similarly, selective separations which are obtained in one type of flotation cell [in which, for example, accelerations exceed gravitational acceleration by factors from 10 to 50—as estimated by Taggart (1957) and Gaudin (1957)] may not be obtainable in a pneumatic cell under conditions approaching gravitational acceleration.

It is also probable that lack of bubble deformation permits flotation under conditions corresponding to the regions of "no contact angle" which had been delineated for numerous static systems by Sutherland and Wark (1955). Hence, tests based on the "pickup" of particles by "captive" bubbles [see Figure 1.5(c)] are likely to be more representative of flotation behavior (particularly if carried out in the presence of frothers as well as collectors) than static contact angle determinations made with flat mineral surfaces.

Figure 2.15. Examples of solid particles attached to two bubbles simultaneously.

However, even the pickup tests do not reproduce all parameters prevailing in a flotation cell; neither does a test in a Hallimond tube reproduce all flotation parameters, particularly if restricted to a single selected mineral species and a collector-acting surfactant only. None of the preceding preliminary techniques can replace the actual flotation test. It must be borne in mind that every auxiliary technique can indicate a particular feature of a complex process but cannot replace it. Each technique has a bias that has to be recognized and allowed for when conclusions are made with respect to the overall process. Similarly, zeta potentials determined in a simplified system containing a single solid may not necessarily reflect the potentials of the same solid in a complex flotation mixture.

Various aspects of dynamic particle–bubble attachment are discussed in Section 9.3. There remain, however, two effects directly concerned with the development of contact angles which should be mentioned at this stage. One involves contact angle developing under conditions of a shear flow of liquid and its relation to the velocity of liquid. The other effect is associated with changes in the magnitude of contact angle resulting from a continued adsorption of surfactants at the bubble surface.

When a drop of a partially wetting liquid ($S < 0$, $\theta > 0$) is deposited on a flat solid surface, there occurs a change in contact angle with time, as depicted in Figures 2.16(a)–(c), after Schwartz and Tejada (1972). This behavior is referred to as self-spreading. It is differentiated from the condition of forced spreading when the solid is moving in the same direction as the spreading liquid at the time of spreading. The work of Hardy, Bangham, and particularly of Bascom *et al.* (1964), who used interferometry and ellipsometry, established two types of patterns associated with the dynamic contact angles observed during spreading, as shown in Figures 2.16(d) and (e). The pattern (d) exhibits only a very thin layer Y of liquid extending a short distance ahead of the first-order interference band. The pattern (e) shows a much thicker (1–2 μm) layer B which follows the Y layer; layer B disappears when the system reaches equilibrium. A number of investigators have developed expressions relating the magnitude of the dynamic contact angle θ_d to its value under static conditions, θ_{eq} and parameters such as the air/liquid surface tension $\gamma_{a/l}$, the viscosity of the liquid η, the forward velocity v of the liquid, the interfacial viscosity of the particular solid–liquid system, etc. Detailed discussion of the dependence between the dynamic contact angle θ_d and the variety of parameters mentioned can be found in the paper by Schwartz and Tejada (1972) and references therein.

With respect to the second effect, namely the change in contact angle resulting from the kinetics of surfactant adsorption, it might be helpful to recall that the action of frother molecules had always been considered to be restricted to the liquid/gas interface in stabilizing air bubbles. The possibility of coadsorption of frothers onto collector-coated solids was discounted for the reason that the magnitude of the contact angle developed in a collector solution was found to be not affected by a subsequent addition of a frother to the solution. Schulman and Leja (1958) discovered that such an absence of a change in contact angle occurs only with those frother acting surfactants which produce gaseous types of films (see Section 6.10) at the air/water interface. However, when the frother-acting molecules are capable of forming more condensed films (liquid or solidified), considerable changes

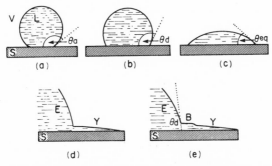

Figure 2.16. Self-spreading of a drop in the absence of movement and during a consonant movement of solid. [From Schwartz and Tejada (1972), in *J. Colloid Interface Sci.* **38**, with the permission of Academic Press, New York.]

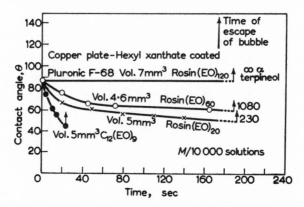

Figure 2.17. The decrease of contact angle with time during adsorption of highly hydrophobic condensing polyoxyethylenes (low EO number) and absence of a change in contact angle for hydrophilic (high EO) noncondensing Pluronic F-68 and Rosin (EO)$_{120}$. [From Schulman and Leja (1958), in *Surface Phenomena in Chemistry and Biology*, J. F. Danielli, K. G. A. Pankhurst, and A. C. Riddiford, eds., with the permission of Pergamon Press, New York.]

in contact angles, initially developed in a solution of collector alone, become evident with time (Figure 2.17). Depending on the kinetics of adsorption and the degree of condensation, the initially developed contact angles (by bubbles attached to a flat surface) decrease until the buoyancy of bubbles forces them to detach from the flat surface.

This phenomenon of condensation of frother or of mixed frother-collector films occurring at the air/bubble interface may be responsible for the occasionally observed nonfloatability of otherwise highly hydrophobic solids. As the condensing film of frother spreads progressively from the surface of the airbubble onto the solid surface, it converts the latter to hydrophilic and allows the bubble to escape from the solid.

2.10.3. *Critical Surface Tension of Low-Energy Solids*

Despite the limitations associated with the contact angle technique, the latter has been found extremely useful for approximate evaluations of parameters that would not be readily determined otherwise. (In the same manner, the BET technique is extensively employed to determine surface areas of solid powders or porous solids despite its obvious theoretical shortcomings, see Section 6.4.5).

Zisman and his co-workers (1941–1953) used the contact angle method to evaluate the $\gamma_{a/s}$ of low-energy solids (natural and synthetic organic compounds such as waxes and polymers). They plotted the surface tension

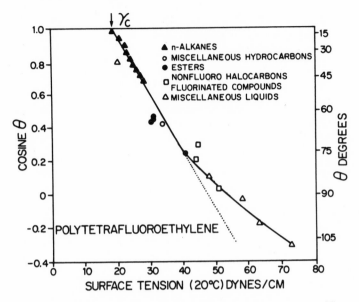

Figure 2.18. Wettability of polytetrafluorothylene by various liquids. [From Zisman (1964), in *Contact Angle, Wettability, and Adhesion*, R. F. Gould, ed., with the permission of the American Chemical Society, Washington, D.C.]

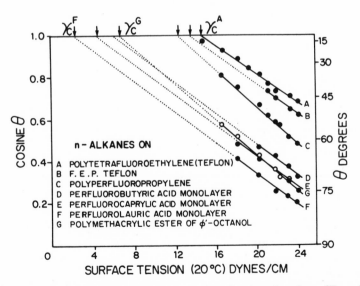

Figure 2.19. Contact angles for *n*-alkanes on various fluorinated surfaces. [From Zisman (1964), in *Contact Angle, Wettability, and Adhesion*, R. F. Gould, ed., with the permission of the American Chemical Society, Washington, D.C.]

of various liquids, which wet the given solid partially, against the value of $\cos \theta$ of the angle measured with these liquids (Figures 2.18 and 2.19). The surface tension for which $\cos \theta = 1$ (or the corresponding spreading coefficient $S = 0$) denotes the so-called *critical surface tension* γ_c of the given solid. A list of critical surface tensions so evaluated for various solid surfaces is given in Table 2.11.

When aqueous (or nonaqueous) solutions of surfactants are used with low-energy solids, the graphs of $\cos \theta$ vs. $\gamma_{a/1}$ give an extrapolated, and approximately identical, value of γ_c as the one obtained with pure liquids in Figures 2.18 or 2.19.

Only those *solutions whose $\gamma_{a/1}$ is less than γ_c spread and wet the solid completely.* Solutions which possess critical micelle concentrations (CMCs)

Table 2.11. Critical Surface Tensions of Low–Energy Surfaces[a]

Surface constitution	γ_c(dyn/cm) at 20°C
Fluorocarbon Surfaces	
—CF$_3$	6
—CF$_2$H	15
—CF$_3$ and —CF$_2$—	17
—CF$_2$—	18
—CH$_2$—CF$_3$	20
—CF$_2$—CFH—	22
—CF$_2$—CH$_2$—	25
—CFH—CH$_2$—	28
Hydrocarbon surfaces	
—CH$_3$ (crystal)	22
—CH$_3$ (monolayer)	24
—CH$_2$—	31
—CH$_2$— and —CH—	33
—CH— (phenyl ring edge)	35
Chlorocarbon surfaces	
—CClH—CH$_2$—	39
—CCl$_2$—CH$_2$—	40
=CCl$_2$	43
Nitrated hydrocarbon surfaces	
—CH$_2$ONO (crystal) [110]	40
—C(NO$_2$)$_3$ (monolayer)	42
—CH$_2$NHNO$_2$ (crystal)	44
—CH$_2$ONO$_2$ (crystal) [101]	45

[a] Source: Zisman (1964).

exhibit a sudden change[†] in slope at those surface tensions corresponding to the CMC. (For discussion of the CMC, see Section 5.7.) Solutions of surfactants which do not form micelles (such as alcohols and 1, 4-dioxane) give a monotonic slope of the $\cos \theta$ vs. $\gamma_{a/l}$ curve, as do pure liquids in Figure 2.19.

The success of Zisman's approach to obtain the surface free energy of the low-energy solids has led to attempts by Good and Girifalco (1960) and Fowkes (1962, 1964) to evaluate the surface energies of other solids with the help of contact angle data.

Girifalco and Good (1960) showed that for "regular" interfaces, that is, systems where the cohesive forces within the phases and the adhesive forces across the interface are of the same type, the solid/liquid interfacial tension $\gamma_{s/l}$ is given by

$$\gamma_{s/l} = \gamma_{a/s} + \gamma_{a/l} - 2\phi(\gamma_{a/s}\gamma_{a/l})^{1/2} \tag{2.110}$$

where $\gamma_{a/s}$ and $\gamma_{a/l}$ are the surface free energies of pure solids in vacuum and of pure liquid phases, respectively, and ϕ is a constant for a given solid and a series of liquids such that if Zisman's γ_c for these systems is known, then

$$\gamma_c = \phi^2\gamma_{a/s} \tag{2.111}$$

Equation (2.110) is combined with the Young expression for contact angle (2.97) to give

$$\cos \theta = 2\phi\left(\frac{\gamma_{a/s}}{\gamma_{a/l}}\right)^{1/2} - 1 \tag{2.112}$$

A comprehensive review of different types of forces—such as dipole, dispersion, and ionic forces—and their effects on estimation of interfacial energies has been published by Good and Elbing (1971).

Fowkes (1962) resolved the surface tension of liquids into constituent contributions such as dispersion forces γ^d and either hydrogen bonding γ^h, so that for water

$$\gamma_{a/l} = \gamma^h + \gamma^d \tag{2.113}$$

or metallic bonding γ^m, so that for mercury

$$\gamma_{a/l} = \gamma^d + \gamma^m \tag{2.113'}$$

The surface tension of hydrocarbons involves only dispersion forces γ^d.

[†] For surfactants possessing CMCs, the value of γ_c should coincide with γ_{CMC} (see Section 8.5) since no further lowering in $\gamma_{a/l}$ occurs past CMC (see Section 5.7) and complete wetting ensues in solutions exceeding CMC.

Further, in analogy with Hildebrand's expression for the solubility of a solute in a solvent, Fowkes introduced the following expression for the interfacial tension:

$$\gamma_{12} = \gamma_1 + \gamma_2 - 2(\gamma_1{}^d\gamma_2{}^d)^{1/2} \tag{2.114}$$

where γ_1 and γ_2 refer to any phases and $\gamma_1{}^d$ and $\gamma_2{}^d$ are the corresponding dispersion contributions.

Again, contact angle measurements are relied upon to evaluate the γ^d values for various phases (Figure 2.20).

An extensive series of evaluations on systems involving numerous

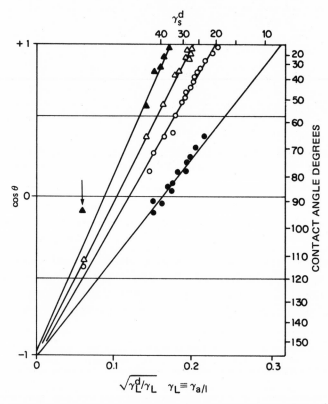

Figure 2.20. Contact angles of a number of liquids in four low energy surfaces. Closed triangles refer to polyethylene, open triangles to paraffin wax, open circles to $C_{36}H_{74}$, and closed circles to fluorododecanoic acid monolayers on platinum. All points below the arrow are contact angles with water. [From Fowkes (1962), in *Chemistry and Physics of Interfaces*, S. Ross, ed., with the permission of the American Chemical Society, Washington, D.C.]

hydrocarbons and water enabled the value of the dispersive component γ^d for water to be determined, namely

$$\gamma^d_{H_2O} = 21.8 \pm 0.7$$

when

$$\gamma_{H_2O} = 72.8 \text{ erg/cm}^2 \qquad \text{at } 20°C$$

Then, for systems comprised of mercury, hydrocarbons, and water the value of the dispersive component for mercury was determined as

$$\gamma^d_{Hg} = 200 \pm 7 \text{ erg/cm}^2 \qquad \text{at } 20°C$$

These values were subsequently used to calculate the interfacial tension of the mercury–water system, and the calculated value was compared with the experimental value, 425 vs. 426 erg/cm², respectively. The agreement suggests that despite the uncertainties associated with the contact angle measurements, the evaluations of the dispersion force contributions to interfacial bonding may be satisfactory.

2.11. *Liquid/Gas and Liquid/Liquid Interfaces*

Our understanding of liquids and their structure is far inferior to that of crystalline solids. However, despite this drawback, the interfacial properties of liquids are easier to evaluate experimentally. This situation is due to the mobility of constituent molecules and ions in the liquid phase which makes the respective interfaces liquid/gas and liquid/liquid *homogeneous* in character on microscale (not on the molecular scale) in contrast to the heterogeneity of solid interfaces. No residual stress or strain is or can be locked in the liquid surface. The molecules at the interface are always in a state of continuous agitation: For example, at 1 atm and 24°C the number of molecules escaping from the liquid surface (and condensing on it) equals 2.9×10^{23} molecules/cm²! Owing to the homogeneity of the liquid interfaces and their mobility, the absolute values of thermodynamic quantities such as surface tension can be experimentally determined for these interfaces if suitable precautions are maintained.

Preferential adsorption of surfactants at the air/solution interface causes a lowering of surface tension in dilute solutions of surfactants. The surface tension–concentration data for various soluble surfactants are given in Sections 5.7–5.14. Shinoda *et al.* (1963), Davies and Rideal (1961),

Harkins (1952), and Adamson (1967) provide comprehensive reviews of surfactant solutions and their properties. Adsorption at highly conducting liquid metal/electrolyte interfaces is also discussed in Chapter 7, Sections 7.4 and 7.5.

Long-chain surfactants which are insoluble in aqueous solutions have provided an extremely valuable means for studying the behavior and chemical interactions of all polar–nonpolar species at the liquid/gas, liquid/liquid, and to some extent liquid/solid interfaces. The discussion of these insoluble surfactant monolayers is deferred to Chapter 6, Section 6.10.

2.12. Selected Readings

2.12.1. Chemical and Molecular Bonding

Barrett, C. S. (1952), *Structure of Metals*, 2nd ed., McGraw-Hill, New York.

Bell, C. F., and Lott, K. A. K. (1963), *Modern Approach to Inorganic Chemistry*, Butterworths, London.

Buscall, R., and Ottewill, R. H. (1975), Thin films, in *Colloid Science*, Vol. 2, D. H. Everett, ed., The Chemical Society, London, pp. 191–245.

Cartmell, E., and Fowles, G. W. A. (1961), *Valency and Molecular Structure*, Butterworths, London.

Cotton, F. A., and Wilkinson, G. (1972), *Advanced Inorganic Chemistry*, Wiley-Interscience, New York.

Debye, P. J. (1929, 1945), *Polar Molecules*, Dover, New York.

Hamilton, W. C., and Ibers, J. A. (1968), *Hydrogen Bonding in Solids*, Benjamin, Reading, Massachusetts.

Israelachvili, J. N., and Tabor, D. (1973), van der Waals forces: Theory and experiment, in *Progress in Surface and Membrane Science*, Vol. 7, J. F. Danielli, M. D. Rosenberg, and D. A. Cadenhead, eds. Academic Press, New York, pp. 1–55.

Parsegian, V. A. (1975), Long range van der Waals forces, in *Physical Chemistry: Enriching Topics from Colloid and Surface Science*, H. van Olphen and K. J. Mysels, eds. Theorex, LaJolla, California, pp. 27–72.

Pauling, L. (1944, 1960), *The Nature of the Chemical Bond*, 3rd ed., Cornell University Press, Ithaca, New York.

Phillips, C. S. G., and Williams, R. J. P. (1965), *Inorganic Chemistry*, Vols. 1 and 2, Oxford University Press, London.

Pimentel, G. C., and McClellan, A. L. (1960), *The Hydrogen Bond*, W. H. Freeman, San Francisco.

Rich, A., and Davidson, N., eds. (1968), *Structural Chemistry and Molecular Biology*, W. H. Freeman, San Francisco.

Richmond, P. (1975), The theory and calculation of van der Waals forces, in *Colloid Science*, Vol. 2, D. H. Everett, ed., Chapter 4, The Chemical Society, London, pp. 130–172.

Syrkin, Y. K., and Dyatkina, M. E. (1964), *Structure of Molecules and the Chemical Bond*, Dover, New York.
Zundel, G. (1970), *Hydration and Intermolecular Interaction*, Academic Press, New York.

2.12.2. *Thermodynamics*

Everett, D. H. (1959), *An Introduction to the Study of Chemical Thermodynamics*, Longmans, Green & Co., Ltd., London.
Gibbs, W. J. (1876, 1948), *Collected Works, Part I—Thermodynamics*, Yale University Press, New Haven, Connecticut.
Guggenheim, E. A. (1950), *Thermodynamics*, North-Holland, Amsterdam.
Hill, T. L. (1960), *Introduction to Statistical Thermodynamics*, Addison-Wesley, Reading, Massachusetts.
Hill, T. L. (1963), *Thermodynamics of Small Systems*, Benjamin, Reading, Massachusetts.
Klotz, I. M. (1950), *Chemical Thermodynamics*, Prentice Hall, Englewood Cliffs, New Jersey.
Lewis, G. H., and Randall, M. (1961), *Thermodynamics*, Rev. by K. S. Pitzer and L. Brewer, McGraw-Hill, New York.
Pourbaix, M. J. N. (1949), *Thermodynamics of Dilute Aqueous Solutions*, trans. J. N. Agar, Edward Arnold & Co., London.
Prigogine, I., and Defay, R. (1954), *Chemical Thermodynamics*, Longmans, Green & Co., Ltd., London.
Zemansky, M. W. (1957), *Heat and Thermodynamics*, McGrawHill, New York.

2.12.3. *Interfacial Relationships*

Adam, N. K. (1941), *The Physics and Chemistry of Surfaces*, 3rd ed., Oxford University Press, London.
Adamson, A. W. (1967), *Physical Chemistry of Surfaces*, 2nd ed., Wiley-Interscience, New York.
Aveyard, R., and Haydon, D. A. (1973), *An Introduction to the Principles of Surface Chemistry*, Cambridge University Press, London.
Everett, D. H., ed. (1973, 1975), *Colloid Science*, Vol. 1 (1973) and Vol. 2 (1975), The Chemical Society, London.
Fowkes, F. M., ed., (1969), *Hydrophobic Surfaces*, Academic Press, New York.
Gould, R. F., ed., (1964), *Contact Angle, Wettability, and Adhesion*, Advances in Chemistry Series No. 43, American Chemical Society, Washington, D.C.
Harkins, W. D. (1952), *The Physical Chemistry of Surface Films*, Reinhold, New York.
Hiemenz, P. C. (1977), *Principles of Colloid and Surface Chemistry*, Marcel Dekker, New York.
Majijevic, E., ed. (1969–1977), *Surface and Colloid Science*, Vols. 1–9, Wiley-Interscience, New York; Vols. 10–12, Plenum Press, New York.
Osipow, L. I. (1962), *Surface Chemistry*, Reinhold, New York.
Van Olphen, H., and Mysels, K. J. (1975), *Physical Chemistry, Enriching Topics from Colloid and Surface Science*, Theorex, LaJolla, California.

3

Structure of Solids

Processes of selective separation of solids by flotation, spherical agglomeration, or selective flocculation exploit *differences in surface characteristics* of the various solid phases. The surface features are primarily determined by the molecular or atomic structure of the solid (whether crystalline or amorphous), the types of bonding involved, and, to a varying degree, by the past history undergone by the solid. In exceptional cases, the physical or mechanical treatment received in the past by different solids may nullify the differentiating effects due to structure or bonding. Invariably, any chemical activity between the environment and the solid modifies the surfaces to some extent, but as long as such an action is not too drastic, the structural features within the interior of the solid are reflected in its surface properties. For more exhaustive treatments of structures and bonding prevailing in various solids (whether minerals or chemical compounds), the reader is referred to the texts listed in Section 3.6. Only a brief review of the salient structural features is undertaken here in order to provide a minimum background for those with no immediate access to those references.

The numerous possibilities of altering the surface characteristics by means of either mechanical or (slight) chemical treatments constitute both a boon and a curse of selective separations. These possibilities are so varied that, at present, no *ab initio* deductions can be made which would predict the influence of all parameters, structural, mechanical, or chemical in nature, on surface properties. Hence, compilation of observations and experiences is relied upon for predictions on the behavior of solid surfaces.

3.1. *Atomic Packing*

Solids consist of either

1. Long-range (near-infinite) arrays of individual ions or atoms held together by *nondirectional bonding*; examples would be NaCl, MgO, and similar ionic solids held by electrostatic forces, metallic solids held by delocalized electrons, or solidified inert gases held by van der Waals bonding; or

2. Arrays of atoms in monoatomic solids, held by *localized covalent bonding*, extending across the whole volume of each crystal; examples are allotropic forms of carbon, silicon, red and black phosphorus, and boron; or

3. Arrays of discrete compounds (molecules) or of complex multiatomic ions held together by nondirectional bonding (van der Waals in the case of molecules and ionic in the case of complex ions) and *locally*, within the aggregate, *by highly directional covalent bonds*; some examples are SiO_2, P_4O_{10}, SiC, CO_3^{2-}, $S_2O_6^{2-}$, CNO^-, and SO_4^{2-}; or

4. Irregular long-range arrangements of atoms as in noncrystalline glasses and polymers, comprising mixtures of bonds.

Minerals represent naturally occurring substances. A classification of crystalline structures represented by different solids, or minerals, may be based on the types of principal interatomic bond: ionic, covalent, metallic, van der Waals. However, in many structures the bonding is often of intermediate character, and different bonding operates between various constituent atoms. Another classification may be based on geometrical considerations. This either associates or separates structures which may be unrelated, or closely related, in other respects. A rigid classification is thus impractical, and it may be best to consider a series of examples representing mineral structures of increasing complexity.

Whenever nondirectional bonding is involved, the crystalline pattern depends on the relative sizes of atoms, ions, or the local directionally bonded aggregates. The criterion for each such structure is the geometry of close packing of the constituent species. If atoms are treated as hard spheres, the three-dimensional arrangement in a nondirectionally bonded solid is obtained by stacking layers of closely packed spheres on top of each other. In the densest packing, the spheres of one layer rest in the declivities of the other layer, forming two types of interstices: tetrahedral and octahedral

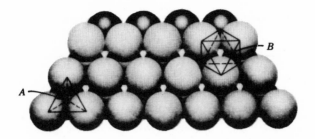

Figure 3.1. One close-packed plane of spheres stacked on top of another one. A tetrahedral interstitial site (A) and an octahedral interstitial site (B) are indicated. [From Moffat *et al.* (1964), *The Structure and Properties of Materials*, Vol. 1, with the permission of John Wiley & Sons, New York.]

(Figure 3.1). The third layer can then be placed so that its atoms are either directly over those of the first layer, forming a so-called hexagonal close packing (hcp), or are slightly displaced, giving a face-centered cubic (fcc) packing. An arrangement not quite as closely packed as fcc or hcp is that of a body-centered cubic (bcc) structure.

The *coordination* number (C.N.) of a given packing arrangement is defined as the number of ions or atoms surrounding the selected ion or atom. If two or more types of atoms are packed together and held by non-directional bonding (ionic, metallic, van der Waals), the geometry of close packing imposes definite limits, critical radius ratios (of cation, r^+ or r_c, to anion, r^- or r_a), on the relative sizes of atoms (or ions) for each of the possible coordination numbers: 2, 3, 4, 6, 8, and 12. These ratios are given in Table 3.1. Thus, a stable configuration of atoms or ions of coordination number 4, that is, the tetrahedral coordination, should occur when their radii ratios are between 0.225 and 0.414; below the lower of the preceding limits the triangular coordination represents a stable arrangement of lower energy, while for the ratios above the upper limit the octahedral coordination represents a state of lower energy.

Table 3.2 shows a comparison between the predicted coordination numbers and the actual, observed ones for a variety of ionic solids and for metals. Generally, the agreement is very satisfactory; the exceptions are due to the presence of some directional bonding.

The three-dimensional structure of a solid may be visualized as stacking of coordination polyhedra, with or without sharing of faces, edges, or corners. Figure 3.2 shows such coordinating octahedra of anions sharing each of their 12 edges with the neighboring octahedra in the simple cubic structure of NaCl. The sharing of corners by the SiO_4 tetrahedra leads to

Table 3.1. Coordination Number Calculated as a Function of Radius Ratio r^+/r^- (or r_c/r_a)

C.N.	Range of radius ratio for which packing is expected to be stable	Coordination polyhedron
2	0–0.155	Line
3	0.155–0.225	Triangle
4	0.225–0.414	Tetrahedron
6	0.414–0.732	Octahedron
8	0.732–1.0	Cube
12	1.0	Cube (hcp or fcc)

Table 3.2. Predicted and Observed Packing for Different Radius Ratios

Compound or metal	Radius ratio of cation/anion, r_c/r_a	Coordination number	
		Predicted	Observed
B_2O_3	0.14	2	3
BeS	0.17	2	4
BeO	0.23	3 or 4	4
SiO_2	0.29	4	4
LiBr	0.31	4	6
MgO	0.47	6	6
MgF_2	0.48	6	6
TiO_2	0.49	6	6
NaCl	0.53	6	6
CaO	0.71	6	6
KCl	0.73	6 or 8	6
CaF_2	0.73	6 or 8	8
CsCl	0.93	8	8
bcc metals	1.0	8 or 12	8
fcc metals	1.0	8 or 12	12
hcp metals	1.0	8 or 12	12

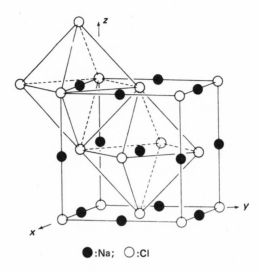

●:Na; ○:Cl

Figure 3.2. The cubic structure of sodium chloride, NaCl, showing the coordinating octahedra of anions round the cations. In addition to chlorides of Na$^+$, K$^+$, Rb$^+$, and Ag$^+$ this structure is also adopted by *galena*, PbS. [From Evans (1966), *An Introduction to Crystal Chemistry*, with the permission of Cambridge University Press.]

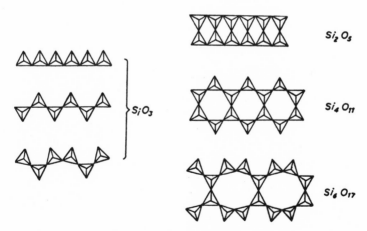

Figure 3.3. Single and double chains formed from SiO$_4$ tetrahedra. [From A. F. Wells (1962), *Structural Inorganic Chemistry*, 3rd ed., with the permission of Oxford University Press.]

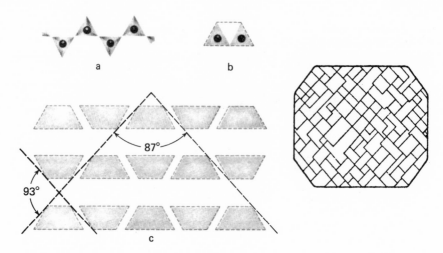

Figure 3.4. Pyroxenes. [Adapted from Bragg and Claringbull (1965) and Moffat *et al.* (1964).]

the single and double chains of silicate minerals such as pyroxenes or amphiboles and to the sheet structure of micas (Figures 3.3–3.6). Other space-filling arrangements of polyhedra are shown in Figure 3.7. The concept of coordinating polyhedra is not limited to ionic structures but applies to all spatial distribution of bonds.

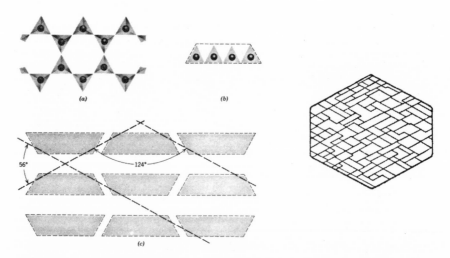

Figure 3.5. Amphiboles. [Adapted from Bragg and Claringbull (1965) and Moffat *et al.* (1964).]

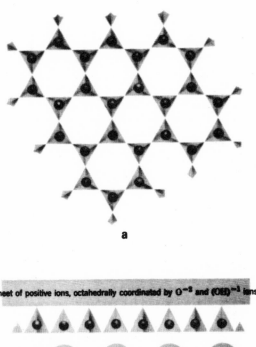

Figure 3.6. Micas. [From Moffat *et al.* (1964), *The Structure and Properties of Materials*, Vol. 1, with the permission of John Wiley & Sons, New York.]

The structure of a solid may also be represented by associating every atom or a group of atoms with a point in one of the Bravais lattices, maintaining the same orientation for a group of atoms at each of the lattice points. The Bravais lattices (Figure 3.8) are defined as three-dimensional arrays of identical points such that every point has surroundings identical to those of every other point. There are 14 such lattices of points possible; each lattice point may be occupied either by one atom or ion or by a group of directionally bonded atoms, forming a discrete unit, a molecule, or a complex ion, that is repeated with the same orientation at every other point.

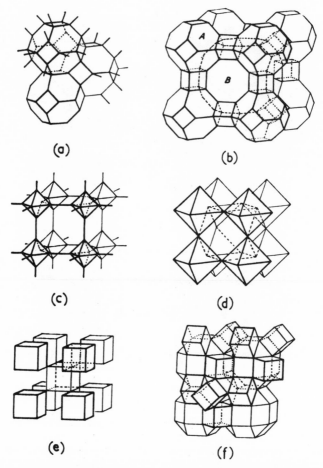

Figure 3.7. Space-filling arrangements of polyhedra. [From A. F. Wells (1962), *Structural Inorganic Chemistry*, 3rd ed., with the permission of Oxford University Press.]

3.2. *Crystal Structures of Simple Compounds*

A perfect crystal represents a regular array of atoms in a space lattice; this array can be described by a repeating unit, called a unit cell. A space lattice may have a number of different unit cells; e.g., the hexagonal Bravais lattice may be described by two primitive unit cells (unique cells possessing only one lattice point per cell when repeatedly shifted in space) (Figure 3.8).

The unit cell shape and the crystalline shape (the external shape of a macrocrystal morphology of the crystal) are not the same, though they are

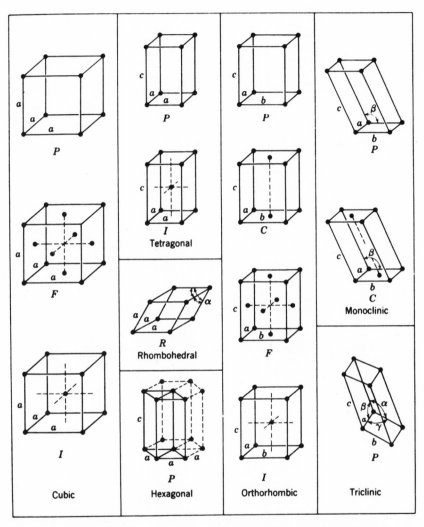

Figure 3.8. Conventional unit cells of the 14 Bravais space lattices: cubic lattices: simple cubic *P*, face-centered cubic *F*, and body-centered cubic *I*; tetragonal lattices: simple *P*, body-centered *I*; orthorhombic lattices: simple *P*, based-centered orthorhombic *C*, face-centered orthorhombic *F*, body-centered orthorhombic *I*; monoclinic: simple mono-clinic *P*, base-centered monoclinic *C*; rhombohedral *R*; triclinic *P*; and the hexagonal Bravais lattice is either the *P* cell shown with solid lines or the *C* cell shown in dashed lines. [From Moffat *et al.* (1964), *The Structure and Properties of Materials*, Vol. 1, with the permission of John Wiley & Sons, New York.]

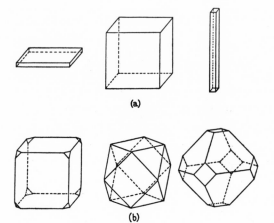

(a)

(b)

Figure 3.9. Variation in habit of crystals. Different relative development of (a) cube faces and (b) cube and octahedron faces. [From A. F. Wells (1962), *Structural Inorganic Chemistry*, 3rd ed., with the permission of Oxford University Press.]

related by symmetry considerations. A wide variation in crystal habit for a given chemical species is possible caused by the conditions under which crystallization occurs, the presence of trace impurities, stress, etc.; see Figure 3.9. According to Mullins (1963), the equilibrium shape of a small crystal is determined by two criteria:

1. The thermodynamic condition that—for a constant volume of a given material—the integral of the product of the surface tension γ for a particular crystallographic orientation and the corresponding area be minimum,

$$\int \gamma \, dA = \text{minimum}$$

2. The least variation in the kinetics of mass transport (volume and surface diffusion) brought about by the differences in the chemical potential within the liquid phase (as caused by the anisotropy of surface tension values, that is, its dependence on the crystallographic orientation of the surface).[†]

[†] For a liquid drop, free from external forces such as gravity, the equilibrium shape is spherical. A solid of constant volume, allowed to crystallize under unrestrained conditions such as, for example, in the vapor phase, also assumes an equilibrium shape that is determined by minimizing the total surface free energy of its faces. A so-called Gibbs–Wulff theorem states that surface tensions of individual faces in an equilibrium solid crystal are proportional to their respective distances from the center of the prism. These distances are obtained by the so-called Wulff's construction of an equilibrium

Trace impurities affect the values of the surface tension γ differently for different orientations (due to the specificity of their adsorption). Hence, new crystal habits are produced (e.g., Figure 3.9). Another indication of the anisotropic nature of the surface tension of crystallographic orientations is the preferential cleavage along certain directions, e.g., cubic cleavage of PbS, NaCl, or KCl, or the rhombohedral cleavage of calcite, or the basal cleavage of graphite, molybdenite, micas, etc. (cf. footnote p. 111).

The unit cells in the Bravais lattices and the external shapes of crystals are grouped into seven symmetry systems (Figure 3.8): cubic, tetragonal, rhombohedral (trigonal), hexagonal, orthorhombic, monoclinic, and triclinic, in a decreasing order of the symmetry elements they possess. The symmetry elements of the external form (shape) of the crystal represent 32 point groups, while the symmetry elements of the unit lattices represent a very much larger number of combinations, i.e., 230 space groups.

The simplest crystal structures are adopted by compounds which involve nondirectional bonding, e.g., in metals (cubic and hcp) and in ionic compounds of compositions AX, AX_2, A_mX_z, $A_mB_nX_z$, and $A_mB_nX_zY_w$, where A and B stand for metallic elements, X and Y for nonmetallic elements, and m, n, z, and w are simple integers. However, compounds with mixed ionic–covalent–van der Waals bonding may also be represented by these structures.

---

shape for any given plot of surface tensions γ. The construction involves drawing a plane perpendicular to the radius vector of the γ plot at the point when these two cross each other; the most inner envelope of all such planes constitutes the equilibrium shape. For details of Wulff's construction, see Mullins (1963), Woodruff (1973), or Murr (1975). Herring (1951) provided several important theorems and proofs relating to γ plots and equilibrium shapes. Solids crystallizing under a restraint (dynamic or spatial) will be geometrically similar to the equilibrium shape in that the corresponding faces will have nearly the same γ's as those of an equilibrium shape, but the distances from the center of the crystal will not be those of the equilibrium.

Depending on the relative balance between the repulsive and attractive force contributions to surface energy, a microscopic observation may reveal curved faces, edges, and corners, but macroscopically the crystal is a polyhedron. The theoretical treatment of the variation of the surface free energy with temperature, as expressed by the surface entropy change $S = -\partial\gamma/\partial T$, shows that there is a critical temperature (for each face of a given crystallographic orientation) below which the surface is planar, free from steps, and this is called a singular surface. Above this critical temperature steps appear, rounding off the surface, which ceases to be singular. Such development of steps during a rise in temperature is frequently observed with metals and ionic solids and is known as thermal etching, leading to faceting or a polymorphic change, giving a new crystal habit.

A. Compounds of the composition AX form the following crystal structures:

1. The *NaCl structure* (Figure 3.2), i.e., the *simple cubic structure,* usually with highly ionic bonds, with the coordination number (C.N.) 6:6, that is, six *anions* surrounding each cation and six cations surrounding each anion. This structure is adopted by most AX halides (e.g., halides of Na, K, Rb, Ag: NaF, NaCl, NaBr, NaI, AgF, AgCl, AgBr, etc.), some oxides (MgO, CaO, SrO, BaO, MnO, FeO), and some sulfides (MgS, BaS, SrS, CaS, MnS, PbS).

2. The *CsCl structure,* i.e., the body-centered cubic (Figure 3.10), with a C.N. of 8:8, adopted by Cs(Cl,Br,I), NH_4Cl, and NH_4Br and low-temperature Rb halides.

3. The *zinc blende* (sphalerite) and the *wurtzite structures*; both are *hexagonal structures* with a C.N. of 4:4; the zinc blende structure [Figure 3.11(a)] is adopted by ZnS CdS, HgS, and BeS and the wurtzite structure by ZnS, CdS, MnS, and ZnO. A gradual replacement of Zn by Fe produces *marmatites* (Zn,Fe)S, with varying amounts of Fe, which adopt the zinc blende (sphalerite) structure. Structurally related to zinc blende are other complex sulfides such as, for example, *chalcopyrite* [Figure 3.11(b)]. When Zn atoms in sphalerite are replaced by alternate Cu and Fe atoms and the unit cell is doubled, the structure of $CuFeS_2$, chalcopyrite, is obtained. (On replacing one-half of the Fe atoms in $CuFeS_2$ by Sn, the structure of another complex sulfide, *stannite,* Cu_2FeSnS_4, is obtained.)

4. The *hexagonal NiAs structure* with a C.N. of 6:6 (Figure 3.12) is highly anisotropic in character due to the covalent bonding between the two atoms A—X; it is adopted by sulfides, selenides, tellurides, arsenides, and antimonides of Ti, V, Fe, Co, and Ni (this structure is not represented

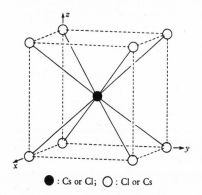

● : Cs or Cl; ○ : Cl or Cs

Figure 3.10. The unit cell of the cubic structure of cesium chloride. CsCl. [From Evans (1966), *An Introduction to Crystal Chemistry,* with the permission of Cambridge University Press.]

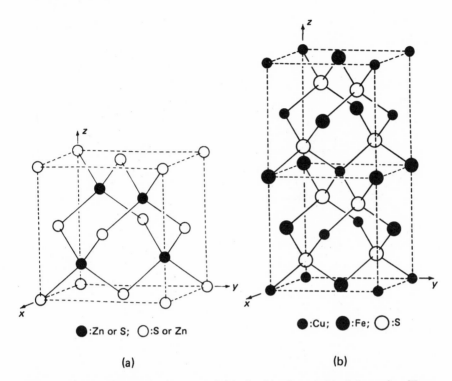

●:Zn or S; ○:S or Zn

●:Cu; ●:Fe; ○:S

(a)

(b)

Figure 3.11. The hexagonal structures of (a) zinc blende and (b) chalcopyrite. [From Evans (1966), *An Introduction to Crystal Chemistry*, with the permission of Cambridge University Press.]

by any oxide). *Pyrrhotite*, $Fe_{1-x}S$, is the most common of all these minerals with NiAs structure. It has a variable composition, shown to be due to the deficiency of iron in the lattice.

5. The *tetragonal PdO or PdS structure* with a C.N. of 4:4 (Figure 3.13) is adopted by CuO, PdO, PtO, PdS, and PtS. Each divalent metal atom is surrounded by four coplanar oxygen (sulfur) atoms at the corners of a square (four planar *dsp* hybrid bonds).

B. Compounds of the composition AX_2 form the following structures:

1. The *fluorite* (CaF_2) *cubic structure* with a C.N. of 8:4 (Figure 3.14) in which the calcium atoms occupy the positions of a fcc unit and the fluorine atoms are in the centers of the eight cubelets into which the cell may be divided. This structure is adopted by some AX_2 halides with ionic radii

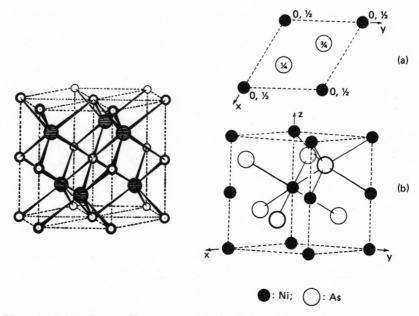

●: Ni; ◯: As

Figure 3.12. The hexagonal structure of NiAs (adopted by pyrrhotite $Fe_{1-x}S$). The numbers in the plan view of the unit cell (top right) denote the elevations of the appropriate atoms within the unit cell. [Adapted from Evans (1966) and Wells (1962).]

ratios, r^+/r^-, greater than 0.7 (e.g., BaF_2, 0.99; PbF_2, 0.88; HgF_2, 0.81; $BaCl_2$, 0.75; CdF_2, 0.71) and some AX_2 oxides (e.g., ZrO_2, 0.57; HfO_2, 0.56; UO_2, 0.64; ThO_2, 0.68; CeO_2, 0.72).

2. The *tetragonal rutile* (TiO_2) *structure* with a C.N. of 6:3 (Figure 3.15), adopted by some halides of divalent metals (MnF_2, FeF_2, CoF_2, NiF_2, ZnF_2, $CaCl_2$, $CaBr_2$, MgF_2) with ionic radii ratios $r^+/r^- = 0.7$–0.3 and the following metal dioxides: PbO_2, 0.60; SnO_2, 0.51; TiO_2, 0.49; WO_2, 0.47; VO_2, 0.43; CrO_2, 0.40; MnO_2, 0.39; GeO_2, 0.38.

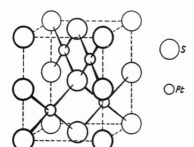

◯ S

◯ Pt

Figure 3.13. The unit cell of the tetragonal structure of PdO, PtO, PdS, and PtS. [From A. F. Wells (1962), *Structural Inorganic Chemistry*, 3rd ed., with the permission of Oxford University Press.]

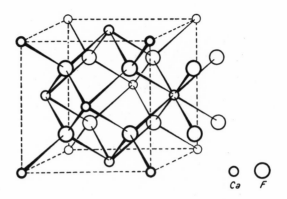

Figure 3.14. The unit cell of the cubic structure of fluorite, CaF_2. [From A. F. Wells (1962), *Structural Inorganic Chemistry*, 3rd ed., with the permission of Oxford University Press.]

O Ca O F

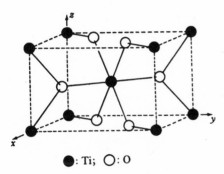

●: Ti; ○: O

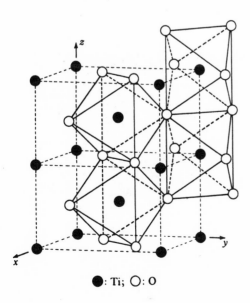

●: Ti; ○: O

Figure 3.15. The tetragonal structure of rutile, TiO_2, showing the coordinating octahedra of anions round the cations and the way in which these octahedra are linked in bands sharing by horizontal edges. The same structure is adopted by SnO_2 (cassiterite), MnO_2, and PbO_2. [From Evans (1966), *An Introduction to Crystal Chemistry*, with the permission of Cambridge University Press.]

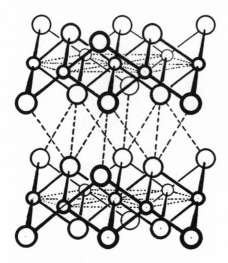

Figure 3.16. Portions of two layers of the CdI₂ structure. The small circles represent metal atoms. The structure is adopted by the hydroxides of Mg, Ca, Mn, Fe, Co, Ni, and Cd. [From A. F. Wells (1962), *Structural Inorganic Chemistry*, 3rd ed., with the permission of Oxford University Press.]

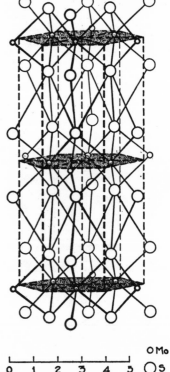

Figure 3.17. The structure of molybdenite, MoS₂. The sheets containing Mo atoms are shaded to emphasize the "layer lattice." The edges of the sheets (resulting from breakage of covalent S—Mo bonds) are ionized and hydrophilic in character. The surfaces of the sheets (resulting from cleavage occurring between two layers of sulfur atoms) are hydrophobic. [From Bragg and Claringbull (1965), *Crystal Structures of Minerals*, Vol. IV, with the permission of Bell and Hyman Ltd., London.]

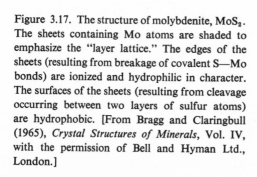

3. The *hexagonal layer structure of cadmium iodide* (CdI_2) with a C.N. of 6:3 (Figure 3.16), with Cd atoms surrounded by six iodine atoms at the corners of an octahedron and three Cd neighbors to each halogen atom. The structure is adopted by Mg, Ca, Mn, Fe, Co, Ni, and Cd *hydroxides*; halides; and also some disulfides, e.g., TiS_2, SnS_2, and PtS_2.

4. The *layer structure of molybdenite*, MoS_2, producing hexagonal crystals in which sheets of molybdenum atoms are sandwiched between two sheets of sulfur atoms. The *three* sheets form a "layer" of the structure and constitute an infinite two-dimensional "molecule"; the sulfur and molybdenum atoms within the layers are strongly covalently bonded, but the successive layers of sulfur atoms are held together by weak van der Waals bonds between the two adjoining sheets. These bonds endow the mineral with excellent cleavage characteristics parallel to the base of the hexagonal crystals (Figure 3.17), producing surfaces hydrophobic in character. The structure is adopted by MoS_2 and WS_2.

5. The *pyrite and marcasite structures* (Figure 3.18), adopted by sulfides, selenides, and tellurides of many transition metals, such as Mn, Fe, Co, and Ni. The pyrite structure resembles that of the simple cubic NaCl structure when Cl is replaced by the S—S group. The marcasite structure is orthorhombic with iron atoms at the corners and the center of the ortho-

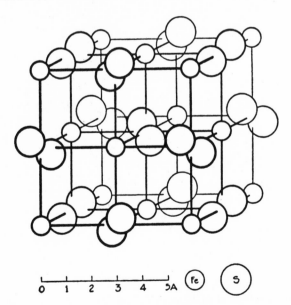

Figure 3.18. The structure of pyrite, FeS_2. [From Bragg and Claringbull (1965), *Crystal Structures of Minerals*, Vol. IV, with the permission of Bell and Hyman Ltd., London.]

rhombic cell and the S—S groups midway between the iron atoms. In both these structures each metal atom is surrounded by six sulfur atoms at the corners of a near-octahedron, while each sulfur is bonded to one other sulfur and three metal atoms. The bonding is wholly *covalent*, with the coordinating octahedra FeS_6 sharing faces. On replacing the S_2 group by the As_2 group or a mixed AsS or SbS group, a number of more complex compounds, crystallizing in structures related to pyrite and marcasite structures (but of lower symmetry), are obtained, for example, CoAsS, NiSbS, and NiAsS.

6. The *silica structure*, SiO_2, with a C.N. of 4:2, has three stable forms (each with α and β modification), i.e., quartz (stable below 870°C), tridymite (870°–1470°C), and cristobalite (above 1470°C). The Si—O bonds possess a considerable degree of covalent character, with oxygen atoms arranged tetrahedrally around Si in such a manner that every oxygen atom is shared by two SiO_4 tetrahedra (Figure 3.19).

C. Compounds of the A_2X composition form the following:

1. The *antifluorite structure* with a C.N. of 4:8; it is a fluorite structure with the positions of the anions and cations interchanged. Many oxides, sulfides, selenides, and tellurides of alkali metals form this structure.

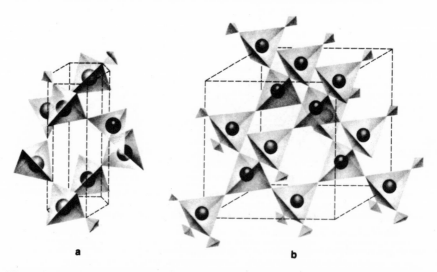

a   b

Figure 3.19. Two of the silica structures: (a) the hexagonal β-quartz, with the SiO_4 tetrahedra arranged in two spiral chains (which can be either left- or right-handed); (b) β-cristobalite, with a more open cubic structure. [From Moffat *et al.* (1964), *The Structure and Properties of Materials*, Vol. 1, with the permission of John Wiley & Sons, New York.]

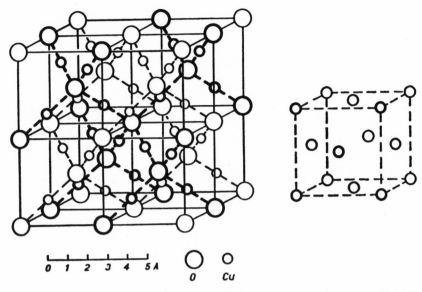

Figure 3.20. The cubic structure of cuprite, Cu_2O. [From Bragg and Claringbull (1965), *Crystal Structures of Minerals*, Vol. IV, with the permission of Bell and Hyman Ltd., London.]

2. The Cu_2O *cuprite structure* with a C.N. of 2:4; the two *sp* bonds of the Cu^+ atom are arranged in line (Figure 3.20).

D. Compounds of the A_mX_z composition (where *m* and *z* are integers 1–3) usually contain negligible ionic but predominantly covalent and van der Waals bonding; their structures become more diverse. The simplest of the structures are the following:

1. The *cubic structure of aluminum fluoride*, AlF_3, with a C.N. of 6:2 (Figure 3.21), adopted by fluorides of trivalent Al, Sc, Fe, Co, Rh, and Pd and by trixides CrO_3, WO_3, and ReO_3.

2. The *hexagonal structure of corundum and hematite* (Figure 3.22) in which the oxygen atoms are arranged in approximately hexagonal close packing; each cation lies between six oxygen atoms, but only two-thirds of the available cation positions are filled. In MgO (periclase), all cation positions are filled. $\alpha\text{-}Al_2O_3$, $\alpha\text{-}Fe_2O_3$, Cr_2O_3, Ti_2O_3, U_2O_3, $\alpha\text{-}Ga_2O_3$, and Rh_2O_3 adopt the corundum structure.

E. Compounds of the $A_mB_nX_z$ composition (where *m*, *n*, and *z* are integers 1–4) comprise a large number of complex oxides, sulfides, and

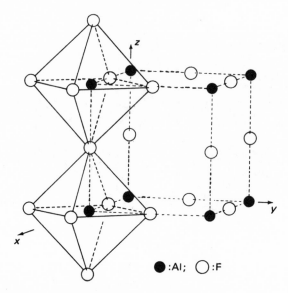

Figure 3.21. The unit cell of the (idealized) cubic structure of aluminium fluoride, AlF_3. Two coordinating [AlF_6] octahedra share corners. [From Evans (1966), *An Introduction to Crystal Chemistry*, with the permission of Cambridge University Press.]

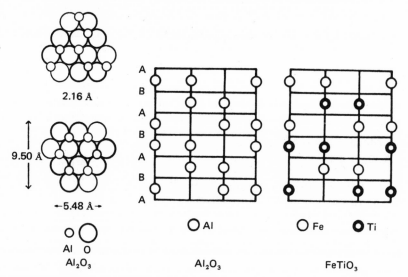

Figure 3.22. The structure of corundum, Al_2O_3, showing successive layers of O—Al atoms. Hematite, Fe_2O_3, adopts the same structure. A comparison of metal atom distribution in corundum and ilmenite, between close-packed layers of oxygen atoms (indicated by horizontal lines *A* and *B*) is shown schematically. [From Bragg and Claringbull (1965), *Crystal Structures of Minerals*, Vol. IV, with the permission of Bell and Hyman Ltd., London.]

halides of two metals and a nonmetal. They adopt structures known as perovskite ($CaTiO_3$), ilmenite ($FeTiO_3$), and spinel ($MgAl_2O_4$).

1. The *ilmenite structure*, $FeTiO_3$, is closely related to that of corundum; the aluminum atoms between two oxygen layers become Fe, and those between the next two layers become Ti, etc., as indicated in Figure 3.22.

2. The *perovskite structure*, $CaTiO_3$, has the arrangement shown in Figure 3.23 with the larger metal atom (or ion) Ca occupying the body-center position of the cubic structure (with a 12 coordination by the oxygen atoms and a 6 coordination by the smaller metal atom, Ti). About 50 complex oxides and fluorides possess the perovskite structure, with the large atoms represented by K, Ca, Sr, Ba, Cd, Pb, and La and the small atoms by Nb, W, Ti, Zr, Sn, Ce, Hf, Mg, and Zn, e.g., $KNbO_3$, $SrTiO_3$,

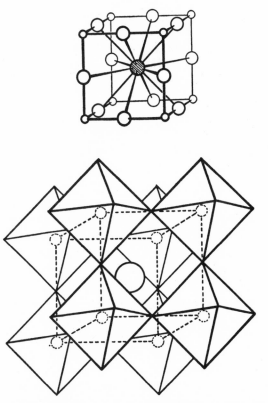

Figure 3.23. The idealized structure of perovskite, $CaTiO_3$. A Ca atom is at the cube center and the Ti—O octahedra are shown diagrammatically. [Adapted from Bragg and Claringbull (1965) and Wells (1962).]

$CaSnO_3$, $PbZrO_3$, $PbTiO_3$, $PbMgF_3$, and $KZnF_3$. Any pair of metallic ions, regardless of valency, can form the perovskite structure provided their radii are appropriate to the preceding coordinations and an aggregate valency of 6 is obtained to maintain electrical neutrality; thus, metal pairs of valencies 1 and 5, 2 and 4, and 3 and 3 and of appropriate radii r_A, r_B, and r_X such that $r_A + r_X \cong (2)^{1/2}(r_B + r_X)$ would form the perovskite structure (A and B represent, respectively, large and small metallic ions and X represents a nonmetallic ion).

3. The *cubic spinel structure* $MgAl_2O_4$ is found in a large number of oxides and some sulfides, fluorides, and cyanides. It is adopted by compounds in which divalent metal ions A and trivalent metal ions B (as in $AX \cdot B_2X_3$, where X is either O or S) form a "normal" spinel structure, as in $MgAl_2O_4$, with A in tetrahedral positions and B in octahedral positions. An alternative "inversed" structure, $MgFe_2O_4$, with the positions occupied by the divalent and the trivalent metals being reversed, is also fairly common and is adopted by the so-called iron spinels: magnetite, $FeFe_2O_4$, and spinels obtained when Mg, Cu, Co, Ti, and Ni replace the divalent Fe in the preceding formula, i.e., $ZnFe_2O_4$, etc. The factors which determine the formation of one or the other of these two spinel structures are not fully understood.

3.3. *Structural Changes and Chemical Relationships*

Almost always the nonmetallic atoms or ions are far larger than those of metals (Figure 3.24). Hence, the greater part of the volume in the structures of the oxides, sulfides, and halides of metallic elements is occupied by the nonmetallic component. The structures of these compounds may then be described as a close-packed framework of nonmetallic atoms or ions with the metallic components in the interstitial positions.

If some of the cationic sites in the structures become vacant, this gives rise to *compounds of variable composition*, e.g., $Fe_{0.9}O$, $Fe_{1-x}S$, etc. This is particularly common when the cation has a lower and a higher valency, so that a vacancy and an ion of a higher valency in the neighborhood can easily satisfy the condition of electrical neutrality within the solid. Another type of change in chemical composition is due to *solid solutions* of two compounds which have the same structures and in which the substituting atoms do not differ by more than about 15% of the size of the smaller atom. For example, Al_2O_3 and Fe_2O_3 are fairly soluble in various spinels because the cation-deficient structure of spinel allows these two oxides to replace

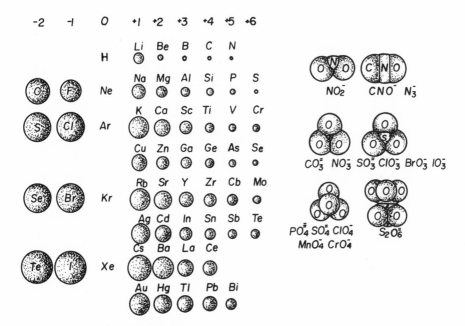

Figure 3.24. The sizes of some ions and the structures of some complex anions. [From Bragg and Claringbull (1965), *Crystal Structures of Minerals*, Vol. IV, with the permission of Bell and Hyman Ltd., London.]

the metallic components in the spinel structure. The size factor is very evident in halides: KCl and KBr form solid solutions of any composition; however, KCl and LiCl are mutually insoluble.

Whenever substances which are chemically unrelated adopt similar structures, so-called *isomorphism* occurs. Thus, for example, NaCl, PbS, NH_4I, and KCN are isomorphous because they all adopt the sodium chloride structure. When the ions are equal in size and charge, as, for example, Mg^{2+} and Fe^{2+}, they can be readily interchanged, giving rise to *isomorphous substitutions* which produce numerous ranges of minerals with composition varying from one extreme set of components to another—but all with the same structure. When ions of different valencies are involved, their ionic size is still the governing parameter, and the condition of electrical neutrality is established by a suitable combination of substitutions and vacancies. Some elements, like rubidium, do not form minerals of their own but are always found as a substituent for K. Mn^{2+} can replace Mg^{2+}, Fe^{2+}, or Ca^{2+}. Li^+ (0.60 Å) does not replace the other alkali ions because it is much smaller $[Na^+$ (0.95 Å), K^+ (1.33 Å)], but it replaces Mg^{2+} (0.65 Å).

Polymeric isomorphism occurs when the similarity of structures is such that groups of several unit cells have to be compared (and not single cells) in order to detect common features. The superposition of two zinc blende cells and their comparison with the unit cell of chalcopyrite, $CuFeS_2$ (Figure 3.11), is necessary to show the isomorphism of these two structures and to obtain a repeating unit. In the case of the stannite (Cu_2FeSnS_4) structure, one-half of the Fe atoms in $CuFeS_2$ are replaced by Sn. Enargite, Cu_3AsS_4, and wolfsbergite, $CuSbS_2$, are isomorphous with the chalcopyrite and zinc blende, while cubanite, $CuFe_2S_3$, and tetrahedrite, Cu_3SbS_3, are isomorphous with the wurtzite structure. Although polymeric isomorphs do not form solutions, close relationships of cell dimensions frequently lead to oriented overgrowth.

Oxygen and sulfur are closely related chemically and yet are sufficiently different in electronegativity to provide appreciable variations in the structures of oxides as compared to those of sulfides. Simple ionic structures are more common in oxides than in sulfides; the latter tend to form surprisingly complex structures. Many common oxides have no counterpart among sulfides and vice versa: For example, PbO_2 is a common oxide, whereas PbS_2 does not exist, FeO_2 does not exist, but FeS_2 is widespread, etc. Sulfides, selenides, and tellurides have luster and electrical conductivity characteristics which are much closer to those of intermetallic compounds than oxides; they also show large variations in composition.

3.4. *Structures Containing Finite Complex Ions*

A number of compounds with $A_mB_nX_z$ composition in which the B atom is less metallic and more strongly electronegative (so that B—X bonds are no longer ionic) form structures in which well-defined $[B_nX_z]$ complex groups exist as anions. Those finite complex anions possess distinct structures of their own (Figure 3.24 and Table 3.3).

The structures formed by compounds containing such complex ions depend on the configuration of the complex ion itself and the coordination of cations and anions. Thus, the plane triangular complexes such as CO_3^{2-} form two structures:

1. The *rhombohedral calcite structure*, $CaCO_3$, with Ca^{2+} coordinated by six CO_3^{2-} and each oxygen atom of these anions coordinated by two Ca^{2+} (Figure 3.25)

2. The *orthorhombic aragonite* structure (related to the hexagonal

Table 3.3. The Structures of Some Finite Complex Anions[a]

Type	Arrangement	Examples
XY	Linear	C_2^{2-}, O_2^{2-}, O_2^{-}, CN^-
XYZ	Linear	N_3^-, CNO^-, CNS^-, I_3^-, ICl_2^-
	Bent	ClO_2^-, NO_2^-
BX_3	Plane equilateral triangle	BO_3^{3-}, CO_3^{2-}, NO_3^-
	Regular trigonal pyramid	PO_3^{3-}, AsO_3^{3-}, SbO_3^{3-}, SO_3^{2-}, SeO_3^{2-}, ClO_3^-, BrO_3^-, IO_3^-
BX_4	Regular tetrahedron	SiO_4^{4-}, PO_4^{3-}, AsO_4^{3-}, VO_4^{3-}, SO_4^{2-}, SeO_4^{2-}, CrO_4^{2-}, ClO_4^-, MnO_4^-; $CoCl_4^{2-}$, BF_4^-; $Cu(CN)_4^{3-}$
	Distorted tetrahedron	MoO_4^{2-}, WO_4^{2-}, ReO_4^-
	Plane square	$PdCl_4^{2-}$, $PtCl_4^{2-}$; $Ni(CN)_4^{2-}$, $Pd(CN)_4^{2-}$, $Pt(CN)_4^{2-}$
BX_6	Regular octahedron	AlF_6^{3-}, TiF_6^{2-}, $ZrCl_6^{2-}$, ReF_6^{2-}, $OsBr_6^{2-}$, $PtCl_6^{2-}$, SiF_6^{2-}, GeF_6^{2-}, $SnCl_6^{2-}$, SnI_6^{2-}, SbF_6^-

[a] Source: Evans (1966).

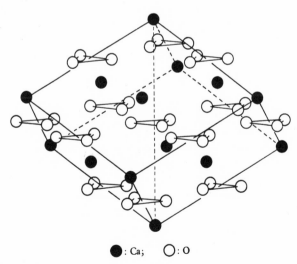

●: Ca;　○: O

Figure 3.25. The structure of calcite. (Its similarity to NaCl structure becomes apparent when the rhombohedron of calcite and the cube of NaCl are aligned with their respective diagonals parallel to each other.) $MgCO_3$, $ZnCO_3$, $FeCO_3$, $MnCO_3$, and $NaNO_3$ adopt the same structure. [From Evans (1966), *An Introduction to Crystal Chemistry*, with the permission of Cambridge University Press.]

NiAs structure, as the calcite structure is related to the cubic NaCl structure) in which each Ca^{2+} is coordinated by nine oxygens and each oxygen by three Ca^{2+} (Figure 3.26).

A number of other carbonates, nitrates, and borates form either the calcite structure (e.g., $MgCo_3$, $ZnCO_3$, $FeCO_3$, $NaNO_3$, $InBO_3$, $MnCO_3$) whenever their metallic ions M^{2+} have their ionic radius smaller than or equal to 0.99 Å or the aragonite structure (e.g., $PbCO_3$, $BaCO_3$, KNO_3, $LaBO_3$) with the M^{2+} radius greater than 0.99 Å.

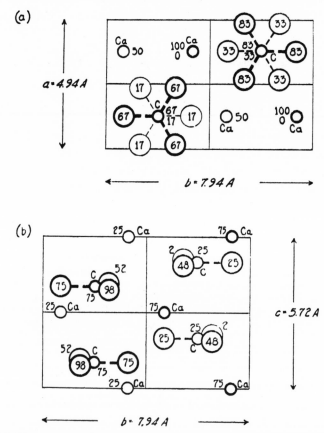

Figure 3.26. The structure of aragonite, $CaCO_3$ [also adopted by $PbCO_3$ (cerussite) and $BaCO_3$ (witherite)]: (a) projection on (001); (b) projection on (100). Superimposed oxygen atoms have been made visible by symmetrical displacement. The numbers denote elevations (from 0 to 100) of the corresponding atoms within the unit cell along the axis perpendicular to the given projection. [From Bragg and Claringbull (1965), *Crystal Structures of Minerals*, Vol. IV, with the permission of Bell and Hyman Ltd., London.]

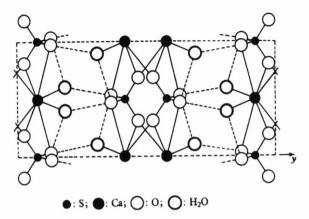

●: S; ●: Ca; ○: O; ○: H₂O

Figure 3.27. The monoclinic structure of gypsum, $CaSO_4 \cdot 2H_2O$, projected on a plane perpendicular to the x axis. Hydrogen bonds are represented by broken lines. [From Evans (1966), *An Introduction to Crystal Chemistry*, with the permission of Cambridge University Press.]

The structures of the basic carbonates of copper, malachite $[Cu_2(OH)_2 CO_3]$, and azurite $[Cu_3(OH)_2(CO_3)_2]$ have rather complex monoclinic cells. Malachite shows two types of Cu ions: One copper has a square planar coordination by two O and two OH ions, and the other has an octahedral coordination. Azurite also has two types of Cu ions, but both have square planar coordination of different neighbors. The triangular CO_3^{2-} complexes are incorporated within the structure.

Several of the tetrahedral complexes, such as SO_4^{2-}, PO_4^{3-}, and AsO_4^{3-}, are substantially the same in form and size, with a distance between the complexing atom and oxygen of ~1.5 Å. Isomorphous substitutions are thus possible and occur. They form a great number of different structures in which the geometrical considerations play a decisive role, as in all ionic structures. Large cations such as Ba^{2+}, Cs^+, Sr^{2+}, Pb^{2+}, NH_4^+, etc., tend to produce

3. The *orthorhombic baryte type of structure*, $BaSO_4$, with Ba atoms coordinated by 12 oxygen atoms

Other compounds possessing the same structure include $SrCrO_4$, $KMnO_4$, $KClO_4$, $(NH_4)ClO_4$, and $BaCrO_4$. Smaller cations, such as Ca^{2+}, form an orthorhombic structure as in anhydrite, $CaSO_4$, in which Ca^{2+} is surrounded by eight oxygens. The more common is the monoclinic structure of calcium sulfate hydrate, *gypsum*, $CaSO_4 \cdot 2H_2O$ (Figure 3.27). It represents a layer structure, each layer containing a sheet of H_2O, two [$Ca^{2+} + SO_4^{2-}$] groupings, and another sheet of H_2O. Within the sheets the cations

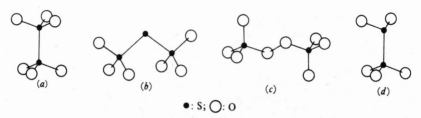

●: S; ○: O

Figure 3.28. The structures of some polynuclear sulfur oxyions: (1) the dithionate ion, $S_2O_6^{2-}$; (b) the trithionate ion, $S_3O_6^{2-}$; (c) the perdisulfate ion, $S_2O_8^{2-}$; (d) the metabisulfite ion, $S_2O_5^{2-}$. [From Evans (1966), *An Introduction to Crystal Chemistry*, with the permission of Cambridge University Press.]

are coordinated by six O of the SO_4 groups and by two water molecules. The perfect cleavage of gypsum (and of barytes) arises from the breakage of hydrogen bonds between the O of the SO_4 groups and H_2O molecules.

Polynuclear complex anions of a general formula B_nX_z form finite structures representing various types of combinations involving simpler groups. Figure 3.28 shows how different combinations of trigonal pyramid groups of SO_3^{2-} result in

1. Dithionate, $S_2O_6^{2-}$ ion
2. Trithionate, $S_3O_6^{2-}$ ion
3. Perdisulfate, $S_2O_8^{2-}$ ion
4. Metabisulfite, $S_2O_5^{2-}$ ion

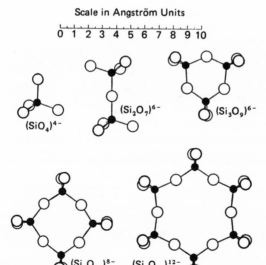

Scale in Angström Units

0 1 2 3 4 5 6 7 8 9 10

$(SiO_4)^{4-}$

$(Si_2O_7)^{6-}$

$(Si_3O_9)^{6-}$

$(Si_4O_{12})^{8-}$

$(Si_6O_{18})^{12-}$

Figure 3.29. Forms of separate silicon–oxygen groups. [From Bragg and Claringbull (1965), *Crystal Structures of Minerals*, Vol. IV, with the permission of Bell and Hyman Ltd., London.]

and Figure 3.29 shows the various structures formed by SiO_4 groups (Section 3.5). Other complex ions, found, for example, in heteropolyacids of molybdenum and tungsten, such as $(As_2Mo_{18}O_{40})^{3-}$ and $(PW_{12}O_{40})^{3-}$, represent an association of sesquioxide octahedra, which form a finite cage or a basket by sharing vertices or edges. The role of complex ions in the formation of surface structures which contribute to changes in surface characteristics of solid particles is discussed in greater detail in Chapters 8 and 10.

3.5. *Silicates*

All silicate structures are based on tetrahedral coordination of silicon by oxygen, SiO_4. The Si—O bond has a considerable degree of ionic character ($\sim 40\%$). Depending on the manner in which the coordinating tetrahedra are linked, the silicates are conveniently classified as follows:

1. *Orthosilicates*, structures in which isolated $(SiO_4)^{4-}$ tetrahedra are linked together through other cations, as, for example, in olivine, Mg_2SiO_4, and garnets, $A_3^{2+}B_2^{3+}Si_3O_{12}$, where $A^{2+} = Ca^{2+}$, Mg^{2+}, Fe^{2+}, Mn^{2+}, and $B^{3+} = Al^{3+}$, Fe^{3+}, Cr^{3+}

2. Structures with $(Si_2O_7)^{6-}$ groups [formed by two SiO_4 tetrahedra linked through one oxygen (Figure 3.29)], as, for example, melilite, $Ca_2MgSi_2O_7$, or hemimorphite, $Zn_4Si_2O_7(OH)_2 \cdot H_2O$

3. *Metasilicates* [closed ring groups of composition $(SiO_3)_n^{2n-}$ containing an indefinite number of members, formed by neighboring SiO_4 tetrahedra sharing two oxygens, shown in Figure 3.29], as, for example, beryl, $Be_3Al_2Si_6O_{18}$

4. *Open single chains* formed by tetrahedra sharing two oxygen atoms, $(SiO_3)_n^{2n-}$ (Figure 3.30), as in pyroxenes, $CaMg(SiO_3)_3$ (Figure 3.4)

5. *Open double chains* formed by SiO_4 tetrahedra sharing two and three oxygen atoms, $(Si_4O_{11})_n^{6n-}$ (Figure 3.30), as in amphiboles, such as tremolite, $(OH, F)_2Ca_2Mg_5Si_8O_{22}$ (Figure 3.5)

6. *Open sheet structures* formed by tetrahedra sharing three oxygen atoms, $(Si_4O_{10})_n^{4n-}$, as in talc, $Mg_3(OH)_2Si_4O_{10}$ (Figure 3.31), and in micas (Figure 3.6)

The essential difference in the structures of talc [and pyrophyllite, $Al_2(Si_4O_{10})(OH)_2$] and micas is that the arrangement of linked tetrahedra in talc and pyrophyllite is represented by uncharged $[Si_4O_{10}—M]^\dagger$ group-

---

† M—metallic ion.

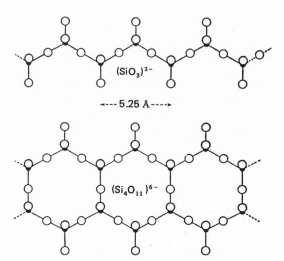

Figure 3.30. Forms of silicon–oxygen chains. [From Bragg and Claringbull (1965), *Crystal Structures of Minerals*, Vol. IV, with the permission of Bell and Hyman Ltd., London.]

ings, whereas in micas the $[AlSi_3O_{10}—M]$ groupings require other ions to balance the charge. The result is that in talc and pyrophyllite, van der Waals bonding exists between oxygen atoms of the neighboring sheets of linked tetrahedra, endowing them with easy cleavage and highly hydrophobic characteristics of cleavage surfaces. In micas, the cleavage surfaces are hydrophilic as a result of broken hydrogen bonds or ionic bonding. For a more detailed discussion on the hydrophobic–hydrophilic differences due to structure modifications, see Forslind and Jacobsson (1975).

Isomorphous substitutions readily occur in micas, pyroxenes, and amphiboles. Thus, Fe^{2+} and Mn^{2+} may replace Mg^{2+}, $2Na^+$ may replace Ca^{2+}, Al^{3+} may replace Si^{4+} and OH^-, and F^- may replace OH^-. A large number of complex sheet silicates exist, all characterized by easy cleavage, flaky appearance, and variable surface characteristics. Micas (muscovite, biotite, etc.), chlorites (vermiculite, etc.), and clays (kaolinite group, montmorillonite group, etc.) represent the most common minerals of sheet silicates. Figure 3.32 shows the distribution of atoms in the sheets of kaolinite, whereas Figure 3.33 shows an analogous distribution for montmorillonite.

A most comprehensive treatment of crystal structures of silicates, clay minerals, and silicate dispersoids is an eight-volume series on silicate science by Eitel [see, e.g., Eitel (1975)].

Clay minerals constitute a field of their own as regards structural differences and industrial applications. Clays are soft and easily become hydrated, disintegrating into minute platelet crystallites; some clays, for example, montmorillonite, swell readily as a result of physically adsorbed multilayers of vapors, especially water vapor, which is capable of penetrating between the individual sheets. For details of clays characteristics, publications such as *Clays and Clay Minerals*, appearing every year or so under the general editorship of Ingerson (1952–1968), should be consulted.

The cleavage in all these silicate sheet structures varies from that of talc (which cleaves readily across the van der Waals bonds holding the sheets together) to less perfect in muscovite mica (where the bonds are ionic through K^+ but still much weaker in comparison with other bonds in the structure) to an imperfect cleavage of sheets more ionically bonded through Ca^{2+} in the layers of the so-called brittle micas.

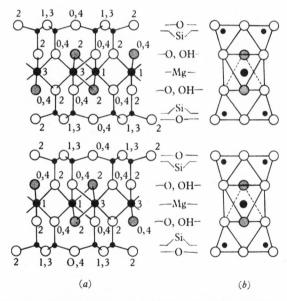

(a)   (b)

Figure 3.31. (a) Plan of the idealized monoclinic structure of talc, $Mg_3(OH)_2Si_4O_{10}$, projected on a plane perpendicular to the x axis. Only one ring of each sheet is drawn, and only some of the magnesium atoms are shown, but the sheets do, of course, extend indefinitely. The numbers indicate the heights of the various atoms above the plane of projection. The heights of the silicon atoms are not marked; they are the same as those of the unshared oxygen to which they are attached. (b) Schematic representation of the same structure showing the coordinating tetrahedra about the silicon atoms and the coordinating octahedra about the magnesium atoms. [From Evans (1966), *An Introduction to Crystal Chemistry*, with the permission of Cambridge University Press.]

The last group of silicates is that possessing the following:

7. The *framework structure* when every oxygen of the SiO_4 tetrahedra is shared, giving a number of distinct families of minerals such as felspars (or feldspars), zeolites, and ultramarines

Feldspars represent about 60% by volume of the earth's crust, amphiboles and pyroxenes $\sim$17%, quartz $\sim$12%, and micas $\sim$4%. Feldspars are the most important rock-forming minerals and are grouped as potassium feldspars, $KAlSi_3O_8$ (orthoclase, microcline, etc.), sodium feldspars (albite, etc.), and barium and calcium feldspars. Details of their structure and properties can be found in Chapter 14 (written by W. H. Taylor) in Bragg and Claringbull (1965).

Zeolites represent a much more open framework than that of feldspars,

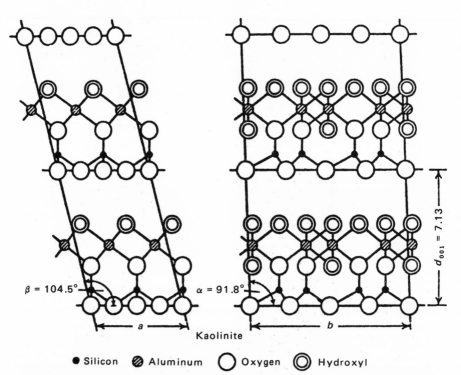

Figure 3.32. Two unit layers of kaolinite. The layers are kept together by weak hydrogen bonds and van der Waals bonding (due to a *cis* coordination of the silica tetrahedra), and these impart a weakly hydrophilic characteristic to the cleaved surface. [From Forslind and Jacobsen (1975), in *Water*, Vol. 5, F. Franks, ed., Plenum Press, New York.]

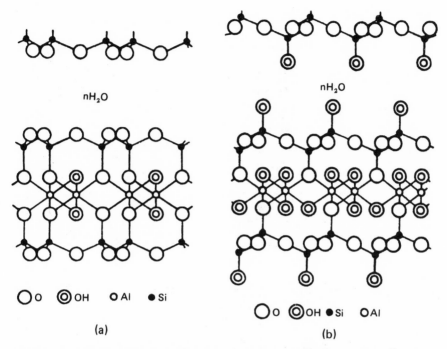

Figure 3.33. Two conceptual modifications of the montmorillonite structure. Structure (a) is due to Hoffman *et al.* (1933), and structure (b) is due to Edelman and Favejee (1940). The modification (b) was introduced to explain the highly hydrophilic characteristics of montmorillonite surfaces with moderate values of cation exchange capacity. Hydrophilic clays react with water by the formation of hydrogen bonds and take on a negative charge in aqueous suspension. [From Forslind and Jacobsson (1975), in *Water*, Vol. 5, F. Franks, ed., Plenum Press, New York.]

with channels in which cations and water molecules are located. Water molecules can be expelled (by an application of heat) without destruction of the silicate framework, and the voids so created can be reversibly replaced by other small molecules such as CH_4, NH_3, etc.; the voids are, however, too small for larger molecules and enable specific zeolites to act as highly selective "molecular sieves." Cations in the framework can be leached out and reversibly replaced by different cations. This forms the basis of utilizing zeolites in the permutite system of water softening.

Other zeolites have the $(Si,Al)O_4$ tetrahedra linked in such a manner that chains of a repeat unit consisting of five tetrahedra give rise to extended fibrous aggregates. The chains of these tetrahedra are cross-linked through the few unshared oxygen atoms, and this accounts for the ease of cleavage parallel to the chains.

3.6. *Selected Readings*

Bragg, L., and Claringbull, G. B. (1965), *Crystal Structures of Minerals*, Bell and Hyman, London.

Eitel, W. (1964), *Silicate Science*, Academic Press, New York.

Evans, R. C. (1966), *An Introduction to Crystal Chemistry*, Cambridge University Press, London.

Forslind, E., and Jacobsson, A. (1975), Clay–water systems, in *Water, A Comprehensive Treatise*, Vol. 5, F. Franks, ed., Plenum Press, New York, pp. 173–248.

Iler, R. K. (1955), *The Colloid Chemistry of Silica and Silicates*, Cornell University Press, Ithaca, New York.

Ingerson, E., ed. (1952–1968), *Clays and Clay Minerals*, Proceedings of the Conferences, 1st–16th, Pergamon Press, London.

Moffatt, E. G., Pearsall, G. W., and Wulff, J. (1964), *The Structure and Properties of Materials*, Vol. 1, *Structure*, Wiley, New York.

Van Olphen, H. (1977), *An Introduction to Clay Colloid Chemistry (for Clay Technologists, Geologists, and Soil Scientists)*, 2nd ed., Wiley, New York.

Wells, H. F. (1962), *Structural Inorganic Chemistry*, 3rd ed., Clarendon Press, Oxford.

4

Water and Aqueous Solutions

Water is the liquid phase used in all industrial mineral processing separations of solids whether by wet-gravity or wet-magnetic techniques, filtration, flotation, flocculation, or spherical agglomeration. The properties of water are, therefore, of great importance to all these processes. The processes exploiting the surface characteristics of solids are particularly affected by the extent of dissociation of water molecules and of other dissolved species, the adsorption of ions dissolved in water, their (ion) hydration, the structure of water in the bulk phase and that at the liquid/gas and liquid/solid interfaces, dielectric constant variation, dynamic behavior on thinning of an aqueous film, etc. A monumental treatise in five volumes, edited by Franks (1972–1975), provides the most comprehensive review on different aspects of the physical chemistry of water and of aqueous solutions. A purely theoretical treatment of water structure and properties is given in the books of Eisenberg and Kauzmann (1969) and Ben-Naim (1974), while Kavanau (1964) and Horne (1972) provide thorough coverages of both theories and properties of water and its solutions. Samoilov (1957) deals also with the spectroscopic evidence on hydration of ions.

In addition to being a component phase in the separation systems, water is essential to life, animal and vegetable. Hence, all effluents derived from mining or mineral separation processes, if they are being returned to the environment, have to be maintained in a form that does not contribute to pollution of natural waters. Therefore, in addition to recognizing the characteristics of water that are needed in industrial processes, the preservation of the quality of water necessary to prevent pollution and to facilitate its biological reusage are the responsibility of mineral processors. A useful text on these aspects is Camp's (1963) *Water and Its Impurities*. The chemistry of seawater is dealt with in books by Horne (1972), Martin (1970), and Church (1975).

4.1. *Structural Models*

Liquids ought to be regarded as disordered solids rather than as highly compressed gases. Water in particular shows a fairly large degree of short-range order, caused by mutual polarization of OH bonds. Numerous models have been proposed in the past for the structure of liquid water; for example, Roentgen (1891) suggested a mixture of ice dissolved in monomeric water; Bernal and Fowler (1933) proposed a tetrahedrally coordinated quartz-like and trydymite-like structure; Samoilov (1946, 1957) introduced a so-called interstitial model in which molecules of water "shaken loose" from their lattice positions occupy interstitial positions within a framework cluster; Eucken (1948) treated water as a mixture of polymers (dimers, tetramers, and octamers); and Pauling (1960) treated water as a clathrate cage structure. [For reviews of these models, see Kavanau (1964, 1965), Drost-Hansen (1967), Conway (1970), or Franks (1972, Vol. I).] The model of Frank and Wen (1957), involving flickering[†] clusters of hydrogen-bonded molecules with free molecules in cavities and surrounded by unstructured free water, is regarded by many as being the most realistic one (Figure 4.1). Nemethy and Scheraga (1962) made extensive thermodynamic calculations based on Frank and Wen's model of clusters. According to these calculations, depending on the extent of hydrogen bonding, the water molecule possesses different relative levels of internal energy; the lowest level of energy is that of a fourfold H-bonded water molecule; it increases with a decrease in H-bonds, so that an unbonded water molecule has the highest level of internal energy. When hydrophobic solutes, such as noble gases, hydrocarbons, and tetraalkylammonium cations, are introduced into water, their presence favors states of lower energy, that is, a higher degree of hydrogen bonding. Thus, hydrophobic solutes promote cluster formation, as indicated by calculations (Table 4.1). Viscosity is a good indicator of the extent of H bonding or structuring in liquid water. The decrease of viscosity with an increase in temperature indicates a decrease in H-bonds. Nemethy and Scheraga (1962) estimated that an average cluster contains about 65 molecules at 0°C and 12 molecules at 100°C [Horne (1978), p. 243].

The molecular dynamical calculations of Rahman and Stillinger (1971) and Stillinger and Rahman (1972, 1974) indicate that an array of all possible energies, skewed by the presence of solutes, should be applicable to

---

[†] Frank and Wen (1957) estimated that the half-life of these clusters is of the order of 10^{-11} sec, which is 10^2–10^3 times longer than the period of molecular vibrations—hence the name *flickering*.

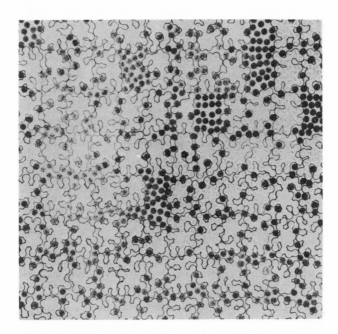

Figure 4.1. Schematic representation of the structure of liquid water as a mixture of hydrogen-bonded clusters (outlined) and monomeric "free" molecules (shaded). [After Courant *et al.* (1972), from Horne (1978) *The Chemistry of Our Environment*, with the permission of John Wiley & Sons, New York.]

liquid water. Frank's flickering clusters could then be regarded as a particular case which gives rise to a two-humped distribution of energies, one hump representing continuous medium–water and the other average-size clusters.

The models in which water is regarded as a mixture of H-bonded clusters and free, unbonded molecules represent one of the possible struc-

Table 4.1. Calculated Mole Fractions of Unbroken H Bonds[a]

T (°C)	Water	Solution of aliphatic hydrocarbon	Solution of aromatic hydrocarbon
0	0.528	0.672	0.633
30	0.434	0.557	0.522
60	0.370	0.462	0.434

[a] Source: Conway (1970).

tures, while the continuum theory of water structure, developed by Pople (1951), represents another. A whole spectrum of hydrogen-bonded water molecules is envisaged by Pople on making an assumption that the hydrogen bonds can be bent into various configurations without becoming dissociated. This model of Pople is capable of interpreting the variations in the properties of water (density, nearest neighbors, dielectric constant) and of explaining some anomalies in the behavior of water with variation in temperature on a par with the mixture models. All models of water suggested so far require further modifications in order to account for the effects of dissolved ions on the local structure and hydrogen bonding and to correlate these features with the experimentally evaluated parameters obtained using X-ray diffraction or spectroscopy (infrared, Raman, NMR, etc.). There is, as yet, no model of water that fits all observations regarding water behavior and its physical characteristics. Most useful information regarding structural interactions between water molecules themselves and those between water and solute molecules can be derived from studies of dielectric properties of water and of aqueous solutions. These dielectric characteristics are extensively discussed by Hasted (1972) in Franks' treatise. The attempts at evaluation of dipoles and of dielectric behavior of substances go back to Debye (1929) (see Sections 2.1 and 2.4). Debye assumed an exponential decay of polarization in a periodic field of angular frequency $\omega = 2\pi\nu$ and introduced the complex dielectric permittivity dependence on frequency ω, $\varepsilon(\omega)$, as a sum of a real term $\varepsilon'(\omega)$ and an imaginary term $\varepsilon''(\omega)$, namely

$$\varepsilon(\omega) = \varepsilon'(\omega) + i\varepsilon''(\omega) \qquad (4.1)$$

[The term $\varepsilon'(\omega)$ represents the out-of-phase term $\sin \omega t$ which reflects the ability of the given dipolar substance to store the applied voltage as a polarization and to create a reverse current after the voltage is removed. The imaginary term $\varepsilon''(\omega)$ represents the dissipated power or the spontaneous fluctuations of the field given by a current in phase with the voltage.] The dependence of ε' and ε'' on ω is given by

$$\varepsilon' - \varepsilon_\infty = \frac{\varepsilon_s - \varepsilon_\infty}{1 + \omega^2\tau^2} \qquad (4.2)$$

and

$$\varepsilon'' = (\varepsilon_s - \varepsilon_\infty) \frac{\omega\tau}{1 + \omega^2\tau^2} \qquad (4.3)$$

where ε_s is the static field dielectric constant, which for water is invariant with frequency from dc to beyond 10^8 Hz; ε_∞ is the dielectric permittivity at a frequency sufficiently high for the intermolecular dipole contributions to be absent but not too high for the induced polarization and vibrations of molecules to be damped out (the choice of appropriate frequency for ε_∞ is not simple); and τ is the dielectric relaxation time, the time for the given permanent dipoles to reorient themselves with the ac field.

When there is a single relaxation time for dipoles in a given substrate, the graphical expression of the Debye equations (4.2) and (4.3) is a semi-circular plot of ε'' (ordinate axis) vs. ε' (abscissa axis), known as the Cole–Cole plot, for corresponding values of ω. When there are two widely differing relaxation times for a given substance, the Cole–Cole plot consists of two rounded hills with a valley in between. As the relaxation times approach each other, the valley disappears, and the hills merge into a semicircle. A distribution of relaxation times gives rise to an arc, or a skewed arc, rather than a semicircle. Any interpretation of the data obtained experimentally regarding the dependence of ε' and ε'' on ω and τ has to start with an assumed model for liquid water structure and its modifications caused by the solutes. Useful information can be obtained about the reorientation energies of dipolar molecules, but any such structural correlations are "model-sensitive" and only approximate at present.

4.2. Hydration of Ions

Ions in aqueous solutions can be obtained either from lattice dissolution of *intrinsic salts* (ionically bonded or true electrolytes), that is, compounds in which there is already complete ionization in their crystal lattices, or from a solvolytic reaction of *potential electrolytes*, such as, for example, dissociation of an acid HA, which is covalently bonded,

$$HA + Sol. \overset{K}{\rightleftharpoons} Sol. H^+ + A^- \tag{4.4}$$

where Sol. denotes water or a proton-accepting solvent and A^- is an anion in solution.

The extent of ionization of an acid is determined by the thermodynamics of the ion-producing reaction, that is, breakage of a bond to create a proton, and its subsequent transfer process. Other ionization reactions involve transfer of a pre-existing charge from one reactant to another, i.e.,

$$BH^+ + Sol. \rightleftharpoons B + Sol. H^+ \tag{4.5}$$

$$HA + B^- \rightleftharpoons HB + A^- \tag{4.6}$$

The main factors determining the ionization equilibrium in ionization reaction (4.4) are the energy of the H—A bond; the solvation energy of the proton in the solvent, Sol. H^+; and the solvation energy of the conjugate base A^-. If the formation of ions in solution occurs as a spontaneous process, the free energy of solvation ΔG is a large negative quantity for ions in both types of reactions: dissolution of intrinsic salts and acid ionization. It is the solvation process that provides energy for rupture of the bonds in acid ionization.

The affinity $\Delta G°$ of the gaseous proton to a gaseous H_2O molecule

$$H^+(g) + H_2O(g) \rightarrow H_3O^+(g) \qquad (4.7)$$

is of the order of -164 ± 4 kcal/mol.

The flat, trigonal pyramidal structure of the protonated species, the H_3O^+ ion, is fairly well known from the NMR studies of solid hydrates.

The hydrated proton H_3O^+ has been identified spectroscopically by Falk and Giguere (1957). It undergoes further hydration to give an ion tetrahedrally coordinated in such a way that three water molecules are hydrogen-bonded to the central hydrated proton $[H_3O^+]$ and a fourth water molecule is electrostatically bonded at the fourth coordination position, forming an ion $H_9O_4^+$ (Figure 4.2).

The overall energy of proton hydration is of the order of -265 ± 7 kcal/mol. The molecules of hydration water are *oriented* and *immobilized* by the ion–dipole forces and form a hydration sheath around the ion.

It is not only the protonated water complex H_3O^+ which is surrounded by four tetrahedrally coordinated, oriented water molecules—every ion orients water molecules.[†] Figure 4.3 shows the different ways in which water can orient itself toward ions. An analysis of the solvation process was carried out thermodynamically by Born (1920) and subsequently by Bernal and Fowler (1933) [see Bockris and Reddy (1970), pp. 80ff]. They showed that a tetrahedral cluster of four water molecules, oriented by a positive ion, is capable of forming 10 H-bonds (out of 12 theoretically possible) with the surrounding water molecules, whereas water molecules surrounding a negative ion are capable of forming only 8 H-bonds out of 12 H-bonds possible.

Since molecules of a liquid are in a state of continuous motion, an ion

---

[†] The neighborhood of an ion is regarded as consisting of three regions: a primary region with completely oriented water molecules, a secondary region with partly oriented water, and a bulk region of unoriented water.

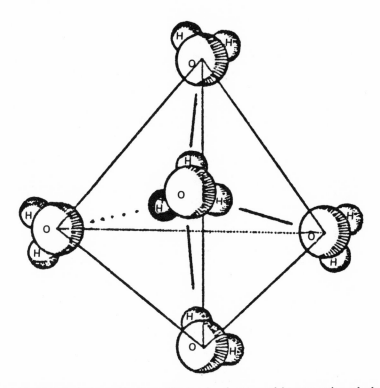

Figure 4.2. Tetrahedral coordination of $H_9O_4{}^+$ ion (not to scale) representing a hydrated H_3O^+ ion in the center, which is hydrogen-bonded to three water molecules and electrostatically bonded to the fourth water molecule. The basic structural data of water molecule and hydrogen bonds are as follows: the O—H distance in the H_2O molecule is 0.96 Å, the O—O distance in hydrogen-bonded H_2O—H_2O is 2.76 Å, the hydrogen bond energy is $\sim$4.5 kcal/mol, the dipole moment of H_2O is $\mu = 1.85$ debye units, and the static dielectric constant $\varepsilon = 79.5$ (78.5 to 80 reported).

with oriented molecules of water moves from site to site at rates depending on the effective number of bound water molecules.

Table 4.2 shows the experimentally determined primary (effective) hydration numbers obtained by a variety of methods; each number is only an approximation owing to numerous assumptions necessary in interpretation of data and to inability of distinguishing between water molecules held less strongly by hydrogen bonds to these primary hydration water sheaths. Large ions, like iodide, cesium, or tetramethylammonium, show the effective number of bound water molecules to be nearly zero, since as the ionic radius increases, the field force which aligns the water dipole decreases.

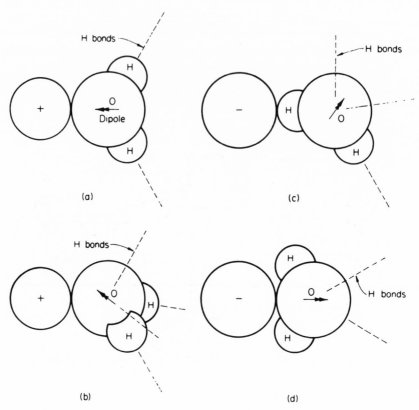

Figure 4.3. Presumed orientations of water molecules toward cations and anions: (a) and (c) after Bernal and Fowler (1933), (b) after Verwey (1941, 1942), and (d) after Halliwell and Nyburg (1963).

The heats of hydration derived theoretically show fairly good agreement with experimental values if water is treated as a quadrupole rather than a dipole [Bockris and Reddy (1970, p. 103)].

The effective radii of ions in solution are different from their radii in the crystalline lattice or the van der Waals radii of the ions in the gaseous state. These differences are revealed by the static dielectric constant ε (Section 2.4). The dielectric constant of any liquid is a function of the dipole moment of its molecules and also of the extent of association of molecules through hydrogen bonding. Figure 4.4 shows these effects for some unassociated and H-bonded liquids.

In ionic solutions, the dipoles of water molecules within the hydration sheath of ions are so firmly fixed in their orientation that they do not con-

Table 4.2. Primary Hydration Numbers[a]

Ion	From compressibility	From entropies	From apparent molal volume	From mobility	Most probable integral value
Li+	5–6	5	2.5	3.5–7	5 ± 1
Na+	6–7	4	4.8	2.4	4 ± 1
K+	6–7	3	1.0	—	3 ± 2
F−	2	5	4.3	—	4 ± 1
Cl−	0–1	3	—	—	2 ± 1
Br−	0	2	—	—	2 ± 1
I−	0	1	—	—	1 ± 1

[a] Source: Bockris and Reddy (1970).

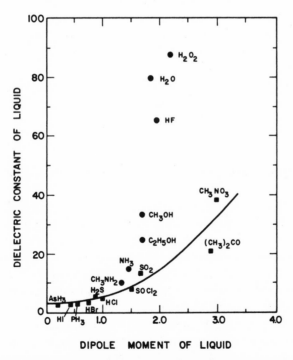

Figure 4.4. The (static) dielectric constants of liquids in relation to their dipole moments. (Squares represent unassociated liquids; circles represent H-bonded liquids.) [From Bockris and Reddy (1970), *Modern Electrochemistry*, Vol. II, Plenum Press, New York.]

tribute to the orientation polarizability of the solutions (only to the deformation polarizability); hence, the static dielectric constant of the primary hydration sheath is much lower than the value in the bulk of water. Measurements using very high alternating field frequencies indicate a dielectric constant of the hydration sheath of about 6. There is, thus, a rather sudden change in dielectric constant within the surroundings of each ion, from approximately $\varepsilon = 6$ within the primary hydration sheath (1–1.5 Å) to $\varepsilon = 80$ (for bulk water) some 6–8 Å away from the central ion.

With an increase in the concentration of ions in solution the proportion of the unoriented water molecules to the strongly oriented water (in hydration sheaths) decreases and so does the dielectric constant of the bulk solution. Thus, for example, solutions of NaCl of concentrations from 0 to 3.5 M show their dielectric constant decreasing from $\varepsilon = 80$ to $\varepsilon = 50$. The extent of ion–solvent interactions is revealed in measurements of volumes of ions in solutions; these volumes can be accurately determined by a variety of experimental techniques, such as pycnometry, dilatometry, a differential buoyancy balance technique, and a magnetic float technique. In any theoretical evaluation of ionic volumes, the real size and shape of ions (with their oriented solvent molecules) have to be taken into account; voids (structural free volume) may represent a considerable proportion of the total volume; for example, the normal molar volume of water at 25°C and 1 atm is 18.07 ml mol^{-1}, while cubically close-packed water molecules would have a volume of $\sim$12.5 ml mol^{-1}, i.e., approximately 70% of the former value.

4.3. *Debye–Hückel Model of Ions in Solution*

In addition to interactions between ions and solvent molecules, there are ion–ion interactions occurring in solutions of all electrolytes, particularly when their concentrations increase. These ion–ion interactions greatly affect the properties of solutions.

Debye and Hückel (1923a, 1923b) proposed a model of ionic solutions which enabled the behavior of charged ions in dilute solutions and the activity coefficients to be theoretically evaluated and compared with the experimentally determined values of activity coefficients. The conclusions based on this model are often extended, unjustifiably, to concentrated ionic solutions. Since the model of Debye and Hückel followed that proposed by Gouy and Chapman for the electrical double layer (see Chapter 7), there is a lot of similarity between the two concepts.

The model of Debye and Hückel selects one ion as a reference ion and replaces the solvent molecules by a continuous medium of dielectric constant ε and the remaining ions by an excess charge density ϱ_r at distance r. Then, by using Poisson's equation for a spherically symmetrical charge distribution in relation to the electrostatic potential ψ_r and Boltzmann distribution for the concentration of ionic species in terms of this electrostatic potential for small ψ_r (that is, when $ze\psi_r \ll kT$), a solution to the Poisson–Boltzmann equation is obtained after expanding the exponential in a Taylor series [for details, see Bockris and Reddy (1970), pp. 180ff].

The spatial distribution of charge density ϱ_r with distance r from the central ion is thus given by

$$\varrho_r = -\frac{ze}{4\pi}\varkappa^2 \frac{e^{-\varkappa r}}{r} \tag{4.8}$$

where z is the valence of the central ion, e is the electronic charge, and $1/\varkappa$ is the effective *radius of the ionic atmosphere* or Debye–Hückel reciprocal length. The physical meaning of equation (4.8) is that a "cloud" of countercharges surrounds the central ion, the excess charge density decaying in an exponential way in such a manner that the maximum value of charge is attained at a distance

$$r = \frac{1}{\varkappa}$$

If the central ion is visualized as a point charge, the electrostatic potential ψ_r is given by

$$\psi_r = \frac{ze}{\varepsilon} \frac{e^{-\varkappa r}}{r} \tag{4.9}$$

However, for a central ion of a definite size $2a$, where a is called the *distance of closest approach* (at which finite-size counterions can approach the central ion), the electrostatic potential is given by

$$\psi_r = \frac{ze}{\varepsilon} \frac{e^{\varkappa a}}{1+\varkappa a} \frac{e^{-\varkappa r}}{r} \tag{4.10}$$

The calculated values of $1/\varkappa$ for various concentrations of salts that give 1:1, 2:2, and 1:3 types of oppositely charged ions are given in Table 4.3.

The mean ionic activity coefficient [which is defined as a square root of the product of individual activity coefficients, $f_\pm = (f_+ f_-)^{1/2}$] can be evaluated as

$$\ln f_\pm = \frac{(ze)^2}{2\varepsilon\varkappa T} \frac{\varkappa}{1+\varkappa a} \tag{4.11}$$

Table 4.3. Thickness of Ionic Atmosphere (in Angstroms) at Various Concentrations and for Various Types of Salts[a]

C (mole liter^{-1})	Type of salt (valence of ions)			
	1:1	1:2	2:2	1:3
10^{-4}	304	176	152	124
10^{-3}	96	55.5	48.1	39.3
10^{-2}	30.4	17.6	15.2	12.4
10^{-1}	9.6	5.5	4.8	3.9

[a] Source: Bockris and Reddy (1970).

if an assumption is made that the distance of closest approach a is much less than the radius of ionic atmosphere $1/\varkappa$; i.e., $\varkappa a \ll 1$. Before Debye and Hückel's treatment, Lewis introduced empirically a quantity called the *ionic strength*, defined as

$$I = \tfrac{1}{2} \sum c_i z_i \qquad (4.12)$$

where c_i is the concentration of ionic species i and z_i is the valence of ionic species i. The concept of ionic strength reflects the quantity of charge in an electrolytic solution. It can be utilized to express the mean ionic activity coefficient as

$$\log f_{\pm} = -A(z_+ z_-) I^{1/2} \qquad (4.13)$$

which for a 1:1 valent electrolyte (e.g., NaCl) reduces to

$$\log f_{\pm} = -A c^{1/2}$$

For water the constant A has a value of 0.5070 at 20°C, 0.5115 at 25°C, and 0.5373 at 50°C.

The preceding approximate theoretical expression for $\log f_{\pm}$ has been repeatedly checked for various electrolytic solutions and was found to be in good agreement with experimental results only for very dilute solutions (Figure 4.5), not for more concentrated ones (Figure 4.6). For solutions more concentrated than $c = \sim 10^{-3}\ M$, an adjustable parameter a, representing the distance of closest approach, has to be employed to obtain

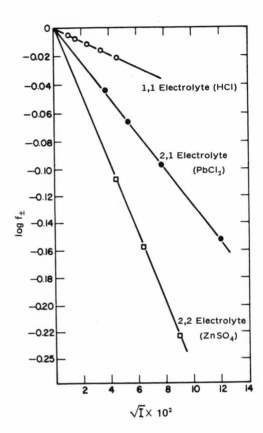

Figure 4.5. The experimental values of log $f_\pm$ (activity coefficient) as linear functions of the ionic strength (raised to the power of one-half) for low concentrations of different electrolytes grouped according to valence type. [From Bockris and Reddy (1970), *Modern Electrochemistry*, Vol. II, Plenum Press, New York.]

a more satisfactory agreement between the experimental results and the theoretical predictions:

$$\log f_\pm = - \frac{A(z_+ z_-) I^{1/2}}{1 + aBI^{1/2}} \tag{4.14}$$

where $BI^{1/2} = \varkappa$.

However, even this adjustment is unable to provide an agreement for concentrations higher than $\sim 0.1\ M$ for which—instead of a continued decrease in $f_\pm$—an actual increase in activity coefficient is frequently observed. A semiquantitative understanding of this behavior is possible if one considers that the amount of water firmly bound to ions as their hydration sheath can, in effect, be subtracted from the amount of solvent to determine the remaining effective solvent available. The three parameters—distance of closest approach, hydration number, and activity coefficient—are mutually interdependent parameters.

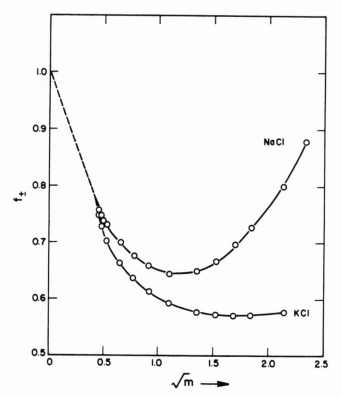

Figure 4.6. The nonlinear dependence of activity coefficients on concentration, m mol/liter, for higher strengths of 1:1 electrolytes in solution. [From Bockris and Reddy (1970), *Modern Electrochemistry*, Vol. II, Plenum Press, New York.]

4.4. Ion–Pair Formation

In addition to the ion-size parameter and the decrease in effective solvent, the formation of neutral ionic pairs by ion–ion association constitutes another effect influencing the behavior of ionic solutions. In 1887 Arrhenius postulated dissociation of *potential* electrolytes (aggregates of atoms held by covalent bonds) into ions, and for a reaction: $AB \rightleftarrows A^+ + B^-$ he defined a dissociation constant:

$$K_D = \frac{a_A a_B}{a_{AB}} \tag{4.15}$$

where a_A, a_B, a_{AB} are activities of species A^+, B^-, AB in solution. Bjerrum (1926) suggested treating ion–pair formation for *true* electrolytes (intrinsic salts M^+Y^-) as an association constant:

$$K_A = \frac{a_{M^+Y^-}}{a_M a_Y} \qquad (4.16)$$

The extent of ion-pair formation, represented by the fraction θ of ions, has to be subtracted from the total ionic concentration to give the correct concentration of free ions. Thus, the Debye–Hückel expression for the activity coefficient, equations (4.13) and (4.14), has to include the term $(1 - \theta)c$ as the concentration of free ions instead of c ions; further, the closest distance of approach a, used in the expression, has to be greater than a corresponding critical value g, or otherwise ion-pair formation (leading to neutral, non-ionized species) occurs. Fuoss (1934) proposed a derivation of the association constant K_A different from that of Bjerrum [for the comparison, see Bockris and Reddy (1970), pp. 251–265], but the experimental results do not justify a choice between the two methods of evaluation of K_A. Table 4.4 shows that, depending on the preceding two parameters a and g (or g/a), the calculated fraction θ of associated ions (ion pairs) already becomes significant (say $> 5\%$) for concentrations greater than ~ 0.02 mol/liter.

Ion pairs are formed by electrostatic (coulombic) forces when their energy exceeds the energy of thermal vibrations that cause dissociation. The distance between the two ions in an ion pair is of the order of 5–10 Å,

Table 4.4. Fraction of Association, θ, of Univalent Ions in Water at 18°C[a]

c(mol/liter)	$a \times 10^8$ cm: 2.82 / g/a: 2.5	2.35 / 3	1.76 / 4	1.01 / 7	0.70 / 10	0.47 / 17
0.0001	—	—	—	—	0.001	0.027
0.0002	—	—	—	—	0.002	0.049
0.0005	—	—	—	0.002	0.006	0.106
0.001	—	0.001	0.001	0.004	0.011	0.177
0.002	0.002	0.002	0.003	0.007	0.021	0.274
0.005	0.002	0.004	0.007	0.016	0.048	0.418
0.01	0.005	0.008	0.012	0.030	0.030	0.529
0.02	0.008	0.013	0.022	0.053	0.137	0.632
0.05	0.017	0.028	0.046	0.105	0.240	0.741
0.1	0.029	0.048	0.072	0.163	0.336	0.804
0.2	0.048	0.079	0.121	0.240	0.437	0.854

[a] Source: Bockris and Reddy (1970).

since both ions are hydrated; if the interposed solvent molecules are removed, bonds stronger than purely electrostatic ones may be created, and a *complex*,[†] as opposed to an ion pair, is formed. Formation of ion pairs and of complexes can be studied by various techniques. Owing to the possibility of interactions between the incident electromagnetic radiation and the various species in solution, it is frequently possible to obtain direct information on the constitution of solutions by studying the radiation which is either transmitted, scattered, or refracted. Thus, the absorption spectra obtained in visible and ultraviolet (VIS and UV) spectroscopy of solutions can be interpreted in terms of free ions or associated ions (if the spectra of these are known) and with respect to charge transfer or crystal-field splitting. Infrared (IR) spectroscopy can provide information on the vibrational and rotational energy levels of a given species, again after absorption, whereas Raman spectroscopy deals with shifts in radiation frequency (due to the polarizability of species) which are detected in scattered radiation. The magnitude of these shifts (with respect to the incident radiation frequency) provides information about the vibrational and rotational levels of the scattering species. Thus, Raman and IR spectroscopy are complementary. Because of strong absorption in the infrared region of radiation by water, IR spectroscopy is not (generally) applicable to the studies of aqueous solutions; however, Raman scattering is, since it is not affected by the aqueous medium. Nuclear magnetic resonance (NMR) spectroscopy can provide useful information on the type of association between an ion and its environment (ion–solvent and ion–ion interactions). The measurements of conductivity of ionic solutions (introduced by Onsager to study ion pairing), the polarographic method, the ion exchange and potentiometric methods, etc., have been extensively employed to study complex formation and ion pairing [for a review of the subject, see Nancollas (1966) and Davies (1962)]. Some of the results, abstracted from Davies' book, are given in Table 4.5 in terms of pK_A values. $pK_A = -\log K_A$, where K_A is given by equation (4.16). The greater the pK_A value, the easier is the formation of ion pairs.

When pK_A values for hydroxide ion pairs are plotted against the parameter $z/(r_M + r_{OH})$ (reflecting coulombic interaction energy) (Figure 4.7), it becomes evident that some metals such as Ag, Pb, Cu, Al, etc., contain contributions other than electrostatic. In fact, many divalent and trivalent metal ions, such as Cr^{3+}, Fe^{3+}, and Co^{2+}, form coordination complexes, in succession, instead of ion pairs. For example, Fe^{3+} forms

---

[†] A more extensive treatment of complexes and of complexing additives is deferred to Section 10.4.

Table 4.5. Selected pK_A Values of Ion Pairs in Aqueous Solutions at $25°C^a$

Hydroxides	Li^+ (-0.08), Na^+ (-0.7), K^+, Rb^+, Cs^+ (no evidence of ion pairs), Ag^+ (2.3), Tl^+ (0.8), Mg^{2+} (2.58), Ca^{2+} (1.30), Sr^{2+} (0.96), Ba^{2+} (0.64), Zn (4.3), Cu (6.5), Fe^{2+} (8.1), Co^{2+} (1.8), Ni^{2+} (3.3), Pb^{2+} (7.8), Al^{3+} (9.0), Fe^{3+} (12.0), Hg^{2+} (11.5), Sn^{2+} (12.1)
Chlorides	Li, Na, K, Mg, Ca, Sr (no evidence of ion pairs), Ag (3.2), Cu^{2+} (0.4), Pb^{2+} (1.5), Fe^{3+} (1.5), Cr^{3+} (0.6)
Sulfates	Li (0.16), Na (0.7), K (1.0), Ag (1.3), Mg (2.23), Ca (2.28), Zn (2.31), Cu (2.36), Mn (2.28), Co (2.47), Ni (2.40), Cd (2.35), Fe^{3+} (4.2)
Thiosulfates	Na (0.6), K (0.9), Ag (8.8), Mg (1.83), Ca (1.95), Ba (2.28), Zn (2.40), Mn (1.95), Co (2.05), Ni (2.06), Cd (3.92), Fe^{3+} (3.25)

^{a} Source: Davies (1962).

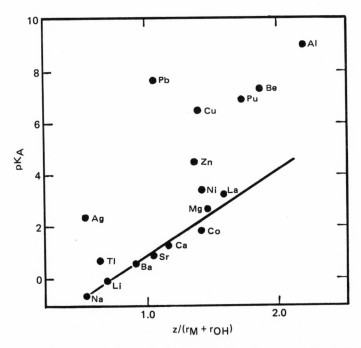

Figure 4.7. Evidence of additional nonelectrostatic interactions occurring in hydroxide ion pairs whose pK_A values do not change linearly with the $z/(r_M + r_{OH})$ parameter.

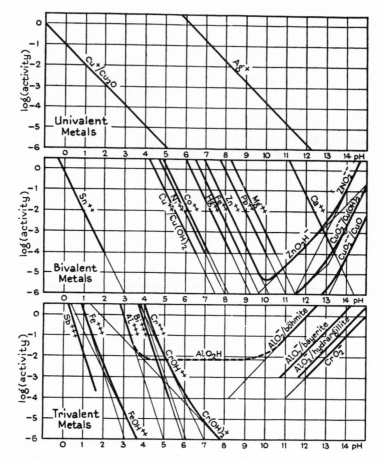

Figure 4.8. Solubility of hydroxides: influence of pH on metal-hydroxide equilibria in heterogeneous systems. [From Pourbaix (1949), with the permission of the author.]

Fe(OH)$^{2+}$ and Fe(OH)$_2{}^+$, the existence of which is associated with a change in color, indicating an inner sphere complex. For further details, see Section 10.4.1 on hydroxy complexes. In systems containing such complexes of multivalent metals, the subsequent precipitation of insoluble metal hydroxides is then delayed (for low metal ion activities) until higher OH$^-$ concentrations are reached; see Figure 4.8.

The formation of ion pairs and, particularly, of complexes is responsible for maintaining in solutions much higher concentrations of metallic ions than those called for by the solubility product relationship:

$$K_{sp} = [M^{+n}][Y^-]^n \qquad (4.17)$$

Theoretically, once the concentrations of ionic species exceed the solubility product K_{sp}, the excess of ions should remove itself from solution as an insoluble precipitate. Ion pairing and complex formation counteract the normal precipitation process (see Section 5.2.3 dealing with precipitation of insoluble metal xanthates).

4.5. *Dissociation of Water and Acids*

In aqueous solutions, the hydrogen ion concentration is often given in terms of pH, defined as $-\log_{10}[H^+]$, where $[H^+]$ is the activity of hydrogen ion (implying H_3O^+ ion, since the proton in solution is always hydrated; similarly, all other ions such as Na^+, and Fe^{2+}, Zn^{2+} are customarily written as such, although the actual species are always hydrated in solution). Water is weakly ionized,

$$H_2O \rightleftarrows H^+ + OH^- \quad \text{or} \quad 2H_2O \rightleftarrows H_3O^+ + OH^- \quad (4.18)$$

and at 24°C the ionic product of dissociation is

$$K_w = [H^+][OH^-] = 10^{-14} \text{ (mol/liter)}^2 \quad (4.19)$$

This value of the dissociation constant for water is significantly temperature-dependent; the pK_w values for different temperatures are as follows: 0°C, 14.9435; 10°C, 14.5346; 30°C, 13.8330; 40°C, 13.5348, 50°C, 13.2617; 60°C, 13.0171 (these values are taken from the *Handbook of Chemistry and Physics*, 46th ed. (1966), The Chemical Rubber Co., Cleveland, Ohio, p. D-80). When $[H^+] = [OH^-]$, i.e., when pH $= 7$, the solution is neutral; solutions of lower pH are acidic, and those of higher pH are alkaline.

Hydrogen compounds HX which ionize in water (and in other polar solvents) act as acids. The extent of ionization depends on the given compound HX and on the solvent. The effect of pH of water on the degree of ionization is shown in Figure 4.9 for some of the most frequently encountered species.

The dissociation constant

$$K_A = \frac{a_H a_X}{a_{HX}}$$

is related to the change in the standard Gibbs free energy (see Section 2.9) by the following expression:

$$\Delta G^\circ = -RT \ln K_A = -2.303 RT \log_{10} K_A \quad (4.20)$$

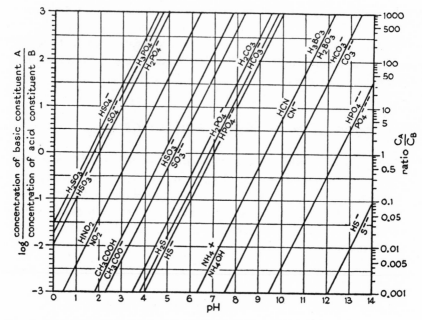

Figure 4.9. Dissociation of acids and bases; influence of pH on equilibria in homogeneous systems. [From Pourbaix (1949), with the permission of the author.]

Fifty percent dissociation occurs at $pH = -\log_{10} K_A$ (9% dissociation at $pH = -\log K_A - 1$; 24% dissociation at $pH = -\log K_A - 0.5$; 91% dissociation at $pH = -\log K_A + 1$, etc.). Dissociation reactions leading to the formation of hydrogen ion, H_3O^+, which becomes further hydrated, can be represented as composed of two steps:

$$\tfrac{1}{2}H_2(g) \rightarrow H^+(g) + e \qquad \text{an oxidation step} \qquad (4.21)$$

and

$$H^+(g) + x\text{-}H_2O \rightarrow H^+(aq) \qquad \text{a solvation step} \qquad (4.22)$$

The reduction of dissolved O_2 provides OH^- ions:

$$\tfrac{1}{2}O_2 + H_2O + 2e \rightarrow 2OH^- \qquad (4.23)$$

The equilibrium potentials of the preceding two ionization reactions, involving hydrogen and oxygen dissolved in water, are represented as straight-line functions of pH [for details, see Pourbaix (1949) or Garrels and Christ (1965)].

4.6. *Solubility of Gases in Water*

Under equilibrium conditions gases from the surrounding atmosphere dissolve in water in direct proportion to their partial pressure; that is, following Henry's law,

$$p_i = K_i X_i$$

where p_i is the partial pressure of the gas, K_i is Henry's law coefficient, and X_i is the mole fraction of the gas dissolved. Since air contains (under moisture-free conditions) 20.95% O_2 by volume, at 20°C pure water dissolves 9.3 mg/liter, that is, 20.95% of the solubility at 1 atm O_2. Table 4.6 gives the solubilities in water of pure gases at 1 atm partial pressure and indicates how readily the solubilities decrease with temperature. Figure 4.10 shows much more dramatically the decrease in solubility of O_2 from air with the rise in temperature.

Henry's law coefficients for the variation of oxygen solubility in water with its partial pressure at different temperature are

$$K_{O_2} = 3.04 \times 10^{-7} \ (20°C) \quad \text{and} \quad 3.33 \times 10^{-7} \ (25°C)$$

whereas those for nitrogen are

$$K_{N_2} = 6 \times 10^{-7} \ (20°C) \quad \text{and} \quad 6.43 \times 10^{-7} \ (25°C)$$

These values reflect the fact that the solubility of oxygen in water is higher than that of nitrogen.

Using interferometry to study dissolution of various gases in solvents, Hyslop (1975) and O'Brien and Hyslop (1975) established two interesting aspects of solubility behavior:

1. Owing to the preferential dissolution of oxygen in water (in comparison with that of nitrogen) when water is injected through an appropriate orifice into a container filled with pressurized air [for example, pressurized to 60–90 psi (4–6 atm)], the water droplets become saturated with oxygen only but not with nitrogen within the few seconds needed for them to fall to the base of the container (of appropriately chosen height).

2. The nucleation of the gas dissolved in water and its release in the form of bubbles require a definite drop in pressure. The results on this aspect are shown in Table 4.7. As yet, no correlation has been established between the pressure drop required for a spontaneous evolution of gas

Table 4.6. Solubility of Gases in Pure Water in Contact with the Pure Gas at a Pressure of 1 atm

Temperature (°C)	Nitrogen,[a] 98.815% N^2 + 1.185% A		Oxygen,[a] O_2		Hydrogen,[a] H_2		Methane,[a] CH_4		Carbon dioxide,[a] CO_2		Hydrogen sulfide,[a] H_2S		Sulfur dioxide,[a] SO_2		Ammonia,[b] NH_3	
	$[N_2]$ $\times 10^3$	mg/ liter[c]	$[O_2]$ $\times 10^3$	mg/ liter	$[H_2]$ $\times 10^3$	mg/ liter	$[CH_4]$ $\times 10^3$	mg/ liter	$[CO_2]$ $\times 10^3$	mg/ liter$_T$[d]	$[H_2S]_T$ $\times 10^3$	mg/ liter$_T$	mT	mg/ liter$_T$	mT	mg/ liter$_T$
0	1.050	29.4	2.18	69.8	0.959	1.93	2.48	39.8	76.4	3360	208.3	7100	3.58	186,700	52.3	471,000
5	0.931	26.1	1.913	61.2	0.912	1.84	2.14	34.3	63.5	2790	177.4	6040	3.03	162,300	45.8	438,000
10	0.830	23.2	1.696	54.3	0.872	1.76	1.864	29.9	53.25	2345	151.6	5160	2.55	140,600	40.1	406,000
15	0.752	21.1	1.523	48.7	0.841	1.70	1.645	26.4	45.45	2000	131.4	4475	2.14	120,500	35.2	375,000
20	0.689	19.3	1.384	44.3	0.812	1.64	1.475	23.6	39.1	1720	115.2	3925	1.80	103,300	30.9	344,000
25	0.639	17.9	1.263	40.4	0.783	1.58	1.342	21.5	33.9	1495	101.8	3470	1.51	88,200	27.9	322,000
30	0.599	16.8	1.163	37.2	0.758	1.53	1.233	19.7	29.65	1305	90.9	3090	1.26	74,700	23.8	288,000
40	0.528	14.8	1.029	32.9	0.734	1.48	1.057	16.9	23.65	1040	74.1	2520	0.91	54,800	19.8	252,000
60	0.456	12.77	0.868	27.8	0.714	1.44	0.872	14.0	16.0	700	53.1	1810	—	—	14.0	192,000
80	0.427	11.96	0.786	25.1	0.714	1.44	0.79	12.7	—	—	40.9	1394	—	—	9.1	134,000
100	0.423	11.85	0.758	24.2	0.714	1.44	0.76	12.2	—	—	36.1	1230	—	—	—	—

[a] From Landolt-Börnstein (1923).
[b] From Landolt-Börnstein (1923); also see Mellor (1922).
[c] Note: 1 mg/liter ≡ 1 ppm.
[d] Subscript T signifies total solubility.

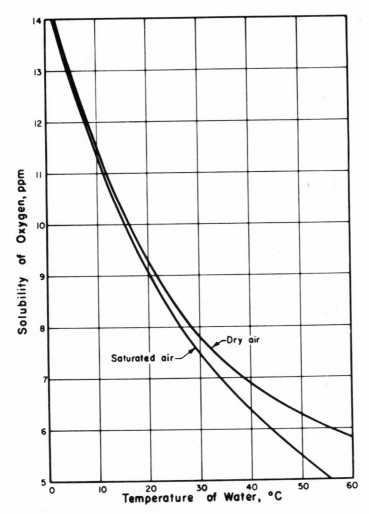

Figure 4.10. Solubility of oxygen in distilled water in contact with air at 760 mm Hg containing 20.95% oxygen. [From Camp (1963), *Water and Its Impurities*, with the permission of Reinhold Book Corporation, New York.]

bubbles and other properties of that gas. However, the data in Table 4.7 show that no N_2-containing bubbles can be nucleated in water as long as the pressure drop occurs from less than 90 psi to atmospheric pressure in a stainless steel vessel. At the same time, when water is saturated with oxygen and the pressure is dropped from above the 60-psi level to atmospheric level, the bubbles released should contain O_2 only.

Table 4.7. Depressurization Conditions Required for Spontaneous Bubble Formation [Values of Pressure (in a stainless steel vessel) from which the System Was Released to Atmospheric Pressure, (in psi)][a]

Oxygen		Nitrogen		Hydrogen		Argon and/or Helium	
Bubbles formed	No bubbles formed	Bubbles formed	No bubbles formed	Bubbles formed	No bubbles formed	Bubbles formed	No bubbles formed
121, 120		120, 121		120		60, 62	
100, 84		100, 95	94	110		58, 63	
81, 74		90	88		105	55, 54	
66, 64			88		90	51, 51	47, 47
61, 59			87				45
	55		86				43
	54						
	54						
	50						

[a] Source: Hyslop (1975).

The preceding findings may help to explain some of the operational and design features of flotation cells used in selective separations of sulfides. One type of mechanical flotation cell may appear distinctly more effective, under otherwise identical conditions, than others. This may be associated with the degree of compression and decompression achieved ahead and behind the impeller blades, correspondingly, and the concomitant action of oxygen-containing bubbles in oxidation and adsorption reactions. Similarly, presaturation of water with compressed air for "vacuum" flotation may confer some significant benefits in those systems where oxygen-nucleated bubbles perform other actions in addition to that of a buoyant medium.

The solubility of gases is reduced not only by a rise in temperature but also by the presence of dissolved salts. This behavior is due to the utilization of water in the hydration of ions and a consequent reduction in available water molecules.

As shown in Figure 4.11, when the dissolved salts reach the concentration of 35,000 ppm, which corresponds to that of seawater, the solubility of O_2 is reduced to approximately 82% of the solubility in pure water.

Carbon dioxide, CO_2, is normally present in atmospheric air at $\sim 0.03\%$ by volume, and the average amount dissolved in water at 25°C would then be ~ 0.45 mg/liter. However, owing to respiration of plants and oxidation of organic matter (within the water, at the expense of dissolved O_2), the dissolved CO_2 may reach 10–20 mg/liter. The solubility of pure CO_2 at 1 atm and 25°C is 1495 mg/liter. (The extent of CO_2 dissociation into HCO_3^- and CO_3^{2-} is indicated in Figure 4.9.) Although atmospheric air is incapable of providing sufficiently high CO_2 partial pressure to reach the saturation level of CO_2 solubility, minerals such as $CaCO_3$, $MgCO_3$, and dolomite, $(Ca,Mg)CO_3$, with which many natural waters are in contact (and the flotation pulps containing the preceding minerals), are usually responsible for an inordinate increase in the HCO_3^- content of such waters to several hundred ppm.

Many of the solutes in aqueous systems are referred to as "structure makers," but the exact definition of this term is rarely made clear. Sometimes a structure maker denotes a species giving rise to a net entropy decrease, without specifying what structure is being "made." Small cations of high surface charge density are capable of holding enough water molecules around them to increase the overall amount of structured water; they are thus structure makers. Tetraalkylammonium salts may act as either structure makers or breakers, depending on their alkyl group and the size and the shape of this group.

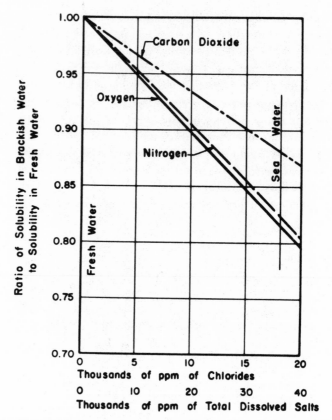

Figure 4.11. Effect of dissolved salts on solubilities of N_2, O_2, and CO_2. [From Camp (1963), *Water and Its Impurities*, with the permission of Reinhold Book Corporation, New York.]

There is, as yet, no agreement as to the manner in which gases become dissolved in water. It is considered most likely that dissolving gas molecules enter cavities in the *clathrate*-type structures[†] to form the so-called gas hydrates. These clathrate hydrate structures have been identified by the

---

[†] Clathrate structures are formed by compounds whenever the host lattice contains cavities that are capable of enclosing other molecules without developing any bonds between the trapped molecule and the host lattice. The best known are quinol (*p*-dihydroxybenzene) clathrates which consist of three molecules, hydrogen bonded, forming an approximately spherical cage of ~4-Å diameter in which numerous small species, e.g., O_2, N_2, CO, NO, CH_4, SO_2, H_2S, etc., can be trapped; CH_3OH can also form a quinol-type clathrate but not C_2H_5OH—this is too large. *Molecular sieves* are solid, inorganic compounds that contain similar cavities for trapping suitable molecules.

X-ray technique after solidification of liquids in the presence of gases. Two types of clathrate clusters, composed of hydrogen-bonded water molecules, are suggested: type I, containing 46 water molecules arranged in (joined) polyhedral cages that effectively produce a cubic cell of $\sim$12 Å in size, and type II, with 136 water molecules arranged in a unit cubic cell of $\sim$17 Å in size. The great majority of gas hydrates are known to conform to one or the other of these structures. Both types contain several cavities (or cages) capable of accommodating species of a given size. Thus, structures of type I accommodate (and are formed by) solute species of sizes less than 5.4 Å in diameter. For example, O_2 and Kr (of van der Waals diameter $\sim$4 Å); N_2, H_2S, and CH_4 ($\sim$4.1-Å diameter); CO_2 (4.7-Å diameter); AsH_3 (4.9-Å diameter); N_2O and SO_2 (5.0-Å diameter); and Cl_2 and SbH_3 (5.1-Å diameter) all form clathrate structures of type I. Solutes whose van der Waals diameters are in the range 5.5–6.6 Å [for example, C_2H_4, $(CH_2)_3$, and BrCl (5.4-Å diameter); COS and C_2H_2 (5.5 Å); C_3H_8, acetone, and furan (6.3 Å); $CHCl_3$ (6.4 Å); and propylene oxide (C_3H_6O) (6.5-Å diameter)] tend to form clathrate structures of type II. It is presumed that the decrease in gas solubility with temperature is a measure of the decreasing availability of clathrate hydrates. Details of various gas clathrate hydrates can be found in papers by Davidson (1973) and Daee *et al.* (1972).

If all available cages were filled with gas molecules, the resultant hydrate would have approximately five to eight molecules of H_2O per gas molecule. Since the solubility of O_2 at 25°C is 1.263×10^{-3} mol/liter, the number of H_2O molecules involved in gas hydrates containing this amount of O_2 constitutes only a very small proportion, approximately 10^{-4}, of all H_2O molecules present in a liter.

As will be seen in Chapter 8, a strict and delicate control of oxygen is indispensable in selective sulfide flotation systems whenever thio compounds are used as collectors. During grinding of ore in rod mills and ball mills, oxygen dissolved in water is being consumed by oxidizable components such as mineral sulfides,[†] steel grinding media, and mill liners. The result is that the oxygen content in the pulp after grinding in rod or ball mills when treating sulfide ores decreases from 9–11 ppm (20°C–10°C) in the feed to approximately 0 ppm in the mill discharge. There is also a con-

---

[†] Sulfides consume oxygen and lower the pH of the pulp due to the formation of acidic oxidation products of sulfur (see Section 10.4). If the sulfides comprise pyrrhotite (or galena) and the lowering of pH is not checked by additions of alkali or lime, H_2S gas may be formed. When this reaction occurs so that the odor of H_2S is detectable, the relevant mineral pulp is without a trace of dissolved oxygen and is acid in character.

comitant rise in temperature of the pulp to between 15° and 30°C due to heating effects during grinding.

The subsequent reoxygenation of the mill discharge is accomplished gradually when the pulp passes through cyclones, conditioners, and flotation cells. In view of the findings of O'Brien and Hyslop (1975), already mentioned, the extent of pressurization occurring in these units of flotation circuit equipment may be of considerable significance for the control of oxidation reactions in a given flotation system. In some sulfide separations with a given combination of thio reagents, the use of a one-stage high-pressure cyclone separation may occasionally produce results different from those obtained with a two-stage cycloning at a lower pressure (due to surface reactions). Similarly, one type of conditioner or a flotation cell may give a suitable degree of pressurization for the subsequent release of O_2 bubbles, whereas other types of cells (or even a different speed of the same cell) may prove distinctly inferior in the ultimate separation effectiveness.

The degree of hydrophobicity of surfaces present in contact with the solution is also an important factor. Air-saturated water solutions in contact with polyethylene or Teflon surfaces will deposit air bubbles on these surfaces on release of pressure less than 5 psi.

4.7. Solubility of Hydrocarbons in Water

As indicated in Section 4.2, there are two types of electrolyte solutes (that is, the intrinsic ionically bonded compounds and the covalently bonded potential electrolytes). There are also two types of nonelectrolyte solutes. One group is represented by amphipatic solutes possessing nonionized polar groups such as —OH, —NH_2, —COOH, and —$CONH_2$ [such nonionized surfactants are also known as amphiphiles]. The other group constitutes the entirely apolar (nonpolar) solutes. (Amphipatic molecules whose polar groups become ionized in a given aqueous environment constitute electrolytic solutes.) Both types of nonelectrolyte solutes are capable of direct interaction with the solvent water.

Figures 4.12 and 4.13 show the extent of molal solubilities of selected nonpolar species. Hydrocarbons possess, generally, low solubilities in water. Methane, ethane, and propane have solubilities nearly the same as that of oxygen, that is, $\sim 1.4 \times 10^{-3} M$ at 20°C (Figure 4.12); benzene shows an appreciably greater solubility, $\sim 1.2 \times 10^{-2} M$, whereas, the solubilities of butane, pentane, hexane, and cyclohexane decrease progressively to levels below $10^{-4} M$. The thermodynamic data and the properties of solutions of

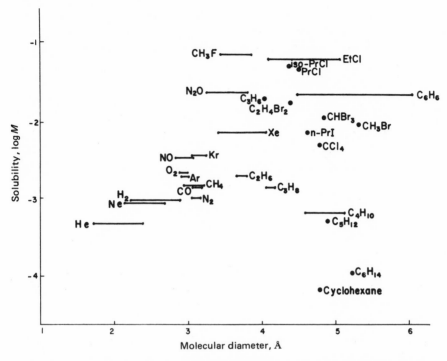

Figure 4.12. Aqueous solubility of selected substances as a function of their molecular size (determined by different techniques, such as gas viscosity, van der Waals *b* coefficient, second virial coefficient, etc.). [From Franks and Reid (1973), in *Water*, Vol. 2, F. Franks, ed., Plenum Press, New York.]

Figure 4.13. Aqueous solubilities of normal alkanes (closed circles) and normal alcohols (open circles). [From Krescheck (1975), in *Water*, Vol. 4, F. Franks, ed., Plenum Press, New York.]

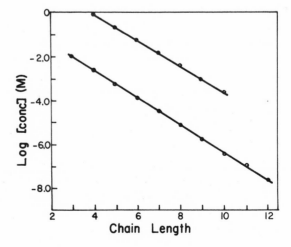

these nonpolar solutes, discussed at length by Franks and Reid (1973), indicate some unusual characteristics; for example, the enthalpies of solutions of alkanes from CH_4 to C_5H_{12} are negative but become positive for hydrocarbons higher than C_5H_{12}. The changes in entropy on dissolution are negative for all hydrocarbons, whereas for polar solutes they are mostly positive.

The dissolution of methane, ethane, and propane, and the thermodynamic properties of their solutions, could be explained on the basis of clathrate hydrate models. However, the otherwise very similar properties of butane solutions cannot be accounted for in the same way, since the molecule of butane is too large to fit into a clathrate cavity, and no clathrate hydrate of butane is known (although distorted clathrate hydrates accommodating butylamines and pentylamines have been suggested). Nemethy and Scheraga (1962) accounted for the properties of solutions of all hydrocarbons by considering the effect of these solutes on the extent of hydrogen bonding and cluster formation in the surrounding water molecules (Table 4.1). Their explanation seems the most satisfactory so far and applies equally to those nonpolar solutes which do form clathrate hydrates and to those which are too large in size. All hydrocarbons cause some structuring of surrounding water molecules and reinforce the stability of existing water clusters, and these effects are consequently similar to stabilization of water in clathrates.

Even the hydrocarbons of the amphiphilic solutes behave in dilute solutions similarly to apolar solutes with respect to the manner in which they affect the solvent water. The polar groups of these amphiphiles provide simply an extra solubilizing interaction. As a result, the thermodynamic data for amphiphilic solutes are parallel to those obtained for analogous nonpolar solutes, as shown in Barclay–Butler plots (Figure 4.14). Each polar group provides an effectively constant increment in both enthalpy and entropy, displacing the line for its homologues parallel to that for alkanes. Thus, solutions of alkyl alcohols up to C_5 can be considered as equivalent to 1:1 mixtures of methanol and the relevant alkanes. Similarly, the changes observed in the heat capacities of amphiphilic solutions (which indicate constant increments per CH_2, independent of the polar group) suggest that the water structure around the nonpolar portion of the amphiphile is only slightly affected by the presence of the polar group.

However, the solubility of alkanes longer than C_{12} ceases to decrease at the rate indicated in the homologous series for shorter chains (Figure 4.13). This behavior is due to chain folding caused by *hydrophobic interaction*; the latter is a common tendency of all nonpolar groups to avoid aqueous environment by presenting a lower number of chain components

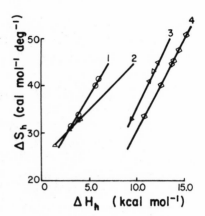

Figure 4.14. Barclay–Butler plots relating entropy and enthalpy of hydration for aqueous solutions of (1) alkanes, (2) rare gases, (3) cyclic ethers, and (4) alkyl alcohols, all at 25°C. [From Franks and Reid (1973), in *Water*, Vol. 2, F. Franks, ed., Plenum Press, New York.]

to the aqueous solvent. [A theoretical treatment of hydrophobic interaction is given by Ben-Naim (1974) in his Chapter 8. The unique features of hydrophobic interaction are that this interaction is very much stronger in water than in any other solvent, polar or nonpolar; further, only for water does the hydrophobic interaction become stronger as the temperature increases.] The solubilities of hydrocarbons in solutions containing other amphipatic solutes are similarly influenced by the hydrophobic interaction. As reported by Somasundaran and Moudgil (1974), a slight decrease in solubility is initially observed (for methane and butane) in solutions of sodium dodecyl sulfonate of less than $10^{-6} M$ concentration. This behavior may be interpreted as due to a "salting out" effect, caused by competition for the same solvent molecules. Thereafter, for solutions of sodium dodecyl sulfonate greater than $10^{-6} M$, an increase in solubility was recorded, caused by hydrophobic interaction.

Hydrophobic interaction is of great consequence not only with respect to increased bulk solubilities of alkanes and surfactants (in the formation of micelles; see Sections 5.6–5.9) but in all biochemical processes, interfacial condensation and agglomeration phenomena of nonpolar groups, stabilization of polymer conformations, etc. Since there are two components in each hydrophobic interaction,

1. An establishment of van der Waals bonding (sometimes accompanied by other bonding as well, such as hydrogen bonding in alcohol or fatty acid dimerization) and

2. A major restructuring of surrounding water molecules,

this type of interaction is not equivalent to a simple development of dispersion-type bonds. In consequence, the effects of branch chains and of molecular structure within the nonpolar groups are highly specific as regards the degree of hydrophobic interaction.

The hydrocarbon solute molecules appear to occupy less volume in an aqueous solution than they do in their own liquid state at the same temperature. There appear to exist unique interactions between hydrophobic solutes and water molecules that result in the partial molar expansion and compression parameters being opposite in sign to the corresponding parameters of polar solutes.

4.8. *Dissolved Mineral Constituents*

Natural surface waters and mine waters contain a large variety of dissolved species derived from minerals that are in contact with these waters. The constituent metallic ions (cations) which react with sodium palmitate to produce soap precipitates (curds) are responsible for the so-called *hardness* of these waters. The hardness is usually expressed in milligrams per liter of equivalent $CaCO_3$. The anions which are neutralized by titration with an acid (i.e., HCO_3^-, CO_3^{2-}, OH^-, $HSiO_3^-$, HPO_4^{2-}, $H_2PO_4^-$, HS^-, NH_3) constitute the so-called *alkalinity* and are also expressed in milligrams per liter of equivalent $CaCO_3$. Although the two concepts, hardness and alkalinity, have developed in connection with the treatment of domestic and boiler feed waters, they have a bearing in flotation systems. High hardness of water means a large concentration of cations that are likely to consume anionic surfactants, whether thio or nonthio in character, added as collectors. High alkalinity usually denotes appreciable quantities of bicarbonate ions. The latter may be utilized in complexing unusually large quantities of metallic ions, such as $Ca(HCO_3)^+$ and $Mg(HCO_3)^+$ or $Cu(HCO_3)^+$. And the hydrogen-bonding propensity of these complexes may be a very important aspect of surfactant adsorption, as discussed in Chapter 8.

For evaluation of bicarbonate content in water, the determination of alkalinity has to be supplemented by separate analyses of OH^- (pH), dissolved silica, phosphates, and sulfides, and the $CaCO_3$ equivalents of all these constituents have to be subtracted from the alkalinity value. All the preceding anions participating in the alkalinity of water may act as activators or depressants with respect to some of the minerals in the flotation system (for more details on this aspect, see Chapter 10). Hence, the ionic composition of local water to be used in flotation of a particular ore is generally a very important parameter; occasionally, it is critical to such an extent that

Table 4.8. Examples of Water Compositions, mg/liter (or ppm) of Dissolved Constituents

Constituent	World Health Organization standards for potable water[a] (mg/liter)	Vancouver Island Stream (Island Copper Mine), British Columbia, Canada (ppm)	Standard seawater (Texada Mine), British Columbia, Canada (ppm)	Broken Hill, Australia (reuse water)[b] (ppm)	El Salado, Chile (process water)[c] (ppm)	Potash mine water (saturated KCl + NaCl) (ppm)
pH	7.0–8.5	7.0	8.2	6.6–6.8	7.6–7.7	—
Total solids	500	25	35,000	—	136,300	—
Ca	75	8	413	1400–1600	2,260	—
Mg	50	2	1,294	130–150	1,890	—
SO_4^{2-} (sulfate)	200	1	2,712	~3500	2,230	—
Cl^- (chloride)	200	—	19,350	~2000	77,200	~105,000 (10.5%)
Zn	5.0	0.001	0.003	10–20	—	—
Mn	0.1	—	—	50–100	—	—
Fe	0.3	0.02	0.01	—	13	—
Na	—	—	10,760	1800–2300	50,200	~ 78,000
K	—	—	387	—	35	~ 58,000

Maximum allowable concentration (mg/liter)

Pb	0.1
As	0.2
Se	0.05
Cr^{6+}	0.05
CN'	0.01
Phenolics	0.001
Hg	?

[a] Lund (1971, Chapter 4, pp. 4-39).
[b] Connor (1977) and Dunkin (1953).
[c] Geisee (1942).

the water has to be especially treated to remove the dissolved ingredient that is harmful in the given flotation system. For example, in the flotation of vanadium minerals in Abenab, South West Africa (Namibia), the bicarbonate ions present in the local water were found to act as a depressant, and the water had to be treated for the complete removal of the bicarbonate ions before it could be employed in flotation [Fleming (1957)]. Even when the water used in the feed to a flotation plant does not initially contain high concentrations of dissolved ions, on being contacted with an ore that possesses some soluble minerals a high content of dissolved minerals may be created as a result of prolonged grinding.

The high content of total dissolved solids (or salinity) of water is per se no deterrent to the use of such water in flotation. In fact, as can be seen from the comparison presented in Table 4.8, the salinity of water used in

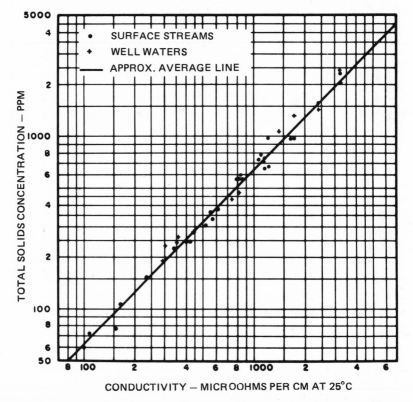

Figure 4.15. Relation between total dissolved solids and electrical conductivity. [From Camp (1963), *Water and Its Impurities*, with the permission of Reinhold Book Corporation, New York.]

the flotation of copper minerals at El Salado in Chile was more than three times that of seawater. And a saturated KCl + NaCl solution used in potash flotation has a still higher concentration of salts. Yet flotation of sulfide minerals of Cu, Pb, and Zn can be, and is being, carried out in seawater [as reported, for example, by Haig-Smillie (1974)] and in the waters of much higher salinity [for example, at El Salado; see Geisee (1942)]. However, the preceding examples do not rule out the possibility that in some flotation systems even a low salinity may not be tolerated for specific separations to be achieved.

In connection with the agricultural use of water for irrigation purposes, a relationship between the total dissolved solids (up to ~ 3000 ppm) and the electrical conductivity (in microohms per centimeter) has been tested; it is shown in Figure 4.15 and indicates that the linearity of the relationship may serve as a useful, quick method for checking changes in the content of dissolved solids.

4.9. Antipollution Treatment of Mill and Mine Effluents

Operators in mineral processing plants have to become conversant with the antipollution treatment procedures necessary for biological reusage of the effluents from the operations they supervise. Toxic ingredients likely to be retained in the effluent have to be eliminated, and any deficiency of oxygen has to be made up by restoring the dissolved oxygen to its normal level. In Table 4.8 maximum allowable concentrations of six toxic constituents are given for potable waters. Many of the national or regional pollution control boards prescribe their own limits for substances which are considered toxic for some aquatic species, and these limits are often even stricter than those for potable waters. For example, the levels of Cu, Zn, and Cd concentrations allowed for waters in which farming of seafood is practiced may be one or two orders of magnitude lower than those listed for potable water in Table 4.8. In fact, Table 4.9 gives levels of concentrations for Zn, Cu, and Pb which are lethal to salmonid fish; as can be seen, in a very soft water 18 ppb of Cu (0.018 ppm) is already considered lethal, whereas the standard for potable water allows 1000 ppb of Cu. Also, some of the trace elements which are known to be extremely toxic, such as Hg, are not even listed among the standard limiting concentrations in Table 4.8.

The procedures to be adopted for the removal of excessive quantities of certain dissolved polluting constituents in effluents from mills are normally

Table 4.9. Incipient Lethal Levels for Zn, Cu, and Pb to Salmonid Fish[a]

Total hardness, mg/liter as CaCO₃	Incipient lethal level, μg/liter (ppb)			Total hardness, g/liter as CaCO₃	Incipient lethal level, μg/liter (ppb)		
	Zn	Cu	Pb		Zn	Cu	Pb
5.0	382	18	575	50	1240	108	1300
10.0	488	28	764	60	1342	126	1402
15.0	587	38	854	100	1750	188	1810
20.0	686	48	943	150	2260	265	2320
25.0	764	58	1033	200	2770	345	2830
30.0	883	68	1096	300	3790	500	3850
40.0	1081	88	1198	—	—	—	—

[a] Source: Hawsley (1972).

based on adjustment of pH, precipitation of insoluble compounds, use of flocculants, and settling of particulates; if found necessary, adsorption on activated carbon or filtration through suitable adsorbent beds, including ion exchange resins, may have to be adopted. All these procedures may have to be followed by reaeration treatment to ensure the requisite level of O_2. It may be of interest to mention that despite a considerable amount of development in the evaluation of oxygen mass-transfer relationships across the water/air interface [see Camp (1963)], the empirically established procedure of spraying aqueous solution into air to replenish its oxygen content is more effective than bubbling massive quantities of air through the same volume of water. When tailings are discharged into lagoons, both settling of solids and reaeration are accomplished over the large surface areas of ponds.

For further details of treatments, specific literature dealing with pollution abatement has to be consulted [for example, Lund (1971)].

4.10. Selected Readings

Ben-Naim, A. (1974), *Water and Aqueous Solutions, Introduction to a Molecular Theory*, Plenum Press, New York.

Camp, T. R. (1963), *Water and Its Impurities*, Reinhold, New York.

Eisenberg, D., and Kauzman, W. (1969), *The Structure and Properties of Water*, Clarendon Press, Oxford.

Franks, F., ed. (1972, 1973, 1974, 1975), *Water, a Comprehensive Treatise*, Vols. I–V, Plenum Press, New York.

Hem, J. D., ed. (1971), *Nonequilibrium Systems in Natural Water Chemistry*, Advances in Chemistry No. 106, American Chemical Society, Washington, D. C.

Horne, R. A. (1969), *Marine Chemistry*, Wiley-Interscience, New York.

Horne, R. A. (1972), *Water and Aqueous Solutions. Structure, Thermodynamics and Transport*, Wiley-Interscience, New York.

Horne, R. A. (1978), *The Chemistry of Our Environment*, Wiley-Interscience, New York.

Jett, C. A., ed. (1977), *Cellulose Chemistry and Technology*, ACS Symposium Series 48, American Chemical Society, Washington, D. C.

Kavanau, J. L. (1964), *Water and Solute–Water Interactions*, Holden-Day, San Francisco.

Martin, D. F. (1970), *Marine Chemistry*, Vols. I and II, Marcel Dekker, New York.

Samoilov, O. Ya (1957), *Structure of Aqueous Electrolyte Solutions and the Hydration of Ions* (trans. by D. J. G. Ives), Consultants Bureau, New York (1965).

5

Flotation Surfactants

Numerous inorganic and organic reagents are employed in flotation for the purpose of controlling the characteristics of interfaces.

Any species, whether organic or inorganic in nature, which has a tendency to concentrate at one of the five possible interfaces (such as liquid/gas, liquid/liquid, solid/liquid, solid/gas, solid/solid) is a *surface active agent*. An ion is thus a surface active agent with respect to an oppositely charged surface site. The name *surfactants* is reserved for surface active *amphipatic* molecules, R—Z (that is, molecules of dual character), represented by a polar group Z and a nonpolar group R. The polar group Z consists of an aggregate of two or more atoms which are covalently bonded but possess a permanent dipole moment; the presence of this dipole makes the group hydrophilic in character. In addition to the dipole, the polar group may (but need not) be ionized. The nonpolar group or radical R is usually represented by a hydrocarbon (but it may be a fluorocarbon or a siloxane). It has no permanent dipole and represents the hydrophobic part of the amphipatic molecule.

Owing to a great variety of polar groupings that are possible and a still greater variety of nonpolar groups, there is a vast number of reagents which act as surfactants. These can be classified either according to the electrical charge associated with their polar group into anionic, cationic, or nonionic, or according to the hydrocarbon structure (alkyl, aryl, phenyl, cyclohexyl, alkylaryl, etc.), or according to the specific type of the polar group. The last classification is by far the most useful in differentiating the action of surfactants and in specifying their characteristics. An overall subdivision of surfactants into one group representing monopolar species and the other comprising multipolar ones is also useful.

For comprehensive reviews of various surfactants, see Schwartz *et al.*

(1966), Schick (1967), Jungermann (1970), and Rao (1971). Extensive listings of commercial surfactants, arranged by trade names and by types of chemical classification, are given in annual editions of McCutcheon's publications on detergents and emulsifiers.

5.1. Classification of Surfactants

Surfactants play a twofold role in flotation: Adsorbing at the solid/liquid interface, they make the surface of selected minerals hydrophobic in character (thus acting as so-called *collectors*), and second, they influence the kinetics of bubble–mineral attachment. The latter surfactants are customarily referred to as *frothers*, although, as will be seen in Chapter 9, their frothing abilities are not the most important characteristic.

Since flotation surfactants are, in general, supplied to the interfaces through the aqueous solution phase, mainly those reagents which are somewhat soluble in water are used in flotation. In some cases it is necessary to use insoluble hydrocarbons or other oils; these liquids are dispersed in the aqueous phase as emulsions, with the help of soluble surfactants, in order to facilitate their reaching the interfaces in a reasonably short time period.

The surfactants of particular importance to flotation may be conveniently grouped as follows: A, monopolar, and B, multipolar, with each group subdivided into three classes in order to facilitate the discussion of their characteristic behavior in aqueous solutions and during adsorption at interfaces:

I. *Thio compounds*, which act primarily as collectors for metallic sulfides

II. *Non-thio, ionizable compounds*, which may act as both collectors and frothers

III. *Nonionic compounds*, some of which act primarily as frothers, while others act as depressants, flocculating agents, and even as activators (or collectors)

A. Monopolar Surfactants

The three classes we have listed contain a number of specific reagents differentiated by the nature of their polar portions. Some of the most frequently encountered polar groups are shown in Figure 5.1, as scale models, in order to emphasize the stereochemical disposition of atoms and the relative sizes of groups.

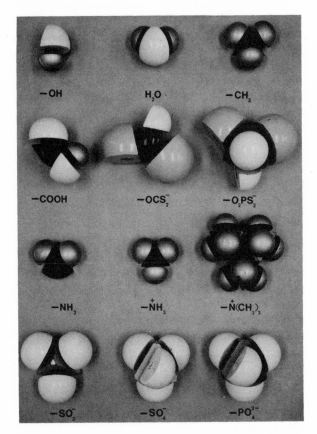

Figure 5.1. Scale models of selected polar groups in surfactants utilized in flotation, detergency, and emulsions. The groups are —OH (alcohol), —COOH (carboxylic acid), —OCS$_2^-$ (xanthate, that is, dithiocarbonate), —O$_2$PS$_2^-$ (Aerofloat, that is, dithiophosphate), —NH$_2$ (amine), —NH$_3^+$ (aminium ion), —N$^+$(CH$_3$)$_3$ (TAB or trimethylammonium ion), —SO$_3^-$ (sulfonate), —SO$_4^-$ (sulfate), and —PO$_4^{2-}$ (phosphate); they are compared with the molecule of water H$_2$O and the terminal hydrocarbon group —CH$_3$. The size of the polar group and the corresponding degree of "packing" or alignment along interfaces play an important (often decisive) role in all adsorption phenomena.

Class I: Thio Compounds

The polar groups of the thio compounds contain at least one sulfur atom *not* bonded to oxygen and are usually derived from an oxygen-bearing parent compound through a substitution of oxygen by sulfur, as shown in Table 5.1. The —SH group is known as the *sulfhydryl group* or, more commonly, as the *mercapto group*.

Table 5.1. Examples of Thio Compounds[a]

Parent compound	Thio derivative, usually a metal (M^+) salt such as Na^+ or K^+		Oxidation product of the thio derivative
1. Alcohols ROH	RSH	Mercaptans	RS—SR (Dialkyl disulfides)
2. Carbonic acid HO—C(=O)OH	R—O—C(=O)S⁻(M^+)	O-alkyl monothiocarbonates	R—O—C(=O)—S—S—C(=O)—O—R (Carbonate disulfide)
	R—O—C(=S)S⁻(M^+)	O-alkyl dithiocarbonates / Alkyl xanthates	R—O—C(=S)—S—S—C(=S)—O—R (Dialkyl dixanthogens)
	R—S—C(=S)S⁻(M^+)	Alkyl trithiocarbonates	
3. Carbamic acid H,H N—C(=O)OH	R,R N—C(=S)S⁻(M^+)	Dialkyl dithiocarbamates	[R_1,R_2 N—C(=S)—S—]$_2$ (Thiurium disulfides)
	R_1,H N—C(=S)O—R_2	Alkyl thionocarbamate esters (e.g., Dow Z-200, where $R_1 = C_3H_7{}^-$, $R_2 = C_2H_5{}^-$)	
4. Phosphoric acid HO,HO P(=O)OH	R—O,H—O P(=S)S⁻(M^+)	Monoalkyl dithiophosphates	Tetraalkyl bis-dithiophosphates
	R—O,R—O P(=S)S⁻(M^+)	Dialkyl dithiophosphates (Aerofloats)	(oxidation products analogous to dixanthogens and disulfides)
5. Urea H_2N,H_2N C=O	(C₆H₅)N(H)—C(=S)—N(H)(C₆H₅)	Diphenyl thiourea (thiocarbanilide)	
	benzothiazole C—SH	Mercaptobenzothiazole (Flotagen, Captax)	

[a] R denotes a nonpolar group, usually a hydrocarbon group: alkyl, aryl, or cyclic.

In addition to the preceding thio compounds there exists a bewildering variety of thioacids (RCOSH), thioamides (RCS · NH$_2$), etc., some of which may be present in commercial flotation agents [Reid (1962)]. The most commonly used flotation thio compounds are xanthates, dithiophosphates (Aerofloats), and Dow Z-200 [Rao (1971)]. These reagents are treated in greater detail in Sections 5.2, 5.3, and 5.5.

The nonpolar groups R of the flotation reagents in the thio compound class are mostly short-chain hydrocarbon groups: ethyl to hexyl ($C_2H_5^-$ to $C_6H_{13}^-$), phenyl ($C_6H_5^-$), cyclohexyl ($C_6H_{11}^-$), and various combinations of alkyl–aryl groups.

The hydrophobic–hydrophilic character of the thiocompounds can change drastically when different metal ions react with the polar group; thus, for example, despite the high dipole moments of many insoluble metal xanthates or dithiophosphates, even the short alkyl homologues of these salts are hydrophobic.

Common characteristics of most thio compounds are the following: their high chemical activity toward acids, oxidizing agents and metal ions, and the absence of surface activity at the liquid/air interface. The latter property may be due to the formation of insoluble oxidation products; in consequence, there are no colloid aggregates (micelles) in solutions of these reagents. For more details on thio compounds, see Sections 5.2–5.5.

Class II: Non-Thio, Ionizable Surfactants

Non-thio, ionizable surfactants are chiefly represented by the following:

1. *Alkyl carboxylate* derivatives of carboxylic acid, such as fatty acids, RCOOH, and their sodium ($RCOO^-Na^+$) or potassium ($RCOO^-K^+$) soaps.

2. *Alkyl sulfates*, $R-O-SO_3^-Na^+,(K^+)$ and *sulfonates*, $R-SO_3^-Na^+,(K^+)$.

3. *Alkyl phosphates*: monoalkyl,

$$R-O \underset{HO}{\overset{}{\diagdown}} P \overset{O}{\underset{O^-Na^+,(K^+)}{\diagup}}$$

or dialkyl,

$$R-O \underset{R-O}{\overset{}{\diagdown}} P \overset{O}{\underset{O^-Na^+,(K^+)}{\diagup}}$$

4. *Amines*, alkyl derivatives of ammonia, NH_3, of which the primary amines RNH_2 are used in flotation in the form of *unsubstituted amine salts* such as acetate, $RNH_3{}^+ CH_3COO^-$, or hydrochloride or hydrobromide, $RNH_3{}^+Cl^-$, (Br^-). The secondary amines, R_1R_2NH,[†] are used in flotation less often but together with the tertiary amines, $R_1R_2NR_3$, are common emulsification agents, for example, dimyristylamine or dimethylmyristylamine. A modification of amine-type surfactants constitute the *substituted amine salts*, e.g., monoalkyl quaternary ammonium salts such as chlorides or bromides, $RN(CH_3)_3{}^+Cl^-,(Br^-)$, or dialkyl quaternary salts, $R_1R_2N(CH_3)_2{}^+Cl^-,(Br^-)$.

Other alkyl or aryl derivatives of amines are

$NH{=}C(NH_2)_2$ $C_5H_{10}NH$ C_5H_5N

guanidine   piperidine   pyridine

$C_6H_{11} \cdot NH_2$ $C_6H_5 \cdot NH_2$

cyclohexylamine   aniline
(aminobenzene)

Of the preceding derivatives the most frequently encountered are alkyl pyridinium salts.

As regards the class of hydrolyzable compounds, only those reagents with R between C_6 and C_{20} are employed in flotation; homologues shorter than C_6 do not show enough surface activity, while homologues longer than about C_{20} become too insoluble for flotation purposes. The solutions of all these compounds are strongly affected by pH, giving rise to hydrolysis or dissociation, which strongly influences the surface activity by providing either the molecular or the ionic species. Also, all long-chain homologues

† *Note*: R_1, R_2, and R_3 represent different alkyl groups and not $1R$, $2R$, or $3R$.

of this class of reagents form aggregates (called micelles) when their solutions reach concentrations higher than a so-called *critical micelle concentration* (CMC) whenever their temperature is above a certain minimum temperature called *Krafft point*. The characteristics of the non-thio ionizable surfactants are discussed in greater detail in Sections 5.6–5.12.

Class III: Nonionic Compounds

Nonionic compounds (amphiphiles) are represented by the following:

1. *Alcohols*, ROH. The C_6–C_{12} isomers, whether alkyl, cyclic, or aryl, are employed in flotation, for example, methyl isobutyl carbinol (MIBC),

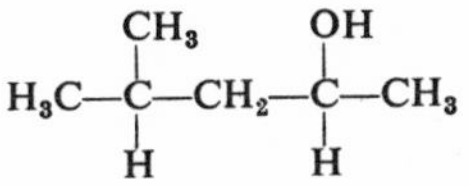

$$H_3C-\underset{\underset{H}{|}}{\overset{\overset{CH_3}{|}}{C}}-CH_2-\underset{\underset{H}{|}}{\overset{\overset{OH}{|}}{C}}-CH_3$$

cyclohexanol ($C_6H_{11}OH$), terpineols, cresylic "acids," and napthols (see Section 5.13).

2. *Aldehydes*,

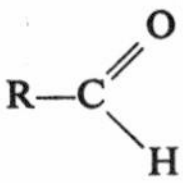

$$R-C\overset{\displaystyle{\nearrow O}}{\underset{\displaystyle{\searrow H}}{}}$$

which are oxidation products of primary alcohols,

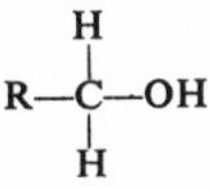

$$R-\underset{\underset{H}{|}}{\overset{\overset{H}{|}}{C}}-OH$$

and *ketones*,

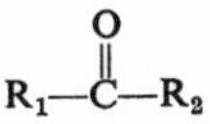

$$R_1-\overset{\overset{O}{\|}}{C}-R_2$$

which are oxidation products of secondary alcohols,

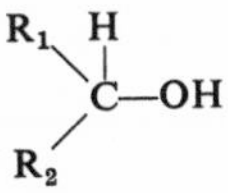

$$\overset{\displaystyle R_1}{\underset{\displaystyle R_2}{>}}\overset{H}{\underset{|}{C}}-OH$$

The polarity of aldehydes and ketones is partway between the nonpolar hydrocarbons (alkanes) and alcohols.

3. *Amides*,

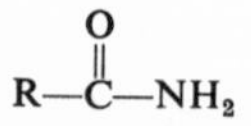

$$R-\overset{\overset{\displaystyle O}{\|}}{C}-NH_2$$

which are reaction products of NH_3 and carboxylic acids.

4. *Esters*,

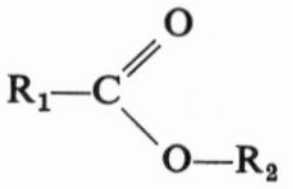

$$R_1-C\overset{\displaystyle O}{\underset{\displaystyle O-R_2}{\diagdown}}$$

reaction products of alcohols and carboxylic acids.

B. Multipolar Surfactants

When more than one polar group is attached to separate carbon atoms of the same hydrocarbon chain, the surfactant is referred to as *multipolar*. A number of surfactants used in flotation are dipolar or tripolar, as, for example, dicarboxylic acids and hydroxy acids (see Section 5.10.1) or triethoxybutane (TEB) used as a flotation frother:

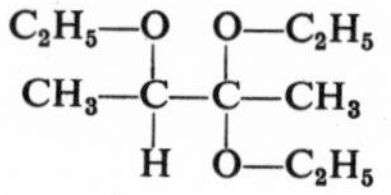

$$\begin{array}{c} C_2H_5-O \quad O-C_2H_5 \\ | \qquad | \\ CH_3-C-C-CH_3 \\ | \qquad | \\ H \quad O-C_2H_5 \end{array}$$

one of the isomers of triethoxybutane (TEB)

These reagents, possessing only a few polar groups attached to a relatively short-chain hydrocarbon, represent the nonpolymeric multipolar surfactants. There are, however, numerous reagents with large numbers of polar groups created by polymerization of one, two, three, or more monopolar compounds. Polymers with molecular weights from 1000 to 2000 (as compared with MW 150–300 for most monopolar and nonpolymeric surfactants) to as high as several million are synthesized to provide surfactants used in flotation, water purification, dispersion of slurries, and detergency.

Multipolar surfactants, whether polymeric or nonpolymeric, are classified as:

1. *Nonionic* if their polar groups consist of either alcohol (—OH), ether (—C—O—C—), aldehyde $\left(-C\overset{\displaystyle O}{\underset{\displaystyle H}{\diagdown}}\right)$, or ketone $\left(\diagup\!\!\!\diagdown C\!=\!O\right)$ groups,

2. *Anionic* with carboxylic, sulfate or sulfonate, phosphate, or any of the thioionized polar groups, and

3. *Cationic* with —NH_2, —N^+H_3 and —$N^+(CH_3)_3$ groups.

When multipolar surfactants possess both anionic and cationic groups, they are known as *amphoteric* surfactants [or *ampholytic*; see Schwartz and Perry (1949, p. 10)]. The characteristics of the various amphoteric agents that can be used in flotation were discussed by Wrobel (1969, 1970) and Smith *et al.* (1973). There are numerous examples of amphoteric reagents among the carboxylic acids listed in Section 5.10.1.

Some natural multipolar polymers such as starch, dextrins, cellulose, and lignins (for structures, see Section 5.15), have occasionally been employed in flotation since its inception, mostly as depressants. However, widespread use of multipolar polymeric surfactants became possible only after the condensation and polymerization techniques were developed sufficiently to provide uniform synthetic products of controlled characteristics. Examples of such polymerized synthetic multipolar surfactants are the following:

1. *Nonionic alkyl (or aryl) polyoxyethylene ether derivatives* of alcohols, carboxylic acids, amines, etc. Any surfactant RY can be converted to such nonionic ether derivatives by condensing a given number n of ethylene oxide molecules onto it:

$$RY(OC_2H_4)_nOH$$

2. *Ionizable derivatives* of any surfactant are obtained by inserting a given number n of ethylene oxide molecules between the nonpolar group R and the polar group Y, that is,

$$R(OC_2H_4)_n \cdot Y$$

To obtain an ionizable polyoxyethylene derivative, the starting material used is a suitable polyoxyethylene alcohol, $R(OC_2H_4)_nOH$, which is then reacted with appropriate reagents to place the polar group Y in the terminal hydrogen position.

3. *Nonionic alkyl (or aryl) polyoxypropylene ether derivatives* of alcohols and surfactants analogous to the oxyethylene nonionic ether derivatives already mentioned, except that instead of $(OC_2H_4)_n$, denoted $(EO)_n$, the polyoxypropylene groups $(OC_3H_6)_n$, denoted $(PO)_n$, are present. The structure of the polymerized oxypropylenes $(PO)_n$ contains a branch CH_3 group in each unit of the polymer:

These branched methyl groups make the propylene polymeric groups much more hydrophobic in character than the analogous oxyethylene polymers, $(EO)_n$:

$$-O-\underset{\underset{H}{|}}{\overset{\overset{H}{|}}{C}}-\underset{\underset{H}{|}}{\overset{\overset{H}{|}}{C}}-\left[O-\underset{\underset{H}{|}}{\overset{\overset{H}{|}}{C}}-\underset{\underset{H}{|}}{\overset{\overset{H}{|}}{C}}-\right]_{n-1}$$

Both types of polymerized ether groups $(EO)_n$ and $(PO)_n$ represent *n* permanent dipoles (centered around oxygen) which are effectively strung in line but have the ability to rotate around the single bonds of the polymer, thus enabling the dipoles to adopt different relative positions.

4. *Ionizable polyoxypropylene derivatives*, analogous to those of poly-oxyethylene polymers.

5. *Block polymers*, consisting of interchanged blocks of polymerized oxyethylene and oxypropylene ethers, such as, for example, $(PO)_m \cdot (EO)_n$ and $(PO)_p(EO)_l(PO)_r$. Block polymers may, in addition, contain some interspaced nonpolar groups R. The oxyethylene and oxypropylene derivatives are discussed in more detail in Section 5.14.1.

6. *Linear polymers* containing amino acid groups. These are represented either by unpurified natural proteins, such as albumin (MW 12,000–25,000), gelatin (MW 10,000–100,000), or casein (MW 75,000–375,000), or are fractions from various stages of purification procedures of proteins.

7. *Linear polymers* of various *alkenes*, such as polymers of acrylic acid, acrylonitrile, or saccharides. For details, see Sections 5.15 and 5.16.

5.1.1. *Hydrocarbon Groups (R) Nomenclature*

The nonpolar hydrophobic group of a surfactant, denoted by R, R_1, R_2, or R_3, usually consists of a hydrocarbon group (see below) which may appear in a large variety of *structural arrangements*, that is, *isomers*, representing compounds of the same molecular formula but differing in structure. The bulk properties and particularly the *surface properties of surfactants* are to a great extent dependent on and determined by the *stereochemistry* of the hydrophobic and the hydrophilic portions of the molecule.

Hydrocarbons are compounds that contain only carbon and hydrogen. There are three classes of aliphatic hydrocarbons: the *alkanes*, in which each carbon has four single bonds; the *alkenes*, in which two carbon atoms are joined by a double bond (two electron pairs); and the *alkynes*, in which two carbon atoms are joined by a triple bond (three electron pairs).

Open-chain alkanes [whether straight-chain, with the prefix *n* (for normal) or branch-chain ones] have the general formula C_nH_{2n+2}; cycloalkanes containing rings of carbon atoms (as in cyclopropane, cyclobutane, cyclopentane, etc.) have the formula C_nH_{2n} if unsubstituted. The possibility of branch-chain compounds isomeric with *n*-hydrocarbons starts with butane and increases very rapidly: Butane has 1 branch-chain isomer; pentane, 2; hexane, 4; heptane, 8; octane, 17; nonane, 34; decane, 74; etc. The group formed when an alkane has one of its hydrogen atoms left out is called an *alkyl group*, e.g., for *n*-hydrocarbons, $-CH_3$, methyl; $-C_2H_5$, ethyl; $-C_3H_7$, propyl; $-C_4H_9$, butyl; $-C_5H_{11}$, pentyl (amyl); $-C_6H_{13}$, hexyl; $-C_7H_{15}$, heptyl; $-C_{14}H_{29}$, tetradecyl (myristyl); $-C_{16}H_{33}$, hexadecyl (cetyl); and $-C_{18}H_{37}$, octadecyl (stearyl).[†]

In naming complex isomers, the longest consecutive chain of carbons is picked out, and the carbons are numbered starting from the end of the chain; the substituent groups are assigned numbers corresponding to their positions in the chain, and the direction of numbering is chosen to give the lowest sum for the numbers of the carbons that carry the substituents. When different substituents are present in a branch-chain isomer, following the practice of Chemical Abstracts Service, they are listed in alphabetical order, i.e.,

4-ethyl–3-methyl–6-propyloctane

In cycloalkanes, the substituents are similarly assigned numbers to keep their sum to a minimum, starting with the substituents in alphabetical order, i.e.,

1-ethyl–3-methylcyclohexane

Small-ring cycloalkanes, such as cyclopropane and cyclobutane, have a high degree of strain in their C—C—C bonding and are more reactive than saturated open-chain hydrocarbons.

An alkyl group is described as *primary* if the carbon at the point of attachment is bonded only to one other carbon, as *secondary* if bonded to two other carbons, and *tertiary* if bonded to three other carbons.

Open-chain alkenes containing one double bond have the general formula C_nH_{2n} (the same as unsubstituted cycloalkanes) and are called

---

[†] The nomenclature of carboxylic acids refers to the total number of carbons C, including that of the polar group COOH. Effectively, alkyl groups of carboxylic acids C_n are C_{n-1} groups.

olefins. The longest continuous chain containing the double bond is given the name of the corresponding alkane, with the ending changed to *-ene*; the alkyl groups derived from alkenes carry the suffix *-enyl*, e.g., pentenyl. Hydrocarbons with a double bond enclosed in a ring, cycloalkenes, are named using the system for open-chain alkenes, but the numbering starts with the carbons of the double bond in the direction that gives the lowest sum for the substituent positions, e.g., 1,3-dimethylcyclopentene (and not 2,5-dimethylcyclopentene). However, most of the natural products extracted from plants, wood, coal, animal fats, etc., are known not by their systematic names but by common names (nonsystematic). Thus, for example, alkenes obtained in the distillation extract derived from pine stumpwood, known as pine oil, are terpineols of which *α-terpineol* is the most common isomer:

$$
\begin{array}{c}
CH_3 \\
| \\
C \\
\diagup \; \backslash\!\!=\\
H_2C \qquad CH \\
| \qquad\qquad | \\
H_2C \qquad CH_2 \\
\backslash \diagup \\
CH \\
| \\
C\!-\!OH \\
\diagup \quad \backslash \\
CH_3 \quad CH_3
\end{array}
$$

α-terpineol ($C_{10}H_{17}OH$)

Other wood extracts are discussed in Section 5.15.3.

Many alkenes contain more than one double bond: *alkadienes* (two double bonds), *alkatrienes* (three double bonds), etc. When two or more double bonds are next to each other, the compound is said to have cumulated double bonds; when the two or three double bonds are separated from each other by one single bond, these are known as conjugated double bonds.

There are four alkenyl groups with 18 carbons (C_{18}) which are frequently found in flotation reagents: The *oleyl* group represents the *cis form* of a monounsaturated hydrocarbon group, $-CH_2(CH_2)_7CH\!=\!CH(CH_2)_7CH_3$, and the *elaidyl* group represents the *trans form* of the same; the *linoleyl* group (from linoleic acid) is a hydrocarbon with two double bonds: *cis*-9, *cis*-12, viz., $-CH_2(CH_2)_7CH\!=\!CHCH_2CH\!=\!CH(CH_2)_4CH_3$; and the linolenyl (from linolenic acid) group is a *cis*-9, *cis*-12, *cis*-15 (three double bonds) hydrocarbon.

It is important to realize that all hydrocarbons or their groups possess freedom of rotation around each single bond but no freedom of rotation around a double or triple bond. Thus, the *paraffinic*[†] alkyl groups are not

---

[†] *Paraffin* means "little affinity" and is applied to saturated hydrocarbons, alkanes, because they are least reactive.

rigid structures but can adopt any *conformation* permitted by the flexibility of the chain. However, the olefinic groups exhibit—in addition to structural isomerism (branch-chain structures)—so-called geometrical isomerism (stereoisomerism), in particular the *cis–trans* isomerism; stereoisomerism is characterized by the *different configurations* adopted by compounds having *the same structure.* Owing to the rigidity of the double bond, the two atoms immediately next to the double-bonded carbon atoms are either at a definite angle, forming a nonlinear configuration (a *cis form* of the molecule), or are in a nearly linear configuration of the molecule (a *trans form*), as shown here and in Figure 5.2:

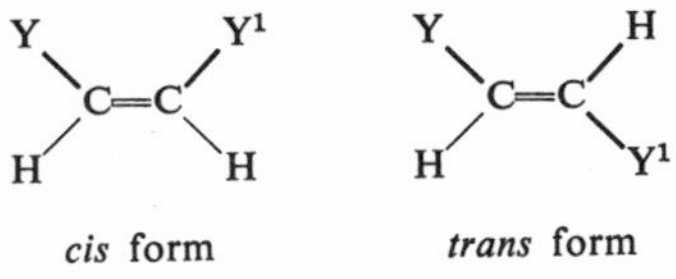

cis form   *trans* form

where Y and Y^1 represent an atom, a hydrocarbon chain, or a group of atoms.

Two conjugate double bonds can also exist in two stereoisomeric *cis* and *trans* configurations, e.g., butadiene,

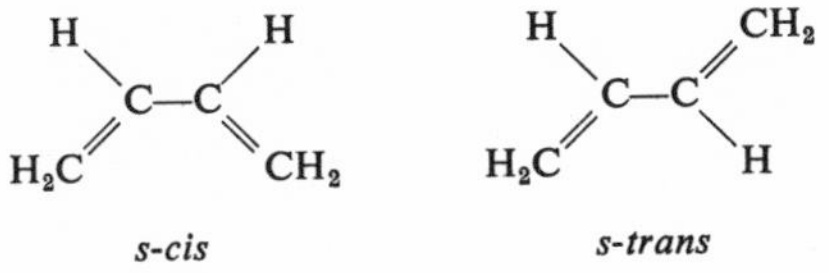

s-cis   s-trans

and of these the *cis* form is stabilized by resonance with the "1,4 π bond":

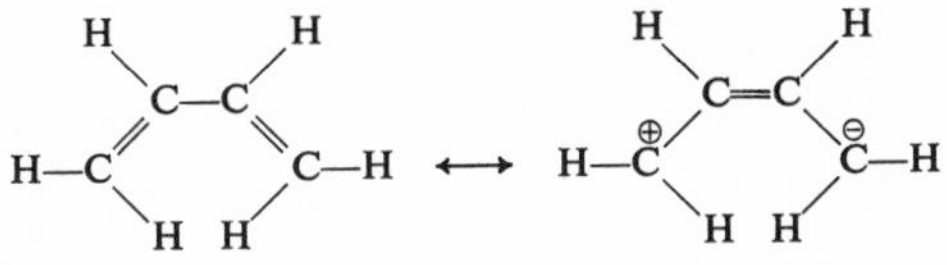

As will be seen later, structural isomers may give rise to different properties of surface films adsorbed at various interfaces (solid/liquid, liquid/gas, solid/gas) and thus may cause pronounced differences in flotation results.

Three conjugate double bonds forming a benzene ring, C_6H_6, represent a special structure, highly stabilized by resonance energy (of ~ 30 kcal/mol, ~ 125 kJ mol^{-1} in SI units).

Benzene gives rise to a very large group of compounds, arenes, or aromatic hydrocarbons. These are formed by numerous substituents at-

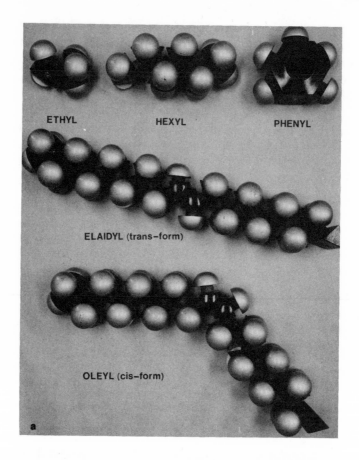

Figure 5.2. Scale models of selected hydrocarbon radicals: (a) ethyl, hexyl, phenyl, elaidyl, and oleyl; (b) dixanthogen, ethyl xanthate ion.

tached to a benzene ring either singly (e.g., phenol, toluene, styrene, aniline, ethyl benzene) or in pairs (e.g., xylene, cresol, etc.):

benzene   phenol   toluene   styrene   aniline   ethylbenzene
(methylbenzene) (vinylbenzene)

Pairs of substituents may occupy different ring positions designated by *ortho*, *meta*, and *para*,

ortho-   meta-   para-

giving rise to three positional isomers, e.g., xylene isomers:

ortho-xylene   meta-xylene   para-xylene
(1,2-dimethylbenzene) (1,3-dimethylbenzene) (1,4-dimethylbenzene)

Another example of positional isomers are *cresols*, $CH_3 \cdot C_6H_4OH$: ortho-cresol (b.p. 141°C), meta-cresol (b.p. 201°C), and para-cresol (b.p. 202.5°C); a mixture of these cresols together with a small amount of phenol, C_6H_5OH (m.p. 43°C, b.p. 182°C), is obtained in coal tar distillation as a heavy oil fraction called *cresylic acid* (a misnomer, since it is a mixture of alcohols), and this has been extensively used as a flotation frother.

Polynuclear aromatic compounds are formed when several benzene rings are "fused" together to form naphthalene (two rings), anthracene (three), phenanthrene (three), and naphthacene (four):

naphthalene   anthracene   phenanthrene

Naphthalene, $C_{10}H_8$, has two isomeric derivative *naphthols*, α-naphthol,

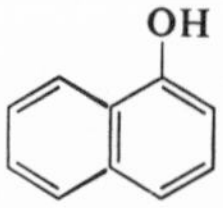

(m.p. 94°C)

and β-naphthol,

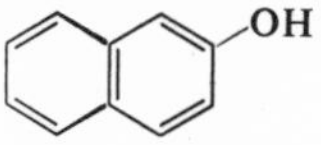

(m.p. 123°C)

and two *naphthenic acids*, α-naphthenic acid,

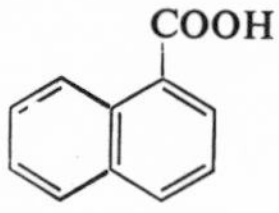

(m.p. 161°C)

and β-naphthenic acid,

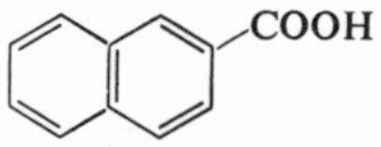

(m.p. 184°C)

The different groups derived from aromatic compounds are

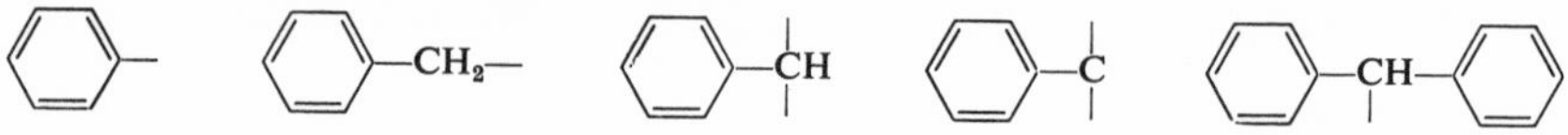

phenyl   benzyl   benzal   benzo   benzhydryl

Alkynes, $R—C\equiv CH$ (where R is either H_3 or an alkyl group), are similar in physical properties to alkenes but do not usually react as rapidly. They are less acidic than water but about 10^{13} more acidic than ammonia. 1-alkynes form insoluble silver salts with silver ammonia solution (a test for detecting terminal triple bonds).

For details on the composition and structure of the numerous organic compounds, the reader is referred to any of the large number of texts on organic chemistry, e.g., Roberts *et al.* (1971), Fieser and Fieser (1961), Karrer (1950), and Noller (1965), Morrison and Boyd (1966).

5.1.2. *Less Common Flotation Surfactants, Mono- and Multipolar*

In addition to the reagents already mentioned that have been extensively used in past flotation practice, there are unlimited possibilities for using less common reagents or creating entirely new surfactants. Examples of reagents with polar groups differing from those we have mentioned are

1. *Alkyl phosphonic acids* and their salts, such as the sodium salt of 2-ethylhexyl phosphonic acid:

$$H_3C(CH_2)_3CHCH_2-O-P \overset{\displaystyle C_2H_5}{\underset{}{}} \begin{matrix} OH \\ O^-Na^+ \end{matrix}$$

2. *Alkyl (aryl) arsonic (or antimonic) acids*, $RAsO_4H_2$ ($RSbO_4H_2$), such as *p*-tolyl arsonic acid:

$$CH_3-\underset{}{\bigcirc}-O-As\overset{\displaystyle O}{\underset{\displaystyle OH}{\diagdown OH}}$$

3. *α-Sulfonated fatty acids*:

$$R-\underset{\displaystyle SO_3H}{\overset{\displaystyle H}{C}}-C\overset{\displaystyle O}{\underset{\displaystyle OH}{}}$$

All three groups of these reagents were extensively tested in flotation of cassiterite [Kirchberg and Wottgen (1964), Trahar (1965), Collins (1967), Töpfer (1967), Collins *et al.* (1968), Wottgen (1969), and Trahar (1970)].

Monosodium salts of α-sulfonated fatty acids [their chemistry has been reviewed by Stirton (1962)] are apparently good collectors for CaF_2, while *p*-tolyl arsonic acid does not float fluorspar [Trahar (1972)].

In addition, there are the following reagents:

4. Alkyl *morpholine* derivatives (used in potash flotation):

$$R-N\overset{\displaystyle CH_2-CH_2}{\underset{\displaystyle CH_2-CH_2}{\diagup \diagdown}}O$$

5. Alkyl *hydroxamates*,

$$R-\underset{\displaystyle OH}{\overset{}{C}}=N-O^-Na^+$$

[used in the flotation of wolframite, $(Fe,Mn)WO_4$; cassiterite, SnO_2; and pyrochlore, $(Ce,Ca)Nb_2O_6 \cdot (Ti,Th)O_3$].

6. *α-Oximes*,

$$R-\underset{\underset{\text{OH}}{|}}{C}-\underset{\underset{\text{NOH}}{\|}}{\overset{\overset{\text{H}}{|}}{C}}-R$$

were also tested [Petersen *et al.* (1965), Bogdanov *et al.* (1974), and Evrard and DeCuyper (1975)].

7. Alkyl derivatives of *taurine* (2-aminoethane sulfonic acid) such as sodium–*N*–methyl–*N*–oleyltaurate,

$$\text{oleyl}-\underset{\underset{\text{CH}_3}{|}}{N}-CH_2-CH_2-SO_3Na$$

which is, apparently, a specific collector for celestite ($SrSO_4$) flotation [Wyslouzil (1970)].

8. Carboxylate derivatives of *alkyl N,N-dimethylformamides*, such as

$$R-\underset{\underset{\text{O}}{\|}}{C}-\underset{\underset{\text{CH}_3}{|}}{N}-CH_2-COONa$$

which are used in the flotation of very fine-size hematite and ilmenite [Wyslouzil (1970)].

9. Derivatives of *quinoline*,

in particular 8-hydroxyquinoline (oxine),

were extensively tested in the flotation of lead and zinc oxide ores by Rinelli and Marabini (1973).

Apart from reagents that incorporate a new polar group or groups, an entirely new family of reagents is created by substituting the hydrocarbon chains with different hydrophobic groups, such as the following:

1. An organosilicon hydrophobic group of the siloxane type:

$$\begin{array}{ccc} & CH_3 & CH_3 \\ & | & | \\ CH_3Si & -O-Si-O- \\ & | & | \\ & CH_3 & CH_3 \end{array}$$

Organosilicon compounds based on a direct C—Si bond represent a major field of recent developments in organic chemistry, with an enormous potential in their application as surfactants. Depending on the number of organic groups R and the remaining silicon–oxygen bonds, there are at least three polymeric groups of compounds based on the three *siloxane* units,

$$(a) \left[-R-\overset{\overset{\displaystyle R_3}{|}}{\underset{\underset{\displaystyle R_2}{|}}{Si}}-O- \right]_n \qquad (b) \left[-O-\overset{\overset{\displaystyle R_3}{|}}{\underset{\underset{\displaystyle R_2}{|}}{Si}}-O- \right]_n \qquad (c) \left[-O-\overset{\overset{\displaystyle R}{|}}{\underset{\underset{\displaystyle O_1}{|}}{Si}}-O- \right]_n$$

and innumerable polymers thereof possible; all these types of compounds constitute the so-called *silicones*. In addition, numerous organic compounds of silicon exist which do *not* contain a Si—C bond (e.g., *silanes*, SiH_4; compounds with oxygen; compounds without oxygen; cyanosilanes; isothiocyanosilanes; etc.). Preliminary flotation tests using polymerized methyl and dimethylsiloxanes—such as the preceding (c) and (b) groups—were reported by Desnoes and Testut (1954). The chemistry of silicones is dealt with by Noll (1968).

2. Similarly, there are compounds in which the hydrocarbon groups are replaced by a *fluorocarbon* (*tetrafluoro*, C_nF_{2n+1}—, or *perfluoro*, $C_nF_{n+1}H_n$—).

The developments in fluorination of organic compounds, by electrolysis in hydrogen fluoride, has made available (since about 1950) numerous fully and partly fluorinated compounds for use as surfactants. The first study on the use of fluorinated collectors, C_9–C_{11} perfluorocarboxylic acids, C_2 and C_3 fluorinated xanthates, fluorinated quaternary compounds, etc., was made by Cooke and Talbot (1955). Subsequent investigations were published by Usui and Iwasaki (1971) and Somasundaran and Kulkarni (1973).

The potential for exploiting these two families of compounds (organosilicon and fluoro–compounds) as surfactants is enormous and has hardly been revealed as yet. It should be kept in mind, however, that some of these surfactants may create environmental hazards.

3. There are compounds created by the introduction of a water-solubilizing, hydrophilic group, such as an ethylene oxide, between the long-chain nonpolar group of either a hydrocarbon or type 1 or 2 compound,

for example, ionizable polyoxyethylene or polyoxypropylene derivatives, (see pp. 213 and 214)

$$R(OC_2H_4)_nZ \quad \text{or} \quad R(OC_3H_6)_nZ$$

where R represents any hydrophobic entity. Depending on the character of the polar group Z and the conditions in flotation, the reagents may behave as ionic or nonionic, collector-acting and/or frother-acting species. Preliminary results with such reagents were reported by Leja and Nixon (1957). Rubio and Kitchener (1976) discussed the mechanism of polyethylene oxide flocculants on silica.

5.1.3. *Commercial Flotation Reagents*

Table 5.2 lists flotation reagents commonly used in industrial practice. In addition to surfactants, a number of inorganic compounds are included. These are either used to regulate pH (lime, soda ash, SO_2, sulfuric acid, etc.) or are surface active at selected solid/liquid interfaces. The latter behave as *activators* (metal ions, sulfide ions) or *depressants* (cyanide ions, sulfide ions, chromate, and silica ions). The role of these inorganic regulating agents is discussed in greater detail in Chapter 10.

Most of the developments of flotation have been carried out empirically. Surfactants employed have been the cheapest ones available at the time and the given locality, and these were usually extracted from vegetable or mineral oils, distilled from coal or wood, or manufactured from complex and impure industrial by-products.

As there were no standards for purity or quality, either with respect to the starting materials or the final products of reagent preparation, the resulting surfactants were distributed (and used) under various proprietary names and trademarks, e.g., Cyanamid's Aero xanthates 301, 322, 325, 343, and 350 or promoters 404 and 444; Dow Chemical's Bear Brand Z-3 to Z-12 xanthates or Z-200 collector; etc. The use of trademark flotation reagents has preserved an attitude to flotation as an art and has hindered the understanding of the process for many years. It is only comparatively recently that the compounds behind the trade names are being disclosed and the utilization of synthetic reagents of higher purity is enabling the mechanism of their action to be investigated and characterized.

Of the raw materials used in the preparation of commercial flotation reagents, the following were (and are) utilized:

1. Products of coal tar distillation—*cresylic acids*, naphthols, and naphthenic acids.

Table 5.2. Most Frequently Employed Flotation Reagents and their Typical Applications

Type	Classification and Composition	Usual form of additions	Typical Applications and Some Properties
Collectors	Xanthates		
	1. Ethyl-Na	10% solution	For selective flotation of Cu–Zn, Cu–Pb–Zn sulfide ores
	Ethyl-K	10% solution	
	2. Isopropyl-Na	10% solution	More powerful collector than ethyl for Cu, Pb, and Zn ores, Au, Ag, Co, Ni, and FeS_2
	Isopropyl-K	10% solution	
	3. Amyl-K secondary	10% solution	Most active collector but not very selective; used for tarnished sulfides (with Na_2S), Co–Ni sulfides
	Amyl-K	10% solution	
	Dithiophosphates, Aerofloats		
	1. Diethyl-Na	5–10% solution	For the Cu and Zn (*but not Pb*) sulfide ores; selective, nonfrothing
	2. Dicresyl, 15% P_2S_5	Undiluted	Ag–Cu–Pb–Zn sulfides, selective, froths
	Dicresyl, 25% P_2S_5	Undiluted	Ag–Cu–Pb–Zn sulfides, selective, froths
	Dicresyl, 31% P_2S_5	Undiluted	Mainly used for PbS and Ag_2S ores
	3. Di-sec-butyl-Na	5–10% solution	Au–Ag–Cu–Zn sulfides; not good for Pb
	Thionocarbamate, ethylisopropyl (Z-200)	Liquid emulsion	Selective collector for Cu sulfides (or Cu-activated ZnS) in the presence of FeS_2
	Mercaptobenzothiazole	Solid	Floats FeS_2 in acid circuits (pH 4–5)

(continued overleaf)

Table 5.2 (continued)

Type	Classification and Composition	Usual form of additions	Typical Applications and Some Properties
	Fatty acids		
	1. Tall oil (mainly oleic acid)	Liquid emulsion	Collectors for fluorspar, iron ore, chromite, scheelite, $CaCO_3$, $MgCO_3$, apatite, ilmenite; readily precipitated by "hard" waters (Ca^{2+} and Mg^{2+})
	2. Refined oleic acid	Liquid emulsion	
	3. Na soap of fatty acids	5–20% solution	
	Alkyl sulfates and sulfonates		
	C_{12}–C_{16} (dodecyl to cetyl)	5–20% solution	Collectors for iron ores, garnet, chromite, barite, copper carbonates, $CaCO_3$, CaF_2, $BaSO_4$, $CaWO_4$
	Cationic reagents		
	1. Primary and secondary amines	In kerosene	Used to separate KCl from NaCl and to float SiO_2
	2. Amine acetates	5–10% solution	Used to float quartz, silicates, chalcopyrite
	3. Quaternary ammonium salts	5–10% solution	Used to float quartz, silicates, chalcopyrite
Frothers	Pine oil (α-terpineol)	Undiluted	Provides most viscous stable froth
	Cresylic "acid" (cresols)	Undiluted	Less viscous but stable froth; some collector action
	Polypropylene glycols		
	MW $\sim$ 200 (D-200)	Solutions in H_2O	Fine, fragile froth; inert to rubber
	MW $\sim$ 250 (D-250)	Solutions in H_2O	Slightly more stable froth than with D-200
	MW $\sim$ 450 (D-450)	Solutions in H_2O	Slightly more stable froth than with D-250
	Polyoxyethylene (nonyl phenol)	Undiluted	Used as calcite dispersant in apatite flotation

Aliphatic alcohols: e.g., MIBC, 2-ethyl hexanol	Undiluted	Fine textured froth; used frequntly with ores containing slimes
Ethers (TEB, etc).	Undiluted	
Regulators		
Lime (CaO) or slaked lime Ca(OH)$_2$	Slurry	pH regulator; depresses FeS$_2$ and pyrrhotite
Soda ash, Na$_2$CO$_3$	Dry	pH regulator; disperses gangue slimes
Caustic soda, NaOH	5–10% solution	pH regulator; disperses gangue slimes
Sulfuric acid, H$_2$SO$_4$	10% solution	pH regulator
Cu^{2+} (CuSO$_4$ · 5H$_2$O)	Sat. solution	Activates ZnS, FeAsS, Fe$_{1-x}$S, Sb$_2$S$_3$
Pb^{2+} (Pb acetate or nitrate)	5–10% solution	Activates Sb$_2$S$_3$
Zn^{2+} (ZnSO$_4$)	10% solution	Depresses ZnS
S^{2-} or HS^{-} (Na$_2$S or NaHS)	5% solution	In sulfidization, activates tarnished (oxidized) sulfide minerals and carbonates
CN^{-} [NaCN or Ca(CN)$_2$]	5% solution	Depresses ZnS, FeS$_2$, and when used in excess, Cu, Sb, and Ni-S
Cr$_2$O$_7{}^{2-}$, CrO$_4{}^{2-}$	10% solution	Depresses PbS
SiO$_2$ (Na$_2$SiO$_3$)	5–10% solution	Disperses siliceous gangue slimes; embrittles froth
Starch, dextrin	5–10% solution	Disperses clay slimes, talc, carbon
Quebracho, tannic acid	5% solution	Depresses CaCO$_3$, (CaMg)CO$_3$
SO$_2$	3% solution	Depresses ZnS and Fe$_{1-x}$S and, with CN^{-}, depresses Cu sulfides

2. Steam distillation products of southern pine trees—*pine oil* with the α- and β-*terpineols* as the main constituents.

3. Extracts derived from the sulfite treatment of wood pulps—*tall oil, sulfonated lignin derivatives*, etc. (see Section 5.15.3).

4. Steam distillates of various seeds (castor, soybean, cottonseed, coconut, etc.). These provide a variety of vegetable oils (see Section 5.10.1), which are subsequently subjected to various treatments such as esterification, sulfonation, hydrogenation, etc.

5. Fractionation products of animal fats and oils (e.g., lard or tallow).

6. Esters of dicarboxylic acids, such as aerosols.

For details of animal and vegetable derivatives, see Section 5.10.1.

CLASS I: THIO COMPOUNDS

5.2. Xanthates

Xanthates are derivatives of carbonic acid, H_2CO_3, in which two oxygens have been replaced by sulfur and one hydrogen by an alkyl (or aryl) group. They are thus O-alkyl (or aryl) dithiocarbonates. They were discovered in 1822 by Zeise, introduced into flotation in 1923, yet despite numerous intensive studies, the details of their behavior under various sets of conditions are still incompletely understood.

An extensive review by Rao (1971) deals with the preparation of xanthates and dixanthogens, their chemical reactivity, and their application in the mineral and cellulose industries, in analytical chemistry, in the vulcanization of rubber, and as pesticides, corrosion inhibitors, and lubricating oil additives. The chemistry of xanthates is dealt with by Reid (1962). The formation and structure of complexes are reviewed by Coucouvanis (1970) and Jorgensen (1962). The reactions of thio compounds with sulfide minerals were recently reviewed by Poling (1976).

5.2.1. Preparations of Alkali Xanthates

The method of preparation of sodium or potassium alkyl xanthate consists of the dissolution of an alkali hydroxide in the appropriate alkyl alcohol, followed by an addition of carbon disulfide to the metal alcoholate. Owing to highly exothermic reactions, the temperature of the mixture should be kept as low as possible to avoid thermal decomposition of the xanthate product.

The reactions are

$$R-OH \quad + \quad NaOH \quad \rightarrow \quad R-ONa \quad + \quad H_2O$$

(alkyl alcohol) + (sodium hydroxide) → (sodium alcoholate) + (water)

$$R-ONa \quad + \quad CS_2 \quad \rightarrow \quad R-O-C\overset{\displaystyle S}{\underset{\displaystyle SNa}{\diagdown}}$$

(sodium alcoholate) + (carbon disulfide) → (sodium alkyl xanthate, solid)

The kinetics of sodium alkyl xanthates formation was studied spectroscopically by Hovenkamp (1963); his results for ethyl xanthate indicated that a direct reaction must occur between CS_2 and the alcoholate. Furthermore, in the absence of ethyl alcohol, CS_2 reacted with NaOH to give an unstable dithiocarbonate ion, CS_2O^{2-}; trithiocarbonate ion, CS_3^{2-}; and hydrosulfide ion, HS^-. In addition, several spectroscopically inactive oxidation products and polysulfides were formed by side reactions. The latter species could be removed by recrystallization from acetone solution and by washing with ethyl ether to give a high-quality sodium ethyl xanthate for laboratory studies. Commercial xanthates are rarely more than 90% pure and frequently only 60–90% pure. In addition to the reaction byproducts we have mentioned, they may contain some residual alcohol or residual alkali hydroxide; the latter, or a metal carbonate, is sometimes added purposely in order to slow down the thermal decomposition of xanthates during storage.

Apart from alkali alkyl xanthates manufactured for flotation purpose, cellulose xanthates are produced and extensively utilized in the making of viscose rayon fibers. Wood pulp and cotton linters are converted into cellulose xanthate by treating them first with sodium hydroxide and then with carbon disulfide. This xanthation procedure solubilizes cellulose (a long-chain polymer of about 300 glucose units, $C_6H_{12}O_6$; see Section 5.15), and when the cellulose xanthate solution is forced through a spinnerette into an acid bath, the xanthate groups are rapidly decomposed, and the cellulose is regenerated in the form of fine fibers (filaments).

Solid alkali alkyl xanthates may become fairly readily decomposed during storage under the combined action of moisture, oxygen, and carbon dioxide. The products of decomposition include dixanthogen; monothiocarbonate; alkali sulfide, oxide, or carbonate; mercaptans and alcohols;

and monosulfides and disulfides. The complex reactions are greatly affected by traces of catalytically active species, photoirradiation, and temperature.

5.2.2. Aqueous Solutions of Xanthates

Alkali xanthates of short hydrocarbon chains are fairly readily soluble in water; however, the solubilities rapidly decrease with chain length (ethyl to hexyl) from about $1.5\ M$ to $\sim 0.1\ M$ at room temperature.

The stability of aqueous solutions of alkyl xanthates and primarily of cellulose xanthates has been the subject of numerous investigations. Of main interest were the reactions involved in highly alkaline solutions (0.1–5 M alkali hydroxide) and highly acid conditions—both being the regions employed in the cellulose industry. The near-neutral and weakly alkaline conditions, pH 6–12, which are of interest to flotation applications of xanthates have been explored less thoroughly.

Ultraviolet spectrophotometry[†] has proved itself to be the most useful technique for investigating some of the decomposition reactions and for monitoring the concentration of xanthate species in industrial flotation circuits. The following UV active species can be identified and quantitatively determined (Figure 5.3):

Xanthate ion, $ROCS_2^-$: Ethyl xanthate ion absorbs at wavelengths of 301 and 226 nm or mμ [1 mμ (millimicron) $= 10\ \text{Å} = 1$ nm (nanometer)] with the respective extinction coefficients (molar absorptivities) ε of $\sim 17{,}500$ (for 301 nm) and 8750 (226-nm band) liter mol^{-1} cm^{-1}. The intensities of these two bands (301 and 226) are in the 2:1 ratio for solutions containing xanthate ion only. With a 1-cm path cell, the concentrations of xanthate solutions within 5×10^{-6} to $5 \times 10^{-4}\ M$ can be readily determined.

Xanthic acid, $ROCS_2H$: Xanthic acid absorbs at ~ 270 nm; the position of the maximum absorption and the molar absorptivity are slightly affected by solvents, for example, in water, 270 nm and $\sim 10{,}700$ liter mol^{-1} cm^{-1}, and in iso-octane, 268 nm and 9772 liter mol^{-1} cm^{-1}. Xanthic acid is readily decomposed, forming an alcohol and a volatile product CS_2, so the UV measurements should be carried out in a closed optical cell (cuvette).

Monothiocarbonate ion,

$$R\!-\!O\!-\!C\underset{\textstyle S^-}{\overset{\textstyle O}{\Big\langle}}$$

[†] For a review of the *UV* spectrometry of xanthate flotation liquors, see Jones and Woodcock (1973a).

Agent	Wavelength, nm	ε Coefficient of Extinction, liter (mole cm)$^{-1}$
CS$_2$	206.5	60,000 - 70,000
Monothiocarbonate	223	12,200 - 13,300
Xanthate Ion	226	8,750
	301	17,500
Dixanthogen	238	17,800
	283	8,600
Xanthic Acid	270	10,700

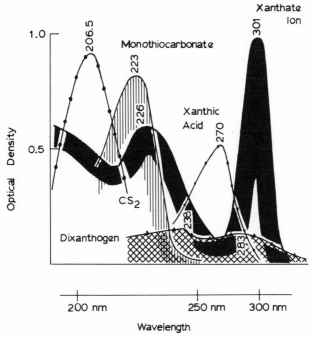

Figure 5.3. Ultraviolet (UV) spectra of species that can be detected in xanthate solutions either before or after oxidation and decomposition reactions. Values of extinction coefficients for characteristic absorption bands are shown in the accompanying table.

(known also as Bender's salt): Monothiocarbonate ion absorbs at the wavelength 221–223 nm with a molar absorptivity $\varepsilon = 12,200$–13,300 liter mol^{-1} cm^{-1}. (Solutions containing HS$^-$ absorb in the same region, 220–230 nm, with $\varepsilon \sim 10,500$ liter mol^{-1} cm^{-1}.)

Carbon disulfide, CS$_2$: Carbon disulfide absorbs at 206.5 nm with a molar absorptivity of 60,000–70,000 liter mol^{-1} cm^{-1} (volatile, as already noted).

Dixanthogen, R—OCS$_2$—S$_2$CO—R: Dixanthogen, an oxidation product of xanthate ion, absorbs at two wavelengths: 283 nm with the molar absorptivity $\varepsilon = 8600$ liter mol^{-1} cm^{-1} and 238 nm with $\varepsilon = 17,800$ liter mol^{-1} cm^{-1}. Owing to the very low solubility of ethyl dixanthogen in water ($\sim$1.25 $\times$ 10^{-5} M), quantitative determinations of dixanthogen can be carried out only after extraction into an organic solvent such as, for example, hexane.

The spectra of various decomposition products that may be encountered in highly alkaline solutions (investigated primarily for the cellulose xanthate chemistry) and the corresponding absorptivities are given in Figure 5.4. The intensity of the weak CS$_2$ band at 315 nm is three orders of magnitude lower than that of the 206.5-nm band; the latter is blanked out in highly alkaline solutions by strong absorption of OH$^-$ in the range of wavelengths less than 240 nm.

5.2.2.1. *Liquid Xanthates*

Whenever large amounts of xanthates (particularly amyl, butyl, and isopropyl) are consumed daily and the cost of labor is high, it may be more economical to deliver to the flotation plants a stock solution of "liquid xanthate" consisting of $\sim$25% wt. xanthate in water at pH $\sim$10. A railway tank-car load or truckload of such a solution is delivered directly from the manufacturer to the consumer, obviating the costs of handling smaller containers such as, for example, drums. The rate of decomposition of such highly concentrated xanthate solutions is (approximately) in the $\frac{1}{2}$%/day range; if the tanker load is consumed within 3 weeks and the transportation distances are reasonable, the overall cost of the reagent at the plant may be much lower than that resulting from shipping dry xanthates in drum quantities.

5.2.2.2. *Reactions of Xanthate Ion Decomposition*

Of the numerous reactions that have been considered by various investigators in the decomposition of aqueous xanthate solutions, there are six reactions which are recognized as pertinent to flotation systems:

Hydrolysis of the xanthate ion

$$\mathrm{K^+ + ROCS_2^- + H_2O} \underset{k_2}{\overset{k_1}{\rightleftharpoons}} \mathrm{K^+ + OH^- + ROCS_2H} \text{ (xanthic acid)}$$

$$\textbf{(I)}$$

Decomposition of xanthic acid

$$\mathrm{ROCS_2H} \overset{k_3}{\rightarrow} \mathrm{ROH + CS_2}$$

$$\textbf{(II)}$$

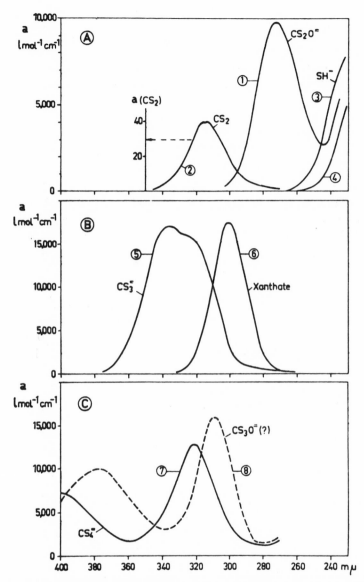

Figure 5.4. UV spectra of sulfur-containing compounds detected in systems comprising CS_2–NaOH–ethyl alcohol. (A): (1) Dithiocarbonate ion, CS_2O^{2-}, in 5 N NaOH solution without an alcohol; (2) carbon disulfide, CS_2 (with its own scale of molecular absorbance—the band at 206.5 nm is outside the abscissa scale shown here); (3) HS$^-$ ion as present in Na_2S solution in 0.1 N NaOH; (4) monothiocarbonate ion or Bender's salt, C_2H_5OCOSK. (B): (5) Trithiocarbonate ion, CS_3^{2-}; (6) ethyl xanthate ion—note that the spectrum does not extend to the lower wavelengths to show the 226-nm band. (C): (7) Tetrathiopercarbonate ion, CS_4^-; (8) trithiopercarbonate ion, CS_3O^- (no molar absorptivities are known; relative values are indicated). [From Hovenkamp (1963), in *J. Polym. Sci. Part C*, No. 2, with the permission of J. Wiley & Sons, New York.]

Hydrolytic decomposition

$$6ROCS_2^- + 3H_2O \rightarrow 6ROH + CO_3^{2-} + 3CS_2 + 2CS_3^{2-}$$

$$\textbf{(III)}$$

(The trithiocarbonate may decompose further into CS_2 and S^{2-}.)

Oxidation to dixanthogen

$$\begin{aligned}
\text{(a)} \quad & 2ROCS_2^- && \rightleftarrows (ROCS_2)_2 + 2e \\
\text{(b)} \quad & 2ROCS_2^- + \tfrac{1}{2}O_2 + H_2O && \rightleftarrows (ROCS_2)_2 + 2OH^-
\end{aligned}$$

$$\textbf{(IV)}$$

Oxidation of monothiocarbonate

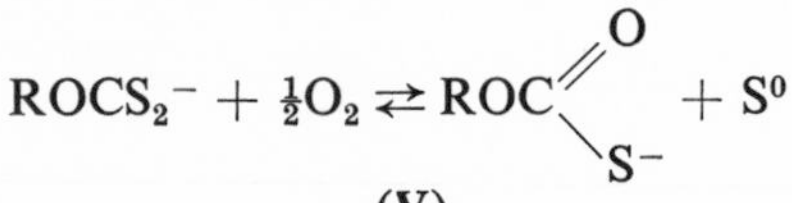

$$ROCS_2^- + \tfrac{1}{2}O_2 \rightleftarrows ROC\overset{O}{\underset{S^-}{\diagup\diagdown}} + S^0$$

$$\textbf{(V)}$$

Oxidation to perxanthate

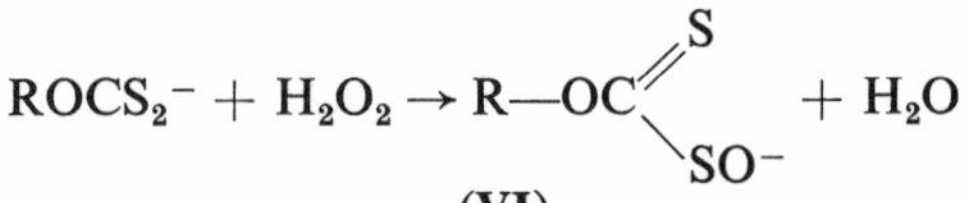

$$ROCS_2^- + H_2O_2 \rightarrow R{-}OC\overset{S}{\underset{SO^-}{\diagup\diagdown}} + H_2O$$

$$\textbf{(VI)}$$

The first two reactions, hydrolysis (**I**) and xanthic acid decomposition (**II**), are the main reactions of decomposition in acidic solutions; they have been extensively studied, and the most complete evaluation of the kinetics is that by Iwasaki and Cooke (1958) (Figure 5.5a). Utilizing UV absorption measurements and assuming that reactions (**I**) and (**II**) take place consecutively, with k_1, k_2, and k_3 denoting the appropriate rate constants, Iwasaki and Cooke derived an expression for the rate equations:

$$\frac{dc}{dt} = -k_3[\text{RXH}] = -\frac{k_3}{1 + K_d/[\text{H}^+]}c \tag{5.1}$$

where c denotes the total concentration of xanthate ion RX^- *and* of xanthic acid RXH (as measured by optical density at the 301-nm wavelength in the UV spectroscopic method) and K_d denotes the dissociation constant for xanthic acid,

$$K_d = \frac{[\text{RX}^-][\text{H}^+]}{[\text{RXH}]} \tag{5.2}$$

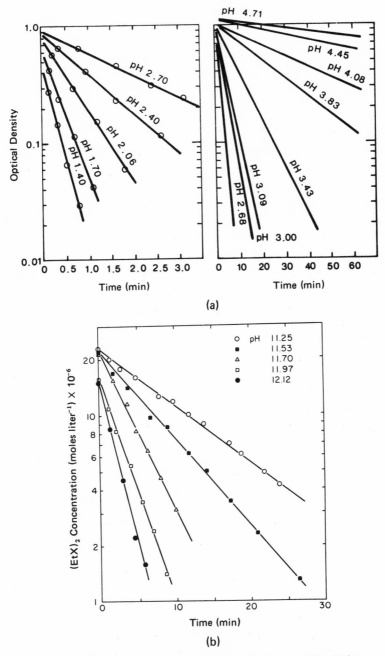

Figure 5.5. (a) Kinetics of decomposition of ethyl xanthate in acid solutions, as indicated by changes in optical density of solutions with time. [Iwasaki and Cooke (1958), reprinted with permission from *J. Am. Chem. Soc.* **80**. Copyright 1958, American Chemical Society, Washington, D.C.]. (b) Kinetics of decomposition of diethyl dixanthogen in highly alkaline solutions. [From Tipman (1970), with the permission of the author.]

The overall logarithmic rate constant can thus be represented by

$$\varkappa = \frac{k_3}{1 + K_d/[\text{H}^+]} = -\frac{d\ln c}{dt} \tag{5.3}$$

and this is found from the plot of $\log \varkappa$ vs. pH.

From these results the values for the monomolecular rate constant ($k_3 = 4.3 \pm 0.21$ min^{-1}) and for the dissociation constant ($K_d = \sim 2 \times 10^{-2}$; hence, p$K_d = -\log K_d = 1.7$) have been derived. These values mean that, at room temperature, about 50% decomposition occurs within 1 hr at $\sim$pH 4.3.

Klein *et al.* (1960) found that in highly acid solutions the rate constant reaches a maximum at 0.4 N and decreases for higher acid concentrations. To account for this change in the rate of decomposition, they postulated the existence of a protonated xanthic acid, RXH_2^+, which was subsequently correlated by Iwasaki and Cooke (1964) with the decomposition mechanism proposed by them earlier.

The new rate constants for reactions

$$\text{K}^+ + \text{ROCS}_2^- \xrightarrow{K_1} \text{ROCS}_2\text{H} \xrightarrow{K_2} \text{ROH} + \text{CS}_2$$

have been compared, after corrections, with the data obtained by Klein *et al.* (1960), and these are shown in Figure 5.6.

In highly alkaline solutions the decomposition reactions yield stable end products: carbonate, hydrosulfide and trithiocarbonate ions, and alcohol, as indicated in reaction (**III**). Philipp and Fichte (1960) suggested that the rate-controlling step is the hydration of the carbon–sulfur double bond. Wronski (1959) derived an expression for the rate of decomposition in terms of NaOH concentration and xanthate concentration c:

$$-\frac{dc}{dt} = k'c + k''[\text{NaOH}]^2 \tag{5.4}$$

where $k'/k'' \cong 3$.

For flotation purposes, the most important reactions are those occurring in xanthate solutions in the pH range 6–12. In the last two decades it has been conclusively demonstrated that some oxidation (caused by an oxidizing agent or oxygen from air) is essential before xanthates (and most thio surfactants) can act as collectors in sulfide systems. Whether it is the oxidation of xanthate ions to dixanthogen, represented by reaction (**IV**), or oxidation of sulfides to provide hydrophobic metal xanthate by-products

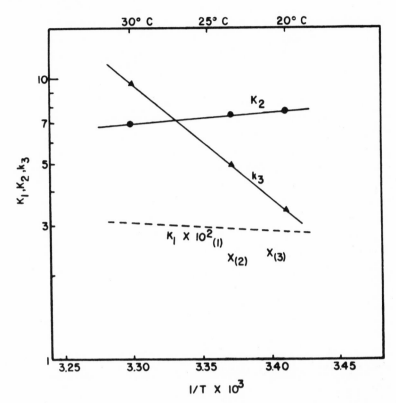

Figure 5.6. A comparison of rate constants as a function of $1/T$ for ethyl xanthate decomposition in acid solutions when the existence of protonated xanthic acid is recognized: (1) $K_1 \times 10^2$ after Klein *et al.* (1960); (2) K_2 is $K_1 \times 10^2$ after Iwasaki and Cooke (1958), and (3) k_3 is $K_1 \times 10^2$ after Iwasaki and Cooke (1959). [Iwasaki and Cooke (1964), reprinted with permission from *J. Phys. Chem.* **68**, Copyright 1964, American Chemical Society, Washington, D.C.]

is still a debatable issue. Oxidation to dixanthogen is readily observed in concentrated xanthate solutions, since the insoluble dixanthogen comes out of solution as an emulsion of dispersed oil droplets (producing a noticeable turbidity of solution) and as oil lenses at the air/water interface. Xanthate oxidation, reaction (**IV**), had been considered irreversible and dixanthogen unreactive. However, spectrophotometric studies on dixanthogen dispersions by Pomianowski and Leja (1963) showed that dixanthogen does react in alkaline solutions and reverts partly to xanthate and decomposes to CS_2 (see studies by Tipman (1970) and Figures 5.5b and 5.7). Rao and Patel (1960, 1961) established that aqueous oxidation of xanthate ion to dixanthogen is catalyzed by multivalent reducible ions such as Cu^{2+} and Fe^{3+}.

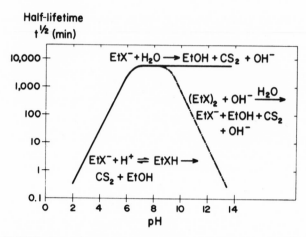

Figure 5.7. Half-life times of decomposition reactions of ethyl xanthate in acid solutions, ethyl xanthate in neutral and alkaline solutions, and diethyl dixanthogen in alkaline solutions at 22°C. [From Tipman (1970), with the permission of the author.]

Golikov and Nagirnyak (1961) found that the surfaces of metallic sulfides can act as oxidation catalysts.

Solozhenkin and Zinchenko (1970) described the use of a complicated 2,2,6,6-tetramethyl-1-iminoxy-4-xanthate,

as a flotation agent for sulfides (galena, chalcopyrite, pyrite, activated sphalerite, and antimonite). They found that oxidation of this xanthate to its dixanthogen form is possible only if metallic cations of variable valence are present but does not occur with molecular oxygen adsorbed on galena.

Finkelstein (1967) observed that the equilibrium constants for reaction (IV) differed by orders of magnitude if calculated from potentiometric data as compared to spectroscopic data. He also established that in contradiction to earlier results, xanthate oxidation to dixanthogen does not take place in pure homogeneous solutions, regardless of the amount of dissolved oxygen. It seems likely that traces of peroxides from ether (used in the purification procedures) were often responsible for inducing oxidation reactions detected previously.

Other researchers also found that every reaction of either xanthate or dixanthogen is influenced by traces of impurities in (or additives to) the system. Thus, for example, Mellgren (1966) found that ethyl xanthate remains quite stable in the presence of thiosulfate ions. Wells *et al.* (1972) proposed that dixanthogen reacts with thiosulfate to form an ionic complex and a xanthate ion.

Tipman (1970) re-examined the reactivity of dixanthogen with OH^- and found that the decomposition of dixanthogen is irreversible and involves a two-step process. The first step is a nucleophilic[†] S_N2 displacement by OH^- (or by any other nucleophilic substituent, in the following approximate order of reactivity: $HS^- > CN^- > OH^- > SO_3^{2-} > SCN^- > I^-$) and results in the formation of a xanthate ion and an intermediate ionic complex between the nucleophilic substituent and the rest of the dixanthogen group.

The second step in the decomposition reaction is a base hydrolysis of the intermediate complex, resulting in an alcohol, carbon disulfide, and OH^- as the end products. Thus, the overall scheme of reactions occurring in solutions of various pH between xanthate species of importance to flotation can be represented as in Figure 5.7. Reactions (**I**) and (**II**) determine the half-life in solutions of pH up to ~6; in the range of pH 6–8 the hydrolytic decomposition occurs, producing alcohol, carbon disulfide, and OH^-; if there are no other reactive species (such as dixanthogen) present in the system to influence this reaction between xanthate ion and water, the half-life of xanthate solutions is approximately constant for pH > 8 up to $\sim$pH 12. Reaction (**III**) does not appear to be significant until the highly alkaline region is reached (pH 13). If, however, catalytically active oxidizing or reducing agents are available, the half-life can be shortened to a fraction of the value shown in Figure 5.7.

When alkyl dixanthogen has been produced as a result of oxidation of a xanthate ion, its stability is determined by the rate of the nucleophilic substitution reaction, which is bimolecular, i.e.,

$$-\frac{d[X_2]}{dt} = k_2[X_2][OH^-] \tag{5.5}$$

[†] The symbol S_N2 stands for *substitution, nucleophilic, bimolecular*. It denotes a mechanism by which a new ligand attacks the original complex, forming a seven-coordinated activated complex; the rate of the reaction is proportional to the concentration of the original complex times that of the ligand; i.e., $v = k[X_2][OH^-]$ in the case of dixanthogen X_2 and substituent ligand OH^-. The other mechanism that is possible between complexes and ligands (substituents) is the S_N1 mechanism, *substitution, nucleophilic, unimolecular*, in which the complex dissociates first (losing the ligand to be replaced) and then reacts with the substituent ligand; the rate law for such a process is $v = k \cdot [\text{complex}]$, and is *independent* of the concentration of the new ligand.

with the rate constant $k_2 = 0.24 \pm 0.005$ liter mol^{-1} sec^{-1} at 23°C. The solubility[†] of dixanthogen is constant in the pH region 2.8–8.4 and equal to 1.25×10^{-5} M.

Oxidation of xanthate ion to dixanthogen can be accomplished by a number of oxidizing agents in aqueous solutions: iodine, chlorine, sodium hypochlorite, nitric acid, etc. The standard dixanthogen–xanthate redox potential has been determined electrochemically [du Rietz (1957), Stepanov *et al.* (1959), Tolun and Kitchener (1964), Woods (1971)], but the values range from -0.037 to -0.081 V.

This wide variation is caused by the factors mentioned earlier, viz. lack of equilibrium, the influence of active impurities in the system (giving a mixed potential), and the incomplete reversibility of the reduction of dixanthogen. The most reasonable value for the redox potential appears to be -0.070 V [Hepel and Pomianowski (1973)]. Theoretically, any oxidizing agent of a potential greater than the standard redox potential can oxidize xanthate ion to dixanthogen, and any agent having a potential lower than the standard redox potential can reduce dixanthogen to xanthate ion. In practice, the reactions are neither uniquely reversible nor limited to the oxidation–reduction cycle. And, as established by Finkelstein (1967), in the case of oxygen dissolved in aqueous xanthate solution, even though thermodynamically the oxidation is possible, it does not occur with any noticeable rate unless catalytically active ions or surfaces are available in the systems. For example, gaseous O_2 was found to be capable of oxidizing xanthate ions in the presence of C_{12} TAB (Finkelstein, 1969b). It should be kept in mind that anything that is added to such systems may influence the measurements obtained; thus, even the buffer additives to keep pH constant may exert an influence on the course of reactions, in the same manner as redox indicators. Tipman (1970) found that an addition of methylene blue (as a redox indicator) increased the rate constant of xanthate decomposition by $\sim 10^6$.

The decomposition of a xanthate ion to a monothiocarbonate ion, reaction (V), was studied by Harris and Finkelstein (1975) in heterogeneous systems containing sulfide minerals (galena and pyrite), oxygen, and xanthate ions. They found that, irrespective of the order in which these components are brought together, as long as metal sulfide, oxygen, and xanthate

---

[†] *Note*: Hamilton and Woods (1979) redetermined the solubility of dixanthogen spectrophotometrically, eliminating the possible effects of reaction products such as xanthate ion and monothiocarbonate by an addition of hydrochloric acid. Their values are 1.14×10^{-5} mol dm^{-3} for ethyl dixanthogen and 1.5×10^{-6} for propyl dixanthogen.

are in the system, the metastable monothiocarbonate ions are formed. The reaction *requires the participation of the sulfide surface* [the rate of formation of monothiocarbonate in homogeneous solution is known to be in orders of magnitude lower than that in a heterogeneous system; Finkelstein (1967)]. Either oxygen or xanthate has to be adsorbed on the surface for the monothiocarbonate to be subsequently generated. If xanthate is preadsorbed, then the reaction is (possibly) as follows:

$$Pb(EtX)_2 + O_2 \rightarrow Pb(MTC)_2 + 2S^0$$
$$Pb(MTC)_2 + 2(EtX^-) \rightarrow Pb(EtX)_2 + 2MTC$$

(A)

(where MTC$^-$ stands for the ethyl monothiocarbonate ion and EtX$^-$ for the ethyl xanthate ion).

If the oxygen had reacted with the sulfide surface then, in the presence of xanthate ions, the following sequence of reactions[†] may possibly have taken place:

$$PbS_2O_3 + 2EtX^- \rightarrow Pb(EtX)_2 + S_2O_3{}^{2-}$$
$$3Pb(EtX)_2 + 2PbS_2O_3 \rightarrow 2PbS + 8S^0 + Pb^{2+} + 6MTC^-$$

(B)

or

$$Pb(OH)_2 + EtX^- \rightarrow Pb(EtX)OH + OH^- \rightarrow PbS + MTC^- + H_2O \quad (C)$$

Whichever is the mechanism, (A), (B), or (C), the overall changes in the residual xanthate, the adsorbed xanthate, and the monothiocarbonate with time were found by Harris and Finkelstein (1975) to follow the outlines indicated in Figure 5.8.

Further, Harris and Finkelstein found that dixanthogen adsorbed on the surface of pyrite is not conducive to the formation of monothiocarbonate and speculated that the intermediate ferric hydroxy xanthate, $Fe(OH)X_2$, may be the surface species that undergoes decomposition to monothiocarbonate.

In acid and alkaline solutions the monothiocarbonate ions undergo decomposition by different mechanisms. In acid solutions first-order decomposition kinetics are obtained; the half-life is pH-dependent. The reactions presumed to occur are analogous to those involved in the xanthate

---

[†] Sheikh (1972) studied the decomposition of $Pb(EtX)_2$ and found evidence for PbS formation, but the results have not been interpreted in a resulting publication [Sheikh and Leja (1973)] in terms of monothiocarbonate formation.

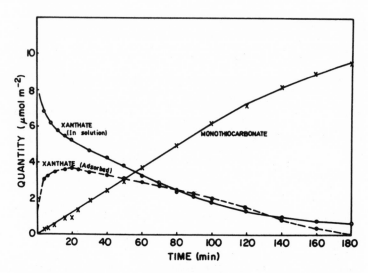

Figure 5.8. The formation of monothiocarbonate from xanthate adsorbed on PbS. [Harris and Finkelstein (1975), reprinted with permission from *Int. J. Miner. Process.* **2**. Copyright Elsevier Publishing Co., Amsterdam.]

ion decomposition under acid conditions, viz.,

$$EtCO_2S^- + H^+ \rightleftharpoons EtCO_2SH$$ (ethyl monothiocarbonic acid formation; $EtCO_2^-$ stands for ethyl monothiocarbonate ion)

$$EtCO_2SH \rightarrow EtOH + COS$$ (decomposition to alcohol and carbonyl sulfide)

$$COS + OH^- \rightleftharpoons CO_2 + HS^-$$

The oxidation of monothiocarbonate ions to the analogue of dixanthogen, i.e.,

$$R-O-C\underset{S-S}{\overset{O\quad O}{\diagdown\diagup}}C-O-R$$

carbonate disulfide

was investigated by Murphy and Winter (1973), and the redox potentials for these carbonate disulfide–monothiocarbonate reactions were found by Winter and Woods (1973) to be more positive (+0.020 to −0.120 V for methyl to hexyl) than those for the corresponding xanthates (−0.004 to −0.158 V for methyl to amyl).

Under specific conditions of alkalinity and oxidation level such as that provided by hydrogen peroxide (H_2O_2), an oxidized species termed *perxanthate* [by Jones and Woodcock (1978)] or *sulphenic acid* [by Tipman and Leja (1975)] is being produced in dilute xanthate (or dixanthogen) solutions, as indicated by reaction (**VI**). Butyl perxanthate is characterized by UV absorption bands at 348 and 215 nm at pH 10 and 298 and 225 nm at pH 2. On solvent extraction from acid solutions these bands shift to various intermediate values depending on the nature of the solvent [data from Jones and Woodcock (1978)]. Reaction (**VI**) was first studied by Garbacik *et al.* (1972).

5.2.2.3. *Soluble (Ionic) Metal-Xanthate Complexes*

Yet another complication arising in the aqueous xanthate systems is due to the formation of soluble metal-xanthate ionic complexes [Majima (1961b) Nanjo and Yamasaki (1966, 1969), Sheikh (1972), Sparrow *et al.* (1977)]. These complexes form in two regions of nonstoichiometric concentrations of metal ions and xanthate ions:

1. When $[M^{n+}] \gg [X^-]$, then cationic $M(X)^{(n-m)+}$ complexes (where $m < n$) may be detected by the spectrophotometric absorption technique.
2. When $[M^{n+}] \ll [X^-]$, anionic $M(X)_m^{(m-n)-}$ complexes may exist. (X^- represents the alkyl xanthate group, such as ethyl xanthate ion.)

Both types of complex species appear under conditions when the solubility product for the appropriate MX_n xanthate is exceeded, even by several orders of magnitude. More details on this aspect are presented in Section 5.2.3 and in Figure 5.11.

Majima (1961b) described the formation of soluble $Pb(X)_2$, $Ni(X)_3^-$, and $Co(X)_3^-$ complexes in 1:1 acetone–water solutions. Nanjo and Yamasaki (1966) reported on the formation and properties of $Cd(X)_4^{2-}$, and Nanjo and Yamasaki (1969) reviewed their studies of 1:1 complexes, $M(X)^{2-}$, of Pb^{2+}, Cd^{2+}, Zn^{2+}, Ni^{2+}, Co^{2+}, and Cu^{2+} (no complex was found for Mg^{2+}). The stability constants[†] were evaluated for these complexes from the spectrophotometric data at 25°C and at an ionic strength of 1 ($NaClO_4$). These are given in Table 5.3.

Absorption spectra could not be used for the evaluation of the stability constant in the case of the copper complex, $Cu(X)^+$, because of the con-

---

[†] The stability constant for the equilibrium $M^{2+} + X^- \rightleftharpoons M(X)^+$ is defined as

$$K = \frac{[M(X)^+]}{[M^{2+}][X^-]}$$

Table 5.3. Stability Constants of 1:1 Complexes, 25°C, $I = 1$ (NaClO$_4$)[a]

Metal ion	K
Cd(II)	$(1.2 \pm 0.1) \times 10^4$
	$(1.8 \pm 0.2) \times 10^{4\,b}$
Zn(II)	$(5.2 \pm 0.1) \times 10^c$
Ni(II)	$(4.4 \pm 0.3) \times 10$
Co(II)	$(1.1 \pm 0.1) \times 10$
Mg(II)	No complex formed

[a] After Nanjo and Yamasaki (1969).
[b] In the presence of 0.02 M acetate buffer.
[c] In the presence of 0.2 M acetate buffer.

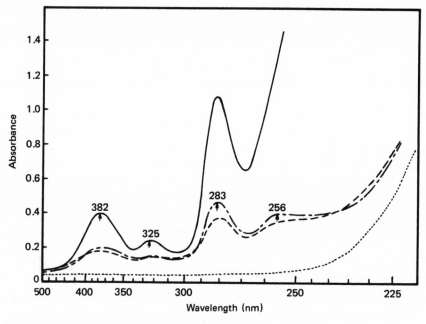

Figure 5.9. The formation of Cu(X)$^+$ complex with characteristic bands 283, 325, and 382 nm. UV spectra of the following solutions (10-cm path length): ——, [Cu^{2+}] = 10^{-2} M, [KEtX] = 10^{-5} M, initial spectrum; — — —, [Cu^{2+}] = 10^{-4} M, [KEtX] = 5×10^{-6} M, initial spectrum; – – –, [Cu^{2+}] = 10^{-4} M, [KEtX] = 5×10^{-6} M, initial spectrum; · · ·, [Cu^{2+}] = 10^{-4} M, [KEtX] = nil. [From Sparrow *et al.* (1977), in *Sep. Sci.* **12(1)**, by courtesy of Marcel Dekker, Inc., New York.]

comitant formation of dixanthogen and (reduction to) cuprous xanthate. Figure 5.9 shows the UV spectra of $Cu(X)^+$ complex formed in solutions of differing excess of Cu^{2+} relative to X^-. The bands 283, 325, and 382 nm are characteristic of this complex. It is seen that in solutions initially containing xanthate ions and a sufficiently high excess of Cu^{2+}, nearly all the xanthate species is preserved in a complex form. An analysis for xanthate, based on UV absorption at 301 nm, would reveal no xanthate ion in such solutions, yet a subsequently formed product (or an adsorbed film) would be a derivative of xanthate. There has not been enough work carried out to assess the role these complex metal-xanthate species may play in adsorption on negatively charged sulfides or, for example, in the mechanism of sphalerite activation by Cu^{2+}. In the latter case, the cupric ions are added to a highly alkaline flotation pulp and are mostly precipitated as $Cu(OH)_2$, yet a subsequent addition of xanthate generates a satisfactory activation and floatability of ZnS. (The sequence of additions, first Cu^{2+} and then X^-, is most important for successful flotation.)

From the spectroscopic data of Sparrow *et al.* (1977), it appears that cupric $Cu(EtX)_2$ is formed first, but it is highly unstable and readily changes to a mixture of cuprous $Cu_2(EtX)_2$ and dixanthogen, $(EtX)_2$.

5.2.3. Insoluble Metal Xanthates

Appreciable oxidation of insoluble metallic sulfides may occur on exposure of ore (after mining) to the atmosphere or during lengthy grinding and aeration of the flotation feed. The oxidation products thus created may become dissolved, producing perceptible concentrations of metallic ions in the flotation pulp. The extent of reactivity between xanthate ions and metal ions has been of great interest, particularly in view of Taggart's hypothesis that flotation is due to precipitation of insoluble heavy metal xanthates. A number of determinations of solubilities and solubility products of metallic xanthates have been made employing several independent methods; among others, the most frequently used were potentiometric titration [du Rietz (1953), Kakovsky (1957), Sheka and Kriss (1959)] and the spectroscopic technique [Majima (1961a,b)]. Tables 5.4–5.6 show some of the results obtained in such determinations.

Kakovsky (1957) also deduced from the thermodynamics of dissolution that in any homologous series of hydrocarbon surfactants the logarithm of solubility, $\log K_{sp}$, must be a linear function of the number m of CH_2 groups; i.e.,

$$\log K_{sp} = a - bm \tag{5.6}$$

Table 5.4. Solubilities of Some Metal Xanthates in Water, at 20°C[a]

Metal ethyl xanthate	Solubility (g/100 ml solution)		Solubility (mol/liter, average)
	By metallic isotope	By sulfur isotope	
$Zn(EtX)_2$	2.96×10^{-2}	2.76×10^{-2}	9.0×10^{-4}
$Ni(EtX)_2$	—	2.07×10^{-2}	6.3×10^{-5}
$As(EtX)_3$	2.60×10^{-3}	2.22×10^{-3}	5.5×10^{-5}
$Cd(EtX)_2$	—	1.18×10^{-3}	3.3×10^{-5}
$Co(EtX)_2$	2.65×10^{-6}	3.89×10^{-6}	7.6×10^{-8}
$Cu_2(EtX)_2$	—	1.54×10^{-6}	4.1×10^{-8}

[a] After Sheka and Kriss (1959).

Table 5.5. Exponents[a] of Solubility Products and of Xanthate Concentrations for Solutions of Metal Ethyl Xanthates at 25°C in Water

Metal M^{n+}	pK_{sp}	pK_{X-}	Reference[b]
Au^+	29.22	14.61	2
Cu^+	19.28	9.64	2
Ag^+	18.1, 18.6	9.1, 9.3	2, 1
Tl^+	7.51	3.76	2
Pb^{2+}	16.1, 16.77	5.5, 5.49	1, 2
Sn^{2+}	~15	~4.9	2
Cd^{2+}	13.59	4.43	2
Co^{2+}	12.25, 13.0	3.98	2
Ni^{2+}	11.85	3.87	2
Zn^{2+}	8.2, 8.31	2.6, 2.67	1, 2
Fe^{2+}	7.1	2.3	2
Bi^{3+}	~30.9	~7.6	2
Sb^{3+}	~24	~5.9	2

[a] Solubility product $K_{sp} = [M^{n+}][X^-]^n$, and the exponent $pK_{sp} = -\log K_{sp} = pK_M + npK_{X-}$, where M^{n+} = metal ion, X^- = xanthate ion, and n = valence of metallic ion.
[b] 1: du Rietz (1953); 2: Kakovsky (1957).

Table 5.6. Solubility Products of Some Alkyl Metal Xanthates[a]

Alkyl xanthate	Au$^+$	Ag$^+$	Cu$^+$	Hg$^+$	Zn^{2+}	Cd^{2+}
Ethyl	6.0×10^{-30}	5×10^{-19}	5.2×10^{-20}	1.7×10^{-38}	4.9×10^{-9}	2.6×10^{-14}
Propyl	—	1.9×10^{-20}	—	1.1×10^{-39}	3.4×10^{-10}	—
Butyl	4.8×10^{-30}	—	4.7×10^{-20}	1.2×10^{-40}	3.7×10^{-11}	2.08×10^{-16}
Amyl	1.03×10^{-31}	1.5×10^{-20}	—	—	1.55×10^{-12}	8.5×10^{-18}
Hexyl	3.5×10^{-32}	3.9×10^{-21}	—	—	1.25×10^{-13}	9.4×10^{-19}
Heptyl	1.05×10^{-32}	—	—	—	1.35×10^{-14}	9.15×10^{-20}
Octyl	—	1.38×10^{-22}	8.8×10^{-24}	—	1.49×10^{-16}	7.2×10^{-22}

[a] After Kakovsky (1957).

It was found that a depends on the polar group but that $b = 4.25$ and is the same for all homologous series regardless of the nature of the polar group.

It appears that the bonding between the metal and sulfur atoms is covalent in all highly insoluble metal xanthates; as their solubility increases, the covalent character of bonding (and the hydrophobic nature of the metal xanthate precipitate) decreases.

The manner of precipitation occurring in a homogeneous system requires some discussion. Generally, it is accepted that the plot of $\log[M^{n+}]$ vs. $\log[X^-]$ should give a straight line from which the value of K_{sp} is derivable. For example, in the case of metal xanthates,

$$\log K_{sp} = \log[M^{n+}] + n \log[X^-] \qquad (5.7)$$

should apply to all systems, regardless of the relative range of concentrations (activities) of the ions present in the system. Once the concentrations $[M^{n+}]$ and $[X^-]$ are such that the solubility product is exceeded, the excess ions should be removed from the homogeneous solution by precipitation of the solid phase. And sooner or later, an equilibrium is established as prescribed by the solubility product (if the latter is correctly evaluated).

However, the linearity of equation (5.7) does *not*, in practice, apply in some systems. It has been established that all such systems contain additional *soluble complex, ionic species*, which are formed between the precipitating component ions. Definite departures from the straight-line relationship are observed and have been reported in the literature for three different types of systems: inorganic salts [Matijevic *et al.* (1961)], metal carboxylates [Matijevic *et al.* (1966), Nemeth and Matijevic (1971)], and metal xanthates [Sheikh (1972)]. Figure 5.10 shows the results obtained in silver bromide precipitation by Matijevic *et al.* (1961). It shows regions of precipitation, of solubility due to the formation of several complex ionic species (such as $AgBr_2^-$, $AgBr_3^{2-}$, etc., in an excess of bromide ions or Ag_2Br^+, Ag_3Br^{2+}, etc., in an excess of silver ions), and of coagulation, all fairly symmetrical with respect to the line of the equivalent concentrations of precipitating ions. The fast precipitation process producing relatively coarse particles occurs in a narrow range of concentrations designated as the equivalent region. On both sides of this equivalent region, mixing of the two appropriate solutions results in the formation of colloidal sols (either positively charged in an excess of Ag^+ or negatively charged in an excess of Br^-) which are stable over a long period of time and show an exceedingly low turbidity. These are denoted on Figure 5.10 as the positive or the negative stability regions. When the solutions undergoing mixing

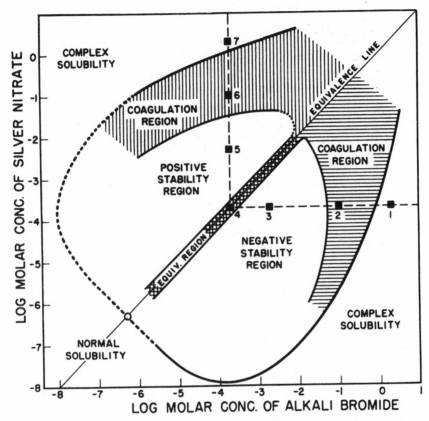

Figure 5.10. Regions of solubility, precipitation, coagulation, and stability of colloidal-size precipitates (sols) obtained by mixing silver nitrate with alkali bromide. Solid lines denote experimental data obtained by measuring the turbidity of solutions 10 min after mixing the precipitating components; they represent boundaries along which a drastic change in turbidity occurs (from 0.0 to $\sim$0.7 cm^{-1} using light of $\lambda = 546$ nm). Both NaBr and KBr give solubility points on the same curve. No data are available for the dotted part of the solubility curve—it has been drawn in analogy to the AgCl solubility line for which complete sets of data are available. [After Matijevic *et al.* (1961), with permission from *J. Chem. Educ.* **38**. Copyright 1961, American Chemical Society, Washington, D.C.]

are more concentrated, again two regions arise, denoted as coagulation. During coagulation the primary sol particles are aggregated, producing coarser particles. However, outside these regions of coagulation, for still more concentrated solutions, two regions of completely homogeneous transparent solutions exist due to the formation of several soluble ionized complexes. Thus, the solubility line in this system departs greatly from the *straight-line relationship* predicted by $\log K_{\mathrm{sp}} = \log[\mathrm{Ag}^+] + \log[\mathrm{Br}^-]$ that would pass at 45° to each axis through the point at equivalent concentrations marked by the circle in Figure 5.10.

For other systems, particularly those involving surfactant ions, the symmetry of behavior, apparent in Figure 5.10, may not be so evident [Matijevic *et al.* (1966)]. Also, the solubility product values reported in the literature may not be too reliable. Figure 5.11 shows data obtained by Sparrow *et al.* (1977), following an initial study of Sheikh (1972), which outline the following:

1. A region of concentrations giving a coarse precipitate (a doubly hatched, near-parabolic region of near-equivalent concentrations)
2. A region in which an initially colloidal precipitate is fairly quickly converted to a coagulated system (marked by a singly hatched area)
3. A transparent sol region which may be stable for several days but is ultimately coagulating to a visible precipitate (between the solid line of solubility and the singly hatched area)
4. Regions of transparent solutions showing no precipitation even after several days of equilibration

Two straight lines, representing the relationship (5.7) for cupric xanthate and cuprous xanthate, are shown. The third heavy solid line delineating the experimentally determined solubility descends below each of these straight lines, indicating that the solubility products may not be reliable. At the same time, the actual solubility line ascends into the areas of nonequivalent concentrations several orders of magnitude above the K_{sp} solubility product lines in a manner not too symmetrical with respect to the line of equivalent concentrations (not shown).

The existence of metal-xanthate ionic complexes, responsible for these deviations of the actual solubility line from the linear predicted line, has been documented so far for the region containing excess M^{n+} ions (see Section 5.2.2.2 and Figure 5.9). In the region containing excess X^- ion,[†] the anionic complexes appear to be present (as indicated by results in Figure 5.11), but, as yet, they have not been characterized or identified spectrophotometrically.

Relationships analogous to that shown in Figure 5.11 for the Cu–X system were shown by Sheikh (1972) and Sheikh and Leja (1973, 1974) to exist for the Pb–X and the Fe–X[†] systems; the extent of regions giving the coarse precipitate, sol formation, or soluble complexes was found to differ for each system.

[†] X denotes alkyl xanthate group.

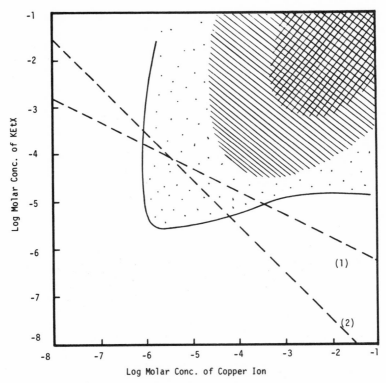

Figure 5.11. Precipitation and complex formation regions for solutions of potassium ethyl xanthate K(EtX) and copper ion. Double-hatched region: *coarse-size precipitate* (formed immediately on mixing). Single-hatched region: very *fine-size precipitate* or *sol*. Region of dots: invisible sol, detected by light scattering. Solid line: solubility line, experimental. Below the solid line: complexes formed. Line (1): theoretical limits of solubility, $K_{sp} = (Cu^{2+})(EtX)^2 = 2 \times 10^{-14}$. Line (2): $K_{sp} = (Cu^+)^2(EtX)^2 = 5.2 \times 10^{-20}$. [From Sparrow *et al.* (1977), in *Sep. Sci.* **12**(1), by courtesy of Marcel Dekker, Inc., New York.]

In a *heterogeneous* system, such as a flotation pulp containing, for example, copper minerals, the different regions shown in Figure 5.11 may become displaced toward lower concentrations. The surfaces of solid phases (from which the metal ions are derived in consequence of oxidation and dissolution reactions) will tend to adsorb xanthate species and will undoubtedly act as nucleating sites for metal-xanthate precipitation. In fact, metal xanthates were identified a long time ago in the adsorbed layers by Gaudin *et al.* (1934). However, the metal-xanthate complexes may still be formed in solution under nonequivalent concentration conditions and may be sufficiently stable to give nonlinear solubility lines even in heterogeneous systems.

Whatever the details of the metal-xanthate precipitation and complexing processes, the extent of these reactions has a definite bearing on the consumption of the collector-acting xanthate species. It may also have a role in the diffusion of xanthate species to the negatively charged surfaces as cationic metal-xanthate complexes. Since the concentrations of metal ions, released into flotation solutions during grinding, have never been (and hardly ever will be) reliably evaluated for any actual flotation system, the optimum addition of the collecter needed for each flotation system has to be established in an empirical manner, by tests. It is known, however, that an addition of an extra amount of the lattice-constituent metallic ions to the flotation system often requires a higher collector addition in order to keep up the recovery of minerals. On the other hand, it has also been encountered in laboratory testing that an addition of metal-xanthate precipitate to the flotation system often reduces the overall mineral recovery. Obviously, the control of the metal and collector ion concentration necessary to achieve optimum recovery is a more complicated procedure than the simple application of the solubility product would seem to indicate.

Metal xanthates can be readily identified by at least two techniques: X-ray diffraction and infrared spectroscopy. Of these, infrared spectroscopy is by far the more versatile and easier to carry out, and, with some limitations, the spectra of solid phases present at interfaces can be obtained *in situ*. The application of infrared spectroscopy enables not only the chemical nature of a metal xanthate to be established, but also the bonding within the polar group and that between the polar group and the adsorbing substrate can be evaluated from the changes in the spectra.

Following an extensive comparison of spectra obtained for a number of thio compounds, Poling (1961) made the following assignment of frequences to specific bonds within the xanthate group: the very strong 1020–1070 cm^{-1} band to the C$=$S stretching mode, two strong bands at 1200 cm^{-1} and 1120 cm^{-1} to vibrations of the C—O—C linkages, and 830–870 cm^{-1} to the asymmetric stretching mode of C—S. A very strong band at 1240–1270 cm^{-1} is characteristic of dixanthogens.

There has been some uncertainty concerning the frequencies of the C$=$S stretching vibrations,[†] but the absence of a band in the region of 1200 cm^{-1} in the spectrum of trithiocarbonates suggests that this high wavenumber cannot be due to the C$=$S group. Goold and Finkelstein (1969) discuss IR spectroscopy of xanthates and provide support for Poling's assignments.

[†] For example, Shankaranarayana and Patel (1965) assigned the 1200-cm^{-1} band to C$=$S.

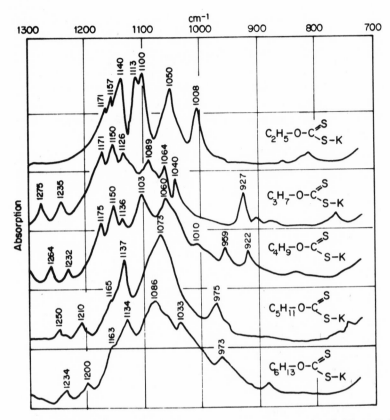

Figure 5.12. Infrared (IR) spectra of potassium xanthates, ethyl to hexyl. [From Poling (1961), with the permission of the author.]

The representative spectra of potassium alkyl xanthates, from methyl to hexyl, are shown in Figure 5.12, and those of dixanthogens and copper xanthates are shown in Figure 5.13.[†] The change in the group frequency of the C=S and the C—O—C bands has been related, for different metal xanthates, to the difference in the electronegativities between sulfur and the metal (Figure 5.14). It is seen from Figure 5.14 that only the ionically

---

[†] It can readily be seen from a comparison of the infrared spectra in Figure 5.13 that the precipitate of copper xanthate (formed by mixing solutions of cupric salt with potassium xanthate) consists of a stoichiometric mixture of cuprous xanthate and dixanthogen (the latter being readily removed by extraction of the precipitate with ether, CCl$_4$, or hexane). The simultaneous reduction of cupric to cuprous xanthate and the oxidation of xanthate to dixanthogen had been recognized already by Zeise and confirmed analytically by Gaudin *et al.* (1934).

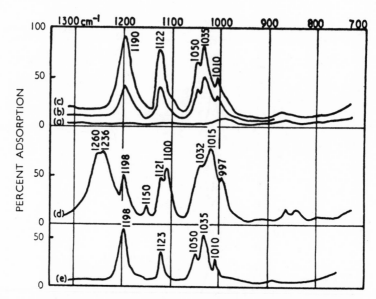

Figure 5.13. Spectra of xanthate species adsorbed on oxidized and sulfidized copper substrates: (a) freshly polished copper plate surface; (b) and (c) cuprous ethyl xanthate formed by adsorption of K ethyl xanthate on oxidized plates, sulfidized plates, and gauze substrates; (d) mixed cuprous xanthate and ethyl dixanthogen on either oxidized or sulfidized copper plates; (e) effect of washing plates (d) in ether. [From Poling (1961), with the permission of the author.]

bonded monovalent metal xanthates (K^+, Na^+, Tl^+) lack the 1200-cm^{-1} band, while the covalently bonded monovalent (Cu^+, Hg^+) and the divalent metals (Sn^{2+}, Pb^{2+}, Ni^{2+}, Zn^{2+}) all show an absorption band at 1200 cm^{-1}, the highest wavenumber being shown by dixanthogen (1270 cm^{-1}) for zero difference in electronegativities. The wavenumbers of the metal–sulfur bonding lie in the range 250–400 cm^{-1}.

Crystal structure determinations [Mazzi and Tadini (1963)] have shown that the ionic alkali metal xanthates consist of isolated cations and xanthate groups. The xanthates of the transition metals [Franzini (1963), Hagihara and Yamashita (1966)] have complex structures in which the sulfurs of xanthate groups have a square planar arrangement around divalent metal ions M^{2+} and a distorted octahedral arrangement around the trivalent metal ions. The two C—S bonds are equivalent (the bond length C—S = 1.68 Å) in some xanthates, such as K, As, and Pb xanthates[†] (on this

---

[†] Note that the IR spectroscopy data (mentioned above) suggest a covalent Pb—S bond in Pb xanthates.

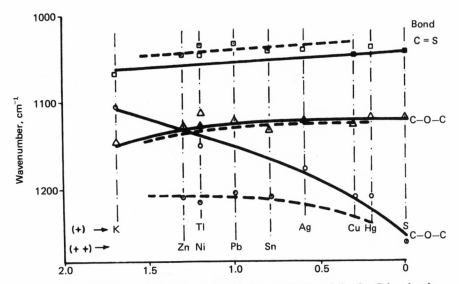

Figure 5.14. Shifts in the infrared wavenumber for the C=S and C—O—C bonds relative to the difference in electronegativities between sulfur and metal for mono- and divalent metal *n*-butyl xanthates. [From Poling (1961), with the permission of the author.]

basis these xanthates are considered to be ionic), while the complex xanthate structures contain one short (1.56–1.64 Å) and one long (1.70–1.79 Å) C—S bond; this differentiation indicated the double-bond character of the shorter of the C—S bonds. All C—S distances in xanthates are shorter than the 1.81-Å single-bond value [Pauling (1960)], indicating delocalization of electrons and resonance structures (a) and (b) in Figure 5.15; the contribution of the resonance structure (c) is negligible in the case of xanthates (but appreciable in the case of dithiocarbamates, —NCS_2^-).

The xanthate group itself, —OCS_2—, is planar in all xanthates.

The dipole moments determined for some metal xanthates and dithiocarbamates are listed in Table 5.7.

It is not often realized that once precipitated, the metal xanthates are subject to further changes. They may undergo various reactions, particularly in open systems like flotation circuits, where pH, oxygen supply, and modifying agents (such as sulfides or cyanides) are continually changing with time.

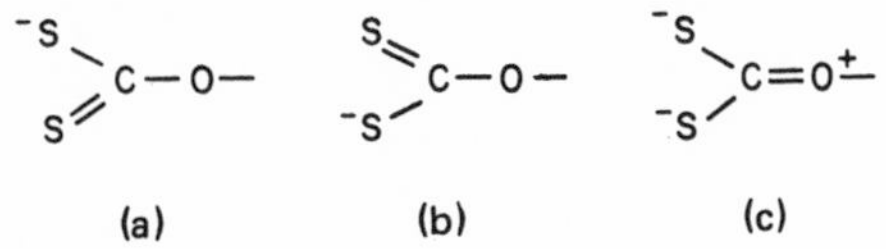

Figure 5.15. Three resonance structures of a dithiocarbonate group.

(a)   (b)   (c)

Table 5.7. Dipole Moments (debye units)

Substance	Xanthates (ethyl–amyl)	Dithiocarbamates (propyl–butyl)	Reference
Ferric	2.5	1.19	Thorn and Ludwig (1962)
Arsenic	1.5	4.5–5.0	Thorn and Ludwig (1962)
		(1.5–1.8)	Malatesta (1940)
Zinc and copper	1.4–1.8	1.2–1.5	Malatesta (1940)
Nickel	1.7		Kakovsky (1957)
Ethyl dixanthogen	1.8		O'Brien and Hyslop (1975)

Initial work on the stability of zinc and lead xanthates in alkaline solutions appears to have been done by McLeod (1934) in the Montana School of Mines and then by Thunaes and Abell (1935) and Fleming (1952). A study by Sheikh (1972) dealt with the stability of Cu, Pb, and Fe xanthates in the pH region 3–12 at 25–60°C. It showed that the decomposition of these xanthates leads to the formation of the corresponding metal sulfides, oxides, and carbonates at rates which are greatly dependent on pH and temperature; see, for example, the decomposition of cuprous xanthate (Figure 5.16). When additional interacting species such as HS^- or CN^- are present, these rates of decomposition become still faster. This behavior is of considerable importance in the selective separation of complex metal sulfides such as those of Cu–Ni and Cu–Mo bulk concentrates or even the bulk Pb–Zn concentrate obtained initially by flotation. The regrinding of these bulk concentrates is followed by chemical treatment which destroys not only the collector coating but also some of the excess of abraded iron.

There are two characteristics of all thio compounds which complicate their chemistry still further:

1. In analogy with the increased solubility of some sulfides in excess HS^- solutions (due to the formation of ionized MS_4^{+z-8}, where $M^{z+} = Ge^{4+}$, As^{5+}, Mo^{4+}, Sb^{5+}, W^{6+}, Hg^{2+}), there is a strong tendency of some metal thio compounds, particularly xanthates and dithiocarbamates, to form complexes with elemental or ionized sulfur species. These sulfur-rich products, referred to in the chemical literature [Coucouvanis (1970)] as perthio derivatives, undergo reversible oxidation and irreversible reduction reactions.

2. The alkyl thio ions (xanthates, dithiocarbamates, dithiophosphates, etc.) constitute ligands which form not only "simple" transition

metal–ligand complexes but also a large number of mixed-ligand complexes, involving additional ligands (adducts) such as NO, CN, pyridine, amines, etc. Metal xanthates in particular tend to react strongly with bases, giving a number of complexes with pyridine and other amines.

Some of the thio complexes are discussed in greater detail in Section 10.4.2.

As examples of such complexing, it may be useful to mention here those likely to be encountered in MoS_2 flotation:

1. The complex $Mo_2O_3(X)_4$ (X stands for xanthate ion, $ROCS_2^-$) reacts with CN^- whereby xanthate is progressively displaced by the CN group; also, the same $Mo_2O_3(X)_4$ reacts with H_2S and RSH (thiols) to give a xanthate of molybdenum sulfide, $MoS_2(X)$, and a substituted molythiol xanthate, $MoS(RS)X$. Another example may be that from the pyrite–xanthate system.

2. FeX_3 (ferric xanthate) is formed only under acid conditions (pH < 3.5), but once the pH changes to pH 5–6, one or more of the xanthate groups in the precipitate begins to be displaced by an OH group, giving $Fe(OH)X_2$ [Sheikh and Leja (1977)].

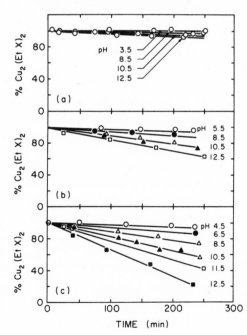

Figure 5.16. The results of estimates (from IR spectroscopic data) of the percentage of cuprous ethyl xanthate precipitate which decomposes with time when agitated in solution of a given pH at 25°, 35°, and 60°C. [From Sheikh (1972), with the permission of the author.]

The chemistry of xanthates and dithiocarbamates has been reviewed by Coucouvanis (1970) and the structural aspects of thio compound complexes by Eisenberg (1970) and Jorgensen (1962).

5.3. *Dithiocarbamates*

Half-amide derivatives of dithiocarbonic acids, known as dithiocarbamates, are formed whenever a primary or secondary amine (in an alcoholic or aqueous solution) is reacted with CS_2 in the presence of an alkali metal hydroxide to form the corresponding salt:

$$RNH_2 + CS_2 + NaOH \rightarrow RNHCSSNa + H_2O \qquad (5.8)$$
$$\text{using a primary amine}$$

$$R_1R_2NH + CS_2 + NaOH \rightarrow R_1R_2NCSSNa + H_2O \qquad (5.9)$$
$$\text{using a secondary amine}$$

With the exception of alkali and alkaline earth metals, all other metallic salts of dithiocarbamates are insoluble, and this characteristic has been extensively utilized in the analysis of metals by colorimetric, titrimetric, gravimetric, and potentiometric methods. Since the mid-1930, some dithiocarbamate salts have found application in agriculture as fungicides, insecticides, and herbicides and in clinical medicine. A comprehensive review of all aspects of dithiocarbamate chemistry, properties, and utilization was published by Thorn and Ludwig (1962).

Dialkyl dithiocarbamates are well known as potential collectors for sulfides (of equal selectivity to that of xanthates), yet they are rarely employed as flotation agents because of much higher production costs. Chatt *et al.* (1956) examined the infrared spectra of dithiocarbamates and came to the conclusion that, in contrast to xanthates, the contribution of the fully ionic resonance structure analogous to (c) in Figure 5.15 is approximately equal to the contributions of the other two resonance structures in the case of dialkyl dithiocarbamates. The two C—S bonds, as determined in crystal structure evaluations, appear to be equivalent (bond length C—S = 1.68–1.71 Å). The higher dipole moments of As dithiocarbamates as compared to those of xanthates (Table 5.7) substantiate this conclusion. The 1438-cm^{-1} band was assigned to the C$=$N stretching mode, while the 1250–1350 cm^{-1} region was assigned to the C—N single bonds. For different metal dithiocarbamates, some shifts are observed in the C—N and C$=$N bands toward lower or higher wavenumbers, depending on the effects of the electron density perturbations involved in each case [Evtushenko (1965), Nakamoto *et al.* (1963), Coucouvanis and Fackler (1967)]. Different assignments of individual bands to different groupings of bonds were made by various workers, and no uniform interpretation appears to have been accepted.

The UV spectra of soluble dithiocarbamates show two strong bands in the regions 275–290 nm and 245–260 nm, both with $\log \varepsilon \cong 4$, and a weaker absorption in the region 325–360 nm ($\log \varepsilon$ about 1.8) [Thorn and Ludwig (1962, pp. 52–56), Koch (1949), Janssen (1960), Zahradnik (1958)]. Infrared spectra of various dithiocarbamate derivatives were published by Randall *et al.* (1949) and Chatt *et al.* (1956), and the different assignments of the 1450–1590 cm^{-1}, 1610 cm^{-1}, and 1435–1415 cm^{-1} absorptions are discussed by Thorn and Ludwig (1962).

The stability of dithiocarbamate solutions in various pH ranges appears to be similar to that of xanthate: In acid solutions a decomposition to amine and CS_2, following a first-order reaction rate, was found to occur. The half-life of decomposition was found by Bode (1954) to range from 0.3 sec at pH 2 to 35 days at pH 9. Eckert (1957) gives the following order of decreasing solubility of metal diethyl dithiocarbamates: Mn^{2+}, As^{3+}, Zn^{2+}, Sn^{2+}, Fe^{3+}, Cd^{2+}, Pb^{2+}, Bi^{2+}, Co^{2+}, Ni^{2+}, Cu^{2+}, Ag^+, Hg^{2+}. It is noteworthy that

1. In contrast to xanthates, cupric dithiocarbamate and not cuprous is the stable salt of copper, and

2. Both cupric and zinc salts exist in two forms in equilibrium as an ionized complex and a neutral salt [Thorn and Ludwig (1962, p. 46)], i.e., $Cu(dtc)^+$ and $Cu(dtc)_2$ (where dtc denotes the dialkyl dithiocarbamate group) and $Zn(dtc)^+$ and $Zn(dtc)_2$.

The UV absorption spectrum of *isopropyl ethyl thionocarbamate* (principal constituent of the Dow Z-200 collector-type reagent) was discussed by Jones and Woodcock (1969). A sharp absorption peak at 242 nm is the only characteristic feature in the spectrum of the pure compound; this absorption peak shifts to 247 nm for a solution of this compound in cyclohexane. The spectrum of Z-200 shows, in addition to 242-nm absorption, a minor shoulder at 282 nm caused by a small proportion (15–20%) of the diethyl dithiocarbamate. When an alkyl xanthate and Z-200 are being used jointly in flotation, the solution can be analyzed spectroscopically for both constituents using an addition of concentrated acid to decompose the xanthate present, as shown in Figure 5.17.

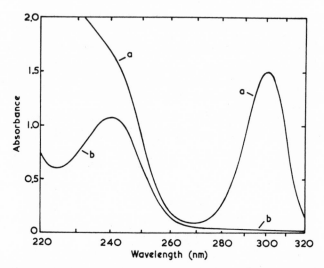

Figure 5.17. The UV spectrum of Z-200 obtained from a mixture containing ethyl xanthate and Z-200 after the xanthate is decomposed by an addition of concentrated HCl. (a) Spectrum of a solution containing 12 ppm Z-200, 14 ppm potassium ethyl xanthate, and 50 ppm sodium thiosulfate; (b) spectrum of the same solution after an addition of a few drops of concentrated HCl. [Jones and Woodcock (1969), with permission from the Australasian Institute of Mining and Metallurgy, Parkville, Victoria, Australia.]

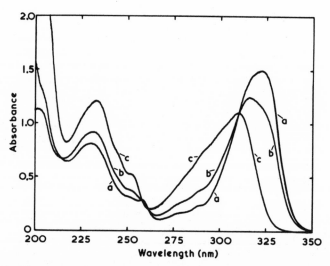

Figure 5.18. UV spectra of solutions containing 10 ppm mercaptobenzothiozole at (a) pH 1.6, (b) pH 6.8, and (c) pH 11.7. [Jones and Woodcock (1973), with permission from the Canadian Institute of Mining and Metallurgy, Montreal, Quebec, Canada.]

The absorption spectra of *mercaptobenzothiazole*[†] at three different pHs (1.6, 6.8, and 11.7) are shown in Figure 5.18, and the molar absorptivities ε of the various peaks were estimated [see Jones and Woodcock (1973)] as

$$\varepsilon230 = 12,600\text{-}13,900; \quad 14,200 \pm 200 \text{ liter mol}^{-1}\text{cm}^{-1}$$
$$\varepsilon235 = 12,600\text{-}25,100; \quad 20,550 \pm 200 \text{ liter mol}^{-1}\text{cm}^{-1}$$
$$\varepsilon310 = 18,650\text{-}21,500; \quad 18,650 \pm 100 \text{ liter mol}^{-1}\text{cm}^{-1}$$
$$\varepsilon323 = 22,500\text{-}29,500; \quad 25,150 \pm 250 \text{ liter mol}^{-1}\text{cm}^{-1}$$

The same authors discussed the effects of various ionic and nonionic additives to flotation systems on the spectroscopic determination of MBT (mercaptobenzothiazole).

5.4. Mercaptans (Thiols)

Mercaptans are the simplest of thio compounds which can be utilized as flotation collectors only if longer than decyl since, owing to the generally pungent odor, the short-chain thiols are unlikely to become adopted in circuits open to the atmosphere. Mercaptans are much less soluble in water than the corresponding alcohols. The S—H bond is weaker than the C—H and has a lower stretching absorption viz. $2600\text{-}2500 \text{ cm}^{-1}$ compared to $2960\text{-}2840 \text{ cm}^{-1}$ for the C—H bond. Thiols are more acidic than alcohols and have lower dipole moments (0.9 debye unit for H_2S vs. 1.8 debye unit for water). Heavy metal salts of thiols (Pb, Hg, Cu, Cd, Ag) are insoluble in water. Thiols oxidize readily to disulfides. The chemistry of thiols is discussed in Gaudin (1957, pp. 190–195) and in Reid (1962, Chapter 2).

5.5. Dithiophosphates

The sodium (or potassium) salts of dialkyl dithiophosphoric acids can be prepared by reacting phosphorous pentasulfide with appropriate alcohols and sodium (potassium) hydroxide [Kosolapoff (1958)]. These salts are employed in flotation under the proprietary name Aerofloats (American Cyanamid Co.). Busev and Ivanyutin (1958) wrote an extensive review of metal dithiophosphates (Pb^{2+}, Hg^{2+}, Fe^{3+}, Ni^{2+}, Cu^{2+}, As^{3+}, Cd^{2+}, Bi^{3+}) and their absorption spectra. Aqueous solutions of Mn^{2+}, Fe^{2+}, Co^{2+}, Zn^{2+},

---

[†] *2-Mercaptobenzothiazole* (formula on p. 208) is formed on heating thiocarbanilide with CS_2 and sulfur. Thiocarbanilide is formed on reacting aniline, $C_6H_5NH_2$, with CS_2 in alcohol. Both reagents are used as rubber accelerators and in flotation.

and Ga^{3+} do not form precipitates with diethyl dithiophosphate ion, though complexes of Co and Zn dithiophosphate can be extracted into organic solvents from such aqueous solutions [Jorgensen (1962, p. 134)]. Ferric dithiophosphate, dark purple in color, is unstable in aqueous solution and tends to form ferrous ion and an oxidized sulfur-bridged bis-dithiophosphate oxidation product (analogous to dixanthogen). Dialkyl dithiophosphoric acids are stronger acids than xanthates (with $pK \sim 0$) and are very stable, not decomposing in an acid aqueous solution. The dithiophosphates of heavy metals dissolve in water much more readily than the corresponding xanthates, as seen from a comparison of Table 5.8(a) with Table 5.6.

Stamboliadis and Salman (1976) extended the range of the solubility product determinations and suggested a model to predict the hydrocarbon chain length required for the formation of an insoluble metal dithiophosphate at a selected pH value. Their data are shown in Table 5.8(b). A comparison of data for cupric dithiophosphates in Tables 5.8(a) and 5.8(b) shows discrepancies of several orders in magnitude, attesting to difficulties in obtaining reliable measurements of solubilities, as already commented upon in connection with Figure 5.11.

The metal dithiophosphates give ultraviolet absorption bands which are characteristic of the central metal ion, forming octahedral $(M^{3+})S_6$ or square planar $(M^{2+})S_4$ structures. Iwasaki and Cooke (1957) reported that solutions of sodium alkyl dithiophosphates have only one UV absorption band at 227 nm, with a molar extinction coefficient of 3700 liter mol^{-1} cm^{-1}.

The infrared spectra (400–1400 cm^{-1}) of various metal dithiophosphates [Zn, Cu, Ni, Co, Mn, presented by Shopov *et al.* (1970), or Cd, Ni, Hg, Ag, Bi, Pb, Cu, Fe, presented by Ripan *et al.* (1963)] showed that both the nature of the metal and the structure of the alkyl (or aryl) radical strongly affect the frequency of the P—S, P=S, and P—O—C absorptions. The P—O—C (alkyl) stretching frequency was found to be at $\sim$1000–1050 cm^{-1} [Rockett (1962)] for phosphates and phosphonates and at $\sim$972–1012 cm^{-1} for the thio derivatives of the same compounds. The branch-chain compounds absorb at a lower wavenumber, 968–972 cm^{-1}, than the straight-chain ones. The assignments of the P=S group range from 720–765 cm^{-1} [Daasch and Smith (1951)] to 800–845 cm^{-1} [Thomas (1957)]. The P=S structure in the highly ionic salts (K, Na) absorbs at a higher wavenumber (shifted by 30–50 cm^{-1}) than it does in the salts with more covalent bonds between the sulfur and the metal (Zn, Cd, Cu, Ni, and H in the alkyl dithiophosphoric salts and acids; unlike the C=S in the thio-carbonates, the P=S structure is preserved in the covalent salts). The P—S— (metal) stretching frequency occurs at 510–575 cm^{-1}, as evaluated by McIvor *et al.* (1956).

Table 5.8a. Solubility Products of Some Alkyl and Aryl Dithiophosphates[a]

Dithiophosphates	Lead	Cadmium	Zinc	Silver	Copper	Thallium
Methyl	5×10^{-11}	—	—	4.2×10^{-16}	8×10^{-16}	—
Ethyl	7.5×10^{-12}	1.50×10^{-10}	1.5×10^{-2}	1.2×10^{-16}	1.4×10^{-16}	1.2×10^{-5}
Propyl	6.0×10^{-14}	4.0×10^{-11}	—	—	—	—
Butyl	6.1×10^{-16}	3.8×10^{-13}	—	5.0×10^{-19}	2.2×10^{-18}	—
Isoamyl	4.2×10^{-18}	—	1.0×10^{-8}	4.0×10^{-20}	—	—
Phenyl	6.8×10^{-16}	1.50×10^{-11}	—	7.4×10^{-19}	—	—
o-Cresyl	1.8×10^{-17}	1.5×10^{-13}	—	3.8×10^{-19}	7×10^{-18}	3.1×10^{-9}
p-Cresyl	7.7×10^{-18}	6.8×10^{-14}	—	7×10^{-20}	4.1×10^{-19}	4.3×10^{-10}
m-Cresyl	—	—	—	5×10^{-20}	2×10^{-19}	2.9×10^{-10}

[a] After Kakovsky (1957).

Table 5.8b. pK_{sp} of Some Metal Dithiophosphates[a]

| Name | Number of carbons | Metal Cation | | | | | |
| | | Cu^{2+} | | Fe^{3+} | | Ni^{2+} | |
		Observed	Estimated	Observed	Estimated	Observed	Estimated
Dimethyl	1	10.87	10.77	[b]	−10.07	[b]	−20.54
Diethyl	2	14.38	14.56	[b]	1.85	[b]	−8.21
Dipropyl	3	16.71	16.77	[b]	8.82	[b]	−1.00
Dibutyl	4	18.48	18.35	13.67	13.77	[b]	4.12
Diamyl	5	[b]	19.60	[b]	17.66	[b]	8.13
Dihexyl	6	[b]	20.70	21.00	20.75	11.33	11.33
Diheptyl	7	[b]	21.60	[b]	23.40	[b]	14.07
Dioctyl	8	[b]	22.30	25.55	25.70	16.45	16.45
Diennyl	9	[b]	23.00	[b]	25.72	[b]	18.53
Didecyl	10	[b]	23.70	[b]	29.54	[b]	20.42

[a] After Stamboliadis and Salman (1976).
[b] Not measured or not forming.

Goold (1972) studied the nature of the reaction products formed by four thio compounds [namely, ethyl dithiophosphate, diethyl dithiocarbamate, sodium mercaptobenzothiazole (Reagent 404), and isopropyl ethyl thionocarbamate (Z-200)] with different sulfide minerals. The latter comprised the minerals that have been floated by the individual reagents; thus, for dithiophosphates (DTP) they constituted molybdenite [Wada and Majima (1963), Chander and Fuerstenau (1975)], pyrite [Mitrofanov and Kushnikova (1958), Popov *et al.* (1971), Kakovsky and Komkov (1970), Fuerstenau *et al.* (1971)], chalcocite and chalcopyrite [Fuerstenau *et al.* (1971), Huiatt (1969), Mitrofanov and Kushnikova (1959)], and galena [Simard *et al.* (1950)]. Goold (1972) reported a number of characteristic IR peaks of the various products obtained in this study: for the oxidation product of diethyl dithiophosphate, $(DTP)_2$, 630 and 495 cm^{-1}; for the cuprous salt $Cu(DTP)$, a single 615-cm^{-1} peak; for the cupric salt $Cu(DTP)_2$, 640, 530, and 510 cm^{-1}; for $Pb(DTP)_2$, 655, 570, and 530 cm^{-1}; for $Fe(DTP)_3$, 640 and 530 cm^{-1}; and for the Mo compound $Mo(DTP)_x$, 640, 615, and 527 cm^{-1}.

CLASS II: NON-THIO IONIZABLE SURFACTANTS

The characteristic features of all reagents belonging to the non-thio ionizable surfactant class (that is, alkyl carboxylates, sulfates and sulfonates, phosphates, amines, etc.) are the following:

1. A proneness to dissociation, ionization, and hydrolysis, the extent of which is governed by the pH of aqueous solutions
2. A pronounced lowering of the air/water and oil/water interfacial tensions in dilute solutions (less than $\sim 10^{-2} M$)
3. A tendency toward the formation of colloidal aggregates, *micelles*, when the concentration exceeds a certain value known as the *critical micelle concentration* (denoted by CMC) and the temperature exceeds a certain minimum level, called the *Krafft point*
4. Solubilization of insoluble hydrocarbons and surfactants within micelles

5.6. *Dissociation and Hydrolysis*

Figure 5.19 and 5.20 show schematically the pH ranges of dissociation (ionization) for different surfactants. In the case of carboxylates it is seen that below $\sim$ pH 4 the predominant species in solution is the undissociated

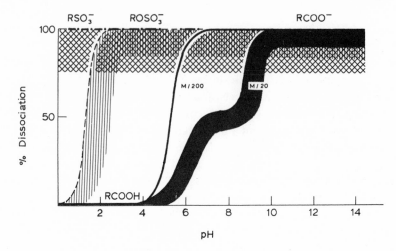

Figure 5.19. Dissociation curves (schematic) of fatty acids, RCOO⁻H⁺, sodium alkyl sulfates, $ROSO_3^-Na^+$, and sodium alkyl sulfonates, $RSO_3^-Na^+$, vs. pH.

fatty acid molecule, whereas above $\sim$pH 10 the carboxylate molecules are fully dissociated as ions:

$$RCOOH + Na^+OH^- \rightleftarrows RCOO^- + Na^+ + H_2O \qquad (5.10)$$

Dissociation constants quoted for fatty acids are of the order of 1.8–1.1

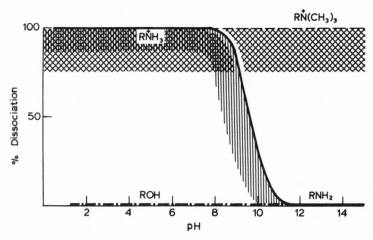

Figure 5.20. Dissociation curves (schematic) of amines (unsubstituted RNH_2 (forming aminium ions, RN^+H_3)), substituted trimethyl ammonium ions, $RN^+(CH_3)_3$ (the positive charge in the amine-type ions is mostly on nitrogen), and alcohols (no dissociation regardless of pH) vs. pH .

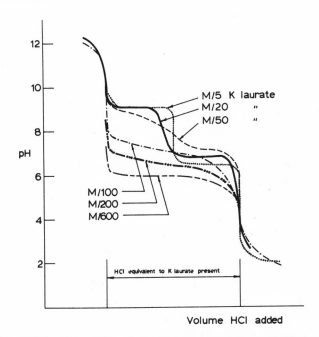

Figure 5.21. Addition of an acid to potassium laurate solutions of different concentrations graphed vs. pH shows the formation of two inflection points for concentrations greater than $M/50$. The development of a "knee" in the titration curve denotes the formation of an acid soap complex (RCOOH · RCOO⁻). [From Rosano *et al.* (1966), in *J. Coll. Interface Sci.* **22**, with the permission of Academic Press, New York.]

$\times 10^{-5}$ for acetic to nonanoic acids (compared with 3.4×10^{-7} for carbonic acid). These values mean that 50% dissociation should occur at pH $\sim$5 and 99% dissociation at pH $\sim$7. Recent titrations of Rosano *et al.* (1966) (Figure 5.21) and of du Rietz (1965) showed that for the longer-chain carboxylates (used in flotation) the dissociation of concentrated solutions is shifted to the pH range 4–10.

Depending on the concentration of, for example, potassium laurate solution ($C_{11}H_{23}COOK$), the shape of the titration curve gradually changes from that of a weak monoprotic acid (like acetic acid, with one inflection point) for low concentrations, less than $M/100$, to that of a diprotic acid—with two inflection points—for concentrations higher than $M/100$; see Figure 5.21. ($M/50$ corresponds approximately to CMC for potassium laurate.) This unusual behavior is due to the formation of an intermediate species, an *acid soap* (RCOOH · RCOO⁻), containing a 1:1 complex of ionized and un-ionized carboxylate. Concentrated solutions of potassium laurate at pH > 12 are perfectly clear (if the temperature is above their

Krafft point), but on addition of HCl the acid soap crystallizes out at pH 9.2, and, on further titration with HCl, this is gradually converted to crystals of the insoluble lauric acid. In solutions of pH > 7.5 the acid soap is in equilibrium with the fully ionized species, while at lower pH, the acid soap and the un-ionized lauric acid are in equilibrium. Below pH 4, the molecules of lauric acid are present in prepondering numbers, while above pH 9.5–10 the fully ionized carboxylate species predominate. Thus, an addition of an alkali carboxylate soap to a flotation circuit at a given pH (or an addition of an equivalent quantity of the carboxylic acid) will result ultimately in the same quantities of un-ionized acid, acid soap, and fully ionized carboxylate species, as indicated by the corresponding titration curve (for the given concentration), provided enough time is allowed for an equilibrium to be established.

Similar intermediate species, analogous to the acid soap in the case of carboxylates, are expected for straight-chain alkyl sulfates, sulfonates, and unsubstituted amines within their appropriate pH ranges of half-dissociation. As will be seen later, the regions of pH where the two species, ionized and un-ionized, occur in appreciable quantities for an ionizable surfactant (within the S curves of dissociation) coincide, for each surfactant, with the region of their dual action as a collector and a frother. Outside such a region the surfactant can act either as a collector only or as a frother only. When it acts as a collector, it always requires an addition of another surfactant acting as a frother for an effective flotation to be achieved. And when it acts as a frother only, there is no floatability until an appropriate collector is used.

Alkyl sulfates are completely dissociated above pH ~ 3; below this pH, hydrolysis of sulfate progressively takes place with the formation of an alcohol and a bisulfate:

$$ROSO_3^-Na^+ + H^+OH^- \rightarrow ROH + SO_2 \Big\langle\begin{matrix}OH\\ONa\end{matrix} \qquad (5.11)$$

Desseigne (1945) found that in boiling 1 N HCl all alkyl sulfates hydrolyze completely within 1–2 hr. At lower temperature and higher pH (lower HCl concentration) the rate of hydrolysis varies as shown in Table 5.9.

In contrast to alkyl sulfates, alkyl sulfonates are stable at low pH; that is, they are not hydrolyzed by strong acids; on the other hand cyclic (aryl) sulfonates do decompose on heating with a dilute HCl under pressure at 150°–200°C.

Unsubstituted amine salts are completely dissociated in acid and neutral

Table 5.9. Rate of Hydrolysis (%/hr) with Relation to Temperature and pH[a]

Concentration of HCl	40°C	60°C	80°C	100°C
$N/1000$ (pH 3)	0.002	0.023	0.27	2.2
$N/1000$ (pH 2)	0.019	0.29	3.4	27.0
$N/10$ (pH 1.05)	0.11	1.7	19.8	
N (pH 0.1)	0.34	5.4	61.5	

[a] After Desseigne (1945).

pHs (Figure 5.20). As the solution is made progressively alkaline, hydrolysis takes place with the formation of undissociated amine:

$$RNH_3{}^+Cl^- + Na^+OH^- \rightarrow RNH_2 + Na^+Cl^- + H_2O \qquad (5.12)$$

The dissociation constant for lauryl (C_{12}) amine is given by Ralston *et al.* (1944) as 2.4×10^{-11}, indicating that at pH 10.6 the *dissolved* amine is half dissociated. Quaternary ammonium salts are stable in both acid and alkaline media and completely dissociated; aqueous solutions of sodium or potassium hydroxide do not decompose the quaternary ammonium salts (however, a methanolic solution of potassium hydroxide reacts with the halide salts giving quaternary ammonium hydroxide). Alkyl alcohols are not affected by pH and remain undissociated in the whole pH range O–14 at room temperature (Figure 5.20).

5.7. *Surface Tension Versus Concentration Relationship*

5.7.1. *Aqueous Solutions of Non-Thio Ionizable Surfactants*

The changes in surface tension γ with increasing concentrations for solutions of sodium dodecyl (lauryl) sulfate, dodecyl amine hydrochloride, myristyl trimethyl ammonium bromide, and sodium laurate at pH 11 are shown in Figure 5.22. The γ–concentration behavior of these solutions is described by the Gibbs adsorption equation:

$$-d\gamma = mRT\Gamma_s \, d\ln c \qquad (2.83)$$

as discussed in Section 2.9. However, the Gibbs equation applies only to

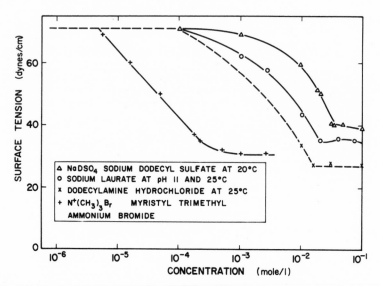

Figure 5.22. Surface tension vs. concentration curves for sodium dodecyl sulfate, sodium laurate, dodecylamine hydrochloride, and myristyl trimethyl ammonium bromide. [Based on data from Smolders (1961), NaDSO₄; Nutting and Long (1941), sodium laurate; Hoffman *et al.* (1942), dodecylamine hydrochloride; Venable and Nauman (1964), myristyltrimethyl ammonium bromide.]

the range of concentrations below that for which γ levels off and attains a constant value thereafter. This concentration is known as the *critical micelle concentration* (CMC). It denotes the concentration starting with which no further adsorption of surfactant species occurs at the air/water interface with continued additions of surfactant to the solution. Instead of adsorbing at the interface, the surfactant species added to the solution in excess of CMC are all utilized in forming colloidal aggregates within the bulk of solution. Such aggregates are known as *Micelles*. At the CMC the slope of the γ–concentration curve, $d\gamma/d \ln c$, changes suddenly from a high value to zero; this sudden falloff signifies that there is no *further change* in the quantity of surfactant adsorbed at the air/water interface; that is, $\Gamma_s = $ constant. The value of Γ_s is that which was attained on approaching γ_{CMC} from lower concentrations, just before γ_{CMC} became constant[†].

[†] There are indications that the transition at CMC is not as abrupt as theoretically defined. Some micelles begin to appear in solutions below CMC, and there is also a very slight decrease in the $\gamma_{\text{CMC}} = $ constant value. For example, Elworthy and Mysels (1966) found for sodium lauryl sulfate solutions a change from $\gamma_{\text{CMC}} = 39.5$ dyn/cm at CMC $= 8 \times 10^{-3} M$ to 37 dyn/cm at $8 \times 10^{-2} M$.

Table 5.10a. Critical Micelle Concentrations (CMCs) of Selected Sodium Alkyl Carboxylates, Sulfates, and Sulfonates[a]

Alkyl group	Carboxylate, Na		Sulfate, Na			Sulfonate, Na		
	MW	CMC (mol/liter)	MW	CMC (mol/liter)	Temperature	MW	CMC (mol/liter)	Temperature
C_7H_{15}	166.2	3.5×10^{-1}						
C_8H_{17}	194.2	9.4×10^{-2}	232.2	1.3×10^{-1}	25°C	216.2	1.6×10^{-1}	25°C
C_9H_{19}	222.3	2.6×10^{-2}	260.3	3.3×10^{-2}	25°C	244.3	4.2×10^{-2}	25°C
$C_{10}H_{21}$	250.3	6.9×10^{-3}	288.3	8.2×10^{-3}	25°C	272.3	9.8×10^{-3}	25°C
$C_{11}H_{23}$	278.4	2.1×10^{-3}	316.4	2.0×10^{-3}	25°C	300.4	2.5×10^{-3}	40°C
$C_{12}H_{25}$	306	1.8×10^{-3}	344.4	2.1×10^{-4}	25°C	328.4	7.0×10^{-4}	50°C
$C_{13}H_{27}$			372.5	3.0×10^{-4}	40°C	356.5	7.5×10^{-4}	57°C
$C_{14}H_{29}$								
$C_{15}H_{31}$								
$C_{16}H_{33}$								
$C_{17}H_{35}$								
$C_{18}H_{37}$								

[a] From Mukerjee and Mysels (1971).

Table 5.10b. Critical Micelle Concentrations, Krafft Points, and Aggregation Numbers for Several n-Alkyl Non-thio Ionizable Surfactants with and without Added Inorganic Electrolyte

Surfactant	Solution	CMC		T°C	Krafft point (T°C)	Aggregation number[a]
		(Moles/liter)	Surface tension (dyn/cm)			
Sulfates						
$C_{12}H_{25}SO_4Na$	Water	8.2[b]	40[b]	25	16,[c] 21[d]	80
	0.01 M NaCl	5.6[e]		21		89
	0.10 M NaCl	1.5[e]		21		112
$C_{14}H_{29}SO_4Na$	Water	2.0[e]		40	30[c]	
$C_{16}H_{33}SO_4Na$	Water	0.58[e]		40[f]	45,[c] 43[d]	
Sulfonates						
$C_{12}H_{25}SO_3Na$	Water	9.7[g]	39.5[g]	40	31.5[h]	54
	0.004 M NaCl				33.0[h]	
	0.008 M NaCl				33.5[h]	
$C_{14}H_{29}SO_3Na$	Water	2.5[e]		40	39.5[i]	80
$C_{16}H_{33}SO_3Na$	Water	0.8[e]		50	47.5,[i] 48[j]	

Primary amines

$C_{12}H_{25}NH_2 \cdot HCl$	Water	$14.8^{e,k}$	28^l	30	28^m	56
	0.01 N NaCl	$11.3^{e,k}$		30		
	0.02 N NaCl	$8.9^{e,k}$		30		
$C_{16}H_{33}NH_2 \cdot HCl$	Water	0.85^a		55		
$C_{18}H_{37}NH_2 \cdot HCl$	Water	0.55^k		60	56^m	

Quaternary amines

$C_{12}H_{25}N(CH_3)_3Br$	Water	14.5^e		25		50
	0.001 M KBr	12.0^n				
	0.1 M NaBr	4.5^e		25		
	0.5 M NaBr	2.0^e		25		
$C_{14}H_{29}N(CH_3)_3Br$	Water	3.5^o	40°	30		75
	0.005 M NaBr	0.42^o	37°	30		
$C_{16}H_{33}N(CH_3)_3Br$	Water	0.92^e		25	21^p	
	0.001 M KBr	0.48^n				

[a] From data compiled by Shinoda et al. (1963).
[b] Elworthy and Mysels (1966).
[c] From data compiled by Rosen (1978).
[d] Raison (1957).
[e] From data compiled by Mukerjee and Mysels (1971).
[f] In this instance, the temperature for CMC measurement is below the available Krafft point values. This apparent contradiction occurs quite frequently in the literature.
[g] Bujake and Goddard (1965).

[h] Tartar and Cadle (1939).
[i] Tartar and Wright (1939).
[j] Murray and Hartley (1935).
[k] Ralston et al. (1949).
[l] Hoffman et al. (1942).
[m] Ralston et al. (1941).
[n] Conner and Ottewill (1971).
[o] Venable and Nauman (1964).
[p] Adam and Pankhurst (1946).

In general, the micelle-forming surfactants behave as indicated in Figure 5.23.

1. For a homologous series of surfactants possessing the same polar group, an increase in chain length $R_1 < R_2 < R_3$ causes a progressive lowering of γ for a given concentration.

2. An addition of a neutral salt causes an overall surface tension lowering of the γ–concentration curve. As discussed in Section 2.9, this decrease in surface tension is associated with a progressive change of the parameter m in equation (2.84) from $m = 2$ to $m = 1$ as the additions of neutral salt increase.

3. The presence of another surfactant in minute quantities (as an impurity) causes the appearance of a minimum around the CMC. Theoretically, such a minimum in the γ–concentration curve is contrary to the Gibbs adsorption process for one surfactant species. Until 1944, numerous attempts to correlate experimental data (showing γ–concentration curves with a minimum) and theoretical predictions of the Gibbs adsorption had been unsuccessful. Careful experiments by Miles and Shedlovsky (1944) proved that traces of alcohols (which could not be detected by chemical analyses) were usually left from the manufacturing and purification procedures used for surfactants and were wholly responsible for the appearance of such minima in γ–concentration curves.

Nowadays, the existence of a minimum in the γ–concentration curve around γ_{CMC} is employed as a most sensitive test for detecting the presence of an impurity–surfactant in a given surfactant sample. If a mimimum is detected, it denotes the presence of an impurity surfactant; when on subsequent purifications of the given surfactant the initially detected minimum in the γ–concentration curve disappears, the surfactant is then considered to be of high purity. The criteria of purity and the techniques for purification of surfactants (and their solutions) are extensively discussed by Mysels and Florence (1970, 1973).

Tables 5.10(a) and (b) list some experimental values of critical micelle concentrations of ionizable surfactants, indicating quantitatively the effects of hydrocarbon chain length and salt additions. A theoretical basis for the preceding effects on γ_{CMC} was established by Shinoda et al. (1963), who treated the formation of micelles in terms of the Derjaguin–Landau–Vervey–Overbeek (DLVO) theory and derived the following relationship:

$$\ln(\text{CMC}) = \frac{(n-1)\omega}{kT} - K_g \ln C_i + K_i \qquad (5.13)$$

where ω is the change in cohesive energy per one methylene group, $-CH_2-$, on transfer from a hydrocarbon environment to an aqueous medium; $\omega = 1.08kT$ at 25°C; $n - 1$ is the number of methylene groups in the hydrocarbon chain of surfactant; K_g is an experimental constant relating to the effective electrical work of introducing a charge into the surface of a micelle; $K_g = 0.4 \div 0.6$; C_i is the concentration of gegenions in gram equivalents per liter; k is Boltzmann's constant; T is the absolute temperature; and K_i is a constant. Previously, the dependence of CMC value on chain length of the alkyl group had been correlated by an empirical relationship of the following type:

$$\log_{10}(CMC) = A - Bn \qquad (5.14)$$

where n is the number of carbons in the homologous chain length and A and B are constants: A is a constant for a given polar group but varies with temperature and concentration of gegenions; $B \cong \log_{10} 2$ for a single polar group, regardless of its nature, but varies with the number of polar groups in the surfactant molecule; see Table 5.11. The value of B may be interpreted as the standard free energy of micelle formation per one CH_2 group, giving $-\Delta G°/CH_2 = 800$ cal.

Either of the equations (5.13) or (5.14) enables the evaluation of the CMC for solutions of a given surfactant with n carbons in its alkyl group.

Table 5.11. Values of Parameters A and B in Equation (5.14) Relating the CMCs with the Type of Surfactant[a]

Surfactant	Temperature (°C)	A	B
R_nCOOK	25	1.63	0.290
R_nCOOK	45	1.74	0.292
R_nSO_3Na	40	1.59	0.294
R_nSO_3Na	50	1.63	0.293
R_nSO_4Na	45	1.42	0.295
R_nNH_3Cl	45	1.79	0.296
$R_nN(CH_3)_3Br$	60	1.77	0.292
Alkyl glucoside	25	2.64	0.530

[a] After Shinoda *et al.* (1963).

5.7.2. *Surface Tension–Concentration Relationships for Surfactants Which Do Not Form Micelles*

When a solute is weakly surface active at the air/water interface, it changes the surface tension of water only to a negligible extent. Examples of such solutes are KOH in Figure 2.6(a) and alkyl xanthates in Figure 5.24. Other surfactants such as fatty acids (in acid pH, less than 4), alcohols, ethers, etc., lower the surface tension of their aqueous solutions more strongly than xanthates until their respective saturation points are reached. When a solution becomes supersaturated, it no longer represents a homogeneous system but forms a colloidal or noncolloidal in size dispersion of an insoluble surfactant phase (solid or liquid) in a saturated solution of surfactant. Figure 5.24 shows schematically a number of γ–concentration curves for

1. Negatively adsorbing inorganic salts [see also Figure 2.6(a) and (b)]
2. Soluble thio compounds such as xanthates
3. Alkyl fatty acids (pH < 4) or alcohols of decreasing solubility and increasing chain length

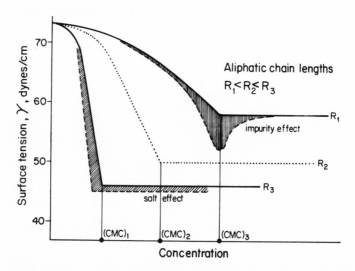

Figure 5.23. Schematic relationships between surface tension and concentration for micelle-forming surfactants of increasing chain lengths $R_1 < R_2 < R_3$ for a surfactant containing small amounts of another surface active species (impurity effect) and for a surfactant solution to which a quantity of a neutral inorganic salt has been added (salt effect).

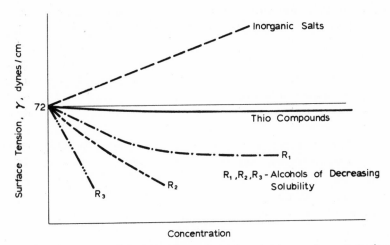

Figure 5.24. Schematic relationships between surface tension and concentration for solutions of most solutes which do not form micelles: inorganic salts (specifically, structure-forming salts), thio compounds, alcohols, ethers, etc.

The effect of increasing the hydrocarbon chain (for surfactants with the same polar group) is analogous to that described earlier for micellar surfactant systems (Figure 5.23); i.e., a greater lowering of surface tension is obtained for the same concentration.

Traube (1891) observed that the surface tension of dilute aqueous solutions in fatty acids, alcohols, and ethers is lowered by the same amount for successive members in a homologous series whenever identical concentrations are compared. Thus, an addition of one CH_2 group (methylene group) to a hydrocarbon chain increases the work of desorption by 710 cal/mol, due mainly to the entropy term ($\cong RT \ln 3$), if the molecules are oriented parallel to the aqueous surface. The effect of the polar group of the surfactant becomes evident for shorter-chain homologues ($n < 7$) and for different substrates, giving other values for the energy of adsorption per methylene group, for example, 345 cal/mol at the formamide/vapor interface. Also, for more concentrated solutions, interactions between adjacent molecules of surfactants and different orientations of molecules with respect to the surface affect the value of Traube's coefficient ($RT \ln 3$).

Extra lowering in surface tension values is also caused by dissociation of the surfactants. The measurements of surface tension of potassium laurate solution during its titration with acid have shown that within the pH range of acid soap formation (Figure 5.21) the surface tension of solutions became practically constant and at a lowest value between 30 and

40 dyn/cm (depending on concentration), whereas solutions of neutral soap, at pH 10, or those containing mainly lauric acid, at pH 4, showed much higher surface tensions, 55–70 dyn/cm. This effect of lowering within the pH range 4–10 is similar to the surfactant–impurity effect, indicated in Figure 5.23, and is caused by coadsorption of the two (undissociated and dissociated) surfactant species at the air/water interface.

5.7.3. *Ideal and Nonideal Mixtures of Soluble Surfactants, Synergistic Effects*

For dilute aqueous solutions of lower fatty acids, Szyszkowski (1908) derived an empirical equation:

$$\frac{\gamma}{\gamma_W} = 1 - \beta \ln\left(\frac{1+c}{\alpha}\right) \qquad (5.15)$$

where γ is the surface tension of solution, γ_W is the surface tension of water, β is the constant characteristic of a given homologous series of surfactant, α is the *capillary constant*, defined as

$$\alpha^2 = \frac{2\gamma}{\Delta\varrho g} \qquad (5.16)$$

$\Delta\varrho$ is the difference in density between the liquid and the gas, c is the concentration of solute, and g is gravitational acceleration.

Equation (5.15) was found applicable over a considerable range of concentrations to *mixtures* of two solutes behaving *ideally* in solution.

Thus, two solutions each of which has the same surface tension γ do not alter their surface tension when they mix *ideally*. If a solution of γ_1 is mixed with a solution of γ_2, the resulting surface tension (if the mixture behaves *ideally*) is obtained directly from equation (5.15), since the ratio of concentrations to the capillary constants is the same: $c_1/\alpha_1 = c_2/\alpha_2 = c_{12}/\alpha_{12}$ (where the subscript 12 denotes the mixture).

However, many mixtures of long-chain surfactants behave *nonideally*, giving a greater lowering of surface tension of their mixture–solutions than that predicted by equation (5.15). The *nonideal* behavior of two or more surfactants in solution is due to molecular *interactions*, consisting of interactions between the respective hydrocarbon chains (due to van der Waals bonding) and also of interactions between the respective polar groups (these may comprise ion–ion, dipole–ion, dipole–dipole, or hydrogen bonding). Such interactions between two or more surfactants may occur not only in bulk solution but preferentially at interfaces: air/liquid, liquid/liquid, and solid/liquid. The surfactants thus interacting are said to act *synergistically* (or to show synergistic effects) toward each other. Evidence of synergistic action at the air/solution interface is provided by the impurity effect (Figure 5.23), in the extra lowering of surface tension in the acid soap region of carboxylates and amine–aminium ion of alkyl amines, and in a pronounced lowering of the whole γ–conc. curves for mixed solutions (as compared with γ–conc. curves of individual components) as in Figure 5.25(a) and (b).

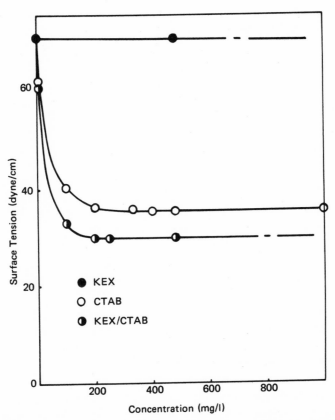

Figure 5.25a. Surface tension vs. concentration curves for solutions of individual surfactants: KEX (potassium ethyl xanthate). $C_{16}TAB$ (cetyl trimethyl ammonium bromide), and their 1:1 mixtures, indicating the occurrence of molecular interactions between the two species. [From Buckenham and Schulman (1963), in *Trans. AIME* **226**, with the permission of the American Institute of Mining and Metallurgical Engineers, Littleton, Colorado.]

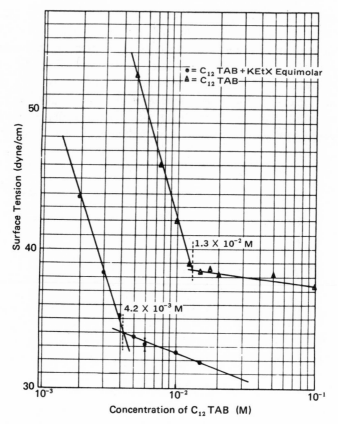

Figure 5.25b. Surface tension curves showing conjoint action of potassium ethyl xanthate and dodecyl trimethyl ammonium bromide ($C_{12}TAB$). [From Finkelstein (1968), in NIM Report No 437, with the permission of National Institute for Metallurgy, Randburg, S. Africa.]

Regardless of the character of γ–conc. curves for individual components, the resultant γ–conc. curve for synergistically interacting systems is always the lowest one. Thus, even the *salt effect* in Figure 5.23 can be considered as evidence of a synergistic action, due to a reduction in repulsion which otherwise exists between the ionic polar groups.

Using radioactive (S^{35}) alkyl sulfates and with radioactive (C^{14}) alcohols and carboxylates, Shinoda (1963) determined relative concentrations of surfactants in different mixtures at the air/solution interface. The results are shown in Table 5.12. In a mixed solution containing lauryl phenyl sulfonate and lauryl sulfate, there are 35–50 times more lauryl phenyl sulfonate species adsorbed at the air/water interface than lauryl sulfate.

Table 5.12. Selective Adsorptivities of Compounds (1) Against
Compounds (2)[a]

Surfactants		Ratio (1)/(2)	
(1)	(2)	Determined	Calculated
C_{12}—$C_6H_4SO_3Na$	C_{12}—OSO_3Na	35–50	25
C_{12}—$C_6H_4SO_3Na$	C_{11}—COOK	180–280	140
C_{15}—COOK	C_{11}—COOK	50–70	55
C_{15}—COOK	C_{13}—COOK	5–8	6
C_{12}—OH	C_{12}—SO_4Na	55 ± 5	
C_{10}—OH	C_{12}—SO_4Na	~ 8	
C_8—OH	C_{12}—SO_4Na	~ 1	

[a] After Shinoda *et al.* (1963).

However, in the case of octyl alcohol and lauryl sulfate (last row of results) there is no preferential adsorption of either surfactant—their mixture behaves ideally. Shinoda suggested that—to a first approximation—the relative adsorptivity of these soluble surfactants is proportional to the inverse ratio of their CMC values. He defined the surface activity of a soluble surfactant at the air/water interface as the reciprocal of its CMC.

Above CMC, a mixture of two surfactants forms *mixed micelles*, whose composition is directly proportional to the concentration of the second (additional) surfactant, according to a general relationship [Shinoda *et al.* (1963)]:

$$(CMC)_{12} = (CMC)_1 - KC_2 \qquad (5.17)$$

where $(CMC)_{12}$ is the CMC of the mixture of surfactants 1 and 2, $(CMC)_1$ is the CMC of solution 1 (alone), C_2 is the concentration of surfactant 2, and K is the relative micelle-forming ability of surfactant 2 with respect to surfactant 1.

The two surfactants in a synergistic system may be either purposely mixed or may coexist as in a natural product; they may represent a product–surfactant and an unreacted feed–surfactant, or the second surfactant may be formed in solution from the original surfactant species by dissociation, oxidation, or reactions with counterions.

As can be seen from Figure 5.26, the extent of lowering the CMC value of a given surfactant such as potassium laurate, due to the presence of alcohols of different chain length, is caused by progressively smaller concentrations of alcohols as their chain length increases.

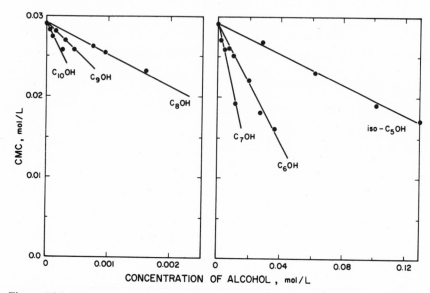

Figure 5.26. The effect of alcohols of different chain lengths on the CMC of potassium laurate (dodecanate) at 10°C. [From Shinoda *et al.* (1963), *Colloid Surfactants*, with the permission of Academic Press, New York.]

5.7.4. *Dynamic Surface Tension*

An adsorption of surface active species at any interface is time-dependent. Equilibrium may be reached within a fraction of a second, in several seconds, or in hours. Different surfactants exhibit fairly wide variations in the kinetics of adsorption; for example, short-chain alcohols adsorb at the air/water interface in fractions of a second, alkyl amines require several seconds, and long-chain sulfates require several hours. Very fast adsorption processes at the liquid/gas interface are best studied by an oscillating jet method [Addison (1943), Sutherland (1954), Hansen *et al.* (1968), Thomas and Potter (1975)]. A bubble pressure technique is used for dynamic surface tension determinations in the 1–300 sec time range [Adam and Schutte (1938), Burcik (1950), Austin *et al.* (1967), Finch and Smith (1972)].

Figure 5.27 shows the dynamic surface tensions of dodecyl amine acetate solutions at concentrations from 4×10^{-5} to $40 \times 10^{-5} M$ [Finch and Smith (1972)]. It is seen clearly that the lower the concentration, the longer is the time of establishing the *equilibrium* value of surface tension. Figure 5.28 shows the effect of pH on the kinetics and the magnitude of the *equilibrium* surface tension established for solutions of one concen-

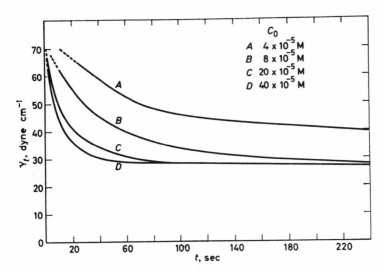

Figure 5.27. Dynamic surface tension of dodecyl amine acetate solutions at pH 9.85 and four different concentrations: $C_0 = 4 \times 10^{-5}$ M to $C_0 = 40 \times 10^{-5}$ M. [From Finch and Smith (1972), in *Trans. IMM* **81**, with the permission of the Institution of Mining and Metallurgy, London.]

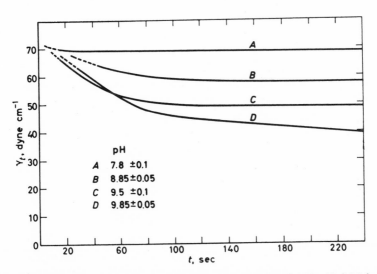

Figure 5.28. The effect of pH on the dynamic surface tension of 4.08×10^{-5} M dodecyl amine acetate solution. [From Finch and Smith (1972), in *Trans. IMM* **81**, with the permission of the Institution of Mining and Metallurgy, London.]

tration, $4.08 \times 10^{-5} M$ dodecyl amine acetate. (At pH 9.85 and 10.8 a precipitate in the form of surface *scum* was observed.)

Synergistic effects due to interaction between ionized and nonionized surfactants, $RN^+H_3-RNH_2$, are apparent, giving distinctly lower surface tensions in the pH range of extensive dissociation (curve D for pH 9.85).

Kulkarni and Somasundaran (1975) have shown that there exists a definite correlation between the dynamic surface tension and the flotation behavior of hematite when potassium oleate is used. They found that optimum flotation occurred at the pH range 7–9, that is, in the region of an appreciable acid soap formation. Smith and Lai (1966) established a similar correlation using an indirect technique, namely contact angle and flotation, for an alkyl amine–quartz system at the pH range 9–12.

5.8. *Krafft Point and Cloud Point*

Solutions of long-chain surfactants (that are capable of forming micelles) will show the typical behavior of a sudden change in the slope of their surface tension vs. concentration curves (Figure 5.23), only when their temperature is above a certain minimum temperature, called the Krafft point. The relationship between the solubility of such surfactants and the temperature is shown in Figure 5.29. It is seen that below a certain temperature T_K, appropriate for each surfactant, the solubility of the surfactant is negligible, and then it increases extremely rapidly once the temperature is raised above T_K. Micelles are not detected in solutions at temperatures below T_K but begin to form once T_K is exceeded, and their formation is responsible for the rapid increase in solubility. Temperatures T_K are called *Krafft points* of respective surfactants (after F. Krafft, who discovered this behavior in surfactant solubility in 1895). As indicated in Figure 5.29(c), Krafft points depend in a linear manner on the chain length and also somewhat on the polar group of the surfactant and the type and the ionic strength of counterions present in solution.

For flotation purposes, the awareness of the value of the Krafft point is useful in preparing highly concentrated stock solutions of long-chain *ionized* surfactants; only above their respective Krafft points will these concentrated solutions be homogeneous and easily dispersed throughout the flotation pulp. Such easy dispersion will greatly reduce the time required for conditioning of the collector-acting surfactant with the given solids.

Solutions of surfactants possessing an alkyl group attached to a nonionic polyoxyethylene and polyoxypropylene group can also form micelles

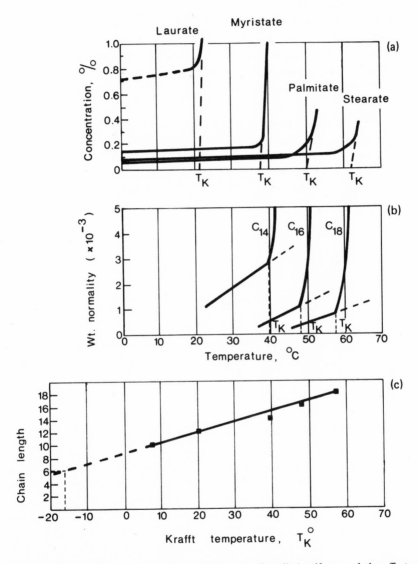

Figure 5.29. Solubility vs. temperature relationships for alkyl sulfates and the effect of chain length on the Krafft point. Based on data from (a) sodium alkyl carboxylates, Preston (1948); (b) sodium alkyl sulfates, Tartar and Wright (1939); and (c) effect of chain length on the Krafft point for sodium alkyl sulfates, Raison (1957).

in appropriate concentrations. However, micellar solutions of such non-ionic surfactants become cloudy or turbid when heated above a certain temperature and often separate into two phases. The temperature of clouding, or the *cloud point*, is not as specific as the Krafft point, but, in contrast to the latter, it indicates a *sharp decrease in solubility* with a rise in temperature. (More details are given in Section 5.14.1.)

5.9. *Micelle Structures*

Numerous properties of ionized surfactant solutions, namely surface tension, detergency, conductivity, osmotic pressure, etc., exhibit sudden changes when their respective CMC values are exceeded and aggregation of the long-chain molecules into micelles occurs. Extensive work by Per

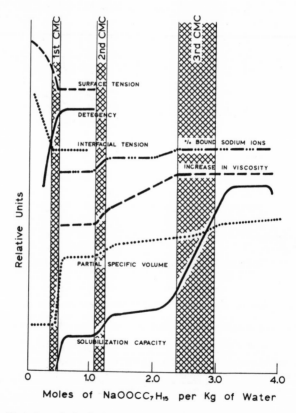

Figure 5.30. Variation of physical properties of solutions with concentration of sodium octanoate. [After Ekwall (1962) and Preston (1948).]

SPHERICAL MICELLE

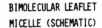

BIMOLECULAR LEAFLET
MICELLE (SCHEMATIC)

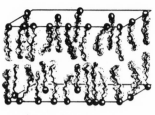

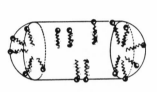

CYLINDRICAL MICELLE

BIMOLECULAR LEAFLET
MICELLE WITH
SOLUBILIZED HYDROCARBON
OIL

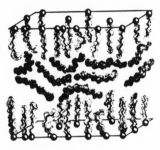

Figure 5.31. Schematic models of micelles: spherical, cylindrical, and bimolecular leaflet.

Ekwall and his associates, carried out on carboxylate soaps, showed that in yet more concentrated solutions, exceeding 1 *M* or so, some further sudden changes occur, and these suggest the existence of a second and a third CMC region (Figure 5.30). Ekwall *et al.* (1972) reviewed all aspects of aggregation in surfactant systems, micelle formation, as well as liquid crystal (mesophase) formation.

Of the many structures proposed initially for micelles, viz., spherical [Hartley (1936)], lamellar [McBain (1939)], cylindrical or rod shape [Debye and Anacker (1951)], disk-shaped (Harkins and Mittelmann (1949)), and hexagonal lamellae (Philippoff (1950)), at least three structures are now recognized as the most likely to exist at differ concentrations.

Spherical micelles are assumed as the most likely to exist between the first and the second CMC regions, cylindrical between the second and third CMC regions, and lamellar (bimolecular leaflet or *sandwich* type) at high concentrations, above the third CMC region (Figure 5.31).

In each of these micellar structures the hydrocarbon chains are assumed to be not as fully adlineated (aligned) as they are in the crystalline form of these surfactants. The extent of van der Waals bonding between the chains within the various micelle structures does not reach the ultimate value attainable theoretically. In spherical and cylindrical micelles the distribution of molecules resembles that in liquid paraffin. In lamellar micelles, the chains are less random but still not fully adlineated. The increase in the heat of solution to form micelles from a pure solid is about 0.9 kcal/methylene group, as compared with 1 kcal/methylene group for the heat of fusion of paraffin chain compounds.

A critical review of micelle formation, and particularly the thermodynamics of micellization, was presented by Pethica (1960); another treatment of association equilibria was prepared by Mukerjee (1967).

X-ray data indicate that in concentrated solutions the micelles consist of layers, the thickness of which corresponds to two fully extended surfactant molecules placed end to end (hence bimolecular leaflet micelle). Concentrated solutions of surfactants are capable of *solubilizing*[†] fairly large quantities of an insoluble hydrocarbon oil; X-ray data indicate that during this process of solubilization the thickness of the lamellar layers increases proportionately to the volume of oil being solubilized.

5.10. Carboxylates

5.10.1. Carboxylic Acids

Fatty acids and carboxylate alkali soaps are used widely as joint flotation collectors–frothers, mostly in the processing of nonsulfide minerals. Caproic acid ($C_5H_{11}COOH$) is the shortest homologue employed, while unsaturated octadecanoic (C_{18}) acids, such as *oleic* (*cis*-9), *linoleic* (*cis*-9 and *cis*-12), and *linolenic* (*cis*-9, *cis*-12, and *cis*-15) acids and their alkali soaps, are the longest ones which are still sufficiently soluble in aqueous systems to be useful in flotation.

Commercial fatty acids are derived from animal and vegetable fats and oils (listed in Table 5.13) and from petroleum fractions by hydrolysis, distillation, hydrogenation, etc. The *commercially pure* C_8, C_{10}, C_{12}, C_{14}, C_{16}, and C_{18} normal fatty acids have 90–92% content of the main component, the rest being represented by the neighboring homologues. In addition to these commercially pure products, there are various mixtures produced, e.g., lauric 70–myristic 30, caprylic 55–capric 45, etc., containing the stated components within ±2–3%. The exceptions are the so-called *single-*, *double-*,

---

[†] The topic is treated extensively in two publications, McBain and Hutchinson (1955) and Elworthy *et al.* (1968).

Table 5.13. Fats and Oils and the Composition of the Fatty Acids Obtained by Hydrolysis[a]

Fat or oil	Composition of fatty acids (%)						% Other components
	Myristic	Palmitic	Stearic	Palmitoleic	Oleic	Linoleic	
Vegetable fats							
Coconut	17–20	4–10	1–5		2–10	0–2	5–10 caprylic, 5–11 capric, and 45–51 lauric acids
Palm	1–3	34–43	3–6		38–40	5–11	3–4 butyric, 1–2 caproic, 1 caprylic, 2–3 capric, and 2–3 lauric acids
Butter	7–9	23–26	10–13	5	30–40	4–5	
Animal fats							
Lard	1–2	28–30	12–18	1–3	41–48	6–7	2 of C_{20} and C_{22} unsaturated fat acids
Tallow	2–3	24–32	14–32	1–3	35–48	2–4	
Vegetable oils							
Nondrying							
Castor	0–1	0–1	1–4		0–9	3–7	80–92 ricinoleic acid
Olive		5–15		0–1	69–84	4–12	
Peanut		6–9	2–6	0–1	50–70	13–26	2–5 arachidic and 1–5 tetracosanoic acids
Semidrying							
Corn	0–2	7–11	3–4	0–2	43–49	34–42	
Cottonseed	0–2	19–24	1–2	0–2	23–33	40–48	
Drying							
Soybean	0–1	6–10	2–4		21–29	50–59	4–8 linolenic acid
Sunflower		10–13			21–39	51–68	
Linseed		4–7	2–5		9–38	3–43	25–58 linolenic acid
Animal oils (Marine)							
Whale	4–6	11–18	2–4	13–18	33–38		11–20 C_{20} and 6–11 C_{22} unsaturated acids
Fish (sardine)	6–8	10–16	1–2	6–15			70 unsaturated acids, C_{16}–C_{22} having one to six double bonds

[a] After Noller (1965).

and *triple-pressed stearic acids* which are *mixtures of 52% palmitic* and 38–44% *stearic acids*, the pressing being used to remove the entrained oleic acid. Also, the *crystallized white oleic* or *crystallized red oils* are mixtures containing 73–80% oleic, 6–4% linoleic, and 4–3% palmitic acids and small amounts of other saturated and unsaturated acids. *Tall oil* is a mixture of fatty acids and resin acids (abietic acids) obtained as a by-product in the kraft process of wood pulp treatment (see Section 5.15.3). Its composition is approximately 40% oleic, 30% linoleic, and 5% palmitic acids and the rest rosins (*abietic* acids).

Fatty acids of hydrocarbon chain length C_{12}–C_{18} (saturated and unsaturated) represent the most important raw materials for all soap-forming and other synthetic surfactants. The alkali soaps of C_{14}–C_{18} fatty acids constitute the oldest and best known washing agents.

Many of the normal carboxylic acids are known under common names, which are listed in Table 5.14.

Some of these common names are used in naming the appropriate alkyl radical C_nH_{2n+1}, regardless of the polar group to which the alkyl

Table 5.14. Common Names and Their Derivation for Selected Carboxylic Acids

Number of all carbon atoms in acid molecule	Name of acid	Derivation of name
C_1	Formic	L. *formic*, ant
C_2	Acetic	L. *acetum*, vinegar
C_3	Propionic	Gr. *proto*, first; *pion*, fat
C_4	Butyric	L. *butyrum*, butter
C_5	Valeric	Valerian root (L. *valere*, to be strong)
C_6	Caproic	L. *caper*, goat
C_8	Caprylic	L. *caper*, goat
C_9	Pelargonic	Pelargonium
C_{10}	Capric	L. *caper*, goat
C_{12}	Lauric	Laurel
C_{14}	Myristic	*Myristica fragrans* (nutmeg)
C_{16}	Palmitic	Palm oil
C_{18}	Stearic	Gr. *stear*, tallow
C_{20}	Arachidic	*Arachis hypogaea* (peanut)
C_{22}	Behenic	Behen oil
C_{26}	Cerotic	L. *cera*, wax

group is attached, e.g., the —C_9H_{19} group is known as pelargonic. Detailed treatments of manufacture and chemistry of fatty acids are given in books by Ralston (1948) and Markley (1960, 1961).

Aliphatic dicarboxylic acids (with two polar groups) are used either directly as depressants or wetting agents or as starting materials for the production of derivatives. The members of the saturated dicarboxylic acid homologue series are as follows: first, C_2, oxalic acid, HOOC—COOH; then C_3, malonic acid, HOOC—CH_2—COOH; C_4, *succinic acid*, HOOC—$(CH_2)_2$—COOH [the sulfonated esters of which, *see below*, are employed (under the proprietary name Aerosols; American Cyanamid Co.) as filtration aids, leaching aids, and depressants]; C_5, glutaric acid, HOOC—$(CH_2)_3$—COOH; C_6, adipic acid, HOOC—$(CH_2)_4$—COOH. The aryl dicarboxylic acid of importance to the flotation reagent industry is *phthalic acid* (benzene-1, 2-dicarboxylic acid):

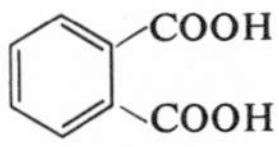

Phthalic acid

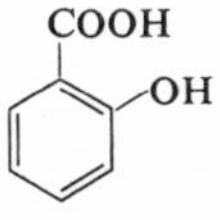

Salicylic acid

The ortho isomer of the hydroxy acid, known as *salicylic acid*, is prepared by the action of CO_2 on sodium phenoxide at 150°C. It is the starting material for the manufacture of aspirin but also for a number of derivatives used as chelating agents.

The unsaturated butenedioic acids exist in two geometric forms, *maleic*, the *cis* form

$$HC—COOH$$
$$HC—COOH$$

and *fumaric*, the *trans* form

$$HC—COOH$$
$$HOOC—CH$$

Aerosols are made by esterification of succinic or maleic acids with an alkyl alcohol, ROH, followed by heating the ester with a concentrated aqueous solution of bisulfite, $NaHSO_3$. The resulting products have the structure

$$CH—COO—R$$
$$NaSO_3—CH—COO—R$$

Aerosols

and aerosol OT is a sodium dioctyl sulfosuccinate, aerosol MA is a sodium dihexyl sulfosuccinate, aerosol AY is a sodium diamyl sulfosuccinate, and aerosol OS is a sodium isopropyl naphthalene sulfonate, while aerosol 18 is a disodium *N*-octadecyl sulfosuccinamate:

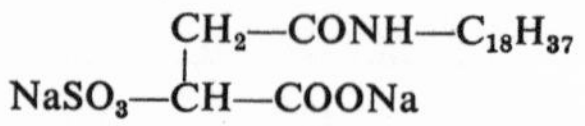

$$CH_2—CONH—C_{18}H_{37}$$
$$NaSO_3—CH—COONa$$

Aerosol 18

The hydroxy acids (one hydroxyl group and one carboxylic acid group) important to flotation are represented by *glycolic acid*,

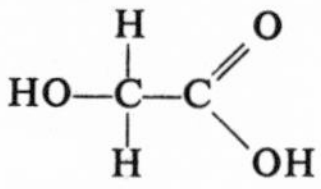

$$HO—C—C \overset{O}{\underset{OH}{\diagup}}$$

Glycolic acid

and its thio derivative, *thioglycolic acid*,

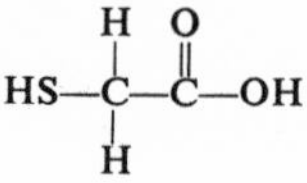

$$HS—C—C—OH$$

and by *lactic acid*,

$$CH_3—C—COOH$$
$$OH$$

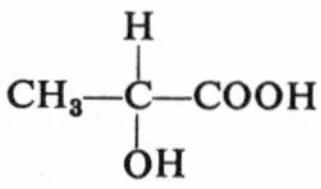

and the unsaturated *ricinoleic acid*,

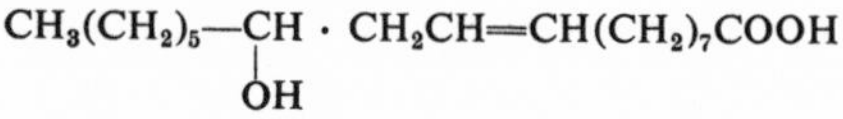

$$CH_3(CH_2)_5—CH \cdot CH_2CH=CH(CH_2)_7COOH$$
$$OH$$

The last is the main constituent of *castor oil*, which is the starting material for the production of a sulfated derivative, i.e., *Turkey-red oil*. (Sulfation of ricinoleic acid takes place either at the hydroxyl group or at the double bond but not twice in the same molecule.)

Next in increasing number of polar groups are hydroxydiacids, e.g., *tartaric acid* (dihydroxysuccinic acid),

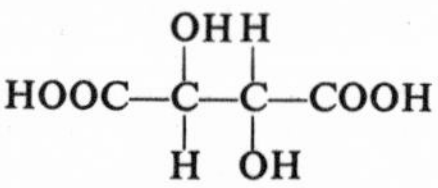

$$HOOC—C—C—COOH$$
$$H \quad OH$$

and hydroxy triacids, e.g., *citric acid,*

$$\underset{\underset{\text{COOH}}{\overset{\overset{\text{OH}}{|}}{\text{HOOC—CH}_2\text{—C—CH}_2\text{—COOH}}}{}$$

2-hydroxy-1,2,3-propanetricarboxylic acids

Tartaric and citric ions form six-valent negatively charged chelating complexes with metal ions, such as Cu^{2+}, decreasing the cupric ion concentration below that required for the precipitation of cupric hydroxide. The two hydroxyl groups of each tartrate enter into the chelate complex in the cupritartrate;

$$\begin{bmatrix} \text{}^-\text{OOC—CH—O} & \text{OCH—COO}^- \\ | & \phantom{:[Cu^{II}]:} & | \\ \text{}^-\text{OOC—CH—O} & \text{OCH—COO}^- \end{bmatrix} + 6\text{Na}^+$$

and in the cupricitrate complex,

$$\begin{bmatrix} \text{}^-\text{OOCCH}_2 & & & \text{CH}_2\text{COO}^- \\ & C & :[Cu^{II}]: & C & \\ \text{}^-\text{OOCCH}_2 & & & \text{CH}_2\text{COO}^- \end{bmatrix} + 6\text{Na}^+$$

one carboxyl and one hydroxyl group of each citrate participate in complexing.

Solubilities of fatty acids in water were determined by Ralston and Hoerr (1942b); Table 5.15 lists some of their values for normal acids at different temperatures and at a single temperature, 25°C, for the three unsaturated acids. The surface tensions of the liquid acids and the interfacial tensions are listed in Table 5.16.

5.10.2. *Metal Carboxylates (Soaps)*

Using a rather complicated potentiometric titration technique (involving metallic ion solutions, carboxylic acid solution, alkali solution, and determination of pH in addition to that of potential), du Rietz (1958) determined a large number of solubility products, given in Table 5.17.

From a study of the kinetics of metal carboxylate precipitation, using light scattering, Matijevic *et al.* (1966) found that the solubility products (as determined by a potentiometric titration technique in a heterogeneous

Table 5.15. Solubilities of Fatty Acids in Water (pH not stated in source)[a]

Acid	Grams of Acid Per 100 g of Water				
	0°C	20°C	30°C	45°C	60°C
Caproic, C_6	0.864	0.968	1.019	1.095	1.171
Caprylic, C_8	0.044	0.068	0.079	0.095	0.113
Capric, C_{10}	0.0095	0.015	0.018	0.023	0.027
Lauric, C_{12}	0.0037	0.0055	0.0063	0.0075	0.0087
Myristic, C_{14}	0.0013	0.0020	0.0024	0.0029	0.0034
Palmitic, C_{16}	0.00046	0.00072	0.00083	0.0010	0.0012
Stearic, C_{18}	0.00018	0.00029	0.00034	0.00042	0.00050

Solubilities of unsaturated acids at 25°C (in water, pH unknown)

Oleic	0.6×10^{-6} mol/liter
Linoleic	0.5×10^{-6} mol/liter
Linolenic	0.2×10^{-6} mol/liter

[a] After Ralston and Hoerr (1942b).

Table 5.16. Surface Tensions of Liquid Acids and Interfacial Tensions of Water–Acid

Fatty acid chain length	m.p. (°C)	Water/acid interfacial tension (dyn/cm) at 75°C	Surface tension of liquid acid (dyn/cm)				
			20°C	30°C	40°C	60°C	80°C
C_6	—3.4	2.1	27.5	26.6	25.8	24.2	22.5
C_8	16.7	5.8	28.3	27.7	26.6	24.6	22.8
C_{10}	31.6	8.0	28.2	27.8	27.3	26.3	24.7
C_{12}	44.2	8.7	—	—	—	27.3	25.8
C_{14}	53.9	9.2	—	—	—	28.4	26.6
C_{16}	63.1	9.2	—	—	—	—	27.5
C_{18}	69.6	9.5	—	—	—	—	28.6

Table 5.17. Exponents of Solubility Products (at 20°C) for Various Fatty Acids[a] and Soaps,[a]

M^{n+}	Palmitate	Stearate	Oleate	Elaidate	Linoleate	Linolenate
H^+	11.9	12.7	10.9	11.2	11.0	11.5
Na^-	5.1	6.0	—	—	—	—
Ag^+	11.1	12.0	9.4	10.5	9.5	9.4
Pb^{2+}	20.7	22.2	16.8	—	—	16.9
Cu^{2+}	19.4	20.8	16.4	—	—	17.0
Zn^{2+}	18.5	20.0	15.1	—	—	15.3
Fe^{2+}	15.6	17.4	12.4 $(15 \cdot 5)^b$	—	—	—
Mn^{2+}	16.2	17.5	12.3	—	—	12.6
Ca^{2+}	15.8	17.4	12.4 $(15 \cdot 6)^b$	14.3	12.4	12.2
Ba^{2+}	15.4	16.9	11.9	—	11.8	11.6
Mg^{2+}	14.3	15.5	10.8 $(15 \cdot 2)^b$	—	—	10.9
Al^{3+}	27.9	30.3	25.5	—	—	25.5

[a] After du Rietz (1958, p. 428). Solubilities of undissociated acids assuming $P^K_{diss} = 4.7$.
[b] Nephelometric determinations of Fuerstenau and Miller (1967) are in parentheses.

system) may be an order of magnitude, or more, higher than those indicated by the process of precipitation in a homogeneous system. A high-turbidity region has been encountered by these authors for systems of equivalent concentrations of Ca^{2+} and oleate ions, exceeding the solubility product, $K_{sp} = 4 \times 10^{-13}$, for calcium oleate by more than one order of magnitude.

In a study on calcite flotation, which involved the determination of solubilities of calcium carboxylate and sulfonate salts, using nephelometry, Fuerstenau and Miller (1967) obtained solubility products one to two orders of magnitude lower then those evaluated previously; see Table 5.18. Solubility products determined by Al Attar and Beck (1970) for Mg, Sr, Ba, and La salts of C_{12}–C_{18} acids are given in Table 5.19.

Dipole moments of carboxylate soaps were the subject of fairly extensive studies. Despite the fact that some long-chain metallic soaps possess fairly high dipole moments in comparison with that for water (1.7 debye units) (see Tables 5.20 and 5.21), they are readily soluble in organic solvents and are used as lubricants because of their high hydrophobic character. The results of dipole moment measurements and calculations given in Table 5.20 are from Banerjee and Palit (1950, 1952). These workers also found

Table 5.18. Solubility Products of Calcium Collector Salts at 23°C[a]

Number of Carbon Atoms	K_{sp} Calcium Sulfonate		K_{sp} Calcium Carboxylate	
	Experimental	Reported[b]	Experimental	Reported[b]
8	6.2×10^{-9}		2.7×10^{-7}	1.4×10^{-6}
9	7.5×10^{-9}		8.0×10^{-9}	1.2×10^{-7}
10	8.5×10^{-9}	1.1×10^{-7}	3.8×10^{-10}	
11	2.8×10^{-9}		2.2×10^{-11}	
12	4.7×10^{-11}	3.4×10^{-11}	8.0×10^{-13}	
14	2.9×10^{-14}	6.1×10^{-14}	1.0×10^{-15}	
16	1.6×10^{-16}	2.4×10^{-15}		1.6×10^{-16}
18		3.6×10^{-15}		4.0×10^{-18}

[a] After Fuerstenau and Miller (1967).
[b] The reported solubility products were calculated from solubilities given by Reed and Tartar (1936), Seidell (1959), and du Rietz (1958).

that the addition of a small quantity of alcohol to the metallic soap solution in benzene increases the dipole moment of Al soaps. The increase in dipole moment of Al soaps with the alkyl chain length of the carboxylate shown in Table 5.21 [after Pilpel (1963)] is fairly pronounced and may reflect the aggregation of the solute, referred to in Section 2.4.

Infrared spectroscopic studies of carboxylic acids and soaps have been carried out by a number of investigators, and the assignments of charac-

Table 5.19. Exponents of Solubility Products of Various Metal Carboxylates[a]

Carbon atoms in molecules	Mg^{2+}	Sr^{2+}	Ba^{2+}	La^{3+}
C_{12}	10.48	11.02	11.46	23.91
C_{14}	13.08	13.37	14.15	27.14
C_{16}	15.77	16.16	17.11	31.03
C_{18}	18.33	18.60	19.60	34.30

[a] After Al Attar and Beck (1970).

Table 5.20. Dipole Moments of Metallic Soaps[a]

	Experimental (debyes)	Calculated by Guggenheim method (debyes)
Divalent metals		
Zn–oleate	0.29	0
Cu–oleate	1.30	1.18
Mg–oleate	1.66	1.60
Pb–oleate	4.29	4.01
Ca–oleate	4.49	4.87
Trivalent metals		
Al–oleate	3.78	3.90
Al–dioleate	3.08	3.11
Cr–oleate	4.36	4.32
Fe–oleate	2.85	2.69
Fe–laurate	2.74	1.70
Fe–stearate	3.46	3.48

[a] After Banerjee and Palit (1950, 1952).

teristic band frequencies are well established. Figure 5.32 shows, for comparison, the spectra of lauric acid, sodium laurate, and cupric laurate. The characteristic bands are assigned as follows:

1. Very strong: 1700-cm^{-1} band due to C=O (carbonyl) stretching mode; 1465-cm^{-1} band due to CH$_2$ deformation.
2. Doublet: 1430- and 1414-cm^{-1} bands and 1300-cm^{-1} band due to C—OH stretching mode coupled with in-plane deformation of the OH.
3. A series of bands 1170–1320 cm^{-1} is due to wagging modes of the CH$_2$ group.

Table 5.21. Dipole Moments of Some Aluminum Soaps (in debyes)[a]

(C$_4$) Al–dibutanoate	3.24	(C$_{14}$) Al–ditetradecanoate	4.42
(C$_5$) Al–dipentanoate	3.44	(C$_{16}$) Al–dihexadecanoate	5.22
(C$_{12}$) Al–didodecanoate	4.06	(C$_{18}$) Al–dioctadecanoate	5.49

[a] After Pilpel (1963).

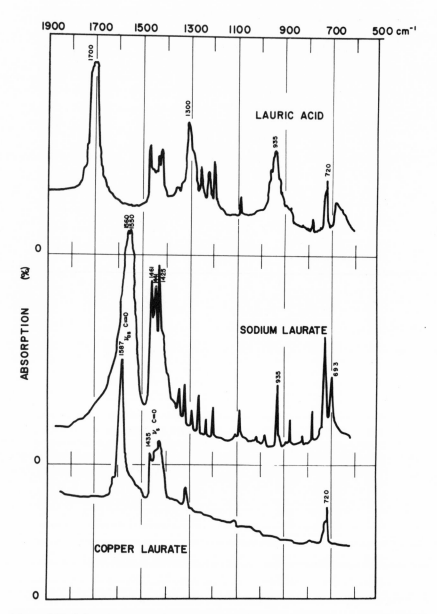

Figure 5.32. Infrared spectra of some lauryl carboxylates. [Scowen and Leja (1967), with the permission of National Research Council, Canada.]

4. A strong 1300-cm^{-1} band observed in the spectra of fatty acids is absent in the soap spectra. The symmetric ν_sC—O, asymmetric ν_{as} C—O stretching modes of the C$\cdots$O groups occur at 1400–1590 cm^{-1} for all metallic soaps; for example, some of the symmetric ν_s and the asymmetric ν_{as} stretching modes are, respectively, Cu^{2+}, 1437 and 1587 cm^{-1}; Ag$^+$, 1424 and 1516; Pb^{2+}, 1423 and 1510; Zn^{2+}, 1402 and 1536; Ca^{2+}, 1433 and 1572; Al^{3+}, 1442 and 1585; Fe^{3+}, 1423 and 1569; and Cr^{3+}, 1428 and 1535 [all data after Scowen (1966)].

5. The $\sim$720-cm^{-1} band exhibited by all compounds containing aliphatic chains is due to a CH$_2$ rocking mode. Some soaps show a strong band between 900 and 1000 cm^{-1} assigned to the stretching vibration of the metal–oxygen bonds of the carboxylate group.

5.11. *Alkyl Sulfates and Sulfonates*

Most of the commercial sulfonates and sulfates are derived from industrial by-products such as petroleum distillates (mahogany oil fractions), coal tar distillates (cresylic and naphthenic derivatives), tall oil, lignin derivatives, etc. Each raw material represents a variety of hydrocarbons, some of which have been oxidized to alcohols, ethers, and carboxylic acids or are unsaturated (contain double bonds). A subsequent sulfation or sulfonation of such raw materials results in a still greater variety of products; in many of these the sulfonate group is attached to the hydrophobic group at different sites or through an intermediate polar group (carboxyl, amide, etc.).

Pure synthetic sodium alkyl sulfates are made by reacting SO$_2$Cl$_2$ with the solution of an appropriate alcohol in chloroform at $-5°$ – $0°$C, followed by a gradual elimination of the evolving HCl, and then neutralizing with NaOH and recrystallization from acetone or petroleum ether.

Alkyl sulfonates are obtained in reacting alkyl halides with aqueous solutions of sodium sulfite at elevated temperatures ($\sim$200°C) according to the reaction

$$RCl + Na_2SO_3 \rightarrow RSO_3^-Na + NaCl$$

In comparison with alkyl sulfates, alkyl sulfonates are stable in hot acid media, owing to a direct C—S bond; calcium and magnesium salts of sulfonates have low solubility, while *lead* and *barium* sulfonate salts are very soluble in water.

Table 5.22. Critical Micelle Concentrations and Krafft Points of Alkyl
Sulfonates[a,b]

Sodium sulfonate of	Critical micelle concentration (mol/liter)	Temperature (°C), Krafft point
C_{10} hydrocarbon	0.040	22.5
C_{12} hydrocarbon	0.0098	31.5
C_{14} hydrocarbon	0.0027	39.5
C_{16} hydrocarbon	0.00105	47.5
C_{18} hydrocarbon	0.00075	57.0

[a] Source: Ralston (1948).
[b] Compare Table 5.10a and Figure 5.29c.

Table 5.22 shows the CMCs and Krafft points of some sodium alkyl
sulfonates [Ralston (1948)].

An infrared study of alkyl sulfates of various metals resulted in the
band assignments shown in Table 5.23 [Scowen (1966)].

Table 5.23. IR Band Assignments for Metal Alkyl Sulfates[a]

Metallic Cation	ν_{as} S=O	ν_s S=O	ν S—OC	ν C—OS
Na^+	1251 1220	1084	833	990–1020
K^+	1275 1215	1071	808	990–1020
Cu^+	1248 1215	1071	811	980–1020
Ag^+	1248 1216	1085	823	990–1015
Ba^{2+}	1276 1232, 1224, 1197	1077	822	950–1030
Pb^{2+}	1203 1152	1060	847	960–980
Cu^{2+}	1248 1172	1067	822	965–990

[a] After Scowen (1966).

The spectra of alkyl sulfonates are, in general, less complex than those of alkyl sulfates, owing to the absence of C—OS and S—OC linkages; their S=O stretching vibrations occur at lower frequencies than in the corresponding sulfates.

Sulfated and sulfonated long-chain ($C_n > C_{12}$) alcohols, introduced around 1930, are still one of the most widely used groups of *synthetic detergents*. Detergents are those ionized surfactants (or nonionic surfactants comprising *polymerized* dipoles) which are capable of displacing a foreign phase—a so-called "dirt"—adhering to a substrate. The substrate may be fibrous, hydrophobic, or hydrophilic in character, or it may be any organic, inorganic, or metallic solid; the dirt may also be inorganic or organic in nature and usually adheres to the substrate because of van der Waals or ionic attraction forces. The action of the detergent species is to adsorb on both the substrate and the dirt to such an extent that the outermost layer on both adsorbents, the substrate and the dirt, consists of identically ionized polar groups. It is the electrostatic repulsion between these identically charged layers on the dirt and the substrate that ultimately overcomes the van der Waals and/or ionic attraction and removes the dirt particulate from the substrate. Alkyl sulfates and sulfonates have been more thoroughly studied than other synthetic detergents during the development of scientific background for surfactant applications [see, for example, Prins (1955)]. Sulfonation of carboxylic acids and esters also resulted in the manufacture of numerous reagents other than detergents, such as wetting agents, dispersants, and lubricant additives.

In flotation, alkyl sulfates and sulfonates are used as collectors for some metal oxides [such as hematite (Fe_2O_3) and alumina (Al_2O_3); see a review paper by Fuerstenau and Palmer (1976)] and for separations of salt-type minerals [such as barite ($BaSO_4$) and scheelite ($CaWO_4$); see a paper by Hanna and Somasundaran (1976)]. Probably owing to the highly heterogeneous nature of commercial sulfated products and the characteristics of their adsorption mechanism, these reagents have not proved sufficiently selective in their flotation applications and, consequently, have not been as popular or effective as fatty acids.

5.12. *Alkyl Amines and Substituted Amines*

Cationic Surfactants, edited by Jungermann (1970), presents a most complete collection and compilation of data on the chemistry (physical and analytical) of fatty amines, cyclic (saturated and unsaturated) alkyl ammonium compounds, amines derived from petroleum constituents, and polymeric cationic surfactants. The application of cationic surfactants to flotation has so far been based on two groups: unsubstituted alkyl amines and quaternary ammonium compounds. These reagents have been used quite extensively in the selective flotation of some oxide minerals, mainly quartz and numerous silicates, and in potash flotation. Gaudin (1957) dealt with the work concerning cationic flotation agents that had been reported up to the mid-1950, Agar (1967) reviewed both the theoretical and practical aspects of cationic flotation, while the most recent reviews of silicate flotation are by Manser (1975) and Smith and Akhtar (1976).

The cationic agents are most frequently employed as salts of hydrochloric and acetic acids (or bromic and iodic acids). The polar groups of these amphipatic molecules ionize in pH ranges indicated in Figure 5.20, with a positive charge residing on the N atom:

1. $RNH_3Cl \rightleftarrows R\overset{+}{N}H_3 + Cl^-$ (primary amine chloride salt, unsubstituted amine)

2. $RN(CH_3)_3Br \rightleftarrows R\overset{+}{N}(CH_3)_3 + Br^-$ (quaternary ammonium bromide salt, substituted amine)

The substituted trimethyl ammonium bromides are frequently referred to as C_n TABs. The four N—C bonds in the quaternary ammonium salts are covalent, both atoms contributing one electron each to this type of bond. This makes the compounds stable in alkaline media, in a fully dissociated form, regardless of pH. On the other hand, the unsubstituted primary amine salts have the proton held by a coordinate bond in which one atom is the source of both electrons, and this arrangement leads to the loss of a proton in alkaline solutions with the formation of the unionized RNH_2. Thus, $RNH_3^+ + OH^- \rightleftarrows H_2O + RNH_2$.

Figure 5.20 shows the extent of dissociation of unsubstituted amines in relation to pH, with 50% dissociation at pH 10.6. Thus, for all n-alkyl amines from C_4 to C_{18} the equilibrium for the dissociation of the RNH_2 base is given by $pK_B = 3.4$. The solubility of the undissociated amine RNH_2 decreases sharply with an increase in chain length. For example, the solubilities are [after Ralston $et\ al.$ (1944)]

$$C_{10}-NH_2: \quad 510 \times 10^{-6}\ M\ (5.2 \times 10^{-4}\ M)^*$$
$$C_{12}-NH_2: \quad 17 \times 10^{-6}\ M\ (2.0 \times 10^{-5}\ M)^*$$
$$C_{14}-NH_2: \quad 1.1 \times 10^{-6}\ M\ (1.2 \times 10^{-6}\ M)^*$$

* Data in parentheses from Schubert (1967).

Despite the preceding excellent agreement, the data available on the properties of amines have to be treated, in general, with caution since it is very difficult to prepare surfactants of a specified hydrocarbon structure in a sufficiently pure state. As has been already stressed, even very small amounts of impurities profoundly affect the physical properties of the surfactants, for example, the existence of a minimum surface tension.

The most common reagent used for flotation of silicate minerals is *n*-dodecylamine acetate, whereas for flotation of KCl, *n*-hexadecylamine is used. These two hydrocarbon chain lengths appear to be sufficient to provide the requisite hydrophobic character after adsorption on the respective minerals. A systematic study of the role played by the structure of the nonpolar portion of amine collectors has been reported by Bleier *et al.* (1976). Although the concept that the structure of the collector molecule is extremely important has been around for a long time, very few quantitative data obtained under rigidly controlled conditions with sufficiently pure reagents have been reported in the past. These data are discussed more thoroughly in Chapter 8. Most of the commercial cationic reagents, as is generally the case, contain mixtures of a number of species. Sometimes such mixtures are beneficial in selective flotation; at other times they may be harmful.

Figure 5.33 shows the infrared spectra of (a) dodecylamine hydrochloride, (b) dodecylamine (undissociated, in liquid form), and (c) dodecyl alcohol.

The assignments of the main bands are as follows:

$3370 \ cm^{-1}$ and $3290 \ cm^{-1}$: ν_{as} and ν_s of hydrogen-bonded NH_2

3200–$3450 \ cm^{-1}$: stretching frequency of OH

$2455 \ cm^{-1}$ and $2670 \ cm^{-1}$: stretching vibration N^+H_2

$1600 \ cm^{-1}$ or $1613 \ cm^{-1}$: deformation vibration of N—H

1628, 1584, and $1512 \ cm^{-1}$: deformation vibration of N^+—H_3

$1070 \ cm^{-1}$: stretching vibration of C—N

$945 \ cm^{-1}$: stretching vibration of C—N^+

$793 \ cm^{-1}$: rocking mode NH_2

1465, 1390, and $720 \ cm^{-1}$: CH_2 and CH_3 deformation and rocking modes

The IR spectra of quaternary ammonium compounds, C_{18}, C_{12}, and C_{10} TAB, exhibit several sharp bands which have been assigned as shown in Table 5.24. Figure 5.34 shows the IR spectrum of C_{12} TAB.

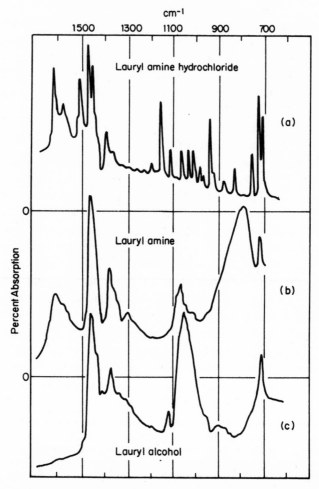

Figure 5.33. IR spectra of dodecyl (lauryl) alcohol, -amine, and -amine hydrochloride. [Scowen and Leja (1967), with the permission of National Research Council, Canada.]

In addition to alkyl amines, alkyl pyridinium salts, and quaternary ammonium salts, there are a number of amino acids (such as glycine, alanine, and valine), ethanolamines, diamines, etc., being introduced to the flotation field. *Glycine* is the lowest member of the neutral amino acids containing an equal number of amino and carboxyl groups. Its formula is H_2NCH_2COOH (that is, amino acetic acid), while *alanine* is α-aminopropionic acid, $CH_3CH(NH_2)COOH$, and valine is $CH_3 \cdot CH_2 \cdot CH(NH_2)COOH$.

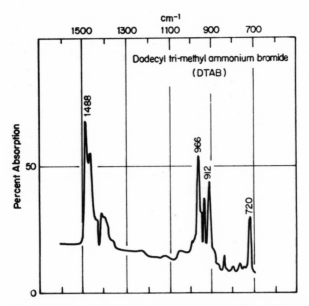

Figure 5.34. IR spectrum of a dodecyl substituted amine (i.e., dodecyltrimethylammonium bromide, DTAB). [Scowen and Leja (1967), with the permission of National Research Council, Canada.]

The reaction of ethylene oxide, $(CH_2)_2O$, with NH_3 gives *ethanolamines* (mono-, di-, and tri-):

$$CH_2\!\!-\!\!CH_2 + NH_3 \longrightarrow HO \cdot CH_2CH_2 \cdot NH_2 \qquad \text{monoethanolamine}$$
$$\diagdown\!\!O\!\!\diagup$$
$$(HO \cdot CH_2CH_2)_2 \cdot NH \qquad \text{diethanolamine}$$
$$(HO \cdot CH_2CH_2)_3 \cdot N \qquad \text{triethanolamine}$$

Salts of these ethanolamines with fatty acids are soluble both in water and hydrocarbons and are good emulsifying agents. Polymerized derivatives

Table 5.24. IR Band Assignment for C_n TABs[a,b]

Compound	ν_{as} C—N$^+$	ν_s C—N$^+$	δ C—N (of N$^+$—CH$_3$)
C_{10} TAB	978 (m), 965 (s), 950 (s)	911 (s)	1487
C_{12} TAB	966 (s), 953 (m), 938 (m)	912 (s)	1488
C_{18} TAB	975 (m), 968 (s), 950 (s)	912 (s)	1488

[a] After Scowen and Leja (1967).
[b] s, strong; m, medium absorption intensity.

of triethanolamines, with a molecular weight less than 5000 and the following structure, are quite effective clay depressants, blocking off the ingress of water between clay platelets [Arsentiev and Gorlovskii (1974), Arsentiev and Leja (1976)]:

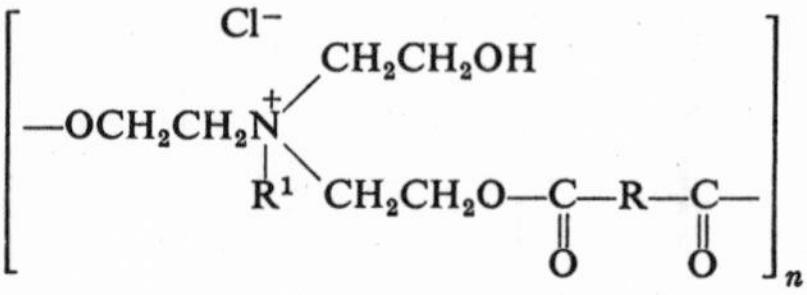

$$\left[\begin{array}{c} \overset{\displaystyle Cl^-}{\underset{\displaystyle R^1}{-OCH_2CH_2\overset{+}{N}}} \overset{\displaystyle CH_2CH_2OH}{\underset{\displaystyle CH_2CH_2O-\underset{\underset{O}{\|}}{C}-R-\underset{\underset{O}{\|}}{C}-}{}} \end{array} \right]_n$$

where $n = 20$–50 and R and R^1 = short hydrocarbon chain. *Ethanolamine* reacting with H_2SO_4 and distilled with NaOH yields *ethylenimine* (or aziridine),

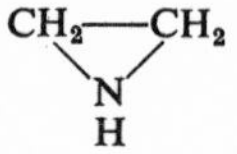

$$\begin{array}{c} CH_2\text{—}CH_2 \\ \diagdown \; N \diagup \\ | \\ H \end{array}$$

which can be polymerized to *polyethylenimines*,

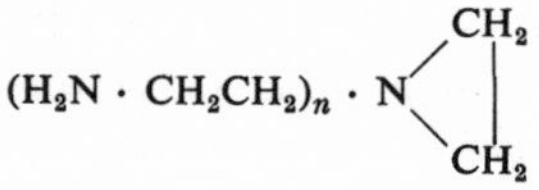

$$(H_2N \cdot CH_2CH_2)_n \cdot N \begin{array}{c} CH_2 \\ \diagup \\ | \\ \diagdown \\ CH_2 \end{array}$$

Ethylenediamine is $H_2N \cdot CH_2CH_2 \cdot NH_2$. The Na salt of ethylenediamine-tetraacetic acid (EDTA, Sequestrene, Versene),

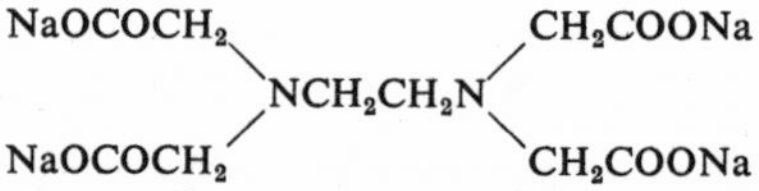

$$\begin{array}{c} NaOCOCH_2 \diagdown \qquad \diagup CH_2COONa \\ \qquad NCH_2CH_2N \\ NaOCOCH_2 \diagup \qquad \diagdown CH_2COONa \end{array}$$

is a strong *chelating agent* for a large number of metallic ions.

Alkyl cyanides (nitriles), $RC{\equiv}N$, for example, acetonitrile, $CH_3C{\equiv}N$, and acrylonitrile, $CH_2{=}CHC{\equiv}N$ (and alkyl isocyanides or isonitriles, isomeric with the nitriles), have dipole moments ~ 4 and ~ 3 debye, respectively. They can be formed on dehydration of amides or on reaction of alkyl halides, sulfates, or sulfonates with metal cyanides. Acyl cyanides,

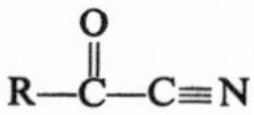

$$R{-}\overset{\overset{\displaystyle O}{\|}}{C}{-}C{\equiv}N$$

can be formed on reaction of acyl halide,

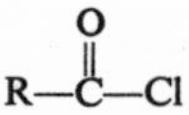

$$R{-}\overset{\overset{\displaystyle O}{\|}}{C}{-}Cl$$

with cyanide ion (CN^-). All of the above compounds are prospective flotation reagents when R is appropriately chosen.

CLASS III: NONIONIC SURFACTANTS

5.13. *Alcohols*

Substitution of one hydrogen in water by an alkyl or an aryl group gives alcohols. Alcohols are classed as primary, secondary, or tertiary according to the number of hydrogen atoms attached to the hydroxylic carbon atom:

$$
\text{RCH}_2\text{OH} \qquad
R{-}\underset{\underset{\displaystyle H}{|}}{\overset{\overset{\displaystyle R_1}{|}}{C}}{-}\text{OH} \qquad
R{-}\underset{\underset{\displaystyle R_2}{|}}{\overset{\overset{\displaystyle R_1}{|}}{C}}{-}\text{OH}
$$

primary   secondary   tertiary

where R_1 and $R_2 \neq R$.

The OH group attached to a carbon chain drastically changes the physical properties of the molecule owing to interactions via hydrogen bonding. The strength of the $[\text{O}{-}\text{H}\cdots\text{O}]$ hydrogen bonds may be as much as one-tenth that of an ordinary carbon–carbon bond. As a result of hydrogen bonding, the water solubility of low-molecular-weight alcohols is high; with the increase in the hydrocarbon length the solubility decreases for the straight-chain homologues (Table 5.25). Nonlinear isomers show an increased solubility in comparison with their linear isomers; for example, the solubility of $n\text{-}C_6$ alcohol is 6.24 g/liter, while that of a branch-chain C_6 isomer, MIBC (methyl isobutyl carbinol), is 17.0 g/liter.

The hydrogen–oxygen bond of the hydroxyl group has a characteristic infrared absorption band, a sharp peak at 3700 cm^{-1} (shown by the vapor phase of C_2H_5OH when no hydrogen bonding exists owing to large intermolecular distances). However, in the spectrum of the same alcohol in, for example, carbon tetrachloride solution, this band is barely visible at 3640 cm^{-1}, but a broad band at 3350 cm^{-1} is shown instead, and this band is characteristic of the hydrogen-bonded hydroxyl groups. The shift of 300 cm^{-1} to a lower absorption frequency indicates that the O—H bond is weakened, and the association band at 3350 cm^{-1} is broad because a whole range of different kinds of hydrogen-bonded aggregates is formed.

Aqueous solutions of alcohols and, in general, all nonionic surfactants are difficult to analyze by spectroscopic techniques. Usually, they require rather complex colorimetric procedures which are not too sensitive. However, owing to an extremely high sensitivity of surface potential to very low concentrations of dipolar surfactants (see Section 7.1 for details on surface potential), the *technique of surface potential measurements* may become a

Table 5.25. Properties of Selected Alcohols

Alcohol	Structure	m.p. (°C)	b.p. (°C)	Water solubility (g/liter, 25°C)	Acid ionization	UV Absorption λ_{max} (nm)	ε [liter (mol · cm)$^{-1}$]
Butyl	n-C$_4$H$_9$OH	−89.8	117.5	Soluble			
Amyl	n-C$_5$H$_{11}$OH	−79	137.3	21.9			
Hexyl	n-C$_6$H$_{13}$OH	−47	158	6.24	~10^{-18}	~200	
Heptyl	n-C$_7$H$_{15}$OH	−35	177	1.81			
Octyl	n-C$_8$H$_{17}$OH	16.7	194–195	Insoluble			
MIBC (methyl isobutyl carbinol) (Aerofroth 70) or 4-methyl-2-pentanol	H$_3$C—CH—CH$_2$—CH—CH$_3$ with CH$_3$ and OH		131.9 (racemic form)	17.0	~10^{-18}		
2-Ethyl hexanol (or 2-octanol)	CH$_3$(CH$_2$)$_3$—CH—C$_2$H$_5$ with CH$_2$ OH	−76	185		~10^{-18}		
Phenol, C$_6$H$_5$OH	phenol structure —OH	43	182	Soluble	1.3 × 10^{-10}	210.5 270.0	6,200 1,450

Compound	Structure						
d-Cresol		34.8	201.9	1.66	1.5×10^{-10}	225.0 / 280.0	7,400 / 1,995
α-Naphthol		93.45	288		4.9×10^{-10}	232.5 / 295.0	33,000 / 5,000
β-Naphthol		122	295	Insoluble	2.8×10^{-10}	226.0 / 273.5	76,000
α-Terpineol, $C_{10}H_{17}OH$		35	220	1.98			
Cyclohexanol, $C_6H_{11}OH$		21.1	161.1	1.98	$\sim 10^{-18}$		
Diacetone alcohol (or 4-hydroxyl-4-methyl-2-pentone)		54-57	164-166			238.5	44.7

reliable procedure for analyzing aqueous solutions for nonionic surfactant content. Faviani *et al.* (1973) report that 10^{-2} ppm of nonionic surfactant concentration may be detected in the presence of up to 1 ppm of ionic surfactant using the surface potential technique. Thus, the method may be particularly suitable for checking the drinking water for nonionic surfactant contamination.

Alcohols of particular usefulness in flotation are those with a chain length between C_6 and C_{10}, preferably branch-chain ones; neither the decreased solubility nor the rate of surface tension lowering is the decisive parameter in the choice of alcohols as flotation frothers, although frequently these features are mentioned as desirable in the frother-acting surfactants. As will be seen later, the structure of the frother molecule appears to play the dominant role but only when analyzed in combination with a particular collector and the given solid onto which the collector is adsorbed. It is speculatively suggested that these structural characteristics determine the relaxation time of the frother's dipole during collision between a collector-coated particle and a (mostly) frother-coated air bubble. (For details, see Section 9.3.)

Alkyl alcohols are practically unaffected by pH, since their acid ionization constant is about 10^{-18}, several orders of magnitude smaller than that of water (10^{-14}). The acid ionization constant of alcohols is defined as

$$K_{\text{ROH}} = \frac{[\text{H}^+][\text{RO}^-]}{[\text{ROH}]}$$

Aromatic alcohols (or aryl alcohols) are compounds containing at least one benzene ring and the OH group on a *side chain*. Phenols are those aromatic compounds which contain the OH group (or groups) bonded directly to a carbon atom of the benzene ring. Phenols are significantly different in their properties from the aromatic alcohols; for example, phenols are distinctly acidic, and their ionization constants are of the order of 10^{-10} (see Table 5.25). Phenols may be monohydric, dihydric, etc., according to the number of OH groups (one, two, etc.) they contain. Aromatic alcohols resemble aliphatic alcohols (they may be considered as aryl derivatives of aliphatic alcohols) and may be either primary (if they contain one benzene ring), or secondary (with two benzene rings), or tertiary (three benzene rings), e.g.,

$C_6H_5 \cdot CH_2OH$	$(C_6H_5)_2 \cdot CHOH$	$(C_6H_5)_3 \cdot COH$
Benzyl alcohol	Diphenyl methanol	Triphenyl carbinol

Phenol and its derivatives, such as cresols, are more polar and capable of forming stronger hydrogen bonds than the corresponding saturated alcohols. The (easier) ionization of phenol gives a phenoxide anion which tends to react with metal cations; thus phenol reacts with ferric chloride in dilute water or alcohol solutions, giving a violet colored ferric–phenoxide complex, while cresols give a blue-colored complex. Phenoxide anion also tends to absorb carbon dioxide, giving phenyl carbonate.

The main constituent of pine oil is α-terpineol, an alicyclic alcohol, $C_{10}H_{17}OH$; the β and γ isomers of terpineol (with different positions of the OH group in the molecule) are not as effective frothers as the α-terpineol.

A synthetic product, diacetone alcohol,

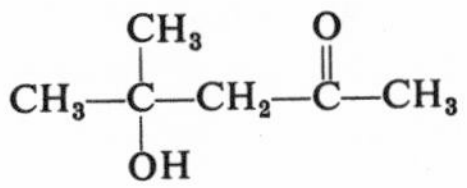

$$CH_3-\underset{\underset{OH}{|}}{\overset{\overset{CH_3}{|}}{C}}-CH_2-\overset{\overset{O}{\|}}{C}-CH_3$$

was tested in the flotation of copper minerals by Oktawiec and Olender (1969) and was found to be a very satisfactory replacement frother for α-terpineol.

1,2-Glycols (diols), dihydric alcohols, are formed by oxidation of olefinic unsaturated compounds. for example, alkenes,

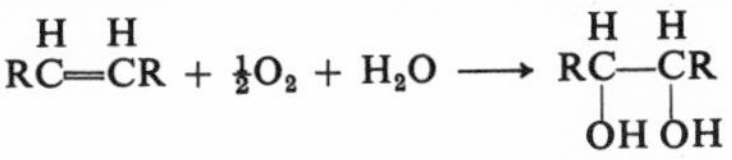

$$\underset{}{\overset{H\ \ H}{RC=CR}} + \tfrac{1}{2}O_2 + H_2O \longrightarrow \underset{\underset{OH\ OH}{|\ \ \ |}}{\overset{H\ \ H}{RC-CR}}$$

Dialkyl glycols

or by hydrolysis of epoxides, for example, ethylene oxide,

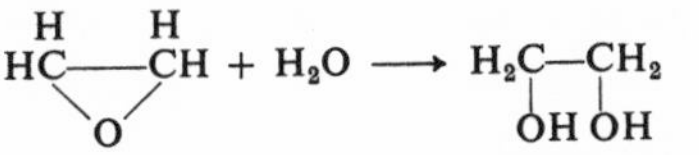

$$\underset{O}{\overset{H\ \ \ \ H}{HC-\!\!-\!\!CH}} + H_2O \longrightarrow \underset{\underset{OH\ OH}{|\ \ \ \ |}}{\overset{}{H_2C-CH_2}}$$

Ethylene glycol
(used in antifreeze)

Glycerol, trihydric alcohol, is a by-product of soap manufacture, used for resins and cellophane; alkyl monoglycerides form important surfactants which act, depending on the length of the alkyl chain, as frothers or emulsifying agents:

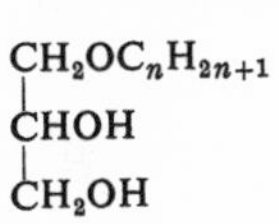

$$\begin{array}{l} CH_2OC_nH_{2n+1} \\ | \\ CHOH \\ | \\ CH_2OH \end{array}$$

Alkyl monoglyceride

(Nitroglycerin is a trinitrate derivative of glycerol and an explosive ingredient in dynamite.)

Sorbitol,

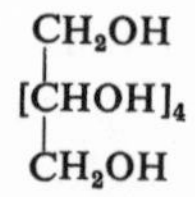

$$CH_2OH$$
$$[CHOH]_4$$
$$CH_2OH$$

is a hexahydric alcohol manufactured by catalytic hydrogenation of glucose. It is used for the synthesis of vitamin C and for the production of cosmetics and of numerous surfactants (by substituting one or more of the H or OH groups with alkyl and other groups).

The alkyl derivatives of the preceding polyhydric alcohols are used primarily as frothers. In the case of glycols, only short-chain alkyl glycols are sufficiently water-soluble to act as frothers, whereas derivatives of sorbitols may comprise long- and/or short-chain hydrocarbons.

5.14. *Ethers*

Substitution of both hydrogens in a water molecule by an alkyl or other hydrocarbon group results in the formation of ethers. They are fairly unreactive compounds, but owing to a permanent dipole (due to oxygen), they are surface active at the air/water interface. Some of them, like TEB (triethoxybutane),[†]

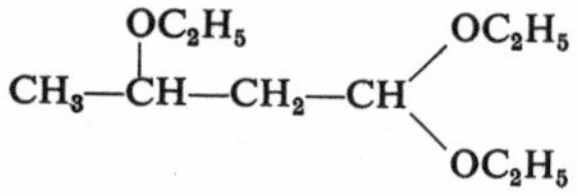

$$OC_2H_5 \qquad OC_2H_5$$
$$CH_3-CH-CH_2-CH$$
$$OC_2H_5$$

(solubility 8 g/liter), or ethers derived from terpenes (e.g., methyl α-terpene), are employed as flotation frothers. The cyclic ethers, such as ethylene oxide,

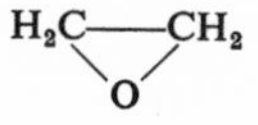

$$H_2C——CH_2$$
$$O$$

and propylene oxide,

$$CH_3$$
$$H_2C——CH$$
$$O$$

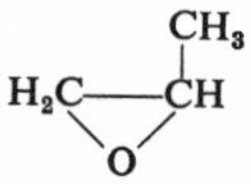

are the most useful raw materials for the manufacture of polymerized derivatives, polyoxyethylene alcohols, and polyoxypropylene ethers; these polymerized reagents recently became the most important nonionic surfactants in several fields of surfactant application.

[†] A different isomer from that shown on p. 212.

Ethylene glycol, distilled with 4% H_2SO_4, yields 1,4-*dioxane*, another cyclic ether which is an important solvent and paint remover (requires ventilation during its use):

$$
\begin{array}{c}
O \\
H_2C \diagup \quad \diagdown CH_2 \\
H_2C \diagdown \quad \diagup CH_2 \\
O
\end{array}
$$

5.14.1. *Polyoxyethylene Nonionic Derivatives*

An insoluble long-chain alcohol can be converted into a soluble nonionic surfactant by an addition of ethylene oxide to the hydroxyl group of the alcohol; the first step is

$$ROH \;+\; H_2C\!\!-\!\!CH_2 \;\longrightarrow\; ROC_2H_4OH$$
$$\overset{}{\underset{O}{\diagdown\diagup}}$$

Alcohol Ethylene oxide Oxyethylene ether

The monoethyl ether ($C_2H_5OC_2H_4OH$) is called *cellosolve* because it is a solvent for cellulose nitrate.

A further reaction with additional ethylene oxide produces polyoxyethylene ethers:

$$ROC_2H_4OH \;+\; n[H_2C\!\!-\!\!CH_2] \;\longrightarrow\; R(OC_2H_4)_n \cdot OC_2H_4OH$$

Ethylene oxide will react with any material containing an active hydrogen; thus, in addition to alcohols, any amphipatic molecule (surfactant) like fatty acid, alkyl phenol, mercaptan, alkyl amide, or alkyl amine can be used as a starting material for condensation with ethylene oxide to give *nonionic surfactants* (with the exception of alkyl amine derivatives which do show some residual cationic character). On the other hand, new ionic surfactants can be synthesized using polymerized nonionic polyoxyethylenes as starting materials for conversion of their terminal hydroxyl group into a sulfate, phosphate, dithiophosphate, carboxylate, xanthate, or quaternary ammonium group, e.g., $R(OC_2H_4)_nOSO_3Na$ (where R is C_7–C_{12} alkyl phenyl hydrocarbon or C_{16}–C_{18} fatty alcohol chain), $R(OC_2H_4)_nOPO_3H_2$, $R(OC_2H_4)_nOCH_2COOH$, or

$$
\left[R \cdot N \overset{\displaystyle (C_2H_4O)_xH}{\underset{\displaystyle (C_2H_4O)_yH}{-C_2H_4-N}} \overset{\displaystyle (C_2H_4O)_zH}{\underset{\displaystyle (C_2H_4O)_wH}{}} \right]^{+} Br^{-}
$$

Since ethylene oxide is flammable and toxic, and forms explosive mixtures with air, the reactions of condensing ethylene oxide with alcohols, etc., have to be carried out under carefully controlled conditions with regard to appropriate temperature and pressure and in the presence of a suitable catalyst. The commercial products are listed under respective trade names (Sterox, Brij, Pluronic, etc.), and in most cases the chemical composition of the product has been disclosed either in the manufacturer's pamphlet or publications such as *Detergents and Emulsifiers* (an annual published by John W. McCutcheon) or in books such as those by Schwartz *et al.* (1966) and Osipow (1962). The large collective volume by Schick (1967) provides an extensive review of all aspects concerning nonionic surfactants: their chemistry, properties, biological behavior, and analytical procedures.

The oxyethylene derivatives of straight-chain primary alcohols became commercially the most important nonionic surfactants because of their easy biodegradability and effectiveness as detergents in household and industrial applications. Depending on the number of moles of ethylene oxide added per mole of a specific alcohol, the nonionic surfactants being produced range in solubility from the completely oil-soluble reagents (e.g., long-chain alcohols C_{12}–C_{20} with 1–3 mol of ethylene oxide) to the completely water-soluble reagents with more than 30 mol of ethylene oxide.

The effect of the increasing number of ethylene oxides on the surface tension values of aqueous solutions of n-C_{12} nonionic derivatives is shown in Figure 5.35. It is seen that as the number of ethylene oxide groups condensed onto C_{12} alcohol increases, the surface tension rises. There is a

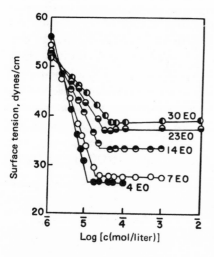

Figure 5.35. Surface tension vs. log concentration for aqueous solutions of *n*-dodecyl polyoxyethylenes (EO = ethylene oxide group). [Reprinted from Schick, ed., (1967), *Nonionic Surfactants*, by courtesy of Marcel Dekker, Inc., New York.]

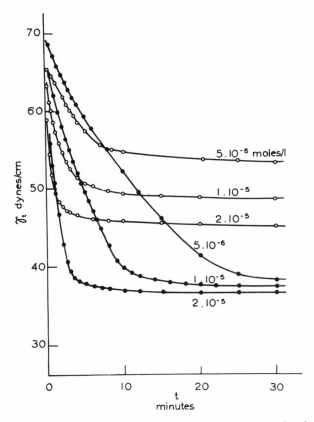

Figure 5.36. Kinetics of surface tension development in solutions of *n*-dodecyl–7 EO and *n*-hexadecyl–10 EO oxyethylene ether alcohols at three different concentrations. (Open symbols refer to 7 EO compound.) [Reprinted from Schick, ed., (1967), *Nonionic Surfactants*, by courtesy of Marcel Dekker, Inc., New York.]

clear indication of a CMC for each oxyethylene derivative. In fact, the micelles tend to form at considerably lower concentrations than with ionic surfactants containing an identical hydrophobic group [compare the 10^{-4}–10^{-5} M range for the polyoxyethylene n-C_{12} derivatives with the value of 9.8×10^{-3} M for the n-C_{12} sulfonate (Table 5.10a), 1.4×10^{-2} M for the n-C_{12} amine, and 6.9×10^{-3} M for the n-C_{13} carboxylate]. Another important feature of the nonionic surfactants is a very slow rate of adsorption at the air/water interface, as indicated by the time dependence of surface tension development (Figure 5.36). The lower the concentration, the longer it takes for the surface tension to reach its equilibrium value, particularly if both hydrophobic and hydrophilic groups are long; for example, the

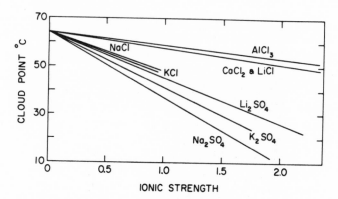

Figure 5.37. Effect of electrolyte additions on the lowering of the cloud point of a 2% triton X (isooctylphenol–100 EO) solution. [Reprinted from Schick, ed., (1967), *Nonionic Surfactants*, by courtesy of Marcel Dekker, Inc., New York.]

5×10^{-6} mol/liter solution of C_{12}–7 EO requires 5 min to reach equilibrium, whereas the same concentration of C_{16}–10 EO requires ~ 25 min.

When the temperature of a nonionic surfactant solution is raised, a sudden onset of turbidity is observed; this is called the *cloud point* (see Section 5.8). It is ascribed to the formation of giant micelles as a result of partial dehydration of the polyoxyethylene-type surfactants; the surfactant molecules (separated at lower temperatures by water-of-hydration molecules) begin to associate, and the aggregates grow progressively in size (with the rise in temperature) until an insoluble phase separates out. An addition of an inorganic salt shifts the cloud point of a given surfactant to lower temperatures, depending on the type of salt and the ionic strength (Figure 5.37); it is seen that for an ionic strength[†] of Na_2SO_4 approaching 2.0, the cloud point of 2% Triton X-100 solution is reduced from $\sim 65°C$ to $\sim 10°C$.

The cloud point is insensitive to the concentration of the nonionic surfactant but depends on the length and the structure of the hydrophobic group and the number of condensed oxyethylenes.

[†] As already mentioned in Section 4.3, ionic strength of a salt solution is defined as half the sum of the products of the molar concentrations of each ion and the square of its valence:

$$I = \tfrac{1}{2}(c_1 z_1{}^2 + c_2 z_2{}^2 + \cdots)$$

$c_{SO_4{}^{2-}} = 0.1$, and $z_{SO_4{}^{2-}} = 2$; hence, for 0.1 M Na_2SO_4, $I = \tfrac{1}{2}(0.2 \times 1^2 + 0.1 \times 2^2)$ $= 0.3$; thus, an ionic strength of 2 would be obtained by 0.66 M addition of Na_2SO_4.

5.14.2. *Polyoxypropylenes and Block-Polymer Nonionic Surfactants*

Polymerized derivatives of propylene oxide,

$$
\begin{array}{c}
CH_3 \\
| \\
H_2C\!\!-\!\!-\!\!CH \\
\diagdown\!\!\diagup \\
O
\end{array}
$$

(often denoted by P, MW 58), condensing on alcohols are obtained in a manner similar to that for ethylene oxide additions. Owing to the presence of a methyl side chain, the hydrophobic character of the polymerized propylene groups is greater than that of the polymerized ethylene groups for the same magnitude of dipole moment represented by each oxygen. Due to this methyl side chain, the solubility of long-chain polypropylene alcohols polymerized on a given hydrocarbon group R does not increase indefinitely with the number of moles of propylene oxide but attains a maximum value. In fact, a polymerized polypropylene glycol (with no hydrocarbon group at all) possessing two alcoholic hydroxyl groups, one at each end of the polymer, becomes completely water-insoluble when the molecular weight exceeds 900. Thus, any polymerized polypropylene glycol with more than $900/58P$ (i.e., $16P$) functions as a hydrophobic compound. If one hydroxyl group at one end of the polymer is replaced by a hydrophobic hydrocarbon chain, the maximum value of the polymerized propylene oxide (P) that would make the compound at least slightly water-soluble would have to be less than $16P$.

The ability of polymerized propylene oxide to act as a hydrophobic group is utilized to produce the so-called *block polymers* with innumerable variations of condensation sequences; the ratios of condensed alkylene oxides (ethylene oxide, propylene oxide, butylene oxide, etc.) can be varied as well. Further, condensation can be carried out with pure oxides in sequence or with mixtures of two oxides in different ratios—the last method of condensation leads to the so-called *heteric* copolymers that contain a random distribution of alkylene oxides within the polymerized chain. The properties of such heteric nonionic surfactants are quite different from their isomers containing identical numbers of the same groups condensed in a sequential manner.

Thus, for example, if the groups that can be condensed are denoted as follows,

Ethylene oxide, OC_2H_4, by E
Propylene oxide, OC_3H_6, by P

and any hydrocarbon group by R, the following types of nonionic block polymers can be obtained:

P glycols: $H(OC_3H_6)_nOH$

PE glycols: $H(OC_3H_6)_a(OH_2H_4)_bOH$

PEP glycols: $H(OC_3H_6)_a(OC_2H_4)_b(OC_3H_6)_cOH$

EPE glycols: $H(OC_2H_4)_a(OC_3H_6)_b(OC_2H_4)_cOH$

RP nonionics: $R(OC_3H_6)_nOH$

RPE nonionics: $R(OC_3H_6)_n(OC_2H_4)_mOH$

ERP glycols: $(OC_2H_4)_nR(OC_3H_6)_p$

$EPRPE$ glycols, etc.: $(OC_2H_4)_s(OC_3H_6)_tR(OC_3H_6)_u(OC_2H_4)_w$

In addition, a large variety of heteric copolymers can also be obtained:

$$(E + P)_x(E + P)_y, \qquad R(E + P)_x(E + P)_yE, \qquad \text{etc.}$$

where $n, a, b, c, m, p, s, t, u, w, x, y =$ integers.

Besides alcohols[†] and polyhydric alcohols (glycols and glycerols), any other starting material such as fatty acids, alkyl amines, amides, alkyl sulfonates and sulfates, etc., can be utilized for condensation of nonionic derivatives.

Most of the nonionic compounds produced commercially find numerous applications in formulations of detergents, emulsifiers, cosmetics, food additives, etc. In flotation, there are at present three polyoxypropylene glycol frothers which find an increasing use, viz. Dowfroth 450, 250, and 200. They represent compounds of approximate molecular weights 450, 250, and 200 and are prepared by reacting propylene glycol, $OH—CH_2—CH_2—CH_2OH$, with the appropriate number of moles of propylene oxide (three for D-250) or condensing propylene oxide on methyl alcohol. The Soviet frothing agent OPSB, representing butyl polyoxypropylenes,

$$C_4H_9—(O—CH_2—\overset{\displaystyle CH_3}{\underset{|}{CH}}—)_nOH$$

where $n = 2$–5, is most useful in flotation systems employing hydrocarbon

---

[†] The so-called *OXO* alcohols are products of synthetic production of alcohols by the addition of carbon monoxide and hydrogen to an olefin in the presence of a cobalt catalyst to form at first an aldehyde, which is hydrogenated in the second step, resulting in the formation of isomeric alcohols. Depending on the olefin used, the mixtures of isomers produced are referred to as *primary amyl alcohol, isohexanol, tridecanol,* and *hexadecyl alcohol.* With the present emphasis on biodegradable detergents, these synthetic alcohol mixtures are used as starting materials for the manufacture of nonionic surfactants.

oil additions to reinforce the collector action [Arsentiev and Leja (1977)]. A comprehensive list of the most probable composition of reagents used under different trade names in the USSR and Eastern Bloc countries is given in an appendix in Manser (1975).

Many other oxyethylene and oxypropylene nonionic reagents have not yet been sufficiently tested in flotation systems to become accepted, although they represent very promising potential frothers with which the hydrophobic–hydrophilic character, degree of frothing, froth persistency, etc., can be varied in a more gradual manner than is presently possible with the alkyl–aryl alcohols now used as frothers.

Owing to the presence of the periodically repeating permanent dipole moment centered around the oxygen atoms of the polymerized alkylene oxide groups, the configuration of the polyoxyethylene or polyoxypropylene chain in solutions of these compounds (and in the bulk phases of pure compounds) differs from the zigzag configuration adopted by the paraffinic compounds. Rotation around single bonds within the chain enables the alkylene groups to adopt a so-called *meander* configuration. This configuration enables interactions of dipoles in the coaxial direction which contact the polyalkylene chain, reducing the distance between the antiparallel dipoles.

In contrast to the polymerized polyoxyethylene ether alcohols (Figure 5.35), the surface tension of solutions of compounds comprising polyoxypropylene ether chains does not increase with an increase in propylene oxide groups but remains constant. On the other hand, the interfacial tension at the oil/water interface decreases with the number of polypropylene oxides for the same strength of solutions but increases for the polyoxyethylene oxides [Schick (1967)].

5.15. *Organic Regulating Agents*

In addition to two types of surfactants used in flotation, one acting as a collector and the other as a frother, numerous regulating or modifying agents are employed to prepare (to condition) the solids and their liquid environment toward a state that would enable the most selective separation to be achieved during flotation. A large number of inorganic reagents serve as regulating agents (these are discussed in Chapter 10); also, there exists an important group of organic regulating agents. These organic compounds are very similar in their constitution to surfactants acting as collectors or frothers. They differ in being invariably multipolar and of such a structure that having adsorbed on selected solids (as do the collectors), they make these solids hydrophilic (instead of hydrophobic), exposing for this purpose some of their polar groups oriented toward the aqueous phase. Because of their structural similarity to the main flotation surfactants, these compounds are discussed here.

In most cases, the organic regulating agents are represented by natural polymerized materials such as extracts or derivatives of starch, cellulose, and tannins. Synthetic polymers are also becoming available, but these are at present used primarily as flocculating agents and only infrequently as depressants and dispersants. (The organic activating agents are known as collectors.)

5.15.1. *Cellulose, Starch, and Their Hydrolysis Products*

Cellulose is the most widely distributed natural polymer, a polysaccharide consisting of long chains of D-glucose units ($C_6H_{12}O_6$, representing one of the naturally occurring carbohydrates), all bonded by a β-glucoside link, formed between C-1 and C-4 of each two neighboring D-glucose monomers:

Cellulose
(linear chain)

It constitutes the fibrous tissue of all plants and trees and is separated on wood repulping; see Section 5.15.3.

Starch, the second most widely distributed polysaccharide, consists of two types of glucose polymers; one is a linear structure (*amylose*) with an α-glucoside linkage between the glucose units, and the other, *amylopectin*, is a branched structure of amylose-like chains in which branch chains form bonds between carbons at positions 1 and 6 of the D-glucose units:

Amylose
(helical chain α-glucoside linkage)

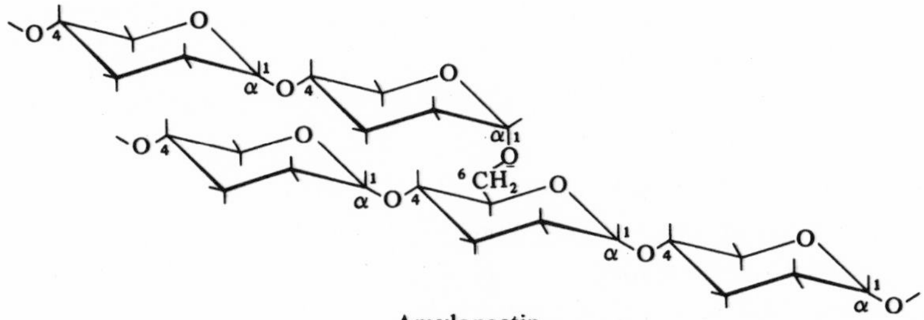

Amylopectin
(branch chain of amylose)

The basic component of both polymers—cellulose and starch—is a mono-saccharide D-glucose; it is one of the four structures of monosaccharides (D-glucose, D-mannose, D-galactose, and D-fructose), $C_6H_{12}O_6$, containing six carbons, five OH groups, and one double-bonded oxygen arranged in cyclic form. Owing to some conformational mobility, there are altogether 16 optically active isomers possible. The two stable crystalline forms of D-glucose, α and β, differ not in structure but in conformation,[†] obtained by rotating the OH and H around the C-1 position by 180°, as shown by both Haworth projection and conformational formulas:

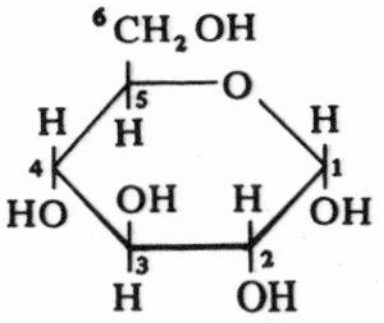

Haworth projection formula 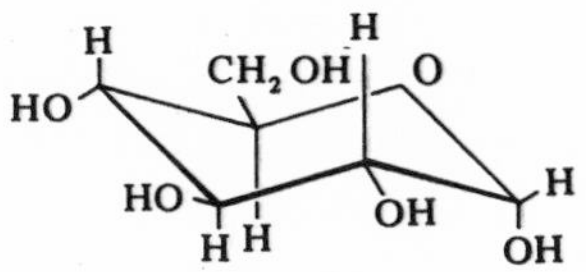   Conformational formula

α-D-glucose

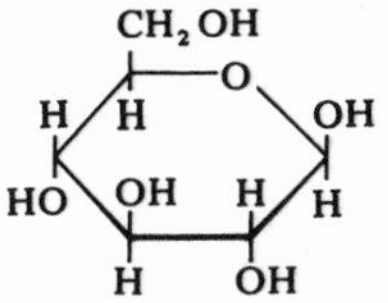

Haworth projection formula 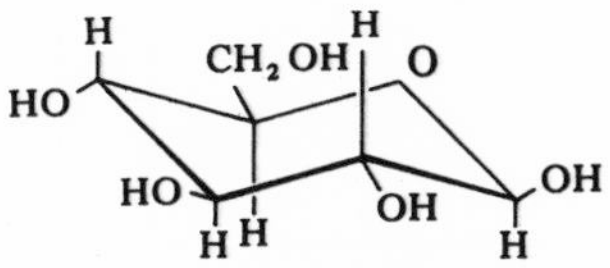   Conformational formula

β-D-glucose

[†] *Structure* designates the order in which the atoms are joined to each other; structural isomers have the same formula, but atoms are linked differently. Rotation about single bonds in a group of atoms can lead to different *conformations* (angular relationship between the atoms of a particular group). If the compound contains a double bond or a ring of atoms, rotation is prevented, and different *configurations* (like *cis* or *trans* stereoisomerism) may be possible (*cf.* p. 217).

In a solution of glucose a mixture of both conformations, 64% of β and 36% of α, exists at equilibrium.

When the monomers of β-D-glucose are joined together in a long-chain polymer, a β-glucoside linkage results, as in cellulose. If the α-D-glucose monomers are joined into a polymer chain, the amylose structure of starch is created; if the amylose-like chains are branch-bonded, the amylopectin structure is formed. In the polymeric structure of either amylose or cellulose, each monomeric unit utilizes two OH groups on the C-1 and C-4 positions to build the polymer chain, thus leaving three OH groups (for each monomeric unit) free. The availability of these OH groups in the polymeric chains of cellulose or starch leads to the formation of micro-crystals (as a result of multiple hydrogen bonding) between some portions of the polymeric chains; the microcrystals are linked to each other by portions of the polymers which are disordered, so that the whole polymeric chain of each cellulose (or starch molecule) participates in several micro-crystals and at the same time in several amorphous or disordered *hinges* linking these microcrystals [Battista (1965)]. The origin of the material (whether cotton of flax cellulose or potato versus corn starch) determines the lengths of polymers, the size of microcrystals, and the number of hinges (Figure 5.38). Cotton cellulose, for example, contains some 3000 glucose units ($C_6H_{12}O_6$, MW 180) per molecule; i.e., its molecular weight is of the order of 500,000. Douglas fir has the longest fiber of all common woods, about 4 mm. Starch is a mixture of amylopectin (75–90%) consisting of thousands of branched-off glucose units with a molecular weight well in excess of 100,000 and linear amylose (10-25%) with several hundreds of glucose units only. Figure 5.39 shows the cellulose structure in a unit crystal.

Granules of starch are extracted from ground grains of cereals (corn, wheat, rice, milo, etc.) or from potatoes and roots [cassava, manion (tapioca), etc.] with water and weak acid (SO_2) or alkali in order to separate them from the noncarbohydrate constituents (proteins, fats, fibrous matter). Changes in the process variables lead to different degrees of granule disruption by hydrolysis.

Controlled acid hydrolysis of either cellulose or starch granules at appropriate temperatures results in H^+ attacking the α- or β-glucose linkages of the *hinge* portions of the molecules and leads to a progressive breakdown of the macrostructure of these polymers. Ultimately with cellulose, the isolation of individual microcrystallites is achieved, giving colloidal-size particles, 25–300 Å wide, 25–300 Å thick, and 50–5000 Å long. This final state is the so-called *level-off degree of polymerization* (D.P.) representing the discrete, unhinged microcrystals of the original polymer. Acid hydro-

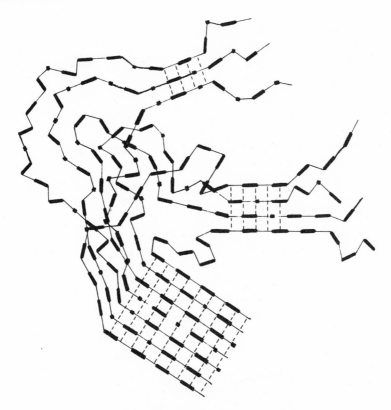

Figure 5.38. Model of high-molecular-weight polymers comprising discrete microcrystals within the polymer matrix separated from each other by disordered regions called hinges.

lysis of starch results in a partial breakdown of polymerized microcrystallites, producing *dextrins*.

In addition to (1) dextrins produced by acid hydrolysis in aqueous media, there are (2) dextrins obtained by the action of enzymes called amylases (α and β), (3) cyclic Schardinger dextrins produced by the action of *Bacillus macerans* enzyme on starch, and (4) pyrodextrins produced by the action of heat (or both heat and acid) on starch. Pyrodextrination of starch appears to involve not only hydrolysis of starch molecules (amylose and allylpectin) to relatively low levels of molecular weight but also a subsequent recombination of these fragments through the formation of $1 \rightarrow 6$ linkages to create highly branched structures. Depending on the time length of heat application, the properties of dextrins change, resulting in a rapidly increasing solubility after ~ 2 hr of heating. Viscosity is simultaneously reduced by nearly an order of magnitude. If the temperature or

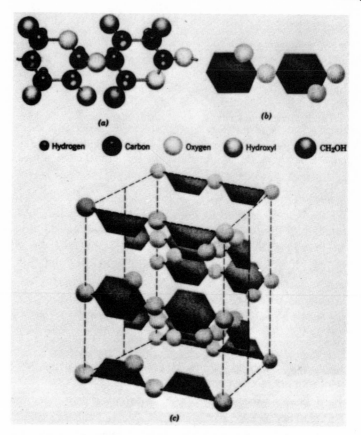

Figure 5.39. Cellulose structure: (a) the atomic arrangement within a chain of cellulose (a cellobiose unit); (b) a schematic representation of the cellulose chain; (c) the arrangement of cellulose chains in a unit cell of the cellulose crystal. [From Moffat *et al.* (1964), *The Structure and Properties of Materials*, Vol. 1, with the permission of John Wiley & Sons, New York.]

acidity of dextrination is varied, numerous types of dextrins are produced, as indicated in Figure 5.40.

Cyclic dextrins, known as Schardinger dextrins, contain six or more D-glucose units bonded by the α-glucoside linkage in a form of a torus. Depending on the number of glucose units in the torus, the dextrins are referred to as α-dextrin (or cyclohexa-amylose), β-dextrin (or cyclohepta-amylose), γ-dextrin (or cycloocta-amylose), etc. All cyclic dextrins are capable of forming numerous inclusion compounds in solution, and, in comparison with flexible open-chain polymers, the cyclodextrins display a large number of *anomalous* chemical properties.

Some enzymes are capable of carrying out the hydrolysis of the poly-saccharide polymers, cellulose or starch, to their respective structural units; the *cellobiose unit* (consisting of two glucose units bonded by a β-glucoside link) is the ultimate product of hydrolysis of cellulose catalyzed by the enzyme *emulsin,* and the *maltose unit* (consisting of two glucose units bonded by an α-glucoside link) is the ultimate product of hydrolysis of starch catalyzed by the enzyme amylase or maltase.

Depending on the size of the ultimate unhinged microcrystallites in the hydrolyzed cellulose or starch product, solutions of these products show unusual rheological characteristics and begin to form gels when a minimum concentration is exceeded. This concentration for the onset of gel formation may be as low as 0.5% wt. solids for very small, near-spherical microcrystallites to 12–15% wt. solids for rod-type or platelet-type microcrystallites.

Three different crystalline forms of amylose have been recognized so far from X-ray spectra: Type A and type B spectra are produced when water-dissolved amylose, which is metastable, recrystallizes (this is the so-called *retrograded* amylose). Spectrum B indicates orthorhombic crystals with eight D-glucose monomers per unit cell. The third form, giving the helical "V" spectrum, corresponds to six D-glucose monomers per each turn of the helix (with a 12.98-Å diameter and 7.91 Å between successive helices); the helices are closely packed to form cylinders that establish the hexagonal nature of the X-ray diffraction patterns. The helical structure of amylose appears to exist in strong solvents, giving rise to inclusion complexes [i.e., with iodine (giving the characteristic blue color), aliphatic and cyclic alcohols, fatty acids, etc.] and to the formation of cyclodextrins. Amylopectin does not react with iodine.

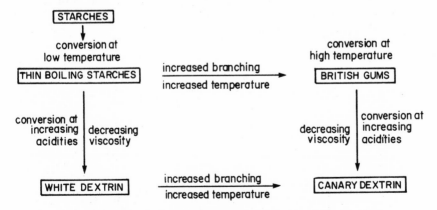

Figure 5.40. Relationships among various starch degradation products.

5.15.2. *Derivatives of Starch and Cellulose*

Treatment of either starch or cellulose with different reagents results in the formation of chemically modified products such as

Methyl cellulose (with O—CH$_3$ groups replacing the hydroxyl group)
Carboxymethyl cellulose or-starch (with a O—CH$_2$COO$^-$Na$^+$ substituent)
Cellulose nitrate (mono-, di-, and trinitrates)
Cellulose diacetate (celanese silk), cationically substituted cellulose or starch with a —O—CH$_2$—CHOH—CH$_2$N$^+$(CH$_3$)$_3$ substituent (at positions 2 and 6), etc.

Some examples of such derivatives are shown in Figure 5.41.

The degree of substitution depends on the intensity of chemical treatment. It may vary in a continuous manner from a small fraction of the theoretically available units of OH groups to the formation of mono-, di-, and triderivatives, e.g., cellulose trinitrate (gun cotton). With the increasing degree of substitution the structure of microcrystallites is progressively destroyed, and the average molecular weight may also be drastically reduced. Thus, the effects of these polymeric compounds should be evaluated with materials whose production procedure gives reproducible products that are clearly identifiable, since under each procedure an enormous variety of polymeric materials can be obtained unless each treatment step is strictly controlled.

Details of chemistry and technology of starches and of cellulose and their derivatives can be found in Whistler and Paschall (1965).

On reacting cellulose with sodium acetyl chloride, Cl · CH$_2$COO$^-$Na$^+$, the hydrogen of the branch-chain hydroxyl group (i.e., the sixth position of the glucose formula) is replaced by the sodium acetyl group, producing the sodium carboxymethyl cellulose:

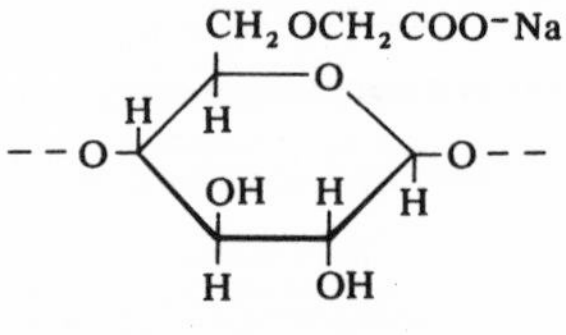

In alkaline solution the above polymeric salt is fully ionized and is capable of being attracted electrostatically to oppositely charged sites. Its adsorption results in selective depression of some solids that would attract anionic collectors, by replacing the positively charged sites on these

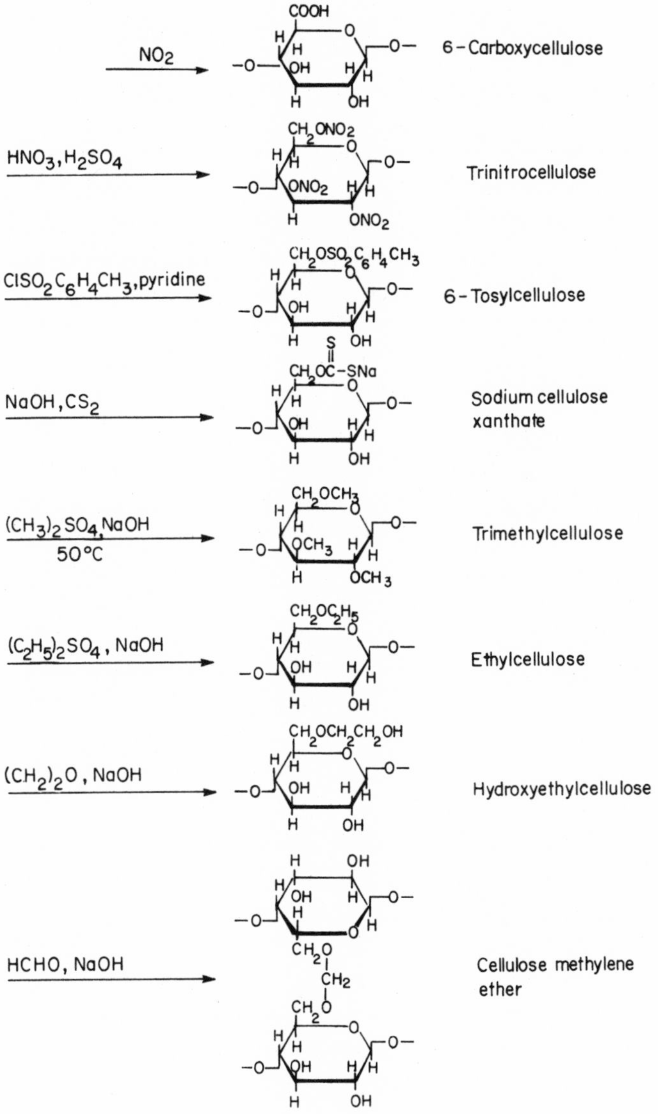

Figure 5.41. Cellulose derivatives. [From Siu (1951).]

solids with the nonionized —OH groups, still highly hydrophilic. To obtain reproducibility of action as depressants, the polymers of the same quality (that is, of the same molecular weight, the same degree of substitution, etc.) must be used, and the manner of preparing the dilute solutions must be standardized. Otherwise any comparison of effects produced by such polymers may be invalid.

Other polymeric derivatives of cellulose or starch such as nitrates, hydroxyethyl or ethyl cellulose, cellulose acetates, etc., may also be employed for depressant action. Modified cellulose derivatives, e.g., carboxymethyl cellulose phosphates, and amphoteric compounds containing cationic as well as anionic groups, etc., can be employed in the control of charge density at solid interfaces.

Hydrocarbon polymers of relatively low molecular weight (that is, less than $\frac{1}{2}$ million) but with multiple nonionic groups such as ether, amide, hydroxyl, ethylene oxide, carboxylic acid in solutions of low pH, or multiple ionized groups such as carboxylate ion, sulfonate, etc., can be dissolved in water to give apparently "clear solutions"; these may still contain colloidal-size aggregates (and/or coiled molecules), which can be detected by light scattering.

Adsorption of such polymers with multiple polar groups occurs through either dipole interactions or electrostatic interaction; multiple bonding takes place, leading to the replacement of the crystalline solid substrate with that of a highly hydrophilic (ionized or dipolar in character) surface of a noncrystalline polymer, extended or coiled. Judicious additions of such polymeric species should, in theory, be capable of differentiating among various solids. Unfortunately, the wide range of mixtures represented in commercial products, their unknown chemical nature, and the empirical manner of testing their efficacy have resulted in a very crude and often nonreproducible application. A great deal of systematic work, with purified reagents of known polymeric types, remains to be done before the full extent of their utilization will be realized.

5.15.3. *Wood Extracts: Quebracho and Wattle Bark, Lignin and Tannin Derivatives*

Trees of forests, native to different regions of the world, provide a large array of chemicals known as silvichemicals. These are either extracted from spent pulping liquors or distilled off individual tree components (needles, heartwood, bark, roots etc.).

Wood consists of $\sim 50\%$ cellulose fibers, $\sim 30\%$ lignin, and 20% car-

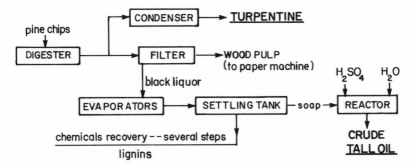

Figure 5.42. Recovery of by-products from the kraft pulping process.

bohydrates, fats, and resins. The silvichemicals obtained from woods represent all types of materials from charcoal, lignin derivatives, gums, rubber, resins, alkaloids, tannins, waxes, essential oils, turpentine, ethanol, and acetic acid. The silvichemicals produced from the pines are referred to as *naval stores* because of the constituent pitch and oils previously used for caulking and preserving boats and wooden ships.

Cellulose is obtained from wood by dissolving lignin carbohydrates on pulping (or digesting) the wood with either sodium sulfite in an acid solution or with sodium sulfide and sodium hydroxide in an alkaline solution. The latter process is called the kraft pulping process (*kraft* means strength in Swedish and German) because it gives strong, unimpaired cellulose fibers. Debarked wood is chipped, the chips are digested in containers, then the wood pulp (cellulose) is filtered off and used to manufacture paper, while the filtrate—*black liquor* (whose color is caused by the lignin and tannin derivatives)—is treated to recover fats, resins, lignins, etc. Volatile matter evolved in digesters is condensed to produce *turpentine*; see Figure 5.42.

The alkaline black liquor is concentrated in multiple evaporators. A mixture of soaps is produced upon saponification of fatty acids (occurring during evaporation of excess water) and rises to the surface of the liquor in settling tanks to be skimmed off. These soaps are then acidified to yield crude *tall oil*. Since about the mid-1950s the crude tall oil has been fractioned, yielding $\sim$ one-third fatty acids, $\sim$ one-third rosin, and the rest pitch fractions. Fatty acids consist of $\sim 75\%$ oleic and linoleic acids, other unsaturated C_{18} and C_{20} acids, and some palmitic acid. They are used in the formulation of protective coatings, of detergents, and as flotation agents.

Rosin consists primarily of resin acids of the *abietic* and *pimaric* types (Figure 5.43). The main use of rosin is in the sizing of paper (to control

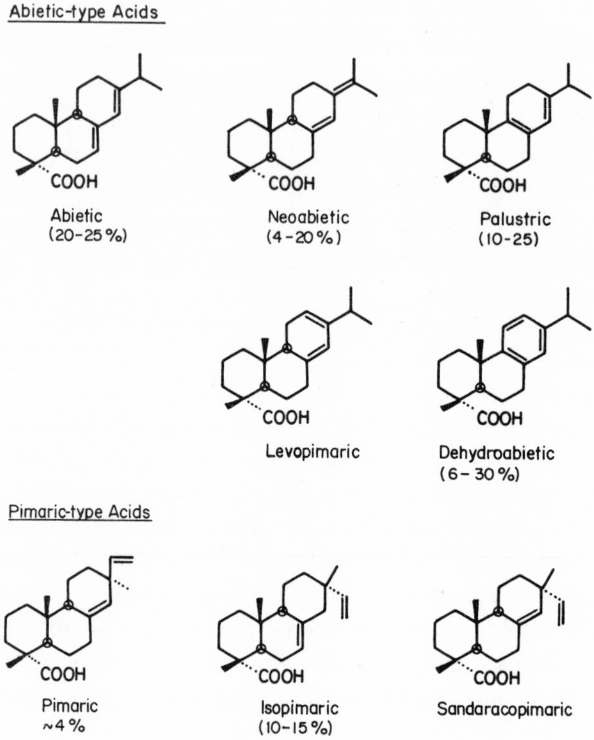
Abietic-type Acids
COOH
Abietic
(20-25%)
COOH
Neoabietic
(4-20%)
COOH
Palustric
(10-25)
COOH
Levopimaric
COOH
Dehydroabietic
(6-30%)
Pimaric-type Acids
COOH
Pimaric
~4%
COOH
Isopimaric
(10-15%)
COOH
Sandaracopimaric

Figure 5.43. Common resin acids of pine.

water absorptivity) and in formulating coatings, chewing gum, and printing inks. Solubilized derivatives of abietic acids have been employed in flotation.†

The volatile components escaping wood chip digesters are condensed to give *turpentine*. This is composed of α-pinene (60–70%), β-pinene (20-25%), and other terpenes. Terpenes obtained by steam distillation or ethe–extraction of the various parts of plants and trees yield the so-called *essential oils*, used widely in perfumes, food flavoring, and medicine. The

† For example, *resorcinol*, 1,3-hydroxybenzene,

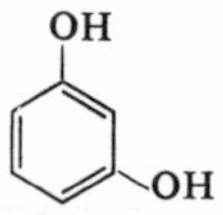
OH
OH

is a product of distillation of certain natural resins; its derivatives (by substitution reactions in position 4) such as 4-alkyl acids are used as multipolar collectors in taconite flotation [Patel *et al.* (1956)].

majority of these compounds are head-to-tail combinations of the C_5 isoprene skeletons that may include hydrocarbons, alcohols, aldehydes, etc. The C_5 isoprene unit,

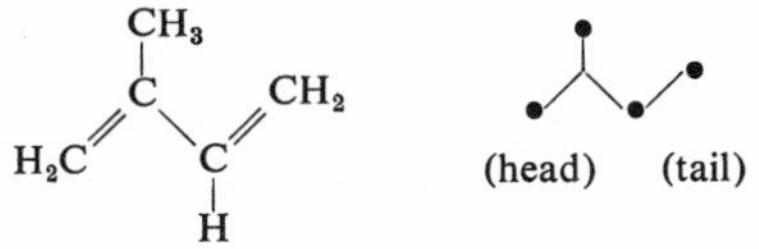

gives rise to C_{10} compounds known as monoterpenes, C_{15} sesquiterpenes, (e.g., farnesol—lily of the valley alcohol), C_{20} diterpenes (abietic acids, vitamin A), C_{30} tri-terpenes (lanosterol wool fat), and C_{40} tetraterpenes. The terpene alcohols shown in Figure 5.44 are important perfume ingredients, while terpene aldehydes are used as flavor chemicals.

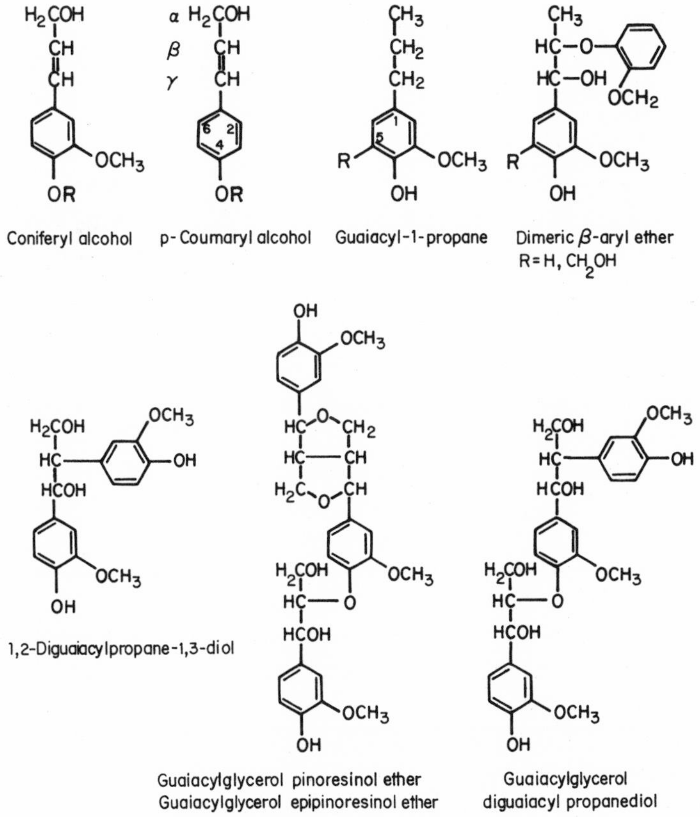

Figure 5.44. Lignin structures.

About one-half of the turpentine produced from pine wood is subsequently converted (on acidification of α-pinene) to the *synthetic pine oil* whose primary constituent is α-terpineol.

Fractionated compounds of turpentine, rosin, and pitch residues are also modified (by esterification, sulfonation, sulfation, condensation of ethylene oxide, etc.) for the surfactant and polymer markets.

Lignin, which binds together the cellulose fibers of wood, represents an extremely complex mixture of monomeric cyclic species with polymers which are both two- and three-dimensional in character. A dioxyphenyl propyl unit,

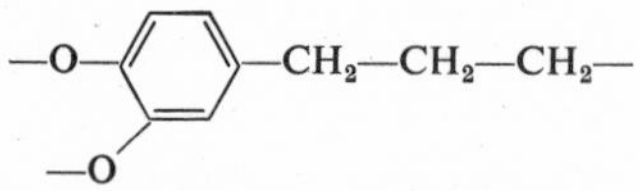

$$-O-\hspace{-2pt}\bigcirc\hspace{-2pt}-CH_2-CH_2-CH_2-$$
$$-O$$

appears to be an important repeating unit of most lignin components. Coniferyl alcohol (Figure 5.44) is considered to be the parent structure for a wide variety of lignin constituents. Freudenberg (1966) reviews the work concerned with the evaluation of lignin structures in terms of over 60 different types of compounds (of increasing complexity) some of which are shown in Figure 5.44.

The *tannin extracts* obtained from two types of trees are well-known flotation agents: *quebracho* and *wattle bark* extract. Both have been used (and are being used) primarily as depressants for calcite in the flotation of fluorite with fatty acids and also, to some extent, with other anionic collectors, as a depressant for sphalerite and carbonaceous gangue in flotation of galena. The following order of quebracho's depressant activity is indicated:

phosphate > calcite > fluorite > barite > scheelite
hematite > calcite > silica
pyrite > pyrrhotite > sphalerite > copper sulfides

Quebracho denotes the extract obtained from the heartwood of the genus *Shinopsis balansae* and *Shinopsis lorentzii* trees, native to a region of South America comprising southern Brazil, western Paraguay, and northwestern Argentina. The name *quebracho* means, in colloquial Spanish, "axe breaker" and refers to the extreme hardness and density of this wood. The extraction is carried out by treating heartwood chips with hot water (220°–240°F) at a slight pressure, followed by vacuum evaporation of the initial 10% solution to obtain a 95% solid, free-flowing powder. This powder is soluble in water above pH 8. To make the extract soluble in cold water at all pHs, it is often treated with sodium bisulfate ($NaHSO_4$). A third form of quebracho is also commercially available, namely quebracho with amine groups introduced into the extract.

Wattle (or mimosa) bark extract is derived by a similar treatment from the bark of *Acacia mollissima* trees and, being more soluble in water, does not require treatment with bisulfates.

King and White (1957) carried out an extensive evaluation of numerous extracts from leaves, twigs, bark, and heartwood of *Shinopsis* trees. They came to the conclusion that the nonwoody parts of plants and trees produce a host of so-called *hydrolyzable tannins*—based on phenols or slightly polymerized phenol structures. The metabolic sequence of reactions starts with acetate → glucose → shikimic acid and quinic and gallic acids (Figure 5.45). The reactions are controlled by enzymes and result in numerous derivatives of the galloyl type (based on gallic acid). On the other hand, the extracts from the heartwood belong to the so-called *condensed tannins*, more highly polymerized, like flavonols, catechins, and derivatives of coumarans (containing three phenyl structures with a variety of hydroxyl and ether groups in different substitution positions).

It is obvious that the commercial tannin extracts are highly complex mixtures of naturally occurring polyphenol compounds and their derivatives formed during heating oxidation and treatment with solubilizing agents. Ionization of the phenolic and carboxylic groups leads to adsorption of these compounds on cationic surface sites, followed by strong lateral hydrogen bonding and conversion of mineral surfaces to negatively charged hydrophilic surfaces. The latter are thus unsuitable for adsorption of anionic collectors but become active adsorption sites for cationic species.

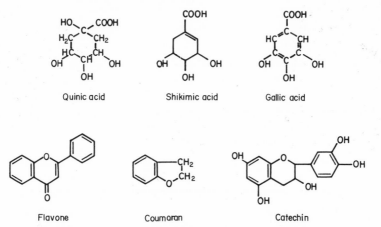

Figure 5.45. Components of hydrolyzable tannins (quinic, shikimic, and gallic acids), and condensed tannins (flavone, catechin, coumaran).

5.16. *Commercial Flocculating Agents*

In addition to reagents which are used specifically to achieve selectivity in flotation separations, there are numerous occasions when other designated reagents have to be used by flotation operators. Examples of such reagents are chelating agents used to remove selected metallic or toxic components from mill effluents or flocculating agents used in the clarification of waters carrying slimes as suspended solids. The future use of flocculating agents is likely to expand considerably. This expansion will be due to an increasing success of selective flocculation (as compared with heteroflocculation) in aggregating individual mineral species of finely subdivided slurries. Selective flocculation may thus become the step preceding flotation of very finely disseminated minerals [Usoni *et al.* (1968) and Osborne (1978)].

Flocculating agents are organic reagents represented by very long-chain molecules, of 10^4–10^7 MW, with recurring structural units, i.e., multi-polar polymers. The polymers may be natural or synthetic. Common natural polymers are polysaccharides such as starch, gums, and cellulose and proteins (amino acids) such as gelatine, casein, and albumin. Natural polymers are inhomogeneous; that is, they represent mixtures of coiled molecules of widely varying weights. For example, the particle weights of gelatin vary from 10,000 to 100,000 and those of casein from 75,000 to 375,000. To control not only the molecular weight of polymers but also the type of charge and its density, derivatives of natural products and synthetic polymers are made, and these are preferred in various applications. Examples of derivatives of natural polymers are dextrins (Figure 5.40) and cellulose derivatives (Figure 5.41). The synthetic polymers which are employed as flocculating agents are *addition* polymers (built up by a repeated addition reaction between monomer molecules) of (1) *alkenes* (unsaturated hydrocarbon species, reacting at the double bond) such as acrylic acids and their esters, acrylonitrile, acrylamide, vinyl alcohol, styrene, etc.,

$CH_2\!\!=\!\!CHCOOH$ Acrylic acid

$$CH_2\!\!=\!\!\overset{\overset{\displaystyle CH_3}{|}}{C}\!\!-\!\!COOH \qquad \text{α-Methyl acrylic acid}$$

$$CH_2\!\!=\!\!C\overset{(H)}{\underset{(CH_3)}{}}COOR \qquad \text{Acrylic esters, where R = methyl, ethyl, butyl, and 2-ethylhexyl groups.}$$

$CH_2\!\!=\!\!CH\!\!-\!\!C\!\!\equiv\!\!N$ Acrylonitrile

$CH_2\!\!=\!\!CH\!\!-\!\!CONH_2$ Acrylamide

or (2) *cyclic ethers* (epoxides) and imines, such as ethylene oxide and ethyl-enimine, reacting by opening their ring structures.

Irrespective of whether the polymers are natural, derivative, or synthetic products, they are either

1. Nonionic: starches, dextrins, polyacrylamides, polyvinyl alcohols, polyoxyethylenes, etc.
2. Anionic: sodium carboxymethyl cellulose, phosphated starch, carboxylated polyacrylamide, etc.
3. Cationic: polyethylenimines, etc.
4. Amphoteric: gelatin, albumin, etc.

If only one monomeric species is used in synthesis of polymers, the products may still represent two types of arrangement:

1. Head-to-tail arrangement:

$$-CH_2-CH-CH_2-CH-CH_2-CH-$$
$$\qquad\quad | \qquad\quad | \qquad\quad |$$
$$\qquad\quad X \qquad\quad X \qquad\quad X$$

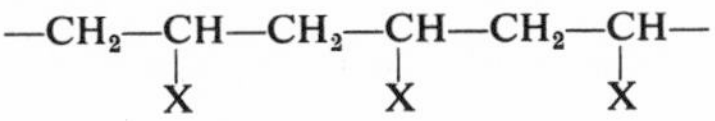

2. Head-to-head (or tail-to-tail) arrangement:

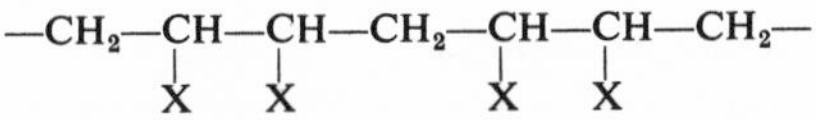

$$-CH_2-CH-CH-CH_2-CH-CH-CH_2-$$
$$\qquad\quad | \quad | \qquad\qquad | \quad |$$
$$\qquad\quad X \quad X \qquad\quad X \quad X$$

Usually, the head-to-tail arrangement predominates. When a mixture of two monomers is used for copolymerization (for example, acrylonitrile and acrylamide), depending on conditions used, either a uniform copolymer with alternating polar groups,

$$R_1X-R_2Y-R_1X-R_2Y \cdots$$

or a *block polymer*,

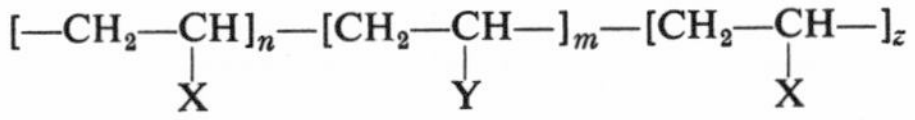

$$[-CH_2-CH]_n-[CH_2-CH-]_m-[CH_2-CH-]_z$$
$$\qquad\quad | \qquad\qquad\quad | \qquad\qquad\quad |$$
$$\qquad\quad X \qquad\qquad\quad Y \qquad\qquad\quad X$$

or a *graft polymer*,

$$CH_2-CH-CH_2-CH-CH_2-CH-CH_2-CH$$
$$\qquad | \qquad\quad | \qquad\quad | \qquad\quad |$$
$$\qquad X \qquad\quad X \qquad\quad X \qquad\quad X$$
$$\begin{bmatrix} CH_2-CH \\ \quad | \\ \quad Y \end{bmatrix}_m \qquad\qquad \begin{bmatrix} CH_2-CH \\ \quad | \\ \quad Y \end{bmatrix}_n$$

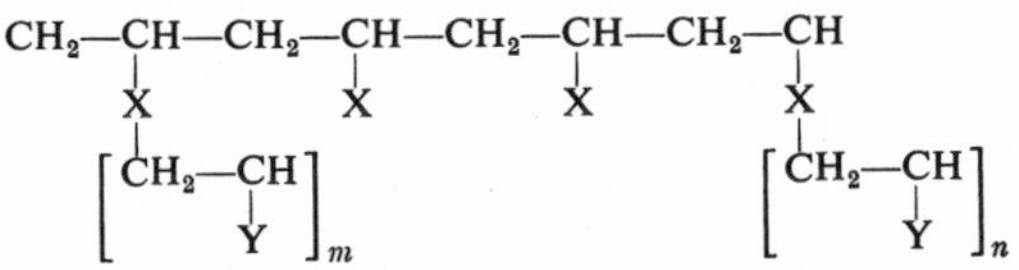

is produced. Obviously, each type of such copolymers has different properties; and these also differ from the properties of mixtures of two individual homopolymers.

(A) In the preceding schematic representations of various polymeric arrangements, the polar groups X and Y may portray

—OH (hydroxyl) as in starch, cellulose, or polyvinyl alcohol

—C≡N (nitrile) as in polyacrylonitrile

—NH$_2$ (amine) as in polyamines (such as Cyanamid Superfloc 310, S3730, and S3645 or Alchem Nalcolyte 600)

—CONH$_2$ (amide) as in polyacrylamides [such as Cyanamid Superfloc 16, 20, 84, and 127, with MW varying from $\sim(3-15) \times 10^6$, or Dow Separan MGL, MW 2×10^6]

(B) The original polymer may be subjected to further treatment, resulting in a partial or complete conversion of the original polar groups to new species. Thus, a polyacrylonitrile may be hydrolyzed to yield polyacrylamide groups which, under still further hydrolysis, may be converted (in part or completely) to carboxylic groups.

Polyimines hydrolyze in a manner similar to polyaldehydes (—CHO). By previous hydrolysis, commercial polyacrylamides may be carboxylated to a different extent, as exemplified by Cyanamid Superfloc 210, 212, and 214 [MW $(12-15) \times 10^6$] which contain increasing proportions of carboxylic groups, or Dow Separan MG 200, MG 300, and MG 700 [MW $(2-3) \times 10^6$]; or Alchem Nalcolyte 672, 673, and 675 (MW $12\times$, $15\times$, and 18×10^6, respectively).

(C) Polymerization can be carried out by several methods:

1. Bulk polymerization of a liquid monomer
2. Solution polymerization—with the monomer in an organic solvent
3. Emulsion polymerization (the monomer is emulsified in an aqueous medium)
4. Suspension polymerization (relatively large drops of an insoluble monomer are dispersed in an aqueous medium)

Unless the conditions of polymerization are strictly controlled, products widely varying in quality may be released under the same designation.

The quality of a polymer depends not only on the overall molecular weight (average, or its distribution curve) but also on the stereochemical configurations in the long-chain molecule and the character of forces acting between the chains.

For a polymer with a side radical or a side polar group, such as styrene

$$n\left[\underset{\overset{|}{H}}{\overset{}{\underset{}{\bigcirc}}}-C=CH_2\right] \longrightarrow \left[-CH_2\overset{\overset{\bigcirc}{|}}{C}H-\right]_n$$

the stereochemical configurations may be either

Atactic: with a random distribution of side groups

Isotactic: side groups distributed only on one side of the chain

Syndiotactic: side groups oriented alternately on one side and then on the other in a repeating, uniform manner

Depending on the stereochemical arrangement, the polymeric solid may be crystalline or amorphous. The crystalline character may be complete or partial, that is, observed only among portions of long polymeric chains or otherwise separated by randomly oriented hinge portions, as shown in Figure 5.38 for cellulose or starch polymers.

The efficacy of flocculating action obtained with long-chain polymers employed for that purpose is also critically influenced by the extent of their dissolution in aqueous media to the state of fully extended molecules. Dissolution of such long molecules (of up to 2×10^7 MW) is an unusually slow process due to penetration by diffusion of solvent molecules among the crystalline and hinge portions of the solid polymer. Irregular coils of incompletely solvated macromolecules, occupying fairly large volumes, are usually left in the *stock solution* unless the solution is sufficiently dilute and enough time has been allowed for gradual solubilization. The volume occupied by the solvated macromolecule depends on the molecular weight of the polymer. For a polymethylmethacrylate of 0.5×10^6 MW, a solution of 5% strength begins to form a gel, denoting lack of any free solvent molecules; for a polymer twice that long, that is, 1×10^6 MW, gelation already starts at 0.2% strength. Thus, stock solutions of flocculating agents when added to a suspension for the purpose of fast flocculation must be sufficiently dilute and *aged* to ensure optimum results.

An evaluation of the composition and flocculating characteristics of some commercial polyacrylamide flocculants has been presented by Rogers and Poling (1978). The principles of the action of polymeric flocculants were discussed by Kitchener (1972, 1978) and the chemical factors involved in flocculation of mineral slurries were treated by Slater *et al.* (1969). Usoni *et al.* (1968), Yarar and Kitchener (1970), and Attia and Kitchener (1975) dealt with selective flocculation of mineral species and a choice of polar groups for tailor-made flocculants. Proceedings of the NATO Advanced Study Institute conference on flocculation have been edited by Ives (1978).

Flocculation involves cooperative action of electrostatic forces, hydrogen bonding, multiple dipole interaction, and hydrophobic bonds. The solvated molecules of flocculating agents have to become strongly adsorbed on the dispersed solids, but only a fraction of a saturation dosage is required for effective flocculation. If more than an optimum quantity of flocculating agent is employed, the system may become stabilized in a dispersed state. Depending on the concentration of the dispersed particles, collisions caused by Brownian motion lead to aggregation of particles at a fairly rapid rate, called *perikinetic* flocculation. At this stage, the rate depends on the temperature and the concentration of particles and is theoretically derivable from the theory of Smoluchowski. To sweep all of the dispersed particles into aggregates, a gentle stirring is necessary to increase collisions by hydrodynamic action; this stage is known as *orthokinetic* flocculation. The rate of collisions in the latter stage depends on shear velocities in the suspension (which are proportional to the gradients of velocity), but when a certain critical value of shear is exceeded, the flocculated aggregates become broken and do not reaggregate on subsequent reduction of the shearing velocity.

A uniform distribution of flocculant, a low rate of shear applied for a long time, and a relatively high solids content are all very important physical parameters contributing to the effectiveness of flocculation. The length of the polymeric flocculant used and the strength of its adsorption determine how successful the application is to a particular system. Adsorption and *bridging* of polymers are referred to in Section 6.7 (Figure 6.36).

5.17. *Selected Readings*

Cross, J., ed. (1977), *Anionic Surfactants—Chemical Analysis*, Marcel Dekker, New York.
Gould, R. F., ed. (1966), *Lignin Structure and Reactions*, Advances in Chemistry Series No. 59, American Chemical Society, Washington, D. C.
Harkins, W. D. (1952), *Physical Chemistry of Surface Films*, Reinhold, New York.
Jungermann, E. (1970), *Cationic Surfactants*, Marcel Dekker, New York.
Kosolapoff, G. M. (1958), *Organophosphorus Compounds*, Wiley, New York.
Linfield, W. M., ed. (1976), *Anionic Surfactants*, (in two parts), Marcel Dekker, New York.
Manser, R. M. (1975), *Handbook of Silicate Flotation*, Warren Spring Laboratories, Stevenage, England.
Markley, K. S. (1960, 1961), *Fatty Acids*, 2nd ed., Parts 1 and 2, Interscience, New York.
McCutcheon, J. W., Inc., *Detergents and Emulsifiers* (published annually), M. C. Publishing Co. Ridgewood, New Jersey.

Mittal, K. L., ed. (1977), Micellization, solubilization and microemulsions, in *Proceedings of the Symposium at the 7th NERM (Northeast Regional Meeting) of the ACS, Albany, New York*, 1976, Plenum Press, New York.

Mittal, K. L., ed. (1979), Solution chemistry of surfactants, in *Proceedings of the 52nd Colloid & Surface Science Symposium, Knoxville, Tennessee*, 1978, Vols. I and II, Plenum Press, New York.

Noll, W. (1968), *Chemistry and Technology of Silicones*, Academic Press, New York.

Noller, C. R. (1965), *Chemistry of Organic Compounds*, Saunders, Philadelphia.

Osipow, L. I. (1962), *Surface Chemistry*, Reinhold, New York.

Ralston, A. W. (1958), *Fatty Acids and Their Derivatives*, Wiley, New York.

Rao, S. R. (1971), *Xanthates and Related Compounds*, Marcel Dekker, New York.

Reid, E. E. (1962), *Organic Chemistry of Bivalent Sulphur in Thiocarbonic Acids and Derivatives*, Vol. IV, Chemical Publishing, New York, Chap. 2.

Rosen, M. J. (1978), *Surfactants and Interfacial Phenomena*, Wiley, New York.

Schick, M. J. (1967), *Nonionic Surfactants*, Marcel Dekker, New York.

Schwartz, A. W., Perry, J. W., and Berch, J. (1966), *Surface Active Agents and Detergents*, Vol. II, Interscience, New York.

Shinoda, K. (1967), *Solvent Properties of Surfactant Solutions*, Marcel Dekker, New York.

Shinoda, K., Nakagawa, T., Tamamushi, B., and Isemura, T. (1963), *Colloidal Surfactants*, Academic Press, New York.

Siu, R. G. H. (1951), *Microbial Decomposition of Cellulose*, Reinhold, New York.

Swisher, R. D. (1967), *Surfactant Biodegradation*, Marcel Dekker, New York.

Thorn, G. D., and Ludwig, R. A. (1962), *The Dithiocarbamates and Related Compounds*, Elsevier, Amsterdam.

Whistler, R. L., and Paschall, E. F., eds. (1965), *Starch, Chemistry and Technology*, Academic Press, New York.

6

Physical Chemistry of Surfaces and Interfaces

Whenever reactions occur in heterogeneous systems (such as flotation pulps), they take place predominantly at one of the interfaces, whether the solid/liquid, the liquid/liquid, or the liquid/gas. Some reactions may occur within the liquid phase, among the dissolved species, but such homogeneous reactions are not the prevailing ones in flotation systems.

In this chapter we shall deal primarily with structural features of solid interfaces, their heterogeneity, and the changes wrought upon the solid surfaces as a result of mechanical and chemical treatments preceding the addition of flotation reagents. Adsorption at interfaces is the first step in all interfacial reactions. The general characteristics of adsorption processes are dealt with in order to provide the background for some of the phenomena leading to alterations of surfaces that are necessary for flotation. A detailed discussion of adsorption of flotation reagents is given in Chapter 8, after an acquaintance with some of the electrical characteristics of interfaces in Chapter 7.

6.1. *Types of Interfaces. Colloids*

Matter exists in three types of phases: solid, liquid, and gaseous.

A *phase* is a region of space in which the chemical composition is uniform throughout and the physical and mechanical properties are the same.

When two phases meet, an extremely thin boundary layer between these phases constitutes an *interface*: This layer is endowed with specific physical and chemical properties which differ from those of the constituent phases.

There are five types of interfaces: solid/gas, solid/liquid, solid/solid, liquid/gas (or liquid/vapor), and liquid/liquid. There is no gas/gas interface since gases always mix spontaneously. Liquid/gas and solid/gas interfaces are generally referred to as *surfaces*.

The physical and chemical properties of matter in bulk are determined by the electronic structure of the constituent atoms and, primarily, by bonds existing between the atoms and molecules within each phase; these properties can be fairly easily evaluated from macroscale experiments conducted on relatively large quantities of matter. The study of interfaces is more complicated because of the very small thickness of interfacial regions. Thickness of interfaces usually ranges from a few angstroms ($1 \text{ Å} = 10^{-8}$ cm) to a few hundred of angstroms, that is, well below the resolution power of optical and even electron microscopes.

Depending on the state of subdivision of a given quantity of matter, the total area of an interface (or of a surface) may be orders of magnitude different from the geometric area of the bulk quantity of this matter. For example, when a cube of a solid with 1-cm sides is subdivided to form cubes that are 1-μm ($= 10^4$ Å) edge length and then the 1-μm cubes are agglomerated together by sintering, the overall surface area of the interface (solid/gas or solid/liquid), for the same volume of the solid, is four orders of magnitude greater, that is, approximately $6 \times 10^4 \text{ cm}^2$ instead of 6 cm^2. To study interfaces, very large areas of the particular type of interface have to be available. An interface of 1-cm^2 area can be completely covered by a single layer of atoms (of say 1-Å diameter) with 10^{16} atoms, that is, by $10^{16}/(6.023 \times 10^{23}) = \frac{1}{6} \times 10^{-7}$ mole. For most atoms or independent species, such a quantity can be barely detected by analytical techniques available nowadays. However, occasionally even a 1% of surface coverage may alter the properties of a particular surface in a drastic manner, and a fraction of 10^{-9} mole would not, usually, be detected (for 1-cm^2 area). Therefore, for studies of interfacial characteristics, the quantities of matter are selected to give total areas commensurate with the sensitivity of the technique employed for monitoring the changes in surface characteristics.

When a very large interfacial area per unit weight (or unit volume) is created for a given phase, the system is said to be in *colloidal state*, and the subdivided phase is referred to as a *colloid*. Depending on the manner in which the bulk phase has been subdivided to give large interfacial area, the following types of colloids can be distinguished:

Laminar colloids: if only one dimension is reduced to submicron (i.e., less than 10^{-6} m) values. Examples might be a thin film of oil spread over the ocean from a tanker, a suspension of any finely dispersed platelet minerals—such as montmorillonite, molybdenite, graphite—when subdivided to contain only a few lattice-cell layers in each platelet, or "dry" foam formed by some detergents when the liquid films separating bubbles are less than ~ 1 μm in thickness ("wet" foam, containing large and nearly spherical bubbles with fairly large amounts of liquid between the bubbles, does not constitute a colloidal system because its interfacial area–weight ratio is too small).

Fibrillar colloids: when two dimensions of the subdivided phase are reduced to submicron range; all fibers, whether inorganic (like asbestos) or organic in character, belong to this type of colloidal state if their diameter is below 1 μm.

Corpuscular colloids: when all three dimensions are reduced to the submicron range. This type is the most frequently encountered colloidal system in smog, fog, aerosols, emulsions, and sols.[†]

The ultimate colloids are micelles, polymer molecules, and the so-called *black films*. For such colloids the ratio of surface area–volume is of the order of 10^7. In general, the colloidal state is characterized by the values of the area–volume ratio between $\sim 10^4$ and $\sim 10^7$. For ratios less than 10^4 (that is, for dimensions larger than ~ 1 μm), the effects specifically due to interfaces no longer override the bulk properties.

6.2. *Characteristics of a Solid Surface*

Solids are crystalline or amorphous. Minerals are mostly crystalline, but a few exceptions in the form of amorphous minerals, such as opal, some fused oxides, and bitumen do occur.

On the atomic scale all solid surfaces representing individual crystal faces are far from perfectly smooth, ideal planes but contain ledges, kinks, and terraces. A terrace–ledge–kink (TLK) model of a heterogeneous surface was suggested by Kossel (1927) and is shown schematically in Figure 6.1. It has been fully confirmed quantitatively by all subsequent electron micrography and interferometric studies for all types of solids, whether metals

---

[†] *Aerosol* denotes a dispersed liquid, or solid, in a gas phase. *Emulsion* denotes a dispersed liquid in another immiscible liquid. *Foam* denotes a dispersed gas in a liquid or in a solid. *Sol* denotes a dispersed solid in liquid (or dispersed solid). All dispersed phases are in the micron and submicron ranges of sizes.

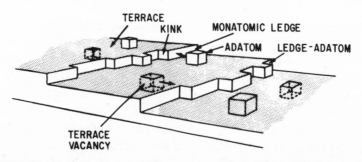

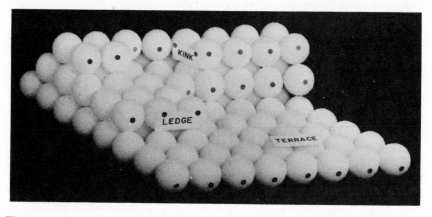

Figure 6.1. Schematic representation of the TLK (terrace–ledge–kink) model of a solid surface.

or covalent or ionic crystals (see Figure 6.2) regardless of the method used for preparing the surface. Only the liquid/gas and liquid/liquid systems can provide homogeneous, uniform, and reasonably clean surfaces that can be obtained experimentally on a sufficiently large macroscopic scale (hundreds of square centimeters) to enable accurate studies to be carried out.

Interfaces involving solids are always highly heterogeneous in character, varying from one submicroscopic speck to another. Because of the relative immobility of atoms in the solid, its surface properties depend greatly on the immediate *past history of the solid*. A freshly formed cleavage surface in a vacuum has different physical and chemical properties from those of ground, abraded, or electropolished surfaces. Any heat treatment or exposure to different environments again produces different surfaces, contaminated by chemical reactions with O_2, CO, H_2O, etc., or by adsorption of organic molecules from the atmosphere. As a result, an unambiguous interpretation of results obtained with solid interfaces is very difficult, and there is always some uncertainty associated with it.

The concept of a clean surface is abstruse—theoretically possible only with (nonexistent) perfect crystals in a complete vacuum (which is unattainable)—and the degree of cleanliness, when measured, depends on the sensitivity of the detector used in evaluating the contamination. Even when we are dealing with surfaces exposed to *vacuum*, a large degree of contamination of the surface by the residual gas molecules can be expected.

In an ultrahigh vacuum of 10^{-13} torr (1 torr = 1 mmHg pressure), which is the best vacuum that can be obtained at present, there is one molecule of residual gas striking every submicroscopic area of 1 (μm)^2 every second (Table 6.3). Under atmospheric conditions of 760 mmHg pressure the number of collisions with the surface at 25°C is 2.9×10^{23}/sec cm^{-2} (of geometric area). Thus, any solid/gas interface under atmospheric conditions will always be covered by numerous layers of adsorbed gas molecules. If the gas phase consists of a mixture, e.g., N_2, O_2, and H_2O vapor, the amounts of individual gases that are adsorbed on the solid depend on their respective partial pressures and on the *sticking coefficient* (see Section 6.4); water molecules or O_2, despite their lower (than N_2) partial pressure may be adsorbed preferentially because of higher sticking coefficients.

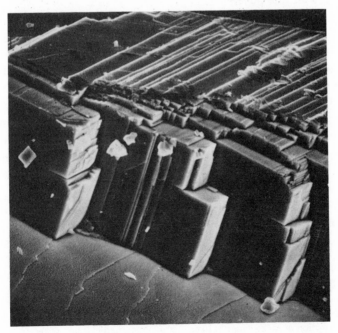

Figure 6.2. Photograph of a galena surface, after cleavage (160 $\times$).

Let us consider the formation of an interface by a cleavage of a perfect crystal in vacuum; a cleavage severs all bonds initially existing across the surface of the cleavage plane between the atoms residing on the opposing sides of this plane. The broken bonds endow the surface atoms with an excess energy, that is, *surface energy*, equal theoretically to half of the *cohesion energy* (sum total of all bond energies) across the plane of cleavage. However, immediately following the fracture of bonds the outermost atoms along the cleavage plane undergo a limited amount of rearrangement, strengthening the bonds between the surface atoms and those directly below the surface; this rearrangement absorbs a certain amount of the excess surface energy, reducing its initial high value. A certain amount of the cohesion energy is also lost in radiation of heat and in electrical discharges. Further, owing to the fact that even in a vacuum there are numerous molecules of residual gas available, adsorption of some of these residual molecules from the gaseous phase partially satisfies an appreciable proportion of the initially broken bonds. The result is a further reduction in the residual excess energy associated with the newly formed surface.[†] However, despite this overall decrease, each portion of the surface (whether represented by a terrace, ledge, or kink) has a definite excess energy, known as surface energy, associated with this portion.

Due to anisotropy of crystalline solids, the surface energy of a cleavage plane depends on its crystallographic orientation. As already mentioned in Section 3.2, a single crystal held at elevated temperature (close to its melting point) ultimately assumes an equilibrium shape bounded by crystallographic planes of minimum surface energy, such as the (111), (100), and (110) orientations expressed by Miller indices. (Miller indices are the numerators in the reciprocals of intercepts for a given crystallographic plane with a set of coordinate axes parallel to the edges of a unit cell; see Figure 6.3.) However, even these low-index crystalline faces are not perfectly flat terraces but contain monoatomic ledges and kinks. A perfect cleavage does not denote a flat surface on the atomic scale.

To give a specific example of the influence exerted by crystallographic orientation, let us consider a crystal of germanium in its diamond cubic structure. The number of broken bonds varies with the orientations, and so do the corresponding work functions (constituting a measure of surface

---

[†] For example, surface energies of mica [quoted by Bowden (1967)] have been determined as: 5000 ergs/cm² —clean in high vacuum; 300 ergs/cm² —exposed to air; 180 ergs/cm² — after adsorption of water vapor; 110 ergs/cm² —immersed in liquid water; 37 dynes/cm² —with deposited monolayer of lauric acid.

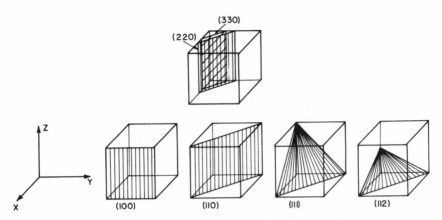

Figure 6.3. Miller indices of planes in cubic crystals.

energy) and the corresponding dissolution rates (i.e., chemical activity), as shown in Table 6.1.

The broken-bond technique can be utilized to evaluate the relative surface energies and to derive relations with the sublimation energy and the fusion energy, using the atomic coordination numbers of the close-packed solid. However, this technique can be applied only to materials which are bonded by short-range forces, such as metallic or covalent bonds. It cannot be applied either to ionic solids where coulomb forces are long range in nature or to mixed, ionic–covalent solids which are represented by the vast majority of minerals.

In the case of ionic crystals, the surface energy can be calculated using the elastic constant E (Young's modulus), the periodic character of the

Table 6.1. Effects of Crystallographic Orientation on Surface Characteristics of a Germanium Crystal[a]

Orientation of cleavage plane	Broken bonds per cm²	Relative work function	Relative dissolution rate
(100)	1.25×10^{15}	1.0	1.0
(110)	8.83×10^{14}	0.95	0.89
(111)	7.2×10^{14}	0.93	0.62

[a] From Gatos (1962).

Table 6.2. Surface Energies (in erg/cm²) of Selected Ionic Crystals[a]

Material	$\gamma_{s/v}$, experimental	$\gamma_{s/v}$, calculated by Eq. (6.1)
NaCl	300	310
MgO	1200	1300
LiF	340	370
CaF$_2$	450	540
BaF$_2$	280	350
CaCO$_3$	230	370

[a] From: Brophy *et al.* (1964).

potential energy of the ionic lattice, and the interatomic distance r_0, resulting in

$$\gamma_{s/v} = \frac{E}{\pi} \int_0^{r_0} \sin\left(\frac{\pi x}{r_0}\right) dx = \frac{E r_0}{\pi^2} \tag{6.1}$$

where $\gamma_{s/v}$ is the surface energy of solid/vapor, E is the elastic constant, and $\sin(\pi x/r_0)$ is the periodicity of the potential energy.

A comparison of calculated and experimental surface energy values for various ionic crystals is given in Table 6.2.

It must be realized, however, that neither the calculated nor the experimental values can be considered truly reliable because of the many approximations adopted in both approaches. Absolute data for solid/gas or solid/liquid interfaces cannot be obtained by any of the experimental techniques available at present. An order of magnitude is the best result that can be hoped for. In both the theoretical and experimental approaches of evaluating the quantitative magnitudes of surface energies or heats of adsorption, the degree of heterogeneity of the solid surface is extremely difficult to account for.

For a cubic structure, a crystalline face other than a low-index orientation can be theoretically repreented by an asrray of ledges at an angle θ to the close-packed low-index face, and its surface energy is then expressed by

$$\gamma(\theta) = \gamma_0 \cos\theta + \frac{\varepsilon}{h} \sin\theta \tag{6.2}$$

where γ_0 is the work (energy) to create a unit area of the close-packed, low-index face, ε is the work to create a ledge, h is the height of the single ledge, and θ is the angle between the crystal face and the close-packed, low-index face.

Equation (6.2) enables some useful predictions to be made about the *relative* dependence of γ on orientation. It shows that the relative tendency to satisfy broken bonds increases for those atoms which are progressively more exposed, that is, for terrace < ledge < kink. Thus, atoms in the kink position represent the most likely sites for adsorption of molecules from the neighboring phase (gaseous or liquid), then the atoms in ledges are next, and the least likely sites for initial adsorption are atoms in terraces.

As shown in Table 6.3, for solid/gas interfaces the number of collisions with each unit area of a surface each second depends on pressure (or vacuum) in the gas phase. (For solid/liquid interfaces, the collisions of given ions depend on their concentrations in solutions.) Collisions leading to adsorption (that satisfy broken bonds) do not necessarily occur preferentially with kinks and ledges but do occur indiscriminately all over the exposed surfaces. Thus terraces are hit by atoms and molecules with the same frequency per atom as the atoms in ledges and kinks. Some of the colliding atoms will diffuse along the terrace to a higher-energy site—kink or ledge—and become adsorbed therein; others will rebound. The proportion of rebounding species is inversely related to the energy associated with the given site.

Table 6.3. Classification of Vacuum

Parameter	Atmospheric conditions	Vacuum[a]			
		Rough	Medium	High	Ultrahigh
Pressure (torr)	760	1	10^{-3}	10^{-8}	10^{-13}
Density (molecules/cm³)	2.5×10^{19}	3.2×10^{16}	3.2×10^{13}	3.2×10^{8}	3.2×10^{3}
Collisions with 1-cm² surface, at 25°C, per second	2.9×10^{23}	3.8×10^{20}	3.8×10^{17}	3.8×10^{12}	3.8×10^{7}
Mean free path at 25°C (cm)	6.6×10^{-6}	5×10^{-3}	5	5×10^{5}	5×10^{10}

[a] The range of conditions from one value to another next to it is described by the heading above the righthand value.

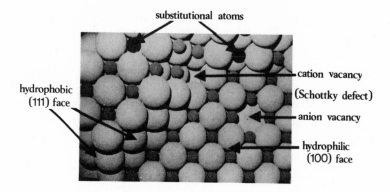

substitutional atoms

hydrophobic
(111) face

cation vacancy

(Schottky defect)

anion vacancy

hydrophilic
(100) face

GALENA SURFACE

Figure 6.4. A model of a galena surface showing a hydrophilic (100) face, initial portions of a hydrophobic (111) face, an anion and an adjoining cation vacancy (constituting a Schottky defect or imperfection), and substitutional impurity atoms replacing Pb^{2+} in the lattice. Not shown is a Frenkel imperfection formed when an interstitial host or impurity atom adjoins a vacancy from which it (presumably) has derived.

With aged solid surfaces, diffusion of surface active impurities from the interior of the solid toward the surface, and relief of stress accompanied by emerging dislocations, tend to decrease the overall excess energy associated with the surface still further in comparison with the surface energy of a newly created surface.

All solids possess imperfection such as point defects and line dislocations distributed throughout the bulk of the crystalline phase, and whenever a surface is created, it incorporates some of these imperfections in addition to ledges and kinks. The types of point defects (very localized interruptions in the regularity of a lattice) which occur in ionic solids are shown in Figure 6.4. A cation vacancy and an associated anion vacancy (the pair is known as a *Schottky imperfection*) occur in different portions of the crystalline lattice; when a cation has moved to a neighboring site in an interstitial position, a so-called *Frenkel imperfection* is created. Also, most of the naturally occurring crystals of many minerals do not have the ideal stoichiometric composition ascribed to them, and, in addition, they contain traces of foreign atoms as impurities. In fact, it is shown by thermodynamic reasoning [see Prigogine and Defay (1954, p. 242)] that slight departures from stoichiometry create a more stable system. The impurity atoms can either occupy the sites in the lattice substituting for the host atoms or become additional *interstitial* atoms. If the crystal contains 0.01% atomic impurity, the number of impurity atoms in a 1-cm² surface would amount to ~10^{12} atoms for each layer of ~10^{16} atoms in this surface if the impurity atoms are not surface active. Surface active impurity atoms would tend to diffuse to the newly created surface and—in time—would establish a surface equilibrium excess in a manner analogous to that for a surfactant adsorbing from the bulk solution at the air/water interface (see Section 5.7).

Two types of line dislocations are distinguished: an edge dislocation [shown in Figure 6.5(a)] and a screw dislocation [shown in Figure 6.5(b)]. The dislocations represent the regions of lattice distortion such that when an extra half plane of atoms has been squeezed out, an *edge dislocation* is formed, and when atoms on one plane are progressively displaced along a helical path to coincide with a plane of atoms lying above or below the original plane, a *screw dislocation* is created.

An edge dislocation causes the surrounding lattice atoms on the side of the extra half plane to be under compressive stress, while the atoms on the side immediately below the slip plane are in tension. A screw dislocation has neither tensile nor compressive strain but only shear strain associated with it. The strain energy of a dislocation leads to a faster rate of chemical reaction at the point of emergence of this dislocation at the free surface. Similarly, because of the strain energy, dislocations act as sites for precipitation of impurity atoms from the solid or the solution phase. These features of increased chemical activity (in dissolution and precipitation) are used extensively as techniques for detecting dislocations.

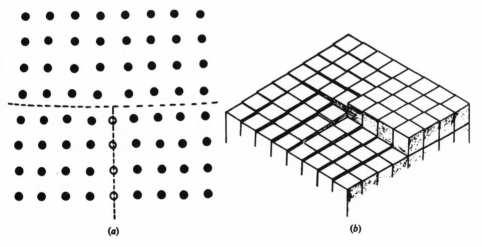

Figure 6.5. (a) Edge dislocation and (b) screw dislocation in a simple cubic crystal.

The helicoidal structure due to a screw dislocation provides the self-perpetuating growth step, particularly in crystallization from solutions of low supersaturation. All minerals derived from hydrothermal deposition (or vapor condensation) contain many dislocations in the as-grown condition, forming dislocation networks. Since the magnitude of either an edge or a screw dislocation is represented by a so-called *Burgers vector* (indicating the length and direction of the shift in lattice due to the dislocation), the density of dislocations is measured in centimeters per cubic centimeter. Highly annealed, strain-free metal may have 10^3–10^4 cm/cm^3 dislocation density; normal crystals deposited from solution or melt have 10^8–10^{10} cm/cm^3 dislocations; highly strained crystals may have 10^{12}–10^{14} cm/cm^3 dislocations.

Alterations of Solid Surfaces Caused by Mechanical Forces (Such as in Grinding)

When solid surfaces are produced by means other than cleavage, for example, by cutting, grinding, abrasion, polishing, etching, electronic bombardment, etc., layers of damaged material extend for some depth below the surface. Such layers may be a few angstroms thick, as for electropolished surfaces, to hundreds of microns thick, as is the case of surfaces obtained by a mechanical saw cut.

The damaged layer, known as the *Beilby layer*, is highly disarrayed; the continuity of crystalline lattices in individual single subgrains is broken, resulting in highly strained material with locked-in dislocations and imperfections.

The prolonged pounding taking place during grinding of ore minerals in ball mills may annihilate all remaining crystallinity in the damaged layer, converting it entirely into an amorphous phase. Such changes from the crystalline toward the amorphous state have been reported in the past by Lin and Somasundaran (1972), Gammage and Glasson (1976), and Lin *et al.* (1975). The last authors reviewed the relevant work on alterations occurring in solids during their comminution by grinding. Lin and Somasundaran (1972) found, for example, that when quartz was ground in air for several hundred hours, a progressive decrease in the density of quartz from 2.65 (for 1000-μm-size particles) down to $\sim$2.37 for 0.1-μm particles occurred. This was interpreted as due to the formation of increasing amounts of amorphous quartz, the density of which is reported in the literature to be 2.203 (while that of the crystalline quartz is 2.650). Wet grinding (that is, grinding in a liquid medium, not necessarily aqueous) produces smaller quantities of the amorphous material for the same expenditure of energy due to a cushioning effect of the liquid, on the one hand, and to a partial disintegration of the freshly wetted surface layer, on the other hand. As reported by Bikerman (1970, p. 229), Gurvich (1915) carried out systematic determinations of the quantity of fines produced by immersion of a known amount of fuller's earth in different liquids. That quantity was 2.82% of the initial weight when the liquid was water, 1.60% when methanol was used, 0.90% for ethyl ether, and 0.22% for hexane. The quantities of fines produced appear to be related to the respective heats of wetting. Calcium carbonate, charcoal, and lead shot provided similar sets of results. Small amounts of *active* solutes such as copper oleate, dioctylsulfosuccinate, lecithin, etc., when added to the liquid, acted as very effective dispersants and increased the quantity of fines produced during wetting of a given amount of a solid.

In an aqueous medium, a partial or a complete removal of the amorphous layer also takes place by dissolution, depending on pH and the presence of other solutes conducive to the formation of soluble complexes. Dissolution is, however, not the only reaction that may be influenced by other solutes since, depending in the nature of the solutes, the initial dissolution may be followed by other reactions such as precipitation and redeposition of the products on the solid surface in place of the initially formed amorphous layer. To the last category belong all surface reactions of metal hydroxide deposition, oxidation of sulfur to sulfate minerals, deposition of insoluble metallic sulfates on initially sulfide minerals, and deposition of insoluble metallic salts of reactive surfactants. The increased chemical reactivity of finely comminuted solids has been well documented [and is reviewed by Lin and Somasundaran (1972)].

The extent of the damaged layer produced on various minerals during grinding depends on the energy input, the type of crystalline lattice of the solids, and the environment. During fracture and abrasion a large amount of heat may be generated, and if this is not dissipated to the environment, the rise in temperature produced locally may be very pronounced. This local change in temperature may cause an alteration of the physical characteristics of the solids varying in intensity from, for example, a transformation in semiconductivity to a recrystallization, solid-state decomposition, or fusion followed by recrystallization. Carta *et al.* (1974) reported that fluorite and calcite changed their semiconducting characteristics from *n* type to *p* type (see Chapter 7 for details on semiconductivity) during a dry grinding but not during an aqueous grinding. Similar *n*- to *p*-type changes were observed by these workers when the preceding minerals were heated to higher temperatures or were wet-ground in the presence of H_2O.

More severe mechanical work can cause recrystallization. Lin and Somasundaran (1972) observed that after $\sim$10 hr of dry grinding nearly a quarter of the calcite had been converted to aragonite (the polymorphic transformation was identified by an X-ray diffraction analysis), while after 100 hr of grinding approximately 70% calcite was so transformed to aragonite. A reverse transformation of aragonite to calcite can be achieved by heating aragonite to 520°C in *differential thermal analysis* (DTA). Polymorphic transformations occurring during grinding have been reported in the literature for wurtzite into sphalerite (ZnS), massicot (PbO) into litharge (PbO), monoclinic tetrametaphosphate of sodium ($Na_4P_4O_{12}$) into the triclinic form, etc. [for references, see Lin and Somasundaran (1972)]. A prolonged dry grinding causes decomposition of some carbonates into oxides (for example, zinc, manganese, and cadmium carbonates) or of highly hydrated into less hydrated salts ($FeSO_4 \cdot 7H_2O \rightarrow FeSO_4 \cdot 4H_2O \rightarrow FeSO_4 \cdot H_2O$; $BaCl_2 \cdot 2H_2O \rightarrow BaCl_2 \cdot H_2O \rightarrow BaCl_2$).

The production of a damaged or an amorphous layer on the solid surface and the physical changes resulting from a local increase in temperature are not the only alterations taking place during mechanical work performed on the solid. Whatever the medium used as a tool or a support in exerting the pressure onto the solid—whether a hammer, a saw, a grinding medium (pebble, iron, or steel), the lining of a mill or a crusher—there is always a certain amount of this medium wearing off and depositing on the surface of the solid. Evidence accumulated in tribophysics and tribology is overwhelming in this respect, and the progressive wear of the grinding media and of the mill liners is a constant reminder to the mineral plant operators of this phenomenon. In many systems concerned with a relatively coarse-size grinding, the extent of wear is not a major problem, but in some fine-grinding systems even the minor quantities of abraded medium (a fraction of a kilogram per tonne of solid treated) may create difficulties, for example, in the manufacture of optical glass or in white paint additives.

With respect to flotation systems, any prolonged grinding of sulfides, necessary for the liberation of highly disseminated ores, creates difficulties in selectivity and recovery, particularly if the grinding is carried out in steel mills. These difficulties have been ascribed by Rao *et al.* (1976) to the changes in the electrochemistry of the surface reactions necessary for flotation. Such changes in reactions are caused by the galvanic action of a greatly increased proportion of the metallic iron phase abraded onto the surface of the semiconducting sulfides (compare Sections 1.5 and 8.2).

From the preceding discussion of surface alterations it becomes obvious that the "past history" undergone by the solid leaves an imprint, more or less indelible, on its surface characteristics. This imprint takes the form of either a damaged semicrystalline layer or an amorphous layer, or redeposited reaction products, or mixed patches composed of a damaged layer covered with flecks of worn-off grinding medium, or highly heterogeneous mixtures of all these alteration "ingredients."

It is rather surprising that despite these numerous possibilities of alteration to the surfaces, the chemical and crystallographic nature of the given mineral is reflected in the surfaces of its coarse particles (that is, particles greater than $\sim$10 μm) obtained by conventional grinding processes and allows their selective separation by flotation. The reason for such behavior may be due to several complementary factors, such as the following:

1. The initial heterogeneity of the mineral surfaces obtained by cleavage is already very high, and its increase due to surface alterations taking place during grinding has no undue ill effects until a very fine grind is required.

2. The extent of the mineral surface that is being exploited in flotation represents only a small fraction of the overall surface of the particles.

3. The aqueous environment counteracts the surface alterations by dissolving a major portion of the highly stressed and amorphous alteration products.

The more prolonged the grinding, the more severe are the alterations suffered by the surfaces of all solids involved (of gangue as well as of valuable minerals) and the lower the likelihood of a highly selective separation. Therefore, for a successful treatment of very finely disseminated ores the present approach of using entirely mechanical grinding for a complete liberation will have to be changed in the future and replaced by a new approach of either a mixed mechanical–chemical liberation or an entirely chemical intergranular disintegration along the grain boundaries. In some instances such an approach has already proved fairly fruitful, for example, in coal disintegration by a treatment in a methanolic solution of NaOH [Aldrich *et al.* (1974)]. A mixed mechanical–chemical treatment is effectively carried out during the heap leaching of coarsely comminuted low-grade ores, sometimes giving a complete removal of the valuable mineral. Whether, in the future, it will be more beneficial to change the process of hydrometallurgical leaching into one restricted to grain boundary dissolution that is followed by disintegration and a subsequent selective concentration step (flotation or spherical agglomeration) will depend on the respective economics of the two processes.

6.3. Adsorption and Its Characteristics

Adsorption denotes the process of concentration of any chemical species i occurring at an interfacial region separating two phases. The species i may be either charged or uncharged (i.e., ionic or nonionized), polar or nonpolar (i.e., with or without a permanent dipole), monoatomic or polyatomic. Diffusion is the main mechanism by which transport of the adsorbing species to the interface occurs. Hence, it is usually the species present in the less dense of the two phases (where the diffusion is faster) that becomes concentrated at the interface more readily.

At equilibrium, the so-called *surface active species* will be concentrated at the interface regardless of whether it is present in the denser or the less dense of the two phases or in low or in high numbers as compared to a nonsurface active species. The adsorbing species is called the *adsorbate*; the denser phase onto which adsorption takes place is the *substrate* or the *adsorbent*.

Absorption involves adsorption followed by an additional penetration of the adsorbing species into the substrate phase. Thus, absorption involves diffusion of the adsorbing species into the bulk of the liquid or the solid representing the substrate phase.

When freshly formed surfaces of solids are exposed to a gaseous phase, adsorption of the gases takes place, depending on the partial pressure of each component gas. Even in a so-called "high" vacuum there are enough molecules of residual gases present to cover completely a solid surface of limited area within a relatively short time—tens of minutes. Table 6.3 shows the relevant data concerning vacuum classification.

When a gas or a vapor is allowed to come to equilibrium with a solid surface, the concentration of gas molecule is *always* greater in the vicinity of the surface than in the gas phase, regardless of the nature of the gas or the surface. When a mixture of several gases is present in the gas phase, the relative concentrations of these gaseous components adsorbed in the interfacial region of a solid will be, in most systems, quite different from the composition of the gas phase.

6.4. *Thermodynamic Models of Isotherms for Physical Adsorption*

An *adsorption isotherm* represents a functional relationship between the amount adsorbed at an interface and the activity of adsorbate (i.e., partial pressure of a gas or concentration of a solute for sufficiently low concentration) obtained under the condition of *constant temperature* in the system. Theoretical expressions for different types of adsorption isotherms are derived when *various assumptions* are made with respect to the behavior of the components in the system. Thus, if the assumption is made that the adsorbing gas behaves as a perfect gas, both before and after adsorption, that is,

$$pV = nRT \quad \text{for the gaseous state}$$

$$\phi A^s = nRT \quad \text{for the adsorbed state}$$

where p is the pressure, V is the volume, ϕ is the surface pressure, A^s is the surface area, n is the number of molecules, T is the absolute temperature, and R is the gas content, the Henry's law linear adsorption isotherm is obtained. Other assumptions yield expressions depicting Langmuir, Frumkin, Freundlich, etc., isotherms (see Sections 6.4 and 6.6).

For dilute solutions of surfactants, the Gibbs adsorption equation,

$$-d\gamma = RT\Gamma_s \, d\ln a_s \tag{2.80}$$

gives an adsorption isotherm for nonionized species at the solution/gas interface in the form

$$\Gamma_s = -\frac{1}{RT} \frac{d\gamma}{d\ln a_s}$$

When ionized surfactants, in the presence of neutral salt additions, are adsorbed at the gas/solution interface, the adsorption isotherm is modified by the salt parameter m (see Section 2.9.2):

$$\Gamma_s = -\frac{1}{mRT} \frac{d\gamma}{d\ln a_s}$$

At the liquid/gas interface the adsorption of surfactants present as a solute in the liquid phase is always physical in nature and fully reversible as long as no chemical alteration of the adsorbed film occurs. However, at solid/gas and solid/liquid interfaces, adsorption may in general be either physical or chemical in nature.

If no transfer of electrons takes place during adsorption, the process is considered to be *physical adsorption*. If, at some stage, electrons are transferred during adsorption, then such adsorption becomes *chemisorption*.

The criterion of charge transfer (or sharing of electrons) is used to differentiate between the physical and the chemical adsorption processes; it is the most favored but by no means the only universally adopted criterion. Numerous criteria were used in the past for the purpose of differentiating the two types of adsorptions, but invariably just as numerous exceptions were encountered. Of the criteria used in the past, such as the relatively lower heat of physical adsorption, the existence of an activation energy for chemisorption, the reversibility of adsorption–desorption, the faster rate of physical adsorption, and, lately, the differences in the entropy of adsorption, many are still frequently utilized as useful guidelines in distinguishing the physical from chemical adsorption. Thus, physical adsorption is usually (but not always) fast and *reversible* and with low heats of adsorption (< 10–15 kcal), while chemisorption is generally slow and *irreversible*, requires an appreciable energy of activation (hence, it is favored at higher temperature), and generates a large exothermic heat of adsorption. However, there are chemisorption processes that are too fast to be measured, with heats of adsorption zero or endothermic and with zero activation energy even at very low temperatures. An example of such chemisorption is the adsorption of H_2 on Ni at $20°$K, which involves dissociation, is very fast, and requires no activation energy.

The distinction between physical and chemical adsorption is not a fundamental one, and many cases of intermediate character of adsorption occur. The analogy with intermediate types of bonding, such as partly covalent and partly ionic or partly covalent and partly van der Waals, is very appropriate.

Since gas adsorption occurs as a natural or spontaneous process and since it involves the loss of at least one degree of freedom for the adsorbing species (its entropy is *decreased*), the process of its physical adsorption is always exothermic.

For the spontaneous process,

$$\Delta G = \Delta H - T \Delta S < 0 \tag{6.3}$$

but since $\Delta S < 0$ for adsorption of gaseous species, $-T \Delta S > 0$. Hence, ΔH must be < 0; that is, the process of physical adsorption of any gas is an exothermic process.

The Lennard-Jones potential energy diagrams, as shown in Figure 6.6

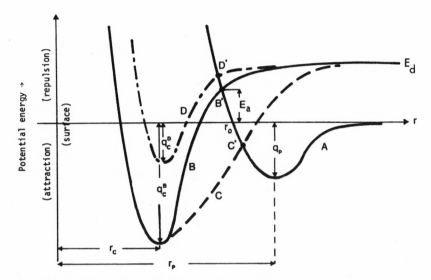

Figure 6.6. Energy diagrams for A: physical adsorption of species XY; q_p is the energy of physical adsorption; B: chemisorption of dissociated X and Y species with activation energy E_a; q_C^B is the energy of chemisorption; C: chemisorption of dissociation X and Y without any energy of activation; D: chemisorption on portions of surface involving lower energy of adsorption q_C^D when increased coverage causes repulsive lateral interactions between adsorbing species.

for an adsorbate species approaching a surface, help to distinguish among the various characteristics of physical and chemical adsorption processes. At a large distance from the surface the potential energy of the adsorbate with respect to the surface is zero by definition. This means that the curve representing the potential energy is asymptotic to the axis of abscissae r, as shown by curve A in Figure 6.6. The potential energy is described by the expression

$$U(r) = -4U_p\left[\left(\frac{r_0}{r}\right)^6 - \left(\frac{r_0}{r}\right)^{12}\right] \tag{6.4}$$

where r_0 is the distance from the surface at $U(r) = 0$ and $U_p = q_p$, which was derived for an inert (monoatomic) gas at low pressure. U_p denotes the depth of the well at r_p. The potential energy curve passes through a minimum at r_p which denotes the equilibrium distance for the physically adsorbed gas atom. As the distance is further decreased, the potential energy curve begins to rise sharply due to a rapid increase in the repulsion forces. The bonding established at r_p by an inert gas atom is a result of van der Waals forces between the adsorbate and the surface, and this type of bonding causes only minor distortion in the structure of the substrate or the adsorbate. However, when electrons are transferred or redistributed, as in chemisorption, the potential energy diagram of a reactive atom, represented by curve B in Figure 6.6, shows a deeper minimum, at $r_c < r_p$. A greater amount of surface–adsorbate perturbation occurs due to the development of chemical bonds stronger than van der Waals. An inert atomic gas is not capable of chemisorption. Reactive adsorbate is therefore either a diatomic or a polyatomic molecule, and the potential energy of individual atoms comprising the adsorbate must take into account the dissociation energy E_d. Dissociation of the molecule would raise the energy (for large distances r) of individual atoms to a level E_d above the axis of abscissae.

Point B′ at which the potential energy curve A for physical adsorption crosses curve B for chemisorption denotes the activation energy E_a required for the transfer from a physically adsorbed to a chemisorbed state. If this crossover point of the two potential energy curves is either on the axis of the abscissae or below it, as shown in the case of curve C with curve A at point C′, then no activation energy is required. Consequently, the molecule would transfer immediately from a physically adsorbed to a chemisorbed state.

The depth of the minimum at r_p or r_c corresponds, approximately, to the heat of adsorption; $U_p = q_p \cong \varDelta H_p$ for the physical adsorption, and $q_c \cong \varDelta H_c$ for the chemisorption. As chemisorption progresses and the coverage increases from zero to unity (for a complete monolayer coverage), the potential energy curve shifts gradually from curve B toward curve D (and beyond), changing the relative values q_c of adsorption energy.

The application of infrared spectroscopy to adsorption systems has proved an extremely useful technique to distinguish whether any transfer of electrons has taken place during adsorption [see Eischens and Pliskin (1958)]. If a given species is physically adsorbed, the spectrum of the adsorbent–adsorbate system shows no new bands but is a composite of individual spectra of adsorbent and of adsorbate. For a chemisorbed species, a definite change in the spectrum (from that representing a composite of adsorbent plus adsorbate spectra) is usually obtained, with new bands and/or definitive shifts in other bands of adsorbate being observed.

As shown in Figure 6.6, physical adsorption is always unactivated, whereas chemisorption may require an activation energy E_a or may also be unactivated. Chemisorption occurs at a shorter distance from the surface than physical adsorption, but the depth of the adsorption well—indicative of the heat of adsorption—may be greater or smaller than that for physical adsorption. Chemisorption is always limited to a monomolecular layer.

In addition to the classification into physical and chemical, adsorption may also be classified as *localized* or *nonlocalized, monolayer* or *multilayer, mobile* or *immobile*.

In *localized adsorption* on a uniform (homogeneous) solid surface, the adsorbate species are considered as residing at discrete sites of minimum potential energy, and the energy for lateral transfer by surface diffusion from one site to another is always equal to or only slightly smaller than the adsorption energy. For a nonuniform heterogeneous surface the diagram of potential energy minima in any direction along the plane of the surface may be represented as in Figure 6.7, where the potential minima are spaced in an irregular pattern over the surface (though in a definite relation to atomic positions of this surface) and vary in magnitude. Nonlocalized adsorption is represented as a limiting case when the relative depths of the potential wells (that is, the energy for lateral transfer) become reduced to zero ($v_0 \rightarrow 0$; Figure 6.7) but the adsorption energy $U_0 \gg kT$. Nonlocalized adsorption is also mobile, but localized adsorption may be mobile or immobile depending on the relative magnitudes of the energy for lateral transfer (surface diffusion) and the energy of adsorption at the potential well U_0. If v_0 or v_0' is small enough, that is, of the order of thermal vibration energy, $v_0 \cong kT$, the adsorbed species may move around on the surface considerable distances during the lifetime of adsorption. The lifetime of an adsorbate gas molecule residing on the surface is related to the heat of

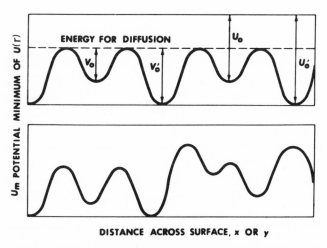

Figure 6.7. Energy diagrams for adsorption on nonuniform surfaces, differentiating between energies of adsorption U_0 or U_0^1 and energies for lateral mobility from one site to another (energy for surface diffusion) v_0 or v_0^1.

Table 6.4. Average Lifetime of Adsorption at 25° and 400°C[a,b]

Heat of adsorption, q (kcal/mol)	t, 25°C (sec)	t, 400°C (sec)
0.1	6.0×10^{-14}	5.0×10^{-14}
1.0	2.7×10^{-13}	1.0×10^{-13}
10.0	1.6×10^{-6}	8.5×10^{-11}
15.0	9.0×10^{-3}	3.5×10^{-9}
20.0	50.0	1.4×10^{-7}
25.0	3.0×10^{5}	1.2×10^{-6}
30.0	2.0×10^{9}	2.5×10^{-4}
40.0		4.0×10^{-1}
50.0		1.4×10^{2}

[a] $t_0 = 5 \times 10^{-14}$ sec.
[b] Source: Clark (1970).

adsorption q by the expression

$$t = t_0 e^{-q/RT} \tag{6.5}$$

where t_0 is related to the period of atomic vibrations in the solid. Table 6.4 shows some calculated average lifetimes at 25° and 400°C.

It is evident from Figure 6.6 that even though in physical adsorption there is no activation energy involved, the *desorption* of a physically adsorbed species always requires an activation energy at least equal to the removal of the adsorbed species from the potential well; that is, E (*desorption*) $\geqq q_p$. For a chemically adsorbed species which underwent dissociation, the activation energy for desorption is at least equal to the sum $q_c + E_d$.

Calculations of the energy involved in physical adsorption have been carried out for a number of gas–solid systems by summation of the whole variety of forces that are possible in the particular system, such as ionic, dipole–dipole, dipole–induced dipole, and dispersion forces. The results of such calculations have then been compared with the experimentally obtained heats of adsorption.

For adsorption of *inert gases on potassium or sodium salts*, such a comparison gives a fairly good agreement. Calculations and experimental data reproduced in Table 6.5 show that the octahedral (111) faces of KCl adsorb argon, Ar, with a higher energy than the cubic (100) faces.

There is a satisfactory agreement in both calculated and experimental results of the two independent investigations. However, the agreement is not so good for N_2, CO_2, and polar molecules.

Table 6.5. Heats of Adsorption of Ar on KCl[a]

Face	Young's Data (1951)		Hayakawa's Data (1957)	
	Experimental (cal/mol)	Calculated (cal/mol)	Experimental (cal/mol)	Calculated (cal/mol)
(111)	2600	2300	2460	2140
(100)	2100	1800	2080	1900

[a] From Young and Crowell (1962, pp. 44, 45.)

Kiselev and Poshkus (1958) calculated the heats of adsorption of benzene (and toluene) vapors on MgO (correcting for lateral interactions at half a monolayer coverage) and compared these with the experimentally determined values, obtaining 9300 cal/mol for calculated vs. 9200 and 9100 cal/mol for experimental results. Further, they estimated that

Dipole–dipole interactions accounted for 81–83% adsorption energy.
Dipole–quadrupole interactions accounted for 12–14% adsorption energy.
Quadrupole–quadrupole interactions accounted for 3% adsorption energy.
Induced dipole interactions accounted for 2% adsorption energy.

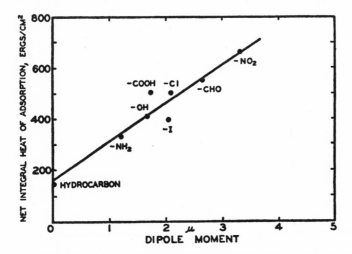

Figure 6.8. Net integral heat of adsorption for various butyl derivatives on rutile (TiO_2) as a function of the dipole moment of the liquid. [From Chessick *et al.* (1955), with the permission of National Research Council, Canada.]

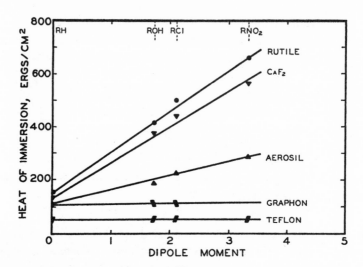

Figure 6.9. The heats of immersion of butyl derivatives on various surfaces as a function of dipole moment of the liquid. [Zettlemoyer *et al.* (1958), reprinted with permission from *J. Phys. Chem.* **62**. Copyright (1958), American Chemical Society, Washington, D.C.]

Adsorption of *polar* molecules on *ionic crystals* involves primarily dipole interactions. Experimental determinations of heats of immersion by Zettlemoyer *et al.* (1958) and Chessick *et al.* (1955) show that the energy of interaction of several polar compounds is linearly proportional to their dipole moments, as shown in Figures 6.8 and 6.9. Figure 6.9 also shows the effect of the substrate itself. The more polar the adsorbent (CaF_2 or TiO_2), the greater is the dependence on the dipole moment of the adsorbate. The nonpolar Graphon and Teflon substrates show no effect of the polarity of adsorbates.

6.4.1. *The Henry's Law Isotherm*

The Henry's law isotherm (obtained in assuming perfect gas behavior in the gaseous and adsorbed states) gives a linear relationship between the amount adsorbed V_a and the equilibrium gas pressure p:

$$V_a = V_m b p \qquad (6.6)$$

where V_m is the amount necessary for a monolayer coverage and b is a constant.

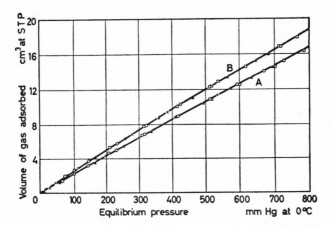

Figure 6.10. Adsorption isotherms of oxygen (A) and nitrogen (B) on silica gel at 0°C. [From Lambert and Peel (1934), *Proc. Roy. Soc. London Ser. A* **144**, with the permission of The Royal Society, London.]

Linear isotherms of this type were obtained experimentally for the adsorption of O_2, N_2, and Ar on charcoal and silica gels at room temperature and pressures up to 1 atm, an example of which is shown in Figure 6.10. In general, Henry's law isotherm applies only to a very small portion of any isotherm during initial adsorption, usually when the fraction adsorbed is less than ∼1% of a monolayer coverage.

6.4.2. *The Langmuir Adsorption Isotherm*

The Langmuir adsorption isotherm has been derived by Langmuir based on the following assumptions:

1. Gas molecules adsorbing on unit area of surface in a unit time represent only a given fraction τ of all those that strike the bare surface, while those striking the already preadsorbed molecules are elastically reflected back to the gas phase.

2. The energy of adsorption is invariant.

From the kinetic theory of gases, the number of molecules striking a unit area of surface every second is given by

$$n_a = p(2\pi m k T)^{-1/2} \tag{6.7}$$

where p is the pressure, m is the mass of the molecule, k is Boltzmann's constant, and T is the absolute temperature. If θ denotes the *fraction of covered surface* and τ the fraction of the inelastic collisions which result in adsorption (τ is called a *sticking coefficient* or a *condensation coefficient*), the rate of adsorption R_a is

$$R_a = n_a\tau(1 - \theta) \tag{6.8}$$

Under equilibrium conditions the rate of adsorbing molecules must equal the rate of desorbing molecules; the number of desorbing molecules is

$$n_d = K_0 e^{-q/kT} \tag{6.9}$$

where q is the activation energy required for desorption and K_0 is a constant. Further, the rate of desorption is

$$R_d = n_d\theta \tag{6.10}$$

and from the equality of rates at equilibrium,

$$n_d\theta = n_a\tau(1 - \theta) \tag{6.11}$$

the expression for the Langmuir isotherm is obtained:

$$\theta = \frac{\tau(n_a/n_d)}{1 + \tau(n_a/n_d)} = \frac{bp}{1 + bp} \tag{6.12}$$

where

$$b = \frac{\tau}{K_0 e^{-q/kT}(2\pi mkT)^{1/2}} \tag{6.13}$$

Langmuir considered b to be a constant; this is equivalent to considering $K_0 e^{-q/kT} = $ constant or to an assumption that the *free energy of adsorption does not vary with the coverage* θ. The parameter b would be constant if adsorption were to take place on an *energetically uniform surface* without any lateral interactions between the adsorbed molecules. Actually, the Langmuir adsorption isotherm is obtained in numerous instances when an interaction between the adsorbed molecules does occur, as is unavoidable. The apparent constant heat of adsorption can be obtained by the compensating effect of two opposing tendencies, one represented by the heat of adsorption decreasing with an increase in the coverage θ, while the other is due to lateral interactions between the adsorbed molecules which increase with θ. The variations of these two parameters occur, and these variations may, and frequently do, produce an overall effect of an apparently constant

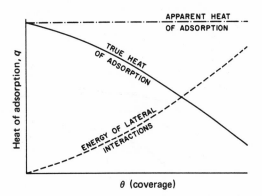

Figure 6.11. An apparently constant heat of adsorption q resulting from compensation of decreasing heat of adsorption and increasing lateral interactions.

heat of adsorption q, as indicated in Figure 6.11. Since all solid surfaces are highly heterogeneous, the prevalence of the Langmuir-type adsorption isotherm among experimental results indicates that such compensating effects play a far greater role than realized. In fact, Ross and Olivier (1964) came to the conclusion that the experimental determination of a Langmuir adsorption isotherm is, in itself, clear evidence of a highly heterogeneous surface with a *mobile* film of adsorbate (see Figure 6.14 and discussion thereof). The fractional coverage by the adsorbate,

$$\theta = \frac{V}{V_m} \tag{6.14}$$

is the ratio of the gas volume V adsorbed at the equilibrium pressure p to the volume of gas V_m necessary to cover the surface with a complete adsorbed monolayer. Equation (6.12) then becomes

$$V = V_m \frac{bp}{1 + bp} \tag{6.15}$$

At very low pressures, the term bp can be neglected in comparison with 1, and the adsorption isotherm becomes a Henry's law isotherm. In the high-pressure region, $1 + bp \cong bp$ (1 may be neglected in comparison with bp), and the isotherm becomes $V_a = V_m$; i.e., the amount adsorbed is independent of pressure, as shown in Figure 6.12.

For the intermediate range of pressure, equation (6.15) can be transformed into a linear form,

$$\frac{p}{V} = \frac{1}{V_m b} + \frac{1}{V_m} p \tag{6.16}$$

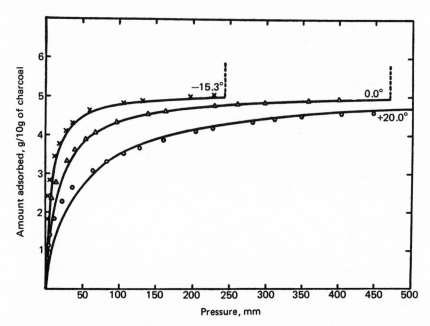

Figure 6.12. Adsorption isotherms of ethyl chloride on charcoal showing decreasing amounts adsorbed at higher temperatures in comparison with the adsorption at lower temperatures. [From Brunauer (1945), *The Adsorption of Gases and Vapors*, Vol. 1, *Physical Adsorption*. Copyright 1945, 1971 by Princeton University Press. Reprinted by permission of Princeton University Press, Princeton, New Jersey.]

and a plot of p/V vs. p should give a straight line. Alternatively, dividing by p,

$$\frac{1}{V} = \frac{1}{V_m b}\,\frac{1}{p} + \frac{1}{V_m} \qquad (6.17)$$

a straight line is obtained on plotting $1/V$ against $1/p$, as in Figure 6.13. Plots like those expressed by equations (6.16) and (6.17) are used for experimental adsorption data to check whether the Langmuir adsorption isotherm is obeyed.

The basic assumptions of Langmuir in the derivation of his adsorption isotherm expression, using dynamic equilibrium considerations, were

1. Localized adsorption on bare surface sites only, thus limiting the adsorption to the formation of a unimolecular layer

2. Constant heat of adsorption

Fowler (1935) used statistical mechanics to derive an equation which

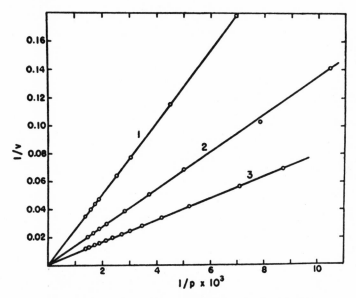

Figure 6.13. Linear Langmuir plots for the adsorption of oxygen at 0°C (curve 1), carbon monoxide at 0°C (curve 2), and carbon dioxide at 100°C (curve 3) on silica. [From Brunauer (1945), *The Adsorption of Gases and Vapors*, Vol. 1, *Physical Adsorption*. Copyright 1945, 1971 by Princeton University Press. Reprinted by permission of Princeton University Press, Princeton, New Jersey.]

is formally similar to that of the Langmuir equation and for which he employed additional assumptions:

3. The adsorbent represents an energetically uniform surface

4. No lateral interactions exist between the adsorbed species

Since then, numerous other derivations have been made for the Langmuir isotherm, all requiring assumptions equivalent to the ones stated.

6.4.3. Adsorption Isotherms for Mobile Adsorbates

Volmer (1932) showed that an adsorbed layer can be extremely mobile (and, as a result, that the transfer of matter through fine pores by surface diffusion of a mobile adsorbed film can be faster than by free diffusion in the gas phase). Subsequently, de Boer (1953) introduced a model of mobile adsorption (as contrasted with localized adsorption in the Langmuir model) taking into account lateral interactions between the adsorbed molecules but maintaining the assumption of an energetically uniform substrate. He used an equation for nonideal behavior of an adsorbate in a two-dimensional adsorbed film (an analogue of the van der Waals equation for a three-dimensional system) in the form

$$\left(\phi + \frac{\alpha}{A^2}\right)(A - \beta) = kT \tag{6.18}$$

where ϕ is the film pressure ($\gamma - \gamma_0$), A is the area of the film, α represents a term due to lateral interactions between adsorbates, β represents the coarea of adsorbate molecule and has the value

$$\beta = \tfrac{1}{2}\pi d^2 = 1.57d^2 \tag{6.19}$$

where d is the diameter of the adsorbate molecule. Then using the Gibbs adsorption theorem, de Boer related the three-dimensional pressure p with the film pressure ϕ:

$$d\phi = \frac{kT}{\beta} d \ln p \tag{6.20}$$

From the two equations (6.18) and (6.20), de Boer and Hill derived the adsorption isotherm equation:

$$p = K \frac{\theta}{1 - \theta} \exp\left(\frac{\theta}{1 - \theta} - \frac{2\alpha\theta}{kT\beta}\right) \tag{6.21}$$

This is now known as the *Hill–de Boer equation* for a two-dimensional adsorption isotherm; it separates two types of interactions taking place during adsorption:

1. The parameter K represents interactions between the adsorbate and the substrate.
2. The term $2\alpha/kT\beta$ represents interactions between the adsorbate molecules themselves.

The Hill–de Boer equation enables predictions to be made on the *two-dimensional phase changes* (*phase transformations*) *in the adsorbed film*, such as condensation, solidification, order–disorder, etc.

Ross and Boyd (1947) were the first to investigate a *reproducible* two-dimensional condensation of ethane adsorbed on the (100) face of NaCl at 90°K. They also showed experimentally that a two-dimensional critical temperature exists above which no condensation could take place in the film. Figure 6.14 shows the results of similar studies by Fisher and McMillan (1958) for adsorption of Kr on NaBr. A two-dimensional condensation is

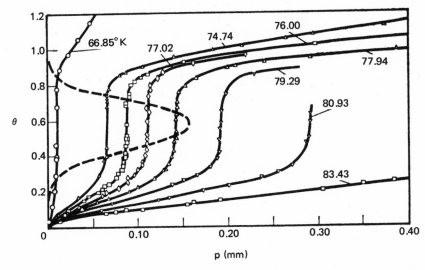

Figure 6.14. Adsorption isotherms of krypton on sodium bromide showing stepwise increases in coverage caused by two-dimensional condensation. [Fisher and McMillan (1958), with permission from *J. Am. Chem. Soc.* **79**. Copyright 1957, American Chemical Society, Washington, D.C.]

indicated by the vertical step in the isotherms obtained for temperatures below approximately 81°K, showing continued increases in coverage at constant pressures of adsorbate. The prerequisites for observing such condensation effects are the following:

1. The substrate must be *energetically homogeneous toward the given adsorbate* (that is, a homotattic substrate). Examples of homotattic substrates are liquid surfaces, perfect cubic crystals showing the same crystal face, or hexagonal layer-lattice crystals with a pronounced cleavage such as graphite, boron nitrate, and molybdenum disulfide.

2. The temperature must be below the *critical temperature* for a two-dimensional film (which is approximately equal to half of the absolute critical temperature for the three-dimensional phase change; see below).

3. The measurement must be carried out in the range of pressures *below that for a saturated monolayer*.

When these conditions obtain, a discontinuity in the form of a sharp rise in adsorption is observed. The vertical steps (on a θ–p diagram) rep-

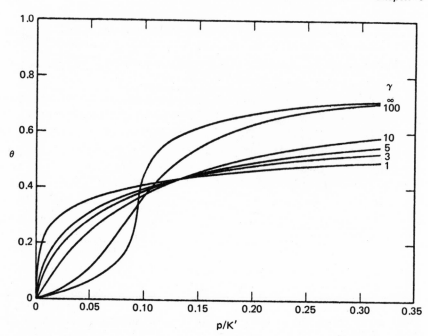

Figure 6.15. Computed model adsorption isotherms for argon adsorbed as a two-dimensional van der Waals gas on a series of substrates of increasing heterogeneity. [From Ross and Olivier (1964), *On Physical Adsorption*, with permission of John Wiley & Sons, New York.]

resenting these discontinuities are due to a two-dimensional condensation of the adsorbate.

Extensive work with near-homotattic solid subtrates has been carried out by Ross *et al.* and has led Ross and Olivier (1964) to suggest a model of a heterogeneous substrate that is capable of providing a large number of computed adsorption isotherms derived for different degrees γ of surface heterogeneity. Figure 6.15 shows some such computed adsorption isotherms of the Ross and Olivier model prepared for adsorption of Ar onto substrates of decreasing heterogeneity, $1 < \gamma < \infty$. When $\gamma = 1$, the surface is highly heterogeneous; when $\gamma \to \infty$, the substrate becomes increasingly more uniform or homotattic. The conclusion of this treatment is that the *Langmuir adsorption isotherm follows directly from a model of a mobile film on a highly heterogeneous substrate* (as well as from *localized* adsorption on an energetically *uniform* substrate, as originally derived).

A further conclusion was that the *empirical adsorption isotherm due to Freundlich,*

$$\theta = cp^{1/n} \qquad (6.22)$$

where n is an integer and c is a constant, which is found frequently in experimental results for a large variety of adsorption systems, describes a highly heterogeneous substrate but less so than that resulting in Langmuir's isotherm.

Subsequent to the preceding theoretical evaluations, experiments have established that a heat treatment applied to an initially heterogeneous substrate of carbon-black gradually alters the character of the substrate and influences the adsorption of Ar on it in the manner predicted by the theoretical model by Ross and Olivier. A progressive decrease in the heterogeneity of the carbon-black surface has resulted from the heat treatment, and this has narrowed the spread of adsorption energies to one single value, resulting in a one-step isotherm. A whole sequence of heat treatments was required to achieve that result.

One limitation of the Ross–Olivier treatment is that with highly heterogeneous substrates the weakly adsorbing patches cannot be distinguished from the nearly completed monolayers on high-energy patches; thus a clear differentiation of a saturation point of the first monolayer is absent.

As indicated, the Hill–de Boer adsorption isotherm, equation (6.21), can be used to predict phase transformations. The critical parameters for a two-dimensional condensation in the adsorbed film have been evaluated theoretically by Hill as follows:

Critical coverage

$$\theta_c = \tfrac{1}{3} \tag{6.23}$$

Critical pressure (p_0 is the saturation vapor pressure of the adsorbate)

$$p_c \cong 10^{-3} p_0 \tag{6.24}$$

Critical surface pressure per molecule

$$\phi_c = 0.361 p_c \left(\frac{V_c}{N} \right)^{1/3} \tag{6.25}$$

Critical surface area per molecule

$$\frac{A_c}{N} = 1.38 \left(\frac{V_c}{N} \right)^{2/3} \tag{6.26}$$

If T_{2c} and T_{3c} denote *critical temperatures* for the two- and three-dimensional phases, then

$$T_{2c} = 0.5 T_{3c} \tag{6.27}$$

All other symbols with subscript c refer to two-dimensional films with N number of adsorbate molecules.

Experimental values of the T_{2c}/T_{3c} ratio for NaCl and carbon-black adsorbents and C_2H_6, CH_4, Kr, and Xe adsorbates varied between 0.39 and 0.47, thus giving even lower T_{2c}'s than the theoretically predicted one.

The single step in the θ–p adsorption isotherm (Figure 6.14) due to a first-order phase transformation in the adsorbate film should not be confused with other, usually multistep, discontinuities. Some of the latter may be due to a nonattainment of equilibrium in experiments, as exemplified in Figure 6.16. Yet others may be stepwise adsorption isotherms obtained under equilibrium conditions but for a substrate consisting of a definite number of uniform crystal faces, each type of face attaining equilibrium at a different rate (Figure 6.17).

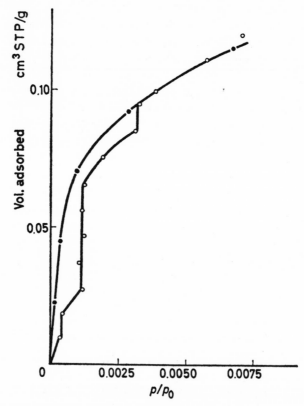

Figure 6.16. An example of a stepwise adsorption isotherm caused by nonattainment of equilibrium. Adsorption of krypton on calcium halophosphate at 77.3°K; ○ when adsorbent is not readily accessible (in a bulb), ● when it is thinly spread on a system of trays. [Corrin and Rutkowski (1954); with permission from *J. Phys. Chem.* **58**. Copyright 1954, American Chemical Society, Washington, D.C.]

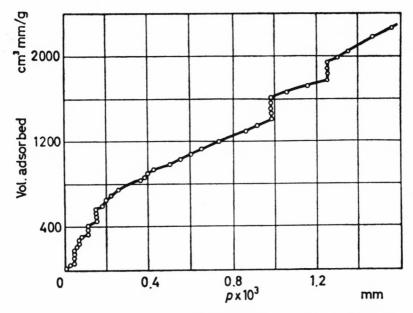

Figure 6.17. An example of a stepwise adsorption isotherm caused by the presence of several types of homogeneous surfaces possessing different adsorption characteristics—ethane adsorbing on CaF_2 at 90°K. [Ross and Winkler (1954), with permission from *J. Am. Chem. Soc.* **76**. Copyright 1954, American Chemical Society, Washington, D.C.]

The examples quoted demonstrate that the same adsorption isotherm models can be derived using sets of different assumptions. Specifically, the shape of a Langmuir isotherm is obtained either for a localized monolayer adsorption with a constant heat of adsorption or for a mobile adsorption on a highly heterogeneous substrate. Therefore, to determine the *true characteristics of the adsorbed layer* (in particular, its thermodynamic properties as functions of temperature and coverage), it is necessary to obtain *adsorption isotherms at several temperatures* and, using appropriate partition functions,[†] to derive the theoretical expressions for parameters such as the

---

[†] The partition function Q^s is derived through statistical mechanics for the various assumptions and is related to the Helmholtz free energy F^s by

$$F^s = -kT \ln Q^s \qquad (6.28)$$

or to the interfacial energy by

$$U^s = kT^2 \left(\frac{\partial \ln Q^s}{\partial T} \right)_{A^s, V^s} \qquad (6.29)$$

entropy functions of the various types of heats of adsorption. These heats can then be compared with the experimental values. A comparison of a single isotherm with the prediction of a model is nearly worthless, since there are usually several models. For details of relationships developed for a variety of models, see Young and Crowell (1962) or Clark (1970). An agreement between the predictions of models and experimental isotherms does not guarantee that the true physical picture of the adsorbed state was obtained. Nevertheless, the various models help to bring some order to the extremely complex picture of adsorption processes.

6.4.4. *Polanyi's Potential Theory Isotherms*

Polanyi's theory, introduced in 1932, provides neither an explicit isotherm equation nor a detailed physical model, yet it is applicable to most adsorption systems. It starts with a general principle that the force

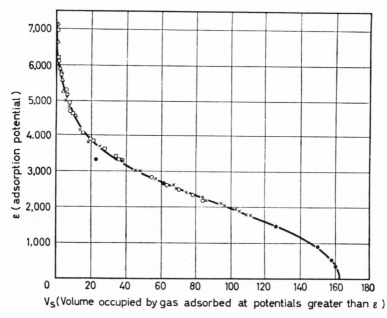

Figure 6.18. Characteristic curve for the adsorption of carbon dioxide on charcoal. The points were determined experimentally by Titoff (1910), the line was calculated by Lowry and Olmstead (1927), Titoff's data: (T in °K): ○, 196.6; ●, 303.1; ×, 273.1; □, 353.1; △, 424.6. [From Brunauer (1945), *The Adsorption of Gases and Vapors*, Vol. 1, *Physical Adsorption.* Copyright 1945, 1971 by Princeton University Press. Reprinted by permission of Princeton University Press, Princeton, New Jersey.]

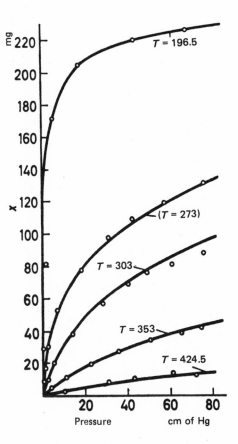

Figure 6.19. The extent of agreement between the experimentally determined adsorption isotherms and the theoretically predicted ones when based on a characteristic curve determined for adsorption at one temperature. Adsorption isotherms of carbon dioxide on charcoal. O, Titoff's experimental points; lines represent isotherms calculated by Berenyi (1920) from the 273 K isotherm. [From Brunauer (1945), *The Adsorption of Gases and Vapors*, Vol. 1, *Physical Adsorption*. Copyright 1945, 1971 by Princeton University Press. Reprinted by permission of Princeton University Press, Princeton, New Jersey.]

of attraction at any point in the adsorbed layer is determined by the adsorption potential ε, defined as the work done by the adsorption forces to bring the molecule from the gas phase to the adsorption site. An interface is thus surrounded by equipotential lines (or surfaces) which give a distribution function $\varepsilon = f(v_s)$ relative to the volume v_s of adsorbate. The adsorption potential is postulated to be independent of temperature; hence $\varepsilon = f(v_s)$ is the same for a given gas *at all temperatures* and is known as the *characteristic curve* (see Figure 6.18). The only way to test the validity of this theory is to calculate a characteristic curve from one experimental isotherm and then predict the remaining isotherms from this curve [see Young and Crowell (1962, pp. 140ff.)]. Figures 6.18 and 6.19 show the sets of relevant curves for CO_2 adsorption on charcoal.

It has been found that once a characteristic curve is obtained for a given substrate and a specific adsorbate, the isotherms for other adsorbates

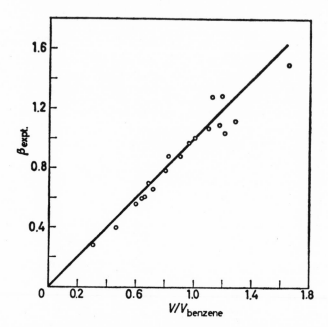

Figure 6.20. Plot of the coefficient of affinity β_{exp} against the molar volume of the adsorbate for several organic vapors on charcoal at 20°C. [From Young and Crowell (1962), *Physical Adsorption of Gases*, with the permission of Butterworths, London.]

adsorbing on the same substrate can be readily obtained by using a suitable coefficient β determined experimentally. Thus, in the adsorption of benzene and other organic vapors on charcoal, the comparison of adsorption isotherms showed that the various characteristic curves could be reduced to a single curve (that for benzene) by the use of the equation

$$\varepsilon_i = \beta_{exp} f(v_i) \tag{6.30}$$

where i denotes any adsorbate. And further, the relative β_{exp} values were found to be very closely approximated by the *ratio of molar volumes*, as indicated in Figure 6.20:

$$\beta_{exp} = \frac{V \ new \ adsorbate}{V \ benzene}$$

Hence, it appears that *molar volume* is the single property of the adsorbate which can be used to predict a priori the adsorption isotherm, at least whenever the adsorbed molecules are bonded to the substrate predominantly by van der Waals bonds.

6.4.5. *BET Theory [Brunauer, Emmett, Teller (1938)]* *of Multilayer Adsorption*

Experimentally obtained isotherms for solid–gas systems were often found to differ from the Langmuir-type adsorption. In fact, relatively few of the vapor adsorption isotherms obey equations (6.12), (6.16), and (6.17), and still fewer adsorptions are restricted to a single layer of adsorbed species. Brunauer *et al.* (1940) introduced a classification of adsorption isotherms, dividing them into five types (Figure 6.21). The first, type I, represents a Langmuir isotherm. Types II and III represent multilayer adsorption on nonporous solids, while types IV and V represent analogous multilayer adsorption on porous solids involving capillary condensation within the pores. However, this classification is not complete since, for example, step-wise isotherms are not included, whether those involving vertical steps due to phase transformations or horizontal steps denoting completion of individual layers. Neither are those isotherms which terminate at a definite multilayer thickness at the saturation point (when $p/p_0 = 1$; see Figure 6.46) included.

Brunauer *et al.* (1938) extended Langmuir's concepts of adsorption by assuming that molecules can adsorb on already preadsorbed adsorbate molecules, forming multilayers. They assumed that the whole area of the adsorbent consists of fractions: S_0 with no adsorbate, S_1 with 1 layer of adsorbate, S_2 with 2 layers, ..., S_i with i layers. At equilibrium, the rate of condensation on the bare surface S_0 must be equal to the rate of evaporation from the first layer S_1; i.e.,

$$a_1 p S_0 = b_1 S_1 e^{-E_1/RT} \tag{6.31}$$

where p is the pressure, E_1 is the heat of adsorption in the first layer, and a_1 and b_1 are constants. Also, they reasoned that at equilibrium the extent

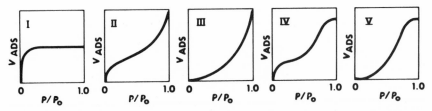

Figure 6.21. Five types of adsorption isotherms according to the classification by Brunauer *et al.* (1940). [Note that this classification does not include the stepwise isotherms due to two-dimensional condensation, as in Figure 6.14, or the *S*-type isotherms (Frumkin's or Frumkin–Fowler's), as in Figure 7.14.]

of monolayer coverage S_1 must remain constant; that is, the condensation on S_1 and the rate of evaporation from S_1 must be equal to the condensation on S_1 plus the rate of evaporation from S_2:

$$a_2 p S_1 + b_1 S_1 e^{-E_1/RT} = a_1 p S_1 + b_2 S_2 e^{-E_2/RT} \tag{6.32}$$

which reduces to

$$a_2 p S_1 = b_2 S_2 e^{-E_2/RT} \tag{6.33}$$

Similarly,

$$a_i p S_{i-1} = b_i S_i e^{-E_i/RT} \tag{6.34}$$

The fractional coverage is

$$\theta = \frac{v}{A v_0} = \frac{v}{v_m}$$

where A is the total surface area of adsorbent, v is the volume of gas actually adsorbed, v_m is the monolayer coverage, and v_0 is the monolayer volume of 1-cm² coverage (specific volume). Before an expression for an isotherm could be derived, two simplifying assumptions had to be made by BET:

1. $E_2 = E_3 = \cdots = E_i = E_{\text{liquifaction}}$; that is, the heats of adsorption in the second and all subsequent layers are equal to the heat of liquefaction.
2. $b_2/a_2 = b_3/a_3 = b_i/a_i = $ constant (that is, the evaporation and condensation characteristics for all layers except the first monolayer are the same as those of the liquid state of adsorbate).

The final form of the BET isotherm is

$$\frac{p}{v(p_0 - p)} = \frac{1}{v_m C} + \frac{C - 1}{v_m C} \frac{p}{p_0} \tag{6.35}$$

where p_0 denotes the saturation pressure of the adsorbate. It should give a straight-line relationship when $p/[v(p_0 - p)]$ is plotted against p/p_0 [the slope $(C - 1)/v_m C$ and the intercept $1/v_m C$ would then enable the values of the terms C and v_m to be evaluated].

For a great majority of vapor adsorption isotherms of type II, the *term C* is constant for p/p_0 values between 0.05 and 0.35, giving excellent straight-line relationships. Such results enable the evaluation of

1. The surface area A^s of the adsorbent per unit mass
2. The approximate heat of adsorption in the first layer

Goates and Hatch (1953) derived the BET equation on thermodynamic grounds and showed that the constant C is related to the heat of adsorption by

$$C = \exp\left(\frac{E_1 - E_L}{RT}\right) \qquad (6.36)$$

where E_1 is the heat of adsorption in the first layer and E_L is the heat of liquefaction. Thus, the enumeration of the terms from the plot representing equation (6.35) also enables the evaluation of the *net heat of adsorption* $(E_1 - E_L)$ by equation (6.36). By analogy, the term a_1b_2/b_1a_2 may be regarded as the *net entropy of adsorption*. Comparisons of heats of adsorption E_1 obtained from the net heats [equation (6.36)] with the experimental values (calorimetric values) of the heats of adsorption show that the calculated values are smaller, sometimes even less than half of the calorimetric values (e.g., Harkins and Boyd's data for adsorption of benzene on TiO_2 at 25°C gave $(E_1 - E_L) = 5.2$ kcal/mol from the heats of immersion and 2.6 kcal/mol from the C value of the BET plot). These differences arise, first, from the fact that the BET theory is applicable to a limited range of p/p_0 (0.05–0.35 and sometimes only to 0.2), and under those conditions the first-layer coverage is far from complete, while the second, third, fourth, etc., layers are already formed on some parts of the substrate. Second, the entropy factor a_1b_2/b_1a_2, assumed to be unity for the purpose of E_1 calculations, should diminish with increasing p/p_0. This decrease should overcompensate the effects of the heats of adsorption in the second, third, etc., layers, which must be greater than the heat of liquefaction of bulk liquid, $E_2 > E_3 \cdots E_L$.

The parameter v_m gives the number of molecules necessary to cover the surface of the adsorbent with a complete monolayer. It is for the evaluation of this parameter, v_m, that the BET isotherms are determined. Assuming a close packing of spherical molecules on the adsorbent, Brunauer and Emmett calculated from the density of the liquid adsorbate that an adsorbed molecule of nitrogen would occupy 16.2 Å², while Livingston calculated from the van der Waals equation that the area occupied should be 15.4 Å²/molecule of adsorbed nitrogen. There is no absolute method available to determine which of these values is correct. The only indication of the correctness of a BET surface area determination is to compare the BET results with the results of determinations by some other method completely independent of BET, such as those of Harkins and Jura's *absolute* method [based on linearity of the adsorption isotherm in the region of condensed film formation; see Harkins (1952, pp. 232ff.)] or an electron microscopic evaluation of particle size and the resultant surface area. For nonporous adsorbents, good agreement was obtained (5–10% difference) when such comparisons were made—as long as the same adsorbate, for example, N_2, was used in the BET method and in the Harkins and Jura method. When different adsorbates are used in surface area determinations, the results may vary by 50–100%, and it is impossible to be certain which value is the correct one. Thus, the whole concept of the surface area on an atomic scale is ill-defined; the surface can be only approximately evaluated even when the "yardstick" used to measure the area is specified.

Reviewing the assignments of areas occupied by adsorbate in the completed monolayer, McClellan and Harnsberger (1967) arrived at the following recommended values: $N_2(77°K)$, 16.2 Å^2 (range $13–20 \text{ Å}^2$); $Ar(77°K)$, 13.8 Å^2 (range $13–17 \text{ Å}^2$); $Kr(77°K)$, 20.2 Å^2 (range $17–22 \text{ Å}^2$); n-$C_4H_{10}(273°K)$, 44.4 Å^2; C_6H_6 $(293°K)$, 43.0 Å^2 (range $30–50 \text{ Å}^2$).

The most frequent use of the BET technique, despite its theoretical shortcomings, is the evaluation of surface areas of fine powders. The determination of several isotherm points is followed by a graphical test of linearity of the $p/v(p_0 - p)$ vs. p/p_0 relationship and the evaluation of v_m and C from the intercept and the slope of the linear relationship. However, for isotherms of type II, that is, when $C > 100$, Emmett and Brunauer suggested an empirical fast method of surface area determination. This method is based on the assumption that the beginning of the straight-line portion of the adsorption isotherm (in the intermediate p/p_0 region), the so-called *point B*, corresponds to the completion of a monolayer. In general, good agreement is obtained between the *point B* and the v_m results.

If the value of C is large, $C > 100$, there is a distinct *knee* (inflection) in the type II isotherm (Figure 6.22), but as C becomes smaller, this inflection occurs progressively nearer the origin, and for a critical value $C = 2$ the inflection coincides with the origin. For $C < 2$, isotherms of type III are obtained. The critical transition from isotherm II to III occurs when $(E_1 - E_L)RT$ is approximately equal to 0.7, that is, when $E_1 - E_L \cong 420$ cal/mol at $T = 300°K$ or $\cong 110$ cal/mol at $T = 80°K$. Qualitatively, this means that if $E_1 \gg E_L$, the second and higher layers will not start adsorbing until the first layer is filled (or nearly filled), and then a sharp knee is observed for $v \to v_m$, and the surface area determinations are most "reliable" for such type II isotherms.

Type IV and V isotherms in Brunauer's classification refer to porous solid adsorbents and denote adsorption in multilayers followed by condensation of adsorbate in pores, resulting in a finite number of layers at saturation pressure. The most general characteristic of adsorption isotherms on porous solids is their *hysteresis*, that is, adsorption is not completely

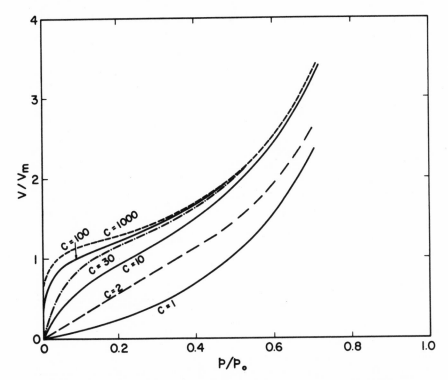

Figure 6.22. The effect of variation in the magnitude of constant C in the BET isotherm on the shape of isotherms, their gradual transition from type III to type II isotherms, and the development of a distinct knee.

reversible, but on desorption greater amounts of adsorbate are retained (for the same p/p_0) than those adsorbed on adsorption steps; see Figure 6.23. For a great majority of adsorbent–adsorbate systems the hysteresis loops are fully reproducible, that is, can be retraced without any changes. An understanding of this behavior is provided by consideration of the vapor pressure of liquid in the fine capillaries. Using thermodynamics, Lord Kelvin (1871) derived the expression

$$\ln\left(\frac{p}{p_0}\right) = -\frac{2\gamma V}{rRT}\cos\theta' \tag{6.37}$$

where γ is the surface tension of the liquid, V is the molecular volume of adsorbate, r is the radius of the capillary, and θ' is the contact angle between the liquid and the wall of the capillary, to describe an equilibrium between the partial vapor pressure and the surface tension of liquid condensing in the

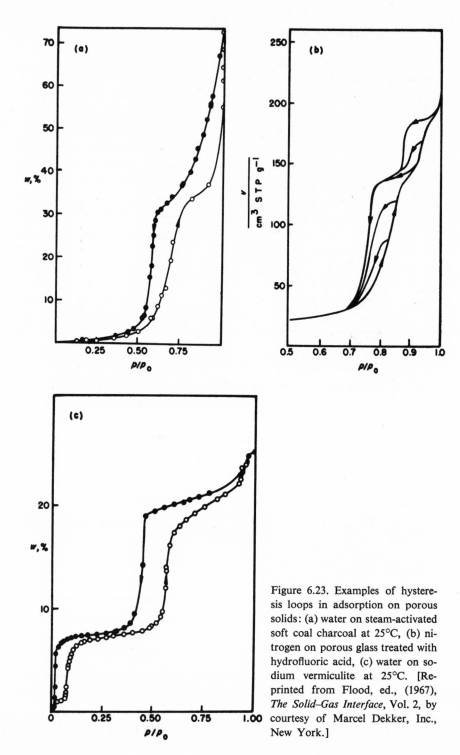

Figure 6.23. Examples of hysteresis loops in adsorption on porous solids: (a) water on steam-activated soft coal charcoal at 25°C, (b) nitrogen on porous glass treated with hydrofluoric acid, (c) water on sodium vermiculite at 25°C. [Reprinted from Flood, ed., (1967), *The Solid–Gas Interface*, Vol. 2, by courtesy of Marcel Dekker, Inc., New York.]

capillary of radius r. For the various points in the *desorption branch* of the isotherm, all pores of radius r and smaller than r should be filled, and those of a larger radius should be empty (as long as $\theta' < 90°$) for any given p/p_0. Assumptions are made about the geometry of pores—usually a set of uniform nonintersecting cylinders is assumed (although the actual pores are probably highly irregular intercommunicating capillaries). The number of pores of different radii is then calculated from the desorption isotherms for the given p, and usually a Gaussian distribution curve is obtained. For details, see Linsen (1970). Porous adsorbents are used extensively as supports for the manufacture of a variety of catalysts. Sintered materials and naturally occurring conglomerates are also porous. Most crystalline minerals—with the exception of zeolites (*molecular sieves*)—are nonporous. Yet, despite a nonporous structure, there are instances of sheet-type minerals, mostly silicates, being capable of adsorbing inordinately large volumes of water vapor (or other vapors) with the resultant *swelling* of the crystalline solid. This behavior results from the ability of the adsorbate to penetrate in between the layers (sheets) of the solid whenever the latter are held by weaker dipole–dipole or van der Waals bonds than those that the adsorbate develops with atoms of the sheet surface. Montmorillonite is the best known example of swelling clays.

Brunauer (1961) described an interesting example of a progressive increase in surface area of tobermorite ($Ca_3Si_2O_7 \cdot H_2O$)—a constituent of hydrated portland cement and concrete—which may be prepared as sheets comprising a single unit cell, or two or three unit cells, rolled into fibers. Nitrogen used as an adsorbate can penetrate in between the rolled sheets, but water vapor cannot.

Prins (1967) reviewed the swelling phenomena encountered in inorganic and organic materials which possess either layer structures or glass structures. Numerous organic polymers swell as a result of adsorption of organic vapors.

6.4.6. *Frenkel–Halsey–Hill Slab Theory of Adsorption*

Despite the wide acceptance of the BET technique for surface area determinations, it gradually became evident that the BET theory suffers from serious limitations due to the simplifying assumptions (1) $E_2 = E_3 = \cdots = E_L$ and (2) $b_2/a_2 = b_3/a_3 =$ constant, which represent a pair of mutually compensating errors. Heats of adsorption measured for various amounts of adsorbed adsorbates indicate that $E_1 > E_2 > E_3 > \cdots > E_L$. Refinements to BET theory introduced since 1948 concern (1) more realistic postulates with respect to E_i energies in successive layers and (2) acceptance of defects superimposed on the uniform crystallographic surfaces. To these refinements belong the Wheeler–Ono approach and the Frenkel–Halsey–Hill slab theory; for details of both, see Young and Crowell (1962).

An interesting corollary of the slab theory is that at low coverages of the adsorbed molecules a stepwise isotherm is obtained which, for real systems, becomes a continuous isotherm because the surface heterogeneity tends to smooth out the steps. Recent experimental work on carbon-blacks confirmed this view, showing that graphitization of the carbon-black increases ordering and results in crystallites with a higher degree of surface uniformity. Isotherms on untreated carbon-blacks are smooth, but as graphitization is progressively achieved (by heating the carbon-blacks for a given length of time at higher temperatures), stepwise isotherms are gradually obtained in the adsorption of Ar or Kr. Calorimetric heats of adsorption show maxima and minima corresponding to the steps in the isotherm.

6.5. *Mechanical Effects of Adsorption at Solid/Gas Interfaces*

The most important application of physical adsorption studies is their use in determining the surface areas of finely divided and porous solids. Many industrial reactions are carried out utilizing interfacial regions, and for those systems large surface areas per unit mass are required. Macroporous materials are those with pore diameter greater than approximately 500 Å and surface areas up to 1 or 2 m²/g; to this group belong *paper, fabrics, building stone, coke, wool,* etc. Microporous materials have surface areas of the order of 50–1000 m²/g and are represented primarily by products of the *cement industry* and *catalysts* and *adsorbents* in the chemical industry.

The shape of pores is indeterminate and not readily controlled. Pores exist as capillaries in tubular form, distorted capillaries, flattened and *ink-bottle*-type pores, etc. The methods of preparing high-area solids are numerous, for example, precipitation (silica gel), leaching of one constituent from an alloy (Ni–Al to give Raney catalyst), sintering of fine precipitates (Fe_2O_3 to give Fe_3O_4), sublimation of salts or metals (CaF_2, SiO_2, metals deposited on Al_2O_3, or SiO_2 support), steam attack to produce active carbons, and dehydration of salts [$Al(OH)_3$ to give Al_2O_3, $MgCO_3$ to give MgO] and of carbohydrates (to give activated carbon-blacks).

Generally, it is very difficult to prepare successive batches of porous solids in a reproducible manner at the same level of purity and of the same surface characteristics between different batches. Even in determinations of surface areas, aging effects take place due to surface diffusion and sintering. For example, fine-size Fe_2O_3 powder shows a progressive decrease in surface area (determined by the same BET technique) with time of storage, even at room temperature.

The development of a bond between the adsorbate and the solid substrate causes a perturbation in the solid surface even when the adsorption is purely physical in character. Of great practical importance are *linear expansion* and *swelling* effects caused by gas or vapor adsorption on rigid adsorbents of *high surface areas*. Swelling and expansions can be of particularly serious consequence in building materials, like cement or mortar, when exposed to varying partial pressures of water vapor. As long as vapor adsorption occurs slowly and uniformly throughout the mass, the structure may be able to accommodate itself to the stresses set up. However, when adsorption is rapid and takes place at different rates in various parts of the structure, *the stresses set up by the adsorbed film may be large enough to crack the material*. A piece of outgassed charcoal exposed to a gradually increasing pressure of methanol vapor swells slowly and remains intact; however, if the same outgassed charcoal is thrown from vacuum into liquid methanol, adsorption causes such an uneven expansion that the whole piece shatters.

The pioneering work on the expansion of rigid adsorbents was carried out by Bangham and his collaborators [Bangham and Fakhoury (1931), Bangham *et al.* (1932, 1934)] on charcoal and coal. Later, silica gels, porous glass (Vycor), kaolin compacts, etc., were also used as adsorbents and H_2O, CO_2, SO_2, N_2, benzene, and CH_4 as adsorbates. The fractional change in length, $\Delta l/l$, of a rod or a prism of the adsorbent was measured at the same time the adsorption isotherm of the material was determined. An expansion, or a contraction followed by an expansion, up to 2% was found to be proportional not to the amount adsorbed but to the *reduction in the surface free energy*, i.e., the surface pressure ϕ, brought about by the adsorption,

$$\frac{\Delta l}{l} = \lambda \phi \tag{6.38}$$

where λ is a constant related to the Young's modulus of elasticity Y, the total surface area A, and the density of adsorbent ϱ by

$$\lambda = \frac{A\varrho}{Y} \tag{6.39}$$

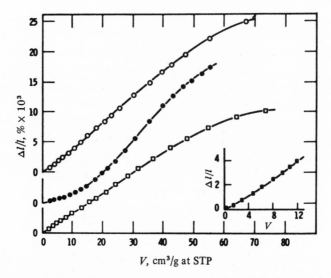

Figure 6.24. Linear expansion (in millipercent) resulting from physical adsorption of gases on Vycor glass at 90°K: □, argon; ●, nitrogen; ○, oxygen; ■, hydrogen. [Reprinted from Flood, ed., (1967), *The Solid–Gas Interface*, Vol. 2, by courtesy of Marcel Dekker, Inc., New York.]

Very extensive investigations were carried out on porous Vycor glass using polar and nonpolar gases and vapors. The results of Yates (1956), shown in Figures 6.24 and 6.25, indicate that nonpolar adsorbates such as Ar, O_2, Kr, and H_2 cause an expansion, while polar adsorbates CO, SO_2, and NH_3 cause, at first, a contraction and then an expansion at higher coverages. Subsequent work by Folman and Yates (1958) has shown that the initial contraction of the sample, due to the adsorption of polar adsorbates on polar substrates, disappears when the substrate is converted from polar to nonpolar in character, as, for example, in methylation of glass, where —OH groups are substituted by —OCH_3 groups in the glass surface (Figure 6.26). With nonrigid solids, such as clays, agar-agar, Fe_2O_3 gels, cotton and wool, etc., adsorption of water vapor (adsorption of benzene vapor on rubber) causes *swelling* which may increase several times the original size of the adsorbent.

In all cases of expansion and swelling, the extent of the effect caused by adsorption depends on the *relative pressure* p/p_0 and not on the absolute pressure p. This aspect is of particular importance in the case of water adsorption. At, say, 50% *relative humidity* the actual partial pressure of water is 2.3 mm at 0°C and 42 mm at 35°C. With a sudden drop in temperature from 35° to 0°C, the absolute humidity (the amount of H_2O vapor) may change very little, while the relative humidity changes severalfold and exceeds saturation by nearly an order of magnitude.

As shown in Figures 6.12 and 6.14, the amounts of vapors adsorbed at lower temperatures are always greater than those adsorbed at higher temperatures. Thus, a sudden drop in temperature would increase slightly the amount of water vapor adsorbed if the relative humidity were maintained unchanged. However, the relative humidity during a drop in temperature changes severalfold into the region of supersaturation ($p/p_0 > 1$), and the fast physical adsorption progresses into multilayers, condensing some of the excess water vapor. The stresses thus produced in the substrate—on the

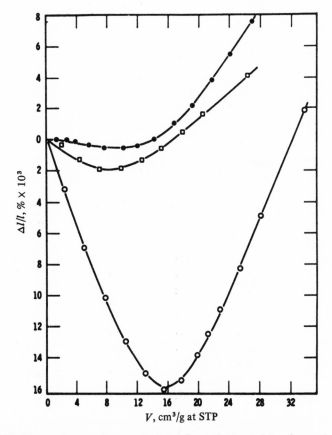

Figure 6.25. Linear changes (contraction and expansion) resulting from physical adsorption of dipolar molecules on porous Vycor glass: ●, carbon monoxide, 90°K; □, sulfur dioxide, 195°K; ○, ammonia, 273°K. [Reprinted from Flood, ed., (1967), *The Solid–Gas Interface*, Vol. 2, by courtesy of Marcel Dekker, Inc., New York.]

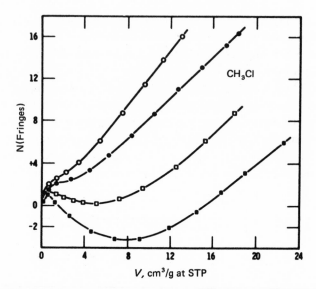

Figure 6.26. The initial contraction caused by adsorption on a polar substrate can be eliminated by converting the surface of the substrate to nonpolar in character. Substitution of the —OH surface groups in glass by —OCH$_3$ (on methylation) produces a change in the type of physical changes due to adsorption of methyl chloride. Unmethylated glass: ■, −78°C; □, 0°C. Methylated glass: ●, −78°C; ○, 0°C. [Reprinted from Flood, ed., (1967), *The Solid-Gas Interface*, Vol. 2, by courtesy of Marcel Dekker, Inc., New York.]

one hand, a contraction due to a temperature change; on the other hand, an expansion caused by the very large volumes of water vapor adsorbed— lead to cracking of the substrate.

In this manner, any rapid transfer of furniture, paintings, or ceramic materials and of all high-surface-area rigid structures from a humid environment at higher temperatures to a much lower temperature (without rigid control of humidity) may lead to serious damage due to cracking and spalling. Numerous occurrences of damage suffered by furniture and paintings transported in winter, by ceramic linings of metallurgical furnaces cooled too suddenly, and by concrete structures subjected to sudden temperature drops in an Arctic environment, etc., have been recorded.

Even during grinding, the sudden wetting of the freshly formed surfaces (created by cleavage during impact) contributes to the production of fines resulting from the contraction–expansion effects produced by sudden adsorption. This mode of production of fines has already been mentioned in Section 6.2.1; such fines should be differentiated from the fines produced by abrasion.

6.6. *Models of Isotherms for Chemisorption*

In Section 6.3 the criteria for differentiating chemisorption from physical adsorption are discussed without emphasizing that chemisorption, as such, has to be limited to a single monolayer. This situation arises from the fact that the same character of an electron exchange (between an adsorbate and an adsorbent) cannot be maintained for more than one layer of adsorbate on one layer of underlying adsorbent. All subsequent exchanges of electrons can occur only between species that have *diffused through* the chemisorbed layer and are thus forming a reaction product and not a chemisorption layer. Therefore, if quantities greater than those required for a monolayer formation are found chemisorbing on a solid, they represent mixtures of a chemisorbed monolayer and reaction products, coadsorbed physically. Multilayer adsorbed films represent either purely physically adsorbed species, or a chemisorbed monolayer with physically coadsorbed adsorbate species or physically coadsorbed reaction products (formed by electron exchange).

When the first chemisorbed layer is localized, there are, in general, threee types of theoretical adsorption isotherms considered, depending on assumptions regarding the variation of adsorption energy with coverage.

The first type is the *Langmuir isotherm* based on the assumption that the heat of localized adsorption q is constant regardless of coverage. Thus, in the expression

$$V = V_m \frac{bp}{1 + bp} \tag{6.14}$$

derived for the Langmuir adsorption isotherm, the term

$$b = \frac{\tau}{K_0 e^{-q/kT}(2\pi mkT)^{1/2}} = \text{constant} \tag{6.40}$$

since $q = $ constant and $K_0 = $ constant.

If, in a given system, mobile chemisorption occurs, the lateral interactions between the adsorbate species give rise to a *Frumkin's adsorption isotherm*:

$$bp = \frac{\theta}{1 - \theta} \exp(-2\alpha\theta) \tag{7.18}$$

producing S-shape curves as in Figure 6.14. This type of isotherm is particularly favored by electrochemists, so its discussion is carried out in greater detail in Section 7.4.

The second type of chemisorption isotherm is derived when an assumption due to Temkin (1938) is made, namely, that the heat of adsorption q varies linearly with coverage:

$$q = q_0(1 - x\theta) \tag{6.41}$$

The expression for the term b is then

$$b = K_T \exp\left|-q_0\left(\frac{1 - x\theta}{RT}\right)\right| \tag{6.42}$$

where x is the proportionality factor.

The preceding type of *adsorption isotherm* is known as *Temkin's*.

The third chemisorption isotherm is based on the assumption of a logarithmic variation of adsorption energy with the coverage,

$$q = q' \ln \theta \tag{6.43}$$

and hence,

$$b = K_F \exp\left(- \frac{q' \ln \theta}{RT}\right) \tag{6.44}$$

which results in a *Freundlich-type isotherm*.

As already stated in Section 6.4, the same isotherm expression can be derived for several different sets of assumptions. Whenever experimental data (for a given system, at one temperature) fit a particular type of theoretical adsorption isotherm, such an agreement cannot be taken as a conclusion or proof that one or another set of assumptions is being fulfilled in a given system. (And this type of "conclusion" is quite frequently encountered in the literature.) For any definitive decision regarding the character of adsorption in a given system, either adsorption isotherms at several temperatures have to be analyzed, or additional thermodynamic and chemical properties have to be taken into account.

6.7. *Characterization of Adsorption from Solutions (in Particular at Solid/Liquid Interfaces)*

Adsorption from solutions differs from that of gaseous or vapor adsorption in that *two* components of widely ranging relative quantities (solute and solvent species) are *competitively* adsorbing at the same interface. The extent of individual adsorption depends on the system, the complete or partial miscibility of the solute and solvent, the degree of

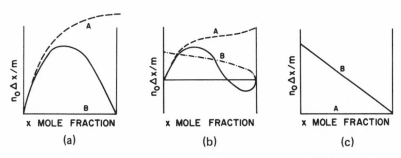

Figure 6.27. Types of composite isotherm—(a) U-shaped, (b) S-shaped, (c) linear—in relation to the corresponding apparent isotherms of solute A and solvent B. [With permission from Kipling (1965), *Adsorption from Solutions of Non-Electrolytes*. Copyright by Academic Press, Inc. (London) Ltd.]

preferential adsorption, etc. For completely miscible liquids some examples of individual isotherms ("apparent" isotherms of solute A and solvent B, represented by dashed lines) and of the composite isotherms (full lines) are given in Figure 6.27. These examples illustrate the following:

1. The occurrence of a maximum in the composite adsorption, that is, a one-branch or U-shaped isotherm. It represents a common type of selective adsorption, both physical and chemisorption, and indicates the existence of definite interactions between the solid adsorbent and the adsorbate, exceeding those between the solid and the solvent molecules. Most polar compounds are more strongly adsorbed by polar solids than are nonpolar solvents.

2. A two-branch S-type or *negative* isotherm, with the two branches usually of different size. This type of adsorption is also common when the adsorbent solid is nonpolar in character, the adsorbate is amphipatic in nature, and the solvent is nonpolar.

3. A linear isotherm representing adsorption of the solvent only. It occurs in sorption by molecular sieves when one component can enter the pores and the other cannot.

Numerous variants of the basic U-shaped and S-shaped isotherms have been encountered and are recognized by more detailed classification schemes. Nagy and Schay (1960) recognized three variants of U-shaped and two of S-shaped types. Giles *et al.* (1960) introduced a classification system consisting of four classes and five subgroups. The most common shape for adsorption on solids from solutions is that which (for low concentrations of adsorbate) can be fitted by a Langmuir or Freundlich type of equation, equations (6.15) or (6.22), respectively.

   A comprehensive monograph by Kipling (1965), dealing with the adsorption of polar and nonpolar solutes in systems comprising different adsorbents (carbon blacks, oxide gels, metals, etc.) and various types of solutions (miscible and partly miscible liquids) provides numerous examples of different isotherms. Of particular interest to flotation are the isotherms of metallic ions adsorbed on solids and of surfactants at the air/water and the solid/solution interfaces for nonpolar (hydrophobic) solids (talc, graphite, etc.) and for polar oxides and sulfides.

   For air/liquid interfaces, surface excess values calculated from the Gibbs equation (p. 357) for dilute solutions outline the individual adsorption isotherms of the solute; selected examples are shown in Figures 6.28 and 6.29. Some of these isotherms reach limiting values [Langmuir-type isotherms] implying, though not proving, the existence of monolayer adsorption; others (such as that for acetic acid) are similar to the type II of Brunauer's classification (Figure 6.21), but any conclusion as regards monolayer vs. multilayer adsorption based on this similarity is unwarranted. Independent evidence is required to ascertain the number of adsorbed layers—particularly if phase separation (due to insolubility) is likely to occur.

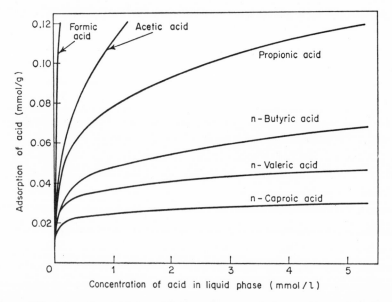

Figure 6.28. Adsorption of fatty acids from aqueous solution on charcoal, under oxidizing conditions. [With permission from Kipling (1965), *Adsorption from Solutions of Non-Electrolytes.* Copyright by Academic Press, Inc. (London) Ltd.]

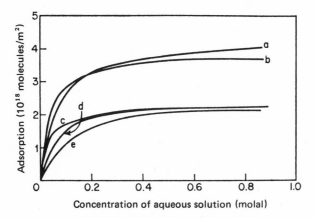

Figure 6.29. Adsorption of *n*-butyl alcohol from aqueous solutions at interfaces with (a) paraffin, (b) air, (c) talc, (d) graphite, and (e) stibnite. [With permission from Kipling (1965), *Adsorption from Solutions of Non-Electrolytes.* Copyright by Academic Press, Inc. (London) Ltd.]

Figure 6.29 shows adsorption isotherms of *n*-butyl alcohol from its aqueous solutions at interfaces of air/liquid, liquid/liquid, and hydrophobic solid/solution.

Adsorption at the hydrophobic solid/solution interface is lower than that at the air/solution or oil/solution interface, implying a different orientation of the molecules; similarly, the continued increase in adsorption at the oil/solution interface (curve a, Figure 6.29) may be caused by partial absorption of alcohol in the oil phase.

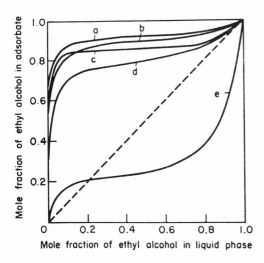

Figure 6.30. Physical adsorption of ethyl alcohol from mixtures with benzene on (a) gibbsite, $Al(OH)_3$; (b) silica gel; (c) boehmite, γ-AlO(OH); (d) γ-alumina; and (e) charcoal. [With permission from Kipling (1965), *Adsorption from Solutions of Non-Electrolytes.* Copyright by Academic Press, Inc. (London) Ltd.]

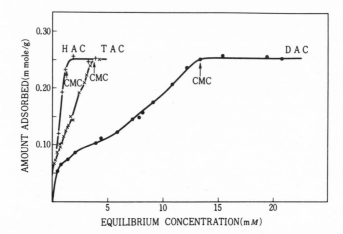

Figure 6.31. Adsorption of surfactants from aqueous solution by alumina: dodecylam-monium chloride (DAC) at 40°C, tetradecylammonium chloride (TAC), and hexadecyl ammonium chloride (HAC). [Tamamushi and Tamaki (1957), with the permission of the authors.]

Adsorption isotherms of an alcohol adsorbing from an organic solvent on polar solids, shown in Figure 6.30, are U-shaped, with no significant solvent adsorption, except for curve e; the latter, representing the adsorp-tion on mostly hydrophobic charcoal, is in the form of an S-shaped iso-

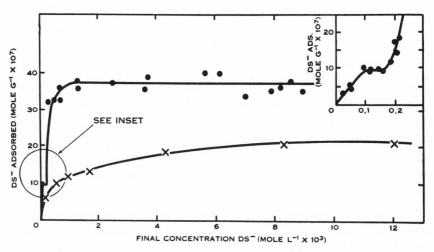

Figure 6.32. Effect of Ba²⁺ ion contamination on the adsorption of dodecyl sulfate (DS⁻) on BaSO₄. ●, Contaminated powder; ×, powder washed free of Ba²⁺. [From Cuming and Schulman (1959), with the permission of *Aust. J. Chem.* CSIRO, Melbourne, Victoria, Australia.]

therm and indicates a pronounced competitive adsorption of the nonpolar solvent.

In systems comprising non-thio ionizable surfactants (carboxylic acids, alkyl sulfates or sulfonates, alkyl amines or quaternary ammonium salts, etc.), adsorption isotherms may exhibit plateaus, steps, and sharp discontinuities, as shown in Figures 6.31 and 6.32. In Figure 6.31 the limiting values of adsorption, in the form of plateaus, are reached around the corresponding CMC for the given surfactant. Initially, these plateaus were interpreted by the authors as due to the completion of four monolayers and later on as due to a two-layer formation of adsorbate.

The adsorption isotherms shown in Figure 6.33 were apparently determined only for concentrations lower than CMC, but the three regions

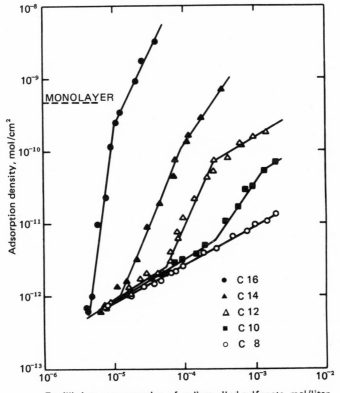

Figure 6.33. Adsorption isotherms for sodium alkyl sulfonates of C_8–C_{16} hydrocarbon chain lengths on alumina at pH 7.2, 25°C, and 2×10^{-3} M ionic strength. [From Wakamatsu and Fuerstenau (1968), in *Adsorption from Aqueous Solution*, W. J. Weber Jr. and E. Matijevic, eds., with the permission of American Chemical Society, Washington, D.C.]

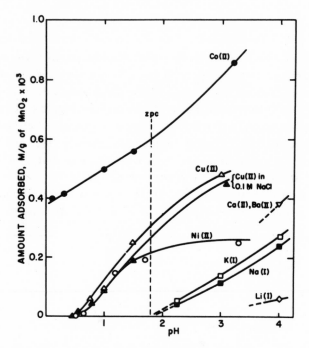

Figure 6.34. The adsorption of cobalt, copper, nickel, potassium, and sodium ions at 5×10^{-2} M concentration on Mn(II) manganite, MnO(OH), at 25°C as a function of pH. (Single data points for calcium, barium, and lithium ions are included.) [From Murray *et al.* (1968), in *Adsorption from Aqueous Solution*, W. J. Weber Jr. and E. Matijevic, eds., with the permission of American Chemical Society, Washington, D.C.]

of different slopes in each isotherm (for a homologous series of alkyl sulfonates) are clearly distinguished (except in the case of C_8). The lowest of the three slopes in the isotherms has been interpreted as denoting electrostatic adsorption of individually isolated surfactant ions, followed by a high-slope region denoting intermolecular association between the hydrocarbon chains of adsorbing species. The uppermost region of each adsorption isotherm has been interpreted by Fuerstenau *et al.* (1964), Wakamatsu and Fuerstenau (1968) as a retarded adsorption due to an electrostatic repulsion.

Results shown in Figure 6.32, obtained by Cuming and Schulman (1959), call attention to the fact that spurious isotherms may be obtained unless the substrate is freed from contaminating counterions (metal ions in the case of anionic surfactants) capable of reacting with the adsorbate. Coadsorption of such counterion surfactant reaction products (and/or of un-ionized hydrolysis products) often tends to conceal the completion of the first monolayer. Only when the system is devoid of secondary surfactant components can the true adsorption isotherm for the main surfactant species at a solid/liquid interface be obtained and unambiguously interpreted.

Adsorption of metallic ions at solid/liquid interfaces is very much pH-dependent, as indicated in Figure 6.34. In some systems metallic ion adsorption may tend to reach a limiting value (plateau level) at a specific pH value, as shown in Figure 6.35 for Co^{2+} on quartz. The progressive formation of hydroxide complexes by polyvalent metal ions, e.g., $Fe(OH)^{2+}$, $Fe(OH)_2^+$, $Fe_2(OH)_2^{4+}$, etc., provides numerous species of a varying degree of polymerization, hydrogen-bonded with each other and hydrogen-bonding extra water molecules. Adsorption of these species at a solid/liquid interface is frequently nonselective (nonspecific) and can vary in extent by an order of magnitude or so, e.g., Li^+ and Co^{2+} in Figure 6.34.

Surfactants containing a single polar group (monopolar or monofunctional) may adsorb at a solid/liquid interface either through the polar group to a polar solid or through the nonpolar hydrocarbon chain onto a nonpolar solid. Thus, the long axis of their molecules may be either parallel

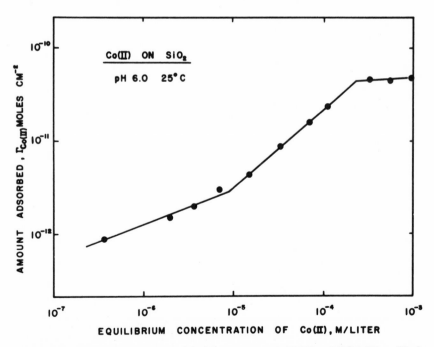

Figure 6.35. Adsorption isotherms of Co(II) on quartz at pH 6.0 and 25°C. [From Healy *et al.* (1968), in *Adsorption from Aqueous Solution*, W. J. Weber Jr. and E. Matijevic, eds., with the permission of American Chemical Society, Washington, D.C.]

to the solid surface or inclined to it at a fairly steep angle. Multipolar and polymeric surfactants possess a large number of polar groups, either of the same type or mixed (ionizable and nonionized dipolar groups), which may interact not only with the solid surface but also with each other. Consequently, the disposition that a polymer molecule may adopt at a solid/ liquid interface depends not only on the reactivity between the solid surface sites and the polymer but also on the possibility and the extent of any additional intramolecular bonding. In the absence of intramolecular bonding, the polar groups of the multipolar surfactant provide multipoint anchoring to the solid surface. When intramolecular bonding exists, a highly coiled configuration of the polymer molecule may produce a complete surface coverage with negligible interaction between the surface sites and the polar groups. Adsorption isotherms for polymers may be of any form discussed so far; their interpretation is extremely difficult and unreliable unless much additional information is obtained by other techniques. The effectiveness of multipolar polymers in flocculation is due to their ability to bridge solid particles (Figure 6.36). When the size of particles is very small, even a single

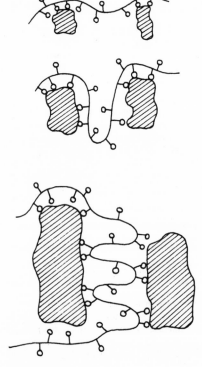

Figure 6.36. Interactions of a polymer molecule with solid particles. Depending on the relative size of particles and the disposition of the groups developing bonds with the solid (polar or nonpolar groups and/or bonds), multipoint bonds may develop between small particles or bridging by intramolecularly bonded segments of polymers may develop between large particles.

anchoring of a particle to an uncoiled polymer suffices for aggregation. Larger particles can provide multipoint anchoring which involves highly coiled polymer molecules.

6.8. Oxidation of Surfaces of Metals and Metallic Sulfides

Physical adsorption of gases or vapors may change into chemisorption (as evidenced by a partial or a complete charge transfer) whenever there exists a characteristic chemical specificity between the adsorbate and the solid substrate. For minerals exposed to air such chemisorption specificity occurs whenever the elements composing the mineral phase react to form either oxides or nitrides. Thus, all sulfides tend to chemisorb oxygen. However, most oxides, silicates, or carbonates show little specificity toward chemisorption of oxygen because they already contain a nearly requisite quantity of this element. Superseding chemisorption, sulfides tend to form additional surface oxidation products, since both of their types of constituent elements (i.e., metals and sulfur) have high affinities[†] for oxygen (Section 10.3).

In view of the heterogeneity of the solid surfaces, neither the initial adsorption nor the subsequent reactions forming surface oxidation products proceed uniformly over the whole exposed surface. Initially, the kink and ledge portions of the surface adsorb oxygen preferentially and may start forming patches of multilayer oxidation products (thus acting as nuclei) before the neighboring terraces become filled with the first chemisorbed monolayer. Different crystal faces (exposed as surfaces) oxidize at different rates. Indeed, careful measurements by Rhodin (1950) and by Young *et al.* (1956) on single crystals of metallic copper have provided striking evidence of large differences in rates (Figure 6.37) for faces with different Miller indices. More recent work on the role of crystalline orientation in electrodissolution (anodic as well as cathodic) and electrodeposition has confirmed the differences in the kinetics of most reactions at different faces. The presence of defects in the solid face, the availability of O_2, and any contamination arising from a variety of sources may have an overriding in-

---

[†] The affinity of a reaction, such as oxidation, is defined in thermodynamics as the derivative of the free energy decrease with respect to the extent of reaction taking place whenever given amounts of reactants are converted into appropriate amounts of products. [For detailed relationships, see Everett (1959, pp. 97ff.).]

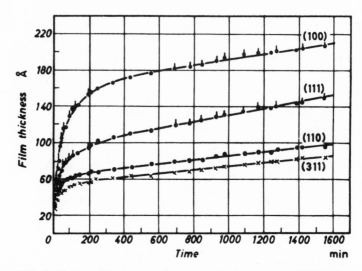

Figure 6.37. Oxidation of four faces of a copper single crystal at 178°C; film thickness vs. time. [Reprinted with permission from Young *et al.* (1956), *Acta Metall.* **4**, p. 145. Copyright 1956, Pergamon Press, Elmsford, New York.]

fluence on the kinetics of reactions at different faces. For details, refer to Jenkins and Stiegler (1962), Jenkins and Bertocci (1965), Bertocci (1966), and Jenkins and Durham (1970).

Depending on the electronic structure of the metal atom incorporated in the surface, the bonds formed on chemisorption at different kinks, ledges, emergent dislocations, etc., may vary considerably in character and energy. Clear evidence for such differences has been amply provided by infrared spectroscopy of chemisorbed gases and is presented by Little (1966). The appearance of new IR adsorption bands at different wavelengths (in comparison with those shown in spectra of free adsorbate and substrate) provides, in general, a direct confirmation of the chemisorption bonds developed on a mineral or solid surface. The energies of the new bonds are directly related to adsorption band wavelengths.

Consequent to the development of a bond during chemisorption, some perturbation of the surface atoms occurs. Independent evidence obtained by Ehrlich (1963a,b) using field-emission microscopy and Farnsworth and Madden (1961) or Germer and MacRae (1962a,b), who used the low-energy electron diffraction technique (LEED), suggests that very drastic structural rearrangement of surface atoms occurs on metallic nickel when traces of oxygen at 10^{-9}–10^{-8} mmHg are allowed to react with the surface; no similar rearrangement is observed with either N_2 or H_2 adsorption, which are physically adsorbed. The conclusion reached from such studies on metals is that in addition to the distortion of crystal spacing taking place among surface atoms during the moment of creation of the surface, a further rearrangement of surface structure must occur during *all* chemisorption processes. No analogous studies have been carried out on surfaces of metallic sulfides, but it is presumed that the conclusion regarding postchemisorption rearrangements applies to most systems involving solid surfaces. The surface perturbation resulting from chemisorption may produce tensile or compressive stresses in different portions of the surface; at some points it may cancel the stress (strain) produced at the time of creation of the surface, or it may increase its magnitude.

The oxidation product forming atop the chemisorbed monolayer patch is frequently characterized by an *epitaxial growth*, whereby the oxidation product continues the crystalline structure of the substrate, as, for example, when a drop of supersaturated solution of $NaNO_3$ (sodium nitrate) crystallizes on calcite ($CaCO_3$), with which it is isomorphous. Numerous metal–metal oxide systems showing epitaxial growth are discussed by Pashley (1970) and Gwathmey and Lawless (1960). The latter authors dispute the validity of some previously held criteria, such as, for example, that the existence of only a small amount of misfit in the close-packed orientations of crystalline lattices of metal substrate and of the epitaxial oxide is a necessary condition for an epitaxial growth or that a thin oxide film is necessarily pseudomorphic with the substrate. Despite the lack of complete understanding of epitaxy, the universal occurrence of epitaxial growth is clearly recognized, as is its importance in the strength of adhesion of the film to the substrate. Generally, whenever the rate of oxidation on a particular crystal face is high, thick oxide films are produced, with numerous orientations caused by increased numbers of nuclei and paths of high diffusion rates. In consequence, thick oxidation films flake off easily. On the other hand, thin films, grown on the faces with low rates of oxidation, adhere tightly.

A surface of polycrystalline freshly fractured metal contains a variety of crystalline faces, each of which chemisorbs oxygen at different rates and reacts subsequently with additional oxygen molecules to give oxidation films of different thicknesses. The overall surface oxidation product is thus nonuniform in thickness and in composition, and each portion of such an oxide film is underlain by a monolayer of chemisorbed oxygen. When the metallic element has several valencies, the overall composition of the oxidation product is much less uniform than that on a metallic element of single valence. Layers of different oxides are formed parallel to the surface, with the lowest valence metal oxide nearest the chemisorbed layer. For example, on iron the sequence of oxidation layers away from the chemisorbed oxygen is FeO, Fe_3O_4, and Fe_2O_3 (outermost).

When a freshly fractured surface of a metallic sulfide undergoes oxidation in air or in aqueous solution, the products formed consist of a still larger variety of species, starting with elemental sulfur, metal hydroxide, metal thiosulfates ($S_2O_3{}^{2-}$), metal polythionates ($S_xO_6{}^{2-}$, $x = 2-6$), and metal sulfates ($SO_4{}^{2-}$) to volatile SO_2 and residual metal oxides. The number of possible species increases rapidly with multiple valence of the metallic element(s) [when the mineral is a complex sulfide, containing more than one metallic element, such as $(Zn,Fe)S$, $CuFeS_2$, etc.]. The conditions under which oxidation reactions occur play a decisive role in establishing the nature of the species present in the oxidation film. The availability of oxygen (as measured by p_{O_2}, the partial pressure of oxygen for the solid–gas reactions, or by μ_{O_2}, the chemical potential of dissolved oxygen in solution, for the solid–liquid reactions), the temperature of the system, the existence of any catalytic agent in the system, and the time of the oxidation reaction all decide the composition of the oxidation film.

Most of the work on the oxidation of sulfide minerals reported in the literature pertains to high-temperature oxidation as in roasting of sulfides to convert them to oxides (before pyrometallurgical smelting) or to water-soluble sulfates and acid-soluble oxides for leaching purposes. Only a very limited amount of published work relates to oxidation of sulfides under conditions preceding their flotation. The classic work is that of Eadington and Prosser (1966, 1969), who carried out extensive determinations of surface oxidation products on synthetic lead sulfide. Their results have resolved the apparent conflict between previous evaluations of surface oxidation products by other workers and also provided clear indications of the role played by additional parameters such as aging, nonstoichiometry, and irradiation by visible light. The previous work on the oxidation of lead sulfide at low temperature consisted of the following: Plante and Sutherland (1949) found that in alkaline solutions both lead thiosulfate and lead sulfate were produced in significant quantities. Reuter and Stein (1957) and Leja *et al.* (1963) concluded that lead thiosulfate was the predominant species of PbS oxidation in aqueous solutions, while Hagihara (1952a) and Greenler (1962) found lead sulfate to be predominant.

Analyzing the oxidation products (after their dissolution from the PbS with ammonium acetate), Eadington and Prosser established the following:

1. The pH of the solution contacting solid PbS and the time of exposure to oxygen were the two major parameters that determined the nature of oxidation products.

2. At pH 1.5, the major products were elemental S and Pb^{2+}, with a small quantity of $PbSO_4$ appearing after about 5 hr of oxidation.

3. At pH 7, the quantity of elemental S produced was very small ($\sim 10^{-6}$ g mol from 2.5×10^{-3} mol of PbS precipitate, with 15.7-m^2/g area); the main product obtained in up to 12 hr was lead thiosulfate, which reached a constant level of $\sim 0.1 \times 10^{-3}$ mol; lead sulfate appeared only after 5 hr, exceeding the thiosulfate after 12 hr, as shown in Figure 6.38(a).

4. At pH 9, lead thiosulfate appeared to be the predominant product, with lead sulfate again appearing only after ~ 5 hr but reaching a constant level of $\sim 0.2 \times 10^{-3}$ mol; see Figure 6.38(b).

A small proportion of polythionate, determined as $S_4O_6^{2-}$, was detected in the oxidation product (approximately 10^{-4} g mol); also, a nearly constant amount ($\sim 0.1 \times 10^{-3}$ mol) of elemental S was found at pH 9, although thermodynamically, judging from the E_h/pH diagram evaluated by Majima (1969), no elemental S should exist at this pH.

The rate of oxidation was found to be nearly constant (to within 10%) for temperatures 25°–50°C and for oxygen pressures between 20 and 760 mm Hg.

The preceding results of Eadington and Prosser are in agreement with the oxidation mechanism proposed by Reuter and Stein (1957), i.e., that the initial products of oxidation are elemental sulfur and $Pb(OH)_2$,

$$2PbS + 2H_2O + O_2 \rightarrow 2Pb^{2+} + 4OH^- + 2S^0 \qquad (6.45)$$

followed by the formation of basic lead thiosulfate and its subsequent gradual disproportionation into lead sulfate and sulfide:

$$4Pb^{2+} + 8OH^- + 4S^0 \rightarrow 2PbS + PbO \cdot PbS_2O_3 \cdot xH_2O + (4-x)H_2O \qquad (6.46)$$

and

$$PbO \cdot PbS_2O_3 \cdot xH_2O \rightarrow PbS + PbSO_4 + xH_2O \qquad (6.47)$$

When $PbSO_4$ is precipitated in quantities approximately equivalent to a monolayer, the oxidation reaction appears to be significantly retarded, as if it were blocking the active sites on the surface of PbS. Other conclusions of Eadington and Prosser were that the stoichiometry of the precipitated

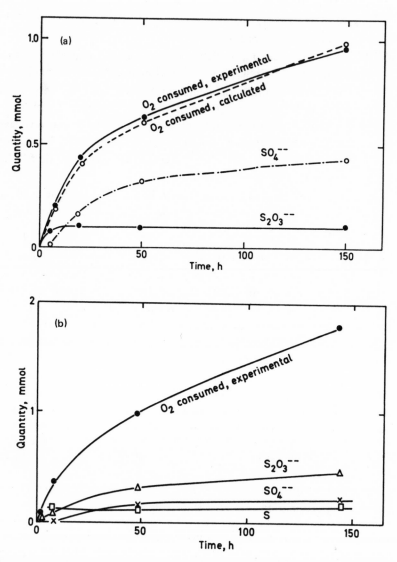

Figure 6.38. (a) Rate of formation of oxidation products from an aged precipitate at pH 7. Total quantity of PbS in the reaction vessel, 2.5 mmol; specific area, 15.7 m²/g. (b) Rate of formation of oxidation products from the same PbS precipitate but at pH 9. Note the reversed positions of SO_4^{2-} and $S_2O_3^{2-}$ formation in comparison with those in part (a) (for pH 7) and the formation of elemental sulfur, which is theoretically not likely to form in a simple system. [From Eadington and Prosser (1969), *Trans. IMM* **78**, with the permission of the Institution of Mining and Metallurgy, London.]

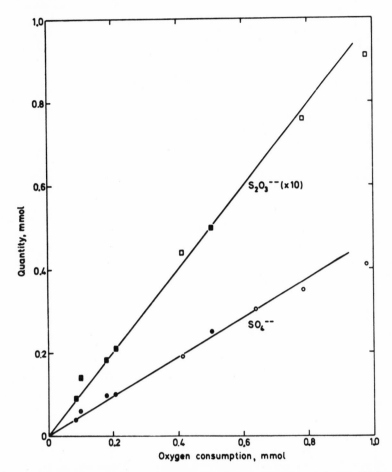

Figure 6.39. Quantities of SO_4^{2-} and $S_2O_3^{2-}$ (note that the $S_2O_3^{2-}$ scale is $\times 10$) formed over long periods of oxidation by different samples; open symbols correspond to aged precipitates and solid symbols to samples of fixed nonstoichiometry. Excess sulfur in PbS caused a higher oxygen consumption and led to SO_4^{2-} formation, and excess lead in PbS to $S_2O_3^{2-}$ formation. [From Eadington and Prosser (1969), *Trans. IMM* **78**, with the permission of the Institution of Mining and Metallurgy, London.]

PbS affected the rate of oxidation: Lead-rich samples were much slower to oxidize than sulfur-rich samples (Figure 6.39). High-intensity visible light also reduced the rate of oxidation, as shown in Figure 6.40. Freshly precipitated PbS and lead-rich PbS samples behaved similarly in that they required an induction period of up to 40 hr before an oxidation at a definite rate would start; the consumption of oxygen during the induction period was below the detection limit, $< 1 \times 10^{-6}$ g mol.

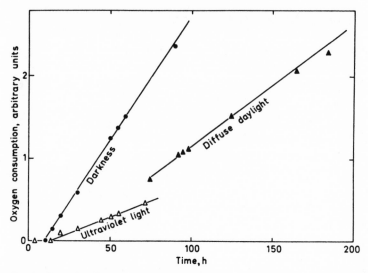

Figure 6.40. Effect of radiation on oxygen consumption in reactions of PbS containing excess S. [From Eadington and Prosser (1969), *Trans. IMM* **78**, with the permission of the Institution of Mining and Metallurgy, London.]

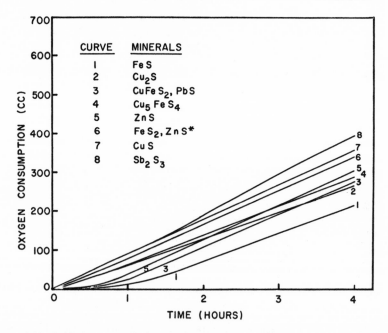

Figure 6.41. Oxidation rates of minerals. Oxygen-consumption curves in *neutral solutions* at 120°C. $P_{O_2} = 124$ psi; pH $= 7.1$ (phosphate buffer). ZnS* denotes sphalerite activated by Cu^{++} ions. [From Majima and Peters (1966), *Trans. Met. Soc. AIME* **236**, with the permission of the American Institute of Mining and Metallurgical Engineers, Littleton, Colorado.]

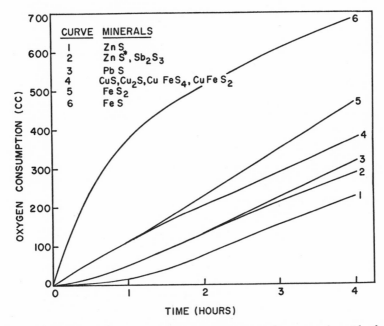

Figure 6.42. Oxidation rates of minerals. Oxygen-consumption curves in *acid solutions* at 120°C. $P_{O_2} = 124$ psi; pH = 2.7 (phosphate buffer). ZnS* = copper-activated sphalerite. [From Majima and Peters (1966), *Trans. Met. Soc. AIME* **236**, with the permission of the American Institute of Mining and Metallurgical Engineers, Littleton, Colorado.]

No similarly detailed investigation appears to have been done regarding oxidation of other sulfides. Majima and Peters (1966) carried out a comparison of oxidation rates at 120°C of various sulfides (each tested individually) using aqueous solutions buffered by phosphate at pH 2.7, 7.1, and 11.2. Their results are shown in Figures 6.41 to 6.43 for the above specified pH values. The oxidation rates were measured in terms of oxygen volume consumed per unit of exposed surface area, employing equivalent-size particles, pulp densities, agitation, and oxygen pressure. The variation in oxidation rates can be explained by a greater or lesser solubility (or insolubility) of the expected oxidation products (the nature of which was not determined) for each individual metallic element.

The catalytic effect of increased oxidation in the case of copper-activated ZnS is clearly shown in Figure 6.43. This effect due to the copper additive is not an isolated effect; other additives were found to act as oxidation catalysts of varying effectiveness, for example, in the oxidation of ZnS using Bi, Ru, Mo, and Fe, as reported by Scott and Dyson (1968), and in the oxidation of CuS- and $CuFeS_2$-activated carbon with additions

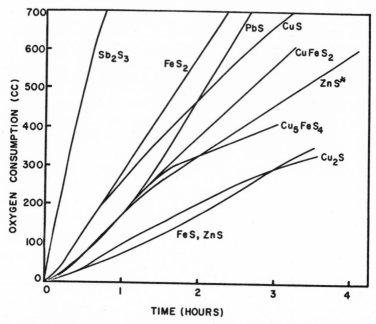

Figure 6.43. Oxidation rates of minerals. Oxygen-consumption curves in *basic* solution at 120°C. $P_{O_2} = 124$ psi; pH = 11.1 (phosphate buffer). ZnS* = copper-activated sphalerite. [From Majima and Peters (1966), *Trans. Met. Soc. AIME* **236**, with the permission of the American Institute of Mining and Metallurgical Engineers, Littleton, Colorado.]

of Fe^{3+}, as reported by Bryner *et al.* (1967). The last authors dealt primarily with the effect of microorganisms, like *Thiobacillus ferrooxidans*, on dissolution at pH < 4 of FeS_2, CuS, and $CuFeS_2$. They found that in a sterile system, without microorganisms, the nonbiological oxidation of pyrite at 35°C was less than 0.6% in 76 days, while the biologically inoculated system showed that over 80% of the iron was oxidized and solubilized. The more complex effect of temperature on the biological oxidation of pyrite and of reagent-grade copper sulfide is shown in Figures 6.44 and 6.45, respectively. The initial pH of the nutrient solution was pH 3.

Although the results of Bryner *et al.* (1967) are of primary interest to hydrometallurgical leaching of residues, as catalyzed by bacterial action, they do have a direct bearing on flotation systems. What they do emphasize is the possibility that microorganisms, likely to be encountered in water which is being recirculated from tailing dams to the mills, *may* strongly affect the oxidation reactions during grinding. As will be discussed in greater detail in Chapter 10, in flotation systems dealing with separations of sulfide minerals using thio compounds, a very close control of the degree of oxidation within the system is indispensable for a successful selectivity of separation. Catalytic action of microorganisms in sulfide oxidation is fostered by an acidic pH (and such a pH is easily generated in uncontrolled aqueous oxidation of sulfides in tailing dams) and an optimum temperature of $\sim 30\text{--}35°C$, which is readily attained (at least locally) during grinding.

Finally, another variable that may affect the nature of the oxidation products and the rates of reactions is that of galvanic action. A galvanic couple is established between two different sulfide phases in direct contact (as in an incompletely liberated particle) or is created by a speck of metallic iron deposited on the surface of a sulfide particle from grinding balls or mill lining by attrition during grinding. The difference in electrical potentials created by such direct contact of two conducting or semiconducting phases immersed in an electrolyte solution leads to additional electrochemical

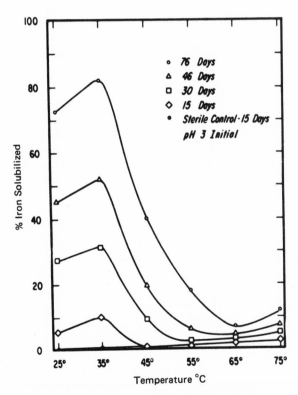

Figure 6.44. The effect of temperature on the biological oxidation of pyrite (effect of *Thiobacillus ferrooxidans* on dissolution at pH 3). [From Bryner *et al.* (1967), *Trans. AIME* **238**, with the permission of the American Institute of Mining and Metallurgical Engineers, Littleton, Colorado.]

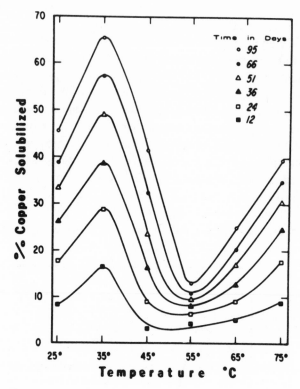

Figure 6.45. The effect of temperature on the biological oxidation of reagent-grade copper sulfide (inoculated with *Thiobacillus ferrooxidans*). [From Bryner *et al.* (1967), *Trans. AIME* **238**, with the permission of the American Institute of Mining and Metallurgical Engineers, Littleton, Colorado.]

reactions, to an increase in the rate of dissolution of the anodic phase, and to a possible change in the character of products that would otherwise be generated in the absence of such galvanic potential. The electrochemical aspects of sulfide dissolution have been discussed by Majima and Peters (1968) in relation to hydrometallurgical applications at acid pH. A thorough evaluation of reactions introduced by a galvanic contact between sulfide minerals and that with metallic iron in a flotation environment remains to be carried out; preliminary results have been reported by Rao *et al.* (1976).

6.9. Hydration of Surfaces. Thin Films

Molecules of water tend to adsorb on all surfaces, but they do so particularly strongly on any freshly formed *polar* solid surfaces; they create thereon a hydration layer that shows some unusual characteristics. Most of the insoluble metal oxides (if not all) adsorb some of the water by a dissociative chemisorption to form surface hydroxyl groups on top of which molecular water is adsorbed. The two species can readily be identified by IR spectroscopy [see Little (1966, pp. 233, 250–267)] when present on minerals that are transparent to IR. The in-plane deformation vibrations of H atoms give an absorption band at 1630 cm^{-1} for liquid water, and the hydroxyl stretching vibrations give a band at 3450 cm^{-1} for liquid water with marked evidence of hydrogen bonding. The surface hydroxyl groups, on the other band, give a narrow band at approximately 3770 cm^{-1} (3740–3790 cm^{-1}) and a low-frequency band of variable intensity at 3650–3680 cm^{-1}.

These differences in bands have been used to distinguish surface hydroxyl groups from adsorbed water molecules. Numerous IR studies on silica gels, porous silica glass (Vycor), silica powder (Cab-O-Sil), alumina, thoria, etc., have established that adsorbed water is not removed even by a procedure more drastic than the one used normally for moisture determinations, namely heating and evacuation at $110°$ or $120°C$. Even after 8 hr of evacuation at $300°C$ some water molecules remain adsorbed on silica. Holmes *et al.* (1968) have shown that the surface of thorium oxide, after high-temperature evacuation at $500°C$, holds adsorbed water irreversibly in quantities far in excess of the amount required to give a simple hydroxylation of the surface; at least an additional monolayer of molecular water, hydrogen-bonded to the surface hydroxyl groups, was presumed to remain on that surface. In spectra of silica that had been treated by evacuation at $800°C$, some surface hydroxyl groups have still been clearly indicated. A complete removal of OH groups (leaving a dehydroxylated surface containing 0^-) was found to occur only with alumina evacuated at $> 650°C$. The so-called active Al_2O_3 is not pure but contains impurities such as alkali oxide, iron oxide and sulfate, and H_2O from a few tenths of 1% to $\sim 5\%$. A large number of compounds are referred to as active aluminas. All of them are obtained from aluminum hydroxide by dehydration at various temperatures; if the temperature is less than $600°C$, the products are known as the γ group of aluminas; if it is $900°$–$1000°C$, they are known as the δ group of aluminas. Water is adsorbed on active aluminas either as hydroxyl ions or as water molecules, depending on temperature and partial pressure. At room temperature aluminas adsorb molecular water bonded with strong hydrogen bonds to the surface. During drying at $120°$–$300°C$, most of the water is removed, but some molecules do not desorb but react to form surface hydroxyl groups. These OH ions on the alumina surface behave as

Brönsted-acid sites. When on dehydration two neighboring OH⁻ ions combine and lose one water molecule, the remaining oxygen bridge between two

atoms in the surface behaves as a *Lewis-acid site.* Lewis and Brönsted acid sites may be distinguished by the differences in the IR spectra of chemisorbed pyridine and ammonia. Extensive studies of alumina surface structure were carried out by Peri, Lippens, Fripiat, and their co-workers; see Clark (1970) or Linsen (1970).

Solids other than metal oxides do not necessarily form surface hydroxyl groups by dissociative chemisorption, but all solids adsorb water to the extent of forming numerous multilayers when the partial pressure approaches saturation.

It is not only water vapor that adsorbs on solids as multilayer films, but all vapors of liquids behave in an analogous manner, producing multilayer films on adsorption.

A similar film, consisting of multilayers of adsorbed liquid molecules, is retained at the solid/liquid interface whenever the excess liquid phase is removed by any mechanical action (such as draining, blowing, application of suction or pressure) to give a de-wetted solid/vapor interface. To determine the thickness of such adsorbed vapor layers of liquid films retained after de-wetting, Derjaguin and co-workers utilized the elliptic polarization of light reflected at an oblique angle from a smooth solid surface.

The results of such measurements, obtained since about 1935, on a number of liquid vapors adsorbing on the surface of glass were reported by Derjaguin and Zorin (1957) and are shown in Figure 6.46.

Two types of behavior can be distinguished among these results:

1. Polar vapors (water, alcohols) adsorb on the polar solid (glass), giving isotherms that intersect the ordinate line at saturation ($p/p_0 = 1$) at a definite thickness (~ 20, ~ 40, or ~ 60 Å).
2. Vapours of nonpolar liquids (CCl_4, benzene) give isotherms (on the polar solid) which approach the ordinate line $p/p_0 = 1$ asymptotically.

[*Note*: The preceding behavior suggests that the glass surfaces used in adsorption experiments were hydrophobic in character, most likely due to an accidental contamination. (Cf. p. 494 and p. 579.)]

Further differences in the behavior of surfaces possessing retained

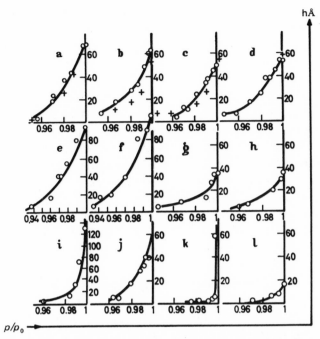

Figure 6.46. Adsorption isotherms on (an apparently hydrophobic) glass: (a) water, (b) ethyl alcohol, (c) propyl alcohol, (d) butyl alcohol, (e) hexyl alcohol, (f) heptyl alcohol, (g) octyl alcohol, (h) nonyl alcohol, (i) carbon tetrachloride, (j) *n*-pentane, (k) benzene, (l) nitrobenzene. [From Derjaguin and Zorin (1957), with the permission of the authors.]

liquid films were noted when these surfaces were viewed under the polarizing microscope while the vapor pressure was changed through the saturation point (by changing the temperature). For example, with vapors of nonpolar liquids the thickness of the film increased in a continuous manner uniformly over the whole surface and resulted in alternated lightening and darkening of the field of view with polarization of light. However, the vapors of polar liquids showed the development of nonuniformly distributed brighter spots in the film as condensation progressed. These spots grew in size and became isolated lenses of condensed liquid phase, scattered over a film of constant thickness, unchanged since saturation.

These results of Derjaguin and co-workers show that an adsorbed layer of a finite film thickness can coexist with the liquid bulk phase. The measured thickness of the film, from ~ 20 to ~ 60 Å, indicates that many layers of adsorbed polar molecules constitute the film, so it is definitely not a monolayer or a double layer.

The thickness of the films remaining between the surface of glass wetted by water (or by electrolyte solutions) and an air bubble pressed against this glass surface was determined by Derjaguin and Kussakov (1939). They found that such wetting films are considerably thicker (e.g., 1150 Å in 10^{-4} N NaCl, 750 Å in 10^{-3} N NaCl) than the thickness of the corresponding electrical double layers. The latter were evaluated by Jones and Wood (1948) as $\sim$300 Å in 10^{-4} M KCl and 96 Å in 10^{-3} M KCl. A redetermination of such residual films underneath air bubbles pressed against silica glass carried out by Read and Kitchener (1967) (see Figure 9.17) has confirmed the order of magnitude of the thickness obtained by Derjaguin and Kussakov for pressures of $\sim$2000 dyn cm^{-2} (1 atm $\equiv 0.980 \times 10^6$ dyn cm^{-2}).

All measurements of this nature are extremely difficult; not only infinitesimal traces of surface active contamination (the removal of which calls for extraordinary precautions) are bound to affect the magnitude of the experimentally determined thickness of the residual films, but also any surface dissolution (from solid surfaces) is likely to have an overriding influence on the properties being measured for these thin liquid films. Examples of such difficulties, due to surface active contaminations and solute contamination, are some of the numerous anomalies frequently reported in the literature [reviewed by Drost-Hansen (1971b)] and, particularly, the protracted controversy surrounding the existence of polywater (water II, anomalous water, orthowater, superwater, etc.).[†]

The existence of thin, interfacial hydration films is beyond dispute; however, the magnitude of the film thickness may vary considerably depending on the surface tension and dissolution from the solid, giving gel-like surface structures. To exist at an interface, thin films must be under the influence of a force that acts normally to the surface or the interface (the films exist at all interfaces involving a liquid phase: solid/liquid, liquid/liquid, and air/liquid). As mentioned in Section 2.9.4, Derjaguin (1955)

[†] In 1965 Derjaguin announced the discovery of highly unusual properties of a new form of water that was obtained by condensation in glass capillaries. Since then, until 1973, scientists in innumerable laboratories all over the world applied the most modern instrumental techniques to produce and to determine the structure and the characteristics of this liquid (obtained only in capillary condensation). The result was that masses of conflicting data were obtained by various groups. Derjaguin, revising his own paper (submitted to a conference in 1971) for publication, see Derjaguin (1973), reversed his long-held stand on this matter and, to his great credit, acknowledged that he and his followers were chasing an illusion. He concluded that excessive dissolution of quartz capillaries takes place during the condensation of vapor on their walls, and the unusual properties of the condensate are due to the presence of minute traces of solutes such as silicon, sodium, and/or boron.

coined the name *disjoining pressure* of the film, Π, defining it as a mechanical pressure that would have to be applied to the bulk liquid in order to bring it into equilibrium with the film of a given thickness h. Theoretically, the disjoining pressure can be evaluated from different contributions, as outlined in Section 9.2.2, provided that the related film thickness h is reliably determined.

The measurements reported by Derjaguin and Zorin (1957), shown in Figure 6.46, demonstrate the existence of positive disjoining pressures in thick wetting films. Films of limited thickness which coexist with microlenses formed during the condensation of vapors indicated negative values of disjoining pressure. A definite contact angle was formed by such lenses, thus manifesting the fact that the film-covered solid is not wetted by the bulk liquid.

Frumkin and Gorodetskaya (1938) observed liquid lenses similar to those formed in the condensation experiments of Derjaguin and Zorin that developed underneath air bubbles adhering to a mercury surface. The lenses were in contact with a residual liquid film of a finite thickness. Mercury was immersed in distilled water or in dilute electrolyte solutions (less than 10^{-3} M Na_2SO_4) and was held at potentials close to the electrocapillary maximum (see Chapter 7). Air bubbles were deposited onto mercury and were found to adhere to the surface. Some time after deposition of bubbles, very numerous and very small lenses began to appear over the mercury–bubble contact area; within 12–48 hr the number of these lenses decreased as they grew in size by coalescence. Frumkin (1938) presented a thermodynamic treatment to explain the instability of the hydration films existing initially under the bubbles and their correlation with contact angles.

This type of behavior whereby bulk liquid does not wet a multilayer film formed by its own vapor adsorbed on a solid is known as *autophobicity*. *Autophobic surfaces* can exist not only at solid/liquid interfaces but also at liquid/liquid interfaces.

Thick layers of organic liquids spread on water may break into an invisible thin film and extra lenses. Water on a mercury surface or hydrocarbon oils on a mercury surface also behave similarly. All autophobic surfaces indicate the existence of negative disjoining pressures.

Sheludko and Platikanov (1961) used an ingenious apparatus to determine the relationship between the disjoining pressure Π and the thickness h of films of benzene on mercury; their results, shown in Figure 6.47, indicate that benzene films between $\sim$240 and 1000 Å in thickness show negative disjoining pressures and are inherently unstable. Analogous negative disjoining pressures (denoting inherent instability) were obtained for

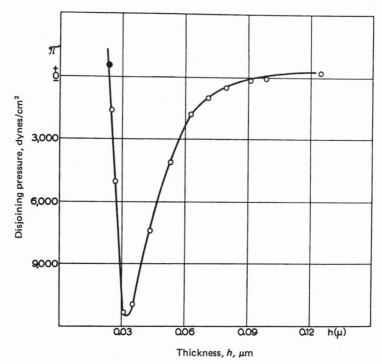

Figure 6.47. Experimentally determined relationship between (negative) disjoining pressure and thickness in benzene films on mercury. [After Sheludko and Platikanov (1961), from Sheludko (1966), *Colloid Chemistry*, with the permission of Elsevier Publishing Corporation, Amsterdam.]

other systems, such as electrolyte (KCl) solutions on mercury, free films of electrolyte solutions between two bubbles, etc.

The thermodynamic instability, indicated by the negative values of the disjoining pressures, and the kinetics of changes associated with the thinning of these films, discussed in Section 9.7, are the basic factors operative in flotation, coagulation, and the stability of foams and emulsions. The existence of a sharp boundary between the multilayer hydration film and the lenses of nonwetting liquid was used by Derjaguin and co-workers as an argument to stress that the structure of molecules near the polar solid must be ordered.[†]

[†] Normally, polished glass is hydrophilic, completely wetted by water and alcohols; thus, it should not be autophobic, as indicated by the results obtained by Derjaguin and Zorin in Figure 6.46. As already suggested, the glass plate used by them must have been contaminated by surfactant species or heated above 650°C, as established by Laskowski and Kitchener (1969) (see p. 494). This does not detract from the generality of their argument about the instability of films possessing negative disjoining pressure. (See Section 9.4.)

There is plenty of evidence that hydration films, existing at polar interfaces, show anomalous characteristics; hence, they must have structural differences with respect to the bulk liquid. If Frank's model of flickering clusters and free water molecules or any more sophisticated model such as that of Rahman and Stillinger (1971) (Section 4.1) is accepted for the bulk liquid, then a three-layer model of water is the most likely for polar solid–liquid interfaces, as discussed by Drost-Hansen (1971b). The first layer consists of water molecules oriented by dipole–dipole interactions. The latter propagate over several molecules away from the polar surface into the liquid. The second layer consists of less oriented free water molecules. Some ions are adsorbed specifically within the first layer, and others, with their hydration sheaths, are distributed within the first and second hydration layers. The third layer, the bulk water containing some flickering clusters, commences at a distance many monolayers away from the solid surface.

The structure of water near a nonpolar (hydrophobic) surface is visualized as consisting primarily of cluster-like entities near the nonpolar surface with a very narrow disordered intermediate region separating the bulk water phase. No specifically adsorbed dehydrated ions are presumed to be present near a nonpolar surface.

6.10. *Insoluble Monolayers of Surfactants at the Air/Water Interface. Interactions Among Surfactants*

The most revealing work carried out on surfactants to evaluate their characteristics has been done with insoluble surfactant homologues at the air/water interface using a Langmuir–Adam surface balance (Figure 6.48). The latter is a device in the form of a trough in which a float F attached to a torsion wire T measures forces exerted by a film of surfactant spread on the surface of water and enclosed between the float F, the barrier B, and the sides of the trough. Detailed accounts of the monolayer studies are given by Adam (1941), Harkins (1952), Davies and Rideal (1961), and Gaines (1966). All the structural and chemical requirements necessary for the molecules in the monolayers of surfactants to behave (in a manner analogous to the three-dimensional states of matter) as a two-dimensional gas, liquid or solid (Figure 6.49), have been evaluated using the trough technique. *Gaseous* or *vapor* surfactant films consist of separate molecules, freely and independently moving along the interface. The alignment of long-chain molecules forming a gaseous film is parallel to the surface of the liquid,

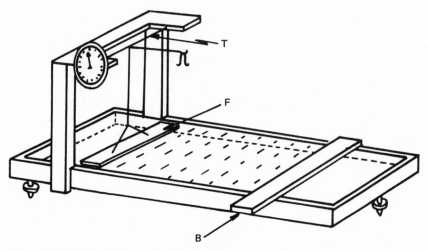

Figure 6.48. Surface balance (Langmuir–Adam trough) for studying insoluble mono-layers. Float F is attached to torsion wire T, which is calibrated to measure the force exerted onto the float by the surfactant film contained between the barrier B and the float F. Flexible strips attached to the edges of the float (F) and the sides of the trough prevent the surfactant from escaping past the float when the area available for the sur-factant molecules in the film is gradually decreased by sweeping the barrier B toward the float F.

with their polar group immersed in the liquid. *Condensed* or *solid films* forming at the air/liquid interface consist of closely packed, steeply oriented long-chain molecules with maximum interactions between the adlineated hydrocarbon chains. Intermediate between gaseous and solid-condensed films are *liquid-expanded* and *liquid-condensed* two-dimensional films, with molecules associated in the form of islands "floating" among single molecules of a gaseous film (Figure 6.50). Transformations somewhat analogous to condensation and solidification in three-dimensional systems take place in two-dimensional films at a definite temperature for each particular sur-factant. For a straight-chain alcohol, carboxylate, or amine, a change in the chain length by one CH_2 (methylene) group corresponds, approximately, to an 8–10°C change in the temperature of transformations. The chain lengths of the homologues which become sufficiently insoluble to be studied at the air/water interface by the Langmuir–Adam surface balance vary in relation to the nature of the polar group. Thus, for alcohols a C_{10} or C_{12} chain length is sufficient to make the molecules insoluble, particularly on acidic substrates. Carboxylic acids become insoluble starting with C_{14} (myristic) if the pH of the substrate is less than 2. Alkyl sulfates and sul-fonates have to be at least C_{20}–C_{22} in aliphatic chain length and require substrates of highly concentrated salt (NaCl or KCl) solutions before they can be studied as insoluble monolayer films. For amines, chain lengths of C_{20}–C_{22} and highly basic substrates are required to enable formation of insoluble films at the air/water interface. The preceding requirements refer to room temperature.

The term *insoluble* really denotes a solubility below the limit of ready detection, that is, less than, say, 10^{-8} or 10^{-9} mol/liter; this still represents a very large number of *molecules* going into the substrate (10^{12}–10^{11} molecules/cm^3).

The behavior of surfactant molecules at an interface is determined by a combination of (1) dispersion forces, i.e., van der Waals bonds developed between induced dipoles; (2) dipole–dipole interactions between the permanent dipoles present in the substrate; (3) ion–dipole and ion–ion interactions; and, occasionally, (4) specific chemical bonds between the polar group and the substrate. Dipole–dipole and dispersion (van der Waals) forces are always present and cooperate in the interactions of all surfactants at any interface. Ionic and chemical bonds may be present under one set of interfacial conditions but not necessarily under another set.

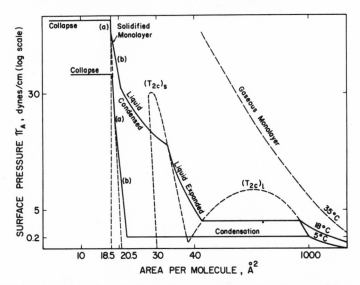

Figure 6.49. Transformations in insoluble monolayers of surfactants from gaseous through liquid to solidified films as shown by pressure–area (π_A–A) isotherms for three different temperatures [A = area per molecule in Å^2 and π_A = surface pressure which equals ($\gamma_{H_2O} - \gamma_{film}$)]. (a) Fully adlineated hydrocarbon chains in a solidified monolayer; (b) incompletely adlineated hydrocarbons in a monolayer. The theoretical cross section of adlineated hydrocarbon chains equals 18.5 Å^2/molecule.

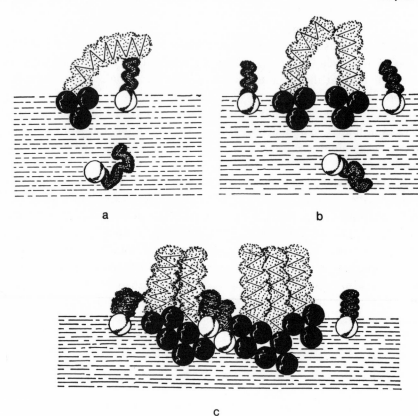

Figure 6.50. Schematic representation of penetration interactions between a soluble surfactant and (a) a gaseous film of an insoluble surfactant monolayer, (b) a liquid expanded (condensed) monolayer, and (c) a solidifying monolayer.

Although in most flotation systems only those surfactants which are soluble (or at least slightly soluble) are employed, the monolayer studies of their insoluble homologues have provided a useful insight into the less evident aspects of their adsorption mechanisms and the associative action of frother molecules.

6.10.1. *Monolayer Penetration, Molecular Association of Surfactants at the Air/Liquid Interface*

The first instance of interactions between *two* different surfactants was detected by Schulman and Hughes (1935) as penetration of monolayers of the insoluble $C_{16}H_{33}OH$ alcohol by the molecules of $C_{16}H_{33}SO_4Na$ dissolved in the aqueous substrate. Their results showed a radical departure of the π_A-A isotherm of the penetrated film from that of the $C_{16}H_{33}OH$ isotherm and also suggested the formation of 1:2 and 1:3 molecular complexes between the alcohol and the sulfate molecules. However, since the rate of compression of the film and the concentration of the soluble surfactant appeared to affect the π_A-A isotherm of the penetrated film, Harkins (1952) objected to the initial interpretation of Schulman that specific molecular complexes are forming on penetration. Using a number of mixed films (such as alcohols–amines, alcohols–sulfates, amines–carboxylates, and alcohols–carboxylates) at varying molecular ratios, Harkins and co-workers, see Harkins (1952), showed that the additive surfactant causes either a condensation or an expansion of the mixed film but not necessarily a specific molecular complex between the components.

Since the time of the Harkins studies, numerous systems of two types of surfactants (one soluble and the other insoluble, or both surfactants soluble in aqueous solutions) have been investigated by a variety of techniques at several interfaces and as bulk mixtures by Dervichian (1958), Goodrich (1957), Shinoda *et al.* (1963), Kung and Goddard (1963), Goddard *et al.* (1968), and Cadenhead and Phillips (1968). These studies have all shown that, in most cases, definite departures in the behavior of mixed systems are observed in comparison with the behavior of individual components or their solutions. All such departures have been interpreted as due to interactions between the component surfactants. The existence of specific stoichiometric complexes between surfactants has been confirmed in some systems but not in all those that were initially suggested.

Bowcott (1957) showed that insoluble monolayers of hexadecyl alcohol, cetyl amine, or palmitic acid are penetrated by an appropriately long alkyl thio-compound, such as lauryl xanthate. The xanthate species penetrating the insoluble monolayer appeared to be stabilized against oxidation when it was complexed with an alkyl alcohol or an alkyl trimethyl ammonium bromide monolayer but was not stabilized when penetrating or complexing an alkyl fatty acid or an alkyl amine (unsubstituted, primary amine) monolayer. The latter surfactant, alkyl amine, caused rapid hydrolysis of xanthate at pH $>$ 7, although in the absence of amine the xanthate species appeared fairly resistant to hydrolysis. Even at pH 10, the amine monolayer penetrated by xanthate molecules initially showed a rapid decay of its surface pressure (within 10–15 min) to a pressure given by the corresponding alcohol–amine monolayer; such behavior has been interpreted by Bowcott as an indication of xanthate degradation to its alcohol derivative.

The preceding experiments with surfactants possessing different polar groups, some ionized and others nonionized, have demonstrated that any two surfactant compounds may interact at the air/water interface to form mixed films which consist of either strongly or less strongly associated molecules of the two components. The interaction may be very weak if it is limited to occasional van der Waals bonds between respective nonpolar (alkyl) groups. It is stronger if the van der Waals bonds become more numerous and/or cooperative bonds between the respective polar groups of the two components are also developing. The latter may be ion–dipole, dipole–dipole, or ion–ion in character. The nature of the surfactants may be such that one may act as a collector in a flotation system, while the other may be acting as a frother.

Interactions between two soluble surfactants have been indicated in studies of surface tension changes (Figures 5.25a and 5.25b) in the lowering of the CMC on the addition of a second surfactant and formation of mixed micelles [equation (5.17)] and in selective adsorptivities (Table 5.12). Changes in froth volume and froth stability are yet another indication of interactions occurring between soluble surfactants. Such changes have been studied by Burcik and Newman (1953). A molecular interaction between two soluble surfactants may be indicated either by an increase in froth volume (or its stability) or by a decrease in frothing to the point of a complete supression of froth or foam (as, for example, in the case of antifoaming agents). Because of multiplicity of bonding involved in interactions between two surfactants (see above) and the variety of effects influencing frothing, only a systematic study of a whole range of surfactants may reveal the degree and the character of interaction. For example, 1:1 mixtures of anionic and cationic surfactants such as cetyl sulfate ($C_{16}H_{33}SO_4Na$) and cetyl amine will produce voluminous froths if the amine is substituted (such as cetyl trimethyl ammonium salt) and will give no froth at all if an unsubstituted amine (cetyl amine salt) is used. A side chain in the alkyl group of the above amine salt will create frothing. Similarly, a change in the length of the alkyl group of the amine or sulfate component in the latter system will also produce a frothing mixture.

6.10.2. Solidification of Ionized Monolayers on Reaction with Counterions

Monolayer penetration studies of Schulman et al. (1935, 1937, 1938, 1953) were followed by investigations of interactions between insoluble surfactant monolayers at the air/water interface and various counterions (metal ions for anionic surfactants, inorganic anions for cationic surfactants) injected into the underlying substrate. These monolayer interaction studies provide an insight into one of the possible mechanisms of collector adsorption in flotation systems. Since all minerals release some quantities of appropriate ions into solution during grinding and agitation preceding flotation, the characteristics of interactions between ionized surfactants and ions in solutions are of interest to flotation.

Insoluble monolayers of alcohols and esters (octadecyl alcohol and methyl stearate) were found completely unaffected by metal ions injected into the substrate over the entire range of pH values. However, monolayers of fatty acids, under suitable conditions, reacted very strongly with metal ions, such as Fe^{3+}, Cu^{2+}, Co^{2+}, Mn^{2+}, Ca^{2+}, Mg^{2+}, and Al^{3+}. The interactions invariably led to solidification of the originally *liquid*-type (expanded or condensed) film whenever the aqueous substrate was held *within a specific range of pH*. This range of pH was closely related to the pH at which formation of metal complexes with hydrogen-bonding species such as OH^- or HCO_3^- occurred, as, for example, $Fe(OH)^{2+}$ or $Fe(OH)_2^+$ or $Cu(HCO_3)^+$, etc. Further, the disposition of molecules in the solidified film, that is, the area per molecule of the initially spread monolayer, was found to be determined primarily by the *size of the metal ion complex* and not by the dimensions of the fatty acid molecules. If a large concentration of complexing species other than OH^- was added to the substrate, for example, $NaHCO_3$, then the area per molecule differed from that obtained when a hydroxide was used. The amount of fatty acid present at the surface of water as a monolayer and other conditions in the preceding experiments were unchanged. As shown in Figure 6.51, when the $Cu(OH)^+$ complex is presumed to form, the solidified monolayer gives an area of 28 $Å^2$/molecule, whereas when the $Cu(HCO_3)^+$ complex is formed, the area is 42 $Å^2$/molecule.

To establish the critical role of metal complexes in surfactant monolayer solidification, Wolstenholme and Schulman (1950, 1951) carried out studies involving various non-hydrogen-bonding and hydrogen-bonding metal complexes with different isomers of carboxylic acids.

The acids chosen represented a wide range of cross-sectional areas per molecule:

1. Myristic acid:

$$C_{12}H_{25}$$
$$|$$
$$CH_2$$
$$|$$
$$COO^-H^+ \qquad \text{20-}Å^2 \text{ area}$$

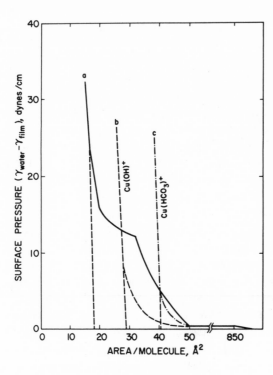

Figure 6.51. Schematic representation of pressure–area curves obtained when myristic acid monolayers are solidified by ionic copper complexes which develop a hydrogen-bonded network underlying the surfactant molecules: (a) Myristic acid on $N/100$ NCl substrate; (b) after injection of $CuCl_2$ at pH 5.2, resulting in $Cu(OH)^+$ complex formation; (c) after injection of $CuCl_2$ and $NaHCO_3$, at pH 5.2, resulting in the formation of the $Cu(HCO_3)^+$ complex occupying a much greater area in the hydrogen-bonded network.

2. Methyl dodecyl acetic acid:

$$C_{12}H_{25}$$
$$|$$
$$CH\ COO^-H^+ \qquad 30\text{-Å}^2\ \text{area}$$
$$|$$
$$CH_3$$

3. Butyl decyl acetic acid:

$$C_{10}H_{21}$$
$$|$$
$$CH\ COO^-H^+ \qquad 46\text{-Å}^2\ \text{area}$$
$$|$$
$$C_4H_9$$

4. Dihexyl acetic acid:

$$C_6H_{13}$$
$$|$$
$$CH\ COO^-H^+ \qquad 55\text{-Å}^2\ \text{area}$$
$$|$$
$$C_6H_{13}$$

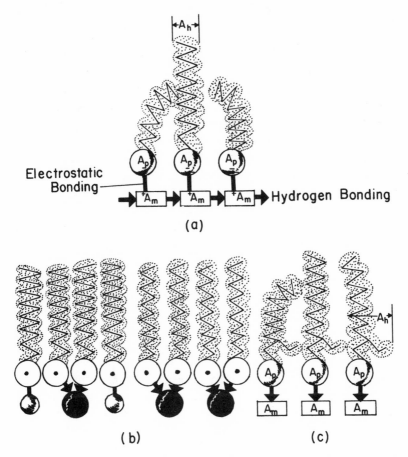

Figure 6.52. Schematic representation of monolayer solidification modes: (a) by electrostatic adsorption to laterally hydrogen-bonded complexes, (b) by van der Waals bonds between hydrocarbons, (c) no solidification due to steric hindrance caused by branched chain hydrocarbons that prevent hydrogen bonding between underlying complexes.

The results obtained have indicated that each metal ion defines a range of pH within which solidification could take place if the cross-sectional area of the metal complex A_m, that of the polar group A_p, and that of the non-polar group A_h are suitably correlated. As shown schematically in Figure 6.52, the solidification of the monolayer may involve primarily:

1. the counterions underlying the polar groups or
2. the hydrocarbon chains.

When counterions are involved, Wolstenholme and Schulman (1950) interpreted the solidification process as follow: At a suitable pH in the substrate strong *lateral* bonds between the neighboring metal complex species in the substrate develop as a result of hydrogen bonding [Figure 6.52(a)]. These hydrogen-bonded complexes provide a rigid network of ionized sites to which the oppositely charged ionized polar groups of the surfactant are attracted by electrostatic forces.

The cooperation between hydrogen bonding and electrostatic bonding leads to solidification, i.e., *immobilization* of surfactant monolayer molecules, whenever both A_h and A_p (the cross-sectional area of the hydrocarbon chain and that of the polar group) are smaller than or equal to the area of the metal complex A_m. Some van der Waals bonds between portions of hydrocarbon chains may also participate. When the network of laterally hydrogen-bonded metal complexes (which exists in the substrate regardless of the presence or absence of the surfactant monolayer) cannot be brought into juxtaposition with the surfactant molecules so that electrostatic bonding is not sufficiently widespread, no solidification occurs; the resultant monolayer is in the liquid expanded state. Similarly, when a dipolar surfactant (nonionized) is present, the dipole–ion interaction (instead of electrostatic ion–ion) is insufficient for solidification.

When the metal ion is incapable of forming a laterally bonded complex (no hydrogen bonding occurs), the film cannot solidify unless its $A_m < A_h$. For example, in the case of Ca^{2+} or Mg^{2+}, the solidification occurs as a result of adlineation of straight-chain hydrocarbons developing a large number of van der Waals bonds between the CH_2 groups of the neighboring hydrocarbon chains. In this instance, the metal ion reduces the repulsive forces between the similarly charged polar groups of the surfactant molecules to allow their adlineation [Figure 6.52(b)].

When the pH of the substrate is raised to a level where precipitation of metal hydroxide occurs at a given concentration of metal ions, the progressive increase of OH ions gradually neutralizes all positive charges on the (cationic) metal complexes, causing a breakdown of the hydrogen-bonded network underlying the monolayer.

In addition to the pioneering solidification studies with carboxylate monolayers [Wolstenholme and Schulman (1950)], the interactions between insoluble monolayers of alkyl sulfates and metallic ions [Thomas and Schulman (1954)] and of alkyl amine hydrochlorides with acetates and silicates [Goddard and Schulman (1953)] have also been investigated. The latter studies confirmed the interpretation of a two-dimensional solidification process developed for carboxylate films, providing further evidence for the influence of pH, the size and shape of the injected metal ion complex (or

Table 6.6. Results of Analytical Determinations of Abstractions of Frothers—Temperature $20°C \pm 2°$[a]

Test No.	Material	Collector			Frother				Abstraction of frother	
		Compound	Concentration	Time of conditioning	Compound	Concentration	Time of conditioning	pH	Amount abstracted	Initial frother concentration (%)
1	50 g of −300 mesh Cu[b]	KAX[a]	$M/1300$ 2:1 S:W	15 min	$C_{12}H_{25}SO_3Na$	$M/7500$ 2:1 S:W final 1:1 S:W	10 min[a]	12.5	0.490 mg/50 g of Cu	18
2	50 g of −100 +200 mesh Cu[b]	KAX[a]	$M/10,000$ 2:1 S:W	2 hr	$C_{12}H_{25}SO_3Na$	$M/10,000$ 2:1 S:W final 1:1 S:W	1 hr[a]	12.5	0.387 mg/50 g	27
3	50 g of −300 mesh Cu[b]	$C_{12}H_{25}SO_3Na$[a]	$M/2500$ 1:1 S:W 26% ads.	10 min[e]	p-Cresol	$M/2500$ 1:1 S:W	10 min[a]	7.5	0.450 mg 50 g	21
4	50 g of −300 mesh Cu[b]	$C_{12}H_{25}SO_3Na$[a]	$M/2500$ 1:1 S:W 28% ads.	10 min[e]	p-Cresol	$M/2500$ 1:1 S:W	5 hr[a]	7.5	0.685 mg 50 g	31
5	20 g of PbS, −60 +200 mesh	KEX cell test	$M/20,000$ 1:1 S:W	30 min	o-Cresol	$M/20,000$ final 1:2.5 S:W	10 min	7.5	0.056 mg/20 g	22
6	50 g of ZnS, −60 +100 mesh[c]	KEX[a]	$M/12,500$ 1:1 S:W	10 min[e]	o-Cresol	$M/1250$ 1:1 S:W	2 hr[a]	7.5	1.54 mg/50 g	35
7	30 g of BaSO₄, −120 +300 mesh	$C_{12}H_{25}SO_3Na$[a]	$M/10,000$ 1:1 S:W 36% ads.	15 min[e]	α-Terpineol	$M/5000$ 1:1 S:W	4 hr[a]	8.0	2.2 mg/30 g	35
8	50 g of Cu, −300[b]	$C_{12}H_{25}SO_3Na$[a]	$M/1000$ 1:1 S:W 46% ads.	20 min[e]	CTAB	$M/10,000$ 1:1 S:W	15 min[a]	7.0	1.092 mg/50 g	48

[a] From Leja and Schulman (1954).
[b] Air atomized Cu powder.
[c] Sphalerite containing 3.1% Cu_2S and 15% pyrite.
[d] Cylinder test.
[e] Solution of collector decanted.

the anion complex in the case of alkyl amine monolayers), the concentration of ions, and the stereochemistry of surfactant molecules. Attempts to study the interactions of monolayer thio compounds with injected metals were unsuccessful. The high oxidation rate of the thio compounds at the air/water interface was presumed to prevent the formation of reactive, stable monolayers.

6.10.3. *Interactions Between Surfactants at the Solid/Liquid Interface*

When one type of two different surfactants is adsorbed at a solid/liquid interface, an addition of the second surfactant (which *normally would not adsorb appreciably* at *this* particular solid/liquid interface *when present in solution by itself*) causes an appreciable coadsorption of the second surfactant. The evidence of a definite coadsorption of frother-acting surfactants on collector-coated solids has been initially presented by Leja and Schulman (1954) (Table 6.6); since then numerous studies of other combinations of surfactants, utilizing a variety of techniques, have shown that this behavior is a general one: As soon as one surfactant is adsorbed at the solid/liquid interface, any other additive, whether a surfactant or a nonpolar oil, will tend to coadsorb at the pretreated solid/liquid interface. The extent of coadsorption depends on the nature of bonds being developed between the two species *and also* on the nature of the solid substrate [Pomianowski and Leja (1963)].

Similar effects of coadsorption of a second surfactant have been observed in corrosion inhibition and in lubrication [Schulman *et al.* (1956)]. Their work on fretting corrosion indicated that the *kinetics of inhibitor adsorption* appears to be particularly strongly affected by the presence of and the coadsorption of the second surfactant.

6.11. *Selected Readings*

Adam, N. K. (1941), *Physics and Chemistry of Surfaces*, Oxford University Press, London.

Adamson, A. W. (1967, 1969, 1973), *Physical Chemistry of Surfaces*, John Wiley, New York.

Aveyard, R., and Haydon, D. A. (1973), *An Introduction to the Principles of Surface Chemistry*, Cambridge University Press, London.

Bikerman, J. J. (1970), *Physical Surfaces*, Academic Press, New York.

Blakely, J. M. (1973), *Introduction to the Properties of Crystal Surfaces*, Pergamon, Elmsford, N.Y.

Brunauer, S. (1945), *The Adsorption of Gases and Vapors*, Princeton University Press, Princeton, N.J.

Danielli, J. F., Pankhurst, K. G. A., and Riddiford, A. C., eds. (1964–1970), *Recent Progress in Surface Science*, Vols. 1–3, Academic Press, New York.

Danielli, J. F., Rosenberg, M. D., and Cadenhead, D. A. (1971–1979), *Progress in Surface and Membrane Science*, Vols. 4–10, Academic Press, New York.

Davies, J. T., and Rideal, E. K. (1961), *Interfacial Phenomena*, Academic Press, New York.

de Boer, J. H. (1953), *The Dynamical Character of Adsorption*, Clarendon Press, Oxford.

Dukhin, S. S., and Derjaguin, B. V. (1974), Equilibrium (and nonequilibrium) double layer and electrokinetic phenomena, in *Surface and Colloid Science*, Vol. 7, ed. E. Matijevic, John Wiley, New York.

Everett, D. H. (1972), Definitions, terminology and symbols in colloid and surface chemistry (adopted by IUPAC), *Pure Appl. Chem. 31* (4) 579–638.

Everett, D. H., ed. (1974), *Colloid Science*, Vols. I and II, The Chemical Society, London.

Flood, E. A., ed. (1967), *The Solid–Gas Interface*, Vols. I and II, Marcel Dekker, New York.

Gaines, G. L., Jr. (1969), *Insoluble Monolayers at Liquid/Gas Interfaces*, Wiley-Interscience, New York.

Gatos, H. C., ed. (1960), *The Surface Chemistry of Metals and Semiconductors*, Proceedings of the symposium on the above topic held in Columbus, Ohio, Oct. 1959, John Wiley & Sons, New York.

Gatos, H. C. (1962), Crystalline Structure and Surface Reactivity, *Science*, **137**, 311–322.

Goddard, E. D., ed. (1975), *Monolayers*, Advances in Chemistry Series No. 144, American Chemical Society, Washington, D. C.

Goldfinger, G., ed. (1970), *Clean Surfaces* (based on a symposium at North Carolina State University at Raleigh), Marcel Dekker, New York.

Gregg, S. J., and Sing, K. S. W. (1967), *Adsorption, Surface Area and Porosity*, Academic Press, New York.

Harkins, W. D. (1952), *Physical Chemistry of Surface Films*, Van Nostrand Reinhold, New York.

Hiemenz, P. C. (1977), *Principles of Colloid and Surface Chemistry*, Marcel Dekker, New York.

Hirth, J. P., and Lothe, J. (1968), *Theory of Dislocations*, McGraw-Hill, New York.

Iler, R. K. (1955), *Colloid Chemistry of Silica and Silicates*, Cornell University Press, Ithaca, N.Y.

Ingerson, E., ed. (1964), *Clays and Clay Minerals*, Pergamon Press, Elmsford, N.Y.

Kane, P. F., and Larrabee, G. R., eds. (1974), *Characterization of Solid Surfaces*, Plenum Press, New York.

Kerker, M., ed. (1977), *Colloid and Interface Science*, Vols. I–V, Proceedings of the International Conference on Colloids and Surfaces, San Juan, Puerto Rico, June 1976, Academic Press, New York.

Kipling, J. J. (1965), *Adsorption from Solutions of Nonelectrolytes*, Academic Press, New York.

Kruyt, H. R. (1952), *Colloid Science*, Elsevier, Amsterdam.

Kubaschewski, O., and Hopkins, B. E. (1962), *Oxidation of Metals and Alloys*, Butterworth's, London.

Kuhn, W. E., ed. (1963), *Ultrafine Particles*, John Wiley & Sons, New York.

Little, L. H. (1966), *Infrared Spectra of Adsorbed Species*, Academic Press, New York.

Matijevic, E., ed. (1969–1979), *Surface and Colloid Science*, Vols. 1 (1969) to 11 (1979), John Wiley (Vols. 1–8), Plenum Press (Vols. 9–11), New York.

Mittal, K. L., ed. (1979), *Surface Contamination*, Proceedings of the symposium on this topic held in Washington, D. C., Sept. 1978, Plenum Press, New York.

Murr, L. E. (1975), *Interfacial Phenomena in Metals and Alloys*, Addison-Wesley, Reading, Mass.

Osipow, L. I. (1962), *Surface Chemistry, Theory and Industrial Applications*, Van Nostrand Reinhold, New York.

Padday, J. F., ed. (1978), *Wetting, Spreading and Adhesion*, Academic Press, New York.

Samorjai, G. A. (1972), *Principles of Surface Chemistry*, Prentice-Hall, Englewood Cliffs, N.J.

Sheludko, A. (1966), *Colloid Chemistry*, Elsevier, Amsterdam.

Shinoda, K., Nakagawa, T., Tamamushi, B., and Isemura, T. (1963), *Colloidal Surfactants*, Academic Press, New York.

van Olphen, H., and Mysels, K. J. (1975), *Physical Chemistry: Enriching Topics from Colloid and Surface Science*, Theorex, La Jolla, Calif.

Young, D. M., and Crowell, A. D. (1962), *Physical Adsorption of Gases*, Butterworth's London.

Electrical Characteristics of Interfaces. Electrical Double Layer and Zeta Potential

The importance of surface charges in establishing electrical characteristics of interfaces, particularly of the solid/liquid, liquid/gas, and liquid/liquid ones, has already been stressed in the opening paragraphs of Chapter 1. Whenever a new solid surface is formed in a gaseous or a liquid environment, as, for example, during dry or wet grinding, it either becomes charged at the moment of rupture of ionic or covalent bonds or picks up a charge by a subsequent adsorption of ions. Freshly cleaved solids remain uncharged only if the cleavage exclusively ruptures van der Waals bonds when the underlying lattice points are occupied by covalently bonded molecules and, in addition, there are in the system no mobile charges such as electrons, ions, or orientable dipoles.

Whenever mobile charges[†] are present, the interfaces become charged; the exceptions are those specific conditions which lead to a mutual compensation of charges, resulting in a point of zero charge (pzc).

[†] The density of charge carriers, electrons, in a metal is of the order of 10^{22} cm^{-3}; the conductivity of most metals is of the order of 10^4 Ω^{-1} m^{-1} and is entirely due to the high concentration of electrons. This high concentration of charge carriers in a metal does not permit the existence of a diffuse excess charge layer within the metal itself, but all the excess charge associated with a metal/gas or metal/solution interface is within an atomic distance or less of the metal surface. Semiconductors, such as metallic oxides or sulfides, have conductivities orders of magnitude lower than those of metals, namely in the range of 10^{-3}–10^{-7} Ω^{-1} m^{-1}. The densities of charge carriers (electrons and holes) in semiconductors are in the range 10^{13}–10^{16} cm^{-3}. Insulators have conductivities lower than 10^{-10} Ω^{-1} m^{-1}.

In this chapter the basis of electrical charge distribution across interfaces is treated. The concept of the equilibrium electrical double layer developed at all interfaces possessing mobile charges or dipoles is dealt with first. Such double layers produce potential barriers which either assist or hinder motion of charges across the interfaces. Only the simplest features of double layers are described for two types of electrodes: reversible and polarizable. An example of a reversible electrode is the calomel electrode $Hg-Hg_2Cl_2$–aqueous solution of KCl. The system metallic Hg–aqueous solution is an example of a polarizable electrode.

Next, electrokinetic phenomena are described; these are due to changes occurring within the electrical double layer when the two adjoining phases (forming the interface) are moving relative to each other. The potential evaluated from experimentally determined parameters with respect to the shear plane, known as the zeta potential, determines the behavior of colloidal particles, their stability in dispersions, or their tendency toward coagulation.

For details of the theory of electrical double-layer and of electrokinetic phenomena, the reader is referred to standard books and reviews on the subject, such as Verwey and Overbeek (1948), Butler (1951), Haydon (1964), Delahay (1965), Vetter (1961), Bockris and Reddy (1970), Payne (1973), Dukhin (1974), and Derjaguin and Dukhin (1974). The kinetics of electrode processes involving continued charge transfer across the interface are dealt with in Conway (1965), Delahay (1965), and Bockris and Reddy (1970).

7.1. *Definitions and Electrochemical Concepts*

Whenever two phases meet, some rearrangement of component species (atoms, aggregates) occurs. If the interface is permeable to ions which are common to both phases (for example, metallic ions in a metal/electrolyte interfacial system), the rearrangement establishes an equilibrium by balancing the diffusive and the electrical effects. As a result, an electrochemical potential (see below) in the system comprising two phases and their interfacial region is established.

The electrical potential at any given point in vacuum is defined as the energy necessary to bring a unit charge from infinity to this point. In a material medium other than vacuum, such as phase α, the energy required to bring the charge through the medium depends on the interactions between the medium α and the charged particle, whether H^+, K^+, or Fe^{2+}. For a species a this work (energy) is called its *electrochemical potential* $\bar{\mu}_a^\alpha$ and is defined by

$$\bar{\mu}_a{}^\alpha = \mu_a{}^\alpha + z_a e \phi^\alpha \tag{7.1}$$

where $\mu_a{}^\alpha$ is known as the *chemical potential* and denotes all nonelectrostatic interactions of species a with the medium α. $z_a e$ denotes the charge carried by a (z_a = valence, e = unit electronic charge). ϕ^α is known as the *inner potential*; it is constant within the phase α and denotes the work done in transporting a unit charge across the interface α/vacuum (of the phase α surrounded by vacuum) which comprises a layer of dipoles [for details, see Parsons (1954)].

The inner potential ϕ^α is a sum of the so-called *outer potential* ψ^α and the surface potential χ^α:

$$\phi^\alpha = \psi^\alpha + \chi^\alpha \tag{7.2}$$

Hence, the electrochemical potential is

$$\bar{\mu}_a{}^\alpha = \mu_a{}^\alpha + z_a e(\psi^\alpha + \chi^\alpha) \tag{7.1'}$$

The outer potential ψ^α denotes the work done to bring a unit charge from infinity to a point just outside the interface α/vacuum. This interface is visualized as carrying a layer of charges q_m and their countercharges q_s, thus representing a layer of dipoles. The work required to cross this layer of dipoles is the surface potential χ^α.

A difference between two outer potentials of phases α and β, $\Delta\psi = \psi^\alpha - \psi^\beta$, can be measured experimentally and is known as the *Volta* or *contact potential difference*, $\Delta\psi$.

The difference $\Delta\phi = \phi^\alpha - \phi^\beta$ of the two inner potentials (which includes contributions from the double layers of phases α and β) is known as the *Galvani potential difference* $\Delta\phi$ and differs from the Volta potential, since it includes the difference of surface potentials χ as well as $\Delta\psi$:

$$\Delta\phi = \phi^\alpha - \phi^\beta = \psi^\alpha - \psi^\beta + \chi^\alpha - \chi^\beta = \Delta\psi + \chi^\alpha - \chi^\beta \tag{7.3}$$

The Galvani potential difference between the solid and liquid phases cannot be measured experimentally in absolute terms, only on relative scale, with respect to the potential of a reference or standard electrode in equilibrium with the electrolyte. To measure a potential difference, an instrument such as a potentiometer or a voltmeter in series with a galvanometer is required. The two terminals of the instrument must be connected in an appropriate manner to allow meaningful interpretation of the measured values. An electrode which is *reversible* to an ion present in the electrolyte solution and in the metallic phase is necessary for measuring potential differences between a metal and an electrolyte. Details of instrumental arrangements for trustworthy potential evaluations can be found in Parsons (1954) or Barlow (1970).

Difficulties are experienced in experimental measurements, mostly due to contamination of surfaces by adsorbed impurities. Even when the impurities are controlled most stringently, the evaluation of, for example, the surface potential χ from Volta potential determinations of dilute electrolyte solutions is not easy or straightforward. The reason for this difficulty is an unknown and variable degree of dipole orientation at the surface. Thus, for χ of the water/air interface values of -0.5 V, -0.3 V, and -0.36 to $+0.4$ V have been obtained by various groups of researchers.

The most frequently measured potential differences are those for metal–electrolyte systems. Two types of electrode–solution systems are recognized: *ideal polarizable* electrode–solution systems and reversible or *ideal nonpolarizable* ones.

The ideal nonpolarizable (or reversible) electrode/solution interfaces allow unhindered ionic exchange between the electrode and the electrolyte. Examples of such nonpolarizable electrode–solution systems are the reference electrodes quoted in Section 2.9.1, such as *calomel electrodes* (containing Hg, Hg_2Cl_2, KCl, and either saturated KCl solution or 1 N or 0.1 N, contacting the test electrolyte solution through a capillary tube) or the *standard hydrogen electrode* (consisting of a platinized Pt wire in contact with H_2 gas at 1 atm pressure in a solution containing H^+ ions at unit activity, pH $= 0$). Other examples are the metal/solution interfaces under conditions of unhindered dissolution $\leftrightarrow$ deposition of metallic ions, that is, whenever the Nernst equation applies to the system (see Section 2.9.1):

$$E = E^\circ + \frac{2.3RT}{nF} \log a_{M^{n+}}$$

Thermodynamically, an ideal polarizable electrode–solution system possesses an extra degree of freedom. This extra degree of freedom is expressed by the Lippmann equation (see Section 7.3). No equation analogous to the Lippmann equation exists for reversible electrodes since for such electrodes it is impossible to vary the potential E at constant T, P, and composition. The classic example of such an ideally polarizable interface is that of pure mercury in an aqueous electrolyte solution. It has provided a particularly useful system for fundamental studies of structure and properties of the electrical double layer.

7.2. *Models of the Electrical Double Layer*

The rearrangement of species occurring during the formation of an interface, such as solid/solution or liquid mercury/solution, results in establishing a layer of charges that has attracted a layer of countercharges. The first model of the distribution of these excess charges originated with Helmholtz (1879) and Perrin (1904), who proposed to treat the total excess of electronic charges in metal as a surface charge with the ionic counter-charges in the electrolyte as if they were forming a charged parallel plate condenser along the surface. This simple model is applicable only to metal-electrolyte system of high salt concentrations, greater than 0.1 M. Some years later, Gouy (1910) and Chapman (1913) independently suggested that the charges form a diffuse continuum of ions in a structureless dielectric. They considered only point charges whose concentration decreases progressively with distance away from the solid into the solution phase. [Debye and Hückel (1923a, 1923b) utilized this approach of Gouy and Chapman to evaluate ion–ion interactions; see Section 4.3.] Stern (1924) modified the Gouy–Chapman model by replacing the point charge approximation with ions of finite size which are capable of approaching the surface of the solid no closer than a minimum distance d. Further, Stern combined the model of a diffuse layer (the Gouy–Chapman layer) with that of a condenser-like compact Helmholtz model. This new (Stern) model, consisting of two layers in series, is referred to as a *compound double layer* or a *triple layer*. If each of these two layers is regarded as a capacitor, then the differential capacity of the double layer C is related to the other two by

$$\frac{1}{C} = \frac{1}{C_H} + \frac{1}{C_{G-C}}$$

where C_H represents the capacitance of the Helmholtz layer and C_{G-C} that of the Gouy–Chapman layer in series.

In addition, Stern introduced the concept of specific adsorption of ions within the Helmholtz portion of the compound double layer at a distance β from the solid surface. The plane of these specifically adsorbed unhydrated ions (at β) is known as the *inner Helmholtz plane* (IHP), while that of the closest approach for the more weakly adsorbed hydrated ions, at d, is known as the *outer Helmholtz plane* (OHP). The three stages of the model development are shown in Figure 7.1(a), (b), and (c).

Depending on the character of specifically adsorbed ions in the IHP and the concentration of ions in the electrolyte solution, two modifications

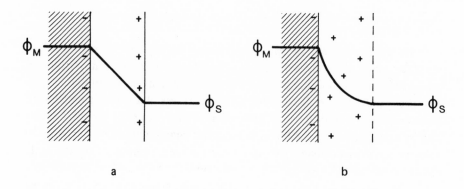

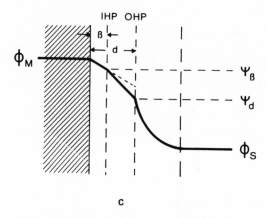

Figure 7.1. Three stages in the development of the model for the electrical double layer: (a) Helmholtz–Perrin compact layer, (b) Gouy–Chapman diffuse layer, and (c) Stern double layer. (IHP = inner Helmholtz plane of adsorption loci for unhydrated specifically adsorbing ions; OHP = outer Helmholtz plane of loci for hydrated adsorbing ions.)

of the Stern compound layer may be encountered in addition to that visualized in Figure 7.1(c). These are shown in Figure 7.2(a), which indicates the occurrence of a charge reversal within the compound layer; Figure 7.2(b) shows the compound double layer in concentrated electrolyte solutions when the diffuse layer disappears.

Within the thickness of the compact layer, the potential changes linearly from the level ϕ_M (inner potential of the metallic phase) to the level ψ_d determined by the excess charge q_M in the compact layer. The potential within the diffuse layer is evaluated starting with Poisson's equation (which relates charge density and potential) and applying Boltzmann's exponential distribution law to the concentration of positive and negative ions within the diffuse layer. By convention, the inner potential of the electrolyte solution is $\phi_s = 0$. The resultant expressions for the various parameters are rather complex functions of sinh and tanh, for example, the following:

(1) the slope of the potential function,

$$\frac{d\phi}{dx} = -\left(\frac{32\pi C_0 kT}{\varepsilon}\right)^{1/2} \sinh\left(\frac{ze\phi}{2kT}\right) \tag{7.4}$$

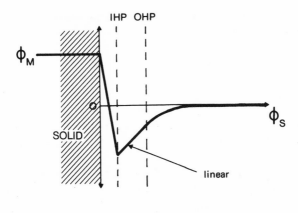

a

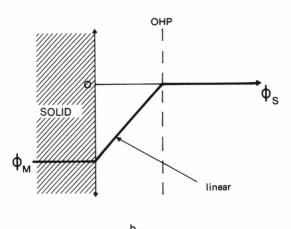

b

Figure 7.2. Modifications of the electric double-layer potential showing (a) charge reversal due to highly specific adsorption of counterions and (b) elimination of the diffuse portion (of the electrical double layer) occurring in concentrated electrolyte solutions.

(2) the potential drop within the diffuse layer[†] $\phi(x)$,

$$\phi(x) = \left(\frac{4kT}{ze}\right) \tanh^{-1}[ze\phi(0)4kT] \exp\left(\frac{-x}{L_D}\right) \qquad (7.5)$$

(3) the potential at $x = 0$ ($x =$ distance from the metallic surface),

$$\phi(0) = \left(\frac{2kT}{ze}\right) \sinh^{-1}\left(\frac{q}{2A}\right) \qquad (7.6)$$

(4) the charge q present on each side of the interface, of equal value but opposite in sign,

$$q = 2A \sinh\left(\frac{ze\phi(0)}{2kT}\right) \qquad (7.7)$$

and (5) the differential capacitance $\bar{C}_2$,

$$\bar{C}_2 = \left(\frac{Aze}{kT}\right) \cosh\left(\frac{ze\phi(0)}{2kT}\right) \qquad (7.8)$$

where $A \equiv kT\varepsilon/4\pi zeL_D$ and $L_D = (kT\varepsilon/8\pi C_0 z^2 e^2)^{1/2}$ (known as the *Debye length*; some physicists call double this value the Debye length; it is also known as the thickness of the diffuse layer, $1/\varkappa$). In equations (7.4)–(7.8), C_0 is the concentration of the ions in the bulk electrolyte, z is the valence, e is the electronic unit charge, and ε is the dielectric constant. For small values of $\phi(0)$, i.e., less than 25 mV, near the electrocapillary maximum (ecm; see below), the diffuse layer acts as a parallel plate capacitor of plate separation L_D filled with a dielectric ε. Similarly, for small values of $\tanh^{-1}$ in equation (7.8), the potential $\phi(x)$ varies as $\exp(-x/L_D)$ [or $\phi(x) = \phi_0 \exp(-\varkappa x)$].

7.3. Experimental Testing of the Double-Layer Theory (*Using an Ideally Polarizable Electrode*)

Early measurements of electrical double-layer characteristics encountered problems because of (1) theoretical misgivings relating to thermodynamics of a metal/electrolyte interface and (2) experimental difficulties of purifying electrode surfaces from traces of surface active impurities.

[†] For a potential difference within the same solution phase, one can write $\phi(x)$ or $\psi(x)$, since there is no surface dipole layer present within the same phase.

Only from the 1940s onward when Grahame, first, clarified the thermodynamic analysis of an ideally polarizable electrode [Grahame and Whitney (1942)] and, second, adopted the dropping mercury electrode for double-layer capacity measurements (thus minimizing the effects of contaminations) did reliable measurements for testing the predictions of double-layer theory become possible.

The complete differential for the Gibbs free energy of the system representing an electrical double layer at an interface is

$$dG = -S\,dT + V\,dP - A^s\,d\gamma - \sum \Gamma_i\,d\bar{\mu}_i \tag{7.9}$$

The absolute value of the Galvani potential difference across the interface, $\phi^\alpha - \phi^\beta$, cannot be determined. However, for a polarizable interface a change in this potential can be measured *provided* the interface is linked to an ideally nonpolarizable interface to form an electrochemical cell. If such a cell is connected to an external source of electricity and the resulting potential E is measured, then an analysis of the thermodynamic relationships in such a circuit [see Bockris and Reddy (1970, Vol. II, pp. 698–701), Conway (1965), Parsons (1954), and Delahay (1965)] leads to the fundamental equation for the thermodynamic treatment of polarizable interfaces:

$$d\gamma = -q_M\,dE - \frac{q_M}{z_j F}\,d\mu_j - \sum \Gamma_i\,d\mu_i \tag{7.10}$$

where q_M is the excess charge *density* on the electrode, Γ_i is the surface excess of species i, F is the faraday gram-mole of ions (of unit charge), j is the particular species that is involved in leakage across the nonpolarizable electrode (Cl^- in the case of the calomel electrode and H^+ in the case of the hydrogen electrode), and μ_i is the chemical potential of all species i, charged and uncharged. For a solution of a fixed composition, $d\mu_j = 0$ and $\sum \Gamma_i\,d\mu_i = 0$; hence, from equation (7.10), at constant T and P,

$$\left(\frac{\partial \gamma}{\partial E}\right)_{\mu_{i,j}} = -q_M \tag{7.11}$$

This is known as the *Lippmann equation.*

Further differentiation gives the capacitance C of the electrical double layer:

$$\left(\frac{\partial q_M}{\partial E}\right)_{\mu_{i,j}} = C = -\left(\frac{\partial^2 \gamma}{\partial E^2}\right)_{\mu_{i,j}} \tag{7.12}$$

Equations (7.11) and (7.12) constitute the basis on which very extensive work relating to the properties and the structure of the electrical double layer has been carried out.

The plots of experimentally determined variations in surface tension γ with the applied potential E for the mercury/electrolyte interfaces are known as the *electrocapillary curves*, since they were usually determined using the capillary electrometer. Unfortunately, the measurements of γ were not very accurate and also could not be applied to solid electrode–electrolyte systems. Hence, it is more useful nowadays to measure the differential capacitance and to derive γ from these measurements. Differential capacity measurements can be made easily using an ac Wheatstone bridge technique.

Since, for a capacitor, the potential difference across the plates is

$$E = \frac{4\pi d}{\varepsilon} q \tag{7.13}$$

where d is the separation distance of the capacitor's plates, ε is the dielectric of the medium, and q is the charge on the plates, integration of the Lippmann equation (7.11),

$$\int d\gamma = -\frac{4\pi d}{\varepsilon} \int q_M \, dq_M \tag{7.14}$$

results in

$$\gamma = \left(-\frac{4\pi d}{\varepsilon}\right) \frac{q_M^2}{2} + \text{constant} \tag{7.15}$$

or

$$\gamma = \left(-\frac{\varepsilon}{4\pi d}\right) \frac{E^2}{2} + \text{constant} \tag{7.16}$$

Equation (7.16) indicates that, theoretically, an electrocapillary curve, γ vs. E, should be an inverted parabola (Figure 7.3). By a long established convention, the negative values of potential are plotted to the right and the positive values to the left on the abscissa axis. The maximum in the electrocapillary curve, γ_{max} or γ_{ecm} (electrocapillary maximum), occurs at a potential at which the charge density [determined by the slope of the capillary curve (equation (7.11)] changes from positive to negative values passing through zero. This potential is known as the *potential of zero charge* (pzc) and is given the symbol E_{pzc}. Experimentally obtained electrocapillary curves are not perfect parabolas but slightly asymmetric near-parabolas

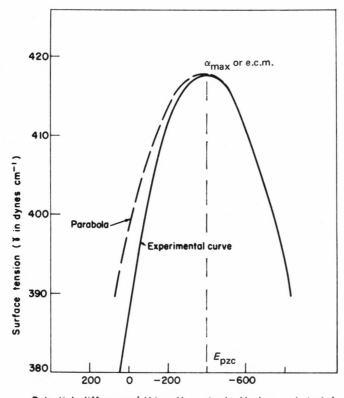

Figure 7.3. The experimental electrocapillary curve for 2.94 N HCl electrolyte concentration is compared with a corresponding (ideal) parabola. [From Bockris and Reddy (1970), *Modern Electrochemistry*, Vol. II, with the permission of Plenum Publishing Corporation, New York.]

(Figure 7.3). For such near-parabolas, the capacity is not constant over the whole potential range [Figure 7.4(b)] but shows a step at E_{pzc}. Also, the charge density is not linear with the potential but changes slope at the E_{pzc} [Figure 7.4(a)]. These departures from the ideal behavior predicted by equations (7.11), (7.12), and (7.16) indicate that the simple Helmholtz–Perrin model does not satisfactorily explain all characteristics of the electrical double layer.

For example, a comparison between the predicted capacitance of the Gouy–Chapman diffuse layer [equation (7.8)] and the capacitance–potential relationships obtained experimentally shows major differences, except in the immediate vicinity of the E_{pzc} (for dilute electrolyte solution). The dif-

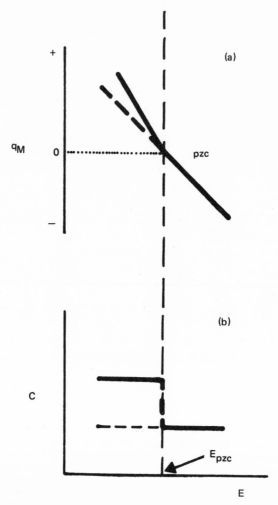

Figure 7.4. The derivative parameters: (a) charge density q_M and (b) capacitance C, relative to the applied potential E, for experimental and ideal (dashed lines) electrocapillary curves, shown schematically.

ferences may be twofold in character: a slight displacement in the potential (at which a sharp minimum in the differential capacitance occurs) with respect to ecm and/or a progressive disappearance of the sharp dip in the capacitance curve as the concentration increases (Figure 7.5). However not all systems may behave the same way. For example, NaF shows no displacement of the minimum in the differential capacitance curve with respect to the theoretically predicted value, whereas other systems show considerable changes in electrocapillary curves, as indicated in Figure 7.6 and 7.7. Hence, ions such as Na^+ and F^- are known as indifferent ions (or electrolyte) and others as potential-determining or specifically adsorbing ions. The position of E_{pzc} is shifted to more negative potentials (that is, to the right on the abscissa scale) in the presence of specifically adsorbing anions and in the opposite direction for specifically adsorbing cations (Figure 7.6). Also, the positive branch of electrocapillary curves (to the left of ecm) changes with the nature and the concentration of specifically adsorbing anions present in the electrolyte (Figure 7.7), while the negative branch may be coincident for some cations used (K^+, Na^+, Ca^{2+}) but may

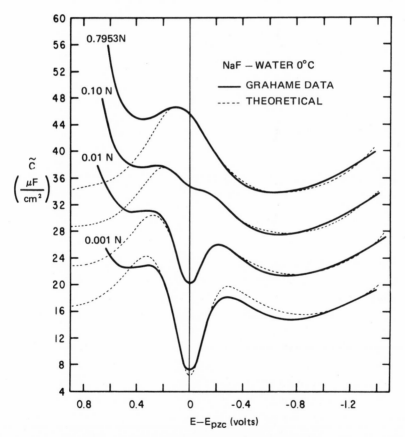

Figure 7.5. A comparison of experimental and theoretical differential capacitance curves for the mercury/electrolyte interfaces in increasing concentrations of an indifferent electrolyte, NaF, 0.001–0.7953 N. [From Bockris and Reddy (1970), *Modern Electrochemistry*, Vol. II, Plenum Press, New York.]

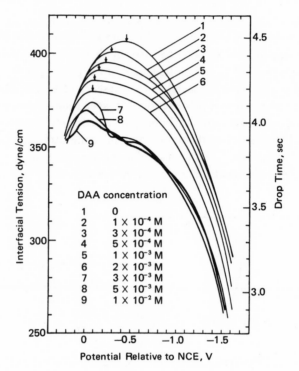

Figure 7.6. Electrocapillary curves for the mercury–aqueous solutions of dodecylam-monium acetate (DAA) at natural pH in 0.1 N KF as supporting electrolyte. Arrows indicate pzc. [From Usui and Iwasaki (1970) *Trans. AIME* **247**, with the permission of the American Institute of Mining and Metallurgical Engineers, Littleton, Colorado.]

change with the concentration (Figure 7.6) and with the nature of specifically adsorbed inorganic cations such as Ba^{2+} or Pb^{2+} or organic dodecylam-monium ions.

The departures from the theoretically predicted relationships have been presumed by Stern (1924) to be caused by specific adsorption of anions and cations in the IHP of the compound double layer. However, the two modifications introduced by Stern (namely, the finite size of charged ions participating in both compact and diffuse portions of the double layer and their specific adsorption in the IHP) have proved insufficient to explain all aspects of the experimental behavior of charged interfaces. Levine and Bell (1962, 1963) introduced a *discrete charge effect*, a concept which to some extent accounts for repulsion between adsorbed ions. Further, following the proposal of Watts-Tobin and Mott (1961), it has been accepted that the first layer of water at the electrode/solution interface consists of oriented water dipoles, some of which are in a *down position*, that is, with hydrogen atoms oriented toward the metal electrode, and some of which are in an *up position* with oxygen toward the metal surface. Figure 7.8 shows schematically a model of a metal/electrolyte interface incorporating these modifications.

The refinements of the latest model and the present understanding of the electrical double-layer characteristics came about through extensive experimental and theoretical work of numerous investigators [for details, see Delahay (1965) or Bockris and Reddy (1970) or the reviews by Payne (1973) and Gerischer *et al.* (1978)].

For very dilute electrolyte solutions, the differential capacity of the mercury/electrolyte interface has a pronounced dip (V-shaped segment) around E_{pzc}, as shown in Figure 7.5. The sharp minimum reflects the predominant influence on the capacity of the diffuse portion in the compound double layer. More concentrated electrolyte solutions no longer show such a dip but, instead, a hump around the E_{pzc} and a constant capacity region at higher potential. Figure 7.9 shows the differential capacity of the mercury–0.1 N HCl solution. In the negative range of potential difference it has a constant capacity portion at $\sim$16–17 μF cm^{-2}. A hump of $\sim$40 μF cm^{-2} occurs in the vicinity of E_{pzc} (i.e., at $E - E_{pzc} = 0$). The constant capacity level is found for all electrolytes irrespective of the size (radii) of ions involved; see Table 7.1. This independence of ionic size has been

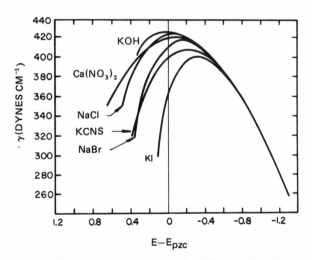

Figure 7.7. Electrocapillary curves for mercury in different electrolytes at 18°C. Potentials referred to E_{pzc} in NaF solution (without specific adsorption). [From Bockris and Reddy (1970), *Modern Electrochemistry*, Vol. II, Plenum Press, New York.]

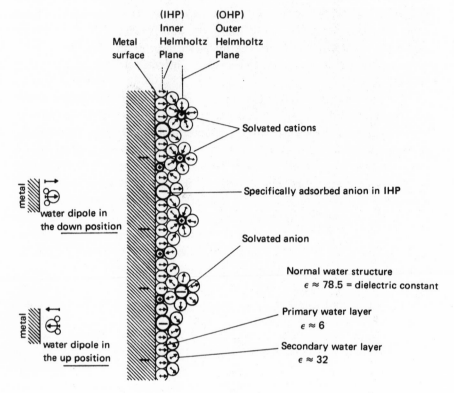

Figure 7.8. Model of a metal/electrolyte interface. [From Bockris and Reddy (1970), *Modern Electrochemistry*, Vol. II, Plenum Press, New York.]

interpreted as evidence that the capacity of the electrical double layer in this highly negative range of potentials is primarily determined by the first layer of oriented water molecules of much lower dielectric constant, $\varepsilon \cong 6$, than that for bulk water, $\varepsilon \cong 80$. Since no contribution from specifically adsorbed ions is apparent, such specifically adsorbed ions are presumed to be nonexistent in this potential range. The formation of a hump in the capacity–potential curve is interpreted as a displacement of the adsorbed water molecules by the specifically adsorbed ions. With an increase in the number of such specifically adsorbing ions, a sharp increase in the capacity is observed from the level of constant value, $\sim 17 \ \mu\text{F cm}^{-2}$. However, the increasing number of charges in the IHP sets up a resistance to further adsorption in the form of lateral repulsion among adsorbed ions (this is the depolarizing effect of the discrete dipoles formed by the ions and metal surface charges). This lateral repulsion is responsible for a slight reduction in capacity, thus creating a hump in the capacity–potential curve (Figure 7.9) [for details, see Bockris and Reddy (1970, pp. 761 ff.) or Barlow (1970)].

The highest coverages of the surface by specifically adsorbed ions were found to be less than 20% of the available area, the rest being occupied by water molecules. Quantitative evaluations of the surface excess of specifically adsorbed ions such as Cl^-, CN^-, I^-, ClO_4^-, BrO_3^-, and CNS^- in different concentrations [Wroblowa *et al.* (1965)] enabled exact locations of the capacitance humps to be predicted. Also, the shifts in E_{pzc} observed for some electrolytes (Figure 7.7) (but not for others, such as NaF) and/or for increased concentrations of the same electrolyte (the latter shifts in E_{pzc} are known as the Esin and Markov effect) (Figure 7.6) have been explained

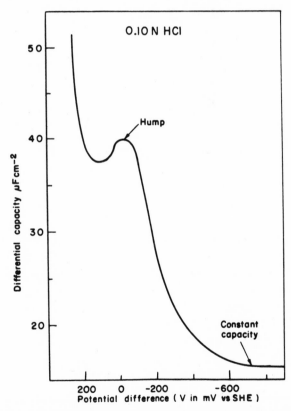

Figure 7.9. The experimental capacitance–potential curves for 0.1 *N* HCl showing the region of constant capacitance, $\sim$17 μF cm^{-2}, and the hump due to a specific adsorption of Cl^-. [From Bockris and Reddy (1970), *Modern Electrochemistry*, Vol. II, Plenum Press, New York.]

Table 7.1. Values of the Constant Capacity 0.1 N Aqueous Chloride
Solutions[a]

Ion	Unhydrated radius (Å)	Estimated hydrated radius, r_i (Å)	Differential Capacity at $q_M = -12 \, \mu C \, cm^{-2}$ ($\mu F \, cm^{-2}$)
H_3O^+	—	—	16.6
Li^+	0.60	3.4	16.2
K^+	1.33	4.1	17.0
Rb^+	1.48	4.3	17.5
Mg^{2+}	0.65	6.3	16.5
Sr^{2+}	1.13	6.7	17.0
Al^{3+}	0.50	6.1	16.5
La^{3+}	1.15	6.8	17.1

[a] Source: Bockris and Reddy (1970).

(and predicted) utilizing the concept of specifically adsorbed species. The evaluations of such shifts in the E_{pzc} position and of surface excesses for specifically adsorbed anions have also brought out the fact that adsorption of such anions is possible (and does invariably occur) at potentials at which the surface of mercury is charged negatively, that is, has the same charge as the adsorbing species.

The occurrence of specific adsorption of anions on a negatively charged surface is due to the multiple nature of adsorption forces. An ion adsorbs in the IHP as a result of electrostatic and van der Waals (dispersion forces) interactions and, in the case of species such as thiourea and S^{2-} ions adsorbing on mercury, a "chemical" bond also appears to be involved. As long as the contribution of the latter two types of interaction exceeds that of the (opposing) electrostatic interaction, adsorption does take place. Further, water molecules of ion solvation sheaths and those which are preadsorbed on the site in the IHP where ion adsorption occurs have to be either displaced or rearranged. Hence, the overall free energy of adsorption must exceed not only the electrostatic contribution but also that necessary for the rearrangement or desorption of the preadsorbed water molecules.

Of the two positions that water molecules may take in the first layer of adsorbed species, the free-energy change of adsorption is greater for the up position than for the down position (since the effective charge is nearer to the electrode when oxygen is oriented toward the metal surface, that is, in the up position). At the E_{pzc} the water molecules are held least strongly to the surface of the electrode. But the number of water molecules in the up position is not equal the number of H_2O molecules in the down position even at the E_{pzc}, because of the (slight) difference in the free energy of adsorption for the two positions.

7.4. *Adsorption of Neutral (Nonionic) Surfactants at the Mercury/Electrolyte Interface*

The work described so far has dealt mainly with electrical double-layer systems comprising a mercury electrode in an electrolyte solution of one inorganic component. Electrocapillary curves obtained for solutions of different concentrations of such single electrolytes enabled the relative surface excess values of positive or negative ions to be evaluated for different potentials E. The presence of an organic compound, such as a surfactant, in an electrolyte solution produces a variety of changes in the electrocapillary curve of the electrolyte, depending on the chemical nature of the compound.

Adsorption of a surfactant at the mercury/solution interface causes a decrease in the interfacial tension and thus always lowers the electrocapillary curve for the solution of the same electrolytes without surfactant. This behavior is indicated in Figures 7.6 and 7.10. On the other hand, an increase in the concentration of the supporting electrolyte may either decrease or increase the adsorption of organic molecules, depending on the relative specific adsorption of electrolyte ions vs. surfactant species. For example, on increasing the concentration of NaI, the specific adsorption of I^- decreases the adsorption of dipolar organic molecules and organic anions (such as alkyl alcohols, phenols, and salicylate anions) but increases the adsorption of organic cations such as $[(C_3H_7)_4N]^+$ [Frumkin and Damaskin (1964)]. On the other hand, the concurrent change in the form of the surfactant (ionized and un-ionized), which occurs on increasing the pH from neutral to alkaline for solutions of unsubstituted alkyl amines, may increase the adsorption of both forms, as shown in Figure 7.11.

With aromatic compounds the contribution of π-bonding may also be involved. Conway and co-workers [reviewed by Conway (1976)] have used rigid organic molecules in order to differentiate the effects of various conformations of the adsorbate at the mercury/solution interface. They found that molecules whose own dipoles do not contribute to the surface potential (created by oriented water dipoles) lie flat and that the change of surface potential is due only to displacement of oriented water [Figure

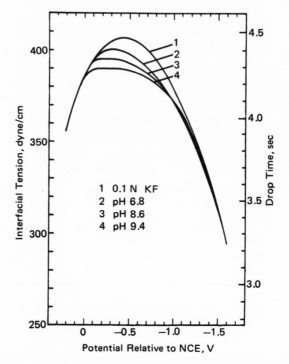

Figure 7.10. Electrocapillary curves as a function of pH for solutions of 10^{-4} M do-decylammonium acetate (DAA); supporting electrolyte, 0.1 N KF; the pH was adjusted with KOH solution. [From Usui and Iwasaki (1970) *Trans. AIME* **247**, with the permission of the American Institute of Mining and Metallurgical Engineers, Littleton, Colorado.]

7.12(a)]. When the same type of organic molecules at higher surface coverage begin to interact with the charges at the interface and become oriented accordingly, they cause a sharp transition in the surface potential change [Figure 7.12(b)].

A variety of adsorption isotherm models may be adopted for organic compounds in electrolyte solutions when interpretation of electrocapillary curves is considered. The Gibbs adsorption isotherm, applicable to the liquid/liquid and liquid/gas interfaces,

$$dy = -\sum \Gamma_i \, d\mu_i = -RT \sum \Gamma_i \, d\ln a_i \qquad (7.17)$$

has to be modified in order to account for the presence of electrolyte ions and for the (possible) varying degree of organic molecule ionization in the electric field. A linear adsorption isotherm (Henry's type) assumed by

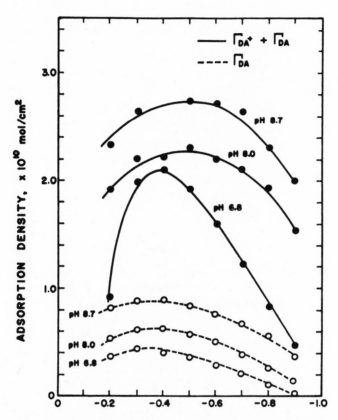

Figure 7.11. Evaluated adsorption densities of dodecylammonium ion (Γ_{DA^+}) and of undissociated amine (Γ_{DA}). Adsorption densities are calculated from the electrocapillary curves by applying the Gibbs adsorption equation in the following forms:

$$-\frac{1}{RT}\left(\frac{\partial \gamma}{\partial \ln a_{\text{DAA}}}\right)_{\text{E,pH}} = \Gamma_{\text{DA}^+} + \Gamma_{\text{DA}}$$

and

$$-\frac{1}{RT}\left(\frac{\partial \gamma}{\partial \ln a_{\text{KOH}}}\right)_{\text{E},\mu_{\text{DAA}}} = -\frac{1}{2.3RT}\left(\frac{\partial \gamma}{\partial \text{pH}}\right)_{\text{E},\mu_{\text{DAA}}} = \Gamma_{\text{DA}}$$

where γ is the interfacial tension, a is the activity, Γ is the adsorption density, E is the applied potential vs. NCE, and μ is the chemical potential. [From Usui and Iwasaki (1970) *Trans. AIME* **247**, with the permission of the American Institute of Mining and Metallurgical Engineers, Littleton, Colorado.]

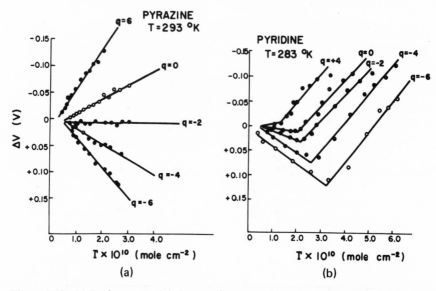

Figure 7.12. (a) Surface potential changes ΔV caused by water dipole reorientation due to displacement of some water molecules by adsorbed pyrazine at various surface charge densities q_M. (b) Surface potential changes ΔV analogous to those in part (a) caused by adsorption of pyridine (C_5H_5N) which itself orients at different surface coverages Γ depending on charge density q_M. [From Conway (1976), with the permission of *Chemistry in Canada*.]

Grahame is of limited applicability. The Langmuir adsorption isotherm does not take lateral interaction between the adsorbed species into account. To correct this deficiency, Frumkin (1926a, 1926b) modified the Langmuir isotherm (Section 6.4),

$$bp = \frac{\theta}{1 - \theta}$$

by introducing a parameter a to account for lateral interactions. The Frumkin adsorption isotherm has the form

$$bp = \frac{\theta}{1 - \theta} \exp(-2a\theta) \tag{7.18}$$

The parameter a in the Frumkin isotherm may be considered to be independent of the potential ϕ (or the charge q_M) at the electrode, and initially it was accepted as such. Alternately, as Damaskin and co-workers have postulated [reviewed by Frumkin and Damaskin (1964)], a may be assumed to depend on the potential ϕ. Experimentally, such a dependence was

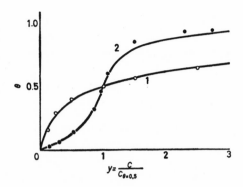

Figure 7.13. Frumkin's adsorption isotherms calculated from equation (7.18) for the parameter of lateral interactions $a = 1$ (curve 1) and $a = 1.6$ (curve 2) as compared with experimental data for adsorption of triethyl aminium ion, $[(C_2H_5)_3HN]^+$ [○ data taken from Lorenz et al. (1960)], and for tertiary amyl alcohol, t-$C_5H_{11}OH$ [● data taken from Lorenz and Müller (1960)].

detected for the first time by Gierst (1958) in the case of pyridine adsorption on mercury. Lorenz et al. (1960) determined a number of adsorption isotherms (from differential capacity data) and found a linear dependence of a on E in the adsorption of t-$C_5H_{11}OH$ (tertiary amyl alcohol) in 1 N KF on mercury (Figure 7.13). A nonlinear relationship (parabolic) was obtained for a vs. E for adsorption of tetraalkyl ammonium ions. When there are attractive lateral forces between the adsorbed molecules, $a > 0$, the coverage θ assumes an S-shape dependence on the concentration (Figure 7.14). When $a < 0$, repulsion between the adsorbates predominates, and the isotherm lies below that of Langmuir (for which $a = 0$), expressing approximately a logarithmic dependence of θ on concentration. When $a = 2$, there appears a vertical section on the adsorption isotherm, and beyond

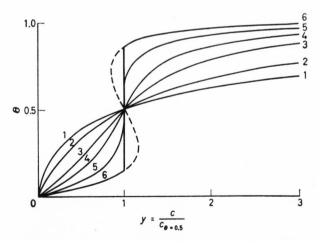

Figure 7.14. Frumkin's adsorption isotherms [equation (7.18)] calculated for different values of a (parameter for lateral interactions).

$a > 2$, unstable states in the adsorbed layer appear (shown by the dashed section of curve 6 in Figure 7.14). The axis of the abscissae in Figures 7.13 and 7.14 represents relative concentration,

$$\frac{c}{c_{\theta=0.5}} = \frac{\theta}{1-\theta} \exp[a(1-2\theta)] \qquad (7.19)$$

which is used instead of equation (7.18). The value of the parameter a can easily be evaluated graphically from experimental data [see Frumkin and Damaskin (1964)] for any S-shape curve, such as in Figure 7.13 or 7.14.

The measurements of the double-layer capacity obtained in the presence of organic molecules provide an extremely useful technique for studying adsorption isotherms at the mercury/solution interface when comparisons with the capacity in a reference electrolyte and Lippmann's equations (7.11) and (7.12) are utilized. A very striking feature of the capacity-potential (C–E) curves obtained in the presence of nonionic surfactants (neutral organic molecules) is the region of low capacity, ~ 5–$1 \ \mu F \ cm^{-2}$, surrounded on both sides by sharp peaks within narrow potential ranges (Figure 7.15c). The region of reduced differential capacity is due to the adsorption of dipolar molecules displacing water. At higher field strengths

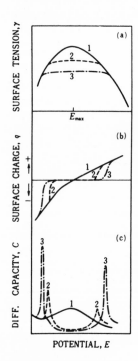

Figure 7.15. Comparison of the effects of nonionic surfactant adsorption at the mercury/electrolyte interface on its surface tension (electrocapillary curve), surface charge q, and differential capacitance, C, shown schematically for increasing concentrations, curves 2 and 3; curve 1, zero concentration of nonionic surfactant. [From Shinoda *et al.* (1963), *Colloid Surfactants*, with the permission of Academic Press, New York.]

(potentials), in either direction, these adsorbed molecules are displaced from the mercury surface by the water molecules, producing a steep rise in the differential capacity (on both sides of the E_{pzc}). The differential capacity curve of the electrical double layer in the presence of the electrolyte alone, without the organic compound, shows only a small hump in the vicinity of the adsorption peak for the given electrolyte. Higher concentrations of the organic adsorbate cause an increase in the height of the peaks (which change linearly with the logarithm of the concentration) and a widening of the low-capacity region (Figure 7.16). Maximum capacity at the peaks is associated with the largest variation in surface coverages. More quantitative data on these aspects can be found in Frumkin and Damaskin (1964).

In a number of cases more complex $C–E$ curves have been obtained. For example, in saturated solutions of heptyl and octyl alcohols the differential capacity curves were obtained with four peaks instead of two. The two inner peaks were interpreted as due to the adsorption–desorption of a second layer of alcohol molecules at the electrode interface. With ionizable surfactants, two low-capacity regions and three peaks may be obtained. One of these two low-capacity regions is interpreted as due to the adsorption of nonionic species and the other as due to ionized adsorbate. Frumkin and Damaskin (1964) and Payne (1973) published comprehensive reviews of adsorption studies at electrode surfaces.

The electrocapillary method is applicable only to liquid electrode–electrolyte systems (mercury and gallium). The differential capacity measurements can be used for studying adsorption of organic compounds on solid electrodes as well. The results are similar to those obtained with liquid mercury.

7.5. *Charge Transfer Across the Electrical Double Layer*

Chemisorption of atoms or small molecules involves charge transfer across the interface. One type of charge transfer is an electronic interaction between the substrate and the adsorbate; the other is an ion transfer followed by its partial or complete neutralization. Electrons can pass the energy barrier existing within the electrical double layer by tunneling. Ions, on the other hand, are initially held at the outer part of the Helmholtz double layer (OHP) until they acquire enough energy to pass the barrier within the Helmholtz layer and become incorporated in the solid at its kink sites.

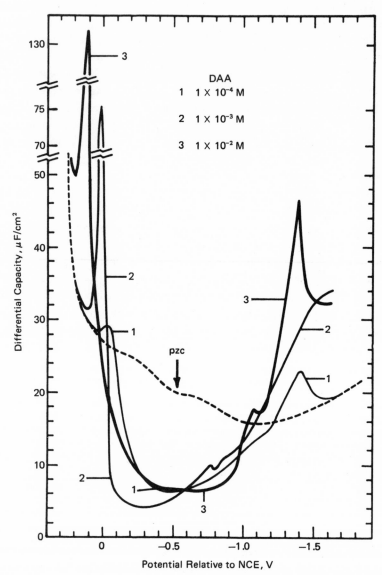

Figure 7.16. The capacitance–potential curves for increasing concentrations of dodecyl-amine acetate solutions at the mercury/electrolyte interface. [From Usui and Iwasaki (1970) *Trans. AIME* **247**, with the permission of the American Institute of Mining and Metallurgical Engineers, Littleton, Colorado.]

The height of the energy barrier depends on the electrode potential difference $\Delta\phi$ for the initial and final states in adsorption. Introducing the so-called *overpotential* η,

$$\eta = \Delta\phi - \Delta\phi_e \tag{7.20}$$

where $\Delta\phi_e$ is the potential difference for the equilibrium state in adsorption, the rate of an electrode reaction can be measured directly as the net current i,

$$i = i_0\left[\exp\left(\frac{\alpha z e\eta}{kT}\right) - \frac{n}{n_0}\exp\left(-\frac{\beta z e\eta}{kT}\right)\right] \tag{7.21}$$

where i_0 is the exchange current at equilibrium, α and β are charge transfer coefficients such that $\alpha + \beta = 1$, n is the concentration of ion (participating in the charge transfer) in solution, n_0 is the concentration of the same ion at equilibrium, z is the valence of the ion, and e is the electronic charge. The expression (7.21) is known as the Butler–Volmer equation, and it relates the kinetics of electrochemical reactions at the electrode/solution interface with the overpotential η. The plot of i vs. η gives a curve similar to a hyperbolic sine function, $\sinh x = (e^x - e^{-x})/2$, especially if $\alpha = \beta = \frac{1}{2}$.

Reactions involving charge transfer and ion exchange between electrodes and electrolytes may represent either reduction (cathodic) processes or oxidation (anodic) processes. With heterogeneous[†] solid surfaces both oxidation and reduction may occur on different portions of the same solid surface. When they occur to a different extent, the net current density represents the difference between the current density due to oxidation and that due to reduction reactions:

$$i = i_{ox} - i_{red}$$

At equilibrium, $i_{ox} = i_{red} = i_0$, there is no net charge transfer across the interface.

When metallic surfaces composed of one element are immersed in solutions of their own ions, the oxidation reaction represents metallic atoms going into solution, $M_{surf} \to M^{2+} + 2e$, and the reduction reaction is the reverse, $M^{2+} + 2e \to M_{surf}$, involving the same species M and M^{2+}. However, with surfaces represented by alloys, conducting oxides, or sulfides, oxidation and reduction may comprise two entirely different reactions, occasionally not even involving the components of the electrode material itself.

[†] Even on faces of single crystals each site must alternate between cathodic and anodic states within a few electron transfers.

For large overpotential values (exceeding 0.1 V) the term $\exp[-(\beta ze\eta/kT)]$ in equation (7.21) tends toward zero, and the whole expression reduces to $i = i_0 \exp(\alpha ze\eta/kT)$. This means that η is changing linearly with $\ln i$. The preceding linear dependence was discovered by Tafel as

$$\eta = a + b \log |i| \qquad (7.22)$$

and the slope b is hence known as the Tafel slope.

When two different reactions, one cathodic and the other anodic, occur on an atomically heterogeneous electrode/electrolyte interface, each reaction establishes a separate relationship between its anodic (or cathodic) current i and anodic (or cathodic) overpotential η. The intersection of the two Tafel slopes (anodic and cathodic) gives the steady-state current density i_0. The corresponding steady-state potential is known as the *mixed potential* (or rest potential) for that particular electrode–electrolyte system (Figure 7.17).

For small values of overpotential η, the current density i is linear with η, and the Butler–Volmer equation (7.21) becomes *Ohm's law*,

$$i = i_0 \frac{F}{RT} \eta \qquad (7.23)$$

where F is the gram-mole of ions, known as the faraday, and R is the gas constant.

In systems consisting of a monoelemental metal electrode in contact with its own ions in an electrolyte solution, the Butler–Volmer equation for equilibrium conditions becomes the Nernst equation for a reversible potential E_r,

$$E_r = E_0 + \frac{RT}{zF} \ln \frac{a_{M^{z+}}}{a_M} \qquad (7.24)$$

where $a_{M^{z+}}$ is the activity of metal ions in the electrolyte and a_M is the activity in the bulk metal phase, taken as unity. On applying potentials positive with respect to the reversible potential (E_r) to the electrode, metallic ions are formed, and dissolution takes place at the electrode surface; at potentials negative to E_r, discharge of metallic ions occurs at the electrode, and they are deposited on the metallic surface until a new equilibrium is reached. (*Cf.* p. 100 regarding potential scale conventions.)

However, when the metal ions in solution are foreign with respect to the metal electrode, deposition often occurs, in apparent violation of the Nernst law, at potentials positive with respect to the reversible Nernst potential, E_r. This effect is known as *underpotential deposition*, and it

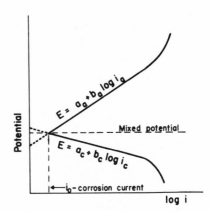

Figure 7.17. Mixed potential and the corresponding anodic and cathodic potentials, with their Tafel slopes, shown schematically.

indicates that foreign atoms may attach to a metal electrode much more strongly than like-metal atoms on the same metal surface.

The technique of varying the potential of a given electrode–electrolyte system within a limited range (linearly with time, in a repetitive manner) and recording the corresponding currents in the anodic and the cathodic direction with respect to E_r is known as cyclic voltametry. If a single anodic or cathodic sweep is made, it is known as a linear potential sweep. In both techniques a peak in a current appears at a potential characteristic of the reaction taking place at the electrode/electrolyte interface. The position and the shape of the peak depend on several factors (sweep rate, material of the electrode, composition of solution, and concentration of reactants). Cyclic voltametry is used primarily to identify various steps in the overall complex reaction, but an approximate value of the rate constant can also be derived from the separation of anodic and cathodic peak potentials. The experiment is conducted in unstirred solutions to eliminate convection. Cyclic voltametry enables detailed investigations to be made of dissolution and deposition reactions at numerous interfaces of conducting solids in electrolyte solutions. It provides information on adsorption and desorption characteristics at the solid/solution interface similar to that provided for desorption alone in gas–solid systems by the flash desorption technique. For details on technique(s), see Gileadi *et al.* (1975) or Yeager and Kuta (1970).

Cyclic voltametry studies carried out with metallic electrodes have established a large number of valuable data, such as the following:

1. The existence of underpotential deposition is indicated for numerous metallic couples but not for all combinations of metals.

2. For a large number of couples the difference in respective work

functions[†] of the substrate metal and the adsorbate metal can be correlated with the difference in potentials of peaks characterizing monolayer and bulk "stripping." The respective work functions for substrate–adsorbate are more appropriate indicators of ionic character of adsorbate–substrate bonding than Pauling's electronegativities. When the work function of the substrate is lower than that of the adsorbate, the given couple does not show underpotential deposition, as, for example, Cu on Ag.

3. Since underpotential is very sensitive to the nature of the surface, the characteristics of monolayer adsorption and desorption may be conveniently studied on composite surfaces or alloys. For example, cyclic voltametry has established that an alloy of 40% Au + 60% Ag bulk composition behaves as an Ag electrode due to preferential segregation of Ag on the surface.

4. Measurements on different crystallographic planes of single-crystal Ag, Au, or Cu electrodes have confirmed the close relationships between pzc (point of zero charge) and work function of each face. Oxygen adsorption begins, for example, on the (100) Au face at potentials more positive than 1.0 V (SCE) and on the (111) face at still more positive potential, $\sim$1.2 V (SCE), but reduction of adsorbed oxygen occurs at the same potential for all low-index faces.

5. Even in the first monolayer adsorption occurs in several steps, each indicated by a separate peak. Detailed analysis of these peaks provides information on the extent of coverage, lateral interactions, and strength of bonding with substrate sites.

Systematic studies with Au have established that when the adsorbing metal atoms are of smaller diameter than that of Au, the maximum number of atoms adsorbed corresponds to the number of substrate atoms for each face. When adsorbate atoms are larger, a close-packed monolayer is formed regardless of the surface structure. Exceptions to these findings occur (e.g., Pb on Cu), with ordered superstructures developing before completion of a monolayer.

Details of studies on solid surfaces and references to the recent papers on cyclic voltametry can be found in a review by Gerischer *et al.* (1978).

[†] A work function is the potential at which electrons can be removed from the highest occupied band within the metal. Since the rate of electron emission is highly temperature-dependent, the term *thermionic work function* is often used. For more information, see Conway (1965).

7.6. *Semiconductor/Solution Interfaces*

Many metal oxides and metal sulfides are semiconductors (see Section 2.7), i.e., substances which possess two types of charge carriers: electrons in the conduction band and holes in the valence band. When electrons from the valence band become sufficiently excited (thermally or by photoillumination) to cross into the conduction band, holes are created in the valence band. Their motion in the direction opposite to that of electrons contributes to charge transfer and to conductivity, as if the holes were positively charged carriers of electricity. Electrons excited into the conduction band are free to occupy large numbers of unfilled energy states and conduct electricity.

The concentrations of charge carriers in semiconductors are much lower than in metals, 10^{13}–10^{16}/cm^3 as compared to $\sim 10^{22}$–10^{23}/cm^3 in metals where only the electrons are charge carriers. For comparison, a 10^{-5} M electrolyte solution possesses $\sim 10^{16}$ ions/cm^3. The mobilities of electrons and holes in some semiconductors and the respective energy gaps between the valence and the conduction bands are indicated in Table 7.2.

Table 7.2. Some Characteristics of Selected Semiconductors[a]

Substance	Mobility (cm^2/V sec)		Energy gap (eV)	Diffusion coefficient of electrons at 300°K (cm^2/sec)	Crystal type
	Electrons	Holes			
Ge	3,900	1,700	0.72	93	Diamond
Si	1,900	425	1.12	31	Diamond
InSb	77,000	1,250	0.26	93	Zinc blende
GaSb	4,000	2,000	0.80	100	Zinc blende
InAs	27,000	280	0.34	14	Zinc blende
ZnS	—	—	3.67	—	
CdS	240	—	2.5	6	
Sb$_2$S$_3$	—	—	2.0	—	Rhombic
PbS	640	350	0.37	16	Cubic
PbTe	2,100	840	0.54	52	
Cu$_2$O	250	57	2.06	—	Cuprite
AgI	30	—	2.8	—	

[a] Data from Teichmann (1964).

In an intrinsic semiconductor (such as germanium, silicon, tellurium, stoichiometric lead sulfide) the energy gap is small, the concentration of electrons excited into the conduction band is equal to the concentration of holes created in the valence band, and the product of these two concentrations is constant at a fixed temperature. This constant depends on the energy gap. Near the surface of a semiconductor, electrons and holes are no longer present in equal numbers, but a so-called space charge layer exists (of specified thickness l_{sc}, which is typically of $\sim 10^{-5}-10^{-3}$ cm) whose capacity depends on the applied potential. The distribution of excess charges in this interfacial space charge region is analogous to the distribution of ions in the diffuse part of the electric double layer in an electrolyte. It contrasts with the distribution of charges in metals which have all interfacial excess charges concentrated in the surface. In semiconductors the space charge may become squeezed onto the surface (with a corresponding decrease in its thickness l_{sc}) when the magnitude of the interfacial excess charge is sufficiently increased (in analogy with a decrease in the thickness of the Gouy–Chapman diffuse layer for an increased ionic strength). Due to the existence of a layer of charges in the OHP in the electrolyte, the bands in the space charge layer become bent, up or down, depending on the ions in the OHP. The potential at which the space charge in a semiconductor electrode attains zero value is called the flat-band potential (no band bending occurs at the surface).

A schematic distribution of charges at a semiconductor/electrolyte interface is shown in Figure 7.18. For a comprehensive treatment of electrochemical behavior of semiconductors, refer to Gerischer (1970).

Nonintrinsic semiconductors are impurity-activated (or doped) semiconductors. They may be n type when an electron donor is added to increase the concentration of electrons or p type when an electron acceptor is added to an intrinsic semiconductor to increase the concentration of holes. Lattice vacancies (Schottky defects) or interstitial atoms and ions (Frenkel defects) (see Figure 6.4) also give rise to appropriate semiconductor behavior in solids.

The extent of nonstoichiometry in compounds possessing semiconducting characteristics is less than 1%. The impurity atoms may ionize to form positive ions (at the donor level, below the conduction band) or negative ions (at the acceptor level, just above the valence band level). These impurity atoms are immobile, but the electrons or holes, resulting from their ionization, contribute to conductivity; the product of electron and hole concentrations is constant (for a given temperature), but the respective concentrations are no longer equal to each other.

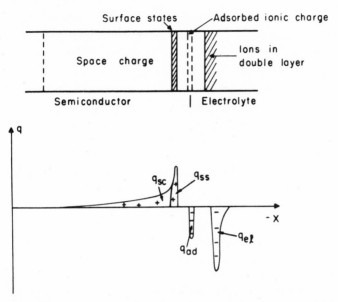

Figure 7.18. Model of a semiconductor/electrolyte interface showing four types of charges: a space charge q_{sc}, a surface state charge q_{ss}, a charge due to specifically adsorbed ions q_{ad}, and electrolyte countercharge q_{el}. [From Gerischer (1970), in *Physical Chemistry*, Vol. IXA, H. Eyring, ed., with the permission of Academic Press, New York.]

As shown in Figure 7.18, the model of a semiconductor/electrolyte interface represents four types of charges:

Space charge, q_{sc}: immobilized donor or acceptor states

Surface state charge, q_{ss}: trapped charges in the surface layer of semiconductor (analogous to specifically adsorbed ions)

Specifically adsorbed ions or ionized groups, q_{ad}: in the IHP of the electrolyte side of interface

Countercharge, q_{el}: in the diffuse layer within the electrolyte

The measurements of differential capacity, illumination effects, and relaxation time effects in surface states have provided useful correlations for comparison with theoretical predictions. The most important conclusions from such studies are repeated here [taken from Gerischer (1970)] to provide a background for relevant studies on semiconductor behavior of metallic oxides and sulfides in flotation environments.

The flat-band potential, that is, the point of zero space charge in the semiconductor, serves as a convenient reference for analysis of space charge behavior. The composition of an electrolyte can greatly influence the flat-band potential. For example, Tyagai (1965) found that the flat-band potential of a CdS single crystal (an n-type, metal excess semiconductor) was shifted from -0.95 to -1.35 V (vs. NHE) by adsorption of S^{2-} from 0.1 M Na_2S solution, whereas both O_2 and Cd^{2+} had only a negligible influence. Adsorption of S^{2-} ions is presumed to occur at a vacancy or an excess Cd site, giving rise to the formation of negatively charged surface states. Specifically, the pH of the electrolyte can change the flat-band potential in a germanium crystal in a (straight) linear manner.

Equilibrium in the space charge layer is established rapidly, but it is achieved much more slowly in surface states. When disturbed by external forces, such as illumination, an increase in charge distribution (concentration of electrons and holes) occurs which is proportional to the light intensity (if the latter is not too high). The resulting increase in differential capacity is used to evaluate the charge of the double layer and the influence of charge carriers on the surface reaction. Frequencies above 50 kHz and rapid sweep techniques are desired for correct results.

Charge transfer reactions at a semiconductor/electrolyte interface differ from those in a homogeneous solution or at a metal/electrolyte interface in that only those electron transfer processes which are on the energy levels of either the conduction or the valence bands are possible, while the acceptor and donor states with levels within the energy gap are excluded from the transfer processes. The rate of the electron transfer is proportional to the density of occupied energy states in one phase and the unoccupied states in the other phase (both of which are on the same energy level across the interface).

If holes are accumulated in the semiconductor surface, the bonds of surface atoms are weakened by the missing electrons in bonding states. Electrons available in nucleophilic reagents in solution, e.g., anions, may react with such surface atoms, forming new bonds and filling the unoccupied states in the valence band. If electrons are injected into the conduction band, the activation energy required is of the order of the energy gap for the given semiconductor, and consequently the anodic processes will occur if the semiconductor is anodically polarized to a sufficient level, or if illumination supplies the energy of activation, or if a suitable redox system (such as $Fe^{2+}-Fe^{3+}$, $H_2O_2-O_2$, $I_3^{-}-I^{-}$, etc.) supplies holes (injects holes into the semiconductor's surface states) through its reduction reaction (electrons from the valence band are consumed in the reduction of redox components).

The mechanism of anodic dissolution is complex even for an elemental semiconductor and more so for a compound one, such as CdS, ZnO, and

Cu_2O. In the photo-oxidation of ZnS the products are Zn^{2+} (going into solution) and S, polymerizing to S_n but remaining on the solid surface.

An addition of a reducing agent to a solution in contact with a semiconductor possessing a wider energy gap causes the photoproduced holes to oxidize the reducing agent. Some reducing agents, viz. those changing their valence by two electrons, are able to double the current across the semiconductor/electrolyte interface, at constant light intensity, in comparison with that obtained for monoequivalent reducing species. Typical current-doubling species are $HCOO^-$, CH_3OH, C_2H_5OH, and As^{3+}, while one-equivalent species are I^-, Br^-, and $Fe(CN)_6^{4-}$. The explanation of this behavior is that the *redox radicals* formed by the single-hole-capture appear to be unstable and on decomposition inject an electron into the conduction band, thus nearly doubling the current, for example,

$$p + HCOO^- \rightarrow HCOO^\circ$$

$$HCOO^\circ \rightarrow e + H^+ + CO_2$$

where p denotes a hole in the semiconductor's valence band.

If oxygen is bubbled through the solution containing the reducing species, $HCOO^-$, the current decreases; this has been interpreted as due to the capture of the electron associated with the redox radical by oxygen (thus reducing the oxygen and preventing the injection of the electron into the conduction band):

$$HCOO^\circ + O_2 \rightarrow H^+ + CO_2 + O_2^-$$

Analogous reactions may take place with xanthate ions and semiconducting sulfides, such as the formation of an adsorbed xanthate radical on reaction with a hole in the semiconductor surface and its subsequent decomposition to an alcohol and CS_2. The possibility of such catalytic decomposition of adsorbates necessitates caution in the interpretation of adsorption isotherms; the chemical nature of the species retained on the solid surface and/or formed in solution after a charge transfer should be determined independently.

The cathodic reduction (decomposition) of semiconductors proceeds via electrons in the conduction band. Unfortunately, the products of the reductive decomposition usually stick to the surface of the semiconductor and change its properties. For example, as a result of an association of two adjacent surface radicals in ZnO reduction, a layer of elemental zinc remains on the surface, representing a new interface with complex (and unknown) structure and properties.

In the presence of some oxidizing agents a semiconductor is likely to undergo a direct chemical oxidation reaction independently of the charge in the surface and additional to electrochemical processes. Only those oxidants which can form two new chemical bonds simultaneously with two identical components on the surface do so. Such oxidants are Cl_2, I_2, Br_2, and H_2O_2 (and in flotation systems, the oxidation products of thio compounds, such as dixanthogen).

7.7. Electrical Double-Layer Studies Relevant to Flotation

Experimental work on the electrical double layer has been carried out mainly on metal electrode–electrolyte systems. Although the quantities of minerals represented in nature by native metals (gold, platinum, silver, copper) and concentrated in part by flotation are very limited, the electrochemical studies on metal–surfactant solution systems have proved extremely useful. In particular, the studies on model flotation of mercury droplets in xanthate solution with the corresponding electrochemical measurements, introduced by Pomianowski (1957, 1967), have enabled very useful correlations to be established. Such studies have been extended to metal particulates (gold, platinum, copper) and to particulates of semiconducting minerals, such as galena and copper sulfide, by Gardner and Woods (1973), Chander and Fuerstenau (1975), Kowal and Pomianowski (1973), etc.

Preceding these direct correlations on model flotation systems, comparisons were made between electrochemical studies of specific solid/liquid interfaces and the contact angle behavior or the flotation behavior of the same solid particulates under otherwise identical conditions. Talmud and Lubman (1930), Frumkin *et al.* (1932), Kamienski (1931), Lintern and Adam (1935), Salamy and Nixon (1953), Iwasaki and de Bruyn (1958), Szeglowski (1960), Tolun and Kitchener (1964), Majima and Takeda (1968), Yarar *et al.* (1969), and a number of investigators following in the wake of these papers dealt with the effect of flotation reagents on the potential of the mineral/solution interface, the nature of electrochemical reactions, and the conditions under which the pzc is established in the given system.

The rest potentials of minerals that were identical but derived from different deposits were found to vary considerably. This behavior has been ascribed in part to a slight modification in the stoichiometry of the mineral composition and in part to local surface heterogeneity. The surface heterogeneity was found to cause potential differences of up to 150 mV in magnitude between neighboring portions of surface only 1 mm apart [Szeglowski (1960)]. A second sulfide phase adhering to the sulfide whose potential is measured (thus representing an incompletely liberated mineral grain) may create in an electrolyte a local galvanic cell with a potential difference amounting to approximately 0.5 V.

Metallic sulfide electrodes immersed in xanthate solutions of varying xanthate concentration have rest potentials linearly dependent on the concentration of xanthate [Majima and Takeda (1968)]. Metal electrodes (Pt, Au, Hg) behave similarly. Tolun and Kitchener (1964) established the reversible character of the xanthate–dixanthogen redox potentials at the Pt and PbS electrodes.

Knowledge of the actual values of the rest potential for any particular solid–electrolyte system does not provide sufficient information regarding the identity of the species and the nature of electrochemical reactions at the electrode. Independent spectroscopic evidence, together with an evaluation of capacitance–potential and/or current–potential (cyclic voltametry) curves, appears to be required to obtain a better understanding of the reactions which precede the particle–bubble attachment. If all such electrochemical measurements were to be made in a system in which flotation is simultaneously carried out, a direct correlation of the important parameters would be possible. Pomianowski (1967) and Pomianowski *et al.* (1968) set out to conduct such determinations of the capacity–potential curves with simultaneous flotation of systems comprising Hg–electrolytes and/or Hg–xanthate. Some of their results are shown in Figure 7.19–7.21.

Curve A in Figure 7.19 represents the capacity–potential relationship in 1 N KCl alone; arrow 0 indicates the pzc of Hg in 1 N KCl; curve B is the capacity–potential curve in 4×10^{-5} M K–ethyl xanthate (KEtX) obtained at 20-Hz frequency, while curve D is that for 710 Hz. There are two humps in each of the curves B and D: hump I at -0.7 V (vs. SCE) and hump II at -0.35 V. As the concentration of KEtX in this system was increased from 1×10^{-5} M, it was found that the intensity of the second hump (-0.35 V) increased with the concentration but only up to 4×10^{-5} M KEtX; then—for higher concentrations—it disappeared, while another hump III at -0.45 V began to appear (Figure 7.20).

At this concentration, 4×10^{-5} M KEtX, the coverage per xanthate ion, calculated from Koryta's (1953) expression for surface excess,

$$\Gamma = \frac{0.627 D^{1/2} m^{2/3} t^{7/6} c}{S} \tag{7.25}$$

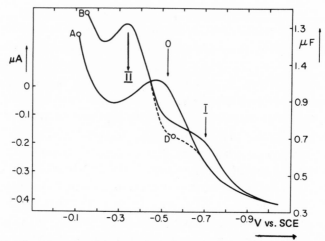

Figure 7.19. Capacity–potential curves for a dropping mercury electrode vs. SCE: curve A, 1 N KCl; curve B, 4×10^{-5} M KEtX (potassium ethyl xanthate) for 20-Hz frequency; curve D, 4×10^{-5} M KEtX for 710-Hz frequency; 0, pzc in 1 N KCl. [After Pomianowski (1967), with the permission of the author.]

where D is the diffusion coefficient of xanthate ion ($D = 8.7 \times 10^{-6}$ cm²/sec), m is the capillary rate (g/sec), t is the drop time (sec), S is the surface area of drop in time t (cm²), and c is the concentration of EtX⁻, was evaluated by Pomianowski to be 44 ± 6 Å²/EtX⁻ ion.

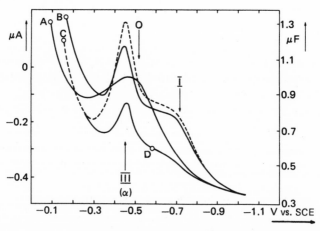

Figure 7.20. Capacity–potential curves for a dropping mercury electrode vs. SCE: curve A, 1 N KCl; curve B, 7.5×10^{-5} M KEtX, 20 Hz frequency; curve C, 9×10^{-5} M KEtX; curve D, 9×10^{-5} M KEtX after correction for pseudocapacity of Faradaic reaction. [After Pomianowski (1967), with the permission of the author.]

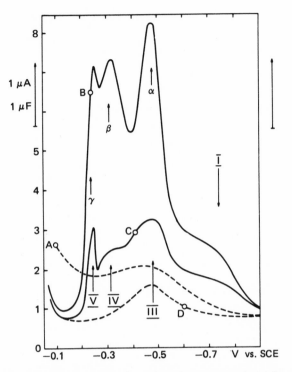

Figure 7.21. Capacity–potential curves for a dropping mercury electrode vs. SCE: curve A, 1 N KCl; curve B, 6×10^{-4} M KEtX, 20 Hz frequency; curve C, 6×10^{-4} M KEtX, 301-Hz frequency; curve D, in KEtX after correction for pseudocapacity. [After Pomianowski (1967), with the permission of the author.]

Figure 7.20 shows hump III at -0.45 V for concentrations of 7.5×10^{-5} M KEtX. This hump (-0.45 V) increased in intensity up to a maximum at 1.2×10^{-4} M KEtX concentration. When the concentration of KEtX was increased above 1.2×10^{-4} M, viz. to 6×10^{-4} M, two new humps appeared in the capacity–potential curves at -0.3 V (β) and -0.25 V (γ) (Figure 7.21). A linear increase in the intensity of the humps at -0.45, -0.3, and -0.25 V with the concentration of KEtX was found to occur, and it showed that the slope for -0.25 V was twice that of the humps at -0.45 and -0.3 V.

The following interpretation of the capacity–potential humps has been made:

Hump I: -0.7 V potential (vs. SCE) is due to a reversible adsorption of EtX$^-$ (ion) on a negatively charged Hg surface.

Hump II: −0.35 V potential is caused by a reorientation of xanthate ion bonded electrostatically to the surface. Such reorientation is possible only when the coverage does not exceed a certain density, that of ∼44 Å²/ion of EtX⁻.

Hump III: −0.45 V (also denoted by α) is the potential corresponding to the maximum of a single electron transfer reaction oxidizing EtX⁻.

Hump IV: −0.3 V (also denoted by β) is the potential of a second single electron transfer reaction resulting in the formation of adsorbed dixanthogen:

$$(EtX–Hg) + EtX^- \rightarrow [(EtX)_2–Hg] + e$$

Hump V: −0.25 V (denoted by γ) is due to a two-electron charge transfer step resulting in the formation of a mercuric xanthate precipitate that accumulates on the mercury surface:

$$2EtX^- + Hg \rightarrow Hg(EtX)_2 + 2e$$

Extending these capacitance studies by voltametric determinations of current–potential sweeps for 0.2×10^{-5} M to 5×10^{-4} M KEtX solutions, Pomianowski confirmed the progressive appearance of the maxima I–IV, identified in the capacity–potential curves, with small differences. Namely, the potentials for the maxima in the capacity–potential relationship were shifted to slightly lower values, and the reaction of a single-electron charge transfer appeared to start at a much lower xanthate concentration. These differences may have been due to adsorption kinetics, since the current–potential work was carried out using hanging Hg drops, allowing the surface to adsorb more xanthate from a weaker solution, while in the capacitance study a dropping Hg surface was exposed to the reagent for a much shorter time.

Impedance measurements on a dropping mercury electrode in xanthate solutions, held at a constant potential, −0.45 V, carried out by Pawlikowska-Czubak and Leja (1972) confirmed the various aspects of surface adsorption processes suggested by Pomianowski. It is clear that even in an apparently simple Hg–EtX system (no sulfide and derivative ions) the process of xanthate adsorption is highly complex and involves numerous steps. Woods (1971) obtained similar indications of stepwise adsorption (and desorption) of xanthate on copper and galena in his current–potential studies. Figure 7.22 shows the current–potential curves obtained by Woods (1971) for a galena electrode in 0.1 M borate supporting electrolyte and 9.5×10^{-3} M ethyl xanthate concentration. The peak in the region −0.1 to +0.1 V was interpreted as indicating adsorption of xanthate ion with a charge transfer.

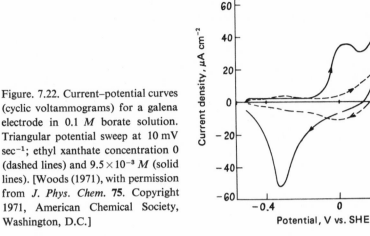

Figure. 7.22. Current–potential curves (cyclic voltammograms) for a galena electrode in 0.1 M borate solution. Triangular potential sweep at 10 mV sec^{-1}; ethyl xanthate concentration 0 (dashed lines) and 9.5×10^{-3} M (solid lines). [Woods (1971), with permission from *J. Phys. Chem.* **75**. Copyright 1971, American Chemical Society, Washington, D.C.]

Kowal and Pomianowski (1973) obtained voltammograms (current–potential sweeps) for natural copper sulfide electrode in 0.1 M NaF solution and various ethyl xanthate concentrations. Two of the peaks obtained in their curves have been interpreted as due to xanthate adsorption, whereas a third peak was assigned to the formation of bulk copper xanthate at the electrode surface.

Chander and Fuerstenau (1975) rightly commented that cyclic voltammetry methods may not provide complete information regarding the reaction mechanisms in complex systems. Not only may the supporting electrolyte influence the course of reactions, but nonelectrochemical reactions may be taking place undetected, and comparisons with a model require identification of products by an independent method.

In studies carried out by Pomianowski (1967), flotation tests were carried out in a cell especially designed for mercury flotation under conditions (as regards xanthate concentration and potential) identical to those in the capacity and cyclic voltammetry studies. In addition, the effects of frother-acting surfactants were also investigated and correlated with the corresponding changes in the electrochemical parameters. The most surprising result of these flotation studies on mixed surfactants was the finding that the improvement in recoveries could not have been predicted from studies of the capacity–potential relationship. [For additional details on electrochemistry of sulfides in flotation systems, reference should be made to Woods' (1976) review paper and to the discussion of reactions between sulfides and thiol reagents by Poling (1976).]

7.8. Electrokinetic Effects

When a phase (solid, liquid, or gas) is in a tangential motion relative to an aqueous solution in contact with this phase, an electric field is induced. The induced field may be interpreted as arising from a charge gradient existing at the plane of slippage (shear plane or slip plane). This plane is situated within the diffuse portion of the electrical double layer[†] (that is, within the solution region of the interface). As a result of the tangential motion along the slip plane, the equilibrium of charges within the interface is disturbed, and this disturbance is counteracted by a flow of charges attempting to restore the electrical balance across the whole interface, on both sides of the slip plane. The overall phenomenon of interfacial flow and the associated charge redistribution is known as the electrokinetic effect. This effect is of great importance in systems involving flow of liquids in capillaries or dispersions of colloidal particulates (solid, liquid, or gaseous, that is, slurries, emulsions, aerosols, etc.) in a fluid under the influence of a velocity gradient or an electrical field gradient.

The charge gradient at the slip plane is treated as a potential difference and is known as the electrokinetic potential (and its synonym is *zeta potential*). This potential difference is *derived* (i.e., calculated) for a *set of assumptions* from a set of *measured and applied* quantities (involving electrical and hydrodynamic parameters appropriate to the technique used in evaluation of the electrokinetic properties of charged interfaces; see below).

[†] The exact position of the slip plane within the diffuse portion of the electrical double layer is not known. In theoretical treatments of electrokinetic effects it is assumed that slippage occurs along, or close to, the OHP; thus the thickness of liquid adhering to the solid is, at the most, a few angstroms. The hydrodynamically derived expression for the volume of a stagnant layer retained by a solid plate from which a liquid of bulk viscosity η is draining in time τ is

$$V_{\text{theor}} = \frac{2wl^{1.5}\eta^{0.5}}{3(g\varrho)^{0.5}\tau^{0.5}}$$

where w is the width of the solid plate, l is the length of the solid plate, η is the viscosity of the bulk liquid, ϱ is the liquid density, g is acceleration, and τ is time.

Experimental values are usually much greater. The usual explanation is as follows: Surface rugosity (roughness), higher viscosity of adhering liquid layer, "swelling" of the wetted surface, etc., may be involved and are responsible for the extra liquid retained [Bikerman (1970, pp. 227ff.)]. In measurements of oils (of various viscosities) retained on the surface of stainless steel, the stagnant layer was evaluated as 0.25–3.6 μm, whereas the diffusion layer on Hg electrodes was measured as 50–350 Å. The thickness of the stagnant liquid layer is of great importance in de-watering of finely divided solids or porous solids and in removing the residual oil from porous oil-bearing strata.

It is regarded as a difference in the potential at the slip plane and the potential in the bulk solution, the latter being taken as the reference, equal to zero. The derivation is carried out in such a way as to make the resulting zeta potential independent of the magnitude of relative motion, that is, independent of the exact position of the slip plane (which, for higher speeds of motion, is presumed to approach the OHP of the electrical double layer). Thus, the derived values of zeta potential represent *relative positions of the potential distribution curves* in the diffuse portion of the electrical double layer, irrespective of the shear plane, for the set of conditions used (pH, concentration, etc.). The absolute value of zeta potential is not known, except at the pzc.

There are four basic types of electrokinetic effects. Theoretically, each can be utilized to evaluate the zeta potential for a given set of conditions. These four types are indicated in Figure 7.23, namely,

1. *Electrophoresis*: dealing with mobility of a *charged colloidal particle* suspended in an electrolyte in which a potential gradient is set up

2. *Streaming potential*: concerned with the magnitude of induced electric potential, generated by a flow of liquid in a single capillary or in a system of capillaries (represented, for example, by a porous plug of compressed solid particles)

3. *Electro-osmosis*: denoting a movement of liquid in a capillary or in a system of capillaries (porous plug) under an applied potential

4. *Sedimentation potential*: determining the magnitude of the potential developed by a dispersion of particles settling under the influence of gravity

Of these four electrokinetic effects, the electrophoretic and the streaming potential are the ones most frequently employed in studies relating to minerals. Laboratory setups for streaming potential measurements have been described by Fuerstenau (1956), Joy *et al.* (1965), and Somasundaran and Kulkarni (1973); commercial equipment models are available for electrophoretic measurements (from Zeta-Meter, Inc., New York, or from Rank Bros., Bottisham, England).

A demountable electrophoretic cell suitable for use with mineral particles was tested by Shergold *et al.* (1968). A laboratory-built electrophoresis apparatus has also been described by Deju and Bhappu (1966). A book on electrophoresis was published by Shaw (1969).

In electrophoretic measurements the mobility v of charged colloidal particles dispersed in solution is measured for different values of applied

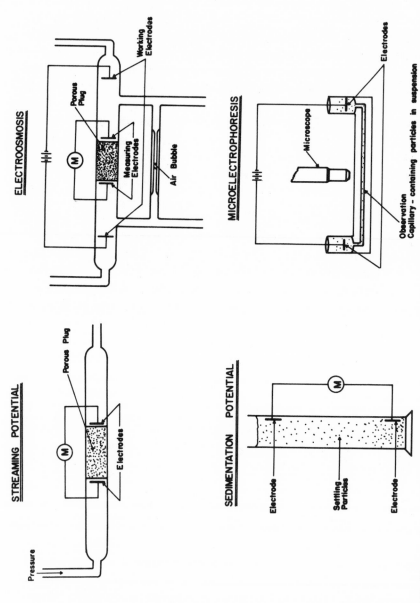

Figure 7.23. Four main techniques used for zeta potential evaluations: streaming potential, electroosmosis, sedimentation potential and microelectrophoresis (shown schematically).

potential gradient, E/l. In streaming potential, different pressures P are applied to the electrolyte, forcing it through a capillary (or a set of pores), and the corresponding streaming potential E is determined.

Smoluchowski (1903) considered laminar flow of an incompressible aqueous electrolyte between two nonconducting plates and derived basic expressions for the zeta potential: (a) in streaming potential,

$$\zeta = \frac{E}{P} \frac{\lambda}{\varepsilon} \eta \qquad (7.26)$$

which becomes zero when the ratio $E/p = 0$ or $E = 0$, and (b) in electrophoresis,

$$\zeta = \frac{4\pi l}{\varepsilon E} \eta v \qquad (7.27)$$

which becomes zero for $v = 0$, where v is the velocity of the electrophoretic colloid, E is the applied potential (in electrophoresis) or measured potential (in the streaming potential technique), P is the pressure applied to the liquid, and l is the distance between electrodes, η is the viscosity, λ is the conductivity, and ε is the dielectric constant of the electrolyte bulk solution.

Let us consider these expressions and the assumptions used in their derivation. First, it is surprising that the properties of the solid comprising one of the two phases in an electrokinetic system are not taken into account at all. Second, the derivations leading to equation (7.26) and (7.27) are based on two broad assumptions:

1. The electrical double layer (along which separation of charges occurs) consists of the Helmholtz condenser-like layers of opposite charges. This assumption would apply only to concentrated electrolyte solutions.

2. When a steady-state condition is reached, the charges carried away by convection in the flowing liquid are supposed to be compensated by an electronic current (the flow of which has never been defined or identified). Such compensation is not likely to take place in every solid–electrolyte system without an extraneous charge buildup on some solids.

All electrokinetic effects are difficult to measure in a reproducible manner. A large number of side effects [such as changes in viscosity caused by the applied potential gradient (resulting in a so-called electrophoretic retardation), surface conductance, "electrode effect," etc.] have been introduced into evaluations of zeta potential for the purpose of overcoming the difficulties of nonreproducibility of measurements. Kruyt (1952), Over-beek (1950), and Mirnik (1970) discussed the significance of such side effects on zeta potential.

Ball and Fuerstenau (1973) reviewed the measurements of streaming potentials from the standpoint of the constancy of E/P ratios. They found that although the plots of E/P are linear, the lines of constant slope hardly ever pass through the origin of the streaming potential vs. pressure axes. They attribute this behavior to an "electrode effect." The latter appears to be a function of the material used, of the treatment and cleaning it received before the measurements, as well as of the electrolyte concentration. Somasundaran *et al.* (1977) suggested that these electrode effects could be eliminated by recording the potential at the instant of stopping the flow of the electrolyte; otherwise the time decay of potential incorporates the polarization effects, producing nonzero intercepts of E/P lines.

In addition to these side effects (caused by nonfulfillment of theoretical assumptions) there are other effects caused by adsorption of impurities, some of which may be very active at interfaces even if present only in trace amounts.

Mackenzie (1966), Hall (1965), Healy *et al.* (1968), and Mackenzie and O'Brien (1969) found very strong effects of trace metal ion adsorption on the zeta potentials of quartz and silica, causing large shifts in zeta potential and even a reversal of charges. Kloubek (1970) found that even minute traces of polyvalent cations present in acids used for preparations of cleaning solutions or in compounds used for pH adjustments and in electrolyte salts affect the measurements of the zeta potential of quartz. A useful review of zeta potential studies in mineral processing has been published by Mackenzie (1971).

In light of the preceding comments, it becomes obvious that the expressions (7.26) and (7.27) are capable of reproducing the value of zeta potential only under a very special set of conditions. Generally, the measurements reflect not only the effects due to zeta potential but also the superimposed effects due to the nonapplicability of the assumptions used in the derivation of equations (7.26) and (7.27) and to specific adsorption of impurities.

In streaming potential measurements, the ionic charges carried away in the flowing liquid may not be completely compensated by the counterflow of electronic charges. Such lack of compensation often results in a buildup of charges at one end of the capillary. The buildup may be reduced to an unknown extent by an external flow of current in the potential-measuring circuit attached to the electrodes.

The difficulties encountered in the reproducibility of zeta potential determinations and the differences between zeta potentials determined by a streaming potential technique and those obtained in electrophoresis for the same solid–electrolyte systems suggest that the values of zeta potential calculated with the expressions (7.26) and (7.27) may not be the true values they purport to represent. In fact, the measured electrophoretic mobilities are lately being reported instead of the calculated zeta potential. The only value that should not be affected by side effects caused by the technique of evaluation is the point of zero charge.[†] Provided that suitable precautions are taken to avoid accidental extraneous contaminations, the pzc should reflect the effects of potential-determining ions adsorbing at the particular crystalline faces of the solid.

7.9. *Examples of Zeta Potential Changes*

Apart from the zero value of the electrokinetic potential (pzc), only the various *trends* in the behavior of the electrical double-layer systems, as indicated by electrokinetic measurements, have significance—not the specific calculated values of zeta potentials, however carefully evaluated. Such trends are briefly discussed below.

An example of zeta potential variation for an ionic solid in solutions of its lattice-constituent ions is shown in Figure 7.24. For AgI in pure water, a negative zeta potential is obtained. Approximately a 10^{-5} M concentration of Ag^+ ($AgNO_3$) is required to reach a zeta potential equaling zero; for

---

[†] *Note*: Three terms are used rather loosely in discussing electrokinetic effects, and some comment is required: the point of zero charge (pzc), the isoelectric point (iep), and the reversal-of-charge concentration (rcc). The *point of zero charge* (pzc) is defined for the electrocapillary curve as the potential E_{pzc} for which $\gamma = \gamma_{max}$, that is, $d\gamma/dE = q = 0$, where q is the charge density. For other systems it should denote the condition under which a *direct measurement of surface charge density* gives a zero net charge on the particle. The measurements of surface charge are carried out by potentiometric titrations. The isoelectric point was originally defined as that value of pH at which a protein or a polyfunctional polymer showed no electrophoretic mobility because the dissociation of the constituent amino and carboxylate groups provided equal numbers of opposite charges, giving a net zero charge. The term isoelectric point should denote the negative logarithm (base 10) of the *concentration* of *potential-determining ions* at which the electrokinetic potential becomes zero (that is, the electrophoretic mobility becomes zero). The term *reversal-of-charge concentration* is defined as the concentration of another additive (at *constant concentration of initially present potential-determining ions*) for which the electrokinetic potential is reduced to zero, and changes sign on further addition.

concentrations of Ag^+ greater than 10^{-5} M the zeta potential progressively increases due to continued adsorption of Ag^+—it is thus marked as positive—while for any addition of KI, the zeta potential is more negative than its value in pure water. Other ionic solids A^+B^-, in solutions of their lattice-constituents ions A^+ and B^-, may have their curve of zeta potentials shifted so that $\zeta = 0$ occurs along the A^+ or the B^- abscissa, but the overall trend of the zeta potential variation will be similar to that indicated in Figure 7.24, as long as the concentrations of added anionic B^- are plotted to the right and those of the cationic A^+ to the left of the point for pure water. The general conclusion is that lattice-constituent ions markedly affect the zeta potential values, causing charge reversal, either in an excess of anions or an excess of cations.

The effects of different electrolytes on zeta potential are well illustrated by the results of Buchanan and Heymann (1949), obtained with $BaSO_4$ (Figure 7.25). It is seen that some electrolytes are, for practical purposes, *indifferent*; their addition shows an insignificant change of zeta potential from the value determined for pure water (about $+26$ mV) as observed with the increasing concentrations of electrolytes such as chlorides of monovalent and of some divalent metals (Pb^{2+}, Ca^{2+}, Mg^{2+}). Other electrolytes,

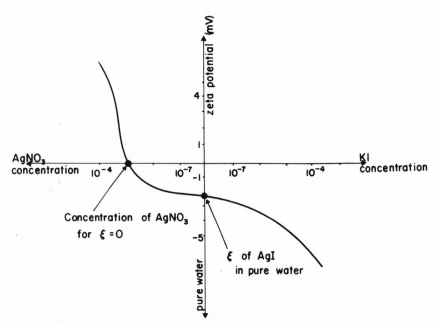

Figure 7.24. Zeta potential curve for an ionic solid, AgI, in solutions of its lattice-constituent ions. [After Lange and Crane (1929).]

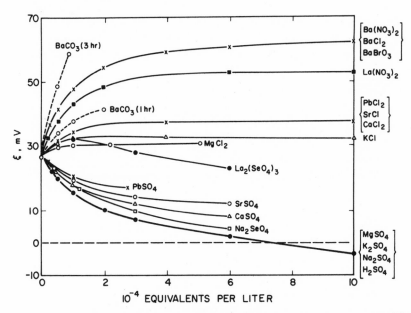

Figure 7.25. The effects of electrolytes (in 10^{-4} equivalent per liter) on zeta potential of recrystallized $BaSO_4$, evaluated from streaming potential determinations, for 20.88% wt. ethyl alcohol–water solution, all at 20°C. All anions which cause a charge reversal by crossing the zeta potential to the negative region are potential-determining ions. [After Buchanan and Heymann (1949), *J. Colloid Sci.* 4, with the permission of Academic Press, New York.]

namely carbonates and nitrates of Ba^{2+}, tend to increase the zeta potential of $BaSO_4$, as expected of the lattice-constituent cation. All sulfates tend to lower the zeta potential developed in pure water, with a clear indication of a reversal of charge achieved by the sulfates of Mg^{2+}, K^+, Na^+, and H^+. These results of Buchanan and Heymann are effectively a confirmation of the result shown in Figure 7.24 (if all Ba^{2+} salts were plotted using concentrations increasing toward the left of zero on the abscissa). In addition, they emphasize the conjoint action of both cation and anion adsorption in the electrical double layer of the solid/liquid interface. Analogous effects are observed in electrocapillary curves.

On the basis of the behavior recorded in Figures 7.24 and 7.25, the following generalization is hereby proposed regarding the zeta potential changes for insoluble inorganic solids in electrolyte solutions. Two types of solids are likely to be encountered, as shown in Figure 7.26. One type will have a positive zeta potential P in pure water and the other type a negative zeta potential N. (Solids having a zeta potential of zero value, i.e., at pzc in

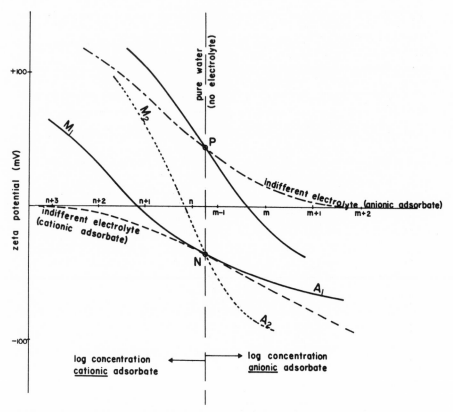

Figure 7.26. Schematic representation of zeta potential changes for two types of solids in electrolyte solutions containing specifically adsorbing cations or anions. One type of solid possesses a positive zeta potential (P) and the other a negative zeta potential (N) in pure water. Dashed lines represent (schematically) the effects of indifferent electrolytes. Increasing concentrations of cationic adsorbates M_1 and M_2 are plotted to the left of pure water and those of anionic adsorbates A_1 and A_2 to the right of pure water.

pure water, are also likely to be encountered, but these will be rare—exceptions rather than the rule.) When electrolytes are added, the concentrations of those which lower the zeta potential are plotted to the right of the line for pure water, and the concentrations of those that increase the zeta potential are plotted to the left of it. Indifferent electrolytes are not capable of causing a charge reversal (and potential reversal) by their adsorption, although they change the magnitude of zeta potential on adsorption. Specifically adsorbing anions cause a charge reversal but only for those solids which show a positive zeta potential in pure water; their specific adsorption cannot be ascertained for solids which show a negative zeta potential in pure water. Specifically adsorbing cations behave in an analogous way toward the solids, showing negative zeta potentials in pure water. An electrolyte M_1A_1 whose cation M_1 is specifically adsorbing to cause a charge reversal and a pzc at a concentration, say, $x \cdot 10^{n+1} M$ may

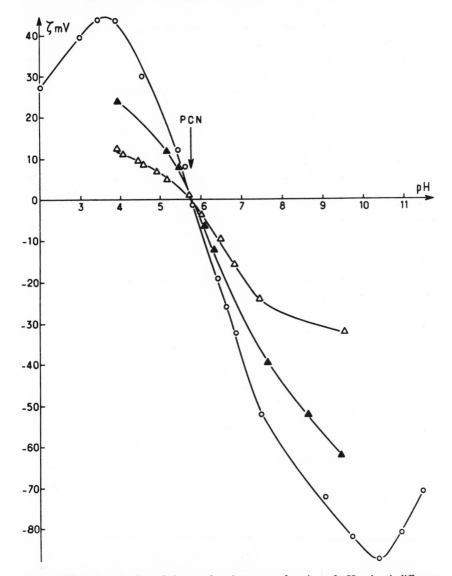

Figure 7.27. Determination of the pzc for zircon as a function of pH using indifferent electrolyte KCl. $\bigcirc$, without KCl; $\blacktriangle$ in $10^{-3} N$ KCl; $\triangle$ in $10^{-2} N$ KCl. PCN $\equiv$ pH$_{pzc}$. [From Cases (1967), with the permission of the author.]

shift the concentration of pzc if it is used as a salt of different anion, A_2, due to a possible conjoint adsorption (not indicated in the graph). A different specifically adsorbing cation M_2 endows the solid with charge reversal characteristics at a different concentration, $y \cdot 10^n$. The preceding effects of specifically adsorbing ions should be contrasted with the conjoint action of indifferent electrolytes, shown in Figure 7.27 for pH_{pzc}. Regardless of the concentration of indifferent electrolyte used, the value of the pH_{pzc} (i.e., the concentration at which the specifically adsorbing H^+ or OH^- ions reverse the charge) is not changed.

Insoluble metal oxides respond to solutions of different pH (different H^+ and OH^- concentrations) in a manner analogous to that of an ionic solid response to solutions of its lattice-constituent ions (because OH^- can be considered an equivalent of a lattice-constituent ion, oxygen, while H^+ may replace the metallic cation). Thus, each metal oxide shows with respect to a pH scale a trend similar to that indicated in Figure 7.24 with respect to the concentration scale of anions and cations, with the pH = 7 corresponding to pure water on the concentration scale. For a pH below a certain value, specific to each metal oxide and denoted as its pH_{pzc}, the zeta potentials are positive due to a preferential adsorption of H^+, while above this pH_{pzc} value the zeta potentials are negative, as indicated in

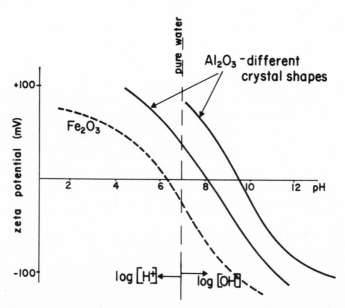

Figure 7.28. Schematic zeta potential curves of pure metallic oxides vs. pH showing the probable effect of different crystalline shapes on the zeta potential values of Al_2O_3.

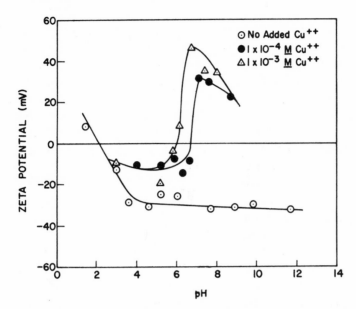

Figure 7.29. Zeta potential curves of crysocollar as a function of pH in the absence and the presence of Cu^{2+} in 10^{-4} and 10^{-3} M concentrations. [From Palmer *et al.* (1975a), *Trans. AIME* **258**, with the permission of the American Institute of Mining and Metallurgical Engineers, Littleton, Colorado.]

Figure 7.28. If the system is free from extraneous ionic contaminants which may specifically adsorb on the solid, there is only one pH_{pzc} for each metal oxide–pH system. In the presence of traces of extraneous metallic ions which adsorb in the Helmholtz layer, a shift in the pH_{pzc} to a higher pH occurs (Figures 7.29 and 7.30). If a surfactant that can specifically adsorb on the solid is present, a similar variation of pH_{pzc} and zeta potential occurs. Thus, in practical evaluations of a metal oxide–pH system whenever more than one pzc point is found, the unavoidable conclusion is that the system is more complex than the single metal oxide–pH or that accidental contamination has taken place during measurements.

Very useful compilations of zeta potentials and pzc data for various oxides and complex oxide minerals were published by Parks (1965, 1967) and Ney (1973). Parks points out that "a distressing multiplicity of observed pzc exists and the reasons for it are not fully understood; e.g., SiO_2 and Al_2O_3 have been the most extensively studied of all oxides and yet there is little agreement between the various determinations." The heterogeneity of solid surfaces, the nonstoichiometry of oxides tested, and any undetected impurities leading to specific adsorption are likely to be the causes of vari-

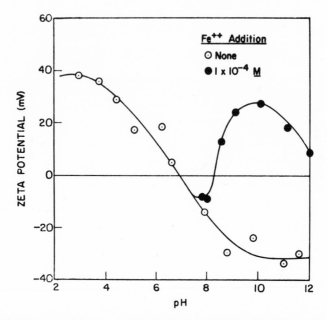

Figure 7.30. Zeta potential curves of chromite(II) as a function of pH in the absence and the presence of 10^{-4} M Fe^{2+} ions. [From Palmer *et al.* (1975b), *Trans. AIME* **258**, with the permission of the American Institute of Mining and Metallurgical Engineers, Littleton, Colorado.]

ation in the pzc determinations. The nonstoichiometry will give a space charge within the solid portion of the interface as well as cause different ions to leach out into the solution portion of the interface (during the so-called "cleaning" of the surface); consequently, the charges and their distributions for a solid with one particular degree of nonstoichiometry will be different (and so will the pzc value) from those for an analogous solid with a different degree of nonstoichiometry. When monodisperse solid particles of different crystalline morphology began to be systematically produced [see the publications by Matijevic *et al.*, reviewed by Matijevic (1978)], it became possible to determine unequivocally the extent of the crystalline face effect on zeta potentials. Two curves shown in Figure 7.28 for Al_2O_3 encompass the spectrum of values reported for the pzc of Al_2O_3 materials tested. The effects on zeta potentials due to various pretreatments (such as leaching, ultrasonic scrubbing, de-sliming) which are commonly applied to solids before electrokinetic measurements have been reviewed by Kulkarni and Somasundaran (1977).

Attempts have been made to predict the pzc values for various oxides either by evaluating the pH of minimum solubility of the given metal oxide or by considering a model of the surface-charging mechanism. Parks (1965, 1967) introduced the methods; Yoon *et al.* (1979) suggested several improvements. The comparisons tabulated by Yoon *et al.* (see Table 7.3) show some good agreement but also large discrepancies, so it is difficult to arrive, at present, at a definitive conclusion.

Some values of the pzc (taken from publications by Ney and Parks) are listed in Tables 7.3 and 7.4. Since many minerals show a range of pzc values,

Table 7.3. pH_{pzc} of Metallic Oxides[a]

Substance	Oxidation state of metal	Experimental pH_{pzc}	Calculated pH_{pzc}
Oxides			
WO_3	+6	0.2	0.34
Sb_2O_5	+5	0.2	—
SiO_2	+4	1.8	−1.85
TiO_2	+4	4.0 (<7)	6.35
MnO_2	+4	4.2	
HgO	+2	7.3	13.14
CuO	+2	9.5 (9.4)	8.06
SnO_2	+4	7.3 (6.6)	6.71
Fe_2O_3	+3	6.6, 8.2 (9.04)	9.45
NiO	+2	10.3	
UO_2	+4	5.8–6.6	
CdO	+2	10.4	
Cr_2O_3	+3	7.0	7.26
ZnO	+2	9.2 (9.3)	9.30
ZrO_2	+4	10.0 (10–11)	12.07
Al_2O_3	+3	5.0–9.2 (9.1)	9.10
BeO	+2	10.2	8.20
ThO_2	+4	9.3 (9–9.3)	10.46
Y_2O_3	+3	9.0	10.12
MgO	+2	12.4	
La_2O_3	+3	10.4 (12.4)	12.4
Sulfides			
$CuFeS_2$ (chalcopyrite)		1.8	
PbS (galena)		?	
FeS_2 (pyrite)		2.0	
FeS (pyrrhotite)		2.0	
ZnS (sphalerite)		2.0–2.3	

[a] After Parks (1967) and Ney (1973) (values in parentheses); calculated values from Yoon *et al.* (1979).

Table 7.4. Experimental pH_{pzc} of Selected Minerals[a]

Substance	Empirical formula	Experimental pH_{pzc}
Silicate group		
Anthophylite	$(Mg,Fe)_7(Si_8O_{22})(OH)_2$	3.8
Antigorite	Serpentine mineral	No pzc detected
Biotite	$K(MgFe)_3(AlSi_3O_{10})(OH)_2$	2.1
Bronzite	$(MgFe)_2Si_2O_6$	3.3
Chlorite	$Mg_3(Si_4O_{10})(OH)_2 Mg_3(OH)_6$	4.0–5.2
Chrysotile	$Mg_3Si_2O_5(OH)_8$	10.5
Cristobalite	Synthetic	1.9
Diopside (pyroxene)	$CaMg(Si_2O_6)$	3.2
Fayerite (fayalite)	Fe_2SiO_4	4, 5
Feldspar, microcline	$K(AlSi_3O_8)$	4.1 (2.4)
Feldspar, orthoclase	$K(AlSi_3O_8)$	4.8
Forsterite	Mg_2SiO_4	3.5–9.0
Glauconite	$K_2(MgFe)_2Al_6(Si_4O_{10})_3(OH)_{12}$	2.0
Hornblende	$Ca_2Na(Mg,Fe^{2+})_4(Al,Fe^{3+},Ti)_3$ $Si_8O_{22}(O,OH)_2$	No pzc detected
Kisselglas	Synthetic	2.5
Lepidolite	$K_2Li_3Al_3(AlSi_3O_{10})_2(OH,F)_4$	1.8
Muscovite	$KAl_2(AlSi_3O_{10})(OH)_2$	4.0
Olivine	$(Mg,Fe)_2SiO_4$	3.2, 11.1 (4.1)
Quartz	SiO_2	1.9
Tremolite	$Ca_2Mg_5(Si_8O_{22})(OH)_2$	No pzc detected
Vermiculite (altered biotite, see above)		2.3
Naturally hydrophobic minerals		
Anthracite		8.1
Graphite	C	2 (Ceylon) 4.2 (Burrowdale)
Molybdenite	MoS_2	1.0
Pyrophyllite	$Al_2(OH)_2Si_4O_{10}$	1.7
Sulfur	S	2.0
Talc	$Mg_3(Si_4O_{10})(OH)_2$	1.9

[a] After Parks (1967) and Ney (1973).

the safest procedure for any system under study is to determine the pzc in that particular system.

An extremely useful, simple, and rapid technique of the pH_{pzc} determination for metal oxides is described by Mular and Roberts (1966). An insoluble metal oxide added to a solution of a given pH causes a change

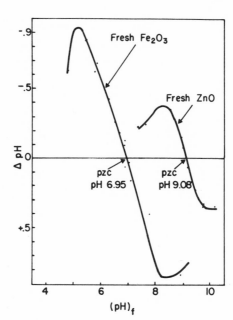

Figure 7.31. Determination of pzc for Fe_2O_3 and ZnO from the shifts of ΔpH vs. pH. Supporting electrolytes KNO_3 for Fe_2O_3 and NaCl for ZnO varied from 10^{-2} to 10^{-1} N. [After Mular and Roberts (1966), with the permission of the Canadian Institute of Mining and Metallurgy, Montreal, Quebec.]

in pH by ΔpH which is negative (due to the specific adsorption of OH^-) if the test solution is below the pH_{pzc} and positive if the test solution is above the pH_{pzc}. On plotting the values of ΔpH in relation to the *final* pH, a graph as in Figure 7.31 is obtained; the intercept for ΔpH = 0 gives the pH_{pzc}. Precautions to avoid contamination from the supporting electrolyte and from additives used to control the pH must be taken.

The foregoing discussion of the zeta potential and the electrical double layer is applicable to solids which do not undergo a continued dissolution or give deposition of a reaction product with an entirely different structure to that of the original solid. Otherwise, pronounced variation of the zeta potentials will be observed with time.

7.10. *Utility of Electrical Double-Layer and Electrokinetic Data*

The electrical double-layer theory has been verified, using electrocapillary and double-layer capacitance techniques, on the ideally polarized mercury/electrolyte interface. Later, Kitchener and Haydon (1959) extended its application to the water/air interface, and Haydon and Taylor, see Haydon (1964), extended its application to the water/hydrocarbon interface.

The significance of the electrical double layer is not, however, limited to the electrochemistry of these three types of interfaces but pertains to all systems that are electrochemical and colloidal in nature. In electrochemical systems the kinetics of electrode reactions are the object of applications and in colloidal systems the stability of dispersions and coagulation phenomena.

Two groups of scientists applied the electrical double-layer theory to explain the coagulation power of electrolytes and the stability of dispersed colloids. It is hence known as the DLVO theory, after Derjaguin and Landau (1941), who comprise one group, and Verwey and Overbeek (1948), who comprise the other group. The DLVO theory explains the stabilization of colloidal dispersions (sols, emulsions, aerosols) on the basis of repulsion caused by overlapping of the double layers surrounding the dispersed particles. At a constant potential on each particle surface, ψ_0, the repulsive energy due to the overlap is directly proportional to the dielectric constant ε, the particle size a, and ψ_0^2. Thus, when (for example) the particle size is 10^{-6} m (1 μm) and $\psi_0 = 14$ mV, the repulsive energy is of the order of $15\,kT$ at the distance of closest approach. For high dispersion stability, repulsion energy greater than $\sim 20\,kT$ is considered to be necessary. This corresponds to surface potentials $\psi_0 > 20\text{–}30$ mV, depending on the size of particles a, and the thickness of the electrical double layer $\varkappa^{-1}$, which is governed by the electrolyte concentration.

The DLVO theory has been developed for a model of electrostatic interactions between spherical particles in relatively low concentrations of electrolytes such that a mathematical treatment of the overlap of electrical double layers deals with potentials in the Gouy–Chapman diffuse layers. In real systems, usually containing a variety of ionic and dipolar species in the electrolyte, specific chemical interactions resulting in strong adsorption of some ions, particularly complexes, may occur on the solid surfaces. A dramatic change in the behavior of the system may ensue. Highly charged complexes may constitute only a small fraction of ionic content but may exert a major effect on dispersion stability or coagulation. The effects of chemical specificity make the quantitative ratios of critical coagulation concentrations for $z:z$ electrolytes inapplicable in all such systems. The critical coagulation concentrations of counterions are the minimum concentrations of counterions necessary to initiate fast coagulation. The DLVO theory predicts that for 1:1, 2:2, and 3:3 electrolytes these concentrations are in the inverse sixth power ratio of their valence; that is, $c_1:c_2:c_3 = 1/1^6:1/2^6:1/3^6$. This prediction quantified the so-called Hardy–Schulze rule, based on observations of a rather dramatic decrease in the concentration of polyvalent counterions needed for coagulation.

Complex formation may also be responsible for a spectacular decrease in the coagulating action of a multivalent ion, known as an *antagonistic effect*. For example, uncomplexed Al^{3+} is very active as a coagulant at $pH < 3$; however, in the presence of SO_4^{2-} ions, a complex $AlSO_4^+$ is formed, reducing the effectiveness of added aluminum ion as a coagulant in a linear proportion to the added SO_4^{2-} ions. Similarly, in the presence of F^- ions, complexes of the type $AlF_m^{(3-m)+}$ are formed, causing a corresponding reduction of free Al^{3+} ions and an alteration of expected effects.

For all aqueous solutions, the ubiquitous OH^- ion is the most important complexing agent. It is capable, on the one hand, of reducing the valence of highly charged cations by forming simple $M^n(OH)_x^{(n-x)+}$ complexes and, on the other hand, of forming *polymerized* complexes [such as $Al_7(OH)_{17}^{4+}$] of higher valence than that of uncomplexed metal ion. For details on their action in colloid stabilization, the review papers by Matijevic (1973a, 1973b, 1979) and the original references therein should be consulted. Some aspects of complex formation are dealt with in Section 10.4.

The electrokinetic phenomena are of primary significance in the stability of colloid dispersions as determined by adsorption of ions. Hence, the structure of the electrical double layer is responsible for the characteristics of the electrokinetic potential. Systems which do not lend themselves to study by electrochemical techniques can have some features of their electrical double layer approximated by electrokinetic techniques. At present, the exact position of the shear plane along which the zeta potential is evaluated in electrokinetic experiments is not known. The electrokinetic measurements are in general of relative value for the semiquantitative estimation of surface charge and adsorption. However, under specific conditions, double-layer polarization of isolated spherical particles should enable evaluations of the thickness of the stagnant layer (bound to the moving particle) and of ψ and ζ to be carried out from electrophoretic measurements.

For a thorough review of all aspects of electrokinetic phenomena, reference must be made to the papers by Dukhin (1974) and Dukhin and Derjaguin (1974).

As already mentioned, the stability and coagulation of colloidal dispersions are considered in terms of electrokinetic data for systems that cannot be tested by electrochemical techniques. In flotation systems, the adsorption of surfactants by electrostatic forces is directly related to electrokinetic characteristics.

7.11. Selected Readings

Bockris, J. O'M., and Conway, B. C., eds. (1954—1979), *Modern Aspects of Electrochemistry*, Vols. 1 and 2, Academic Press, New York, Vol. 3, Butterworth's, London, Vols. 4–13, Plenum Press, New York.

Bockris, J. O'M., and Reddy, A. K. N. (1970), *Modern Electrochemistry*, Vols. I and II Plenum, New York; see, in particular, Chapters 7–10.

Butler, J. A. V. (1951), *Electrical Phenomena at Interfaces*, Methuen, London.

Conway, B. E. (1965), *Theory and Principles of Electrode Processes*, Ronald, New York.

Delahay, P. (1965), *Double Layer and Electrode Kinetics*, Wiley-Interscience, New York.

Derjaguin, B. V., and Dukhin, S. S. (1974), Equilibrium (and nonequilibrium) double layer and electrokinetic phenomena, in *Surface and Colloid Science*, Vol. 7, ed. E. Matijevic, Wiley, New York, pp. 49–335.

Dukhin, S. S. (1974), Development of notions as to the mechanism of electrokinetic phenomena and the structure of the colloid micelle, in *Surface and Colloid Science*, Vol. 7, ed. E. Matijevic, Wiley, New York, pp. 1–47.

Frumkin, A. N., and Damaskin, B. C. (1964), Adsorption of organic compounds at electrodes, in *Modern Aspects of Electrochemistry*, Vol. 3, eds. J. O'M. Bockris and B. C. Conway, Butterworths, London, pp. 149–223.

Gerischer, H. (1970), Semiconductor electrochemistry, in *Physical Chemistry*, Vol. IXA, Electrochemistry, ed. H. Eyring, Academic Press, New York, pp. 463–542.

Ney, P. (1973), *Zeta Potentiale und Flotierbarkeit von Mineralen*, Springer, Berlin.

Parks, G. A. (1965), The isoelectric points of solid oxides, solid hydroxides, and aqueous hydrox. complex systems, *Chem. Rev.* **65**, 177–198.

Parks, G. A. (1967), Aqueous Surface Chemistry of oxides and complex oxide minerals, Advances in Chemistry Series No 67, *Equilibrium Concepts in Natural Water Systems*, ACS, Washington, D.C.

Parsons, R. (1954), Equilibrium properties of electrified interphases, in *Modern Aspects of Electrochemistry*, Vol. I, eds. J. O'M. Bockris and B. C. Conway, Butterworth's, London. pp. 103–179.

Payne, R. (1973), Double layer at the mercury–solution interface, in *Progress in Surface and Membrane Science*, Vol. 6, eds. J. F. Danielli, M. D. Rosenberg, and D. A. Cadenhead, Academic Press, New York, pp. 51–123.

Verwey, E. W., and Overbeek, J. T. H. G. (1948), *Theory of the Stability of Hydrophobic Colloids*, Elsevier, Amsterdam.

8

Adsorption of Flotation Collectors

A hydrophobic solid such as talc, graphite, paraffin, molybdenite, sulfur, high-rank unoxidized coal, etc., does not require a collector for its separation by flotation unless contaminated by polar species so strongly that its surface turns hydrophilic. To improve the kinetics of flotation, a hydrophobic solid does require an addition of a frother-acting dipolar surfactant; occasionally, an addition of a nonpolar oil is also needed to improve the solid's hydrophobicity if it becomes partially reduced by contamination.

Qualitatively, a hydrophilic solid is described as being completely wetted by water and nonwetted by nonpolar oils, not attaching air bubbles when submerged in water. A hydrophilic solid is polar in nature, endowed with dipoles within its surface layer and, usually, possessing an excess charge on its surface. However, lack of an excess charge (as under conditions established at the pzc) is not *per se* sufficient to furnish the surface with a hydrophobic character. The latter is determined thermodynamically by the relative values of the free surface energies for the three participating interfaces. In practice, only semiquantitative evaluations of hydrophobicity are provided by the expressions for the negative spreading coefficient (work of nonwetting) or the contact angle—since these are usually modified by energy barriers pertinent to the system and the technique employed for evaluations.

Splitting the work of adhesion of water to a solid into three terms $W_A = W^d + W^h + W^i$, the dispersion term W^d, the hydration of nonionic sites W^h, and the contribution from ionic sites W^i, Laskowski and Kitchener (1969) analyzed their data on the contact angle and zeta potential of methylated silica (and of pure silica) and came to the following conclusions:

1. The hydrophilic state of methylated silica, re-established after slow rehydration, i.e. after lengthy contact with water, is due to water molecules which are only physically bound to the surface.

2. Heating pure (unmethylated) silica to 650°–850°C in order to remove the chemisorbed water converts silica to a markedly hydrophobic solid, and several hours of contact with water are needed for a complete rehydration of such silica, indicating that the Si—O—Si (siloxane) group is intrinsically hydrophobic.

3. The presence of the electrical double layers does not exclude hydrophobicity (though it affects the thickness of the wetting film on a silica surface) since methylation of the silica surface had no effect on the zeta potential values under all pH conditions in comparison with the zeta potential of pure (unmethylated) silica. The authors estimated that 54% of the silica surface was covered with the —$Si(CH_3)_3$ groups; utilizing their own contact angle data and the dispersion force contributions of Fowkes (1964), they concluded that the hydration term W^h is responsible for wettability, while hydrophobicity arises when W^h and W^i are small in relation to W^d; thus, the hydrophobic character of solids arises from the instability of a water film (which is deficient of hydrogen bonding in comparison with liquid water) adjoining a layer of highly structured water adsorbed onto the solid. [*Cf.* p. 168—hydrophobic solutes promote cluster formation in liquid water; and cf. p. 419—a three-layer model of water near a solid.]

To convert a hydrophilic solid into hydrophobic, the polarity of its surface must be reduced below the value of the polarity of water. This situation may be realized in some instances by a special rearrangement of surface atoms or by adsorption of suitable adsorbate species. The most dependable method is the adsorption of an appropriate surfactant, which replaces some portions of the polar substrate with a layer of a nonpolar hydrocarbon-like surface. Surfactants capable of an adsorption such that the solid surface is converted from hydrophilic to hydrophobic in character are referred to in flotation systems as collectors.

8.1. *Requirements for Collector Adsorption*

From past experience it is well known that not every surfactant can act as a collector for a given solid. What are the requirements that have to be fulfilled before collector action by the particular surfactant is assured?

The criterion of partial wetting of the hydrophobic surface connotes a resistance to surface displacement of the adsorbate at the line of contact between the three phases solid–liquid–gas. A surfactant that converts a hydrophilic surface to a hydrophobic one must be adsorbed in such a way that it is *relatively* immobilized. This immobilization denotes the ability to withstand movement of the liquid along the shear plane adjacent to the solid without displacement; otherwise a complete wetting of the solid would ensue.[†] In addition, the initially high attraction of the surface (over a specific patch or over several scattered patches on the surface) toward water molecules must be reduced below a certain critical value.

It appears that there are at least three ways of immobilizing a collector-acting surfactant at a hydrophilic solid/liquid interface:

1. Chemisorption involving a charge transfer (but not causing an undue perturbation in the substrate)
2. Specific adsorption in the IHP layer (without charge transfer)
3. Electrostatic adsorption over an immobilized network of hydrogen-bonded counterions at OHP or at a shear plane within the Gouy–Chapman layer

Each of these constitutes a localized adsorption. However, there are considerable differences in relative energies of adsorption and in the manner in which different forms of adsorption may affect the residual free interfacial energy of scattered patches on the solid substrate.

With heterogeneous solid surfaces, localized chemisorption takes place on active sites: kinks and ledges, emerging dislocations, and lattice defects. High energy of adsorption may lead to a perturbation of the surfaces[‡]; it may produce a definite distortion and, occasionally, may even break the chemical bonds between the atomic species constituting the adsorption site and the rest of the solid. The latter aftermath of chemisorption may have significant consequences in dynamic systems such as flotation.

[†] The requirement of relative immobilization of a collector species at the solid surface appears to conflict with the finding by Hassialis and Myers (1951) of unexpected mobility of xanthate preadsorbed on a galena surface and its transfer onto a fresh galena surface introduced into the system. Such mobility can readily occur with physically coadsorbed secondary adsorbates whenever the primary adsorption bonds between the polar groups and the substrate are stronger than the bonds of physically coadsorbed entities.

[‡] *Note*: As already referred to in Section 6.8, Germer and MacRae (1962a,b) found with LEED (low-energy electron diffraction) a complete *restructuring* of the Ni surface as a result of H_2 adsorption—with a return to the original surface structure once H_2 is desorbed on raising the temperature. The condition of minimum energy to be attained by the adsorbate and adsorbent atoms may lead to novel, highly complex configurations. New *surface states*, with distinct properties, are theoretically possible, and some evidence for their existence has been obtained with chemisorbed gases on metals using the flash desorption technique [Ehrlich (1963a, 1963b), Samorjai (1972), Murr (1975), Prutton (1975)].

An example of the breakage of bonds on chemisorption is the catastrophic oxidation of Mo, W, V, and their alloys at temperatures greater than 725°C. Liquefied oxides are formed—MoO_3 (m.p. 795°C), V_2O_5 (m.p. 675°C), WO_3 (m.p. 1470°C)—which readily volatilize when certain minimum temperatures are exceeded [see Kubaschewski and Hopkins (1962, pp. 226, 278)]. Another example is the removal of C from the surface of W by adsorption of O_2; similarly, adsorption of O_2 on coal and its reaction with C and H to give volatile CO and H_2O provides clear evidence for breakage of bonds between the adsorption site and the rest of the solid. Any dissolution process at a solid/liquid interface taking place on addition of a specific solvent or a solute is, effectively, an adsorption process leading to breakage of bonds between the adsorption site and the rest of the solid. This type of bond breakage differs from a true desorption in that the *adsorption bond* itself is not broken during the removal of the surface complex. Bonding by electrostatic forces is generally stronger than dipole–dipole bonding on a *per bond* or *per mole of bonds* basis. Yet multiple dipole–dipole interactions of small water molecules often break individually stronger ionic bonds of various solids, causing their dissolution, because of relative differences in the number of bonds *per unit area* of solids.

Hence, the concept of *relative immobilization* of collector molecules imposes a condition such that of the various adsorption "states," as exemplified in Figure 8.1, only the states within the intermediate energy levels, that is, within the range denoted by F, may be appropriate for flotation purposes. A very strong adsorption, as in the chemisorbed state at $\theta \cong 0$, may lead to an excessive weakening or even a breakage of residual bonds between the adsorption site and the solid. Some species, either physically adsorbed or those in the chemisorbed state within the range, say, $\theta = 0.8$ and $\theta = 1$, may be too weakly adsorbed to withstand the external forces imposed during attachment to bubbles. The immobilization implies that the energy for translation from one adsorption site to another is of appropriate value with respect to F in Figure 8.1. In this situation, a nonpolar molecule may become relatively immobilized when a *sufficient number* of —CH_2— groups of this molecule develops multiple van der Waals bonds with a nonpolar substrate even though each individual —CH_2— group is very weakly bonding by itself. Similarly, development of any lateral bonding, whether van der Waals bonds between the molecularly associated hydrocarbon chains or lateral hydrogen bonding (between the hydrated polar groups or hydroxylated counterions acting as adsorption sites), may lead to an immobilization of the adsorbate species.

In view of the major differences in the properties of surfactants, their

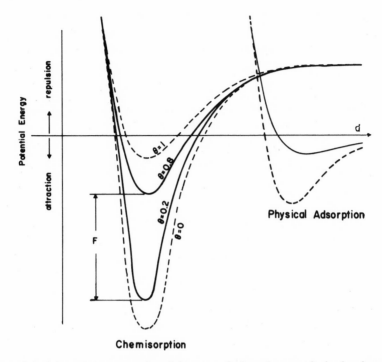

Figure 8.1. Schematic representation of the range of adsorption energies in chemisorption and physical adsorption of surfactants conducive to flotation.

adsorption may be discussed relative to their classification, i.e., adsorption of thio compounds, adsorption of non-thio ionizable surfactants, etc., or—as is customary in the flotation literature—relative to the group of solids being floated such as sulfides, oxides, silicates, soluble salts, etc. Due to the complexity of flotation systems, no classification approach is sufficiently unique to pertain to a particular group chosen for discussion. For example, clearly hydrophobic sulfides such as molybdenite, MoS_2, and stibnite, Sb_2S_3, or a hydrophobic silicate such as talc, $Mg_6(Si_8O_{20})(OH)_4$, do not adsorb surfactants through the polar group, as do all hydrophilic sulfides or silicates, and do not require collectors unless contaminated.

The approach adopted here for the discussion of collector adsorption combines the manner of adsorption, the requirements of the flotation process (imposed by the dynamics of the system), and some applications to solids most frequently floated in the mineral industry.

The following requirements for adsorption of a surfactant to act as a collector are considered:

1. Relative immobilization of the surfactant species (as already discussed).

2. Development of a sufficiently hydrophobic character of the solid particles to be floated such that it can "survive" the mechanical and dynamic effects of the particular flotation system

3. Selection of a surfactant–modifying agents combination that would permit control of adequate selectivity

4. Suitable choice of a collector–frother combination that would permit satisfactory kinetics of separation to be achieved

The first two requirements are basic for the selection of an appropriate collector-acting surfactant. They impose conditions on the polar group which develops bonding with the solid and on the nonpolar group that regulates subsequent development of hydrophobicity. These two requirements may be met by selecting a surfactant species incorporating appropriate polar and nonpolar groups or by a choice (intentional or accidental) of two species, one fulfilling the requirement of developing the primary bond with the substrate and the other producing a sufficient degree of hydrophobicity by secondary (mostly van der Waals) bonding. Requirement 3 determines which of the available modifiers can be used to improve selectivity; if no suitable modifiers can be found, the surfactant is not a useful collector, however successfully it meets the first two requirements. Similarly, as regards requirement 4, if a collector causes such hydrophobicity that solid particles agglomerate preferentially and no appropriate frother can be found to foster their attachment to air bubbles, that reagent is also not a suitable collector (at least temporarily).

8.2. Mechanisms of Xanthate Adsorption

From the discussion on aqueous xanthate systems in Section 5.2 it is apparent that—depending on conditions—a large number of xanthate-derived species may be present in the pulp in various relative proportions viz.,

1. Xanthate ion, $ROCSS^-$ (or X^-)
2. Xanthic acid, $ROCSSH$ (in acidic conditons)
3. Dixanthogen, $ROCSS$—$SSCOR$
4. Monothiocarbonate ion, $ROCOS^-$, and its oxidation product (5)
5. Carbonate disulfide, $ROCOS$—$SOCOR$

   6. Perxanthate ion, $ROCSSO^-$, and if divalent M^{2+} (or trivalent M^{3+}) metal ions are present, dissolved molecules of undissociated metal perxanthates

   7. Metal xanthates, $ROCSS-M-SSCOR$

   8. Metal monothiocarbonates, $M(SOCOR)_2$

In addition, under the conditions of highly nonstoichiometric concentrations of metal M^{2+} ions and xanthate X^- ions, metal xanthate complexes may be present:

   9. $M(X)^+$

   10. $M(X)_3^-$

The relative proportions of the preceding species will be determined by the kinetics of various oxidation and complex formation reactions, which, in turn, are governed by the following:

   1. pH of the pulp

   2. Oxidation level—dissolved O_2, oxidizing and reducing species in the pulp

   3. Metal ions available in the bulk of solution and particularly in the vicinity of particle surfaces

   4. Presence of any catalytic species fostering a specific reaction

   In view of such an unusually large variety of species that may be available in any xanthate flotation system, it is very unlikely that a single mechanism of adsorption is operative. It is not surprising, therefore, that numerous studies carried out by many investigators since about 1930 have been throwing progressively more and more light on the complexity of the xanthate systems without producing a general agreement on the mechanism of adsorption.

   A detailed treatment of earlier work on the role of xanthate in flotation and on its adsorption mechanisms is available in the texts by Sutherland and Wark (1955), Gaudin (1957), and Klassen and Mokrousov (1963). The reviews dealing with later developments are those of Rogers (1962), Joy and Robinson (1964), Wottgen (1967), Hopstock (1969), Rao (1969) Granville *et al.* (1972), and Gutierrez (1973). The review by Granville *et al.* (1972) is the most exhaustive and authoritative and should be consulted in conjunction with the paper by Gutierrez (1973) for details of studies carried out since 1957 on the chemistry of xanthates, oxidation of galena, thermodynamic, electrochemical, and spectroscopic evidence of galena–

xanthate interactions (207 references). A later paper by Leja (1973) is somewhat complementary to the reviews of Granville *et al.* and Gutierrez, as it deals with chemical and electrochemical work that was not accessible to these reviewers.

Basically, three different theories on the mechanism of the action of xanthates and, generally, collectors in flotation have been advanced:

1. Adsorption theory [Langmuir (1920), Taggart and Gaudin (1923), Gaudin (1928), and Wark and Cox (1934)]
2. Ion exchange theory [Taggart *et al.* (1930)]
3. Neutral molecule adsorption theory proposed by Cook and Nixon (1950).

The ion exchange theory of Taggart *et al.* was based on the determinations of a near-stoichiometric balance between xanthate ions abstracted by galena from solutions and the combined sulfate, carbonate, thiosulfate, and thionate ions released by galena surfaces into solution. The extent of initial oxidation of galena surfaces controlled the amounts of lead salts available for exchange with xanthate ions.

Cook and Nixon (1950) pointed out that all results regarding xanthate adsorption isotherms, contact angles on polished sulfides, and flotation results appeared to be equally well explained by postulating adsorption of nonionic collector species. The advantage of a nonionic adsorbate was that its adsorption was not hindered by excess charges in the electrical double layers surrounding solid particles.

As a result of studies carried out between the 1940s and the 1960s, xanthate adsorption became generally accepted as being a chemisorption and yet leading to multilayer formation (this constitutes an apparent contradiction until the multicomponent nature of the adsorbed film and physical coadsorption of secondary adsorbates are recognized). Using autoradiography, Plaksin *et al.* (1957) clearly established that a nonuniform, patchlike distribution of adsorbates is created on the mineral surface (Figure 8.2). Previous to this finding, the hydrophobic character of the surface covered with the xanthate species was judged by the contact angle technique, and a monolayer coverage was inferred from the fact that a constant value of contact angle was obtained for all concentrations exceeding a critical (very low) concentration. Wark and Cox (1934) have shown that maximum contact angle was determined by the length of the alkyl chain and was independent of the nature of the substrate. Gaudin and Schuhmann (1936) established that chalcocite abstracts amounts of xanthate equivalent to

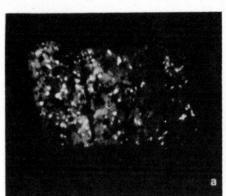

Figure 8.2. Autoradiography of adsorbed surfactants: (a) xanthate (white spots) on the face of a sphalerite particle, 500×, [Plaksin *et al.* (1957), from Klassen and Mokrousov (1963), *An Introduction to the Theory of Flotation*, Trans. by J. Leja and G. W. Poling, with permission of Butterworths, London], and (b) xanthate (dark spots) on the surface of a galena particle, approximately 400-μm size, both showing a nonuniform distribution of adsorbates aggregated in multilayer patches. [Plaksin *et al.* (1956), from Klassen and Mokrousov (1963), *An Introduction to the Theory of Flotation*, Trans. by J. Leja and G. W. Poling, with permission of Butterworths, London.]

several multilayers and that cuprous xanthate can be leached out from this adsorbed layer, leaving an unleachable species equivalent to a monolayer.

Eyring and Wadsworth (1956) were the first to apply the IR spectrophotometry technique to study adsorption of surfactants at the solid/liquid interface (after emersion from the surfactant solution) and found that hexanethiol chemisorbs on the surfaces of zincite (ZnO) and willemite ($ZnSiO_4$).

As the results, obtained by various research groups using different techniques, began to multiply, more controversies were generated. The nature of the adsorbate, the role of oxygen, the effects of excessive oxidation, and the role of metal xanthates were all questioned and argued, and many still remain controversial to a greater or lesser extent. The views to be expressed here are therefore primarily the views of the author.

8.2.1. *The Nature of the Adsorbate Species*

Of the 10 (or so) possible surfactant species in a xanthate–sulfide flotation system, the ones which are most likely to be present under conditions of air access to an alkaline pulp are

Xanthate ion, ROCSS⁻

Dixanthogen, ROCSS—SSCOR

Monothiocarbonate, ROCOS⁻

Metal xanthates, MX_2

and, occasionally, metal–xanthate complexes, $M(X)^+$ or $M(X)_3^-$. Under acidic conditions, xanthic acid may also be formed.

The work of Pomianowski (1967) using capacity–potential determinations in a mercury–xanthate system (Figure 7.19, Section 7.5) suggests that electrostatic (physical) adsorption of *xanthate* ion takes place first, even on a negatively charged metallic surface (mercury), without any charge transfer being initially involved. Then, at a higher potential, a single charge transfer reaction (denoted by III in Figure 7.20) takes place, signifying chemisorption of xanthate ion. This is followed by a second charge transfer, reaction IV, interpreted by Pomianowski as the formation of adsorbed dixanthogen. A subsequent two-electron transfer step is due to the formation of metal xanthate (in the case of mercury, mercuric xanthate).

The spectroscopic evidence of Poling (1963) (Figure 8.3) showed that in the absence of oxygen in the system, xanthate ions [X⁻] cannot chemisorb (though they may adsorb physically without a change in spectrum) on an evaporated PbS film. In contrast, dixanthogen can chemisorb under these conditions. Chemisorption of dixanthogen results in the formation of a Pb—X in the first layer of adsorbate, and in multilayers of coadsorbed dixanthogen X_2 if this species is present in sufficient excess relative to the available surface area of PbS. The spectrum of the first Pb—X layer is identical with the spectrum left on the surface of PbS (onto which xanthate ion was allowed to adsorb from an *oxygen-containing* system) after dissolution of multilayers of PbX_2.

The electrochemical studies, starting with those of Frumkin *et al.* (1932), Kamienski (1931), Salamy and Nixon (1953, 1954), Pomianowski (1957), Nixon (1957), Tolun and Kitchener (1964), Majima and Takeda (1968), Toperi and Tolun (1969), Woods (1971, 1972a and b), and Gardner and Woods (1973), have established a number of important relationships among the electrode potential, pH of solution, concentrations of xanthate and oxygen, etc. The more significant of these findings are the following:

1. Reactions at interfaces in the absence of oxygen are appreciably different from those in the presence of oxygen in the system.

2. Owing to the heterogeneity of surfaces, areas of different electrochemical potential coexist next to each other. Szeglowski (1960) found

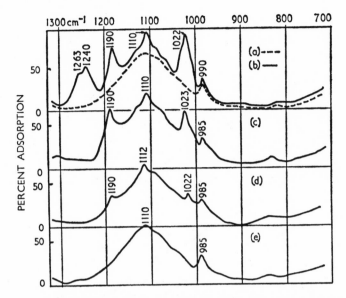

Figure 8.3. Infrared spectra indicating ethyl dixanthogen adsorption onto evaporated PbS film. (a) Spectrum of evaporated PbS film showing the presence of oxidation products on its surface. (b) The same film after treatment with an aqueous emulsion of ethyl dixanthogen. (c) The preceding film with adsorbed dixanthogen after washing in ether. (d) The preceding after washing in boiling benzene. (e) The preceding after washing in pyridine. [From Poling (1963), with the permission of the author.]

differences in potential on a pure galena surface ranging from 10 to 150 mV; if incompletely liberated mineral phases coexist in the surface in a galvanic contact, local potential differences of up to ~0.5 V can develop.

3. Using a small pneumatic flotation cell adapted for simultaneous electrochemical measurements and control, Gardner and Woods (1973) confirmed for solid particles (metal and sulfide) the correlations between electrochemical potentials and floatability established previously by Pomianowski (1967) for liquid mercury in xanthate solutions. They established that flotation of metallic Au or galena powder is not induced until the potential of the solid/liquid interface is above that for PbX_2 formation (in the case of galena flotation) or above the reversible potential for oxidation of xanthate ions to dixanthogen (Figure 8.4).

This result corroborated the IR spectroscopic evidence that the adsorbed Pb—X and PbX_2 are the main species on PbS. Allison and Finkelstein (1971) detected only ~1% dixanthogen in the PbX_2 extracted from a galena surface (and this dixanthogen may have been formed during extraction).

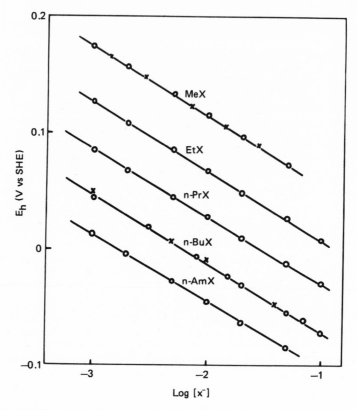

Figure 8.4. Redox potentials of the alkyl xanthate–dialkyl dixanthogen couples for methyl-*n*-amyl xanthates. Pt electrode, He bubbling, 25°C, pH 9. [From Majima and Takeda (1968), *Trans. AIME* **241**, with the permission of American Institute of Mining and Metallurgical Engineers, Littleton, Colorado.]

   In light of the preceding electrochemical evidence, it appears clear that chemisorption of xanthate species per se does not suffice for flotation. Either a metal xanthate or dixanthogen, formed at the interface, has to coadsorb on the first layer of chemisorbed species, Pb—X or Cu—X (or presumably Au—X), before flotation is induced. Thus, the degree of hydrophobic character of the adsorbed xanthate is furnished and controlled by its metal xanthate or dixanthogen, physically coadsorbed, and not by the chemisorbed Pb—X alone (for the short chain xanthates used).

   Further, the nature of the adsorbate species is not restricted to dixanthogen alone or to xanthate ion, but either of these two species can and does act as adsorbate, depending on conditions and availability. If the sulfide surfaces are too highly charged negatively and dixanthogen is formed or is added as such to the system, it will act as adsorbate, whereas xanthate ion *may* not be capable of adsorption under these conditions. However, it does not appear necessary to oxidize xanthate ion to dixanthogen in order to initiate its adsorption, so dixanthogen formation is not a prerequisite of adsorption. Flotation of galena powder was observed by Gardner and Woods (1973) at potentials about 100 mV lower than the reversible potential for dixanthogen formation—but this potential was above the potential required for lead xanthate formation.

When xanthate ion is available and can chemisorb, it will do so without being converted to dixanthogen. However, with short-chain alkyl xanthates the surface will not become hydrophobic until the chemisorbed species Pb—X acquires enough PbX_2 to improve the water-repellant characteristics of the first layer. The requisite amount of PbX_2 necessary to make the particles of galena hydrophobic will not be available if the potential is too low and/or the conditions are such that other products are formed in preference to PbX_2.

Toperi and Tolun (1969) prepared an electrochemical phase diagram, E_h vs. pH, representing regions for equilibria of reactions between galena and ethyl xanthate, reactions that can be evaluated from available thermodynamic data (Figure 8.5). [For details of the preparation and interpretation of such E_h–pH diagrams, the reader is referred to Pourbaix (1949), who introduced the graphical method of representing thermodynamic data for metal–solution systems, and to Garrels and Christ (1965), who applied this technique to mineral–solution systems.] Depending on the concentration of xanthate ions, at pH > ~10.5 the formation of PbX_2 should cease (thermodynamically), and $HPbO_2^-$ and $S_2O_3^{2-}$ ions should form preferentially. Indeed, all flotation data in simple galena–xanthate systems indicate an abrupt cessation of floatability, contact angle development, etc., at such a pH. Galena flotation may still be possible at higher potentials (and pH > 10.5) when X_2 formation provides the necessary replacement for PbX_2 to make the surface hydrophobic or at potentials below X_2 formation and pH > 10.5 in complex systems when other hydrophobic species may be available to replace PbX_2 in its hydrophobicity-making function. Unfortunately, no detailed and carefully executed experiments seem to have been performed as yet to distinguish unambiguously the regions and the conditions under which adsorption to form a partial monolayer of Pb—X can be clearly differentiated from the regions where lack of flotation is due to an absence of hydrophobic supplementary species.

Despite the ease with which xanthate ion species is decomposed under acidic conditions (Section 5.2), adsorption and flotation in the acid region

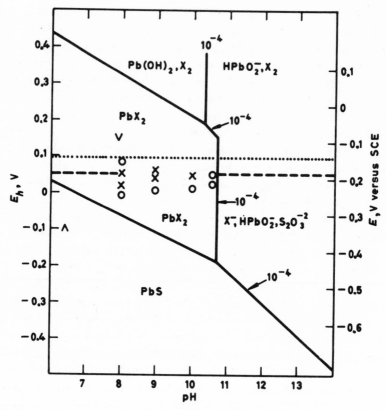

Figure 8.5. Potential Eh–pH relationships for a galena–xanthate–oxygen system. ×, Strong bubble contact; ○, feeble bubble contact; ∨, oxidizing depression by potassium chromate; ∧, reducing depression by sodium sulfite–ferrous sulfate mixture; ···, xanthate (10^{-3} M)–dixanthogen (saturated); – – –, galena–dixanthogen system at equilibrium with 10^{-3} M xanthate ion. [From Toperi and Tolun (1969), *Trans. IMM* **78**, with the permission of the Institution of Mining and Metallurgy, London.]

of pH do occur because of differences in kinetics of decomposition and of adsorption. Fuerstenau *et al.* (1968) have provided data on the recoveries of pyrite in microflotation experiments showing the effects of pH, of different xanthate concentrations, and of an addition of Fe^{3+} (Figures 8.6–8.8). These authors conclude that dixanthogen adsorption is primarily responsible for flotation of pyrite. Indeed, an addition of dixanthogen does lead to flotation—a complete recovery in a low pH region and a partial recovery under mildly alkaline conditions. However, regardless of the nature of the species used, whether dixanthogen or xanthate ion, when low concentrations are employed, two humps in the recovery curve vs. pH are observed (Figures

8.6 and 8.7). These two humps suggest that either two distinct mechanisms of adsorption (for example, chemisorption and specific adsorption in the IHP) or two separate species supplementing the hydrophobicity of pyrite surfaces may be involved in the two respective regions of pH (or a combination of both these possibilities). Lacking additional detailed information on the state of oxidation of the pyrite system, freedom from traces of dixanthogen in xanthate solutions used, etc., any explanation of the humps appearing in the recovery curves is speculative. On the basis of Sheikh's (1972) work on Mossbauer spectroscopy of iron xanthates, it would appear that, in Figure 8.6, FeX_3 may be responsible for the improved recoveries at pH ~ 2 and for the shift of the recovery curve toward lower pH, whereas the formation of $Fe(OH)X_2$ may be the cause of lower hydrophobicity and lower recoveries at pH 8–10. A complete lack of recovery (depression) of pyrite between pH 4.5 and 6.5 in Figure 8.8 may be due to the elimination of the particular xanthate species which is capable of primary adsorption in the first layer or that which improves hydrophobicity. This lack of recovery

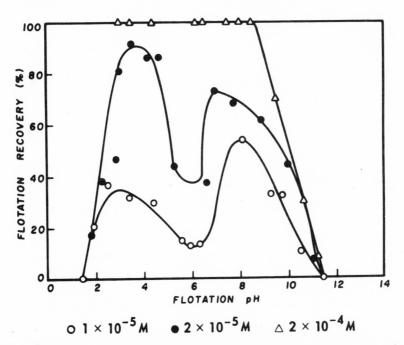

Figure 8.6. Flotation recovery of pyrite in relation to pH using $1 \times 10^{-5}\ M$, $2 \times 10^{-5}\ M$ and $2 \times 10^{-4}\ M$ solutions of potassium ethyl xanthate. [From Fuerstenau *et al.* (1968), *Trans. AIME* **241**, with the permission of the American Institute of Mining and Metallurgical Engineers, Littleton, Colorado.]

is obviously highly dependent on relative quantities of Fe^{3+} and X^-, so it may be associated with $Fe(OH)X^+$ complex formation. Until more details become available, it is impossible to determine the cause of depression.

It is suggested, however, that the mechanism of adsorption in the acid region is that of chemisorption of xanthate ion and/or dissociative chemisorption of dixanthogen, whichever of these two species is present. In the alkaline region, the physical adsorption of the ionized complex $Fe(OH)X^+$ on the negatively charged pyrite surface appears to be involved, reinforced by lateral hydrogen bonding and the physically coadsorbed $Fe(OH)X_2$ species. Due to similarities of IR spectra of X_2 and different $Fe—X$ compounds, no clear differentiation of adsorbed species is possible with the help of the IR technique alone—a number of additional techniques will have to be resorted to.

Generally, no direct evidence is available at present as to adsorption characteristics of metal xanthate complexes, $M(X)^+$ and/or $M(X)_3^-$. It can only be speculated that, when available, these species are likely to participate in depositing xanthate on sulfides.

Monothiocarbonate ions present in the pulp are derived primarily from

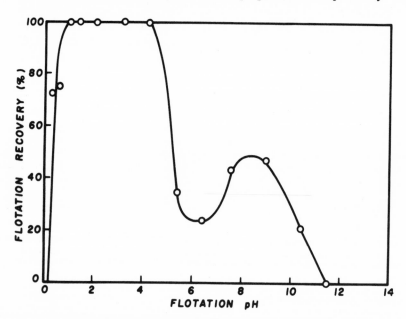

Figure 8.7. Flotation recovery of pyrite as a function of pH in solutions of $1.3 \times 10^{-5}\ M$ diethyl dixanthogen. [From Fuerstenau et al. (1968), Trans. AIME **241**, with the permission of the American Institute of Mining and Metallurgical Engineers, Littleton, Colorado.]

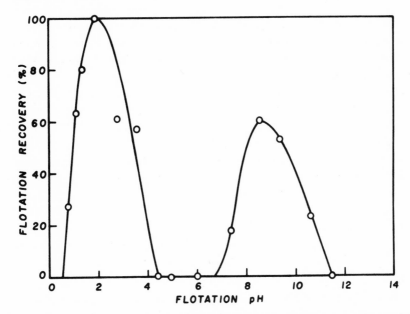

Figure 8.8. Flotation recovery of pyrite as a function of pH using solutions of 2×10^{-5} *M* potassium ethyl xanthate and 4×10^{-5} *M* FeCl$_3$. [From Fuerstenau *et al.* (1968), *Trans. AIME* **241**, with the permission of the American Institute of Mining and Metallurgical Engineers, Littleton, Colorado.]

metal xanthate decomposition; their readsorption on solid sulfides has not been investigated. They are undoubtedly more hydrophilic than the corresponding dithiocarbonate (xanthate ions); therefore only monothiocarbonates with longer alkyl chains may find some applications in future if their polar group is capable of specific adsorption on selected solids.

8.2.2. The Critical Role of Oxygen in Sulfide Flotation Systems

Once it is recognized that coadsorption of metal xanthates or dixanthogen is necessary to establish the hydrophobicity of a short-chain xanthate monolayer, the critical role of oxygen needed for flotation becomes understood. Oxygen is necessary to provide the requisite quantities of metal ions for metal–xanthate formation (at or near the surface of sulfide particles) by oxidation reactions that require a suitable electron acceptor. In other terms, oxygen allows a sufficient extent of charge transfer to take place on the sulfide surface. Without oxygen being available in the system, the charge transfer is stifled, and the particles with sparsely chemisorbed short-chain xanthate (a limited exchange of charge transfer may occur at the surface) possess an insufficient degree of hydrophobic character. Only when a continued charge transfer has occurred to release enough M^{2+} ions in the immediate vicinity of surfaces are these surfaces likely to become hydrophobic by coadsorption of metal xanthates formed near the interface.

Unfortunately, due to a mistaken notion regarding the hydrophilic nature of metal-xanthate precipitates, the true role of these species in converting some mineral surfaces into hydrophobic (by coadsorption) had not been previously recognized. Only recently the hydrophobic nature of metal-xanthate precipitates has been definitely confirmed [see Poling (1976)] and their role in improving the hydrophobic character of the chemisorbed xanthate appreciated.

The metal-xanthate group itself (if the alkyl group is temporarily disregarded) can be considered to have a varying degree of "hydrophobic" character depending on the difference in electronegativities between the sulfur and the metal ions. Poling (1963) related the IR group frequency shifts for the $C{=}S$ and $C{-}O{-}C$ bands to electronegativities, as shown in Figure 5.14; the same correlation may be used for gauging the hydrophobic–hydrophilic character of the metal–xanthate grouping. Dixanthogen, for which the difference in electronegativities is zero, is at one end of the scale as the most "hydrophobic" of the xanthate species (despite its apparently high dipole value). On the other end of the scale are the metals with the greatest difference in electronegativities, that is, the highly soluble and hydrophilic alkali xanthates. The heavy metal xanthates Hg_2X_2, Cu_2X_2, PbX_2, and ZnX_2 (X denotes the alkyl + xanthate group) show progressively increasing differences in electronegativities between S and the metal ion; accordingly, the hydrophobic character of these species progressively decreases: $Hg_2X_2 > Cu_2X_2 > PbX_2 > ZnX_2$. For all practical purposes ZnX_2 is hydrophilic unless the alkyl group of the xanthate exceeds amyl (Figure 8.9). The increase in solubility of metal xanthates parallels their change from hydrophobic to hydrophilic character.

If the conditions in the flotation system favor catalytic formation of dixanthogen (in addition to metal-xanthate precipitation), it will also physically coadsorb and will supplement the hydrophobic character of the sulfide (or metal) surface.

Metal xanthates and/or dixanthogen, formed at the interface, are not the only species that can improve the hydrophobic character of the first chemisorbed xanthate layer. Taggart and Arbiter (1943), as well as Klassen and Plaksin (1954), had already observed that additions of judicious quantities of appropriate nonpolar hydrocarbon oils could greatly improve

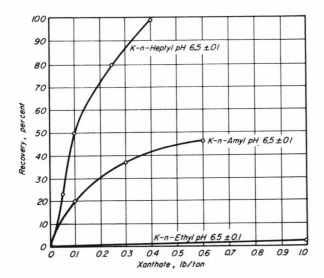

Figure 8.9. Flotation of sphalerite at pH 6.5, using increasing concentrations of progressively more hydrophobic xanthates, ethyl, n-amyl and n-heptyl. [From Gaudin (1957), *Flotation*. Copyright 1957, by the McGraw-Hill Book Company, Inc. Used with the permission of McGraw-Hill Book Company.]

flotation recoveries in systems with threshold hydrophobicity. An additional parameter in controlling the hydrophobic character is a change in the length of the alkyl group (Figure 8.9).

Heyes and Trahar (1977) have definitely established that a limited extent of sulfide oxidation does produce sufficiently hydrophobic surfaces on some sulfides (chalcopyrite, sphalerite) for their flotation in the absence of xanthate. In these instances, the hydrophobic state may be generated by an appropriate combination of M—S—O atoms on the surface or by the formation of polymerized elemental sulfur. Evidence is not available at present to ascertain which of such speculations is valid.

In the absence of oxygen in the system, the oxidized form of xanthate, dixanthogen, when added in sufficient quantity, can fulfill both roles, that of initial adsorption at the surface and that of improvement in the hydrophobic character of the adsorbed layer by physical coadsorption. Dixanthogen can dissociate at surfaces of some sulfides (such as PbS and Cu_2S but not necessarily at surfaces of all sulfides) to provide locally the first chemisorbed monolayer (not necessarily covering the whole particle surface), and, in addition, it then physically coadsorbs onto this partial monolayer (forming multilayer patches) to improve the hydrophobic character for flotation.

Thus, the chemisorption of xanthate onto a sulfide surface may or may not require oxygen depending on the type of species available—an ionic xanthate or a nonionized dixanthogen—and the type of sulfide mineral. However, to establish the requisite degree of hydrophobicity on the initially hydrophilic surface, a definite oxidation level is required whenever an ionic xanthate species is employed. This oxidation level is required to provide a sufficient quantity of metallic salts that can subsequently transform to metal xanthates by an ion exchange mechanism.

In this respect, the most relevant are the determinations carried out by Bushell and Malnarich (1956) and Bushell (1958, 1964) on the amounts of xanthate abstracted in operating circuits of Cominco's Sullivan Concentrator (at Kimberley, B.C., Canada). They found that the amounts of xanthate abstracted in lead roughing were proportional to $m^2[H^+]$ and the corresponding recoveries of lead to $m[H^+]$, where m denotes the xanthate concentrations used and $[H^+]$ the hydrogen ion concentration. For the lead roughing circuit a typical value of $m[H^+]$ was found to be 10×10^{-16}, and for the lead cleaning circuit, from 1.1×10^{-16} to 5×10^{-16}. Bushell (1964) suggested that the $m^2[H^+]$ relationship [or the $m^n[H^+]$ relationship, developed by Gaudin (1957) from contact angle results] is determined by the nature of surface oxidation compounds undergoing metathesis. This suggestion appears to be the most appropriate one. The fact that different relationships are established for xanthate adsorption and for recoveries in separations may be an indication that the development of hydrophobic characteristics does not parallel adsorption isotherms.

Equally relevant and significant are the metal-xanthate solubility products, frequently paralleling the floatabilities of the corresponding metal sulfides [du Rietz (1953, 1965, 1975)]. The latter correlations are, however, not unequivocal because of the influence of other parameters affecting either the adsorption in the first layer or the improvement in hydrophobicity.

As rightly pointed out by Granville *et al.* (1972), the numerous reactions occurring in xanthate flotation systems, in practice, do not reach their respective equilibria. They are all governed by the extent of preoxidation of sulfide surfaces and/or diffusion of the slowest species across the respective interfaces. In an open system (with a continued supply of oxygen from air) the "adsorption" reactions will proceed until all xanthate is consumed— if sufficient time is allowed. However, if (as happens in practical flotation operations) the time is limited, some xanthate ions may remain temporarily unconsumed. In a closed system, with limited quantities of oxygen and xanthate species, the reactions will proceed until the constituent which is in lesser quantity is exhausted if sufficient time is allowed; if the time is curtailed, both oxygen and xanthate may be left in detectable quantities.

Cusack (1967) found that the floatability of galena is affected both by the degree of oxidation of the mineral surface and the partial pressure of oxygen in the system. The optimum floatability was achieved when the liquid of the flotation pulp was saturated with oxygen but the mineral surface was preoxidized only to a limited extent, below a certain threshold value. Gowans (1972) established that the oxygen deficiency, resulting from oxidation reactions during grinding in ball mills, must be replenished through an appropriate length of conditioning time before the sulfide minerals become readily floatable.

The preceding findings testify to the establishment of a balance in flotation circuits between the kinetics of mostly electrochemical reactions and the extent of products formation, specifically those products that determine the requisite degree of hydrophobicity. In a simple system such as Pt in xanthate solution (Figure 8.4) or pyrite or galena in solutions of various electrolytes (Figure 8.10), the potential of the given electrode is

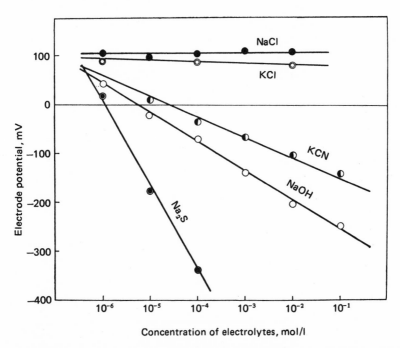

Figure 8.10. Mixed potentials of galena electrode in increasing concentrations of indifferent electrolytes (NaCl, KCl) and specifically adsorbing electrolytes. [From Mukai and Wakamatsu (1962), with the permission of the authors.]

uniquely determined by the concentration of the individual electrolyte, except for indifferent electrolytes such as NaCl and KCl. As the concentration changes due to a reaction consuming the given electrolyte, the potential varies in a linear manner on a logarithmic scale, and the change in potential is a measure of the extent of reaction. For a redox reaction—such as oxidation of metallic species to corresponding ions or oxidation of xanthate ion to dixanthogen—the redox potential varies with the extent of oxidation (Figures 2.5 and 8.4). So, theoretically, monitoring of the E_h potential in a flotation pulp should provide adequate information about the extent of redox reactions. However, it appears that because of a multiplicity of reactions, polarization effects, and a possible poisoning of the Pt electrode (or any specific ion electrode), the utility of redox potential measurements to control flotation systems is at present still insufficiently established. Papers by Woodcock and Jones (1969), and Natarajan and Iwasaki (1973, 1974) should be consulted for a more detailed examination of the E_h monitoring in flotation systems.

An example of complexity already encountered in a flotation system as simple as that represented by metallic copper in ethyl xanthate solutions is the paper by Hepel and Pomianowski (1977). Despite the fact that 44 equilibria equations had been considered, additional assumptions about adsorption, hydrophobicity, etc., are required to utilize the E_h–pH diagrams constructed with these 44 equations. The difficulties envisaged in constructing corresponding E_h–pH diagrams for metallic sulfide–xanthate systems, followed by modifications due to depressants and activators, appear too great to tackle such constructions at present.

8.2.3. *Flotation Problems Caused by Excessive Surface Oxidation*

It is well established that excessive oxidation of sulfide minerals causes a deterioration of their floatability when using thiocompounds. Not only are much higher collector additions generally required, but both the recovery and the selectivity of separations become greatly impaired or altogether inhibited. Equally, many metal oxides and oxidation compounds of metals (carbonates, sulfates, etc.) are not readily floated using thio surfactants.

Two possible causes responsible for this behavior are usually cited. One is the failure of the surfactant employed to adsorb; the other is the failure to make the adsorbed layer sufficiently hydrophobic. Yet determinations of adsorption isotherms for heavily oxidized sulfides frequently show an ever-increasing abstraction of xanthate species from solutions contacting such sulfides. Further, it is found (in some systems at least) that if unusually large additions of xanthates are made, greatly improved recoveries of minerals may be obtained. For example, in the flotation of vanadium minerals approximately 3–5 lb/ton additions of xanthates were found necessary to achieve satisfactory separations [Fleming (1957)].

The impairment or even a complete inhibition of flotation (taking place despite the removal of xanthate species from solution by chemisorption) may be caused by flaking off of unevenly distributed thick patches of metal xanthates formed by metathesis. As a result of excessively strong chemisorption bonds developing between the metal ions and the xanthate ions, the lattice bonds within the oxidized layer surrounding the sulfide particles are disturbed to such an extent that adhesion of the covalently bonded metal xanthate molecules to the underlying substrate representing the oxidized layer is minimal. This type of weakening of bonds between the adsorption product and the substrate, referred to in Section 8.1, is also encountered in other systems, for example, during rapid air oxidation of metallic substrates such as Ca, Ba, Mg, etc., and is explained by nonregistry of lattices formed by the metallic oxide in relation to that of the substrate metal.

An indication that such peeling off of the adsorbed layers, due to rupture of adhesion bonds, does occur in flotation systems was obtained by Leja (1956) in the flotation of malachite. In attempts to establish optimum conditions for the flotation of pure malachite particles (using xanthates and frothers in a laboratory cell) when the xanthate added was found to be completely extracted from the solution, flotation was attempted, and an apparently barren froth was collected. After the froth had collapsed, a small amount of residue was found, which consisted of Cu_2X_2 and dixanthogen.

In light of the preceding indications concerning extensive chemisorption of xanthates by heavily oxidized sulfides and the flotation of reaction products formed with collectors, together with the increased floatability using excessively large collector additions, the following interpretation is suggested: When an addition of a xanthate-type collector is such that the given oxidized mineral reacts to produce only isolated multilayer patches of collector chemisorption products, their adhesion to the heavily disturbed substrate is insufficient to keep the whole mineral particle attached to the bubbles, and only the collector reaction products may be floated. However, when the collector addition is increased to such an extent that the multilayer patches of collector reaction products are no longer isolated but a continuous layer of these products (of uneven thickness) envelops each particle, the floatability of the particles improves. The reason for this is that the initially very weak adhesion to the substrate (of the chemisorption and reaction products) is now strengthened by lateral bonds existing within the continuous layer surrounding the particle. Consequently, the peeling-off occurrence is prevented. Despite the fact that these lateral bonds are only van der Waals bonds, they are sufficiently numerous to be capable of withstanding the disruptive forces acting on the particle–bubble aggregate. The two aspects of this interpretation are schematically presented in Figures 8.11 and 8.12. Lapidot and Mellgren (1968) postulated that a quasi-continuous reagent film (of ~ 370-Å thickness, consisting of oleate and fuel oil) was formed on ilmenite particles. If this film became broken by attrition, the recovery of ilmenite particles decreased.

The interpretation we have suggested for the cause of nonfloatability or impaired floatability of heavily oxidized sulfides (which float readily when only threshold oxidation has taken place) would not apply to those oxidation products in which metal atoms are held in the solid lattice by bonds stronger than the adsorption bonds or metal–collector bonds. Thus, the two causes cited at the beginning of this section as possible explanations for nonflotation may legitimately apply in a number of situations. Unfortunately, no detailed studies seem to have been carried out on specific

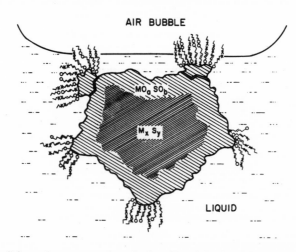

Figure 8.11. Schematic representation of a heavily oxidized sulfide particle M_xS_y, where M stands for a metal such as: Pb, Cu, Fe, etc., made partially hydrophobic by strong chemisorption of collector at *isolated* sites. The particle may be severed from an attaching air bubble as a result of breakage of bonds (weakened by excessively high chemisorption energy) immediately underlying the adsorption sites. The breakage of the underlying bonds depends on the dynamic forces in the flotation system.

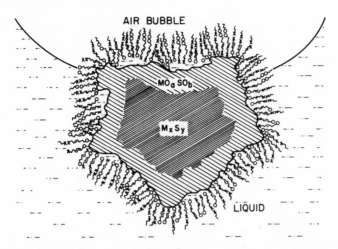

Figure 8.12. Schematic representation of the beneficial role exerted by a multilayer continuous coating produced by large doses of a collector and/or molecularly associated coadsorbates (hydrophobic precipitates of collector, undissociated species of collector, hydrocarbon oil). Instead of breakage of the bonds underlying the adsorption sites, as indicated in Figure 8.11, the continuity of the adsorbate coating acts as an enveloping "bag" for the particle and thus provides additional bonds in particle–bubble adhesion.

minerals which are difficult to float (such as CuO, CaO, $PbSO_4$, etc.) in order to document either the lack of adsorption or the lack of hydrophobicity despite adsorption.

8.3. Mechanisms of Adsorption of Non-Thio Collectors

All non-thio compounds employed in flotation as collectors (whether in separation of sulfides or nonsulfides), that is, alkyl carboxylates, alkyl sulfates and sulfonates, alkyl amines, and alkyl-substituted amines, etc., are surfactants with C_{10}–C_{18} hydrocarbon chains, whereas thio collectors employed generally have C_2–C_6 hydrocarbon chains. This difference in alkyl groups is due to the relative insolubility of longer-chain thio compounds in water and the apparent efficacy of the short-chain thio collectors. Xanthates, and thio compound collectors in general, become immobilized at solid/liquid interfaces mainly by the chemisorption mechanism. Some non-thio compounds, such as carboxylates, appear to be capable of becoming immobilized for the purpose of flotation by all three mechanisms mentioned in Section 8.1, that is, (1) chemisorption, (2) specific adsorption in the IHP without charge transfer, and (3) electrostatic adsorption over a network of laterally bonded complex counterions. Others, such as sulfonates and amines, appear to adsorb mainly by the latter two mechanisms. There has been no sufficiently systematic study carried out to state unequivocally under what conditions the three mechanisms may take place in a particular surfactant–flotation system, but there exists evidence of the individual mechanisms occurring in different surfactant systems.

8.3.1. *Chemisorption of Carboxylates Accompanied by Physical Coadsorption*

Wadsworth and his co-workers provided definitive IR spectroscopic evidence for the chemisorption mechanism in the hematite–oleic acid system [Peck *et al.* (1966)] (Figure 8.13). The $C=O$ stretching band of oleic acid, at 1705 cm^{-1}, is shifted on ionization to higher wavelengths, that is, lower wavenumbers, 1520–1540 cm^{-1} for sodium oleate and 1590 cm^{-1} for ferric oleate. The chemisorption of oleate onto hematite to form the first layer of Fe–oleate species is indicated by the $C \cdots O$ band appearing at 1565 cm^{-1}, a shift in the wavenumber from the 1590 cm^{-1} for ferric oleate. A similar shift was observed by Poling (1963) for the Pb—X chemisorption relative to the PbX$_2$ spectrum, namely from 1210 to 1195 cm^{-1} for the C—O—C band in PbX$_2$ and Pb—X, respectively. A shift of an absorption band such as the preceding is associated with a change (usually an increase) in the covalent bonding formed on adsorption relative to a mostly covalent bond in the corresponding compound. The physical coadsorption of ionized sodium oleate is indicated by the presence of 1520–1540 cm^{-1} bands and that of un-ionized oleic acid by the 1705-cm^{-1} band.

A concurrent physical coadsorption of sodium oleate on fluorite and barite was found by Peck and Wadsworth (1964) to occur at alkaline pHs, whereas the coadsorption of oleic acid occurred at acidic and neutral pHs. This finding agrees with the predominance of oleate and oleic species in the respective pH regions (Figure 5.19). Using absorbances (after suitable calibration), the authors evaluated the proportions of the chemisorbed oleate species to the physically coadsorbed species and found a strong dependence on the type of the solid–inorganic anion combination (Figures 8.14 and 8.15). The anions added appear to compete in their adsorption with the oleate ion to a different extent at fluorite surfaces and at barite surfaces; the lattice constituent anions, F$^-$ and SO$_4^{2-}$, respectively, lower the proportion of chemisorbed oleate to a much greater extent than other anions.

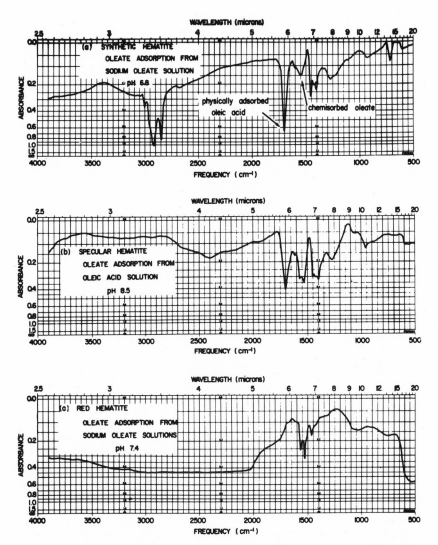

Figure 8.13. Infrared spectra indicating the presence of ferric oleate and oleic acid as distinct adsorbate species during adsorption of oleic acid on hematite at different pH. [From Peck *et al.* (1966), *Trans. AIME* **235**, with the permission of the American Institute of Mining and Metallurgical Engineers, Littleton, Colorado.]

The physical coadsorption of nonionized carboxylic acid may be interpreted as contributing to an increase in the overall hydrophobicity of the adsorbed layer. However, the physical coadsorption of an excess ionized carboxylate species should, in general, cause a deterioration in the hydro-

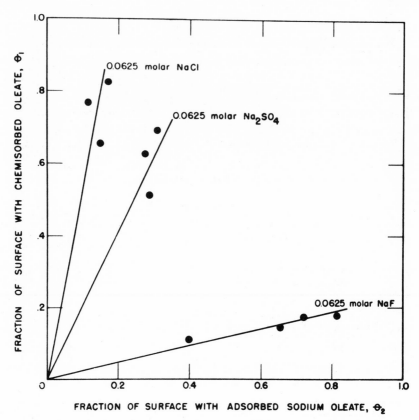

Figure 8.14. The effect of different inorganic salts present in 0.0625 M concentrations on the relative proportion of chemisorbed and physically coadsorbed oleate species in the system fluorite–sodium oleate at pH 9.5; a complete coverage occurred at $\sim$55 mg of oleate per 1 g of fluorite. [From Peck and Wadsworth (1964), *U.S. Bureau of Mines R.I. 6412*, with permission of the U.S. Department of the Interior, Bureau of Mines, Washington, D.C.]

phobic characteristics. Indeed, Taggart and Arbiter (1943, p. 503) stated that "the depressing effect of an excess of alkali soap...may be corrected by use of a mineral oil." They explained this effect as displacement of the second layer of adsorbate by the oil phase.

It may be helpful to realise and to emphasize that depending on the types of bonds, their relative strengths and/or the respective steric parameters, the concurrent adsorption of two surfactant species may result in a variety of conformations of the two molecules, and a consequent increase or a decrease in hydrophobicity of the adsorbed film. If the species which is chemisorbed forms a *diffuse* monolayer covering the surface, then the physically adsorbing species may associate with the pre-adsorbed molecule either in a *parallel* mode (i.e., the hydrocarbon groups interacting by van der Waals forces, and the polar groups developing some dipole–dipole interaction) or in an *inverted* mode (i.e., the respective hydrocarbons are interacting with each other, and the polar groups are not associating, but facing in opposite directions). This situation is likely to occur when the polar group of the physically co-adsorbing species is electrostatically repelled from the site it would otherwise occupy (if it were able to adsorb in a *parallel* mode). As, on continued adsorption, the diffuse monolayer of chemisorbing species becomes more condensed, even the *parallel* mode of phys-

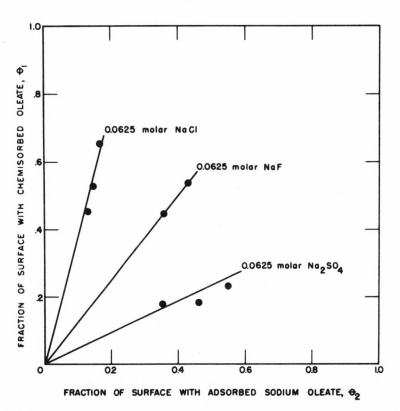

Figure 8.15. The effect of the same inorganic salts as in Figure 8.14 (0.0625 M concentration) on the relative proportion of chemisorbed and physically coadsorbed oleate species in adsorption on barite; a complete coverage at $\sim$25 mg of oleate per 1 g of barite; pH 9.5. Note the effect of lattice constituent anion SO_4^{2-} in lowering the proportion of chemisorbed oleate—analogous to the effect of F^- in respect to fluorite in Figure 8.14. [From Peck and Wadsworth (1964), *U.S. Bureau of Mines R.I. 6412*, with permission of the U.S. Department of the Interior, Bureau of Mines, Washington, D.C.]

ically co-adsorbing molecules may become restricted by steric parameters; consequently, in solutions approaching CMC, the limiting arrangement of the *inverted* mode of adsorption is reached regardless of the initial mode of co-adsorption. The second layer is adsorbed through van der Waals bonds of the terminal CH_3 groups of the respective hydrocarbon chains and the polar groups are in opposite directions.

Since, at present, there is no means available to determine precisely the arrangements of molecules adsorbed at interfaces, the speculations about the different modes possibly cannot be restricted to one or two arrangements for all different structures of surfactants.

8.3.2. *Molecular Interactions Among Surfactants at Solid/Liquid Interfaces*

In analogy with the condensation of adsorbates at solid/gas interfaces (Figure 6.14), regions of increased adsorption density occur in solid–solution systems, as already indicated, for example, in Figures 6.32–6.35. Surfactants which are preadsorbed at kink positions of a solid/liquid interface tend to act as nuclei and exert an increasing attraction toward all surfactant species present in the surroundings, developing the so-called *molecular interactions*. These denote associations which involve both constituent groups of the participating surfactants, namely associations between

1. The respective nonpolar portions (hydrocarbon chains), resulting from van der Waals bonds, and
2. The respective polar portion, due to ion–dipole, ion–ion, or dipole–dipole bonding.

Initial evidence for molecular interactions at the liquid/gas interface was provided by the penetration studies of Schulman *et al.* (i.e., Section 6.10), and for interactions at solid/liquid interfaces by Leja and Schulman (1954). Subsequently, Gaudin and Fuerstenau (1955) coined the word *hemi-micelle* to describe patches of associated surfactant ions "perhaps containing also some associated uncharged molecules."

Analyzing the results of the adsorption of dodecylammonium ions on quartz (through changes of zeta potential of quartz) as a function of dodecylammonium ion concentration, Gaudin and Fuerstenau (1955) came to the conclusion that at a certain concentration an association into patches of ions must take place. The zeta potential not only decreased in value but reversed its sign. This change did not result from a compression of the

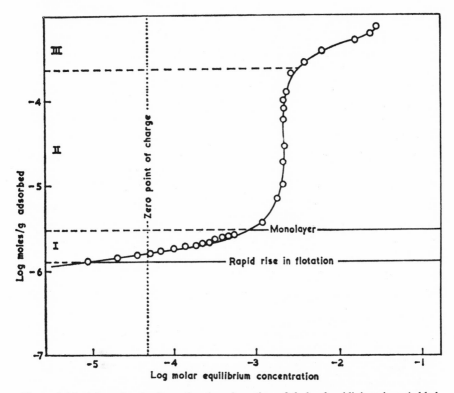

Figure 8.16. Adsorption isotherm for the adsorption of dodecylpyridinium ions (added as a bromide salt) on silver iodide particles suspended in $10^{-4}\,M$ KI solution. [From Jaycock and Ottewill (1963), *Trans. IMM* **72**, with the permission of the Institution of Mining and Metallurgy, London.]

double layer since an addition of large amounts of an indifferent electrolyte, such as NaCl, did not cause a charge reversal in zeta potential but only its decrease in absolute value toward zero.

Associations of surfactants showing nearly vertical steps in adsorption isotherms have been reported by Ottewill *et al.* [Ottewill and Rastogi (1960), Jaycock *et al.* (1960), Jaycock and Ottewill (1963)] (Figure 8.16). Such adsorption isotherms indicated multilayer formation of adsorbates, presumably in patches of two-component adsorbates (adsorbed dodecylpyridinium species and hydrophobic in character silver dodecylpyridinium reaction product). Analogous adsorption isotherms were obtained by Fuerstenau *et al.* [Somasundaran and Fuerstenau (1966), Wakamatsu and Fuerstenau (1968)] in the adsorption of alkyl sulfonates on alumina (see Figure 6.33) and of dodecylammonium acetate on quartz (Figure 8.17). As

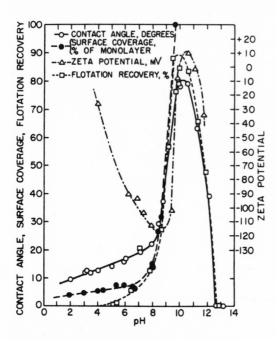

Figure 8.17. A correlation of surface coverage, contact angle, zeta potential, and flotation recovery for the system dodecylammonium acetate–quartz. [From Fuerstenau (1957), *Trans. AIME* **208**, with the permission of the American Institute of Mining and Metallurgical Engineers, Littleton, Colorado.]

shown in Figure 8.17, the isotherms were found to be closely correlated with changes in contact angle and zeta potential determinations. In all these correlations three regions could be distinguished:

1. An initial region representing the onset of primary adsorption and of contact angle development
2. A region of a marked increase in adsorption (accompanied by an increase in contact angle and a change in zeta potential toward and past the zero value) caused by associations of surfactants with initially preadsorbed species
3. A region of less rapid adsorption

Flotation recoveries were found to parallel these curves of adsorption, contact angle, and zeta potential—within the range of concentrations tested. It is significant, however, that a steep increase in flotation recovery becomes evident with a rise in associations. This increase may be taken as an indication that it is the change in the hydrophobicity of the adsorbed layer, and not the occurrence of the adsorption bond itself, which determines the behavior in floatability of mineral particles. It has been known for a long time that an increase in the alkyl chain of a given collector improves floatability and lowers the threshold concentration needed for flotation. The effects have been well documented by the contact angle studies of Wark *et al.* and the flotation recovery correlations of Gaudin and Fuerstenau *et al.*

A systematic study of the role played by hydrocarbon chain lengths in floatability, carried out by Somasundaran *et al.* (1964), Somasundaran and Fuerstenau (1968), Cases (1968), and Predali (1968), showed that a linear dependence exists between the concentration and the corresponding chain lenght of collector species needed for flotation or for a reversal of charge. The slope of the line enabled the evaluation of interactions between CH_2 groups of associating hydrocarbon chains. The value of $0.9\ kT$ so obtained compares well with the $1.03\ kT$ value obtained in solidification of monolayers at the liquid/gas interface.

The benefits of using branched-chain alkyl groups (for creating hydrophobicity) and/or aromatic and cyclic hydrocarbon groups (except in the systems where KCl and other soluble salts are floated; see Section 8.4) have not been well documented so far. Contact angle studies indicate that higher contact angles are created by isoalkyl xanthates in comparison with normal alkyl xanthates. Bleier *et al.* (1976) studied the influence of the structure of amine-type collectors on the flotation of quartz; their data seem to indicate the beneficial role of branching and N substitution in improving the recoveries of quartz and silicates.

It is important to realize that molecular associations between the preadsorbed collector and other species of surfactants (which *per se* may not be capable of adsorbing as collectors) also influence the overall degree of hydrophobicity. Thus, molecular associations extend the degree of hydrophobic character by lateral bonds, along the surface, without contributing an equal degree of change to the adsorption bond with the solid, in the direction perpendicular to the solid surface.

The number of hydrocarbon chains per unit area of the adsorbent surface is greatly increased as a result of molecular associations without a concurrent increase in the primary adhesion bonds. Therefore, the requirements of geometric or steric correspondence (juxtaposition) between the polar groups in the collector layer and the available substrate sites to achieve a given degree of hydrophobicity are not as strict whenever molecular associations occur with surfactants that are nonionized (*neutral species*) or completely nonpolar in comparison with systems devoid of molecularly associating species.

Modi and Fuerstenau (1960) showed that excellent recoveries of corundum can be obtained at pH 6.0 with joint use of sodium laurate and sodium dodecyl xanthate; these two collectors, when used jointly, are very effective at considerably lower concentrations than when dodecyl sulfate alone is used. Modi and Fuerstenau explained this as "probably due to the presence of un-ionized lauric acid or dodecyl xanthic acid molecules," i.e., neutral molecules coadsorbing with the collector ions. Fuerstenau and Yamada (1962) (Figure 8.18) showed that the recovery curve in the Hallimond tube flotation of corundum can be shifted to lower concentrations by employing neutral species "such as decyl alcohol in near equivalent concentrations." (Their claim that decyl alcohol is not a frothing agent does not prevent this nonionized species from acting as a flotation "frother.") Similarly, additions of appropriate nonpolar oils, such as kerosene or fuel oil, were found by Taggart and Arbiter (1943), Klassen and Plaksin (1954), and Lapidot and Mellgren (1968) to improve floatability, although each group of researchers provided different explanations of the effects. Klassen and Plaksin (1954) considered that kerosene aligns itself at the border of contact between the three phases solid–liquid–gas and in this way improves the tenacity of attachment between the solid and the gas bubble. A most convincing piece of evidence of the cooperative effect between oxine and free oil in the flotation of smithsonite was presented by Rinelli and Marabini (1973), as shown in Figure 8.19.

Molecular associations are of prime importance in the frother–collector interactions taking place during the attachment of particles to air bubbles, and these are discussed in greater detail in Section 9.8. What is important to

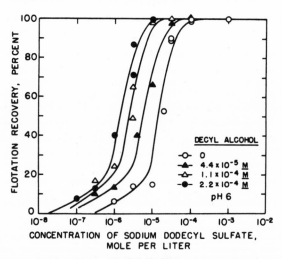

Figure 8.18. The effect of nonionic cosurfactant additions on the flotation recovery of alumina using sodium dodecylsulfate as a collector at pH 6. [From Fuerstenau and Yamada (1962), *Trans. AIME* **223**, with the permission of the American Institute of Mining and Metallurgical Engineers, Littleton, Colorado.]

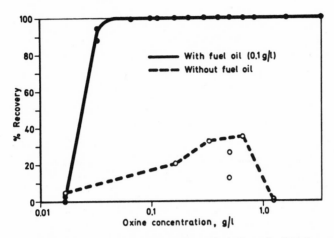

Figure 8.19. The effect of an addition (0.1 g/liter) of fuel oil on the flotation recovery of smithsonite (ZnCO₃) using increasing oxine concentrations at pH 7. [From Rinelli and Marabini (1973), in *Proceedings of the Tenth International Mineral Processing Congress*, with the permission of the Institution of Mining and Metallurgy, London.]

realize, however, is the fact that, owing to molecular interactions, the hydrophobic character of the particle is also greatly affected *before* the moment of attachment. The result is an increased degree of hydrophobicity if tested under near-static or mildly dynamic conditions. However, eventually, with excessively associated systems a deterioration in adhesion may occur if the proportion of the molecularly associated noncollector (species not bonded strongly to the substrate) greatly exceeds a certain value in relation to the primary collector-acting species which is directly adsorbed to the substrate.

In surfactant systems which provide near-equivalent proportions of ionized and nonionized species in appropriate pH regions (see Figures 5.19–5.21), molecular associations prevail around the middissociation points (pK values). Whenever one of the species is capable of adsorbing as a primary collector, optimum flotation conditions result in this pH region of the S curve—hence the importance of acid soaps in flotation with carboxylates and of the corresponding amine–aminium associations in the case of cationic collectors.

8.3.3. *Specific Adsorption of Ionized Collector Species in the IHP (Without Charge Transfer)*

Extensive studies of the zeta potential of quartz, corundum, goethite, etc., in various nonsurfactant (inorganic) electrolytes and in surfactant

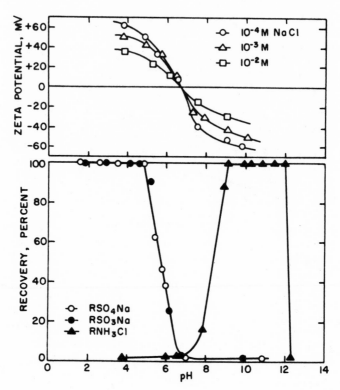

Figure 8.20. The dependence of the flotation recovery of goethite (FeOOH) on the zeta potential (surface charge) of the solid/liquid interface and the charge on the polar group of the collector used. The zeta potential curves for different pH (upper curves) were obtained at 10^{-4}, 10^{-3}, and 10^{-2} M NaCl concentrations, indicating pzc at pH 6.7. The recovery curves were obtained for 10^{-3} M solutions of dodecylammonium chloride (RNH$_3$Cl), sodium dodecylsulfate (RSO$_4$Na), and sodium dodecyl sulfonate (RSO$_3$Na). [From Iwasaki *et al.* (1960b), *U.S. Bureau of Mines R.I. 5593*, with permission of the U.S. Department of the Interior, Bureau of Mines, Washington, D.C.]

electrolytes (carried out by Gaudin, D. W. Fuerstenau, Iwasaki, S. R. B. Cooke, and their co-workers) have established the existence of a direct correlation between the zeta potential and the adsorption of oppositely charged surfactants. Examples of such correlations are given in Figures 8.20 and 8.21. These examples show clearly that flotation of the particular solid is possible only when an anionic surfactant is used in the pH conditions of positive zeta potentials and a cationic surfactant in the negative zeta potential region.

Such correlations suggest that the basic adsorption onto the solid is electrostatic in nature, involves the ionized form of the surfactant, and,

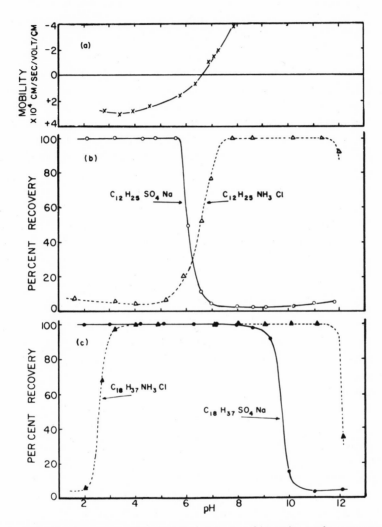

Figure 8.21. The dependence of the flotation recovery of hematite on the zeta potential of its solid/liquid interface, the charge on the polar group, and the chain length of the collector used: (a) electrophoretic mobility as a function of pH, indicating the isoelectric point at pH 6.7; (b) flotation recoveries of hematite vs. pH using 10^{-4} M dodecylammonium chloride and sodium dodecyl sulfate solutions; (c) flotation recoveries vs. pH using 10^{-4} M solutions of octadecyl analogous compounds. Note the dramatic expansion of the floatability into the pH region of opposite surface charges. [From Iwasaki *et al.* (1960b), *U.S. Bureau of Mines R.I. 5593*, with permission of the U.S. Department of the Interior, Bureau of Mines, Washington, D.C.].

since it causes a reversal of zeta potential, must take place in the Helmholtz plane. The fact that surfactants with longer alkyl chains are capable of crossing the electrostatic barrier and cause flotation in the neighborhood of the pzc but in the zeta potential region of similar charges to those carried by the polar group, Figure 8.21 may be due to either (1) the heterogeneity of the solid surfaces being so high that sites of opposite charge to that indicated by the mean value evaluated in the zeta potential function exist on the solid and cause initial adsorption by electrostatic forces, reinforced by molecular associations, or (2) the existence of some covalent bonding (or partial chemisorption) in the first, primary adsorption layer. (This effect is similar to the underpotential effect discussed in Section 7.5.)

There is no reason why, with heterogeneous solid surfaces, the character of adsorption should be uniform over the whole extent of the solid surface. As established in the case of carboxylates studied so far, their adsorption may involve chemisorption in some systems, and this may be indicated by a shift in the IR spectrum of the molecules adsorbed in the first layer. However, the absence of such a shift does not necessarily negate the existence of a charge transfer—as documented by the lack of an IR shift in the cuprous xanthate formation and its chemisorption. The majority of available evidence collected so far indicates that in numerous non-thio ionizable surfactant systems the zeta potential—adsorption correlations suggest a charge reversal by the surfactant and its adsorption in the IHP. However, when the adsorption of charged surfactants is extended to the region of identical charges (indicated by the zeta potential curve), a possibility of some covalent characteristic in the primary bonding exists. And this possibility becomes particularly strong if the adsorption occurs in the pH region of the S curve of dissociation, where the un-ionized surfactant species become predominant. Unfortunately, there appear to be few systematic investigations of IR changes during adsorption of ionizable surfactants on solids. The correlations between zeta potential and adsorption are the only evidence for the existence of specific adsorption in the IHP without charge transfer.

Adsorption in the IHP always precedes chemisorption; the fact that charge transfer has not taken place (when adsorption is limited to the IHP adsorption without charge transfer) may signify that the system either does not possess a sufficient energy of activation (for charge transfer to take place) or does not possess appropriate acceptors or donors under the given set of conditions. Therefore, a change from adsorption in the IHP to chemisorption may occur imperceptibly, and there may be regions for some variables when both forms of adsorption will participate.

Direct and indirect evidence that adsorption in the IHP is involved in some metal oxide–ionized surfactant systems has been provided by the studies of Coelho (1972) (Figure 8.22), Using IR spectroscopy, Coelho found that, in the system (CuO) tenorite–lauric acid, below the pH_{pzc} for tenorite (pH $\sim$ 8) the adsorbed carboxylate species consisted of laurate ions and coadsorbed lauric acid molecules. In the presence of copper ions added to the system, cupric laurate and lauric acid were found adsorbed. No adsorption was detected above the pH_{pzc} in the region of negative zeta potential. The flotation tests paralleled the adsorption findings. When sodium sulfide was added to the system, the curve for flotation recovery (and, presumably, adsorption of laurate) shifted to lower pH. However, more significantly, the curve for flotation using laurylamine hydrochloride has also shifted to lower pH. This behavior indicates that the competitive adsorption of HS^- ions in the electrical double layer occurs and affects both types of surfactants equally. If sodium sulfide were reacting with CuO to provide a sulfide surface, only the curve for one surfactant, lauric acid, would be expected to shift but not that for amine hydrochloride. The hydrosulfide ion, HS^-, shows a depressing action when lauric acid is employed but an activating action with laurylamine ions.

Similar effects of the competitive adsorption of potential-determining ions were observed by Modi and Fuerstenau (1960). On adding SO_4^{2-}

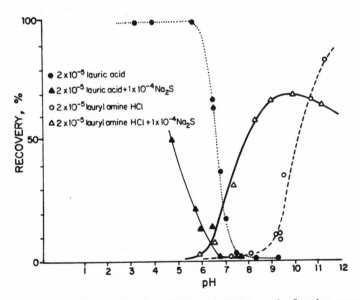

Figure 8.22. The influence of HS^- ion addition (as Na₂S) on the flotation recovery of tenorite (CuO) in 2×10^{-5} M lauric acid and laurylamine hydrochloride solutions, respectively. [After Coelho (1972), with the permission of the author.]

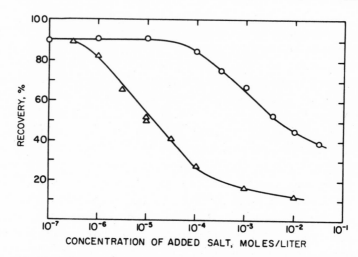

Figure 8.23. The influence of increasing concentrations of inorganic electrolytes (○, NaCl; △, Na₂SO₄) on the flotation recovery of alumina in 4×10^{-5} M sodium dodecyl sulfate solutions at pH 6. [From Modi and Fuerstenau (1960), *Trans. AIME* **217**, with the permission of the American Institute of Mining and Metallurgical Engineers, Littleton, Colorado.]

(Figure 8.23), the recovery of corundum began to be reduced by concentrations nearly two orders of magnitude lower than that of the collector used.

8.3.4. *Adsorption of Surfactants to a Network of Polymerized Counterion Complexes, Activating Agents*

Solidification studies on surfactant monolayers (carried out by Schulman and co-workers, Section 6.10) have provided evidence that immobilization of ionized surfactants may take place at the solution/air interface when oppositely charged complex ions polymerize by hydrogen bonding to form a network of adsorption sites. Subsequent flotation tests by Schulman and Smith (1953) confirmed that the conditions under which such immobilization occurs correlate well with floatability of minerals which can release ions required for the necessary complexing counterions. The polymerizing ionic complexes establish at the solid/liquid interface an entirely new solidified substrate. This new substrate may represent a three-dimensional layer of bonded complexes formed within the original Gouy–Chapman layer of the solid, or it may be a two-dimensional film formed at the original OHP. The formation of such a new solidified substrate leads to the establishment of a new electrical double layer. The adsorption of ionized surfactants takes place at the new IHP as the specific adsorption of counterions which could not specifically adsorb at the original IHP of the solid. But the presence of surfactants is not a prerequisite for the formation of the solidified network of polymerized complexes. When the conditions in the neighborhood of the solid particle (that is, within the diffuse Gouy–Chapman layer) favor formation of ionized complexes that can hydrogen-bond, a polymerized network of such complexes is created regardless of the presence or the absence of surfactants in the solution phase. When excessive quantities of complexes become available, the polymerized network may also start forming in the bulk of the solution (flotation pulp) and may build new substrates around any solid particle present in the pulp. Therefore, flotation resulting from adsorption of surfactants onto such a network of polymerized complexes is likely to be nonselective unless suitable precautions are used to eliminate excess complexing ions. For this reason de-cantation or de-sliming may be necessary before the addition of surfactant is made. Also, the adhesion of the complexing network to the original solid surface and the cohesion within the polymerized layer are likely to be generally weak. The magnitudes of both cohesion and adhesion are governed by the densities of hydrogen bonds and of dipole–dipole or dipole–ion bonds developed in the system; hence, the steric parameters of the complexing inorganic species play a major role in determining selectivity. The steric parameters of the complexes also determine the disposition of the adsorption sites to which surfactants attach by electrostatic bonds. Hence, the length of the hydrocarbon chain in the surfactant molecule may have to be adjusted accordingly in order to develop a sufficient degree of hydrophobicity. (An increase in hydrophobicity by molecular associations with nonpolar oils may unduly weaken the adhesion.)

The characteristics of complexes formed in aqueous systems with different inorganic ligands (sulfate, fluoride, nitrate, silicate, etc.) and, particularly, of the most prevailing ones, hydroxy complexes of transition metals, are of the utmost importance in developing conditions for selective flotation whenever this third mechanism of surfactant adsorption is involved. Some aspects of such complexes are discussed in Section 10.4. Cotton and Wilkinson (1972) have devoted several chapters in their text to different types of complexes and have provided most comprehensive lists of relevant references. A more recent review on the adsorption of aquometal ions is that by Dalang and Stumm (1976). Reviews on the effects of various complex species on stability of dispersions, coagulation, etc., have been published by Matijevic (1973a, 1976). Stability constants of metal complexes for inorganic and organic ligands have been compiled by Sillén and Martell (1964, 1971).

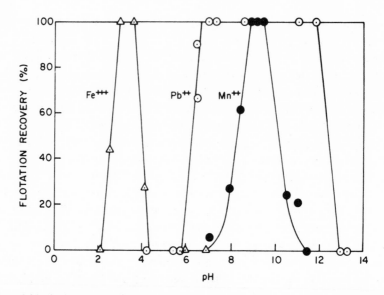

Figure 8.24. Activation of quartz by metallic ions in relation to pH in the flotation of quartz with sodium dodecyl sulfonate at $1 \times 10^{-4}\ M$ (metallic ion concentration also $1 \times 10^{-4}\ M$). [From Fuerstenau and Palmer (1976), *Flotation, A.M. Gaudin Memorial Volume*, with the permission of the American Institute of Mining and Metallurgical Engineers, Littleton, Colorado.]

M. C. Fuerstenau and co-workers have carried out an extensive research program on several systems comprising a metal oxide and a collector-acting surfactant. Many of their findings on the action of activating metallic ions are relevant to the topic under discussion here. The quartz–sulfonate system and the action of activating ions such as Fe^{3+}, Al^{3+}, Pb^{3+}, Mn^{2+}, Mg^{2+}, and Ca^{2+} were described by Fuerstenau *et al.* (1963). Figure 8.24 shows the regions of pH where flotation of quartz occurs as a result of activation by the given metal ion. Flotation begins with the onset of ionized metal hydroxy complex formation and ceases when neutral metal hydroxide starts to precipitate. The quartz–xanthate system, activated by Pb^{2+} and Zn^{2+}, was described by Fuerstenau *et al.* (1965). The quartz–sulfonate system, activated by Pb^{2+}, and the quartz–carboxylate system, activated by Ca^{2+}, were described, respectively, by Fuerstenau and Atak (1965) and Fuerstenau and Cummins (1967). A general review paper on the activation mechanism by hydrolyzed species was published by Fuerstenau *et al.* (1970).

The conclusion reached was that, in all cases, adsorption of surfactants leading to flotation occurred only when the corresponding hydroxylated complexes were present. Two types of complexes have been recognized:

an ionized metal hydroxyl complex, $M(OH)^{(n-1)+}$, and a neutral hydroxyl complex of metal–collector, $M(OH)YR$, where R represents the hydrocarbon chain and Y the polar group. The latter complex is identical with that formed by the electrostatically bonded surfactant and the basic unit of the network (as suggested by Schulman), which polymerizes and in this manner immobilizes the surfactant. The results of M. C. Fuerstenau and co-workers provide an independent confirmation of the original hypothesis of Schulman on the special role of complexes in the immobilization of surfactant ions by hydrogen bonding as a prerequisite to flotation.

The concept developed by Schulman encompasses much more than the activating action of metal hydroxy complexes alone. It applies not only to cationic complexes immobilizing anionic surfactants but equally well to anionic complexes adsorbing amine-type collectors. With respect to the latter, the most comprehensive source of information are publications from Warren Spring Laboratory on flotation of silicates by Read and Manser (1975) and Manser (1975). Activation of feldspar by fluoride ion additions results in optimum separation from quartz by flotation at pH ~ 2.5. The bond $H \cdots F$ is by far the strongest hydrogen bond, as shown in Table 2.7, and the maximum in feldspar flotation is "clearly associated with the presence of HF as a stable species in solution." Yet none of the five explanations of fluoride activation mechanisms, listed by Read and Manser (1975, p. 9), includes HF_2^- complex as an activator. Neither is there any mention of tests carried out with complexes that have been suggested as activators (viz. SiF_6^{2-} AlF_4^-, etc.) to ascertain whether these are capable of solidifying monolayers of amine-type surfactants.[†]

In view of numerous roles that any metallic cation or anionic ligand may play in multicomponent flotation systems, it is extremely difficult to ascertain the exact mechanism involved in activation or adsorption of surfactants. The apparent abundance of a species may not necessarily be the determining factor in deciding its participation in adsorption at interfaces.

[†] An interesting example of activation by complex ions is provided by the suggestion of Shergold *et al.*, (1968) for the flotation of hematite with dodecylamine at pH 1.5 in HCl solution. A negatively charged site is created by Fe^{3+} in the solid by complexing with Cl^- ions and $FeCl_3$ in solution to give

$$\left[\begin{array}{c} Cl \quad OH \quad Cl \\ Fe \diagdown \quad Fe \diagdown \\ \diagdown Cl \quad Cl \quad \diagdown Cl \end{array} \right]^{\ominus} \cdots H_3N^{\oplus}-R$$

It would be instructive to confirm or to deny the existence of hydrogen bonding between the neighboring activating complexes.

8.4. Adsorption of Surfactants to Highly Soluble Salts in Their Saturated Salt Solutions

It is accepted that in flotation systems dealing with sparingly soluble solids the collector-acting surfactants *dissolve* in water, diffuse to the solid/ liquid interface of minerals, and *adsorb* on selected solids to make them hydrophobic in character. During the actual flotation process the already *hydrophobic* solid particles attach to the oscillating air bubbles with the help of frothers.

The separation of constituent solids from naturally occurring potash deposits, trona, and other highly soluble salts is carried out in saturated solutions which represent about 7–10 M concentrations of electrolytes (K^+, Na^+, Cl^-, SO_4^{2-}, CO_3^{2-}, etc.). Long-chain surfactants such as amines, carboxylates, and sulfonates employed in the flotation of salt components become highly insoluble in such concentrations of electrolytes. And yet flotation in saturated brines is kinetically analogous to flotation with soluble surfactants despite the high viscosities of solutions which lower the rates of surfactant diffusion.

Numerous hypotheses have been put forth in the last two decades to explain the mechanism of soluble salt flotation, with no agreement. Gaudin (1957) postulated that ion exchange is involved in the adsorption of surfactants on salts and that the difference in floatability of sylvite and halite, which are so similar chemically and crystallographically, is caused by the size relations among the K^+, Na^+, and RNH_3^+ ions. Fuerstenau and Fuerstenau (1956) extended the ionic size considerations to salts of the whole range of alkali and ammonium halides. Using vacuum flotation tests and contact angle measurements, they concluded from their studies that whenever the collector ion (i.e., the primary alkyl amine) can fit reasonably well into the crystal lattice of the salt surface in place of the constituent cation, that particular halide can be floated by alkyl amines.

Bachmann (1951, 1955) compared the structures of salt lattices with those of solidified alkyl amines and found that halides with lattice constants differing from the lattice constants of solidified alkyl amines by less than $\sim 20\%$ are relatively readily floated. However, he also found that the condition of lattice similarity was a necessary but not a sufficient condition. Although the three alums (NH_4^+, Na^+, and K^+ alum) have nearly identical lattices, only two of the alums, K^+ and NH_4^+, are floated with alkyl amines, not the Na^+ alum. Other salts with noncubic lattices such as K_2SO_4,

Table 8.1. Heats of Solution (ΔH) of Alkali Halides at 18°C in Kcal Mol^{-1}

	Li$^+$	Na$^+$	K$^+$	Rb$^+$	Cs$^+$	NH$_4^+$
			No flotation			
F$^-$	−1.0	−0.4	2.2	5.9	8.5	−1.5
Cl$^-$	4.9	−0.5	−3.8	−4.2	−3.9	−3.4
Br$^-$	7.7	−4.6^a	−4.1	−5.9	−6.6	−4.5
I$^-$	14.8	−3.9^a	−3.4	−6.3	−8.1	−3.6
	No flotation		Flotation			

^{a} Hydrate. The flotation of lithium fluoride was not tested.

$(NH_4)_2SO_4$, $K_2Mg(SO_4)_2 \cdot 6H_2O$, KNO_3, $CaCO_3$ (calcite),[†] etc., show lattice differences within the permissible limits and are reported to float with alkyl amines. Bachmann considered the lattice energy of the salt crystal—in addition to the similarity of lattices—as the other parameter that influences floatability. However, this still does not explain the effectiveness of other collector-acting surfactants. For example, alkyl sulfates float $BaSO_4$, KCl, and $MgSO_4 \cdot H_2O$ (kieserite), whereas alkyl amines float only KCl. Neither is the effectiveness of NaCl flotation with fatty acids nor its flotation after Pb^{2+} activation using alkyl amine as a collector explained by the criteria of lattice energy and lattice similarity.

Rogers and Schulman (1957) came to the conclusion that whenever the heat of solution of a given salt is positive, no adsorption or flotation is possible, but when it is sufficiently negative, adsorption and flotation can occur; see Table 8.1. A soluble salt with a negative heat of solution adsorbs (and can be floated by) only those surfactants whose polar groups are below a certain size. If the surfactant becomes preferentially precipitated, too few molecules (or ions) are left to make the salt surface sufficiently hydrophobic. Alkyl carboxylic and alkyl phosphoric acid surfactants can float salts with small negative heats of solution, while alkyl sulfonates, alkyl sulfates, and alkyl amines can float salts with larger negative heats of solution. Singewald (1961) critically reviewed these theories.

[†] For details on calcite floatability with alkyl amines, oleate, and sulphate ions, see the paper by Smani *et al.* (1975).

Schubert (1965, 1966) distinguished several contributions to the overall energy of collector adsorption, viz. that of van der Waals association of nonpolar (hydrocarbon) chains, that of hydration of the salt surface, and that of hydration of the polar group of surfactant in addition to the electrostatic interactions. The primary region of hydration affects the thermal motion of the nearest water molecules, either decreasing their frequency as in the case of Mg^{2+}, Ca^{2+}, Li^+, and Na^+ ions or increasing this frequency as in the case of K^+, Cs^+, Cl^-, Br^-, and I^-. These effects are reflected in the phenomenon of a *positive* and a *negative* viscosity change in electrolytic solutions [Schubert (1965)]. Thus, according to the interpretation of Schubert, the salt which is not strongly hydrated (such as KCl) can be floated with *n*-alkyl amines, *n*-alkyl sulfates, or *n*-alkyl sulfonates, whereas the more hydrated salt (such as NaCl) requires collectors which are capable of additional hydrogen bonding with the strongly attached water molecules of the primary region, e.g., fatty acids or *n*-alkyl morpholine.

Another parameter was introduced into the considerations of soluble salt flotation mechanisms by Roman *et al.* (1968), viz. the surface charge and its role in the adsorption of the collector. Although the type and the magnitude of the surface charge on the different salt/solution interfaces were inferred rather than experimentally determined, there is no doubt that differences in surface charge distribution must exist among the various salt surfaces. Under some conditions these surface charge differences may be important and may play a critical role in differentiating between flotation and nonflotation. Seidel *et al.* (1968) also determined the effect of temperature and of particle size on the selective flotation of potash. The effect observed was that an increase in temperature lowers the recovery and that the finer sizes of a nonfloating salt component become mechanically occluded in the froth, lowering the grade of the concentrate.

Arsentiev and Leja (1977) carried out investigations with monolayers of amines and carboxylic acids spread on saturated salt solutions. A single crystal of NaCl or KCl was then raised from underneath to establish contact with the spread monolayer. Subsequently, the crystal was pulled downwards, and the force necessary to separate it from the monolayer was determined. The tests were carried out with monolayers compressed to a state represented by a different area per molecule each time the separation force was determined. The results obtained are indicated in Figure 8.25. When a comparison is made with flotation tests of KCl or NaCl by respective surfactants, these results of force determinations suggest that yet another mechanism of surfactant adsorption may be operative in salt flotation. Due to the fact that the surfactants used are highly insoluble in concentrated brine solutions,

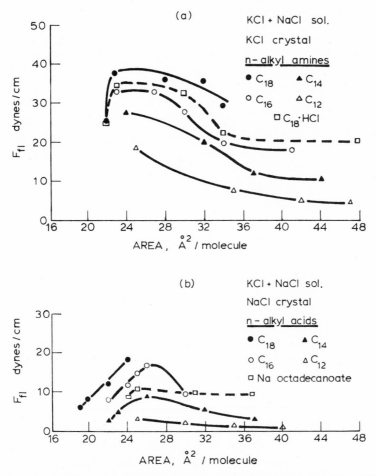

Figure 8.25. Adhesion force between a KCl or NaCl crystal and a surfactant film in relation to the area per molecule of insoluble (a) *n*-alkyl amines and (b) *n*-alkyl acids spread on a saturated (KCl + NaCl) solution in a Langmuir–Adam trough. Note that the force is lower for the corresponding surfactant salt in comparison with that for the un-ionized surfactant film. [From Arsentiev and Leja (1977), in *Colloid and Interface Science*, M. Kerker, ed., with the permission of Academic Press, New York.]

they cannot readily diffuse through the solution to the solid to make it hydrophobic. However, they can spread readily at the air/liquid interface the moment a contact is made between the insoluble surfactant particulate and a generated air bubble. The spread monolayer at the air bubble surface is compressed to a definite area per molecule depending on the quantity of surfactant added for a given rate of bubble generation.[†] A contact between a *hydrophilic* salt particle colliding with an air bubble which is covered with a spread surfactant leads to an attachment if a sufficiently high degree of adhesion is developed on contact. The degree of adhesion (as indicated by the magnitude of the separation force determined in monolayer studies; see Figure 8.25) depends on the *steric (geometrical) correlation* between the structure of the monolayer as decided by the polar groups and the hydrocarbon chains (Figure 8.26) and the hydration energy of the salt surface, which must be overcome by the adhesion. Branch-chain amines do not float KCl. A transfer of a surfactant (which adsorbs preferentially at the air/liquid interface) from the surface of the bubble onto the solid/liquid interface has been suggested by Schulman and Leja (1958); that suggestion dealt with conditions when a three-phase contact had been previously established. Digré and Sandvik (1968) suggested that in adsorption of amine on quartz a transfer of a surfactant from the air/liquid interface onto the solid/liquid interface may be involved at the moment of collision. The measurements of Arsentiev (1977) provide quantitative data to support the possibility of such a transfer mechanism operating under specific conditions.

The tendency to use salts (of fatty acids and of alkyl amines) as collectors is mainly due to their ability to be more easily dispersed in water. Flotation experiments showed that if a suitable solvent is used for amines (for example, isopropyl alcohol), they could be successfully used as collectors without being converted to salts. Experiments in the flotation of langbeinite, $K_2Mg_2(SO_4)_3$, showed that better flotation results are achieved when fatty acids are used in the acid form instead of their salts.

The various explanations proposed to account for the selectivity in soluble salt flotation provide an interesting example of the intricate balance that must be struck between numerous factors likely to contribute to the overall phenomenon of selective flotation. Crystal lattice, ionic size, hydration of polar groups and of salt surfaces, solubility of surfactants and their complexes, surface charge, temperature, etc., all contribute to the chemical aspect of flotation. In addition there are *physical* parameters due

---

[†] Under some operating conditions a second surfactant, *frother*, is needed. Its role may consist, primarily, in adjusting the state of the collector monolayer at the air/solution interface to the requisite degree of area per molecule. Such adjustment is needed if the initial monolayer is too condensed, or too diffuse.

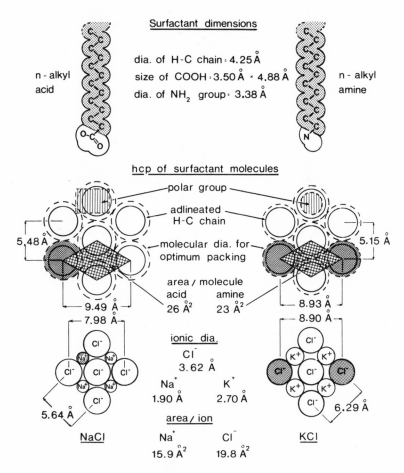

Figure 8.26. The importance of minor differences in steric parameters determining the structure of the surfactant film at the liquid/gas interface and those of the crystalline lattices of KCl and NaCl is highlighted by a comparison between the (probable) hpc arrangement of surfactant molecules in relation to the fcc lattices of NaCl and KCl. The greater degree of coincidence between the amine film and KCl lattice (shown on the right by the juxtaposed patterns) is (most likely) responsible for the higher adhesion forces, shown in Figure 8.25, for this system in comparison with the NaCl–alkyl acid system. [From Arsentiev and Leja (1977), in *Colloid and Interface Science*, M. Kerker, ed., with the permission of Academic Press, New York.]

to hydrodynamics, agitation, aeration, probability of encounter between air bubbles and particles, etc. Each parameter, whether physical or chemical, may come to the fore and become of critical importance in a particular set of conditions but otherwise may not play a major role.

8.5. *Adsorption of Surfactants Leading to Suppression of Floatability*

Many of the non-thio surfactants employed in flotation also act as detergents. Although the formulations of detergents comprise numerous components classed as builders, antideposition agents, and surfactants, the last ones are the primary materials that remove *dirt* or *soil* from a variety of substrates. Of the various mechanisms involved in detergency, the adsorption of surfactants on the substrate and on the soil surfaces in such a form as to break the adhesive bond between these two phases by electrostatic repulsion constitutes the primary action of a detergent surfactant. The condition under which this primary action of a detergent surfactant is most likely to be achieved is the use of residual surfactant concentrations exceeding the respective CMCs. It is only under such a condition that each of the surfaces in the system, regardless of its initial hydrophilic or hydrophobic state, is likely to be covered by an outer layer of ionized or dipole-oriented, similarly charged surfactants repelling one phase from the other, thus removing soil particulates from the substrates. For details of discussions on detergents, see Schwartz *et al.* (1966) or Schwartz (1972), and regarding techniques in studying detergency, see Schwartz (1979).

Whenever ionized non-thio surfactants used in flotation systems are added in excessive amounts, exceeding CMC concentrations, regardless of the mechanism of adsorption, a second inverted layer of surfactants is likely to be adsorbed. All solids then become hydrophilic and charged. Such situations are clearly indicated by adsorption isotherms in Figures 6.32, 8.17, and 8.27. Any air/liquid interface of air bubbles generated in such pulps becomes immediately aligned with ionized surfactants, making the bubbles highly charged, with charges of the same sign as those on the outermost layer of surfactants adsorbed on solid phases. Therefore, regardless of the initial character of solids in the flotation system, all solid particles dispersed in solutions of surfactants above CMC are charged with excess charges due to the polar group of the surfactant used, and, similarly, all air bubbles are charged with charges due to the same polar group of the same surfactants. On collision, the identical charges will repel the two particulates—particles and colliding air bubbles; no attachment is likely to occur if concentrations exceed the CMC sufficiently to ensure that, whatever the kinetics of adsorption are on the solid and on the air bubbles, these two types of interfaces will be charged identically.

This action of non-thio collectors can easily be demonstrated in test-tube experiments with a given solid particle. For example, a small quantity

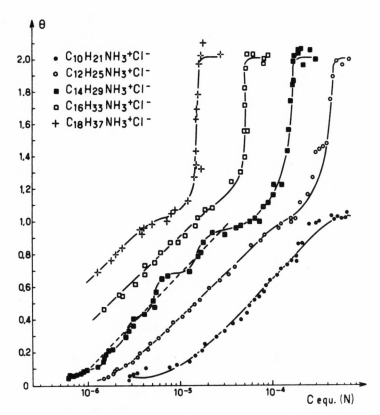

Figure 8.27. Adsorption isotherms of alkyl amine hydrochlorides (C_{10}–C_{18} hydrocarbons). Concentrations at which the double-layer coverage is complete have been evaluated as C_{12}, 5.40×10^{-4} M; C_{14}, 1.70×10^{-4} M; C_{16}, 5.50×10^{-5} M; and C_{18}, 1.72×10^{-5} M. [From Cases (1968), with the permission of the author.]

of quartz (say, $-100 + 270$ mesh size) added to a test tube containing a solution of n-C_{18} amine hydrochloride in 10^{-1} M concentration will show (on shaking the contents) a fully hydrophilic quartz, not showing any tendency to adhere to (voluminous) air bubbles. After one or more decantations and dilutions with water the same quartz particles will behave as hydrophobic, attaching to air bubbles and indicating high floatability.

A complete cessation of flotation has been indicated in a number of systems whenever residual concentrations of the collectors used exceeded their respective CMCs. Figure 8.28 gives the correlation of data confirming this statement.

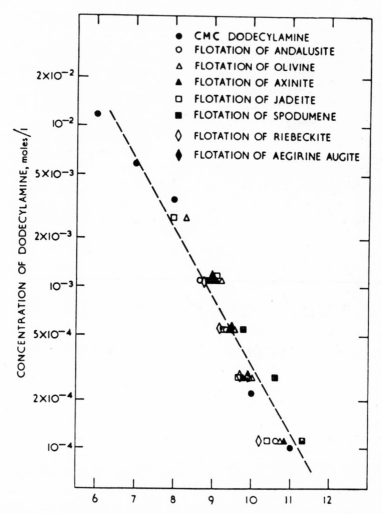

Figure 8.28. Correlation between the critical micelle concentrations of dodecylamine (as affected by changes in the pH of solutions) and the cessation of floatability for various silicates. Flotation ceases to the right of the line. [From Watson and Manser (1968), *Trans. IMM* 77, with the permission of the Institution of Mining and Metallurgy, London.]

8.6. *Concluding Remarks on Adsorption Mechanisms*

Different mechanisms of surfactant adsorption onto a given solid may operate under different conditions of pH, e.g., absence or participation of complexing ions, the type of solid surface exposed to surroundings, etc.

Examples of such variations have been provided by Fuerstenau and Rice (1968) in the flotation of pyrolusite and by Fuerstenau *et al.* (1968) in the flotation of pyrite (Figure 8.6). The interpretations may differ, and the correct assignment of the adsorption mechanisms may only be possible after exhaustive additional determinations.

The unavoidable conclusion is, however, that the action of a surfactant as a collector for a given solid cannot be ascribed to a single adsorption mechanism—a single type of bond—regardless of conditions imposed on the system. There is always a cooperative effect of two or more types of bonds needed to develop the required degree of immobilization of surfactant and of hydrophobicity on the solid surface. Whether a charge transfer has taken place or not (i.e., whether chemisorption or physical adsorption is involved) is less important than the cooperative effect of bonds between

1. The polar group and the surface site
2. The surface site and the rest of the solid
3. The adjoining polar groups
4. The hydrocarbon portions of the surfactant(s)

All these bonds exert their effect with respect to bonds developed by water molecules at the shear plane of the solid surface that is covered by surfactant(s) (and also at the initial, uncovered surface). Steric parameters, i.e., the relative densities of respective bonds, become important and may play a decisive role. Also, since a shear plane is established during a particle-to-bubble attachment, dynamic forces within the pulp may play, and do play, an important role in improving the selectivity or reducing the recovery in flotation.

Because of the multiplicity of bonding involved in establishing the necessary hydrophobicity of particles under dynamic flotation conditions, it is possible to achieve relatively high selectivity in separations. However, any improvements will come about by exertion of a more intricate control over the variety of parameters, which is possible only if their correct role in the process is adequately understood.

8.7. *Selected Readings*

Aplan, F. F., and Fuerstenau, D. W. (1962), Principles of non-metallic mineral flotation, in *Froth Flotation—50th Anniversary Volume*, ed. D. W. Fuerstenau, AIME, New York, pp. 170–214.
Ball, B., and Rickard, R. S. (1976), The chemistry of pyrite flotation and depression, in

Flotation, A. M. Gaudin Memorial Volume, Vol. I, ed. M. C. Fuerstenau, AIME, New York, pp. 458–484.

Bleier, A., Goddard, E. D., and Kulkarni, R. D. (1976), The structural effects of amine collectors on the flotation of quartz, in *Flotation, A. M. Gaudin Memorial Volume*, Vol. I, ed. M. C. Fuerstenau, AIME, New York, pp. 117–147.

Cook, M. A. (1968), Hydrophobicity control of surfaces by hydrolytic adsorption, *J. Colloid Interface Sci.* **28**, 547–556.

Cook, M. A., and Nixon, J. C. (1950), Theory of water repellant films on solids formed by adsorption from aqueous solutions of heterpolar compounds, *J. Phys. Colloid Chem.* **54**, 445–459.

du Rietz, C. (1976), Chemisorption of collectors in flotation, in *Proceedings 11th IMPC*, 1975, Instituto di Arte Mineraria, Università di Cagliari, Cagliari, Italy, pp. 375–403.

Finkelstein, N. P., and Allison, S. A. (1976), The chemistry of activation, dewatering and depression in the flotation of zinc sulphide, a review, in *Flotation, A. M. Gaudin Memorial Volume*, Vol. I, ed. M. C. Fuerstenau, AIME, New York, pp. 414–457.

Finkelstein, N. P., and Poling, G. W. (1977), The role of dithiolates in the flotation of sulphide minerals, *Miner. Sci. Eng.* **9**, 177–197.

Fuerstenau, D. W., and Raghavan, S. (1976), Some aspects of the thermodynamics of flotation, in *Flotation, A. M. Gaudin Memorial Volume*, Vol. I, ed. M. C. Fuerstenau, AIME, New York, pp. 21–65.

Fuerstenau, M. C., and Palmer, B. R. (1976), Anionic flotation of oxides and silicates, in *Flotation, A. M. Gaudin, Memorial Volume*, Vol. I, ed. M. C. Fuerstenau, pp. 148–196.

Garrels, R. M., and Christ, C. L. (1965), *Solutions, Minerals and Equilibria*, Harper & Row, New York.

Gaudin, A. M. (1957), *Flotation*, McGraw-Hill, New York.

Glembotskii, V. A., Klassen, V. I., and Plaksin, I. N. (1963), *Flotation* (trans. from Russian), Primary Sources, New York.

Granville, A., Finkelstein, N. P., and Allison, S. A. (1972), Review of reactions in the flotation galena–xanthate–oxygen, *Trans. IMM* **81**, C1–C30.

Gutierrez, C. (1973), The mechanism of flotation of galena by xanthates, *Miner. Sci. Eng.* **5**, 108–118.

Hanna, H. S., and Somasundaran, P. (1976), Flotation of salt-type minerals, in *Flotation, A. M. Gaudin Memorial Volume*, Vol. 1, ed. M. C. Fuerstenau, AIME, New York, pp. 197–272.

Healy, T. W., and Moignard, M. S. (1976), A review of electrokinetic studies of metal sulphides, in *Flotation, A. M. Gaudin Memorial Volume*, Vol. 1, ed. M. C. Fuerstenau, AIME, New York, pp. 275–297.

Joy, A. S., and Robinson, A. J. (1964), Flotation, in *Recent Progress in Surface Science*, Vol. 2, ed. J. F. Danielli *et al.*, Academic Press, New York, pp. 169–260.

Klassen, V. I., and Mokrousov, V. A. (1963), *An Introduction to the Theory of Flotation* (trans. from Russian by J. Leja and G. W. Poling), Butterworths, London.

Kubaschewski, O., and Hopkins, B. E. (1962), *Oxidation of Metals and Alloys*, Butterworth's, London.

Leja, J. (1968), On the mechanisms of surfactant adsorption, in *Proceedings, VIII International Mineral Processing Congress, Leningrad, 1968*.

Leja, J. (1973), Some electrochemical and chemical studies related to froth flotation with xanthates, *Miner. Sci. Eng.* **5**, 278–286.

Little, L. H. (1966), *Infrared Spectra of Adsorbed Species*, Academic Press, New York.

Palmer, B. R., Fuerstenau, M. C., and Aplan, F. F. (1975), Mechanisms involved in the flotation of oxides and silicates with anionic collectors, Part II, *Trans. AIME* **258**, 261–263.

Poling, G. W. (1976), Reactions between thiol reagents and sulphide minerals, in *Flotation, A. M. Gaudin Memorial Volume*, Vol. 1, ed. M. C. Fuerstenau, AIME, New York, pp. 334–363.

Polkin, S. I., and Berger, G. S. (1968), On forms of fixation, and flotation action of long chain collectors, in *Proceedings, VIII International Mineral Processing Congress, Leningrad, 1968*, Mekhanobr Institute, Leningrad, Paper S-7.

Pourbaix, M. J. N. (1949), *Thermodynamics of Dilute Aqueous Solutions*, Edward Arnold & Co., London.

Predali, J. J., and Cases, J. M. (1974), Thermodynamics of the adsorption of collectors, in *Proceedings 10th IMPC, London 1973*, ed. M. J. Jones, Institute of Mining and Metallurgy, London, pp. 473–492.

Rao, S. R. (1969), Collector mechanism in flotation, *Sep. Sci.* **4**, 357–411.

Rao, S. R. (1971), *Xanthates and Related Compounds*, Marcel Dekker, New York.

Richardson, P. E., and Maust, E. E., Jr. (1976), Surface stoichiometry of galena in aqueous electrolytes and its effect on xanthate interactions, in *Flotation, A. M. Gaudin Memorial Volume*, Vol. 1, ed. M. C. Fuerstenau, AIME, New York, pp. 364–392.

Rochester, C. H. (1976), Infrared spectroscopic studies of powder surfaces and surface-adsorbate interactions including the solid/liquid interface, the review, *Powder Technol.* **13**, 157–176.

Rogers, J. (1962), Principles of sulphide mineral flotation, in *Froth Flotation—50th Anniversary Volume*, ed. D. W. Fuerstenau, AIME, New York, pp. 139–169.

Schubert, H., and Schneider, W. (1968), Role of non-polar groups association in collector adsorption, in *Proceedings, VIII International Mineral Processing Congress, Leningrad, 1968*, Vol. 2, Mekhanobr Institute, Leningrad, Paper S-9, pp. 315-324.

Sheludko, A. (1963), Zur theorie der flotation, *Kolloid Z.* **191**, 52–58.

Shimoiizaka, J., Usui, S., Matsuoka, I., and Sasaki, H. (1976), Depression of galena flotation by sulphite or chromate ion, in *Flotation, A. M. Gaudin Memorial Volume*, Vol. I, ed. M. C. Fuerstenau, AIME, New York, pp. 393–413.

Smith, R. S., and Akhtar, S. (1976), Cationic flotation of oxides and silicates, in *Flotation, A. M. Gaudin Memorial Volume*, Vol. I, ed. M. C. Fuerstenau, AIME, New York, pp. 87–116.

Sutherland, K. L. and Wark, I. W. (1955), *Principles of Flotation*, Australasian Institute of Mining and Metallurgy, Melbourne.

Taggart, A. F., Taylor, T. C., and Knoll, A. F. (1930), Chemical reactions in flotation, *Trans. AIME* **87**, 217–260.

Taggart, A. F., del Giudice, G. R. M., and Ziehl, O. A. (1934), The case for the chemical theory of flotation, *Trans. AIME* **112**, 348–381.

Usui, S., and Iwasaki, I. (1970a), Adsorption studies of dodecylamine at the mercury/solutions interface through differential capacity and electrocapillary measurements and their implication in flotation, *Trans. AIME* **247**, 214–219.

Usui, S., and Iwasaki, I. (1970b), Effect of pH on the adsorption of dodecylamine at the mercury/solution interface, *Trans. AIME* **247**, 220–225.

Woods, R. (1976), Electrochemistry of sulphide flotation, in *Flotation, A. M. Gaudin Memorial Volume*, Vol. 1, ed. M. C. Fuerstenau, AIME, New York, pp. 298–333.

9

Flotation Froths and Foams

Foams and froths are of considerable practical importance not only in froth flotation of minerals or in adsorptive bubble-separation techniques applied to water purification but also in the food industry. Protein-stabilized foams have to be carefully controlled in the boiling and the fermentation stages of food preparation. Also, fire-fighting foams must meet exacting specifications as regards drainage, high temperature stability, speed of generation, etc. Obnoxious foams in boilers and in various stages of oil production and processing must be effectively combated. There have been several books and numerous reviews published on the subject of foams, for example, Bikerman *et al.* (1953), Manegold (1953), de Vries (1957), Kitchener and Cooper (1959), and Akers (1976). The last is the proceedings of a symposium, comprising 18 papers dealing with different aspects of foams.

9.1. *Single Gas Bubbles in Liquid*

Bubbles can be generated in liquids by several different methods, for example,

1. An increase in temperature to cause boiling; at first, dissolved gases are released, and ultimately vapor-filled bubbles are generated.
2. A decrease in pressure to cause *precipitation of bubbles* (only dissolved gases are released).
3. A mechanical agitation to cause gas entrapment.
4. An injection of pressurized gas through an orifice or a porous membrane (plug).

Depending on their diameter, the bubbles may be classed as macrobubbles [of diameters greater than $\sim$0.1 mm (subgroups: small or large)], microbubbles (of diameters 1–1000 μm), or submicrobubbles (< 1 μm).

Once formed within a pure liquid, bubbles tend to coalesce[†] (to reduce the extent of total surface area and thus the overall amount of the free energy in the system) and—if free—to rise in the liquid under the effect of buoyancy. Both the coalescence and the rise of bubbles can be prevented, the coalescence by surface active impurities and the rise by attachment to a hydrophobic solid surface.

In pure liquids bubbles tend to coalesce regardless of the character of the liquid, whether polar or nonpolar. When a liquid contains an impurity, a dissolved surfactant, the coalescence of bubbles may be strongly retarded or altogether prevented, giving rise to a foam or a froth. A single bubble in a liquid, under equilibrium conditions, assumes the shape of a minimum free energy in the system, i.e., the smallest (near-spherical) surface area of the gas/liquid interface. The fundamental equation of capillarity, known as the equation of Young and Laplace, relates the difference in pressure, $p_1 - p_2$, between the pressure on the concave side, i.e., within the bubble, p_1, and that on the convex side, p_2, with the interfacial tension $\gamma_{a/1}$ and the two principal radii of curvature R_1 and R_2,

$$p_1 - p_2 = \gamma_{a/1}\left(\frac{1}{R_1} + \frac{1}{R_2}\right) \qquad (9.1)$$

for a system in which gravitational and other forces are absent. In the presence of gravitational forces, the shape of the liquid surface can—in theory—be calculated from a more general relationship:

$$p_1 - p_2 = \gamma_{a/1}\left(\frac{1}{R_1} + \frac{1}{R_2}\right)g(\varrho_1 - \varrho_2)Z \qquad (9.2)$$

where g is the gravitation acceleration, ϱ_1, ϱ_2 are the densities of the two phases, and Z is the level of the meniscus from a reference level (on the concave side of the meniscus). Since the radii of curvature cannot be readily measured, approximate solutions have been provided by Bashforth and Adams (1883) for figures of revolution about a vertical axis.

Free gas bubbles ascend (rise) in the liquid to its surface under the effect

---

[†] The mechanism of coalescence of liquid drops at liquid/liquid interfaces is discussed by Charles and Mason (1960a,b). Sagert *et al.* (1976) investigated bubble coalescence in aqueous solutions of *n*-alcohols.

of buoyancy, which is determined by Archimedes' law:

$$F = \frac{\pi d^3}{6} (\varrho_1 - \varrho_2)g \qquad (9.3)$$

where d is the diameter of the bubbles, ϱ_1 is the density of liquid, and ϱ_2 is the density of gas. Small bubbles, less than about 0.2-mm diameter, behave in pure water essentially like rigid solid spheres, and the relationship due to Stokes (1880) between the terminal velocity v and the radius r, (η is the viscosity of liquid)

$$v = \frac{2}{9} \frac{gr^2(\varrho_1 - \varrho_2)}{\eta} \qquad (9.4)$$

is found to apply. Ascending bubbles larger than ~ 0.2 mm begin to change their shape, at first to spheroidal and then to various convoluted and distorted cups (as a result of slippage and flexibility of their interface). The theoretical formula, due to Rybczynski (1911), for the terminal velocity of such flexible bubbles modified the velocity given by the preceding relationship of Stokes by a factor

$$\frac{3\eta_1 + 3\eta_2}{3\eta_1 + 2\eta_2} \simeq 1.5$$

where η_1 is the viscosity of air and η_2 is the viscosity of water. Levich (1949) derived an approximate equation for the velocity of single bubbles in the absence of deformation:

$$v = \frac{1}{36} \frac{d^2 g(\varrho_1 - \varrho_2)}{\eta} \qquad (9.5)$$

This equation was found to apply to bubbles of approximately $d \cong 1$ mm, i.e., Reynolds number[†] Re ~ 200.

[†] *Note:* The Reynolds number is defined as the ratio of the forces of inertia to the forces of viscosity:

$$\text{Re} = \frac{U_0 L \varrho}{\eta} \qquad (9.6)$$

where U_0 is the characteristic velocity of the fluid motion, L is the characteristic dimension (such as the radius of the tube in which the fluid flows, or the diameter of the orifice, or the diameter of the bubble, etc.), and ϱ is the density of fluid. The Reynolds number represents a dimensionless parameter which is developed by the similarity theory for the treatment of hydrodynamics. For a steady flow of an incompressible liquid the Reynolds number represents the only controlling parameter in terms of which all other hydrodynamic variables may be expressed. For an unsteady flow or a flow in the presence of external forces, other controlling parameters are needed as well to

For other sizes of bubbles, the experimental terminal velocities vary as shown in Figure 9.1. There appears to be a fairly wide spread of velocity values. Gorodetskaya's (1949) results reproduced in Figure 9.2 bring out clearly the major influence of traces of surfactants introduced into water:

1. The terminal velocity of bubbles is progressively decreased by a factor of 1.5–2.5, particularly for intermediate sizes of bubbles.
2. The average size of bubbles issuing from the same orifice or formed in a flotation cell is also decreased, as observed by Bogdanov *et al.* (1950) and Miagkova (1955) (Figure 9.3).

Gas bubbles in a liquid are unstable; if the liquid is undersaturated with gas, the bubbles diminish in size as the gas diffuses into the liquid; if the liquid is oversaturated, the bubbles increase in size. Thermal, viscous, and elastic properties of a liquid may be considerably influenced by the existence of bubbles within the liquid. Freely rising bubbles in an undersaturated liquid exhibit a dissolution rate more than twice that of stationary (trapped or attached) bubbles. For an air–water system, the coefficient of diffusivity was found by Liebermann (1957) to be 1.7×10^{-5} cm²/sec at 8°C and 2.9×10^{-5} cm²/sec at 27°C for attached bubbles; for a freely rising bubble, it was determined as 6.3×10^{-5} cm²/sec. The observed rate of gas transfer from a freely rising bubble is apparently independent of velocity.[†]

Brown *et al.* (1953) [quoted in detail by Davies and Rideal (1961, pp. 404–408)] evaluated the permeability coefficients in tests on single bubbles of air released beneath the surface of fairly concentrated solutions, 0.1%, of different surfactants by determining the changes in diameter of the bubble with time. The values of permeability obtained varied from 0.2×10^{-2} to 1.3×10^{-2} cm/sec.

The permeability coefficient k for mass transfer of, for example, a gas across a plane surface is defined as the coefficient in the differential equation

$$\frac{dq}{dt} = kA \, \Delta c \tag{9.7}$$

---

describe the hydrodynamic relationships of the system. A small Reynolds number means that viscous forces predominate, and a large one that the forces of inertia predominate. The unit of viscosity η is newton-second per square meter (N · sec m⁻²) in the SI system (1 poise = 1 g/cm · sec in the cgs system). The ratio $\eta/\varrho = \nu$ is called the kinematic viscosity, and its unit is square meter per second (m² sec⁻¹) (1 stoke = 1 cm/sec in the cgs system).

[†] A detailed assessment of data on absorption rates of CO_2, C_2H_4, C_4H_8, and N_2 from a single rising bubble has been published by Johnson *et al.* (1969).

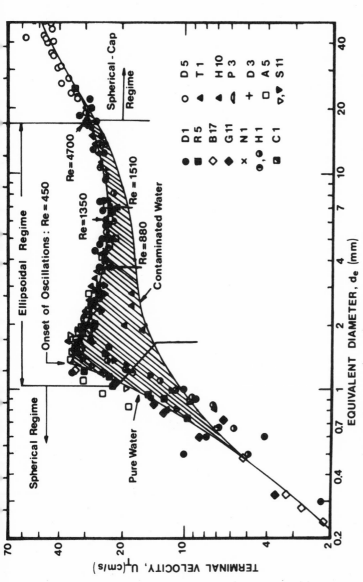

Figure 9.1. Experimental terminal velocities of gas bubbles in pure and contaminated water. Reynolds numbers and regimes where bubbles behave as spherical, ellipsoidal, or spherical cap shapes are indicated. ●, Datta, R. L., *et al.* (1950), *Trans. Inst. Chem. Eng.* **28**, pp. 14–26. ○, Davies, R. M., and Taylor, Sir. G. I. (1950), *Proc. Roy. Soc.*, Ser. **A200**, pp. 375–390. ■, Rosenberg, B. (1950), *David Taylor Model Basin Rep. No. 727.* ▲, Tadaki, T., and Maeda, S. (1961), *Kagaku Kogaku* **25**, pp. 254–264. ◇, Bryn, T. (1949), *David Taylor Model Basin Rep. No. 132.* ▲, Houghton, G., *et al.* (1957), *Chem. Eng. Sci.* **7**, pp. 111–112. ◆, Gorodetskaya, A. (1949), *Zh. Fiz. Khim.* **23**, p. 71–77. ◓, Peebles, F. N., and Garber, H. J. (1953), *Chem. Eng. Prog.* **49**(2), 88–97. ×, Napier, D. H., *et al.* (1950), *Trans. Inst. Chem. Eng.* **28**, 14–31. +, Davenport, W. G., *et al.* (1967), *Chem. Eng. Sci.* **22**, 1221–1235. ◑ and ◐, Haberman, W. L., and Morton, R. K. (1953), *David Taylor Model Basin Rep. No. 802.* □, Aybers, N. M., and Tapucu, A. (1969), *Wärme-Stoffübertrag.* **2**, 171–177. ◪, Calderbank, P. H., *et al.* (1970), *Chem. Eng. Sci.* **25**, 235–256. ▽ and ▼, Sumner, B. S., and Moore, F. K. (1970), *NASA Contract Rep. CR-1669.* [From Clift *et al.* (1978), *Bubbles, Drops and Particles*, with the permission of Academic Press, New York.]

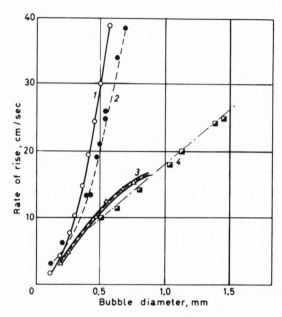

Figure 9.2. A comparison of terminal velocities of bubbles and hollow spheres in water: (1) according to Levich's equation, (2) measured in pure water, (3) measured in solution of 10^{-4} M n-amyl alcohol, and (4) hollow spheres with a correction for density. [After Gorodetskaya (1949), from Klassen and Mokrousov, 1963, *An Introduction to the Theory of Flotation*, Trans. by J. Leja & G. W. Poling, with permission of Butterworths, London.]

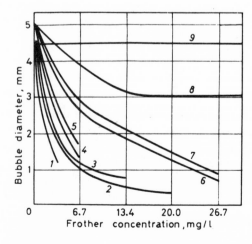

Figure 9.3. Effects of surfactant's concentration on the average size of bubbles: (1) octyl alcohol, (2) terpineol, (3) pine oil, (4) hexyl alcohol, (5) amyl alcohol, (6) cresol, (7) laurylamine, (8) oleic acid, and (9) xanthate. [After Miagkova (1955), from Klassen and Mokrousov, 1963, *An Introduction to the Theory of Flotation*, Trans. by J. Leja & G. W. Poling, with permission of Butterworths, London.]

where q is the moles of material transferring in time t, A is the surface area across which the transfer occurs, and Δc is the difference in concentration in moles per cubic centimeter. The diffusion coefficient D is defined by

$$\frac{dq}{dt} = DA\,\frac{\Delta c}{\Delta x} \tag{9.8}$$

where Δx is the thickness of the region through which transfer occurs; all other terms have the same meaning as in equation (9.7), and $\Delta c/\Delta x$ is the concentration gradient. Thus,

$$k = \frac{D}{\Delta x} \tag{9.9}$$

Two plots of a^2 (where a is the radius of the bubble released under the surface) against time, obtained by Brown *et al.* (1953), are shown in Figure 9.4; from these, $k = 1.3 \times 10^{-2}$ cm/sec for the initial portion of the plot for 0.1% NaLS (sodium lauryl sulfate) solution, and $k = 0.5 \times 10^{-2}$ cm/sec for a mixed 0.1% NaLS and 0.002% LOH (lauryl alcohol) solution. Different permeabilities are interpreted as an indication of a different thickness Δx through which the transport of gases occurs—and not as a variation in diffusion coefficient D.

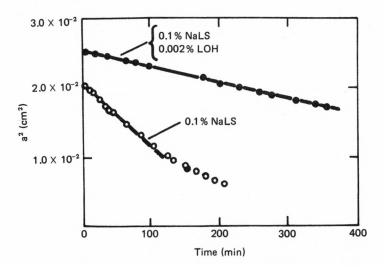

Figure 9.4. Data plot for an evaluation of permeability coefficients [in equation (9.7)] for 0.1% sodium lauryl sulfate (NaLS) solution and a mixed 0.1% NaLS + 0.002% lauryl alcohol (LOH) solution. The addition of alcohol reduces permeability (as indicated by the decrease in the slope). [From Brown *et al.* (1953), *J. Colloid Sci.* **8**, with the permission of Academic Press, New York.]

Another aspect of importance is the relative movement of layers adjoining air/liquid or liquid/liquid interfaces. Linton and Sutherland (1957) analyzed the internal circulation within a moving drop (of an insoluble liquid rising under buoyancy in another liquid). When low interfacial liquid–liquid tension exists and there are no solutes in either of the liquid phases, internal circulation occurs readily and gives high rates of mass transfer between the two liquids. When surfactants are present in the system, they tend to concentrate at the rear portion of the moving drop; this gives rise to a lower interfacial tension at the rear of the drop and thus a higher surface pressure, which tends to spread the surfactant to the front of the moving drop and so oppose the flow along the surface. The movement of the surfactant film causes a considerable movement of the underlying liquid, particularly if the adsorption–desorption of surfactants is not as fast as the transport over the surface. Garner and Hammerton (1954) found that a vigorous toroidal circulation sets in in rising bubbles of > 0.3-mm diameter.

The presence of contaminants was found by Liebermann (1957) to affect the dissolution of bubbles to a negligible extent. Bubbles of size less than 1 μm, submicrobubbles, if they are attached to hydrophobic particles, appear to be stable, are not soluble, and can exist indefinitely. The surface tension pressures opposing the hydrostatic pressure of the liquid for such submicrobubbles are of the order of $2\gamma/R \cong 10^6$ dyn/cm^2. These submicrobubbles appear to act as nuclei for bubble growth (when the degree of saturation is increased) and in boiling and cavitation.

Freely rising bubbles produced by an escape from an orifice or by boiling acquire a vibrational motion, oscillate, or pulsate; depending on the frequency of such oscillations, even audible sounds are produced by radial pulsation of bubbles. Spedden and Hannan (1948) used high-speed photography to study particle–bubble encounter in a collector solution (galena particles in xanthate solution) and found that bubbles escaping from a nozzle undergo vibrations with a frequency of $\sim$1000 cycles/sec. Fuerstenau and Wayman (1958) observed that the size at which bubbles begin to oscillate, and thus lose their sphericity, depends on the amount of surfactant (terpineol) added. An indication of both the frequency of oscillations and the variety of shapes adopted during bubble ascent is provided by a sequence of frames (see Figure 9.5) from high-speed photography of bubble attachment carried out by Schulman and Leja (1958). The time interval between successive frames (as assembled) represents 0.0025 sec. Elastic vibrations and contortions of bubbles were found to be much greater in pure water (Figure 9.5) than in a 10^{-3} M α-terpineol solution. Frother additions dampened oscillations and reduced the velocity of bubble ascent.

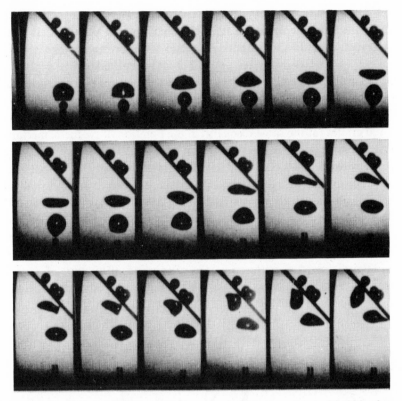

Figure 9.5. Examples of bubble oscillations acquired during release of bubbles from a near-capillary glass tube (by displacement of air using 10–12 cm of water head). Sequence of frames: top left progressing to the right in each row. Time interval between each two frames is 0.0025 sec. Bubbles are colliding with an inclined glass slide made hydrophobic by coating it with wax. [Schulman and Leja (1958), from *Surface Phenomena in Chemistry and Biology*, J. F. Danielli, K. G. A. Pankhurst, and A. C. Riddiford, eds., with permission of Pergamon Press, Oxford.]

The oscillations of bubbles affect the rate of bubble dissolution and bubble coalescence and, at certain frequencies, can have a destructive effect on vegetation and living matter. Harbaum and Houghton (1960) found that the absorption of carbon dioxide from bubbles by water is strongly affected by the frequency generated within the liquid by a vibrator. The absorption rate showed peaks at certain frequencies. The most significant increase in the absorption rate at 70–80 cycles/sec was also associated with an increase in turbulence and coalescence in the bubble bed. At some frequencies, such as 250 cycles/sec, the coalescence of bubbles decreased below the level for no vibration. These findings suggest that oscillations of bubbles in flotation systems may occur over a wide range of frequencies not only within the audible range (20–20,000 cycles/sec) but also in the ultrasonic range. In fact, Stoev and Pirinkov (1966) found that when frequencies of 50–100 cycles/sec and 0.1–2 mm amplitude were applied to a froth in a flotation cell, a better grade of a concentrate was obtained due to the removal of the mechanically entrained gangue minerals. Stoev and Watson (1967) obtained improved recoveries (at the same grade) and improved grades (at the same recovery) in the flotation of synthetic mixtures of barytes–quartz (see Figure 9.6) using an ultrasonic vibrator of 20-kcycle/sec frequency with 200-, 400-, and 600-W output intensity. A systematic investigation on the effects of vibration frequency, size of bubbles, and other parameters seems to be warranted.

A beam of high-intensity ultrasound causes cavitation, i.e., formation of micro- and submicrobubbles in a liquid. If the liquid contains dissolved air or other gases, the rapid changes in pressure produced by the sound cause the dissolved gases to come out of solution. (This type of cavitation

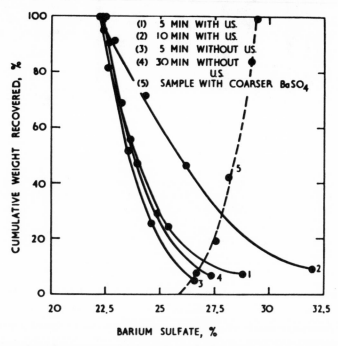

Figure 9.6. Effects of ultrasonic vibrator (U.S.) applications to mineralized froth on the barytes content of the froth (0 on ordinate axis represents bottom of froth). [From Stoev and Watson (1967), *Trans. IMM* **76**, with the permission of the Institution of Mining and Metallurgy, London.]

is sometimes called pseudocavitation.) Pseudocavitation can accelerate certain types of chemical reactions: For example, H_2O_2 is formed in water treated by ultrasonic vibration when dissolved oxygen is present in water. True cavitation occurs when the liquid is completely degassed and freed from contaminants capable of producing bubble nuclei and the change in accoustic pressure is great enough to overcome the cohesional strength of the liquid.[†] The change in pressure necessary to produce true cavitation depends on viscosity; it is about 40 atm for ether, about 30 atm for water, and about 200 atm for plant sap [from Dean (1944)].

9.2. *Foams and Froths (Two-Phase Systems)*

When two bubbles come in contact with each other, the liquid film between them thins and breaks, causing bubbles to coalesce, unless excess surfactant is present at their interfaces or the time of contact is too short. Bubble coalescence can be studied either on two bubbles grown in solution from adjacent nozzles or on a column of foam or froth generated in solution by blowing compressed gas through a fine sintered-glass frit. Sagert *et al.* (1976) determined experimentally the times for coalescence of two bubbles emerging from adjacent nozzles as a function of alcohol concentration (Figure 9.7). Using high-speed photography, they could determine times down to 0.5 msec. Their results show that even very small concentrations ($1 \text{ ppb} = 10^{-9}$) appear to slow (lengthen) the coalescence time in the case of $C_6H_{13}OH$.

When coalescence of bubbles formed in a liquid (or released into it) does not take place in fractions of seconds, the bubbles rise to the surface and aggregate, forming a foam or a froth. In a review on foaming, Kitchener and Cooper (1959) provided a sequence of photographs showing a gradual transition of structure from that of a nondrained foam with spherical bubbles as, for example, in Figure 9.8, to a relatively well-drained froth with polyhedral bubbles, separated by very thin lamellae, as in Figure 9.9. The terms *foam* and *froth* are sometimes used interchangeably; more frequently, however, the imperfectly drained liquid–bubble systems, generally unstable,

---

[†] *Note*: The reduction in pressure necessary to overcome the cohesional strength of the liquid is often referred to as the *negative pressure*; a liquid in contact with a gas phase can rise under suction only to a height not greater than the barometric height for that liquid; its negative pressure is 1 atm at the most. However, a liquid which is completely degassed and is not exposed to any gas phase can withstand very much higher negative pressures without rupturing.

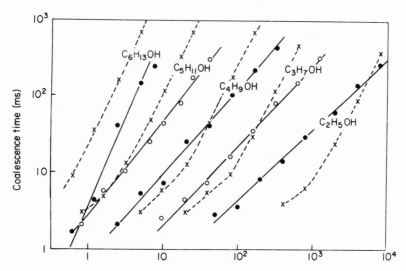

Figure 9.7. Double logarithmic plots of both experimental and calculated coalescence times against solute concentration for the five alcohols investigated; ●, experimental; ×, theoretical. [From Sagert *et al.* (1976), in *Foams*, R. J. Akers, ed., with the permission of the Society of Chemical Industry, London.]

are called foams, while the more persistent, better drained bubble systems are referred to as froths.

Pure liquids of low viscosity do not foam or froth[†] and neither do the solutions containing hydrophilic solutes (purified inorganic salts, sucrose, or glycerol). Unstable, transient foams are obtained with short-chain surfactants, such as alcohols or carboxylic acids, which lower the surface tension of their aqueous solutions to a moderate extent; persistent two-phase froths are obtained with solutions of ionized surfactants, particularly those which lower the surface tension rather strongly even in highly dilute solutions.

There is a strong correlation between the viscosity of the liquid (and particularly the surface viscosity) and the stability of foam. The more viscous the liquid into which the bubbles are injected, the slower is the drainage of liquid from layers between the bubbles. A comparison of stability of foams built up in pure viscous liquids of different surface tensions (glycerol, liquid paraffin, and silicone) showed that the rate of collapse was the same for all systems when their viscosities were made equal by keeping each system at an appropriate temperature. The value of surface tension was not a critical factor in these foams.

[†] Except, probably, viscoelastic liquids composed of coiled, chain-like molecules.

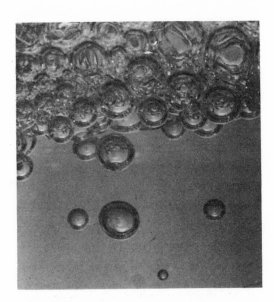

Figure 9.8. A foam, nondrained, with near-spherical bubbles.

Bikerman *et al.* (1953) distinguished two classes of foams:

1. Foams whose stability is due to a high viscosity or rigidity of films, as, for example, solid foams of rubber latex, expanded polystyrene foam, glass foam, polyurethane foam, and protein foams in food products (whipped cream, etc.).

Figure 9.9. A relatively well-drained froth with polyhedral bubbles.

2. Foams whose persistence is due to the Marangoni effect (see below).
Kitchener (1964) distinguished three types of foam:

1. A temporary dispersion of near-spherical bubbles in a viscous
liquid (Kugelschaum)
2. A transient foam obtained with dilute solutions of short-chain,
mostly nonionic, and weakly surface active surfactants
3. A persistent foam (froth) with polyhedral bubbles of well-drained,
very thin walls (lamellae) produced by highly active ionic and nonionic surfactants (soaps, detergents, and proteins)

Foams and froths are thermodynamically unstable, since the free
energy of the system decreases when they collapse. The apparent "stability"
or lifetime ranging from a few seconds to many hours and days can be
attained only if all external disturbances such as mechanical vibrations,
radiant heat, evaporation, temperature differences, etc., are avoided. In the
absence of these extraneous "disturbing influences" the lifetime of foam or
froth depends on a number of surface chemical and physical parameters,
namely concentration of surfactants, surface viscosity and elasticity, bulk
viscosity and transport (by diffusion, drainage, or surface transport); all
of these, in turn, are determined by the molecular structure of the surfactant used and interactions between surfactants, solutes, and water molecules.

Two processes are responsible for the ultimate collapse of foams or
froths:

1. The slow diffusion of gas from bubbles into surrounding liquid and
from small bubbles into larger ones
2. The thinning by drainage and, ultimately, the rupture of the liquid
lamellae separating the bubbles

The diffusion of air across the lamellae separating bubbles is responsible
for the growth of larger bubbles at the expense of smaller ones; the driving
force for this diffusion is a high Laplace pressure $(2\gamma/r)$ within the smaller
bubbles. This diffusion process may reduce the total number of bubbles to
$\sim 10\%$ of the initial number without any rupture of the liquid lamellae [de
Vries (1957)]. Kitchener and Cooper (1959) pointed out that the growth of
large bubbles may be followed by a mechanical rearrangement in their
packing—and then any slight shock may lead to a sudden rupture of thin
lamellae and a collapse of froth.

9.3. *Kinetics of Drainage in Single Films Supported on Frames*

The thinning of the interbubble layer of liquid occurs at first by drainage under gravitational forces and is then followed by movement of the liquid within the lamella (in between the two air/liquid interfaces) caused by suction into the Plateau borders. The suction is created by the curvature of the meniscus in the Plateau border regions. This border suction is very effective in *mobile* interfacial films, where it normally is the predominant drainage mechanism. Mysels *et al.* (1959) published a monograph on their studies into the mechanism of thinning of soap films. Analysis of their colored cine photographs of interference bands, formed during the progress of thinning of liquid films, enabled them to calculate the thickness of the film.

A large number of nearly invisible, so-called *black* films, ranging from ~ 300 to 50 Å in thickness, have been observed with solutions above the CMC (but not in solutions below the CMC); the most commonly observed are the so-called Perrin's or Johonott's "first" and "second" black films.[†] Johonott (1906), and then Perrin (1918), noticed that the first thicker black films formed in the absence of evaporation, while the second films were developing only when the temperature was high enough to cause moderate evaporation. Closer visual observation suggested that the second black film (of 50–60 Å) represents a two-dimensional phase of bimolecular leaflet thickness. The second black film could be obtained only with commercial (impure) surfactants or with mixed surfactants (sodium lauryl sulfate + lauryl alcohol) but not with solutions of a highly purified sodium laurate sulfate, which gave only the first black film, of ~ 90-Å thickness, before any rupture occurred. Superficially, the black film appears as a black hole in the lamella of liquid supported by the frame. Corkill *et al.* (1961, 1963) used radioactive mixed surfactants (sodium dodecyl sulfate with S^{35} and dodecanol with C^{14}) and confirmed that a mixed film, anionic–nonionic (or anionic–cationic in the 1963 paper), determines the structure and the limiting thickness, about 60 Å, of the black film.

If the negative pressure in the Plateau borders (existing at the contact line of the film with the supporting frame) sucks out the liquid without breaking the interfacial film, the lamella thins down and acts as a bottleneck, preventing any further thinning of the lamella (Figure 9.10). This type of

---

[†] Recommended IUPAC nomenclature: CB film (common black film) and NB film (Newton black film); see Section 9.7.

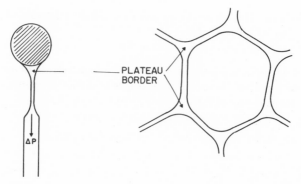

Figure 9.10. The Plateau border is the seat of a suction ΔP which tends to draw the intralamellar liquid out of the film. If the liquid begins to flow out, the film collapses and prevents further flow.

necking occurs with rigid, viscous films and is responsible for a much slower rate of thinning of such rigid films by gravity drainage alone.

The thin films of liquid in a partly drained froth, of the type shown in Figures 9.9 and 9.11, behave similarly to the films supported on a frame, with which these films form a continuous Plateau border.

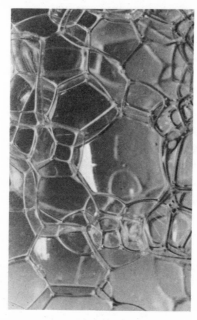

Figure 9.11. A well-drained froth with very thin lamellae walls separating bubbles.

9.4. *Characteristics of Single Thin Liquid Layers (Thin Films). Disjoining Pressure*

The thickness of films between two air bubbles, the corresponding pressures within these films, and the kinetics of thinning were first investigated by Derjaguin and Titievskaya (1953, 1957). These studies were expanded by Sheludko and co-workers (1956–1967), Read and Kitchener (1967, 1969), Barclay and Ottewill (1970) and Barclay *et al.* (1972).

Since the introduction of the concept of *disjoining pressure* by Derjaguin, it became recognized that at least four types of forces operate in any thin layer of liquid:

1. Laplace capillary suction (or pressure), $\Delta P = 2\gamma/R$, for a spherical surface of radius R.
2. Electrical double-layer repulsion or attraction, Π_{el}, caused by an overlap of the two electrical double layers associated with the interfacial boundaries on each side of the thin liquid layer. When the thin layer is formed by two bubbles in a froth, it is the overlap of the two identical electrical double layers of the liquid/gas interface; when the thin layer is formed between a bubble and a solid/liquid interface, two *different* electrical double layers interact to give Π_{el}.
3. Long-range van der Waals pressure, Π_W.
4. *Steric hindrance* in oriented and packed layers, Π_{SH}.

The thin layers may be one of the following: the liquid film wetting a solid near the meniscus (as the solid emerges from the liquid), or the liquid lamella between two air bubbles, or that between a bubble and a solid pressed against it, or that separating dispersed drops in an emulsion or in dispersion of solids (undergoing coagulation). All such layers comprise thin films. The Laplace capillary pressure exists in a film of any liquid, polar or nonpolar, pure or a solution, whenever the interface (liquid/gas or liquid/liquid) is curved; its magnitude is determined entirely by the curvature of the surface and the surface tension of the liquid. Derjaguin's term disjoining pressure encompasses the other three contributions Π_{el}, Π_W, and Π_{SH}:

$$\Pi = \Pi_{el} + \Pi_W + \Pi_{SH} \tag{9.10}$$

These three types of forces arise only when the liquid layer becomes sufficiently thin, so that the electrical and van der Waals forces begin to cooperate to a greater or lesser extent depending on the content of electrolytes and of surfactants. Steric hindrance Π_{SH} becomes evident only in the oriented

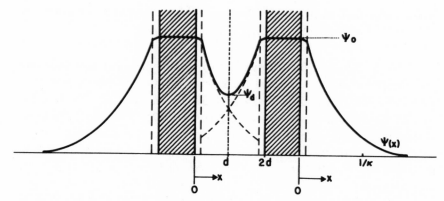

Figure 9.12. Distribution of the outer potential ψ in a thick liquid layer (outside the capillary) and in a thin liquid layer (within the capillary) showing the higher value of ψ_d in the latter (thin film) due to an overlap of two electrical double layers, shown schematically.

hydration layers very close to the solid phase, or in oriented monolayers of long-chain surfactants, as an extra resistance to separation of molecules from these layers; the extent of its action in thin layers is, as yet, completely unknown, and no technique has been devised to evaluate its contribution. So far, it has been assumed that $\Pi_{\mathrm{SH}} = 0$, and all theoretical analyses of disjoining pressure Π apply only to films thicker than the bimolecular films.

Frumkin (1938) and, independently, Langmuir (1938) were the first to derive expressions for the Π_{el} pressure in aqueous wetting films in the case of high surface potentials, $\psi \gg kT/e$, while Derjaguin (1937) derived an analogous expression for small potentials, $\psi < 25$ mV. The Π_{el} contribution proved to be of paramount importance in the stability of *lyophobic* sols and colloids, that is, dispersions of insoluble particles, which are easily influenced by electrolytes.

In a thin liquid film constrained between two interfaces (solid/liquid, liquid/gas, or liquid/liquid), there is an overlap of the electrical double-layer potentials developed in each interface. When the liquid has a very low concentration of electrolytes, the overlap of the potentials ψ in the respective diffuse (Gouy–Chapman) layers can be represented as in Figure 9.12. Such an overlap produces the electrostatic component Π_{el} of the disjoining pressure, a force acting per square centimeter area of the thin film. From electrostatics, this Π_{el} pressure can be evaluated as

$$\Pi_{\mathrm{el}} = \frac{2\pi}{\varepsilon}\,(q_a^2 - q_h^2) \qquad (9.11)$$

where ε is the dielectric constant of the thin layer phase, q_a is the charge density for the layer in contact with the bulk phase, and q_h is the charge density in the layer of thickness h. For symmetrical thin films, that is, films constrained between two identical interfaces, the electrostatic contribution Π_{el} to the disjoining pressure is always positive; that is, it represents a repulsion.

Sheludko and Exerova (1959), using the DLVO theory, showed that Π_{el} can be expressed by

$$\Pi_{el} = 64nkTy^2 \exp(-\varkappa h) \tag{9.12}$$

where

$$y = \frac{[\exp(e\psi_\delta/2kT)] - 1}{[\exp(e\psi_\delta/2kT)] + 1}$$

and e is the electronic charge, ψ_δ is the surface potential (potential at OHP), k is Boltzmann's constant, n is the concentration of 1:1 electrolyte in molecules per centimeter, $\varkappa$ is the reciprocal of electrical double-layer thickness, and h is the thickness of the thin film.

London (1930) derived an expression for the interaction energy between two molecules [Section 2.6, equation (2.17)]. de Boer (1936) and Hamaker (1937) suggested that the London–van der Waals forces are the source of attraction between two *macroscopic* bodies, e.g., parallel plates, or two spheres. de Vries (1957) pointed out that an attraction between two cavities (bubbles) in a liquid is the same as the attraction between two masses of the same liquid separated by the same air gap h, and it can be expressed by an analogous de Boer formula [equation (2.29)]:

$$\Pi_W = -\frac{A_h}{6\pi h^3} \tag{9.13}$$

where A_h is the Hamaker constant of the liquid.

Kitchener (1964) and Sheludko (1966a, 1966b, 1967) have derived expressions for the van der Waals attraction, Π_W. Lifshits (1955) and Dzyaloshinski *et al.* (1959) used the generalized electromagnetic theory and obtained a differentiation in expressions for Π_W: For small gaps or film thicknesses h, Π_W is a function of $1/h^3$, whereas for larger thicknesses it is a function of $1/h^4$. For more details, see Section 2.6.1. and Table 2.9.

The experimental results obtained by Sheludko *et al.* [and reported by Sheludko (1962)] confirmed the linear dependence of Π_W with $1/h^3$ for a number of free liquid films (both nonaqueous and aqueous solutions sta-

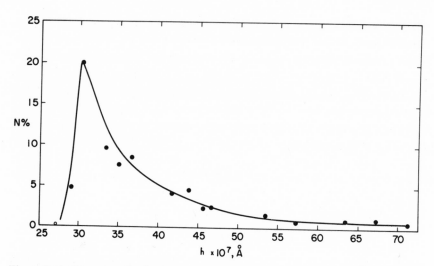

Figure 9.13. Frequency distribution curve for the thickness of rupture of free water films. [From Sheludko (1967), *Advances in Colloid Interface Sci.* **1**, with the permission of Elsevier Publishing Co., Amsterdam.]

bilized by surfactants). The slope of the line enabled the Hamaker constant for the particular liquid to be evaluated.

The sum of Π_{el} and Π_W can be either positive, zero, or negative (assuming $\Pi_{SH} = 0$). When the sum approaches zero and just compensates the capillary pressure, the so-called *equilibrium* free film is obtained; this can be a thick film if the electrolyte concentration is low, and then Π_W can be neglected. Usually, both Π_{el} and Π_W act simultaneously and can be investigated only by suppressing Π_{el} with the use of high electrolyte concentrations. However, this leads to nonequilibrium films [possessing negative disjoining pressure Π_W (Figure 9.13)] that require dynamic methods for their study.

It has been found that thin liquid films rupture on attaining a certain critical thickness h_{cr} unless

1. A state of metastability is achieved first at a thickness $h > h_{cr}$ as a result of a high positive value of Π_{el}, or

2. A jump-like transformation to a metastable structure of thickness $h < h_{cr}$ occurs.

The critical thickness for rupture of free films formed by aqueous solutions of short-chain alcohols, e.g., isopropyl of 0.143 M concentration with 0.1 M KCl, was found by Sheludko (1962) to be ~300 Å. For other solutions e.g., isoamyl alcohol and propionic and butyric acids, the critical thickness was found to be 270–280 Å, while for a nonaqueous solution of decyl alcohol in aniline, $h_{cr} \cong 420$ Å.

If the "equilibrium" thickness is greater than h_{cr}, the film thins smoothly, and a metastable equilibrium persists until thermal or mechanical fluctuations cause rupture. If the equilibrium thickness is smaller than h_{cr}, a jump-like transition through the unstable h_{cr} state is necessary and becomes evident by the development of black spots (or black film holes) in the film. In Sheludko's experiments, when the radius of the free thin film was $r = 10^{-2}$ cm and $h_{cr} \cong 275$ Å, the concentration of 1:1 electrolyte that caused the transition from thick films (without black spots) to very thin ones (with black spots) was $\sim 2 \times 10^{-2}$ M KCl; below this KCl concentration, the equilibrium thickness of the film was reached smoothly without black spot formation. Films with radii larger than 10^{-2} cm ruptured at thicknesses much greater than ~ 300 Å.

An order of magnitude difference in determinations of the thickness of films rupturing (~ 300 Å for microscopic-size films of $r = 10^{-2}$ cm vs. 1700, 3400, and 6700 Å in the case of free bubbles blown from surfactant solutions) demonstrates that not all parameters that affect the thinning phenomenon

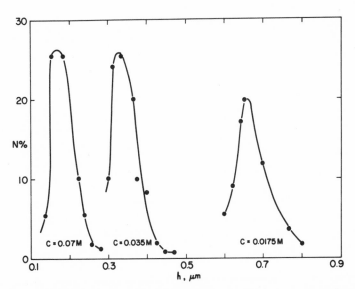

Figure 9.14. Distribution curves showing frequencies of rupture at a given thickness of soap bubbles at various concentrations of sodium oleate: 0.0175, 0.035, and 0.07 M. [From Sheludko (1967), *Advances in Colloid Interface Sci.* **1**, with the permission of Elsevier Publishing Co., Amsterdam.]

are equally evident in different techniques. The detailed evaluation of conditions, under which the tests reported in Figure 9.14 were conducted, showed that approximately 6×10^{-10} mol/cm^2 (i.e., ~ 28 Å^2/molecule) of surfactant was adsorbed at the air/liquid interface of bubbles (regardless of the concentration of solutions used) just before the bubble burst.

The concentration for the initial formation of black spots varies, as shown in Table 9.1, from 1.3×10^{-6} M for sodium laurate to 1.25×10^{-3} M for sodium hexyl sulfate; the value C_{b1} is much lower than the corresponding CMC of the surfactant. As yet, however, no clear correlation has been established between the chemical nature of the surfactant used to stabilize the thin layer and the four types of forces known to operate in thin layers.

Attachment of hydrophobic particles to air bubbles is preceded by thinning and rupture of the liquid layer between the solid/liquid and liquid/gas interfaces. The basic ideas on drainage and thinning discussed in the preceding sections for the free liquid layers between two air/liquid interfaces apply—with modifications—whenever thin layers between two

Table 9.1. Concentrations of Surfactants, C_{b1}, for Initial Formation of Black Spots in Froth Films Compared with Some CMC Values

Surfactant	$C_{b1} \times 10^6$ (mol dm^{-3})[a]	CMC (mol dm^{-3})[b]
Lauryl (EO)$_4$[c]	9.7	5×10^{-5} [d]
Lauryl (EO)$_6$	4.9	9×10^{-5}
Lauryl (EO)$_{18}$	1.3	13×10^{-5} [d]
Sodium octyl sulfonate	650.0	16×10^{-1}
Sodium hexyl sulfate	1250.0	?
Sodium octyl sulfate	134.0	1.3×10^{-1}
Sodium decyl sulfate	13.0	3.3×10^{-2}
Sodium dodecyl sulfate	1.6 (3.5)[d]	8.2×10^{-3}
Sodium tetradecyl sulfate	1.4	2.0×10^{-3}
Sodium laurate	1.3	2.6×10^{-2}
Sodium oleate	40.0	15×10^{-4} [e]
Cetyl pyridinium chloride	14.0	9×10^{-4} [e]
Cetyl trimethyl ammonium bromide (CTAB)	8.5	0.92×10^{-3}

[a] C_{b1} data from Exerowa *et al.* (1976).
[b] CMC data from Mukerjee and Mysels (1971).
[c] (EO) = ethyleneoxide [—OC$_2$H$_4$—] group.
[d] Value quoted in a subsequent tabulation by Exerowa *et al.* (1976).
[e] Data from Shinoda *et al.* (1963).

types of interfaces are formed. The characteristics of thin layers formed between two approaching solid/liquid interfaces are of decisive importance in coagulation and flocculation of dispersed fine solid particles. The stability of emulsions is also governed by the behavior of thin layers between two liquid/liquid interfaces.

9.5. *Theories of Foam and Froth Stability*

The most important factor in the formation of a foam or of a froth is the presence of a surfactant at the liquid/gas interface. The effectiveness of different surfactants is assessed by measuring the so-called foam stability or foam lifetime. This is done by one of the several methods, such as pneumatic, when air is injected at a given rate into a solution through a porous plug of a given pore size distribution, or mechanical, when the solution is agitated with a propeller of a specific design, etc. In each instance, the air is dispersed for a given length of time, at the end of which the volume of froth is estimated and its decay with time measured.

As already indicated, two types of foams appear to be easily distinguished:

1. Transient foams, obtained with dilute solutions of short-chain alcohols and carboxylic acids, with lifetimes ranging from a few seconds to approximately 20 sec
2. Persistent froths, lasting for hours and days, obtained with longer-chain surfactants of detergent character.

The characteristic values of lifetime and the optimum concentration for the first group of transient-foaming agents are given in Table 9.2. and the concentration dependence of their lifetime is graphed in Figure 9.15. A clear maximum is evident in the lifetime–concentration curves. The second group of persistent frothers gives, in contrast with transient foams, a foamlife increasing with concentration initially in an exponential manner (Figure 9.16) until at a certain concentration it levels off and remains constant for higher concentrations (S curve). Measurements by numerous investigators revealed that the surface concentration of surfactants, at the maximum foam life, τ_{max} (Figure 9.16), approaches that of a close-packed monolayer. The lamellae separating the bubbles in the froths of τ_{max} are of the black film thickness (less than ~ 300 Å).

At least four distinct theories to explain the stability of foams and froths have been proposed in the past.

Table 9.2. Characteristic Values of Unstable Foams[a]

Aqueous solutions	Maximum foam persistence, τ_{max} (sec)	Optimum concentration (mol/liter)
Alcohols		
Ethyl	5.0	0.28
Propyl	11.0	0.34
Isobutyl	12.0	0.09
Isoamyl	17.0	0.036
Tert-amyl	10.0	0.034
Heptyl	8.0	0.00007
Octyl	5.0	0.0003
Acids		
Formic	4.0	0.45
Acetic	8.0	0.20
Propionic	11.0	0.25
Butyric	18.0	1.00
Valeric	9.0	0.015
Caproic	13.0	0.0075
Heptanoic	16.0	0.0015
Caprylic	12.0	0.00025
Pelargonic	5.0	0.00007

[a] Source: Sheludko (1966a).

1. Plateau (1873) assumed that the drainage of liquid from the interbubble layers was determined by the viscosity of the adsorbed surface film and that this viscosity, in turn, determined the foam stability. Surface viscosity of surfactant solutions was determined, and indeed a definite correlation between the surface viscosity and the foam life was found to exist for some systems—but not in all cases. Oldroyd (1955) has shown that a complete description of rheological properties of an interface requires *two viscosity coefficients* (one of which is a surface analogue of Newtonian shear viscosity at a constant area; the other represents the resistance of the interface to changes in area and is called a *surface dilational viscosity*) and *two elasticity coefficients* (a surface shear modulus and a surface dilational elastic modulus). With two viscosity coefficients and a non-Newtonian behavior of those surface layers which are not freely mobile, the difficulties in establishing correct experimental procedures and reliable interpretations of results are enormous.

The rheological properties of adsorbed surface films are particularly readily changed by an addition of a second (usually nonionic) surfactant to the solution of a foaming or a frothing agent. The experiments on penetration of a soluble surfactant into an insoluble surfactant monolayer by Schulman *et al.* (see Section 6.10) have shown not only that the structure of the monolayer film is changed by the penetrating species but also that the kinetics of adsorption is affected. Numerous examples of the effects produced by these nonionic additives (*foam builders*) are described by Miles (1945), Miles *et al.* (1950), Epstein *et al.* (1954a,b), Burcik *et al.* (1954), Davies (1957), Kaertkemeyer (1957), Schick and Fowkes (1957), etc.

2. Marangoni (1871) explained foam stability by the ability of a surfactant solution to withstand a variation in its surface tension (within 0.1–0.001 sec). He suggested that a higher surface tension (than its static value) is created in the expanding portion of a bubble–liquid surface and a lower temporary surface tension in a compressed region. Because of this variation in surface tension, a restoring force instantly counteracts the disturbance; this becomes evident in the damping of waves (without changing the wavelength of ripples) on dilute solutions of pure surfactants, which do not impart significant surface viscosities.

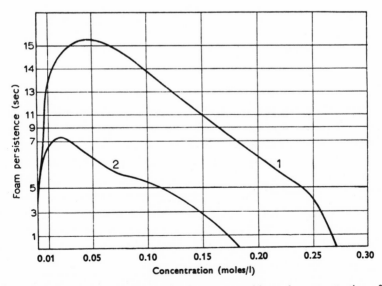

Figure 9.15. Dependence of foam persistence (in seconds) on the concentration of the foaming agent: (1) aqueous solution of isoamyl alcohol and (2) aqueous solution of *m*-cresol. [From Bartsch (1924), in Sheludko (1966), *Colloid Chemistry*, with the permission of Elsevier Publishing Co., Amsterdam.]

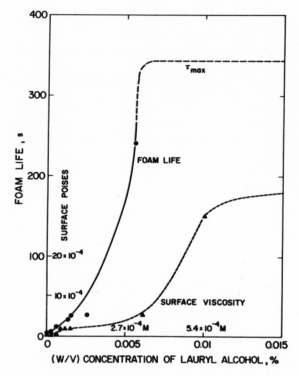

Figure 9.16. The effect of increasing additions of lauryl alcohol on foam stability (foam life) and surface viscosity of an 0.1% sodium laurate solution at pH 10. [From Davies (1957), with the permission of the author.]

There are no clear-cut techniques to measure the magnitude of the Marangoni effect, and the theoretical treatment is still incomplete. Sterling and Scriven (1959) analyzed the extent of Marangoni effect on interfacial turbulence, which raises the rates of mass transfer.

3. Gibbs (1878) suggested that during stretching of foam lamellae a local rise in surface tension occurs in thin layers as a result of surfactant depletion by adsorption into the stretched portion, producing a change in surface elasticity. Gibbs defined the coefficient of static surface elasticity E as the stress divided by strain for a unit area. For an area A of a two-face lamella when the surface tension rises to $\gamma + d\gamma$, the thickness of the lamella decreases to $h - dh$, and the surface elasticity is given by

$$E = \frac{2d\gamma}{dA/A} = \frac{2d\gamma}{d\ln A} = -\frac{2d\gamma}{\ln h} \qquad (9.14)$$

Kitchener and Cooper (1959) discuss some results of surface elasticity calculations for solutions of sodium dodecyl sulfate. These results indicate that surface elasticity could be of influence in foam stability for films of 10^{-3}–10^{-4}-cm thickness when the concentration is 0.01 M and for films of 10^{-5}-cm thickness when 0.1 M sodium dodecyl sulfate concentration is used.

Effectively, both the theories of Gibbs and of Marangoni deal with surface elasticity effects caused by different mechanisms.

4. In 1953 Derjaguin and Titievskaya (1953) proposed another explanation of foam stability, viz., that a positive value of disjoining pressure inhibits the thinning of lamellae. As already discussed (under the characteristic forces within thin layers), the only positive component (repulsion) of the disjoining pressure is Π_{el}, evident especially in dilute surfactant solutions and in the presence of small concentrations of electrolytes when Π_W is small. Additions of electrolyte reduce the value of Π_{el} and thus should sharply reduce the stability of foam or froth. This conclusion is, however, contrary to practical experience since, generally, no special sensitivity to electrolytes is detected either in transient foams or in persistent froth systems.[†]

Sheludko (1966a) differentiates three types of foam stability:

1. Transient foams formed in concentrations below C_{bl} (the concentration for which the initial black film formation occurs); the values of the critical thickness h_{cr} are determined experimentally.

2. Highly persistent froths obtained from solutions of concentrations greater than C_{bl}, close to the CMC, whose rupture is assumed to be due to surface fluctuations (under mechanical stimuli).

3. Froths of intermediate stability when the surfactant concentration is close to C_{bl}, giving a sharp transition region from low to high stability; both the process of thinning and of black spot formation play a part in these froths.

It is obvious that despite a considerable amount of attention being paid in the last 100 years to the characteristics of bubbles, of thin layers separating these bubbles in foams and froths, and of the lifetime of froths, there is no unanimity among the research workers regarding the parameters

---

[†] There are a few exceptions, for example, solutions of carboxylic acids, whose froths show thinning and collapse in 10^{-3} N HCl but produce stable froth in 10^{-3} M NaCl. When an NaCl concentration is increased, the stability decreases, and the froth finally collapses at sufficiently high concentration: 10^{-2} M NaCl for undecanoic acid and 10^{-1} M NaCl for oleic acid of 10^{-3} M concentration.

that govern the stability and the collapse of froths. The chemical nature of the surfactant has not been taken into consideration, yet it appears to be a significant parameter in some situations; numerous examples of synergistic effects recorded whenever mixed surfactants are used indicate chemical specificity which is not entirely reflected by all the changes in the physical parameters considered hitherto. Synergistic effects of beneficial or antagonistic character have often been recorded for pairs of surfactants when one of the pair differs only slightly in the molecular structure, e.g., the *cis* form of oleic acid as compared with the *trans* form of elaidic acid, or the use of an unsubstituted vs. a substituted amine of the same chain length, or the use of sodium laurate vs. calcium laurate (an addition of sodium laurate stabilizes the froth of sodium *p*-dodecylbenzenesulfonate much better than an addition of calcium laurate). The chemical effect of a surfactant is coming to the fore as the (previously mentioned) steric hindrance type of the equilibrium forces active in thin layers. The steric hindrance force represents a cooperative effect of van der Waals interactions between the hydrocarbon chains of the two surfactant molecules, dipole–dipole or ion–dipole interaction between the polar groups, and the interactions of associated counterions, water molecules of the hydration sheath, and hydrogen bonding. As yet no theoretical attempt has been made to correlate these various interactions with their spatial distribution along the interfaces of thin layers. As long as the chemical structure and specificity are not included in the theory of foaming and frothing, our understanding of this phenomenon will be incomplete even in the case of a two-phase frothing system.

9.6. *Destruction (Collapse) of Foams and Froths*

Frequently, it is necessary to destroy a persistent foam or froth, both in a two-phase and in a three-phase system (the latter is a solid-stabilized froth). If the parameter which is primarily responsible for the foam stability in the given system is drastically altered, the existing foam may be destroyed. For example, an additive that eliminates *surface elasticity* either by neutralization of the surfactant responsible for froth formation or by its displacement will cause a collapse of the existing froth. When *viscosity* or *drainage* are mainly responsible for froth stability, a change in the conditions such that either the viscosity is reduced or the drainage is facilitated will often collapse the froth.

The removal of an ionized surfactant from the air/liquid interface may be achieved by an addition of an oppositely charged inorganic species in a near stoichiometric quantity whenever the resultant product of reaction is not appreciably surface active. Thus, an addition of suitable metallic ions to produce insoluble salts with anionic surfactants may be effective in froth suppression whenever cosurfactants such as alcohols and nonionic polyoxyethylenes are absent in the system. Equally effective may be an addition of an oppositely charged surfactant which gives a reaction product with the foam-forming surfactant in the form of an insoluble precipitate or mixed micelles with extremely low residual concentrations of both surfactants. Insoluble fatty acid esters, alkyl phosphate (esters), and alkyl amides probably act in this manner, forming mixed micelles. The exact mechanism of removing the surfactant from the interface is immaterial provided the interface is left bare of surface active species, and the coalescence of bubbles (separated by the thin film) is thus facilitated.

The second means of eliminating the surface elasticity relies on displacing the adsorbed surfactant from the interface with a nearly insoluble film of oil, which itself facilitates coalescence. For an insoluble oil to displace a surfactant, it must have a positive spreading coefficient, which, in the absence of electrical (excess) charges along the interfaces, is expressed by

$$S = \gamma(\text{air/solution}) - \gamma(\text{air/oil}) - \gamma(\text{oil/solution}) > 0$$

Since the $\gamma(\text{air/solution})$ for a foam or froth system containing a hydrocarbon surfactant is of the order of 35–45 dyn/cm and the $\gamma(\text{oil/solution})$ is likely to be of the order of 5–10 dyn/cm, the surface tension of the spreading oil would have to be less than 25 dyn/cm for an effective spreading to occur. Mobile silicone oils have a surprisingly low surface tension, 15.7 dyn/cm for hexamethyldisiloxane, rising to about 20 dyn/cm for the increasingly more viscous methylsiloxane polymers. Perfluorohydrocarbon oils have a still lower surface tension, down to about 10 dyn/cm. These two types of oils, silicone and perfluorohydrocarbons, are very effective foam inhibitors if dispersed in aqueous solutions as an emulsion, with or without a diluent. They are also effective in suppressing foams in hydrocarbon oils. The variable effectiveness of silicone antifoam oils can be strongly influenced by dispersibility and long-range electrical forces, as discussed in detail by Kulkarni *et al.* (1977).

Destruction of a troublesome, persistent froth should be carried out in a container separated from the main supply of the liquid so that the liquid does not become contaminated with the antifoam additive. [For details on patented additives and procedures, refer to Kerner (1976).]

Drainage is promoted by additives such as traces of tributyl phosphate (which reduce the viscosity of surface layers in relation to that of water), whereas additions of 1,3-dimethylbutyl alcohol cause rupture of thick lamellae between the bubbles before drainage can occur [Ross and Cutillas (1955), Ross and Butler (1956), Ross and Haak (1958)].

9.7. Theoretical Conditions (*from Analysis of Film Thinning*) for Particle–Bubble Attachment (*Mineralization of Bubbles*)

The *selective* attachment of a hydrophobic particle to an air bubble in a flotation cell constitutes the most important act of the flotation process and is its primary objective. It contrasts sharply with the nonattachment of hydrophilic particles which are subjected to the same action of all mechanical and hydrodynamic forces within the cell. The chemical aspects of the process come into play in the mechanism of rendering the selected mineral particles hydrophobic and in exploiting this characteristic under the conditions created within the flotation cell.

Through the application of mechanical and gravity forces the mineral particles and air bubbles are repeatedly brought close together within the cell into positions of frequent near collisions or actual collisions, regardless of their hydrophobic or hydrophilic character. It is in the subsequent stages of the particle–bubble interaction that a differentiation between the hydrophobic and the hydrophilic particles occurs. Three stages are of critical importance.

1. The thinning of the intervening liquid layer separating the two bodies—the particle and the air bubble
2. The rupture of the thinning layer to its black film thickness followed by establishment of a *bound* particle–bubble
3. The withdrawal of water, receding from the black film, thus expanding the *hole*

Numerous attempts were made in the past to analyze these parameters. Frumkin (1933) was one of the earliest researchers who suggested that the thinning of the intervening liquid layer was decisive in the attachment of particles to air bubbles in flotation. The process of thinning of the layer formed between a hydrophobic particle and an air bubble during encounters, created mechanically in the pulp, may be represented as follows: As long as the two bodies are several microns ($> 10^4$ Å) apart, there are no special attraction or repulsion forces acting between them. When the hydrodynamic flow brings them together much closer so that the capillary force ($2\gamma/R$, where R is the radius of the bubble) can exert its action, the thinning of the intervening layer starts to take place, but it proceeds relatively slowly and is soon counteracted by the disjoining pressure. Under static conditions, a nearly spherical air bubble lightly pressed against a solid can approach it to a definite distance h_0; this distance is determined by the equilibration of external pressure and capillary forces with the resistance (repulsion) stemming from the disjoining pressure. Under dynamic conditions the distance h_0 may be much smaller than the static value, depending on the relative momentum of the two bodies. If this distance h_0 is within the range of positive values of disjoining pressure, no further thinning of the intervening layer can take place, and the solid behaves as a hydrophilic solid. However, if the hydrodynamic flow (or an external pressure) and the initial capillary thinning bring the two bodies within the range of a negative disjoining pressure, a rupture of the intervening liquid may occur, giving a very thin residual *adsorbed* film, *black hole*. The rest of the bulk liquid forms a contact angle with this thin film and is receding from this black hole at a certain rate.

Since an overlap of the two electrical double layers (encompassing the liquid layer between a solid and a bubble) is of primary importance in thinning, Read and Kitchener (1969) measured the thickness of wetting films between the surface of polished silica plate and a small bubble pressed against it in relation to the concentration of pure mono-, di-, and trivalent electrolytes (Figure 9.17). A fairly good agreement between the experimental and the theoretical thickness values [calculated with the help of computerized tables for surface potentials prepared by Devereux and de Bruyn (1963)] is evident for all three electrolytes. It shows clearly the reduction in the thickness of the intervening layer with an increase in the concentration of electrolytes. No rupture of the thinning film on the silica plate was observed (even when the zeta potential was reduced close to zero by additions of trivalent La^{3+}). In contrast to the previous observations of Derjaguin and Zorin (1957), the surface of silica remained hydrophilic regardless of the electrolyte and its concentration. Pashley and Kitchener (1979) also confirmed the hydrophilic character of quartz surfaces.

For thin liquid layers formed between two nonsymmetrical interfaces (such as the solid/liquid interface of a particle colliding with the liquid/gas of an air bubble), the existence of a higher surface potential ψ_1 on the solid side of the thin liquid layer and a lower surface potential ψ_2 on the liquid side (bubble) may result in an actual *attraction* between the charges on the

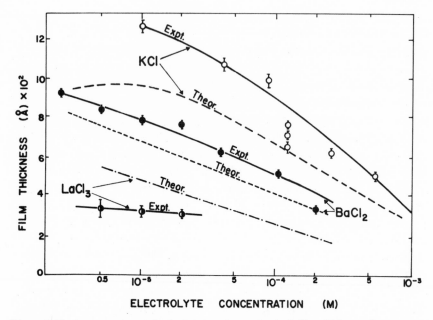

Figure 9.17. A comparison of calculated and experimental film thickness (between an
air bubble at a pressure of 1450 dyn cm^{-2} and a silica plate) as a function of electrolyte
concentration for a mono-, di- and trivalent salt. [From Read and Kitchener (1969),
J. Colloid Interface Sci. **30**(3), with the permission of Academic Press, New York.]

two interfaces if the midpoint potential ψ_d (Figure 9.12) is higher than the
actual potential on the liquid/gas interface of the air bubble. Such a negative
Π_{el} value may lead to a spontaneous rupture, particularly if it happens to
occur for sufficiently small h_0 so that it becomes reinforced by the van der
Waals Π_W negative contribution.

A theoretical analysis of possible variations in the disjoining pressure
Π in relation to the thickness h of the liquid layer separating a solid and a
bubble has been carried out by Derjaguin and Dukhin (1961). Subsequently,
Sheludko *et al.* (1968) and Laskowski and Kitchener (1969) also discussed
these $\Pi(h)$ relationships. These analyses deal with a generalized function,

$$\Pi(h) = -\left(\frac{\partial f^s}{\partial h}\right)_{T,A^s} \tag{9.15}$$

defined as the derivative of the specific interfacial free energy of the two
associated interfaces, $f^s = \gamma_{s/l} + \gamma_{l/a}$ (where $\gamma_{s/l}$ is the interfacial tension of
the solid/liquid and $\gamma_{l/a}$ is that of the liquid/air), with respect to the thickness
h of the liquid layer enclosed between these interfaces at a constant T and

A^s (area of the layer). When Π is positive, the liquid layer is stable and resists any thinning below the thickness corresponding to the external pressure $P = \Pi_H$ (Figure 9.18). If $\Pi = 0$, the layer is unstable, and there is no longer any resistance toward thinning; if $\Pi < 0$, an attraction exists between the interfaces enclosing the liquid layer, this attraction accelerates the thinning process, and a spontaneous rupture occurs, forming the stable layer of thickness h_a.

It appears from the analyses (of Derjaguin, Sheludko, Kitchener, and their co-workers) that, in general, five types of $\Pi(h)$ functions are possible, depending on the relative values of ψ_1, ψ_2, Π_{el}, and Π_W; these functions are indicated schematically in Figures 9.18(a) and (b). A function representing positive values of Π with a monotonic decrease of the slope, $d\Pi/dh = d^2f^s/dh^2 < 0$, indicated by curve A in Figure 9.18(a), denotes a liquid layer between, say, a hydrophilic solid and an air bubble. Any liquid film between two highly charged interfaces (such as those existing in dispersed sols, stabilized emulsions, or stabilized well-drained froths) will be represented by the relationship of curve A. Curve C represents an opposite extreme—such as a layer of liquid between *a highly hydrophobic solid and an air bubble*; the layer is completely unstable and will rupture spontaneously, forming a very thin adsorption layer of thickness h_a and residual drops or lenses of bulk liquid with a definite contact angle. Mysels *et al.* (1966) measured such an angle formed in an 0.05% sodium dodecyl sulfate solution containing 0.4 M NaCl as being $8°50' \pm 0.5°$. Sheludko *et al.* (1968) discussed two different techniques of measuring such angles between the thin film and the bulk liquid in equilibrium. De Feijter and Vrij (1975) employed contact angle determinations in thin films to evaluate thermodynamic excess quantities associated with such films and interaction energies.

In addition to curves A and C, three intermediate types of films, shown schematically as curves B, D, and E in Figures 9.18(a) and (b), are possible, depending on the relative values of $\Pi(\Pi < 0$, $\Pi = 0$, and $d\Pi/dh = d^2f^s/dh^2 = 0$), as governed by ψ_1 and ψ_2 and the magnitude of Hamaker's constant A_h [which determines $\Pi_W = -(A_h/6\pi)(1/h^3)$, one of the two main contributions to Π]. If the long-range van der Waals forces are dominant in the range of thickness h of the layer greater than $1/\varkappa$ (the thickness of the electrical double layer), that is, whenever the concentration of electrolytes is increased above a certain minimum value, then the disjoining pressure of the layer assumes the negative values indicated by curve B. When the electrical contribution Π_{el} is greater in the thicker layers and becomes overcompensated by the long-range van der Waals forces within a relatively narrow range of h values, curve D reproduces the variation of Π with h.

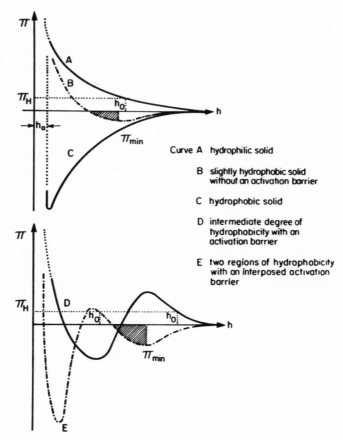

Curve A   hydrophilic solid

B   slightly hydrophobic solid
without an activation barrier

C   hydrophobic solid

D   intermediate degree of
hydrophobicity with an
activation barrier

E   two regions of hydrophobicity
with an interposed activation
barrier

Figure 9.18. Schematic plots of possible relationships between the disjoining pressure Π existing within a thin film and the thickness h of this thin film separating two interfaces; depending on the relative magnitudes of Π_{el} (electrical double-layer contribution) and Π_W (van der Waals contribution) and the magnitude of Hamaker constants and those of outer potentials ψ_i, there appear to be five types of relationships $\Pi(h)$ discernible.

Finally, when an intermediate concentration of electrolyte is present in the system containing a hydrophobic solid, a range of positive values Π may be created as a result of electrostatic repulsion within a given thickness range, providing an energy barrier toward thinning of the intervening liquid layer, as shown by curve E. As the thickness h diminishes, curve E shows at first a relatively shallow minimum (similar to that shown by curve B). This is associated with an appearance of a so-called *common black film* (CB film) or *first black film*. On continued thinning of the film, a second, much deeper minimum is indicated by curve E. This second minimum is associated with the appearance of a second black film, also known as Perrin film. A recent IUPAC recommendation is to use the term *Newton black film* or NB film instead of Perrin film [Everett (1972)].

A combined action of the thermodynamic, mechanical and capillary forces is equivalent to a pressure Π_H and is capable of forcing the two bodies—a particle and an air bubble—to approach each other to a distance h_0; for a hydrophobic solid without a barrier, this distance h_0 may be well within the range of van der Waals forces, and a final h_a thickness will be obtained by a highly accelerated local thinning with the establishment of an appropriate contact angle. For solids with an energy barrier, curves D and E, the combined external energy ($P = \Pi_H$) may be sufficient to overcome the electrostatic barrier; a change in the concentration of electrolytes or an increase in the Π_W contribution may be necessary to reduce these thermodynamic barriers. [Additional variants of the two curves D and E are possible whenever one of the maxima or minima is reduced to $\Pi = 0$; also, the function $\Pi(h)$ represents a combined effect of Π_{el} and Π_W (with the assumption that $\Pi_{SH} = 0$) of which Π_{el} may be positive or negative, while Π_W is mostly negative, representing attraction forces. (*Cf.* p. 85).]

Sheludko (1967) presented a detailed analysis of requirements leading to

1. The thinning of the liquid layer to a critical thickness h_{cr} at which a rupture occurs
2. The actual rupture with an establishment of a stable contact angle[†]

The induction time τ necessary for thinning to the critical thickness h_{cr} can be evaluated using the Reynolds relationship for viscous flow between parallel rigid walls. Sheludko considered that the total induction time consists of two terms: the time τ_0 taken to thin from macrothickness, $h = \infty$, to h_0 (determined by the hydrodynamic and capillary forces). Applying the Reynolds equation, we obtain

$$\tau_0 = \frac{1}{h_0{}^2 a P_0} \qquad (9.16)$$

where

$$a = \frac{4}{3\eta r^2} \quad \text{or} \quad \frac{16}{3\eta r^2}$$

(η is the viscosity of the bulk liquid and r is the radius of the thin film) and

[†] Sheludko *et al.* (1976) analyzed attachment of spherical particles to planar liquid surfaces, concluding that a lower limit for particle size which can be floated is ~ 1 μm. The maximum size of particles which can attach, and the time of impact between a particle and a bubble were also evaluated.

$h_0 \cong 1000$ Å, while P_0 is the capillary pressure $2\gamma/R$. The second term of the total induction time of thinning, τ_{cr}, is the time taken for thinning from h_0 to h_{cr} and is given by

$$\tau_{cr} = \frac{2}{aK} (h_0 - h_{cr}) \qquad (9.17)$$

where K is the van der Waals–Hamaker constant for retarded forces.

It is the second term, τ_{cr}, which is the characteristic parameter for a solid in a given flotation system. The determinations of h_{cr} for a free liquid (symmetrical) film and for the same liquid layer but resting on a substrate (such as a film of benzene on quartz, for which no rupture occurred even at $h = 200$ Å) indicated to Sheludko that for hydrophilic solids the rupture of supported films is much more difficult unless helped by adsorption. He analyzed the effect of an adsorbed layer on the disjoining pressure in terms of the corresponding van der Waals–Hamaker constants:

1. K_{11}, the constant for the liquid phase
2. K_{12}, that for the solid/liquid interface
3. K_{12}^{α}, that for the adsorbed film/liquid interface

Thus, for example,

$$\Pi = \frac{K_{12} - K_{11}}{h^3} \qquad (9.18)$$

for the disjoining pressure of a liquid film 1 on a solid substrate 2, and

$$\Pi = \frac{K_{12} - K_{12}^{\alpha}}{(h + \Delta)^3} - \frac{K_{11} - K_{12}^{\alpha}}{h^3} \qquad (9.19)$$

for the disjoining pressure of an adsorption film 2 of thickness Δ interposed between the solid substrate 2 and the liquid 1 of thickness h. For the relative values of constants $K_{12} > K_{11} > K_{12}^{\alpha}$, the $\Pi(h)$ curve obtained by Sheludko (1967) was analogous to curve D [in Figure 9.18(b)], and its interpretation is that flotation is possible whenever Π is negative and the slope $d\Pi/dh$ exceeds the value of a given surface tension gradient for thin films [the second condition arises from the considerations proposed by de Vries (1957) regarding the role of surface tension in a *deflection* mechanism leading to local thinning, as contrasted with the *hole formation* rupture mechanism]. For the case of $K_{12} < K_{11}$ and in the absence of an adsorption layer 2^{α}, flotation is possible only when Π is negative and $d\Pi/dh$ is greater than a given surface tension gradient, but it becomes greatly improved by an adsorption layer 2^{α} for which $K_{12}^{\alpha} > K_{11}$; the $\Pi(h)$ curve under these conditions resembles that of curve B in Figure 9.18(a) (e.g., between Π_{min} and $\Pi = 0$).

The van der Waals–Hamaker constants for the different adsorbing species, i.e., flotation reagents, activators, or depressants, would have to be evaluated, and thus the (as yet) qualitative predictions of Sheludko's theory of flotation await quantitative confirmation.

9.7.1. Effects of Surfactants on Induction Time in Particle–Bubble Attachment

The existence of induction time and its importance in the kinetics of flotation were first realized by Sven-Nillson (1934), and since then several contact-time measuring units have been devised by Eigeles (1950), Eigeles and Volova (1960), Glembotsky (1953), and Evans and Ewers (1953). The results of extensive measurements carried out by Eigeles and co-workers may be summarized as follows:

1. Additions of collectors to the solution decrease the time required to establish particle–bubble contact from infinity (for strongly hydrophilic minerals) to the range of 0.01–0.001 sec.

2. The use of mixtures of surfactants sharply increases the velocity of attachment; i.e., the induction time decreases by an order of magnitude or more, as first observed by Taggart and Hassialis (1946).

3. The effects of temperature, size of particle, and concentration of the collector species are shown in Figures 9.19(a)–(c). It is apparent that an increase in temperature reduces the contact time in an exponential manner, that is,

$$\log \tau = \frac{A}{T} + B \qquad (9.20)$$

whereas the size of particles affects only the constant B, not the exponent (A).

No systematic study of induction time appears to have been carried out to evaluate the effect of increasing the chain length of frother molecules used with a given hydrophobic mineral or the effect of the structure of the two surfactants used as a collector and a frother.

The physical parameters involved in the probability of collision between a rising air bubble and a falling mineral particle, or in the contact time after collision, have been taken into consideration in calculations of Sutherland (1948), Philippoff (1952), and Evans (1954). Table 9.3 shows the values of contact time and of the required velocity of film recedence; contact time is

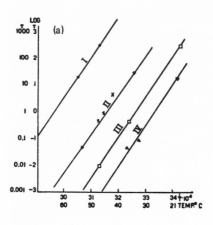

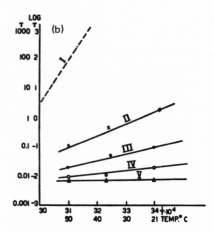

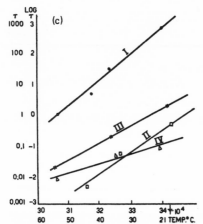

Figure 9.19. Experimentally determined relationships between induction time τ and temperature for (a) different sizes of quartz particles: I, $-0.3 + 0.25$ mm; II, $-0.25 + 0.21$ mm; III, $-0.21 + 0.175$ mm; IV, $-0.175 + 0.147$ mm; (b) increasing concentrations of a collector-acting surfactant, dodecylamine: I, no collector; II, dodecylamine, 0.7×10^{-5} g; III, dodecylamine, 1.4×10^{-5} mol/liter, accelerated retraction; IV, dodecylamine, 2.1×10^{-5} mol/liter, accelerated retraction; V, dodecylamine, 2.8×10^{-5} mol/liter, accelerated retraction; (c) different nature of mineral particles of $-0.175 + 0.147$ mm size: I, calcite; II, opatite; III, barite; IV, galena. In all tests (a) and (c) mineral particles (without cleaning the surfaces) in pure water were used. [From Eigeles and Volova (1960), in *Proceedings of the Fifth International Mineral Processing Congress, London*, with the permission of the Institution of Mining and Metallurgy, London.]

calculated by Philippoff for different sizes of particles on the basis of weight–buoyancy, elasticity, and the frequency of particle vibration, i.e.,

$$t_c = \frac{2}{\text{frequency}} = \frac{\pi}{m^{1/2}/E} \tag{9.21}$$

where m is the mass of the particle in milligrams, E is the elasticity of the bubble surface (representing a restoring force resisting a mechanical deformation) in dynes per centimeter, t_c is the time of contact in milliseconds,

Table 9.3. Calculated Parameters Relating to Particle–Bubble Attachment[a]

Diameter of particle, D (μ)	Contact time, t_c (msec)	Velocity of recedence, V_0 (cm/sec)	Final film thickness, h_0 (μ)	Minimum contact angle for attachment (deg)	
				Dynamic	Static
1000	15.3	3.27	9.8	0.675	12.0
500	6.33	3.95	6.9	0.675	3.1
200	1.88	5.32	4.4	0.726	0.5
100	0.74	6.75	3.1	0.787	0.13
50	0.28	8.90	2.2	0.865	0.032
20	0.077	13.0	1.4	1.000	0.0050
10	0.029	17.3	1.0	1.121	0.0013
5	0.0117	21.4	0.7	1.265	0.00032
2	0.0029	34.5	0.44	1.500	0.000050
1	0.0011	45.4	0.31	1.715	0.000013

[a] After Philippoff (1952).

and the speed of radial recedence of the liquid from the surface of the particle (in the form of a cylinder with a diameter D) is

$$V_0 = \frac{D}{2t_c} \qquad (9.22)$$

Also, the final film thickness (using Stefan's equation) and the minimum required dynamic contact angle to give the necessary velocity of radial recedence have been calculated and are listed in Table 9.3 for the different sizes of particles. High-speed photography experiments enabled the contact times to be determined for different particles; these contact times proved to be $\sim15\%$ longer than the calculated ones. Measured velocities of recedence were found to be only a fraction ($\sim1/10$) of those calculated by equation (9.22).

Using high-speed photography and an ingeneous optical system permitting photographs of colliding particles and bubbles to be taken from two directions simultaneously, Whelan and Brown (1956) obtained useful information on the trajectories, collision velocities, and times of contact. A comparison of the time of contact calculated by the methods of Philippoff and Evans and the ones established experimentally by Whelan and Brown (Table 9.4) shows a quite satisfactory agreement.

An important conclusion from the results of both Philippoff and Whelan and Brown was the finding that the more strongly a particle collides

Table 9.4. Comparison of Calculated and Experimentally Determined
Times of Contact in Particle–Bubble Attachment[a]

Mineral	Time of Contact (sec)	
	Calculated	Experimental
Galena	0.0013–0.0027	0.0008–0.0015
Pyrite	0.0011–0.0021	0.0004–0.0015
Coal	0.0005–0.0010	0.0004–0.0008

[a] After Whelan and Brown (1956).

with a bubble, the more likely is the attachment; in fact, at a sufficiently
high momentum the particle is able to pierce the surface of the bubble.

9.7.2. The Role of Surfactants and Their Diffusion in Particle–Bubble Attachment

Analyzing the process of the particle-to-bubble approach, Derjaguin
and Dukhin (1961) distinguished three zones in the overall thickness of
liquid surrounding an ascending air bubble:

1. An outer zone in which only the hydrodynamic forces operate,
   giving streamlines of liquid flowing past the bubble
2. An intermediate zone of thickness of the order of 10^{-4}–10^{-3} cm,
   called a diffusophoretic zone, in which diffusion acts and electrophoretic potentials are created
3. An inner zone which denotes a thin layer ($\sim 10^{-5}$ cm thick) in which
   the electrical double-layer and the long-range van der Waals forces
   begin to exert their effects

The paper of Derjaguin and Dukhin was the first theoretical approach to
deal with the interplay of the diffusion zones (around the particle and the
bubble) and the effect of a concentration gradient of a surfactant across
the intervening layer of liquid in particle–bubble attachment. Some of the
assumptions made in this paper do not appear entirely reasonable, for
example, the assumption that when a particle "sinks into the diffusion
boundary layer (of the bubble) the flotation reagent is desorbed off its
surface, diffuses in the dissociated form to the surface of the bubble and is
adsorbed on it." The time required for thinning of this diffusion zone to a

limiting thickness (imposed by the theory), based on the preceding assumption, was "some tenths of a second," whereas induction times smaller by two orders of magnitude are involved under actual flotation conditions. Also, two types or forms of surfactants are generally employed (except in flotation of already hydrophobic minerals). The assumption that the flotation reagent desorbs off the particle surface, although likely to take place under special conditions, does not represent the usual occurrence in systems with strongly adsorbed, immobilized collector species. And the omission of recognizing the existence of the second type of surfactant may be the cause of the two-orders-of-magnitude difference in the necessarily very approximate calculations of the thinning times.

As clearly shown by Schulman *et al.* (1935, 1937, 1938, 1949) in monolayer penetration experiments, two species of surfactants tend to interact. The simplest interaction is mainly electrostatic, leading to appreciable changes in surface potential with only minor changes in surface pressure; generally, however, large changes in both surface pressure and surface potential are indicated. Schulman and Leja (1954, 1958) showed that the collector and the frother-acting surfactants interact strongly at the solid/liquid as well as the liquid/gas interfaces and (by inference) also at the solid/gas interface; they suggested that the attachment of predominantly collector-coated particles to predominantly frother-coated air bubbles occurs as a result of penetration of the two interfaces (solid/liquid and liquid/gas) to form a solid/gas interface, possessing a higher density of surfactant coverage than either the gas/liquid or liquid/gas interface. The energy of this interaction is composed of contributions due to van der Waals bonding between the corresponding nonpolar groups, ion–dipole or dipole–dipole interactions between the polar groups, the energy of overlap between the electrical double layers, and the energy of partial dehydration (due to a lower dielectric constant) in the residual thin liquid layer remaining on the solid/gas interface.

The treatment of Derjaguin and Dukhin (1961) may well describe the events taking place on attachment under static conditions, when a weakly collector-coated particle and a bubble approach each other in the absence of frother molecules. However, under dynamic conditions in a flotation cell, in the presence of a frother-acting surfactant, the conditions of convective diffusion for two types of *interacting* surfactants obtain. The driving force for a chemical reaction between surfactants is not only the chemical potential gradient responsible for the diffusion of surfactant species but, in addition, the free-energy change of the reaction taking place between the two surfactants at the two interfaces on penetration. The situation of two overlapping diffusion zones comprising interacting surfactants implies that, first, the thickness h of the thinning liquid film in which van der Waals bonds of diffusing surfactants are able to exert their action (of providing a negative Π_W contribution) may be effectively much greater than that predicted for a noninteracting system. And, second, an additional energy of interaction is available to overcome any energy barriers in the system [as represented by curves D and E in Figure 9.18(b)]. A modified mathematical treatment of such an interacting system of two surfactants diffusing in an intervening liquid layer between a particle and a bubble, as outlined by Levich (1962b, Chapter II), may provide more realistic solutions for the various boundary conditions. A more detailed experimental work is, however, also required to evaluate the correct thermodynamic and kinetic parameters for the chemical and structural aspects of the two reacting surfactants.

High-speed photography of particle–bubble encounters [Whelan and Brown (1956), Hornsby (1975)] showed that particles colliding with the expanding portions of oscillating air bubbles frequently rebound from them, whereas particles in the vicinity of receding portions of the bubbles appeared to be sucked in and become attached. These events can be explained by changes in the local concentration of surfactants, C_x, at the liquid/gas interface during an expansion and contraction cycle in bubble oscillations. The changes in concentration from an equilibrium value C_0 to the value $C_0 - x$ on denudation during expansion of the interface or to $C_0 + x$ on recession of the bubble cause pronounced changes in the rate of molecular interactions for the two surfactants, C_0 and C_0^1:

$$k(C_0 - x)^n (C_0^1 - x^1)^m \ll k(C_0 + x)^n (C_0^1 + x^1)^m \qquad (9.23)$$

At the same time, the free energy of interaction ΔG_0 increases the rate constant k in an exponential manner:

$$k = k_0 \exp\left(\frac{-U}{RT} + \frac{\Delta G_0}{RT}\right) \qquad (9.24)$$

where U/RT is the activation energy for diffusion.

When one of the surfactants is replaced by an entirely nonpolar oil (of a limited solubility but, nevertheless, capable of diffusion particularly when solubilized by the polar frother surfactant), the interaction still occurs; it is now limited to the van der Waals contributions only between the hydrocarbon chains. Extensive use is made of nonpolar oils in the flotation of inherently hydrophobic solids (MoS_2, talc, graphite, coal, sulfur, antimonite) in order to improve the kinetics of attachment by extending the range of action of van der Waals forces without the introduction of opposing electrical charges. A judicious use of nonpolar oils (reasonably short-chain species) in flotation of oxidized ores of copper, lead, and zinc is of considerable assistance, as shown in Figure 8.19. It introduces diffusing species capable of van der Waals bonding, with no buildup of electrical charges to provide an increase in the energy barrier preventing attachment.

Similarly, attachment between a particle containing microbubbles precipitated on its hydrophobic surface and a macrosize bubble is greatly facilitated. The explanation is that a lower energy barrier exists in the thinning layer between the two bubbles because of a decreased Π_{el} contribution in comparison with the Π_{el} value of a thinning layer between the solid/liquid and liquid/gas interfaces. Klassen (1948a,b, 1960, 1963) reported extensively on the beneficial effects of vacuum precipitated bubbles being used jointly with the mechanically dispersed ones. Figure 9.20 shows schematically the degree of improvement obtained by a joint action of precipitated and dispersed bubbles in flotation of fluoride ($-147\ \mu m + 74\ \mu m$ size fraction) using sodium oleate. Dzienisiewicz and Pryor (1950) found that the rate of attachment of rising bubbles to polished specimens of sulfide mineral surfaces held in xanthate solutions was higher if the surfaces were first "frosted" with tiny bubbles.

It is frequently observed that under the conditions of optimum flotation a pronounced flocculation of the hydrophobic mineral particles invariably occurs. The same parameters that have been considered in the thinning and in the attachment of a particle–bubble system apply—with slight modifications—to flocculation. The latter is but an attachment in the same system,

Figure 9.20. Improvements in flotation recovery caused by a combined action of mechanically generated and precipitating microbubbles: 1, flotation solely by bubbles precipitating from solution; 2, flotation solely by bubbles produced by mechanical means; 3, combined flotation by air precipitating from solution and by bubbles produced by mechanical means. [Klassen (1948), from Klassen and Mokrousov (1963), *An Introduction to the Theory of Flotation*, Trans. J. Leja and G. W. Poling, with the permission of Butterworths, London.]

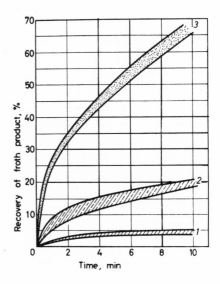

particle to particle. The modifications concern the symmetry of the interfaces on both sides of the intervening liquid layer, both being the solid/liquid interface. A different shape of the $\Pi(h)$ curve (Figure 9.18) from that which applies in the particle–bubble attachment is obtained. Again, the diffusion of frother molecules and their ability to react with the collector coatings on the solids provides distinct advantages over the purely physical condition presented in the absence of these mobile species.

9.7.3. *Withdrawal of Water To Establish the Area of Particle–Bubble Contact*

The third stage of the attachment mechanism consists of the withdrawal of water from the spot with the black film (hole) thus to expand the area of the solid/gas interface and to increase the strength of the particle–bubble adhesion. This stage was investigated by Sheludko *et al.* (1970). The method consisted of photorecording the velocity of expansion of a contact angle formed with a planar solid surface by a column of air held in a tube so close to the solid surface that two related contact angles were formed: one within the tube with its walls and the other with the planar surface. From the geometry of the system the rate of recession of the liquid was evaluated. The results obtained for a silica surface in a solution of 1.35 $\times 10^{-5}$ M dodecylamine hydrochloride are shown in Figure 9.21. A startling variation of nearly four orders of magnitude was obtained in the mobility of liquid for different radii of contact but for the same surfactant solution. However, Sheludko's conclusions that a decrease in the rate of contact area expansion may be the controlling process in floatability of very small particles, and that it is responsible for a sharp decrease in silica recovery at pH > 10, are not convincing in the light of an optimum recovery of silica at pH ~ 9.5 (*Cf.* Figure 8.17, p. 524) and a small difference in Sheludko's curves 2 and 3, Figure 9.21.

Undoubtedly, the rate of liquid withdrawal from the black film spot formed on rupture of the thinning liquid layer may be a controlling parameter under some conditions. In the case of silica, which rapidly dissolves in alkaline pH, and a dodecylamine hydrochloride solution containing different proportions of dissociated and undissociated species at pHs used, another explanation for the sharp decrease in the rate of water withdrawal and changes in silica recovery may be more appropriate.

The preceding study of Sheludko *et al.* does not appear to make allowance for the effects of surface roughness and surface heterogeneity. Oliver and Mason (1977) used scanning electron microscopy to study

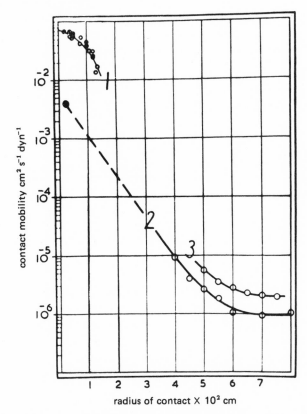

Figure 9.21. Contact area expansion (contact mobility) in relation to the radius of bubble contact with a flat solid surface: curve 1, for low pH, ● pH 4.6, and ○ pH 5.3; curve 2, for pH 11 (the first point ● is obtained from τ_m); curve 3, for pH 8.1. [From Sheludko *et al.* (1970), in *Special Discussions of the Faraday Society No. 1*, with permissions from the Royal Society of Chemistry and the authors.]

spreading of liquids on surfaces of different textures and orientations of roughness. They found that sharp edges of step heights less than 0.05 μm (500 Å, invisible under optical microscopy conditions) influenced spreading to a significant extent. Since withdrawal of water from the black film spot is the reverse of spreading, sharp edges are likely to influence the withdrawal to an equal degree.

Oliver *et al.* (1977) have presented a most comprehensive account of the influence of sharp edges in the spreading of liquids. They examined theoretically and confirmed experimentally the various conditions for drop stability on solid disks with solid edges of different edge-angles ϕ (Figure 9.22). On supplying the liquid through a central hole in the disk, they were

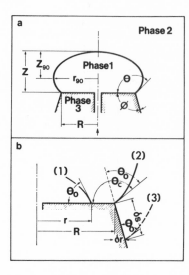

Figure 9.22. Equilibrium of a nonspreading liquid drop ($0 < \theta < 180°$) at the sharp edge of a solid ($0 < \phi < 180°$). (a) Dimensions of an axisymmetric sessile drop (i.e., when there is an appreciable gravity effect) formed on a horizontal surface with a circular edge. (b) Detail at the moment of crossing the edge showing a drop approaching the edge (location 1), in the critical position θ_c (location 2), and after crossing the edge (location 3) and creeping a distance ds down the side. [From Oliver *et al.* (1977), in *J. Colloid Interface Sci.* **59**(3), with the permission of Academic Press, New York.]

able to record, using photography at speeds from 1 to 10 frames/sec, the spreading of the liquid on the surface of the disk [position 1, Figure 9.22(b)] to the critical position 2 just before the liquid crossed the edge to position 3. A parameter $E = (V_2 - V_1)/V_1$, based on the maximum drop volume that can be contained by the edge with the angle ϕ, was introduced. The changes in this parameter E in relation to the advancing contact angle θ_0 (assumed equal to an equilibrium contact angle) have then been evaluated for different ratios of the disk radius R to the capillary constant $a \equiv \gamma_{a/l}/g \, \Delta\varrho$. Figure 9.23 and 9.24 show the curves for $E–\theta_0$ obtained theoretically compared with the values measured experimentally; for details a reference to the paper should be made.

The authors found that the use of the parameter E has advantages over the Gibbs equation for the critical angle θ_c necessary for the spreading across the edge of angle ϕ to occur, viz.,

$$\theta_c = (180 - \phi) + \theta_0 \tag{9.25}$$

In addition, their findings are of paramount importance to the understanding of numerous puzzling occurrences in solid separations. Among such are the following: Nonspherical particles appear to attach to air bubbles in flotation at sharp edges rather than along their planar faces; an appropriate distribution of bubble sizes to suit the particle size distributions appears to be necessary for optimum recoveries; some highly hydrophobic particles give in flotation a "dry" mineralized froth, with particles projecting above the liquid/gas interface into the gas phase; similarly, in spherical agglomeration the sharp-edged particles must be incorporated (entrained) into the oil phase before a satisfactory agglomeration and subsequent separation from water-entrained hydrophilic particles can be achieved. For a thorough assessment of the implications that the work of Oliver *et al.* (1977) has in flotation, their results should be presented in a form that replaces the liquid drop (phase 1) by an air bubble (phase 2) and the spreading of liquid by its withdrawal. The parameter E would incorporate the relative volume change of bubbles, and the ratios of R/a represent two very important flotation parameters: $\gamma_{a/1}$ and the (gravitational) acceleration g, which in mechanically agitated cells may be increased by an order of magnitude or more. The significance of the role played by the mechanical variables in flotation, in relation to surface chemical parameters, becomes more comprehensible if they are analyzed jointly.

The effect of surface heterogeneity with respect to hydrophobic–hydrophilic patches has been the subject of theoretical considerations, and some experimental testings, by Cassie and Baxter (1945), Dettre and Johnson (1967), and Smith and Lindberg (1978). Using patterned surfaces (litho-

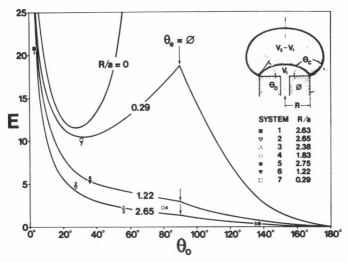

Figure 9.23. Edge effect $E = (V_2 - V_1)/V_1$ of sessile drops in systems with various values of θ_0 on a sapphire disk ($R = 0.4978$ cm and $\phi = 90°$). The curves were calculated for various values of R/a using the Bashforth–Adams tables, and the points correspond to experimental values for the various systems (inset). Very good agreement between the theoretical and the experimental values is indicated. The cusps at $\theta_e = \phi$ occur because V_2 remains constant for $\theta_0 > \phi$ but V_1 increases, so that E subsequently decreases. [From Oliver *et al.* (1977), in *J. Colloid Interface Sci.* **59**(3), with the permission of Academic Press, New York.]

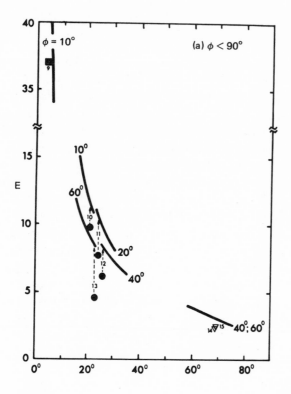

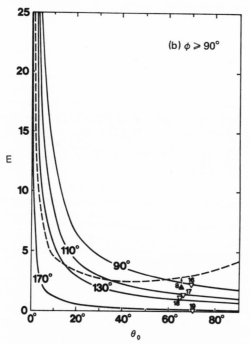

Figure 9.24. Edge effect based on the drop volumes of different liquid systems for various aluminum edges; acute [part (a)] and obtuse [part (b)]. Experimental (points) and corresponding theoretical (solid curves) values are given for various R/a. For simplicity only portions of the curves are shown for systems with $\phi < 90°$. Measured values deviating appreciably from the theory are denoted by broken arrows pointing toward the appropriate curve. The numbers accompanying the experimental points refer to the systems listed in Table III of the original paper by Oliver *et al.* (1977). [From Oliver *et al.* (1977), in *J. Colloid Interface Sci.* **59**(3), with the permission of Academic Press, New York.]

graphic plates) composed of different ratios of hydrophobic ($\theta_a = \sim 81°$) and near-hydrophilic ($\theta_a = \sim 3°$) patterns, the last authors established that a linear relationship exists between $\cos \theta_a$ vs. fraction of hydrophobic surface. The receding contact angles were approximately zero regardless of the fraction of hydrophobic patterns. Further, they established that on supplying acoustic energy to the drop resting on patterned surfaces the contact angle decreased by ~ 10–25% for relatively small doses of extra energy, whereas larger doses expanded the drop periphery too much and established a receding contact angle.

By using surfaces of porous hydrophobic solids, or various patterns of square, rectangular, or circular openings in screens made of hydrophobic materials, it can be readily shown that—when these substrates are submerged and bubbles are made to attach to them—the thinning of films occurs above the hydrophobic patches (asperities) regardless of the pattern used. The withdrawal of water to establish the maximum area of solid–bubble contact may be important for contact on smooth surfaces but not for heterogeneous surfaces composed of randomly distributed hydrophobic patches with a high degree of roughness present. On the latter contact is established with isolated asperities of hydrophobic patches, leaving water trapped in between asperities (as indicated, for example, by attachment of bubbles to wax-coated wire-mesh surfaces).

9.8. *Joint Action of Collectors and Frothers in Flotation*

The association of frothers and collectors [suggested by Christman (1930) and demonstrated by Leja and Schulman (1954) by means of analytical determinations] influences some parameters of a flotation system quite strongly, leaving others completely unaffected. Thus, for example,

1. As long as the frother molecule forms a gaseous type of a monolayer at the air/water interface, the magnitude of the contact angle established by the adsorbed collector is unaffected. However, when the frother molecules tend to form a condensed film, the contact angle is altered as seen in Figure (2.17) [Schulman and Leja (1958)].

2. The surface tension of the mixed solution is decreased in comparison with the values predicted on the basis of additivity (ideal behavior) for the surfactants employed, indicating an increase in adsorption [Buckenham and Schulman (1963), Figure 5.25a; Shinoda *et al.* (1963)]. The interfacial

tension at a liquid/liquid interface may also be affected, e.g., mercury–aqueous solution. An increase or a decrease in the adsorption of a collector at the solid/liquid interface may take place, depending on the relative proportions of the frother and collector and their relative extent of interactions at interfaces and in the bulk of solution.

3. Spectroscopic evidence regarding collector–frother interactions in bulk solution is inconclusive; in some systems appreciable shifts in absorption bands of mixed surfactants were observed [Finkelstein (1968); Mukai *et al.* (1972)]; in other systems a shift in spectral bands can be observed for concentrations higher than the CMC [Szymonska and Czarnecki (1978)].

4. Definite changes in the induction time are shown, for example, in Figure 9.19b for a single surfactant (acting as both a collector and a frother,

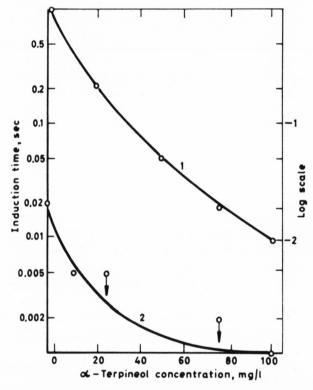

Figure 9.25. Dependence of induction time (seconds) of chalcocite particles (−0.2 + 0.15 mm size) after 1 min of conditioning in increasing concentrations of α-terpineol at pH 9.7: curve 1, for a 0.2-mg ethyl xanthate addition per g chalcocite; curve 2, a 1-mg/g ethyl xanthate addition. [From Lekki and Laskowski (1971), *Trans IMM* **80**, with the permission of the Institution of Mining & Metallurgy, London.]

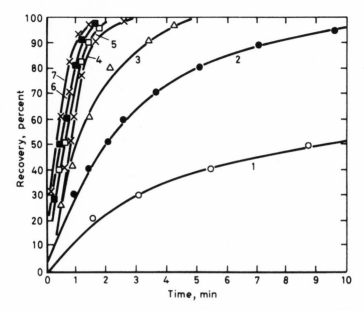

Figure 9.26. Effect of α-terpineol additions on the flotation recovery of chalcocite at a 0.7-mg/g ethyl xanthate addition at pH 9.7. Additions of α-terpineol in mg/liter: curve 1, 0, curve 2, 10; curve 3, 20; curve 4, 50; curve 5, 60; curve 6, 75; curve 7, 100 (mg/liter). [From Lekki and Laskowski (1971), *Trans IMM* **80**, with the permission of the Institution of Mining & Metallurgy, London.]

owing to the presence of dissociated and undissociated forms) and in Figure 9.25 for a two-surfactant system. An addition of a frother is capable of decreasing the induction time by more than one to two orders of magnitude, and this effect is closely related to the amount of collector. Lekki and Laskowski (1971) carried out a systematic investigation on the joint action of α-terpineol and xanthate in the flotation of chalcocite; Bansal and Biswas (1975) provided evidence for coadsorption of tripropylene glycol and tetrapropylene glycol monomethyl ethers (active components in Dowfroth 200 and 250, respectively) with sodium oleate on rutile surfaces and at the air/water interface.

5. There are indications that the zeta potential of a solid, whether a sulfide in a xanthate solution or an oxide in a carboxylate or amine solution, undergoes a change on adsorption of frother molecules, Figures 7.10 and 8.17, but no systematic study of such changes appears to have been made.

6. For practical purposes the most important of all effects is the change in flotation recovery resulting from frother additions. Figure 9.26 shows a progressive improvement in the recovery (obtained under constant additions

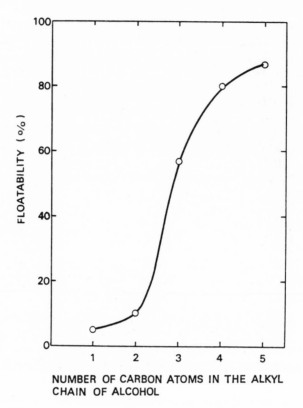

Figure 9.27. Effect of an increase in chain length of the frother-acting alcohols on the flotation recovery of galena at constant additions of the collector and the frother. [Mukai *et al.* (1972), with the permission of the authors.]

of a collector at constant pH) caused by a gradual increase in frother additions. Figure 9.27 shows the effect of increasing the chain length of the frother-acting alcohols on the recovery at constant additions of the collector and the frother.

Instances of improved recoveries obtained on replacing one frother by another are quoted in the literature, e.g., Eigeles and Donchenko (1966) and Plaksin and Khaginskaya (1956, 1957). In plant experience, it is a well-established fact that some types of minerals present in the ore and the type of the collector used impose, occasionally, definite limitations on the type of frothers that can be employed due to strange antagonistic effects. Beneficial synergistic effects are needed and are being exploited.

As mentioned earlier, a minor change in the structure of the surfactant can make all the difference between an antagonistic and a beneficial synergistic effect in frothing; for example, $C_{16}H_{33}N^+(CH_3)_3:C_{16}H_{33}SO_4^-$ solutions froth very well, even at a 1:1 ratio, whereas the unsubstituted amine $C_{16}H_{33}N^+H_3:C_{16}H_{33}SO_4^-$ solutions, in a 1:1 ratio of surfactants, give a complete suppression of frothing.

9.9. *Flotation Frothers: Their Requirements and Mechanisms of Their Action*

On comparing two characteristics of all flotation systems, viz., frothing and selective recovery, it becomes obvious that additives and conditions which lead to good recoveries of selected minerals do not coincide with optimum frothing. In fact, with non-thio surfactants the conditions for maintaining best froths are those in excess of the CMCs—when all recoveries are bound to drop to zero. Similarly, an additive that is unable to generate any frothing on its own is occasionally a very good flotation frother in some special systems. Even when frothing characteristics of solutions (without solids) containing both collector and frother surfactants (used in a given separation) are compared with mineralized froths produced during actual flotation tests, it is obvious that no parallel relationship exists between the two. Two interacting surfactants which give very poor two-phase froths may be, and frequently are, an excellent combination in some specific flotation separations. On the other hand, invariably, good stable froths in two-phase systems are completely unsuitable for flotation purposes.

What, then, are the requirements for a surfactant to serve as a good flotation frother? The primary purpose of a frother addition is to change drastically the kinetics of particle–bubble attachment. It achieves this by ensuring that the thinning of the liquid film between the two colliding particulates, the particle and the bubble, and the rupture of the thinned film (with the necessary establishment of the particle–bubble attachment area) can take place within the collision time. Secondary requirements relate to ensuring a proper degree of stability of the aggregate particle–bubble and allowing most (or preferably all) of the mechanically entrapped (and weakly adhering) particles of unwanted minerals to escape with the draining liquid. The stability of the particle–bubble aggregate and the stability of the mineralized froth formed during the draining process atop the pulp in the cell are not synonymous—one is concerned with adhesion and the other with cohesional characteristics within a mixed liquid–particle film.

A significant feature of all frothers used in flotation systems is their nonionic polar group. (The best two-phase frothing systems are composed of ionized or mixed ionized–nonionic surfactants.) Another characteristic of the better-acting frothers is the nonadlineating structure of their nonpolar groups. All *n*-alkyl hydrocarbon chains of surfactants readily develop multiple van der Waals bonds, that is, become adlineated. This adlineation occurs, particularly, if the polar groups associated with the chains are not much larger than the cross section of the hydrocarbon chain itself. Such multiple bonding gives rise to solidification of films, with many undesirable characteristics, insofar as dynamic flotation systems are concerned.

Another feature that needs to be taken into account is that, owing to the high degree of heterogeneity of particle surfaces that are being dealt with in flotation systems, the condition of no residual charge being present on the solid surface is not likely to exist at the moment of particle–bubble collision. The bubbles themselves are always charged [see Dibbs *et al.* (1974), Huddleston and Smith (1976), and Collins *et al.* (1978)], whether in pure water, in solutions of inorganic electrolytes, or in solutions of surfactants. Owing to a preferential adsorption of OH^-, there is a definite pH dependence of their charge, whether in water or in nonionic surfactant solutions.

In view of these features, noted with respect to residual charges being present on solid particles and on air bubbles, there is a high probability that similarly charged hydrophobic particles and bubbles will be colliding with each other within a flotation pulp. A definite energy barrier due to an electrostatic repulsion is thus likely to be encountered, as indicated by curve D in Figure 9.18, before thinning and rupture of the intervening film can occur.

It is postulated hereby that the main role of a frother-acting surfactant in flotation is to provide means of *replacing repulsive forces*, whenever necessary, with *attraction*, owing to the ability of such surfactant molecules to *align their dipoles* appropriately and instantaneously at the moment of particle–bubble collision.

It is from this ability to align their dipoles that nonionic surfactants with nonadlineating hydrocarbon groups usually prove themselves the better frothers. The interaction with the collector species is beneficial, but not necessarily always so. When it is so strong that it prevents dipole realignment, collisions may result in no attachment (the respective polar groups are then locked together). On the other hand, even a strong interaction between collector and frother species may allow an attachment to take place (dipole realignment is not affected) but no froth is generated because the nonpolar groups of the reacting frother–collector molecules form an adlineated, solidified film. For optimum characteristics of mineralized froths, only such interactions between collector and frother-acting surfactants are required that the relaxation times of their dipoles remain shorter than the collision times and the respective nonpolar groups do not form a mixed solidified film.

Surfactants with charged polar groups may still act as frothers but only in situations under which solid particles are oppositely charged with respect to the charge on the bubbles. Such conditions are not likely to be met in practice over an extensive range of concentrations or pH. It is more likely, however, that some charged surfactants, such as alkyl quaternary ammonium compounds, are capable of acting in a dual capacity as a collector and a frother by virtue of providing a sufficient quantity of neutral products through reactions with counterions (which effectively replace a repulsion by an attraction).

There is, as yet, no direct experimental evidence to confirm the variation of a frother's relaxation time τ in different states (as gaseous, condensed, and solified films) either before or after an interaction with a collector species. Indirect evidence on the change of relaxation times for water adsorbed on kaolinite from 10^{-11} sec to $\sim 10^{-4}$ sec [Hall and Rose (1978)] suggests that similar times could be involved in adlineation of alcohols and ethers (polyoxyethylenes). The collision times estimated from high-speed photography of particle–bubble encounters appear to be about 10^{-3} sec. It must be realized, however, that this estimate may be an order of magnitude, or more, too long. The inhibitive influence of similarly charged interfaces, solid/liquid and liquid/gas, attempting to establish a contact will be eliminated if the concentration of nonionic surfactant is sufficient (to neutralize the excess charges involved) and, at the same time, the relaxation time remains shorter than the collision time.

Although in flotation practice the additions of a frother are selected experimentally after an evaluation of some three to five readily available and most popular products (pine oil, cresylic acid, Dowfroth 250, MIBC, etc.), this does not guarantee that an optimum type of frother has been chosen. A more systematic approach to evaluate the optimum frother combination for the particular flotation system is desirable. The difference in performance (on replacing one frother by another) is frequently only a matter of, say, 1–2% (in unusual cases, up to 5%) higher recovery or a similar improvement in grade, yet sometimes even a fraction of a percent may represent an enormous economic advantage.

On reviewing the topic of froth and frothing agents, Booth and Freyberger (1962) have provided very useful, pertinent information on practical applications of frothers in the flotation of different minerals (Table 9.5).

Table 9.5. Frother Usage (percentage of total flotation plants or mills)[a]

Frother	Gold ores,[b] 52 mills	Simple copper ores,[c] 66 mills	Complex copper ores,[a] 35 mills	Lead ores,[e] 95 mills	Zinc ores,[f] 104 mills	Pyrite ores, 14 mills	Bulk sulfide ores,[g] 23 mills	Coal,[h] 15 mills	Amine flotation, 21 mills	Fatty acid, soap, sulfonate flotation, 61 mills
Pine oil	15	17	23	4	13	50	17	27	30 (mica; feldspar)	8
Cresylic acid	8	11	8	27	17	22	13	—	—	5
Pine oil and cresylic acid	19	5	3	2	6	—	9	—	—	—
Methylisobutylcarbinol (MIBC)	—	14	26	29	33	—	25	53	25 (KCl)	11
Other alcohols	4	3	3	—	4	7	4	7	5 (KCl)	5
Polyglycol types	13	21	8	9	16	7	9	13	5 (feldspar)	7
Triethoxybutane (TEB)	15	8	3	—	—	7	—	—	—	—
None	—	—	—	9	—	—	—	—	30 (quartz)	64

Combinations (other than pine oil and cresylic acid):

Polyglycols and

a. Pine oil	4	5	8	2	2	—	—
b. Cresylic acid	8	1	3	2	2	—	—
c. Pine oil and cresylic acid	6	—	—	—	—	—	—

MIBC and

a. Pine oil	2	5	6	2	1	9	—
b. Cresylic acid	4	3	6	12	3	5	—
c. Pine oil and cresylic acid	—	—	—	—	7	—	—

Miscellaneous	2	7	3	2	3	9	5

[a] Source: Booth and Freyberger (1962).

[b] Ores from which gold is the only important product recovered by flotation, probably in most cases in pyrite.

[c] Ores from which copper is the product of major importance; 49 ores classed as containing copper only, 17 ores classed as containing some gold and/or silver.

[d] Ores from which two important products are obtained, generally by selective flotation; 14 Cu–MoS_2 ores; 13 Cu–Zn ores; 2 Cu–Co ores; 6 Cu–Ni ores.

[e] Four lead ores; 71 Pb–Zn ores; 15 Pb–Cu–Zn ores. In the last named subgroup it was assumed that a lead–copper float was aimed primarily at recovering the lead.

[f] Eight zinc ores; 70 Pb–Zn ores, 13 Pb–Cu–Zn ores; 13 Cu–Zn ores. (Zinc frother unidentified in one Pb–Zn ore and two Pb–Cu–Zn ores.)

[g] Bulk sulfide flotation from ore, not followed by differential flotation. Eight instances of flotation of sulfides from tungsten ores.

[h] Survey limited to domestic mills (U. S.) in this group.

9.10. *Mineralized Flotation Froths and Their Stability*

Once a particle–bubble attachment is achieved, the adhesion between the two phases must be strong enough to survive the disruptive centrifugal forces generated within the mechanical cell, capable of attaining ~ 50 g. On reaching the top of the pulp in the cell, a definite thickness of froth has to be maintained to allow "cleaning" of the floated material to take place. Mechanically entrained gangue particles will tend to escape with the draining liquid. With "stiff," rigid, and stable froths an additional spray of water over the top of the froth may facilitate the cleaning action by drainage.

The presence of solid particles, incorporated within the froth structure, has a profound effect on the characteristics of the froth and its stability. A most informative account of the role played by hydrophobicity of the particles (as measured by contact angles on particles and on plates of the same solid kept in the same solutions), the shape of the particles, particle roughness, their size, the type of frother, etc., has been prepared by Dippenaar (1978). He found that highly hydrophobic particles with contact angles greater than 90° will destabilize froth as a result of facilitated thinning of the interbubble liquid bridged by the particle. Rounded or spherical particles (beads) accomplished the thinning and ruptured the liquid film in ~ 0.1 sec, whereas sharp-edged galena particles ruptured the film in ~ 0.02 sec, at the perimeter of the three-phase contact (Figure 9.28). With less hydrophobic particles (contact angle $\sim 74°$) the thinning occurred in the symmetric liquid film, outside the three-phase perimeter of contact, and required ~ 0.5 sec in the case of spherical particles.

Froth becomes stabilized by hydrophobic solids when they adhere to the air/water interface so closely together that the draining of the liquid is restricted. If, in addition, extensive flocculation of particles occurs, fairly thick layers of interlocked particles are formed at each of the two air/water interfaces separating the bubbles. The froth becomes stable against mechanical handling and does not collapse on overflowing the weir of the flotation cell into the launder, as does the froth with thin layers of solids.

The ability to build a mineralized froth of a given height atop the pulp, its draining characteristics, and its capacity to support particles of a given maximum size in a flotation system are features which depend entirely on the collector–frother interactions at participating interfaces: solid/liquid and liquid/gas. Very little systematic work has been carried out to specify and to quantify these aspects.

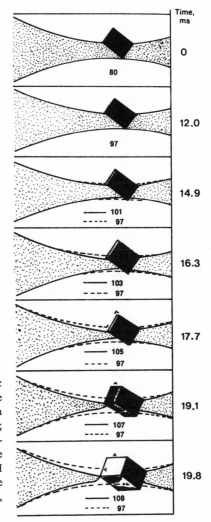

Figure 9.28. Reproduction of a photographic sequence (numbers underneath the particle denote the frames in the cinephotograph) of a particle of galena rupturing a liquid film; numbers to the right denote the time in milliseconds elapsed in the sequence from the first frame. [From Dippenaar (1978), NIM Report No. 1988, with the permission of the National Institute for Metallurgy, Randburg, S. Africa.]

9.11. *Bubble Generation in Mechanical Flotation Cells*

A comprehensive evaluation of the hydrodynamic aspects of flotation cells (Denver, Agitair, Fagergren, Wedag) was presented by Arbiter *et al.* (1969), following an extensive period of investigation since 1960. A number of equations were established to correlate the various studies of operating variables; for details of these a reference to the paper itself is necessary.

The important conclusions were as follows: Aeration affects the power consumption of the cell as well as the ability of the machine to keep the solids in suspension. When a critical value of the airflow number (defined as $N_Q = Q_A/ND^3$, where Q_A is the volumetric flow rate of air at STP, N is the impeller rotational speed, and D is the impeller diameter) is exceeded, a drastic sedimention occurs with closely sized particles and a less drastic one with a wide range of particle sizes. The presence of a frother and solids reduces the air intake (self-aerating capacity) of the cell, reduces the average size of bubbles and their ascent velocity, and diminishes the power consupmtion.

Detailed examination of the manner in which air is drawn in by a mechanical flotation cell and of the mechanism of bubble generation was carried out by Grainger-Allen (1970). There is a linear relationship between the suction (in cm Hg) generated behind an impeller blade (in the absence of air) and the impeller speed; for example, a 2 cm Hg suction is generated at 750 revolutions/min and a 4 cm Hg suction at 1000 revolutions/min. When no air is admitted and the water is not degassed (saturated or near-saturated), cavitation occurs in the region of reduced pressure, with vapor-filled cavitation bubbles. If air is admitted (either through a hollow shaft or by an external tube—a so-called *sparger*), an air cavity is formed behind the revolving impeller.

The use of baffles reduces the relative velocity between the impeller and the pulp in the vicinity of air cavity—these baffles increase the number of bubbles generated behind the impeller but not their size, and change their flow away from the impeller. Depending on the disposition and the form of the impeller blades (or cylindrical stubs) and of the baffles, the air cavity is found to be separated into individual small cavities following each blade or forming a large central cavity for the whole impeller. The air cavity forms small air bubbles by the vortex-shedding mechanism, identical with that in which a large rising bubble splits fine bubbles in its wake (Figure 9.29). Additions of a surface active agent decreased the sizes of bubbles but did not appear to affect the air cavity. Striations were noted on numerous bubbles immediately downstream of the cavity, and these were interpreted as due to vibrations of the bubble surface. When cryolite (a mineral with a refractive index close to that of water) was added to the cell in order to observe the effect of solids on bubble generation, it was concluded that fine solids, up to 40% weight addition, did not materially affect the air cavity shape or the bubble generation process. However, it was clear that there was a very limited particle–bubble collision; the air and the mineral appeared to occupy separate regions in the impeller zone, the air in the low-pressure zone and

Figure 9.29. Subdivision of a large rising bubble into fine bubbles by splitting. [From Kampe and Tredmers (1952).]

the pump containing the solid mineral in the high-pressure zone; they were not intimately mixed as was previously visualized. Obviously, more experimental work is needed to clarify the role of the geometric factors (particularly for the large, modern flotation cells of 400–700-ft³ volume) with the hydrodynamics, collision attachment, etc., peculiar to the cell.

9.12. *Selected Readings*

Akers, R. J., ed. (1976), *Foams* (proceedings of a symposium by the Society of Chemical Industry at Brunel University, Sept. 1975), Academic Press, New York.

Bikerman, J. J., Perri, J. M., Booth, R. B., and Currie, C. C. (1953), *Foams, Theory and Industrial Applications*, Van Nostrand Reinhold, New York.

Booth, R. B., and Freyberger, W. L. (1962), Froths and frothing agents, in *Froth Flotation—50th Anniversary Volume*, ed. D. W. Fuerstenau, AIME, New York, pp. 258–276.

Clift, R., Grace, J. R., and Weber, M. E. (1978), *Bubbles, Drops and Particles*, Academic Press, New York.

Davies, J. T., and Rideal, E. K. (1961), *Interfacial Phenomena*, Academic Press, New York.

de Vries, A. J. (1957), *Foam Stability*, Rubber-Stichting, Delft.

Kitchener, J. A. (1964), Foams and free liquid films, in *Recent Progress in Surface Science*, Vol. 1, eds. J. F. Danielli, K. G. A. Pankhurst, and A. C. Riddiford, Academic Press, New York, pp. 51–93.

Kitchener, J. A., and Cooper, C. F. (1959), Current concepts in the theory of foaming, *Q. Rev. Chem. Soc.* **13**, 71–97.

Klassen, V. I., and Mokrousov, V. A. (1963), *An Introduction to the Theory of Flotation*, Butterworth's, London.

Lovell, V. M. (1976), Froth characteristics in phosphate flotation, in *Flotation, A. M. Gaudin Memorial Volume*, Vol. I, AIME, New York, ed. M. C. Fuerstenau, pp. 597–621.

Manegold, E. (1953), *Schaum*, Strassenbau, Heidelberg.

Mysels, K. J., Shinoda, K., and Frankel, S. (1959), *Soap Films*, Pergamon Press, London.

Read, A. D., and Kitchener, J. A. (1969), Wetting films on silica, *J. Colloid Interface Sci.* **30**(30), 391–398.

Sebba, F. (1962), *Ion Flotation*, Elsevier, Amsterdam.

Sheludko, A. (1966), *Colloid Chemistry*, Elsevier, Amsterdam.

10

Inorganic Regulating Agents, Activators, Depressants

The purpose of making an addition of any reagent to a flotation system is to improve floatability of the desired solid at the desired moment, i.e., to achieve selectivity of separation. Depending on the main aspect of their action, the additives are grouped under different headings. Thus, the regulating agents are those additives, inorganic or organic in nature, which

1. *Control the pH of the system* (i.e., inorganic acids and/or alkali, various salts such as lime, soda ash, etc.).

2. *Adjust the charge density* at the various solid/liquid interfaces (i.e., polyvalent inorganic salts of sufficient solubility to provide high electrolyte concentration, capable of dispersing the solid particles which otherwise tend to coagulate or, if fine, deposit on larger solids).

3. *Regulate the oxidation states* of ions in solution, of surface species on solids, or of surfactants used as collectors. These additives include the oxygen dissolved in water and diffusing from the gas phase during aeration, the purposely added oxidizing or reducing agents—MnO_2, nitrates, As_2O_3, CN^-, HS^-—or the conditions which determine the potential in the system under which the oxidizing or reducing reactions can proceed, i.e., galvanic contact between metallic iron and sulfides or among different nonliberated grains of sulfides. The extent and the intensity of these oxidizing or reducing conditions cause different responses among the various sulfides involved in a flotation system.

4. *Control the concentration of metallic ions* in solution and at interfaces through precipitation or complexing with HS^-, CN^-, OH^-, phosphate ions, etc.

The various types of organic additives are dealt with in Chapter 5. Here only the inorganic agents are considered.

Additives which are more specific in their action, e.g., they facilitate explicitly the adsorption of a collector-type surfactant on a given mineral, are called *activators.* These may modify either the surface of the wanted mineral (thus helping to immobilize the surfactant) or the surfactant itself to let it act as a collector. Reagents which have a specific opposite effect to that of activators are called *depressants.*

It is obvious that the same chemical species may act as a regulatory agent or an activator or a depressant, depending on the character of the solid or the flotation system employed. For example, HS^- can act as an activator in the flotation of oxidized heavy metal sulfides or as a depressant in the flotation of unoxidized sulfides. At the same time, depending on the concentration employed, it is a reducing agent and a highly selective charge density regulator. Similarly, oxygen can be considered a regulator for sulfide minerals or an activator for flotation with xanthate ions and any thio-type collector.

10.1. *pH Control*

An adjustment of pH by an addition of an acid or an alkali affects:

1. *The surfaces of the solids,* the extent of their hydration, the excess charge density, the zeta potential and the types of ions specifically adsorbed in the IHP, and, generally, the formation or the destruction of a crystalline (or, more frequently, a noncrystalline) surface coating. Thus, for example, the destruction of a carbonate or a thio-sulfate surface layer (some of which may be cementing other particles to the surfaces of the wanted solids) and the formation of a noncrystalline coating of hydrogen-bonded metallic complexes like hydroxides, bicarbonates, etc., are determined by the acidic or alkaline range of pH in the pulp.

2. *The extent of ionization or solubility of a surfactant* and its participation in the formation of complexes with inorganic counterions.

3. *The concentrations of inorganic ions,* whether complexed (by OH^-) or uncomplexed, remaining in the liquid phase of the pulp.

4. *The level of the electrochemical (E_h) potential* and of the corresponding oxidation–reduction potential.

The complexity of all these effects, particularly the kinetics and incomplete reversibility involved in the reactions producing these effects, makes it important to establish the correct sequence of pH adjustment procedures, the optimum points, and the levels of additions. The type of additive used to control the pH is often of paramount significance. Although an acidic pH can be obtained by HCl, H_2SO_4, HNO_3, etc., the economic aspect alone, i.e., the price, is obviously not always the sole criterion to follow in deciding which additive to use, since the side effects of Cl^-, SO_4^{2-}, and NO_3^- ions may play a very significant role in some systems. Similarly, $NaOH$, $Ca(OH)_2$, or $Na_2CO_3 \cdot 9H_2O$ may have to be used in some systems because of their special role in side reactions, e.g., CO_2^{2-} in precipitating Pb^{2+} or Ca^{2+} ions and in depressing iron sulfide minerals.

The solution pH determines the extent of ionization and hydrolysis of surfactants. This, in turn, either helps or hinders the adsorption of the surfactant at the various ionized solid/liquid interfaces, contributing to greater or lesser selectivity of flotation. The very intricate balance that has to be achieved between the solid surfaces of the various minerals, the surfactant acting as a collector, and the appropriate concentrations of ions in solution and at the solid/liquid interfaces is determined, in the first instances, by the pH. In a few systems, the range of the favorable (or permissible) pH which enables the required floatability to be achieved is a wide one, but in many complex and difficult separations, the narrowness of the requisite pH range and the difficulty of achieving a steady pH value under the dynamic and kinetic conditions in a flotation cell are chiefly responsible for the poor results.

The incomplete reversibility of dissolution–precipitation reactions (e.g., see Figure 5.11) indicates that, particularly in exploratory laboratory testing, the overshooting of pH should be avoided and any concomitant results treated with suspicion in systems in which some solids are susceptible to large changes in zeta potential around the pH value aimed for if their value of zeta potential plays a critical role in achieving selectivity. The action of pH in the complexing of cations is discussed in Section 10.5.

10.2. *Control of Charge Density at the Solid/Liquid Interface*

The control of charge density at the various solid/liquid interfaces is accomplished by ionized additives, mostly polyvalent in character (in agreement with the Hardy–Schulze rule, Chapter 7). The inorganic additives employed for this purpose are sodium silicates, calcium polyphosphates, aluminum salts, polysulfides, etc. The organic regulating agents are represented by polysaccharides such as starch and dextrins, cellulose derivatives, tannin, block polymers of ethylene oxides and propylene oxides, etc. These reagents are discussed in Chapter 5.

10.2.1. Sodium Silicate

Sodium silicate (or water glass) is the most common additive used for dispersing colloidal particles, *slimes*, produced during grinding from the more friable ore components. According to Vail (1952), there are 29 liquid products manufactured under the name of sodium silicate with densities varying from 1.318 to 1.871, the ratios of SiO_2 to Na_2O varying from 1.6 to 3.75 (this ratio is known as the modulus of the given sodium silicate), and the viscosities up to 0.5×10^6 poise (p. 552). Figure 10.1, representing the triaxial diagram of the Na_2O–SiO_2–H_2O system, gives the types of products available commercially: Areas denoted 9 and 10, shown in this diagram, give the approximate composition of ordinary liquid sodium silicates produced commercially.

When the commercial liquid sodium silicate is diluted in an aqueous solution containing a supporting electrolyte, it undergoes hydrolysis, resulting in the production of hydroxylated species of a general formula: $Na_mH_{4m}SiO_4$, where m denotes the average number of OH^- bound per Si atom and varies between ~ 0.5 and ~ 1.6. Potentiometric titrations carried out by Lagerström (1959) and Ingri (1959) led to suggestions that the silicate components in solution were three monomeric species—$Si(OH)_4$ (uncharged), $[SiO(OH)_3]^-$, and $[SiO_2(OH_4)]^{4-}$—together with a dimeric species, $[Si_2O_3(OH)_4]^{2-}$, and a tetrameric species, $[Si_4O_8(OH)_4]^{4-}$.

A more recent study by Aveston (1965) indicates that, in addition to the preceding species, some polymerized polyvalent aggregates must exist, particularly when m (the average OH^- per Si) decreases from ~ 1.0 to ~ 0.5. The existence of such polyvalent aggregates of silicic acid, silica micelles, was suggested, according to Klassen and Mokrousov (1963), by Peskov (1940). Their structure is, at present, only speculated upon. They are, however, highly charged species.

It is clear that the presence of such polyvalent polymeric species would exert a drastic effect on all oppositely charged solid surfaces, leading to a very rapid reversal of charge and a highly negative zeta potential. However, even the tetravalent monomeric and tetrameric species, already listed, possess sufficiently high activity with respect to the reversal of charge on

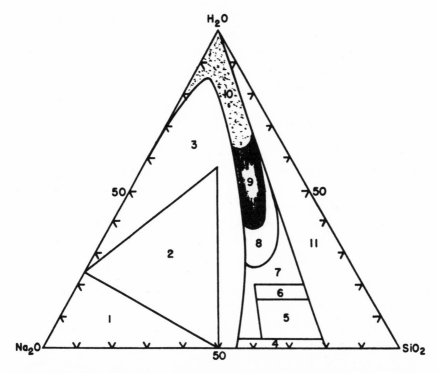

Figure 10.1. Phase diagrams of sodium silicates; commercial production is restricted to areas denoted by 1, anhydrous sodium silicate and mixtures with NaOH; 2, crystalline alkaline silicates; 4, glasses; 9, ordinary sodium silicate liquids; and 10, dilute sodium silicate liquids. Other compositions (marked 3, 5, 6, 7, 8, 11) are either uneconomical or have no applications. [After Vail (1952).]

solids with a positive zeta potential. The polymeric, polyvalent aggregates are mainly present in more concentrated solutions, which are less hydrolyzed.

The kinetics of depolymerization taking place on making a dilute solution or on adding a concentrated solution to a flotation cell may result in widely differing effects being produced in a given flotation system. Therefore, the modulus of the particular sodium silicate used, the manner in which the solution was prepared, and its strength and dosage adopted in testing have to be clearly specified before a comparison of the effects produced by sodium silicate additions can be valid. It is often exasperating to attempt a duplication of the published sodium silicate effects in a given flotation system if the total dosage of silicate is the only information that is given.

The role played by sodium silicate additions can easily be perceived if reference is made to Figure 7.26, which shows the effects of various electrolytes on the zeta potential.

Two solids, such as metallic oxides or SiO_2 and Al_2O_3, usually show a different behavior in aqueous systems containing only one dispersed solid from that when both solids are dispersed in the same solutions. Traces of dissolved ions or molecules derived from each of the solids tend to adsorb on the other solid, thus altering their characteristics. Changes in zeta potential values are the most readily detected alterations in such characteristics. The zeta potential curve of each individual solid (when determined for the solid alone in solution) is shifted toward the zeta potential curve of the accompanying solid when present in the mixture. The extent of this shift in pH_{pzc} is determined by the concentration of ions or complexes released into solution [Mackenzie (1966, 1971), Mackenzie and O'Brien (1969), Healy et al. (1968)] and the *degree of their hydrolysis* [Matijevic (1973a, 1973b)].

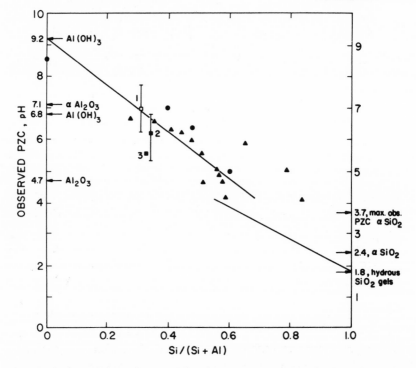

Figure 10.2. Coprecipitates in the Al_2O_3–SiO_2 system; variations in their observed pzc. [After Parks (1967), from *Equilibrium Concepts in Natural Water Systems*, R. F. Gould, ed., with the permission of the American Chemical Society, Washington, D.C.]

Healy *et al.* (1973) discuss examples of shifts in zeta potentials of oxides SnO_2, TiO_2, Al_2O_3, and SiO_2 in mixed three-component systems (two solids in a solution) as compared with two-component solid–solution systems in respect to heterocoagulation of colloidal suspensions. Parks (1967) discusses variations in the pzc of aluminum and silicon hydrous oxides which are either coprecipitated or formed by an amorphous substitution (Figure 10.2).

An addition of a suitable cation, specifically adsorbing on one mineral, is often necessary before the addition of sodium silicate in order to differentiate between two minerals that show the common tendency to shift the respective zeta potential curves and merge their respective pzcs. Such additions of polyvalent cations of aluminum, iron, chromium, etc., are made to improve the action of the subsequently added sodium silicate ions in the flotation of nonsulfide minerals such as fluorite, scheelite, and barite. The action of silicate ions is affected also by the presence of carbonate ions. For example, if silicate and carbonate ions are present together in the system, a selective desorption of oleic acid from calcite and apatite may result but not from scheelite. The amount of each additive (polyvalent ion, carbonate and/or silicate) is critical for each system. Very small additions of sodium silicate act in a highly selective manner. When additions exceed approximately 2–2.5 lb/ton, there is a corresponding loss of selectivity and, for higher additions, a complete dispersion of the highly negatively charged colloids ensues.

10.2.2. *Sodium Polyphosphates*

Sodium polyphosphates constitute another group of important multivalent anions. Three different groups of polyphosphates are distinguished:

1. Linear-chain polyphosphates containing 2–10 P atoms
2. Cyclic metaphosphates containing 3–7 P atoms
3. Infinite-chain metaphosphates

Linear polyphosphates form anions of the general formula

$$[P_nO_{3n+1}]^{(n+2)-}$$

which give soluble complexes with some metals. An orthophosphate ion, a dipolyphosphate ion, $P_2O_7^{4-}$ (a *pyrophosphate*), and a tripolyphosphate ion, $P_3O_{10}^{5-}$, are shown below:

Orthophosphate:

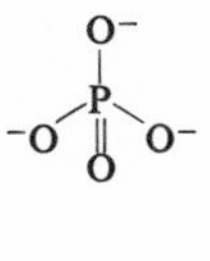

$$PO_4^{3-}$$

Pyrophosphate: $P_2O_7^{4-}$

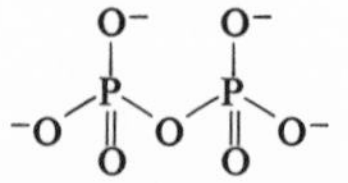

Trimetaphosphate: $P_3O_{10}^{5-}$

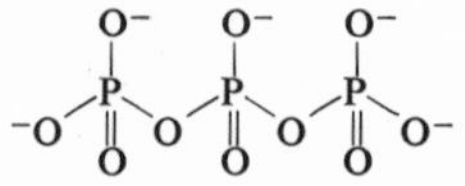

The sodium salts of these polyphosphoric acids are highly soluble, while Ca^{2+}, Mg^{2+}, and heavy metal ions are readily precipitated. Cyclic polyphosphates form anions of the general formula $[P_nO_{3n}]^{n-}$. The cyclic tetrametaphosphate, $Na_4P_4O_{12}$, is shown below together with a cyclic hexametaphosphate ($Na_6P_6O_{18}$). These are the two most important polyvalent species used extensively in detergency, coagulation–dispersion, and flotation for the control of charge density.

Cyclic Polyphosphates—Cyclic Metaphosphates:

Tetrametaphosphate: $P_4O_{12}^{4-} + 4Na^+$

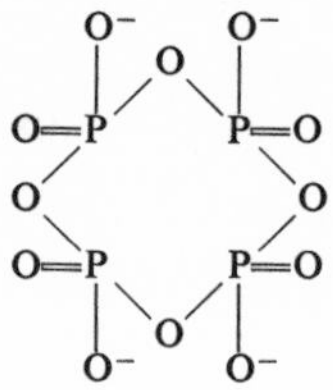

Hexametaphosphate: $P_6O_{18}^{6-} + 6Na^+$

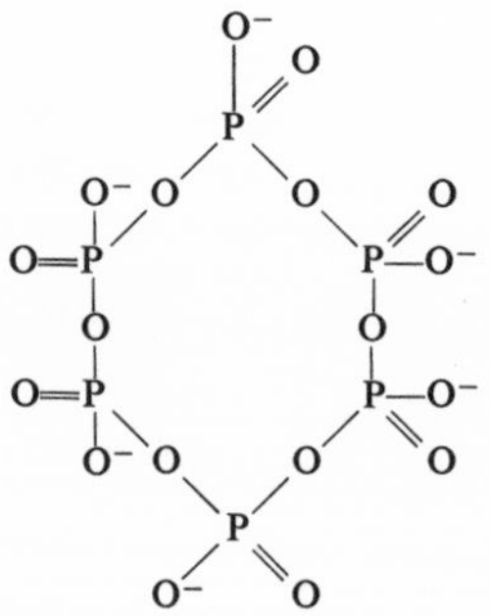

The action of the polyphosphate ions may differ from that of polyvalent silicate ions. The latter are adsorbed purely electrostatically, and a water wash or a dilution of the flotation pulp would bring about a complete or partial removal of the adsorbed silicate species from the solids. The phosphate ions can adsorb in the same manner as silicate ions, i.e., electrostatically, but in addition they are also capable of forming strong covalent bonds with a number of metallic cations. Thus, they can become either incorporated in the electrical double layer or can react covalently through one or two of their ionized oxygens, leaving all other unreacted groups to charge the surface with excess negative charges. Conversely, they can react strongly with the adsorbed metallic cations, removing these away the solid surface (as a precipitate dispersed within the liquid) and thus returning the solid surface to the charge it possessed before the metal cations adsorbed on it.

A mixture of sodium salts of cyclic hexametaphosphoric acid and linear dimeric pyrophosphoric acid ($H_4P_2O_7$) is produced commercially under the trade name Calgon; it is used in water softening to precipitate any residual Ca^{2+} and Mg^{2+} ions (solutions are buffered at pH 8.1).

10.3. *Additives Regulating the Oxidation States of Various Components in the Pulp*

These additives are of critical importance in the flotation of metallic sulfides with thio-compound-type collectors. In the flotation of nonsulfide solids with ionizable non-thio surfactants, the oxidation state of the solid surface may be important (but not to the same extent as with thio collectors and sulfides) but, generally, is not critical.

The most important additive is the oxygen dissolved in the water, both that used in grinding and that used to dilute the mineral *pulp* to a requisite density. Numerous determinations of oxygen content at the various stages of a flotation circuit [Bessonov (1954), Gowans (1972), Woodcock and Jones (1970)] have shown that during grinding of a sulfide ore the level of dissolved oxygen is lowered from near saturation in the feed to the rod mill (at $\sim$8 ppm of O_2 dissolved) to 0.1–0.5 ppm in the ball mill discharge, increases to $\sim$4 ppm in the classifier (or cyclone) overflow (feed to the flotation circuit), and finally reaches the near-saturation point after 10 to 20 min of conditioning or aeration in the flotation cells (Figure 10.3). These measurements indicate that oxygen dissolved in water is rapidly consumed during grinding whenever a large extent of fresh sulfide surface is produced. Oxidation of sulfides and of abraded iron is primarily responsible for the consumption of oxygen.

The various oxidation states for sulfur and for oxygen are shown in Figure 10.4 and 10.5 in terms of *volt equivalents* VEs at 25°C.

The oxidation state of any element, e.g., S or O, in a particular compound or ion is defined as the formal charge per atom of the element, assigning a charge of -2 to every oxygen and $+1$ to every hydrogen atom.

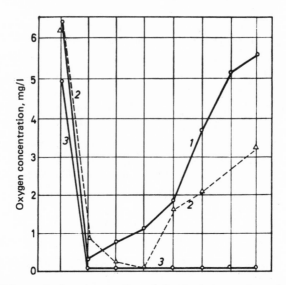

Figure 10.3. Changes in the concentration of dissolved oxygen determined at different stages of flotation circuits [(a), water in the ball mill, (b), grinding product, (c), classifier overflow, (d), collecting tank, (e), tank after transfer of the pulp by pump; in the last cell of each flotation circuit: (f), rougher flotation, (g), cleaning flotation, (h), scavenger flotation] treating the following: 1, a covellite–chalcocite ore with no pyrite; 2, a chalcopyrite ore with a slight amount of pyrite; 3, a highly pyritic ore ($\sim$70% pyrite; note a complete denudation of oxygen in all treatment stages from grinding until the last stages of flotation). [Bessonov (1954), from Klassen and Mokrousov (1963), *An Introduction to the Theory of Flotation*, Transl. by J. Leja and G. W. Poling, with the permission of Butterworths, London.]

Thus, in the $S_2O_4^{2-}$ ion the sulfur atom is in the $+3$ oxidation state, while in the H_2S molecule it is in the -2 oxidation state. The volt equivalent VE of an ion or a compound is the product of its oxidation state and its redox (oxidation–reduction) potential *relative to the element in its standard state*. All VE values are referred to the SHE (standard hydrogen electrode) scale (potential zero is assigned to the SHE; the SHE consists of a Pt electrode in 1 atm of H_2 gas bubbling through a solution of unit activity of H^+, i.e., 1 mol/liter of H^+). In any oxidation state diagram (Figures 10.4–10.7) the gradient of the line joining *any* two points is equal to the oxidation–reduction potential of the couple formed by the two species. The *more positive* the gradient, the *more oxidizing the couple*. Conversely, the *more negative* the gradient, *the more strongly reducing* is the given couple.

The reversible electrode potentials $E^\dagger$ and the corresponding free ener-

---

$\dagger$ See p. 100 regarding two conventions used.

gies of chemical reactions ΔG are directly related by

$$-\Delta G = zFE \tag{10.1}$$

where z is the number of electrons involved and F is Faraday constant, and for the reaction, e.g.,

$$2S^2 + O_2 + 4H^+ \rightarrow S + 2H_2O \tag{10.2}$$

the free energy change is given by

$$-\Delta G = -\Delta G° + RT \ln \frac{a_{S^2} - P_{O_2} a_{H^+}^4}{a_S a_{H_2O}^2} \tag{10.3}$$

and the corresponding e.m.f. is

$$E = E° + \frac{RT}{zF} \ln \frac{a_{S^2} - P_{O_2} a_{H^+}^4}{a_S a_{H_2O}^2} \tag{10.4}$$

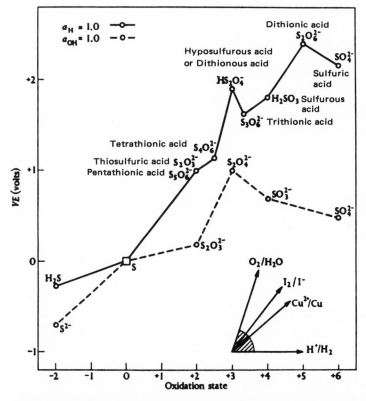

Figure 10.4. Diagram of oxidation states for sulfur for pH = 0 and pH = 14. [From Phillips and Williams (1966), *Inorganic Chemistry*, Vol. II, copyright 1966, Oxford University Press, Oxford, with permission.]

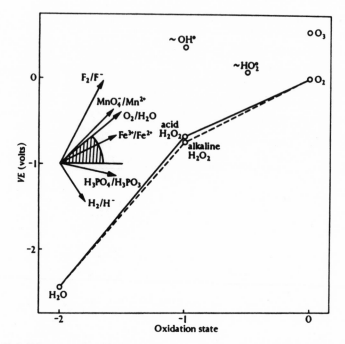

Figure 10.5. Diagram of oxidation states for oxygen. [Reproduced with permission from Phillips and Williams (1966), *Inorganic Chemistry*, Vol. II, copyright 1966, Oxford University Press, Oxford.]

where a_Y denotes the activity of species Y; *in the absence of strong complexing*, the activities can be replaced by the corresponding concentrations. For an oxidation → reduction reaction,

$$\text{oxidized species} + ze^- \rightleftharpoons \text{reduced species} \qquad (10.5)$$

Equation (10.4) becomes

$$E = E^\circ + 0.059 \log_{10} \frac{[\text{oxid}]}{[\text{red}]} \qquad (10.6)$$

The graphs in Figure 10.4–10.7 are for reactions in acid solutions of $a_{H^+} = 1.0$ i.e., pH = 0 (represented by full lines), or in alkaline solutions of $a_{OH^-} = 1.0$, i.e., pH = 14 (represented by dashed lines). The insert scales in Figures 10.4–10.7 show the equivalent free-energy change per atom of the element at 1 eV = 23.06 kcal/mol = 96.5 kJ mol⁻¹.

The oxidation state diagrams show the potentials (or the free-energy changes) involved in the reaction between the two species, a couple, represented by the points giving the particular gradient. When two different couples, four species, are considered, the comparison of the two slopes representing the two couples will show immediately the differentiation between the thermodynamically possible reactions (or combinations of couples) and those which are thermodynamically impossible. The exact course

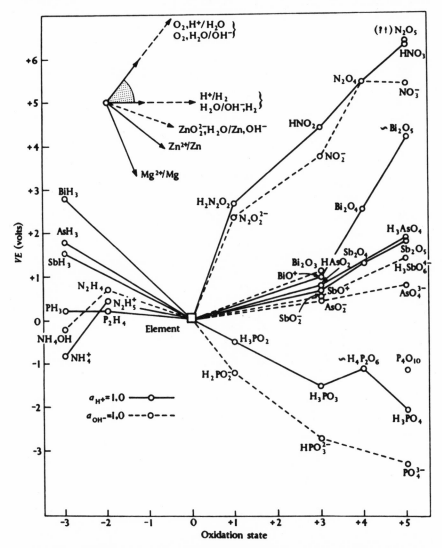

Figure 10.6. Oxidation state diagrams for N, P, As, Sb, and Bi. [Reproduced with permission from Phillips and Williams (1966), *Inorganic Chemistry*, Vol. II, copyright 1966, Oxford University Press, Oxford.]

of the reaction is, however, determined by *kinetic factors*, and these are relatively unknown. The presence in solution of some ionic species which can act as *homogeneous* catalysts (or inhibitors) may facilitate (or inhibit) the given reaction which is possible thermodynamically. More frequently, a heterogeneous catalyst is available in a flotation system owing to the presence of numerous solid phases. If a reaction is not possible on thermo-

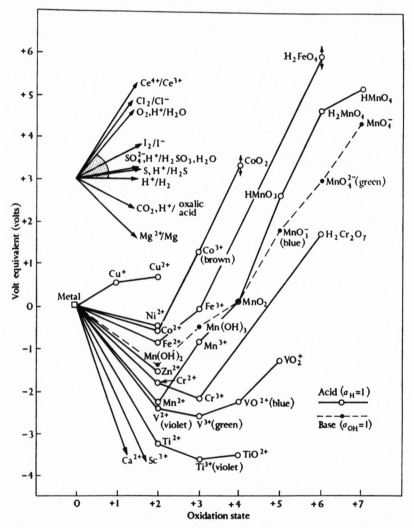

Figure 10.7. Oxidation state diagrams for the transition metals Ti, V, Cr, Mn, Fe, Co, Ni, Cu, Ca, Sc, and Zn. [Reproduced with permission from Phillips and Williams (1966), *Inorganic Chemistry*, Vol. II, copyright 1966, Oxford University Press, Oxford.]

dynamic grounds, it will not occur until the whole system is transferred to a different level of free energy by an addition of a more powerful oxidant or a reducing agent.

Figure 10.5 shows the oxidation states of oxygen: In its elemental state the value of the oxidation state is zero, but in H_2O_2 and H_2O, oxygen has an oxidation state of -1 and -2, respectively. The potential of the oxygen–water (O_2–H_2O) couple is $+1.23$ V in acid or alkaline solution. (There is only a slight change in the H_2O_2 position for these two sets of conditions but none in the *VE* values of O_2 or H_2O.) Any couple which gives a higher potential than $+1.23$ V should attack water to liberate O_2, while a couple with a negative potential should liberate H_2. The inserts in oxidation state diagrams show the potentials for other couples, with the thermodynamically stable range of the oxygen–water couple denoted by shading. Within the shaded area water is thermodynamically stable; however, the range of stability is frequently extended beyond the shaded area by kinetic factors. Thus, for example, sulfur or phosphorus oxyacids such as $S_2O_4^{2-}$, $S_2O_3^{2-}$ (Figure 10.4), or $H_2PO_2^-$, HPO_3^{2-}, PO_4^{3-} (Figure 10.6) couples should act as powerful reducing agents but are kinetically stable, especially in neutral solutions.

Similarly, within the shaded area, O_2 dissolved in water should readily oxidize Fe^{2+} to Fe^{3+} (insert in Figure 10.5) since the gradient of Fe^{3+}–Fe^{2+} couple is well within the gradient of the O_2–H_2O couple. However, this oxidation is kinetically very slow in the acid region, and only when the precipitated $Fe(OH)_3$ is formed in sufficient quantity (above pH ~ 6) does the rate constant of Fe^{2+} oxidation show a rapid increase [Figure 10.8, Singer and Stumm (1969)]. Figure 10.9, after Morgan (1967), shows the comparison of oxidation rates for Fe^{2+} and Mn^{2+} in relation to pH. The rapid oxidation by dissolved oxygen (oxygenation) of Mn^{2+} occurs at a considerably higher pH range than that of Fe^{2+}.

Even a couple involving the same element but in a different oxidation state may act as a reducing couple for the species in a higher oxidation state. Thus, for example, H_2S in acid solutions should reduce (unless hindered kinetically) any of the sulfur acids to the elemental sulfur (Figure 10.4) since its gradient, S–H_2S, is smaller than the gradient of any other couple involving species in oxidation states from $+2$ to $+6$.

In aqueous alkaline solutions sulfur itself is unstable and is liable to disproportionate into sulfide and thiosulfate (Figure 10.4). Further, although oxidations to the thiosulfate and the sulfate species (from the elemental zero oxidation state) require the same free-energy change, the intermediate step of oxidation to dithionate, i.e., $S_2O_3^{2-} \rightarrow S_2O_4^{2-}$, has a

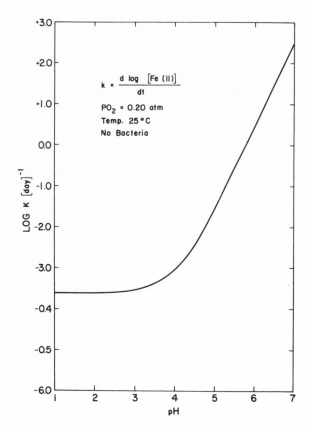

Figure 10.8. Oxidation rate of ferrous ion as a function of pH. [From Singer and Stumm (1969), with permission from the Environmental Protection Agency, Washington, D.C.]

gradient nearly that of the $O_2 \rightarrow H_2O$ reduction and may be the rate-determining step, particularly for activities of O_2 less than 1.0. (A reduction of activity by an order of magnitude corresponds to a decrease in gradient by 0.059 [equation (10.6)].) This high slope of an intermediate couple explains the preponderance of thiosulfate ions in the oxidation of mineral sulfides (in solution and on the solid surfaces) instead of the sulfate species, the final oxidation state, which is considered to be the main oxidation product.

Finally, a comparison of the negative gradient for the couple metallic Fe–Fe^{2+} (Figure 10.7) with the gradients for S^{2-} or S oxidizing to $S_2O_3^{2-}$ shows clearly that *the presence of metallic iron* exerts a *reducing action in a given system*—whether it is during grinding of the ore in a steel mill with steel rods or balls or in a flotation cell after a large degree of abrasion has deposited excessive amounts of metallic specks on the finely ground ore particles. The degree of this reducing action depends, *in the presence of Fe*, on the concentration (activity) of Fe^{2+} regardless of whether this is derived from the oxidation of metallic iron itself or from the oxidation of massive pyrite or pyrrhotite accompanying the heavy metal sulfides. Once the metallic iron is oxidized—even on the surfaces only, but in sufficient extent to screen off the metallic phase—the concentration of Fe^{2+} no longer controls the Fe–Fe^{2+} couple but the Fe^{2+}–Fe^{3+} couple.

Although the oxidation state diagrams, Figures 10.4–10.7, give only the thermodynamic criterion of the behavior (which is far from complete without the knowledge of the kinetic aspects for each pair of reactions under comparison), the information provided by these diagrams is nevertheless extremely useful. Reactions which cannot occur under the given set of conditions (determining the overall potential of the system) can be readily excluded; those which involve activation energy barriers in their kinetics may be the subject of suitable catalytic effects, operative under one set of conditions but not under a different set. The restraints imposed by the kinetics of the thermodynamically feasible reactions are modified by the influence of the pH, by the reduced activity coefficients, or by the higher gradients for any intermediate oxidation states.

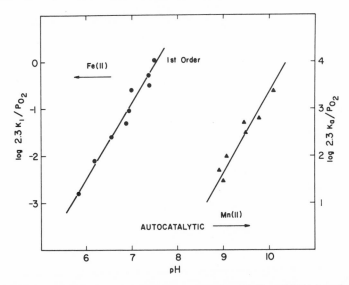

Figure 10.9. A comparison of oxidation rates of divalent Mn^{2+} and Fe^{2+} in solutions at 25°C in relation to pH. [Morgan (1967), from *Equilibrium Concepts in Natural Water Systems*, R. F. Gould, ed., with permission of the American Chemical Society, Washington, D.C.]

10.4. *Complexing Additives*

Any combination of cations with neutral molecules or anions containing free pairs of electrons gives a *coordination complex*. The anions or molecules with which the central cation is coordinated are referred to as *ligands*. The bonding between the central cation and ligands can be electrostatic, covalent, or a mixture of both. If the ligand is composed of several atoms, the one responsible for the nucleophilic nature of the ligand is the ligand atom. And if the complex contains one, two, three, etc., ligand atoms coordinated with the central cation, it is referred to as a *unidentate, bidentate, tridentate*, etc., complex, respectively. If the ligand grouping of atoms is multidentate, e.g., ethylenediamine (a bidentate ligand) or EDTA (ethylenediaminetetracetate, a hexadentate ligand), then the complex is referred to as a *chelated complex* or a *chelate*.

The most important ligands in flotation systems can be grouped according to the ligand atom into the following categories:

1. Oxygen donors such as hydro complexes (H_2O and OH^-) and inorganic ions: CO_3^{2-}, PO_4^{3-} and polyphosphates, NO_3^-, SO_4^{2-}, and ClO_4^{2-}. The organic oxygen donors consist of alcohols, phenols, ethers, ketones, and carboxylates.

2. Sulfur donors such as HS^-, S^{2-}, SO_3^{2-}, $S_2O_3^{2-}$, CNS^-, thioethers (R—S—R), mercaptans (RSH), thiocarboxylate groups

$$\left(R-C\overset{\displaystyle O}{\underset{\displaystyle S^-}{\Big\backslash}} \ ; \ R-C\overset{\displaystyle S}{\underset{\displaystyle S^-}{\Big\backslash}} \right),$$

thiocarbonate groups (mono-, di-, and trithiocarbonates), thiophosphates, thiocarbamates, etc.

3. Nitrogen donors such as NH_3, NO_2^-, SCN^-, primary amines (RNH_2), secondary amines (R_2NH), tertiary amines (R_3N), nitroso groups (R—N=O), oxime groups C—N · OH, etc.

4. A carbon donor (the only one), CN^-.

5. Halogen donors, primarily as halide complexes (fluoro, chloro, bromo, and iodo complexes).

The parameters which influence the formation of metal complexes and determine the character of the metal ion–ligand bonding consist of the following:

1. The *electronegativities* of atoms acting as electron donors (these can be arranged in the series $F > O > N > Cl > Br > I \sim S \sim Se \sim C > Te > P > As > Sb$)

2. The *total dipole moment* of the ligand (i.e., the sum of the *permanent dipole moment* μ and the induced dipole moment μ_i, the latter being the product of the polarizability α and the inducing electrostatic field E; hence, $\mu_{\text{total}} = \mu_{\text{permanent}} + \alpha E$)

3. The ionization potential of participating metal ions and the *electronic structure of cations* thus formed

4. The extent of π bonding

5. The steric parameter of ligand species

Due to the multiplicity of parameters, there is no simple rule determining the preferential complex formation or stability. The relative bond energies of complexes are indicated by a comparison of heats of formation or, preferably, the free energies of formation (given by the stability constants). The various theories put forth to interpret complex formation are concisely treated in Cotton and Wilkinson (1972, Chapter 20), i.e., electrostatic crystal field theory, ligand field theory, molecular orbital theory for transition metal complexes, etc. Phillips and Williams (1966) and Hughes (1974) also provide a comprehensive coverage of complex formation and most instructive comparisons of appropriate sequences of ions or ligands. The main feature of most ligand theories is the splitting of d-orbital energies of the metal ion and the ligand which occurs on complexing; the manner of splitting is determined by the coordination of ligands, whether tetrahedral, square planar, octahedral, or trigonal bipyramid. The actual value of splitting is obtained from electronic spectra of various complexes, which range from 7500 to 12,500 cm^{-1} for divalent metal ions and from 14,000 to 25,000 cm^{-1} for trivalent ions. When arranged in order of increasing splitting values, the ligands form the spectrochemical series

$$I^- < Br^- < Cl^- < SCN^- \text{ (S-bonded)} < F^- < OH^- < \text{oxalate}$$
$$< H_2O < NCS^- \text{ (N-bonded)} < NH_3 < NO_2^- < CN^- \sim CO$$

If the bonding between the metal ion and the ligand involves covalency, i.e., a sharing of electrons occurs, appropriate absorption bands appear in the UV or visible region of the complex spectrum. In addition, bands in the visible and near-infrared regions of the spectrum appear due to various $d \rightarrow d$ electronic transitions; all these spectra provide extremely useful information regarding the structure of the complex. The interpretation of complexes formed by transition metal ions relies heavily on their magnetic properties; these are determined by NMR (nuclear magnetic resonance), EPR (electron paramagnetic resonance, particularly useful in the case of copper and iron complexes), and, to some extent, Mössbauer spectroscopy (hyperfine magnetic splitting).

The electronic structure of metal cations plays a paramount role in complexing. To make some sensible predictions concerning their behavior in complexing, metals have been classified in several different ways. Phillips and Williams (1966) use the classification indicated in Table 10.1, which distinguishes four main classes:

1. The pretransition metals. These occur in Groups IA and IIA of the Periodic Table, and for a number of purposes it is also convenient to include with these the Group III metals, Al, Sc, and Y.

2. The lanthanides and actinides.

3. The transition metals. These occur from Group IIIA to Group IB, although transitional character is not very marked in Sc, Y, and La, while the metals of Group IB also show B-metal character.

4. The B metals. These occur from Group IB onwards into the non-metals.

Another classification, introduced by Ahrland *et al.* (1958), distinguishes *a* and *b* class metals in addition to transition metals.

The *a class metals* consist of cations with a rare-gas configuration, d°, i.e., Be^{2+}, Mg^{2+}, Ca^{2+}, Sr^{2+}, Ba^{2+}, Al^{3+}, Sc^{3+}, Y^{3+}, lanthanides^{3+}, Ti^{4+}, Zr^{4+}, Hf^{4+}, Th^{4+}, Nb^{5+}, Ta^{5+}. All of these are characterized by the fact that in aqueous solution they are able to form complexes only with F^- and *oxygen* as donor atoms.

The important organic complexing agents for these *a* metals are *oxygen donors* (tartrates, citrates, aminopolycarboxylic acids, etc.). Alkali sulfide additions result in hydroxide formation, mostly precipitation. *Dithiocarbamates, xanthates,* and dithiozone—all sulfur donors—*do not react with a metals.*

Cyanide and ammonia additions also result in hydroxide precipitation. Regarding the stability of complexes, anionic ligands give much more stable complexes than uncharged ligands; the more basic[†] the ligand's oxygen, the more stable the complex; hence,

$$OH^- > \text{phenolate} > \text{carboxylate} > F^-$$

[†] Basicity is defined as the ability to accept a proton.

Table 10.1. The Classification of Metals Used by Phillips and Williams (1966)

Pretransition Metals		Transition Metals									B Metals					
Li	Be															
Na	Mg											Al				
K	Ca	Sc	Ti	V	Cr	Mn	Fe	Co	Ni	Cu	Zn	Ga				
Rb	Sr	Y	Zr	Nb	Mo	Tc	Ru	Rh	Pd	Ag	Cd	In	Sn	Sb		
Cs	Ba	+	Hf	Ta	W	Re	Os	Ir	Pt	Au	Hg	Tl	Pb	Bi	Po	
Fr	Ra	++														

Lanthanides and Actinides

+	La	Ce	Pr	Nd	Pm	Sm	Eu	Gd	Tb	Dy	Ho	Er	Tm	Yb	Lu
++	Ac	Th	Pa	U	Np	Pu	Am	Cm	Bk	Cf	Es	Fm	Md	No ?	Lw

and

$$CO_3{}^{2-} \gg NO_3{}^-; \qquad PO_4{}^{3-} \gg SO_4{}^{2-} \gg ClO_4{}^-$$

(more stable ≡ more insoluble)

No complexes with Cl^-, Br^-, I^-, S, N, or C are formed by a metals.

The *b class metals* are shown in Table 10.2. (Some confusion arises since the *b* metals in the two classifications quoted so far are not all identical.)

The *b* metals form strong complexes in water with most halides. The stability of these complexes decreases in the sequence $I > Br > Cl > F$. π bonding and crystal-field effects become more important for these metals, particularly with ligands such as CO and CN^-. They also form very insoluble sulfides, give complexes with sulfur and phosphorus donors, and bond to S in thiocyanate or sulfite complexes (but not to N or O). In aqueous solutions some of the *b* metals form strong amine or cyanide complexes; others do not.

The *transition metals* (those cations with partly filled *d* and *f* shells in oxidation states as well) consist of the following:

The *first transition series*: Sc, Ti, V, Cr, Mn, Fe, Co, Ni, and W (partly filled 3*d* shell)

The *second transition series*: Zr, Nb, Mo, Tc, Ru, Rh, Pd, and Ag (partly filled 4*d* shell)

The *third transition series*: Hf, Ta, W, Re, Os, Ir, Pb, and Au (partly filled 5*d* shell)

The *rare earths*: [lanthanide series of elements (4*f*)] and [actinide series of elements (5*f*)]

Electrons occupying *d* shells are strongly influenced by the surroundings of the ion and, in turn, influence the environment, whereas 4*f* electrons are screened off from the surroundings.

Yet another classification of increasing importance is that of Pearson (1963) (Table 10.3) (R— represents an alkyl group CH_3—, C_2H_5—, etc.).

Pearson introduced the concept of *soft and hard acids and bases* in an attempt to correlate chemical reactions and stabilities of many compounds. The *hard acid* type of metallic ions corresponds roughly to *a* metals and the *soft acid* type of ions to *b* metals of the previous classification; however, Pearson's classification encompasses, in addition, many nonmetallic groups which act as ligands, e.g., $NO_2{}^-$, RS^-, CN^-, etc.

Metallic ions of the hard *acid type*—most of which have an outer shell of eight electrons and are characterized by small size, low polarizability,

Table 10.2. The *b* Class Metals[a]

Li	Be											B	C	N	O	F
Na	Mg											Al	Si	P	S	Cl
K	Ca	Sc	Ti	V	Cr	Mn	Fe	Co	Ni	Cu	Zn	Ga	Ge	As	Se	Br
Rb	Sr	Y	Zr	Nb	Mo	Tc	Ru	Rh	Pd	Ag	Cd	In	Sn	Sb	Te	I
Cs	Ba	La	Hf	Ta	W	Re	Os	Ir	Pt	Au	Hg	Tl	Pb	Bi	Po	At
Fr	Ra	Ac	Th	Pa	U											

[a] See Ahrland *et al.* (1958).

Table 10.3. Hard and Soft Acids and Bases[a]

	Acids			Bases		
	Hard	Intermediate	Soft	Hard	Intermediate	Soft
	H^+	Zn^{2+}	Cu^+	Pd^{2+}	H_2O	RNH_2	Pyridine	—SCN$^-$	R_3P
	Li^+	Cu^{2+}	Ag^+	$Pt^{2+,4+}$	ROH	—NCS$^-$	Br$^-$	—CN$^-$	R_3As
	Na^+	Ni^{2+}	Au^+	Cd^{2+}	R_2O	Cl$^-$	N_3^-	RSH	$S_2O_3^{2-}$
	K^+	Fe^{2+}	Tl^+	CH_3Hg^+	OH$^-$	PO_4^{3-}	NO_2^-	R_2S	
		SO_2	Hg^+	Tl^{3+}	OR$^-$	SO_4^{2-}		RS$^-$	
	Mn^{2+}	NO^+		Au^{3+}	NH_3				
	Cr^{3+}								
	Fe^{3+}	Sn^{2+}	I^+, Br^+						
	Co^{3+}	Pb^{2+}	ICN, HO^+						
		Co^{2+}	I_2, Br_2						
	Group IIA								
	Be^{2+}	Group V							
	Mg^{2+}	V_A—V, Nb, Ta							
	Ca^{2+}	V_B—N, P, As^{3+},							
	Sr^{2+}	Sb^{3+}, Bi^{3+}							
	Ba^{2+}								
	Group IIIA								
	Sc^{3+}								
	Y								
	La^{3+}								
	Ac								

[a] After Pearson (1963).

and a high oxidation state—form complexes with ligands having oxygen as a donor (whose bonds are more ionic in character than, e.g., those of nitrogen). Thus, water is more strongly attracted to these hard acid types of metal ions than ammonia or cyanide; no sulfide complexes or precipitates are formed because OH^- is more firmly bonded than HS^- or S^{2-}. On the other hand, metallic ions belonging to the soft *acid type* bond ammonia, NH_3, more strongly than H_2O, bond CN^- instead of OH^-, and also form stable I^- and Cl^- complexes. The soft metals, as well as those grouped in the intermediate category in Table 10.3, form insoluble sulfides, and soluble complexes with HS^- and S^{2-}, of primary importance in biological complexes [Hughes (1974)]. The latter reactions are often catalyzed by metal ion, e.g., hydrolysis of phosphate esters by Zn^{2+}. In flotation systems the reactions of the ligands are exemplified by transformations of cyanide complexes into thiocyanate complexes, dissolution of insoluble sulfide precipitates by alkaline sulfide solutions through the formation of polysulfide complexes, etc.

10.5. *Hydro Complexes*

Metal cations (mono- or divalent) exist in aqueous solutions primarily as complexes of H_2O molecules and OH^- ions; when the oxidation state of the metal cation is higher (tri- and tetravalent), in addition to hydroxyl complexes some OXO complexes (containing O^{2-} as a ligand) appear in the pH range 4–10.

Before 1930 Brönsted postulated that hydrated multivalent metal ions participate in a series of consecutive proton[†] transfers, e.g.,

$$Fe(H_2O)_6^{3+} \rightarrow Fe(H_2O)_5OH^{2+} + H^+ \rightarrow Fe(H_2O)_4(OH)_2^+ + 2H^+$$
$$\rightarrow Fe(H_2O)_3(OH)_3 + 3H^+ \rightarrow Fe(H_2O)_2(OH)_4^- + 4H^+ \qquad (10.7)$$

Because of the limited pH range of aqueous solutions, not all of the possible metal complexes can exist as anionic hydroxyl or OXO complexes. Polymerization of hydrolysis species results in multinuclear products (dimers, trimers, etc.) and in the formation of colloidal hydroxyl polymers. These multinuclear species are bonded through OH bridges (the formation

---

[†] Brönsted defined acids as proton donors, while Lewis proposed a more generalized definition of an acid as any species (mono- or multiatomic) with an unfilled orbital capable of accepting a pair of electrons. Examples of Lewis acids are H^+, metal cations, SO_2, $AlCl_3$, $SOCl_2$, etc.

Table 10.4. Polymerized Oxyanions and Cationic Hydroxides

Oxyanions	Cationic Hydroxides
$[SiO_2(OH)_2]^{2-}$, $[Si_4O_6(OH)_6]^{2-}$, etc.	$[Al_6(OH)_{15}]^{3+}$
$[P_4O_{12}]^{4-}$, $[P_3O_9]^{3-}$, etc.	$[Fe_2(OH)_2]^{4+}$
$[Mo_7O_{24}]^{6-}$, $[Mo_{12}O_{42}]^{12-}$	$[Cu_2(OH)_2]^{2+}$
$[V_{10}O_{28}]^{6-}$, $[V_3O_9]^{3-}$	$[Pb_4(OH)_4]^{4+}$
$[W_{12}O_{42}]^{12-}$, $[W_7O_{24}]^{6-}$	$[Zr_4(OH)_8]^{8+}$
	$[Bi_6O_6]^{6+}$

of which is then referred to as *olation*) or through O bridges (*oxolation*), e.g.,

$$\left[(H_2O)_4Fe\underset{OH}{\overset{OH}{\diagup\diagdown}}Fe(H_2O)_4\right]^{4+}$$

dimers, usually written as $Fe_2(OH)_2^{4+}$. Very strongly acidic or basic species

Table 10.5. Examples of Hydrolysis Species[a]

Be^{2+}	$Be_3(OH)_3^{3+}$, Be_2OH^{3+}, $Be(OH)_2$
Mg^{2+}	$MgOH^+$
Sc^{3+}	$Sc[Sc(OH)_2]_n^{(3+n)+}$, $n = 1, 2, 3, \ldots$
Cr^{3+}	$CrOH^{2+}$, $Cr_2(OH)_2^{4+}$, $Cr_6(OH)_{12}^{6+}$, $Cr(OH)_4^-$
UO_2^{2+}	$(UO_2)_2OH^{3+}$, $(UO_2)_2(OH)_2^{2+}$, $(UO_2)_3(OH)_5^+$, $(UO_2)_3(OH)_4^{2+}$, $(UO_2)_4(OH)_6^{2+}$, $(UO_2)_4(OH)_7^+$
Mn^{2+}	$Mn(OH)^+$, $Mn(OH)_3^-$
Fe^{2+}	$Fe(OH)^+$, $Fe(OH)_3^-$
Fe^{3+}	$Fe(OH)^{2+}$, $Fe_2(OH)_2^{4+}$, $Fe(OH)_2^+$, $Fe(OH)_4^-$, and multinuclear intermediates in precipitation of colloidal $Fe(OH)_3$
Cu^{2+}	$Cu(OH)^+$, $Cu_2(OH)_2^{2+}$, $Cu(OH)_3^-$, $Cu(OH)_4^{2-}$
Ag^+	$AgOH$, Ag_2OH^+, $Ag_2(OH)_2$, $Ag(OH)_2^-$
Hg^{2+}	$HgOH^+$, Hg_2OH^{3+}, $Hg_2(OH)_2^{2+}$
Al^{3+}	$Al_8(OH)_{20}^{4+}$, $Al_7(OH)_{17}^{4+}$, $Al_{13}(OH)_{34}^{5+}$, $Al(OH)_4^-$
Pb^{2+}	$Pb_4(OH)_4^{4+}$, $Pb_6(OH)_8^{4+}$, $Pb_3(OH)_4^{2+}$, $Pb(OH)_3^-$
Si^{4+}	$SiO(OH)_3^-$, $SiO_2(OH)_2^{2-}$, $Si_4O_6(OH)_6^{2-}$

[a] After Stumm (1967).

do not polymerize; only the weak acids or bases do. Examples are given in Table 10.4.

Studies of Sillén and co-workers (1959) have shown that multinuclear hydrolysis products of metallic cations are of almost universal occurrence in aqueous solutions. Table 10.5, after Stumm (1967), gives examples of complex hydroxyl species, the stability constants of which have been reported by Sillén and Martell (1964).

The extent of hydrolysis depends on the pH and the total concentration of metallic ion. Figure 10.10 shows the distribution of the various species for three different total concentrations of Fe^{3+}: 10^{-4}, 10^{-3}, and 10^{-2} M in the pH range 1–5.

Two ways of graphing the distribution of complexes are shown, either as percent mole fraction of individual species at each pH [e.g., Figure 10.10(b)] or as relative intermediate S curves [Figures 10.10(a) and (c)]. Figure 10.11 shows a distribution of individual hydrolysis species of Pb^{2+}.

A third way of graphing the relationship between pH and hydrolysis species is shown in Figures 10.12 (for Cu^{2+}) and 10.13 (for Al^{3+}). In these graphs the log concentration of each hydrolysis species is plotted with respect to the pH for a thermodynamic equilibrium between species, giving straight-line relationships. When the solution becomes oversaturated with respect to $M^{n+}(OH)_n$—depending on the kinetics of precipitation—a colloidal hydroxide or a hydrated oxide product is present in addition to the hydrated complexes in solution.

The *coagulation* of any colloidal particles is effected by the *hydroxylated complex* species and not by the free uncomplexed multivalent ions. Matijevic (1967) emphasized that nonhydroxylated species are, generally, unable to reverse the charge and to stabilize the lyophobic colloids, while the adsorption of hydroxylated products gives charge reversal on numerous types of solids (clays, silica, ionic solids such as AgBr, polystyrene, bacteria, etc.).

The formation of a colloidal metal hydroxide precipitate can be considered to be the final stage of polynuclear complexing. Titrations of hydrous metal oxides with acids or alkali indicate that H^+ and OH^- are the potential determining ions for these precipitates. Further, at pH below the pzc the precipitates are positively charged and readily interact with OH^- anions due to a competitive ligand exchange; similarly, the interaction of the hydrous oxides with other cations can also be interpreted as an exchange leading to a complex formation. This cation adsorption on hydroxide precipitates is comparable to a cation adsorption on clays—and is referred to as the cation exchange capacity of hydrous oxides. Colloidal precipitates of ferric hydroxide or MnO_2 possess very appreciable cation exchange capacities. Further

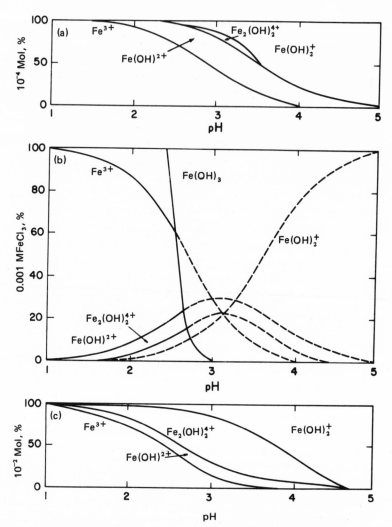

Figure 10.10. Hydrolysis species for solutions of $FeCl_3$ in concentrations of 10^{-4}, 10^{-3}, and 10^{-2} M total $[Fe^{2+}]$ in relation to pH. [From Stumm (1967), in *Principles and Applications of Water Chemistry*, S. D. Faust and J. V. Hunter, eds., with permission of J. Wiley and Sons, New York.]

changes in the surface characteristics of these metal hydroxide precipitates are created by isomorphic replacements or substitutions of the metal ion in the network (e.g., Al in Si tetrahedra) or an introduction of anions such as bicarbonate, sulfide, phosphate, silicate, molybdate, etc., into the structure of these precipitates. These new anions successfully compete with OH^-

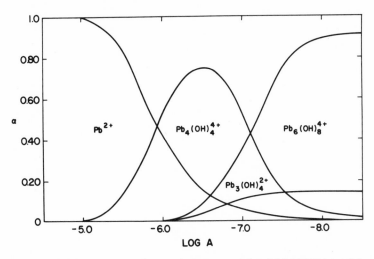

Figure 10.11. Hydrolysis species for a solution containing 0.04 M Pb^{2+} total ion concentration. Log $A = \log[H^+]$. [After Olin (1961) in Stumm (1967), in *Principles and Applications of Water Chemistry*, S. D. Faust and J. V. Hunter, eds., with permission of J. Wiley and Sons, New York.]

in the coordination of divalent metal ions and may greatly influence the solubility of the hydroxide precipitates. Thus, for example, $Ca(OH)_2$ is rendered soluble in CO_2^- containing water due to $Ca(HCO_3)^+$ formation; similarly, $Cu(OH)_2$ becomes soluble due to $Cu(CO_3)_2^{2-}$ formation (see the appropriate lines in Figure 10.12).

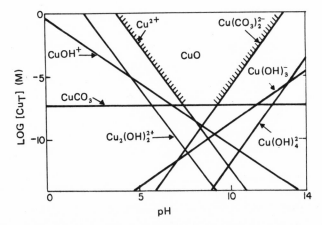

Figure 10.12. Equilibrium diagram for solubilization of CuO showing complexes in relation to pH at 25°C for a total Cu^{2+} ion concentration of 1 M and $P_{CO_2} = 3.52$.

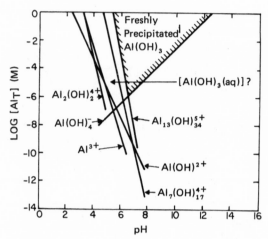

Figure 10.13. Equilibrium diagram for monomeric and polymeric Al hydroxide species in contact with freshly precipitated Al(OH)$_3$ (log K_{sp} = −33).

10.5.1. Selective Flotation Using Control of pH

When simplified systems containing individual minerals in suspension are tested in collector solutions at different pH, the comparison of results may indicate definite differentiation in collector adsorption. This would suggest that subsequent flotation of individual mineral particles should occur in different pH ranges, giving rise to selectivity controlled by pH only. However, as soon as mixed two-mineral (or highly complex multi-mineral ground ore) suspensions are tested, the *theoretical* differentiation derived from simplified systems frequently does not hold true without either minor or major modifications.

In the period 1930–1950 a great deal of research on the fundamentals of flotation was carried out using a captive-bubble technique on simple systems containing polished surfaces of individual minerals. An assumption was made that the conditions under which a cling of a bubble to the polished mineral surface was observed (or a measurable contact angle was obtained) represented the necessary and sufficient requirements for the flotation of this particular mineral from ores. This assumption is not necessarily valid on two accounts:

1. Attachment of a static air bubble to a flat, polished surface does not completely correspond to an attachment of an irregularly shaped particle to an oscillating bubble in a flotation cell.

2. The presence of other minerals may drastically affect the surface

character of the selected mineral—as already discussed on p. 616 (action of silicates).

The relationships between the collector concentration and the pH differentiating the region of no contact angle from that where a cling or a contact angle was observed were established initially by Wark and Cox (1934) for xanthate, dithiocarbamate, and dithiophosphate solutions and a range of sulfide minerals—galena, chalcopyrite, pyrite, etc.—as shown, for example, in Figures 10.14 and 10.15. Barsky (1934) suggested that the plot

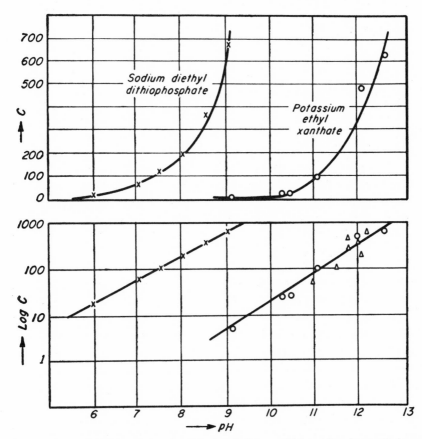

Figure 10.14. Critical contact angle curves in relation to pH for galena in an ethyl xanthate or a diethyl dithiophosphate solution (concentrations C in mg/liter). Contact angles develop under conditions to the left of the curves; no contact angles are possible to the right of the curves. [Based on data from Wark (1938) and Wadsworth (1951), from Gaudin (1957), *Flotation*. Copyright 1957 by the McGraw-Hill Book Company, Inc. Used with the permission of McGraw-Hill Book Co.]

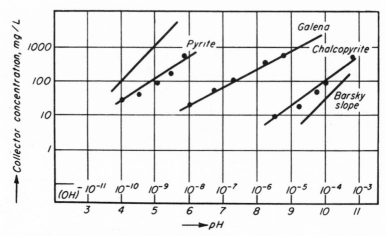

Figure 10.15. Critical contact angle curves for pyrite, galena, and chalcopyrite in solutions of sodium diethyl dithiophosphate being compared with the theoretical Barsky unit slope. [Based on data given in Wark (1938), from Gaudin (1957), *Flotation.* Copyright 1957 by the McGraw-Hill Book Company, Inc. Used with the permission of McGraw-Hill Book Co.]

of log concentration of collector vs. pH should give a straight line of unit slope. As shown in Figure 10.15, the relationships actually obtained are

$$[X^-] = m[OH^-]^n \qquad (10.8)$$

where X^- denotes xanthate ions (or other thiocollectors) and m and n are constants and where n varies, being equal to 0.52 for diethyl dithiophosphate and galena, 0.65 for ethyl xanthate and galena, 0.75 for diamyl dithiocarbamate and sphalerite, etc. It is significant that n is always less than unity.

Bushell (1958) considered similar types of relationship in attempts to control reagent concentrations in flotation solutions; see Section 8.2.2., p. 512.

It would appear that this so-called Barsky relationship depends on other parameters in the system in addition to $[X^-]$ and OH^- of H^+.

10.6. *Complexing Additives Containing Sulfide, Sulfite, or Thiosulfate Ions*

Whenever sulfide minerals are encountered in a system, the sulfur-containing ligands play a major role. Metallic ions belonging to the soft acid class in Pearson's classification have preference for sulfur ligands (rather than for oxygen ligands). However, the soft acid type of ions are not the only ones that comprise the naturally occurring metallic sulfides. Numerous metals of the intermediate class (Fe, Co, Ni, Pb, Zn, Bi, etc.) are also well represented among sulfide deposits, because of the influence due to the participating S (or HS) ligand itself. The two sequences of compounds (one consisting of hydrogen oxide, alcohols, and ethers and the other of hydrogen sulfide, thiols, and thioethers) show the following characteristic features:

1. The permanent dipole *decreases* in the oxygen sequence,

$$\mu_{H_2O} > \mu_{ROH} > \mu_{R_2O}$$

but *increases* for the analogous sulfur compounds,

$$\mu_{H_2S} < \mu_{RSH} < \mu_{R_2S}$$

2. The polarizability is *decreased* by $\sim 5\%$ in going from H_2S to R_2S and much more, $\sim 24\%$, in going from H_2O to R_2O.

H_2S is more polarizable than water and coordinates metallic ions of high field strength (e.g., Hg^{2+}, Ag^+, Cu^+, etc.) more strongly than H_2O does, forming insoluble sulfides.

H_2S is a weak acid ($pK_1 = 6.88$, $pK_2 = 14.15$) the oxidation of which yields elemental sulfur, S, except when strong oxidants are used to give S^{4+} or S^{6+} states. It dissolves in liquid sulfur to produce high-molecular-weight sulfanes, $H_2S_n (n = 2, 3, 4, 5, 6-8)$. Alkali polysulfide solution with an average chain length less than $S_{4.5}$ can be prepared by heating $Na_2S \cdot 9H_2O$ with a sulfur suspension in water. Acid decomposition of thiosulfate solution produces a yellow nonviscous oil which is a mixture of sulfanes.

Sulfides obtained by precipitation from aqueous solution frequently are almost amorphous, nucleation being much faster than the subsequent growth of crystallites. On standing or heating, the precipitates often produce mixtures of stable phases. The variations in the solubility products (as determined by different techniques) may be due to such changes which follow precipitation. Table 10.6 [after Sillén and Martell (1964)] gives the K_{sp} values (the logarithms of solubility products) for some sulfides in water at 25°C.

Sulfides of cations with an s^2p^6 electronic configuration and low charges do not form thio-anions because they are hydrolyzed by water. Cations with s^2p^6 configuration and high positive charges (Mo^{6+}, Re^{7+}, W^{6+}, etc.) form almost insoluble sulfides; however, these are soluble in ammonium or

Table 10.6. Logarithm of Solubility Products of Selected Sulfides:
$M_xS_y \rightleftarrows xM + yS$; $K_{sp} = [M]^x[S]^y$

Sulfide	Log K_{sp}
HgS	-47.0 to -53.5
Ag_2S	-47.75 to -51.22
Cu_2S	-47.6 to -49.44
CuS	-35.1 to -41.9
PbS	-27.47 to -29.37
CdS	-26.1 to -28.3
ZnS	-23.8 to -26.13 (sphalerite)
FeS	-16.38 to -19.42
MnS	-9.6 to -12.6

potassium sulfide, giving thio-anions which are stable in solution—MoS_4^{2-}, WS_4^{2-} ($C^{4+} \rightarrow CS_3^{2-}$)—while cations such as Cr^{6+} and Mn^{7+} oxidize sulfide ions. Some highly charged cations, e.g., Si^{4+}, P^{5+}, V^{5+}, Nb^{5+}, Ta^{5+}, and Re^{7+}, form thio anions, which are readily decomposed by hydrolysis.

The cations with a d^{10} configuration (18-electron shell, referred to as b cations, i.e., Cu^+, Ag^+, Zn^{2+}, Cd^{2+}, Hg^{2+}, Ga^{3+}, In^{3+}, Tl^{3+}, Sn^{4+}) form sulfides which are insoluble in water but give stable thio-anions in excess sulfide solutions: Ge^{4+}, Sn^{4+}, As^{5+} and Sb^{5+}, and Zn^{2+}—these cations are always tetrahedrally coordinated, e.g., SnS_4^{4-} and AsS_4^{3-}. Cu^+ gives poly-sulfide complexes, e.g., $[CuS_4]^-$.

The cations with an s^2 configuration also form sulfides insoluble in water, but these are readily converted to thio-anions, AsS_3^{3-} and $Sb_1S_3^{3-}$, and trimers, $As_3S_6^{3-}$ and $Sb_3S_6^{3-}$. There are no known thio-anions of Sn^{2+} or Pb^{2+}.

Alkali and alkaline earth metal sulfides are almost completely hydro-lyzed in solution:

$$Na_2S + 2H_2O \rightarrow Na_2O + H_2S + H_2O \rightarrow 2Na(OH) + H_2S\uparrow$$

Molybdenum gives a series of thio salts:

$$K_2[MoO_3S], \quad K_2[MoO_2S_2], \quad K_2[MoOS_3], \quad \text{and} \quad K_2MoS_4$$

The spectra of complex sulfide solutions have been little studied.

Sulfite ion, SO_3^{2-}, forms complexes with Mn, Fe, Co, Ni, Pd, Cu, Ag, Au, Zn, Cd, and Hg. In many complexes the sulfite ion is unidentate, but in others it occupies a two-coordination position. The ligand can be bonded

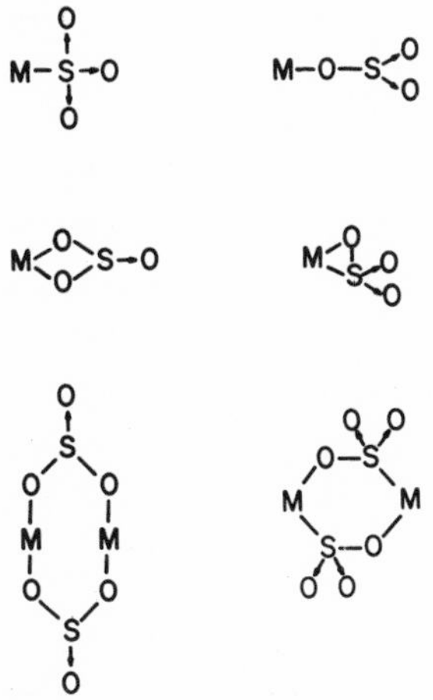

Figure 10.16. Different modes of M (metal) bonding, through O (oxygen) or S (sulfur), in complexes with sulfite ions.

in a number of ways, through O or S, as indicated in Figure 10.16. Mercury sulfite complex is exceptionally stable, and this suggests that the bonding is Hg—S and not Hg—O. The infrared spectra of sulfite complexes of Co^{2+}, Rh^{3+}, Ir^{3+}, Pd^{3+}, and Pt^{3+} are all similar to the spectrum of $(NH_4)_2[Hg(SO_3)_2]$, again suggesting the S—metal bonding.

The Cu–Tl double salt, $TlCu(SO_3)_2$, appears to be O-bonded, while the ammonium double salts, $(NH_4)_2[M^{2+}(SO_3)_2]$, where M^{2+} = Mg, Mn, Fe, Co, Ni, Zn and Cd, and NH_4CuSO_3, are all similar to Na_2SO_3.

Thiosulfate ion, $S_2O_3^{2-}$, forms strong complexes with monovalent ions Cu^+, Ag^+, and Au^+; divalent Cu^{2+} is reduced to Cu^+ complex. Alkaline thiosulfate solution dissolves many insoluble salts of Pb^{2+}, Hg^{2+}, Cu^+, and Ag^+. The unidentate complexes are mostly S-bonded, while bidentate ones may be S- or O-bonded.

As discussed in Chapter 6 (Section 6.8), the main product of PbS oxidation appears to be thiosulfate. Rao *et al.* (1976) found that many other heavy metal sulfides (chalcopyrite, pyrite, pyrrhotite, marmatite) release thiosulfate into solution as the principal soluble product of their oxidation.

Thiocyanate ion, SCN^- or NCS^-, is treated in Section 10.7.

10.6.1. *Use of Sulfur-Containing Inorganic Additives in Selective Flotation*

Sodium sulfide additions are used in selective flotation for three distinct purposes:

1. To sulfidize the heavily oxidized metallic sulfides (or oxide or carbonate minerals) in preparation for their flotation with a thio collector surfactant.

2. To improve selectivity in flotation of sulfide minerals by a combined control of pH and sulfide ion concentration. The latter may be acting as a reducing agent (controlling the level of oxygenation and of oxidation) or as a highly effective thiophilic agent neutralizing the collective properties of elemental sulfur (formed on the surfaces of some sulfides during their oxidation).

3. To decompose the xanthate-type collector coating on the bulk-floated sulfide minerals (either selectively on a few minerals or non-selectively on all floated sulfides) ahead of their subsequent selective reflotation (see the flowsheet in Figure 1.5).

A paper on the kinetics of sulfidization of oxidized minerals (cerussite, malachite, and chrysocolla) was published by Mitrofanov et al. (1955). These authors established the isotherms of sulfidization (under simplified conditions of a single mineral–solution system) using radioactive sulfur determinations; the effects of pH and temperature on the progress of sulfidization were also evaluated. The nature of the sulfide coating depositing on the oxide mineral (whether crystalline or amorphous) was not determined; however, the adhesion of the sulfidized coating to the substrate was found to increase with an increase in pH for cerussite and to decrease with an increase in pH for malachite. The tenacity of adhesion was evaluated by measuring the loss of radioactive sulfide left on the sulfidized minerals after agitation for a given time with a given quantity of quartz particles. This loss of sulfidized coating by attrition is counteracted in flotation practice by making stage additions of $Na_2S \cdot 9H_2O$ instead of a single large addition.

Other studies on the mechanism of sulfidization and on oxidation of the added sulfide are due to Abramov (1969), Alferova and Titova (1969), and Castro et al. (1974). The effects of sulfidization on the recovery of individual oxidized minerals by flotation are treated by Fleming (1953), Glembotsky et al. (1963), Mitrofanov (1967), and Ser et al. (1970).

The use of sodium sulfide as a depressant in selective flotation of heavy metal sulfides is quite distinct from that of sulfidization. Each sulfide mineral appears to have a different tolerance toward the level of sulfide ion that would inhibit flotation of that particular sulfide mineral. Three parameters are very closely involved: the pH, the collector concentration, and the sulfide ion concentration (or Na_2S addition level). Using the static contact angle technique, Wark and Cox (1934) showed the extent to which these three parameters affect the development of a cling (or attachment) between a static bubble and a polished mineral surface and thus, to a rough approximation, the relative floatability of the corresponding mineral particles. Galena was found to be much more sensitive to slight additions of sodium sulfide than chalcopyrite, the latter more sensitive than pyrite, etc. Last (1951) established that the higher the collector concentration, the greater is the corresponding critical sulfide ion concentration that separates the region of floatability from that of nonfloatability, as indicated in Figure 10.17.

These findings obtained in simplified (one mineral–solution) systems cannot apply, without modifications, to the floatability of sulfides in actual ores. The attendant complexity caused by the multitude of associated minerals is bound to exert an effect on the relative values of the main parameters: HS^-, OH^-, and the collector ion concentration. Even the latterly very popular Hallimond tube flotation experiments are unlikely to represent reliably the true behavior of the given mineral in an ore flotation (with respect to, e.g., HS^- additions) since, generally, the associated minerals are absent.

Owing to the difficulties in analytical procedures as regards determinations of the various forms of sulfur oxidation species present on the surfaces of different sulfide minerals and the sulfur oxidation species present in solution, the true role of $Na_2S \cdot 9H_2O$ additions as a depressant may, as yet, be ill-understood. The presence of elemental sulfur and of thiosulfate on the surfaces of sulfides and in solution has been documented, but it is not known with certainty whether the added $Na_2S \cdot 9H_2O$ (or NaHS) prevents or destroys the hydrophobic character on individual sulfides by forming thio anions or sulfanes or other compounds. The difficulties in flotation systems are still further compounded by the kinetics of the various reactions involved and the influence thereon of the possible catalytic agents that may be present in the system. It becomes obvious that in achieving selective flotation through sulfide additions each complex sulfide ore may require a different set of necessary parameters: pH and sulfide and collector additions.

The third mode of utilizing (sodium or ammonium) sulfide or hydrosulfide additions is to destroy the hydrophobic collector coating on the bulk-floated minerals. As the complex sulfide ores available for treatment become more finely disseminated, bulk flotation of sulfides becomes the preferable treatment, as outlined in the flowsheet in Figure 1.5. Any subsequent retreatment of the bulk concentrate necessitates a partial or a

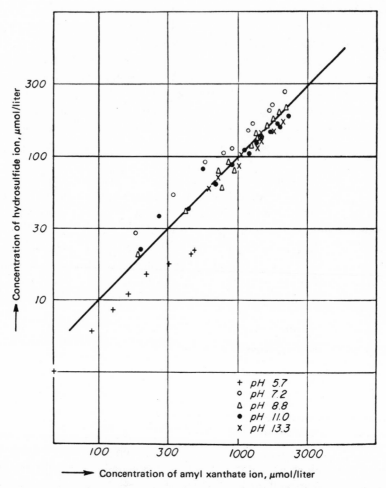

Figure 10.17. Critical relationship between the concentrations of HS⁻ (depressant) and amyl xanthate ion (collector) for the flotation of galena. No flotation to the left of the relationship band and flotation to the right. [Based on data by Last (1951), from Gaudin (1957), *Flotation*. Copyright 1957 by the McGraw-Hill Book Company, Inc. Used with the permission of McGraw-Hill Book Co.]

complete destruction of the initially induced hydrophobic coating, composed of adsorbed thio compounds, their oxidation products, and/or their metal salts. A preliminary work on the evaluation of the kinetics of such decomposition and the effects of temperature, pH, and the level of sodium sulfide additions has been carried out by Sheikh (1972) on metal xanthates (Cu, Pb, Fe) and their oxidation products. The decomposition technique is employed in industrial practice for the separation of Cu–Mo minerals and Cu–Ni minerals. Little investigative work appears to have been done on evaluating all the parameters involved in this procedure.

10.7. *Cyanides, Cyanates, Thiocyanates, and Ferrocyanates*

The cyanide ion is of great importance as a ligand, and numerous cyano complexes of transition metals are known: Zn, Cd, Hg, Cu, Ag, Au, Co, Ni, and Fe. The stability of cyanide complexes is due in a significant measure to π bonding.

Most cyano complexes are anionic and have the general formula $[M^{n+}(CN)_x]^{(x-n)-}$ where $x = 2$, 4, 6 or 8. For example, $Cu(CN)_2^-$, $Cu(CN)_4^{2-}$, $Fe(CN)_6^{3-}$, $Fe(CN)_6^{4-}$, $Mn(CN)_6^{4-}$, $Mn(CN)_6^{5-}$, $Mo(CN)_8^{3-}$, $Mo(CN)_8^{4-}$, $Mo(CN)_8^{5-}$, etc. Some mixed complexes of the type $[M(CN)_5Y]^{n-}$, derivatives of hexacyanides, where Y may be H_2O, NH_3, CO, NO, or a halogen, are also known. Cyanide groups can form linear bridges between metal atoms, giving numerous polymeric complexes in the form of infinite chains, e.g., AuCN, $Zn(CN)_2$, and $Cd(CN)_2$. In several instance, cyanide complexes of metal ions possessing two or three successive oxidation states are known, e.g., $[M(CN)_x]^{n-}$, $[M(CN)_x]^{(n+1)-}$, and $[M(CN)_x]^{(n+2)-}$.

The cyanide complexes of importance in flotation of sulfide minerals are the following:

1. *Iron.* Divalent $Fe^{2+} \rightarrow K_4Fe(CN)_6 \cdot 3H_2O$, *ferrocyanide*, is strongly hydrated. Solutions of $Fe(CN)_6^{4-}$ do not respond to reagents for either Fe^{2+} or CN^-. Trivalent $Fe^{3+} \rightarrow K_3Fe(CN)_6$, *ferricyanide*, is weakly hydrated and not as stable as $Fe(CN)_6^{4-}$. Its solutions react with CN^-. It is formed on oxidation of ferrocyanide by Cl_2 or other strong oxidizing agents.

Ferrocyanide is formed on addition of KCN to solutions containing Fe^{2+}. Its alkaline solutions are stable; acid solutions are rapidly oxidized by air. Heavy metal ions give precipitates with ferrocyanide ions. The solubility products of ferrocyanates vary considerably, e.g., $\log K_{sp}$ of $Cu^{2+}[Fe(CN)_6^{4-}] = -14$ to -16, of $Pb^{2+} = -14$ to -17, and of $Zn^{2+} = -14$ to -16 [data from Sillén and Martell (1964)]. With the exception of the silver salt, the composition of the precipitated ferrocyanide salts of heavy metals varies with the conditions of precipitation, producing a wide range of colors of the metallic compounds.

The insoluble metallic ferrocyanides are amorphous and combined with a considerable quantity of water, which can only partly be driven off by heating the salt at 100°C, but on cooling all water lost will be reabsorbed.

A large number of double salts of ferrocyanides are known. The double salts of alkali ferrocyanides and the ferrocyanides of alkaline earths are sparingly soluble, while the double salts obtained by adding a heavy metal salt to a solution of a soluble ferrocyanide are insoluble. By varying the method of precipitation, compounds of differing compositions can be prepared of the types $M_3^{2+}K_2[Fe(CN)_6]_2$, $M_4^{2+}K_4[Fe(CN)_6]_3$, or $M_6^{2+}K_8[Fe(CN)_6]_5$, in addition to the simplest, $M^{2+}K_2Fe(CN)_6$, where M^{2+} represents a divalent metal (when using potassium ferrocyanide solution in precipitation).

Ammonium salts have a greater tendency to form double ferrocyanide salts than potassium, and sodium less than potassium—if the three types of ions coexist in solution. Many of the insoluble metal ferrocyanides form additional compounds with ammonia, e.g., $Zn_2Fe(CN)_6 \cdot 7NH_3$, $Co_2Fe(CN)_6 \cdot 3NH_3$, etc. One of the most important mixed complex cyanides of Fe^{2+} is the ruby-red potassium nitroprusside, $K_2Fe(CN)_5NO$, which is readily soluble in water but is not stable in solution. It is used as a reagent for detecting S^{2-} and HS^- with which it forms a complex with an intense violet color [probably $Fe(CN)_5(NSO)^{4-}$]. The solutions of nitroprusside and its complex with sulfides are readily decomposed to ferrocyanide and by-products.

Complex salts of Fe^{2+}—CN—Y, where Y stands for H_2O, NH_3, NO, NO_2, CO, SO_3, AsO_2, phenyl, etc., are known. (Fe^{3+} complexes are not as abundant as those of Fe^{2+}.)

2. *Copper.* Cuprous cyanide complexes include CuCN, $Cu(CN)_3^{2-}$, and $Cu(CN)_4^{3-}$. When KCN is added to a solution containing Cu^{2+}, at first $Cu(CN)_2$ is precipitated, but it rapidly decomposes into CuCN (log $K_{sp} = -19.49$) and cyanogen $(CN)_2$, which is slowly hydrolyzed by water. In alkaline solution cyanogen reacts with OH^-:

$$(CN)_2 + 2OH^- \rightarrow CN^- + OCN^- + H_2O$$

$(CN)_2$ also reacts with a large number of substances, e.g., with H_2S, giving thiocyanoformamide,

$$\overset{\displaystyle S}{\underset{\displaystyle N \equiv C\overset{\|}{C}NH_2}{}}$$

or dithio-oxamide,

$$\left[\underset{\displaystyle H_2N\overset{\|}{C}-\overset{\|}{C}NH_2}{\overset{\displaystyle S \quad S}{}} \right]$$

CuCN is extremely stable in aqueous solution but is soluble in excess KCN to give complex cyanides $Cu(CN)_3^{2-}$ and $Cu(CN)_4^{3-}$. Cyanide replaces ammonia from $Cu(NH_3)_4^{2+}$ to the extent of 89% at pH 9.

Divalent $Cu^{2+} \rightarrow Cu(CN)_2$, precipitated at first in solution, is very unstable and decomposes to CuCN and $(CN)_2$ or $2CuCN \cdot Cu(CN)_2$, giving transitory intermediates: $Cu(CN)_4^{2-}$ and $Cu(CN)_3^{-}$.

3. *Zinc.* Solutions containing Zn^{2+} treated with CN^- yield $Zn(CN)_2$, which, in excess CN, produces complex ions $Zn(CN)_3^{-}$, $Zn(CN)_4^{2-}$ and, possibly $Zn(CN)_5^{3-}$ and $Zn(CN)_6^{4-}$; $Zn(CN)_2$ itself is practically insoluble in water ($\log K_{sp} = -12.59$).

4. *Molybdenum.* KCN added at room temperature, and in the absence of oxygen, to a solution containing Mo^{3+} (e.g., K_3MoCl_8) would produce red-brown $K_4Mo(CN)_7 \cdot 2H_2O$, which is very soluble in water but readily oxidizes to Mo^{4+} and Mo^{5+} on exposure to air and is precipitated with other heavy metal cations. Complex Mo^{4+} cyanide is obtained when excess KCN is used in the presence of oxygen (at raised temperatures) with either Mo^{3+} or Mo^{5+} compounds as $K_4Mo(CN)_8 \cdot 2H_2O$; this is readily soluble in water $[Mo(CN)_8^{4-}]$ and also is oxidized (by, e.g., $KMnO_4$ or electrolytically), giving Mo^{5+}, i.e., $Mo(CN)_8^{3-}$. The latter is photosensitive and readily reduced to Mo^{4+}, the most stable Mo–cyanide complex, $Mo(CN)_8^{4-}$. Mixed cyanide complexes of molybdenum with OH, $[Mo(CN)_4(OH)_4^{4-}]$, S, SO, NH_3, N_2H_4, and NO exist and form a large number of double salts such as $K_4Mo(CN)_5(OH)_2(NO)$ or $(Fe^{2+})_3[MoSOH(CN)_4(H_2O)_2^{3-}]_2$ with other divalent ions (Ni, Cu, Zn, Pb, etc.) replacing Fe^{2+} in the last formula.

5. *Rhenium.* Rhenium forms a greater variety of complex cyanides than any other element. Re^+ forms the complex $K_5Re(CN)_5 \cdot 3H_2O$; Re^{2+} forms two complexes, $K_4Re(CN)_6$ and $K_3Re(CN)_5 \cdot H_2O$; Re^{3+} also gives $K_3Re(CN)_6$ and $Re(CN)_5 \cdot H_2O^{2-}$ complexes. Initially tetravalent Re^{4+} gives a number of not too well-defined complexes, of which dimeric $Re_2(CN)_8^{2-}$ and $Re_2(CN)_8^{+}$ have been determined by the polarographic technique; Re^{5+} forms $K_3Re(CN)_8$, $K_3ReO_2(CN)_4$, and $Re(CN)_4(OH)_4^{3-}$ complexes. Finally, Re^{6+} gives $Re(CN)_8^{2-}$, and Re^{7+} gives $Re(CN)_8(OH)^{2-}$ ionized complexes under suitable oxidizing conditions. Numerous mixed hydroxy- and chlorocyanide complexes have also been reported [Ford-Smith (1964)]. Chadwick and Sharpe (1966) review the cyanides and their complexes of transition metals.

6. *Lead.* Pb^{2+} forms an insoluble complex in water, $Pb(CN)_2$ (decomposed by acids), which gives $Pb(CN)_4^{2-}$ anion in excess KCN.

Table 10.7. Dissociation Constants of Metal–Cyanide Complexes[a]

Metal–Cyanide Complex	K_d
$Ag(CN)_2^-$	1.8×10^{-19}
$Au(CN)_2^-$	5×10^{-39}
$Cd(CN)_4^{2-}$	1.4×10^{-19}
$Cu(CN)_2^-$	2×10^{-24}
$Fe(CN)_6^{3-}$ (ic)	10^{-42}
$Fe(CN)_6^{4-}$ (ous)	10^{-35}
$Hg(CN)_4^{2-}$	4×10^{-42}
$Ni(CN)_4^{2-}$	10^{-22}
$Zn(CN)_4^{2-}$	1.2×10^{-18}

[a] The values quoted for the dissociation constants of ferrocyanide (10^{-35}) and ferricyanide (10^{-42}) do not correlate with the statement given earlier [after Ford-Smith (1964)] that ferricyanide solutions give a reaction for CN^- while those of ferrocyanide do not.

The dissociation constants of metal–cyanide complexes are given in Table 10.7 [after Gaudin (1957), from data in Latimer (1952) and by Vladimirova and Kakovsky (1950)].

Hydrogen cyanide, HCN, is an extremely poisonous (though less so than H_2S) colorless gas; in aqueous solution it is capable of dissociation, and also of polymerization, which is induced by ultraviolet radiation passing through the solution, giving trimers, tetramers, etc. In aqueous solutions HCN is a very weak acid, $pK_{25°} = 9.21$, and solutions of soluble cyanides are extensively hydrolyzed, e.g.,

$$NaCN + H_2O \rightleftarrows HCN\uparrow + NaOH \tag{10.9}$$

Mild oxidation of aqueous CN^- solutions produces linear cyanate ion, OCN^-, e.g.,

$$PbO(s) + NaCN(aq) \rightarrow Pb(s) + NaOCN(aq) \tag{10.10}$$

The free *cyanic acid, HOCN*, decomposes in aqueous solution to NH_3, H_2O, and CO_2 (its stability constant $K = 1.2 \times 10^{-4}$). Alkali cyanates also undergo slow hydrolysis to yield ammonium carbonate. Metallic salts of cyanic acid (cyanates) are mostly soluble in water, except for Cu^+ and Ag^+ cyanates (log K_{sp} for Ag^+ cyanate $= -6.64$). Some double salts, Co–K cyanates, Cd–pyridine cyanates, mercuric iodide–potassium cyanate, etc., are also formed. In complexes containing cyanate ion, OCN^-, as a ligand, the bonding to other elements (metals and nonmetals, such as P) takes place either through the O or the N atom.

The cyanide ion reacts with any allotropic form of sulfur, converting it quantitatively to the *thiocyanate ion, SCN⁻* (the only difference being the rate at which each allotropic form reacts). The reaction is free from by-products and represents a suitable analytical procedure for elemental sulfur [Scott (1964)]:

$$S_8 + 8CN^- \rightarrow 8SCN^- \tag{10.11}$$

Similarly, cyanide ion reacts with thiosulfate to form *thiocyanate*:

$$CN^- + S_2O_3^{2-} \rightarrow CNS^- + SO_3^{2-} \tag{10.12}$$

Dixanthogen reacts with CN^-, forming, at first, an alkyl thiocyanate as an intermediate product; this is very unstable and decomposes to dialkyl monosulfide and thiocyanate [Cambron (1930)]; i.e.,

$$[ROCSS]_2 + KCN \rightarrow (ROCS)_2S + KSCN \tag{10.13}$$

Cyanic acid added to the solution containing alkali polysulfides yields thiocyanate:

$$2HCN + Na_2S_3 \rightarrow 2NaSCN + H_2S\uparrow \tag{10.14}$$

A solution of ferrocyanide heated with a polysulfide or a thiosulfate, under pressure, also produces thiocyanate:

$$Na_4Fe(CN)_6 + 2Na_2S_4 \rightarrow 6NaCNS + FeS + Na_2S \tag{10.15}$$

and

$$Na_4Fe(CN)_6 + 8Na_2S_2O_3 \rightarrow 6NaCNS + Na_2S + 6Na_2SO_4 + FeS + S_2 \tag{10.16}$$

Equations (10.11)–(10.16) show that certain nucleophiles can either displace thio anions by stronger thio anions or add a sulfur atom to an anion to form a thio anion. (A thio anion is an anion which bears a negative charge on the sulfur.) Foss (1950, 1961) suggested that the thermodynamic criterion for such displacement reactions be considered in the form of the difference of the dimerization potentials[†] for the reactant nucleophile and the thio anion produced. Edwards (1954) extended this suggestion by includ-

---

[†] The oxidation dimerization potential E_d is defined as the potential for the oxidation reaction (at unit activities): $2YS^- \rightarrow YS\text{—}SY + 2e^-$.

Table 10.8. Oxibase Scale Parameters in Water at 25°C[a]

Nucleophile	E_d	H_b	Nucleophile	E_d	H_b
ClO_4^-	−0.73	(−9)	OH^-	1.65	17.48
F^-	−0.27	4.9	NO_2^-	1.73	5.09
H_2O	0.000	0.000	$C_6H_5NH_2$	1.78	6.28
NO_3^-	0.29	(0.4–0.3)	SCN^-	1.83	(1)
SO_4^{2-}	0.59	3.74	NH_3	1.84	11.22
$ClCH_2COO^-$	0.79	4.54	$(CH_3O)_2POS^-$	2.04	(4)
CO_3^{2-}	0.91	12.1	I^-	2.06	(−5)
CH_3COO^-	0.95	6.46	$(H_2N)_2CS$	(2.18)	0.80
C_5H_5N	1.20	7.04	$S_2O_3^{2-}$	2.52	3.60
Cl^-	1.24	(−3)	SO_3^{2-}	2.57	9.00
$C_5H_6O^-$	1.46	11.74	SH^-	2.60	8.70
Br^-	1.51	(−6)	CN^-	2.79	10.88
N_3^-	1.58	6.46	S^{2-}	3.08	14.66

[a] From Davis (1968) (approximate values in parentheses).

ing the oxidation dimerization potential and the basicity of the nucleophiles and introduced an empirical so-called *oxibase scale* (Table 10.8); the latter is in the form of an equation for the relative rate of reaction k/k_{H_2O}, of displacement of Y^- compared to water, i.e.,

$$Y^- + AZ \xrightarrow{k} YA + Z^-$$

$$H_2O + AZ \xrightarrow{k_{H_2O}} AOH + H^+ + Z^-$$

viz.

$$\ln \frac{k}{k_{H_2O}} = \alpha E_d + \beta H_b \qquad (10.17)$$

in which E_d is the oxidative dimerization potential of the nucleophile less the dimerization potential of water and H_b is the basicity of the nucleophile relative to water.[†] Davis (1968) gives a detailed explanation of the terms and discusses the utility of the oxibase scale. By using the values given in Table 10.8 for the product nucleophile in relation to the reactant nucleophile, when ΔE_d and ΔH_b are negative, the reaction is thermodynamically possible; when they are positive, it is thermodynamically impossible, e.g., for the

[†] The parameters α and β refer to the ease of reduction and the acidity of the reactant in which the given nucleophile displaces or adds the thio anion.

reaction of sulfite with sulfur:

$$\tfrac{1}{8}S_8 + SO_3{}^{2-} \rightarrow S_2O_3{}^{2-} \tag{10.18}$$

$$\Delta E_d = E_{d(S_2O_3)} - E_{d(SO_3)} = 2.52 - 2.57 = -0.05$$

$$\Delta H_b = 3.60 - 9.0 = -5.40$$

Both values indicate the free energy change $\Delta G < 0$.

Since the reaction (10.18) is thermodynamically favored, the opposite reaction between the thiosulfate and sulfur is not possible, as it would indicate $\Delta G > 0$; this is confirmed in practice, as thiosulfate does not react with sulfur. Similarly, OH^- is incapable of attacking the sulfur–sulfur bond of a disulfide,

$$2OH^- + RS—SR \nrightarrow H_2O_2 + 2RS^- \tag{10.19}$$

since $\Delta E_d = E_{d(RS)} - E_{d(OH)} = 2.9 - 1.65 > 0$. The oxibase values for the ferrocyanide ion $Fe(CN)_6{}^{4-}$ are not indicated in Table 10.8, but the reactions (10.15) and (10.16) suggest that the oxidative dimerization potential E_d for ferrocyanide should be greater than that for thiosulfate (i.e., 2.52), since it is a good thiophile. Table 10.8 shows that all nucleophiles with $E_d > 2.5$ and $H_b > 3.5$ (i.e., $S_2O_3{}^{2-}$, $SO_3{}^{2-}$, CN^-, HS^-, S^{2-}) are good thiophiles. And of these, CN^- is one of the most highly reactive with sulfur [equation (10.11)], thiosulfate [equation (10.12)], polysulfides [equation (10.13)], and polythionates:

$$S_xO_6{}^{2-} + (x-1)CN^- + H_2O \rightarrow 2HCN + S_2O_3{}^{2-} + SO_4{}^{2-} + (x-3)CNS^- \tag{10.20}$$

again producing thiocyanates. This high degree of thiophilic reactivity may be the main reason for the depressing action of the CN^- ion.

Alkali and alkaline earth thiocyanates, including some heavy metal thiocyanates (Fe^{3+}, Mn^{2+}, Co^{2+}, Ni^{2+}, etc.), are very soluble in water. Other divalent metal thiocyanates (Cu^{2+}, Cd^{2+}, Pb^{2+}, Hg^{2+}) are sparingly soluble, while cuprous CuCNS ($\log K_{sp} = -14.32$) and AgCNS ($\log K_{sp} = -11.93$) are insoluble. Cuprous thiocyanate (white precipitate) is used as an antifouling paint in marine applications. Oxidation of aqueous SCN^- with MnO_2 produces thiocyanogen, $(SCN)_2$, which is rapidly decomposed by water.

Ferric thiocyanate, $Fe(CNS)_3 \cdot 3H_2O$, highly soluble in water and red in color, is quantitatively extractable by ether from aqueous solutions. It forms numerous double salts with other thiocyanates (ammonium, alkali, guanidine, pyridine, etc.)

Ferrous thiocyanate, $Fe(CNS)_2 \cdot 3H_2O$, forms very soluble, pale-green crystals and is rapidly oxidized to the ferric state (red thiocyanate).

Lead thiocyanate, $Pb(CNS)_2$, soluble to $\sim 1\%$ in water, forms a number of double salts with lead halides.

The thiocyanate ion forms complexes with most transition metals; there are four ways of coordination (except in ionic compounds such as KSCN) possible (M stands for metal atom)

$$M—S—C\equiv N \qquad\qquad M—N=C=S$$
$$\text{(a)} \qquad\qquad\qquad \text{(b)}$$

$$\begin{array}{c} N^+ \\ \parallel \\ C \\ \parallel \end{array}$$
$$M—S^-—C=N^+ \rightarrow M \qquad M—S^- \rightarrow M$$
$$\text{(c)} \qquad\qquad\qquad \text{(d)}$$

In Hg, Cu, and Ag (as in most class B metals) the metal is S-bonded to the ligand and the C—S stretching frequency is used to distinguish between S and N bonding: In S-bonded complexes the C—S stretching occurs at $690-720 \text{ cm}^{-1}$, and in N-bonded complexes at $780-860 \text{ cm}^{-1}$; the M—SCN linkage is always angular,

$$\underset{M}{\diagup}^{S—C\equiv N} \quad \text{or} \quad \underset{M}{\diagup}^{S^-=C=N^+}$$

while the M—NCS linkage can be collinear or angular. NMR spectroscopy has also been used to distinguish between N- and S-bonded thiocyanates.

Strong π-electron acceptors (such as tertiary phosphines) can make the metal orbitals less available for binding with π orbitals of the sulfur atoms; then the effective charge on the metal plays an important role: Complexes with bases weaker than tertiary phosphines (e.g., CO) are S-bonded, whereas stronger bases such as amines are N-bonded.

Thioethers, R_2S, do not coordinate very strongly to metals apart from Hg^{2+}, Pt^{2+}, Ir^{2+}, Pd^{2+}, and Rh^{3+}.

There are numerous double salts of alkali and alkaline earths with the thiocyanates of heavy metals (Cr, Co, Fe, Mn, Hg, Ni, Zn) which are very soluble in water and stable in solution. In addition, a large number of double salts exist containing two heavy metal thiocyanates; these are sparingly soluble; e.g., the solubility product of (Cu^{2+},Hg^{2+}) thiocyanate $[CuHg(CNS)_4]$ is $\log K_{sp} = -7.48$ and of $ZnHg(CNS)_4$ is $\log K_{sp} = -6.66$. The double salts of cuprous and silver thiocyanates are not so stable in solution; on dilution the complex thiocyanate ion decomposes to precipitate the insoluble cuprous or silver thiocyanate (single) salts.

In addition to double salts containing two metallic ions, there are single-salt thiocyanates of organic bases such as aniline, pyridine, guanidine, and double salts containing heavy metal ions and organic bases; these are mostly soluble in water and are used as insecticides.

Molybdenum forms a large number of such double salts with ammonium salts or organic bases, e.g.,

$$`(NH_4)_2MoO(SCN)_5$$

or pyridine salts,

$$(C_5H_6N)_2 \cdot Mo(OH)_2 \cdot (SCN)_3$$

or

$$(C_5H_6N)_4 \begin{bmatrix} (CNS)_3 & & O & & (CNS)_3 \\ & \diagdown Mo \diagup & & \diagdown Mo \diagup & \\ O \diagup & & \diagdown SO_4 \diagup & & \diagdown O \end{bmatrix}$$

or alkyl salts,

$$R_4 \begin{bmatrix} (CNS)_3 & & O & & (CNS)_3 \\ & \diagdown Mo \diagup & & \diagdown Mo \diagup & \\ O \diagup & & CNS{-}SNC & & \diagdown O \end{bmatrix}$$

For a review of the enormous number of various thiocyanates, Chapters XI and XII in Williams (1948) should be consulted.

In addition to thiocyanates of metallic ions, organic bases, and alkyl thiocyanates, there are also phosphoryl thiocyanate, $[POCl_2(SCN)]$; sulfur monothiocyanate, $[S_2(CNS)_2]$; thionyl thiocyanate $[SO(CNS)_2]_x$; sulfuryl thiocyanate, $[SO_2(CNS)_2]_x$; and silicon trichlorothiocyanate, $SiCl_3(CNS)$. Isocyanides, RNC, known also as alkyl or aryl carbylamines or isonitriles, readily form numerous complexes with metal ions and other π-bonding ligands. Malatesta and Bonati (1969) deal extensively with these isocyanide complexes of metals.

10.8. *Use of Cyanide Ion as a Depressant*

Using the captive-bubble technique with polished mineral samples in a solution of a thio collector, Wark and Cox (1934) established the relationships between the critical CN^- concentration for bubble attachment and the pH (Figure 10.18) for a number of sulfide minerals. The effect of increasing the concentration of the collector is to shift the line of critical

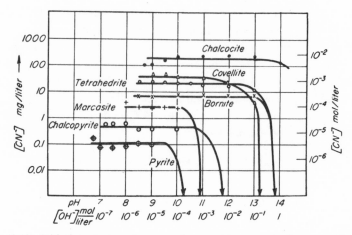

Figure 10.18. Critical contact angle curves for cyanide concentrations vs. pH for different minerals; no contact angle development occurs to the right and above the individual curve for a mineral, only below and to the left is the contact angle detected. [Based on data given in Wark (1938), from Gaudin (1957), *Flotation*. Copyright 1957 by the McGraw-Hill Book Company, Inc. Used with the permission of McGraw-Hill Book Co.]

CN⁻ concentration to a higher value. The same limitations as discussed in Section 10.6.1 apply in transferring these findings to the actual flotation system; i.e., the critical values of CN⁻ additions for nonflotation of a particular sulfide may vary over a much wider pH range and collector concentration range than indicated by the simple system used in the captive-bubble testing. The multiplicity of reactions entered into by CN⁻ in a multicomponent system represented by an ore makes it quite evident that variations in critical levels of additions are to be expected in flotation of ores with varying mineral contents.

Regardless of the possibility of variation in critical additions, flotation experience and practice show that cyanide is often an indispensable additive, required to achieve the desired selectivity, particularly when the depression of pyrite, pyrrhotite, marcasite, arsenopyrite, and marmatite is involved. Despite an extensive usage of cyanide in the past, the exact mechanism of its action has not been thoroughly investigated. Sometimes it is found necessary in flotation practice to add cyanide directly to the ball mill in order to achieve the desired selective action, while on other occasions conditioning time of a few minutes to an hour may be required. It is commonly assumed that complexing of the metal ion (whether Fe^{2+}, Fe^{3+}, or Cu^{2+}, Ni^{2+}, Pb^{2+}, Zn^{2+}, etc.) by the cyanide ion prevents the adsorption of collector ions that utilize these metallic ions. However, the details of what actually happens in the flotation system when cyanide is added have not been unraveled. This problem of detailed mechanism becomes relevant in view of a much stricter control of mill effluents by the pollution control authorities and an apparent need to seek an equally effective replacement for the effective action of cyanide ion. When the mechanism of action is understood, the task of finding a suitable replacement may on longer depend on a purely empirical testing alone. Attempts to replace cyanide by thiourea, as a depressant, have been made.

10.9. *Halide Ligands*

As with other ligands, the properties of metal halides and halide complexes are determined by an interplay of a large number of factors: ionization energy, electronegativity, polarizability, size of ionic and covalent radii, etc.

Although many metal halides are mainly ionic in character, there is a uniform transition through intermediate to essentially covalent halide salts, particularly as the ratio of charge to radius for the metal ion increases. For example, the sequence KCl, $CaCl_2$, $ScCl_3$, $TiCl_4$ shows a change from the completely ionic KCl to the essentially covalent $TiCl_4$. Similarly, a metal in a lower oxidation state forms ionic compounds, e.g., $PbCl_2$, while the same metal in a higher oxidation state gives a covalent one, $PbCl_4$. Polarizability of the halide ion and the size are also important; e.g., AlF_3 is ionic, while AlI_3 is covalent.

The halide ions, of which fluoride is the most reactive, have the ability to form complexes with various metal ions and/or covalent halides or mixed complexes with other ligands. The stability of these complexes in aqueous solutions depends on the metal–halide bond relative to the bonding of metal–water of hydration. Generally, the stability of halide complexes decreases in the series $F > Cl > Br > I$, though with some metals the order may be reversed. The heats of halide ion hydration (Table 10.9) follow the same order, $F^- > Cl^- > Br^- > I^-$.

Many of the less ionic metallic halides and the nonmetal halides are soluble in organic solvents, e.g., $FeCl_3$ in ether. The solvent extraction of some metal halides, e.g., Co^{2+} from Co–Ni solutions in HCl, is based on anionic complexes $CoCl_3^-$ and $CoCl_4^-$ adsorbing on anion–exchange resins. Frequently, a whole series of complexes exists:

$$M^{n+}_{(aq)}, \quad [MY]^{(n-1)+}_{(aq)}, \quad [MY_2]^{(n-2)+}_{(aq)}, \quad \ldots \quad [MY_x]^{(n-x)+}_{(aq)}$$

where Y is a halide ion and x is the maximum coordination number of the

Table 10.9. Heats of Ion Hydration[a]

	$\Delta H_{\text{hydration}}$ (kcal mol^{-1})			$\Delta H_{\text{hydration}}$ (kcal mol^{-1})	
	Standard[b]	Single Ion		Standard	Single Ion
H^+	0.0	-257.0	Ni^{2+}	18.1	-495.9
Li^+	137.674	-119.3	Cu^{2+}	19.45	-494.6
Na^+	163.794	-93.2	Cd^{2+}	89.76	-424.2
K^+	183.98	-73.0	Zn^{2+}	32.84	-481.2
Rb^+	189.9	-67.1	Pb^{2+}	167.68	-346.3
Cs^+	197.8	-59.2	Sc^{3+}	-164.4	-935.4
Tl^+	182.83	-74.2	Y^{3+}	-83.1	-854.1
Be^{2+}	-73	-587	La^{3+}	-2.5	-773.5
Mg^{2+}	61.98	-452.0	Ga^{3+}	-337.6	-1108.6
Ca^{2+}	140.77	-373.2	In^{3+}	-199.9	-970.9
Sr^{2+}	176.05	-337.9	Tl^{3+}	-217.9	-988.9
Ba^{2+}	209.80	-304.2	Al^{3+}	-331.6	-1102.6
Ra^{2+}	220	-294	Fe^{3+}	-264	-1035
Cr^{2+}	79.3	-434.7	F^-	-366.3	-109.3
Mn^{2+}	80.4	-433.6	Cl^-	-348.8	-91.8
Fe^{2+}	62.5	-451.5	Br^-	-340.7	-83.9
Co^{2+}	30.4	-483.6	I^-	-330.3	-73.3

[a] Source: Phillips and Williams (1965).
[b] Arbitrary standard of zero for the H^+ ion hydration.

metal ion M^{n+}. The halides which are predominantly ionic conduct electricity in the fused state but not so the less ionic halides; e.g., the equivalent conductivities are NaCl, 134 (Ω^{-1} cm^2); KCl, 104; $CaCl_2$, 52; and $MgCl_2$, 29; but HCl, 10^{-6}; $AlCl_3$, 1.5×10^{-5}; $TiCl_4$, 0; and $NbCl_5$, 2×10^{-7}. Some interhalogen and nonmetal halides also conduct electricity, particularly those with elements of odd valence: BrF_3, ICl_3, PCl_5, $POCl_3$, SO_2Cl_2, etc. Solid PCl_5 is known to consist of PCl_4^+ and PCl_6^- units; the conductivity of other nonmetal halides and interhalogen compounds shown in Table 10.10 is explained on the basis of similar ionization processes, i.e., $2BrF_3 \rightleftarrows BrF_2^+ + BrF_4^-$. In fact, potentiometric measurements show that various halides can be arranged in an acid–base order (e.g., BrF_2^+ would be the acid and BrF_4^- the base). In nonmetallic halides used as solvents, these acid–base halides precipitate as mixed halides. Such acid–base reactions occur readily in halide solvents and also in the absence of any solvents; numerous mixed halides exist in which one of the halides may be regarded as a Lewis base (i.e., an electron donor), while the other is a Lewis acid

Table 10.10. Interhalogen Anions and Cations

AB_2^-	AB_4^-	AB_6^-	AB_8^-	AB_2^+	AB_4^+	ABC^-
I_3^-	ICl_4^-	IF_6^-	I_9^-	BrF_2^+	IF_4^+	$IBrCl^-$
ICl_2^-	I_5^-	I_7^-		ICl_2^+		
IBr_2^-	BrF_4^-					
$BrCl_2^-$						
Br_3^-						
I_2Cl^-						
I_2Br^-						
Br_2Cl^-						
Cl_3						

The stability constants for the association equilibria

$$AB + B^- \rightleftharpoons AB_2^-$$

determined at 25°C,

	Cl_3^-	Br_3^-	ICl_2^-	IBr_2^-	I_3^-	
$K =$	0.01	17.8	167.0	370.0	725.0	mol^{-1} liter

[a] Source: Phillips and Williams (1965).

(i.e., an electron acceptor), e.g.,

Na_3AlF_6 (cryolite, with AlF_6^{3-} anions)
Na_2SiF_6 (SiF_6^{2-} anions)
H_2SnCl_6 ($SnCl_6^{2-}$)   etc.
$KICl_2$ (ICl_2^-)

For a review of different crystal structures formed by the numerous complex halides, see Wells (1962, Chapters VIII and IX).

The anionic and cationic halide species form a large number of complexes not only with inorganic species but also with organic solvents, e.g., iodine–pyridine complexes with fatty acids (IpyOCOR), etc. Also, carbon–halogen and hydrocarbon–halogen compounds are formed, and of these the organic compounds of fluorine find particularly numerous commercial applications owing to their low reactivity.

A general characteristic of the halide compounds is their extensive range of colors, from white, or colorless, to black, e.g., AgCl (white); AgBr (yellow); $BiBr_3$ (orange); TiI_4 (red); WCl_6 (violet); $PtBr_4$, PtI_2, and PtI_4 (black); or H_2PbCl_6 (yellow), H_2PbBr_6 (red), H_2PbI_6 (black); etc.

Many of the complex polyanions and polycations are more important as short-lived kinetic species rather than as stable ions. The kinetics of various reactions involving halides and their complexes frequently overrides all other factors; e.g., all the positive oxidation states of halogens (perchlorates, iodates, etc.) would not be encountered in practice if it were not for the slowness of their decomposition reactions.

Extensive use of the ionic halogen and halide complexes is made in selective separations of metallic ions by ion exchange (in water purification, recovery of metals from hydrometallurgical effluents, etc.), in ion flotation, and in the concentration of naturally occurring potash deposits.

Extensive literature is available on each of these applications, e.g., Marcus and Kertes (1969), Sebba (1962), and Lemlich (1972). A brief review of various foam separation processes was presented by Mahne (1971), and an overview of foam separation methods, authored by Somasundaran (1972), provides an extensive list of pertinent references.

The concentration of components from naturally occurring potash deposits represents one application of froth flotation which is unusual in that it is carried out in saturated brine solutions. These solutions represent about 7–10 M concentrations of electrolyte (K^+, Na^+, Cl^-, SO_4^{2-} etc.,) ions.

10.10. *Closing Remarks—Problems Created by Fine Particles*

It is evident from the variety of topics discussed in the preceding chapters that froth flotation is influenced by an exceedingly large number of factors. Of these, the parameters which are of critical importance to a particular system have to be optimized for each feed material individually. The size of the liberated particles undergoing flotation is one of those all important factors (see Figure 1.13). Very fine solids (ultrafines, less than 10 μm) and, particularly, colloidal materials (less than 1 μm) give rise to many problems not only in selective flotation separations but also in dewatering, transportation handling, waste disposal, etc. In order to deal with these troublesome features effectively, it is necessary to know the characteristics of fines.

It is the variation in the character of solid surfaces, represented by particles of different sizes, that is chiefly responsible for such problems. The surface chemical considerations apply equally to coarse or to fine materials once the true nature of their surfaces (or the interfaces) is taken into account. The size, or rather the extent of surfaces, affects the relative magnitudes of physical parameters involved in practical operations.

The characteristics of the surfaces are determined by the manner in which the specified fine particles have been generated. The different mechanisms may be considered under the following categories:

1. mechanical—by impact and, primarily, by attrition. Very large residual surface stresses are acquired by the fines, causing rearrangement of atoms in surface layers or even phase changes if the heat generated in bond-fracturing is not dissipated. Amorphous layers, with different hydration and absorption characteristics, are also formed in various proportions, depending on the intensity of mechanical forces used and bulk solid characteristics.

2. thermal—fines produced by heat decrepitation. Heat may be supplied to the system, or generated within the system, as heat of wetting of surfaces (old or newly created), heat of adsorption, etc. The intensity of heat supplied, or generated, and heat conductivities of bulk phases in the system determine the extent and the nature of surfaces in the fines of this type. New phases, hydrates, recrystallization, etc. may take place, but the level of surface stresses is presumably low (due to stress relief by heat).

3. electrical, magnetic, or ultrasonic means—some materials can be disintegrated using high frequency or high intensity electrical fields. [Maroudas (1967), Carley-Macauly (1968), Yidit *et al.* (1969)]. High residual charges cause stable dispersion of such fines (sols). Magnetic and ultrasonic fields are also employed for dispersion of weak or weakly cemented solids.

4. chemical—fines produced by condensation of vapours, crystallization from saturated solutions, solvent penetration into cracks and grain boundaries (as in chemical comminution).

Depending on the fines-generation mechanism, there may be large differences in surface characteristics and, consequently, in the behaviour of fines. And what is also important to realise, in a given system there may be different proportions (in a mixture) of fines generated by various mechanisms, not just one type of fines. All fines are characterised by small mass and large surface area; however, the truly differentiating and troublesome characteristics are those of high or low surface stresses, amorphous or new-phase surface layers, and high or low surface charges of similar or opposite sign.

In consequence, the reactivities of fines produced by different mechanisms from the same solid phase may, and do, differ. Hence, treatment of fine materials may require different procedures not only in comparison with coarser particles of the same phase but is also dependent on the mechanism by which the given portion of fines had been generated.

Owing to the ever increasing necessity of dealing with lean and finely disseminated ores, the interest in the topic represented by fines is high. Several publications appeared recently on the subject. Instead of duplicating their coverage unnecessarily it seems only appropriate to quote here the relevant references. The most comprehensive recent publication on the subject of fine particles consists of two volumes edited by Somasundaran (1980). It comprises some 91 papers (presented at an international symposium in Las Vegas, Feb. 1980) dealing with all aspects of fines production, classification, surface chemistry, agglomeration, flocculation, flotation, solid/liquid separation, and waste disposal. The individual papers (listed separately) have to be consulted for details. Preceding this publication, there have been numerous papers (referred to in the symposium papers) published on isolated topics related to fines, and three specific reviews: Trahar and Warren (1976), Shergold (1976), Collins and Read (1971). The review of Trahar and Warren deals with the theories of size effects in flotation (rate of collision, probability of adhesion, chemical effects of adsorption, solubility, hydration, etc.) and the techniques for improving flotation of fines. Shergold's review deals specifically with a flotation process modifed by additions of nonpolar oils, and briefly with spherical agglomeration. Collins and Read (1971) reviewed the problems of material handling as well as special processes for recovery of fine materials, such as "piggy-back" flotation, spherical agglomeration, selective flocculation, electrophoresis, and electrostatic separation. Selective flocculation and selective spherical agglomeration are two of the most promising (in this author's opinion) techniques for separation of fines from mixtures. Relevant references are Yarar and Kitchener (1970), Attia and Kitchener (1975), Read and Hollick (1976), Ives (1978), and Bratby (1980) as regards flocculation. Farnand *et al.* (1961), Sirianni *et al.* (1969), Capes *et al.* (1974) deal with spherical agglomeration. In addition, papers in the proceedings of recent International Mineral Processing Congresses should also be consulted on these topics. These also follow, listed separately.

10.11. *Selected Readings*

Canterford, J. H., and Colton, R. (1968, 1969), *Halides of Transition Metals*, Wiley-Interscience, New York.

Cotton, F. A., and Wilkinson, G. (1972), *Advanced Inorganic Chemistry*, Wiley-Interscience, New York.

Davis, R. E. (1968), Mechanism of sulphur reactions, in *Inorganic Sulphur Chemistry*, ed. G. Nickless, Elsevier, Amsterdam, pp. 85–135.

Gutman, V., ed. (1967), *Halogen Chemistry*, Vols. 1, 2, and 3, Academic Press, New York.

Hunt, J. P. (1963), *Metal Ions in Aqueous Solution*, Benjamin, Reading, Mass.

Malatesta, L., and Bonati, F. (1969), *Isocyanide Complexes of Metals*, Wiley-Interscience, New York.

Mann, F. G. (1970), *The Heterocyclic Derivatives of Phosphorus, Arsenic, Antimony and Bismuth*, 2nd ed., Wiley-Interscience, New York.

Matijevic, E. (1973), Colloid stability and complex chemistry, *J. Colloid Interface Sci.* 43(2), 217–245.

Nickless, G., ed. (1968), *Inorganic Sulphur Chemistry*, Elsevier, Amsterdam.

Phillips, C. S. G., and Williams, R. Y. P. (1966), *Inorganic Chemistry*, Vols. I and II, Oxford University Press, London.

Price, C. C., and Oae, S. (1962), *Sulphur Bonding*, Ronald, New York.

Reid, E. E. (1963), *Organic Chemistry of Bivalent Sulphur*, Vol. V, Chemical Publishing, New York.

Sharpe, A. G., and Chadwick, B. M. (1966), Transition metalcyanides and their complexes, *Adv. Inorg. Chem. Radiochem.* 8, 83–176.

Sillén, L. G., and Martell, A. E. (1964), *Stability Constants of Metal-Ion Complexes*, Special Publication No. 17, The Chemical Society, London.

Singer, P. C., and Stumm, W. (1969), Oxygenation of ferrous ion, reproduced in R. D. Hill and R. C. Wilmoth, Limestone treatment of acid mine drainage, *Trans. AIME*, 250, June 1971, 162–166.

Stumm, W. (1967), Metal ions in aqueous solutions, in *Principles and Applications of Water Chemistry*, eds. S. D. Faust and J. V. Hunter, Wiley, New York, pp. 520–560.

Taube, H. (1970), *Electron Transfer Reactions of Complex Ions in Solution*, Academic Press, New York.

Vail, J. G. (1952), *Soluble Silicates*, Vol. 1, Van Nostrand Reinhold, New York.

van Wazer, J. R. (1958), *Phosphorus and Its Compounds*, Vol. 1, Wiley-Interscience, New York.

van Wazer, J. R., and Callis, C. F. (1958), Complexing of metals by phosphate, *Chem. Rev.* 58, 10–11.

Wells, A. F. (1962), *Structural Inorganic Chemistry*, 3rd ed., Oxford University Press, London.

Williams, H. E. (1948), *Cyanogen Compounds*, Edward Arnold, London.

List of papers in Flotation—A. M. Gaudin Memorial Volume (Fuerstenau, 1976)

Antoine Marc Gaudin

1. *Spedden, H. Rush*. Antoine M. Gaudin. His Life and His Influence on People.
2. *Arbiter, Nathaniel*. Antoine M. Gaudin. His Influence on Mineral Processing.

Fundamentals of Nonmetallic Flotation

3. *Fuerstenau, D. W. and Raghavan, S*. Some Aspects of the Thermodynamics of Flotation.

4. *Miller, J. D., and Calera, J. V.* Analysis of the Surface Potential Developed by Non-Reactive Ionic Solids.
5. *Smith, Ross W., and Salim Akhtar.* Cationic Flotation of Oxides and Silicates.
6. *Bleier, A., Goddard, E. D., and Kulkarni, R. D.* The Structural Effects of Amine Collectors on the Flotation of Quartz.
7. *Fuerstenau, M. C., and Palmer, B. R.* Anionic Flotation of Oxides and Silicates.
8. *Hanna, H. S., and Somasundaran, P.* Flotation of Salt-Type Minerals.

Fundamentals of Sulfide Flotation

9. *Healy, T. W., and Moignard, M. S.* A Review of Electrokinetic Studies of Metal Sulphides.
10. *Woods, R.* Electrochemistry of Sulfide Flotation.
11. *Poling, G. W.* Reactions Between Thiol Reagents and Sulphide Minerals.
12. *Richardson, Paul E., and Maust, Edwin E., Jr.* Surface Stoichiometry of Galena in Aqueous Electrolytes and Its Effect on Xanthate Interactions.
13. *Shimoiizaka, J., Usui, S., Matsuoka, I., and Sasaki, H.* Depression of Galena Flotation by Sulfite or Chromate Ion.
14. *Finkelstein, N. P., and Allison, S. A.* The Chemistry of Activation, Deactivation and Depression in the Flotation of Zinc Sulfide: A Review.
15. *Ball, B., and Rickard, R. S.* The Chemistry of Pyrite Flotation and Depression.
16. *Hoover, R. M., and Malhotra, D.* Emulsion Flotation of Molybdenite.

Flotation Fundamentals

17. *Rao, S. R., Moon, K. S., and Leja, J.* Effect of Grinding Media on the Surface Reactions and Flotation of Heavy Metal Sulphides.
18. *Tipman, N. R., Agar, G. E., and Paré, L.* Flotation Chemistry of the Inco Matte Separation Process.
19. *Rinelli, G., Marabini, A. M., and Alesse, V.* Flotation of Cassiterite with Salicylaldehyde as a Collector.
20. *Parkins, E. J., and Shergold, H. L.* The Effect of Temperature on the Conditioning and Flotation of an Ilmenite Ore.
21. *Vazquez, L. A., Ramachandaran, S., and Grauerholz, N. L.* Selective Flotation of Scheelite.
22. *Lovell, V. M.* Froth Characteristics in Phosphate Flotation.

Kinetics and Cell Design

23. *Anfruns, J. P., and Kitchener, J. A.* The Absolute Rate of Capture of Single Particles by Single Bubbles.
24. *Woodburn, E. T., Kropholler, H. W., Greene, J. C. A., and Cramer, L. A.* The Utility and Limitations of Mathematical Modelling in the Prediction of the Properties of Flotation Networks.
25. *Huber-Panu, I., Ene-Danalache, E., and Cojocariu, Dan G.* Mathematical Models of Batch and Continuous Flotation.

List of papers in Fine Particles Processing (*Somasundaran, 1980*)

Production of Fines and Ultrafines

1. *Randolph, A. D.* Generation and Measurement of Fine-Crystal Distributions.
2. *Ramjilal, and Ramakrishnan, P.* Preparation and Properties of Fine Lead Zirconate Titanate Powders.
3. *Paulus, M.* Freeze-Drying—A Method for the Preparation of Fine Sinterable Powders and Low Temperature Solid State Reaction.
4. *Gardner, R. P., Verghese, K., and Rousseau, R. W.* On The Feasibility of Using a Radioactive Tracer Method and a Size-Discretized Mass Balance Model for the Measurement of Growth Rates in MSMPR Crystallizers.
5. *Davis, E. G., Hansen, J. P., and Sullivan, G. V.* Attrition Microgrinding.
6. *Trass, O.* The Szego Grinding Mill.
7. *Skelton, R., Khayyat, A. N., and Temple, R. G.* Fluid Energy Milling—An Investigation of Micronizer Performance.
8. *Inoue, T., and Imaizumi, T.* Simulation of Closed Circuit Wet Grinding Process.
9. *Pitchumani, B., and Venkateswarlu, D.* Matrix Representation of Batch and Continuous Size Reduction Processes.

Classification, Mixing, and Flow

10. *Luckie, P., Hogg, R., and Schaller, R.* A Review of Two Fine Particle Processing Unit Operations—Classification and Mixing.
11. *Hukki, R. T., and Airaksinen, T.* A Study of the Effect of Improved Pneumatic Classification on Micropowder Production.
12. *Etkin, B., and Haasz, A. A.* Performance Characteristics of the University of Toronto Infrasizer MK III.
13. *Woodcock, C. R., and Mason, J. S.* Air-Float Conveying of Particulate Bulk Solids.
14. *Sakamoto, H., Moro, T., Yamamoto, M., Kokubo, T., Masuda, K., and Sekimoto, T.* Development of the Choking-Free Cyclone and Its Classification Performance.
15. *Link, J. M., Pouska, G. A., and Kirshenbaum, N. W.* Mineral Slurry Transport—An Update.
16. *Dogan, M. Z., Metin Ger, A., and Yucel, O.* Slurry Pipeline Transport of Concentrate Processed from a Turkish Low Grade Magnetite Ore.
17. *Clark, S., and Komardi.* Concentrate Slurry Properties and Their Effect on the Operation of the Freeport Indonesia, Incorporated Pipeline.

Characterization of Particles and Particle Systems

18. *Sresty, G. C., and Venkateswar, R.* Particle Size Analysis—A Review.
19. *Svarovsky, L., and Hadi, R. S.* A New Simple and Sensitive On-Line Particle Size Analyser for Fine Powders Suspended in Gases.
20. *Bayvel, L. P.* Particle Size Distribution Measurements in the Submicron Radius Range by the Method of Spectral Transparency.

Surface Chemical Aspects

Fine Particles Flotation

46. *Lynch, A. J., and Whiten, W. J.* The Control of Fine Particle Beneficiation Processes.
47. *Gardner, R. P., Lee, H. M., and Yu, B.* Development of Radioactive Tracer Methods for Applying the Mechanistic Approach to Continuous Multi-phase Particle Flotation Processes.

Flocculation, Dispersion, and Selective Flocculation

48. *Somasundaran, P.* Principles of Flocculation, Dispersion, and Selective Flocculation.
49. *Rao, T. C., and Bandopadhyay, P.* Processing of Indian Coal-Fines.
50. *Hogg, R.* Flocculation Problems in the Coal Industry.
51. *Onoda, G. Y., Jr., Deason, D. M., and Chhatre, R. M.* Flocculation and Dispersion Phenomena Affecting Phosphate Slime Dewatering.
52. *Rinelli, G., and Marabini, A. M.* Dispersing Properties of Tanning Agents and Possibilities of Their Use in Flotation of Fine Minerals.
53. *Colombo, A. F.* Selective Flocculation and Flotation of Iron-Bearing Materials.
54. *Iwasaki, I., Smith, K. A., Lipp, R. J., and Sato, H.* Effect of Calcium and Magnesium Ions on Selective Desliming and Cationic Flotation of Quartz from Iron Ores.
55. *Sadowski, Z., and Laskowski, J.* Selective Coagulation and Selective Flocculation of the Quartz-Carbonate Mineral (Calcite, Magnezite, Dolomite) Binary Suspensions.
56. *Banks, A. F.* Selective Flocculation—Flotation of Slimes from Sylvinite Ores.
57. *Wang, Y. H., and Somasundaran, P.* A Study of Carrier Flotation of Clay.
58. *Klimpel, R. R.* The Engineering Analysis of Dispersion Effects in Selected Mineral Processing Operations.

High Intensity Magnetics and Electrostatics

59. *Zimmels, Y., Lin, I. J., and Yaniv, I.* Advances in Application of Magnetic and Electric Techniques for Separation of Fine Particles.
60. *Oberteuffer, J. A., and Wechsler, I.* Recent Advances in High Gradient Magnetic Separation.
61. *Nesset, J. E., and Finch, J. A.* A Loading Equation for High Gradient Magnetic Separators and Application in Identifying the Fine Size Limit of Recovery.
62. *Hopstock, D. M., and Colombo, A. F.* Processing Finely Ground Oxidized Taconite by Wet High-Intensity Magnetic Separation.
63. *Petrakis, L., Ahner, P. F., and Kiviat, F. E.* High Gradient Magnetic Separations of Fine Particles from Industrial Streams.
64. *Roath, S., Paul, F., and Melville, D.* High Gradient Magnetic Separation of Red Blood Cells: Current Status.
65. *Corrans, I. J.* A Development in the Application of Wet High Intensity Magnetic Separation.
66. *Shimoiizaka, J., Nakatsuka, K., Fuijta, T., and Kounosu, A.* Preparation of Magnetic Fluids with Polar Solvent Carriers.
67. *Revnivtsev, V. I., and Khopunov, E. A.* Fundamentals of Triboelectric Separation of Fine Particles.
68. *Simkovich, G., and Aplan, F. F.* The Effect of Solid State Dopants Upon Electrostatic Separation.

Gravity and Hydrometallurgy

69. *Burt, R. O.* Slime Recovery by Gravity Concentration—A Viable Alternative?
70. *Blaschke, W., and Malysa, E.* Gravitational Beneficiation of Ultrafine Grains of Zinc–Lead Ores from Olkusz Region.
71. *Warren, G. W., and Wadsworth, M. E.* Hydrometallurgical Processing of Fine Mineral Particles.

Agglomeration

72. *Holley, C. A.* Agglomeration—The State of the Art.
73. *Capes, C. E.* Principles and Applications of Size Enlargement in Liquid Systems.
74. *Sastry, K. V. S., and Cross, M.* Basic and Applied Aspects of Pelletizing of Fine Particles.
75. *Kestner, D. W.* Industrial Trends in the Pelletizing and Sintering of Metallic Ores, Concentrates and Reverts.
76. *Takamori, T., Hirajima, T., and Tsunekawa, M.* An Experimental Study on the Mechanism of Spherical Agglomeration in Water.
77. *Bemer, G. G., and Zuiderweg, F. J.* Growth Regimes in the Spherical Agglomeration Process.

Solid/Liquid Separation

78. *Tiller, F. M., Crump, J. R., and Ville, F.* A Revised Approach to the Theory of Cake Filtration.
79. *Tadros, M. E., and Mayes, I.* Effects of Particle Properties on Filtration of Aqueous Suspensions.
80. *Kos, P.* Review of Sedimentation and Thickening.
81. *Pearse, M. J.* Factors Affecting the Laboratory Sizing of Thickeners.
82. *Sen, P.* Separation of Subsieve Size Particles from Suspension.
83. *Nicol, S. K., Day, J. C., and Swanson, A. R.* Oil Assisted Dewatering of Fine Coal.
84. *Szczypa, J., Neczaj-Hruzewicz, J., Janusz, W., and Sprycha, R.* New Technique for Coal Fines Dewatering.
85. *Parekh, B. K., and Goldberger, W. M.* Removal of Suspended Solids from Coal Liquefaction Oils.

Waste Treatment

86. *Hanna, H. S., and Rampacek, C.* Resources Potential of Mineral and Metallurgical Wastes.
87. *Hill, R. D., and Auerbach, J. L.* Solid Waste Disposal in the Mining Industry.
88. *Moudgil, B. M.* Handling and Disposal of Coal Preparation Plant Refuse.
89. *Oxford, T. P., and Bromwell, L. G.* Dewatering Florida Phosphate Waste Clays.
90. *Nakahiro, Y.* A Study on the Separation of Cadmium from Copper in Synthesized Waste Cyanide Water by Precipitation–Flotation Method.
91. *Sprute, R. H., Kelsh, D. J.* Dewatering Fine-Particle Suspensions with Direct Current.

List of papers presented at the XIIIth International Mineral Processing Congress (*Laskowski, 1981*)

Flotation

Derjaguin, B. V., Dukhin, S. S. Kinetic Theory of the Flotation of Fine Particles (Invited Lecture).

Doren Gattas, A., van Lierde, A., de Cuyper, J. Influence of Non-ionic Surfactants on the Flotation of Cassiterite.

Laskowski, T. Mineral Processing in Poland (Inaugural Lecture).

Mielczarski, J., Nowak, P., Strojek, J. W., Pomianowski, A. Infra-red Internal Reflection Spectrophotometric Investigations of the Products of Potassium Ethyl Xanthate Sorption on Sulphide Minerals.

Predali, J. J., Brion, D., Hayer, J., Pelletier, B. Characterization by ESCA Spectroscopy of Surface Compounds on the Fine Sulphide Minerals in the Flotation Process.

Schulze, H. J., Gottschalk, G. Investigations of the Hydrodynamic Interaction Between a Gas Bubble and Mineral Particles in Flotation.

Wakamatsu, T., Numata, Y., Sugimoto, Y., The Effect of Amino Acid Addition on the Flotation of Sulfide Ore.

Fine particle technology

Gräsberg, M., Mattsson, K. Novel Process at Yxsjöberg, a Pointer Towards Future More Sophisticated Flotation Methods.

Gururaj, B., Prasad, N., Ramachandran, T. R., Biswas, A. K. Studies on Composition and Beneficiation of a Fine-Grained Alumina-rich Indian Iron Ore.

Heiskanen, K., Laapas, H., On the Effects of the Fluid Rheological and Flow Properties in the Wet Gravitational Classification.

Hencl, V., Svoboda, J. The Possibility of Magnetic Flocculation of Weakly Magnetic Minerals.

Koh, P. T. L., Warren, L. J. Flotation of an Ultrafine Scheelite Ore and the Effect of Shear-Flocculation.

Kuznetsov, V. P., Volova, M. L., Lyubimova, E. I., Shishkova, L. M., Lifirenko, V. E., Sokolov, Yu. F. Beneficiation of Finely Dispersed Ores by Selective Flocculation and Investigation of Its Interrelation with Physicochemical Dispersing of Pulp at High Concentrations of Salts.

Rhodes, M. K. The Effects of the Physical Variables of Carboxymethyl Cellulose Reagents on the Depression of Magnesia Bearing Minerals in Western Australian Nickel Sulphide Ores.

Rinelli, G., Marabini, A. M. A New Reagents System for the Selective Flocculation of Rutile.

Somasundaran, P. Thickening or Dewatering of Slow-Settling Mineral Suspensions (Invited Lecture).

Steiner, H. J. Motion and Interaction of Particles in a Polydispersed Suspension.

Comminution—Fundamentals and models

Barbery, G., Huyet, G., Gateau, G. Liberation Analysis with The Help of Image Analysis: Theory and Applications.

Bodziony, J., Kraj, W. Stereological Analysis as One of Methods in Modern Granulometry.

Denev, S. I., Stoitsova, R. V. Effect of Mechanical Stress on the Structure Alterations of Minerals.

El-Shall, H., Gorken, A., Somasundaran, P. Effect of Chemical Additives on Wet Grinding of Iron Ore Minerals.

Jones, M. P. Automatic Mineralogical Measurements in Mineral Processing (Invited Lecture).

Manlapig, E. V., Seitz, R. A., Spottiswood, D. J. Analysis of the Breakage Mechanisms in Autogenous Grinding.

Tkáčova, K., Sekula, F., Hocmanová, I., Krúpa, V. The Determination of Energy Irreversibly Accumulated in the Grinding Process.

Hydrometallurgy

Ammou-Chokroum, M., Sen, P. K., Fouques, F. Electrooxidation of Chalcopyrite in Acid Chloride Medium; Kinetics, Stoichiometry and Reaction Mechanism.

Bertram, R., Hillrichs, E., Galitis, N., Müller, R., Greulich, H. Electrochemical Leaching of Copper Sulphide Ores.

Bolton, G. L., Zubryckyj, N., Veltman, H. Pressure Leaching Process for Complex Zinc–Lead Concentrates.

Habashi, F., Recent Advances in Hydrometallurgy.

Łetowski, F. Acid Hydrometallurgical Winning of Copper and Other Metals From Polish Sulphide Concentrates.

Majima, H., Awakura, Y. Non-oxidative Leaching of Base Metal Sulphide Ores.

Miller, J. D., Portillo, H. Q. Silver Catalysis in Ferric Sulfate Leaching of Chalcopyrite.

Raghavan, S., Harris, I., Fuerstenau, D. W. Lyometallurgical Treatment of Porous Oxidized Ores.

New trends in technology

Chanturia, V. A., Filinova, V. V., Lunin, V. D., Gorencov, N. L., Bochcarev, G. R. Electrochemical Method of Water-Preparation before Flotation of Manycomponent Polimetal Ores.

Hoberg, H., Breuer, H., Schneider, F. U. Investigations into the Improvement of Cassiterite Flotation by Surface Reduction.

Klassen, V. I. Magnetic Treatment of Water Systems in Mineral Processing.

Lin, I. J., Yaniv, I., Zimmels, Y. On the Separation of Minerals in High Gradient Electric Field.

Lupa, Z., Laskowski, J. Dry Gravity Concentration in the Fluidizing Separators.

Nagata, K., Bolsaitis, P. Selective Sulfidization of Iron Oxides with Mixtures of Carbon Monoxide and Sulfur Dioxide.

Revnivtsev, V. I. Recent Trends in the Preparation of Ores For Beneficiation Processes (Invited Lecture).

Tu, T. Y., Yan, C. H. A Combined Beneficiation Method for the Treatment of Refractory Eluvial Tin Ore.

New machines

Blising, U., Varbanov, R., Schmidt, M., Weber, H., Nikolov, A. Flotation with Intermittent Air Supply—A New Energy-Saving and Flow-rate Increasing Method.

Chin, P. C., Wang, Y. T., Sun, Y. P. A New Slime Concentrator—The Rocking-Shaking Vanner.

Fallenius, K. A New Set of Equations for the Scale-up of Flotations Cells.

Forssberg, K. S. E., Sandström, E. Operational Characteristics of the Reichert Cone in Ore Processing.

Komorowski, J., Myczkowski, Z., Nawrocki, J., Ogaza, H. Hydrodynamical and Technological Investigation of Large Volume Flotation Machines of Iz Type.

Plouf, T. M. Development of Large Volume Flotation Machines Design and Development Considerations for Denver Equipment Division's, Single Mechanism 1275 ft³ (36 m³) DR Flotation Machine.

Schubert, H., Bischofberger, C. On the Optimization of Hydrodynamics in Flotation Processes.

Optimization and automatization

Andersen, R. W., Grönli, B., Olsen, T. O., Kaggerud, I., Romslo, K., Sandvik, K. L. An Optimal Control System of the Rougher Flotation at the Folldal Verk Concentrator, Norway.

Fewings, J. H., Slaughter, P. J., Manlapig, E., Lynch, A. J. The Dynamic Behaviour and Automatic Control of the Chalcopyrite Flotation Circuit at Mount Isa Mines Limited.

Herbst, J. A., Rajamani, K. Evaluation of Optimizing Control Strategies for Closed Circuit Grinding.

Hołyńska, B., Lankosz, M., Ostachowicz, J., Cynien, B., Jaskulski, Z., Widera, K. New Analyzers for Continuous Particle-size Measurements of Copper Ore Slurries.

Kallioinen, J. D. Trends in the Development of Optimal Flotation Circuits Strategies for the Treatment of Complex Ores in Outokumpu Oy, Finland.

Meloy, T. P., Durney, T. E., Eppler, D. T. Sophisticated Shape Analysis of Green Balls from a Balling Drum.

Mular, A. L. Off-line Optimization of a Grinding Circuit.

Woodcock, J. T. Automatic Control of Chemical Environment in Flotation Plants (Invited Lecture).

Treatment of raw materials with total utilization of all constituents

Betz, E. W. Beneficiation of Brazilian Phosphates.

Clerici, C., Mancini, A., Mancini, R., Morandini, A., Occella, E., Protto, C. Recovery of Rutile from an Eclogite Rock.

Cohen, E., Hammoud, N. S. Low-Temperature Roasting in Upgrading The Non-oxidized Phosphorites of Abu Tartur Plateau (Western Desert, Egypt).

Detienne, J. L., Houot, R., Larribau, E., Vestier, D. Beneficiation of a Low-Grade Scheelite Ore.

Gulaikhin, E. V., Kotlarov, V. G., Eropkin, Yu. I., Pieskov, V. V., Novikov, Ya. V. Research and Practice of Integrated Raw Material Treatment in Processing Sulphide–Tin Ores of Complex Mineralogical Composition.

Ustinov, V. S., Snurnikov, A. P., Revnivtsev, V. I., Bogdanov, O. S., Barsukov, V. T., Tishenko, A. G. Basic Trends in Integrated Treatment of Non-ferrous Ores at Concentrators.

Waste treatment and ecology

Evans, J. B., Ellis, D. V., Leja, J., Poling, G. W., Pelletier, C. A. Environmental Monitoring of Porphyry Copper Tailing Discharged into a Marine Environment.

Khavskij, N. N., Tokarev, V. D. Intensification of the Clarification of Sewage and the Mineral Processing and Metallurgy Effluents by a Floto-Flocculation Method.

Neczaj-Hruzewicz, J., Sprycha, R., Janusz, W., Szczypa, J., Monies, A. Treatment of Flotation Tailings by the Spherical Agglomeration of Flocculated Slimes.

Sazonov, G. T., Evdokimov, P. D., Vitoshkin, Yu. K., Ivanov, P. L., Antonov, V. N. Tailing Transportation, Storage and Treatment at Concentrators Under More Stringent Environmental Regulations.

Taylor, R. K. Liquefaction Characteristics of Coal-Mine Tailings With Respect to Storage and Use.

Processing of oxidized and mixed oxide-sulfide lead-zinc ores

Abramov, A. A., Leonov, S. B., Kulikov, I. M., Zapov, V. Z., Kiseleva, M. A. Technological Peculiarities of Flotation Dressing of Oxidized and Mixed Lead–Zinc Ores.

Azefor, A., Kesler, S., Shergold, H. L. The Hydrometallurgical Treatment of Zinc Oxide Minerals by Solvents which Form Soluble Zinc Complexes.

Caproni, G., Ciccu, R., Ghiani, M., Trudu, I. The Processing of Oxidized Lead and Zinc Ores in the Campo Pisano and San Giovanni Plants (Sardinia).

Cases, J. M., Trabelsi, K., Predali, J. J., Brion, D. Flotation Enrichment of Oxidized Lead and Zinc Ore.

Ciccu, R., Curreli, L., Ghiani, M. The Beneficiation of Lean Semioxidized Lead–Zinc Ores. Technical and Economical Evaluation of Different Flotation Flowsheets Applied to Sardinian Ores.

Rinelli, G., Abbruzzese, C. Ammoniacal Leaching of Oxidized Lead–Zinc Ores.

Slusarek, M., Molicka-Haniawetz, A., Zawiślak, L., Oktawiec, M. A Mineralogical Characterization and the Attempts of Upgrading the Concentrates from Polish Zinc–Lead Oxidized Ores.

Stachurski, J., Sanak, S., Zdybiewska, K., Wicher, S. Combined (Thermochemical Treatment—Flotation) Enrichment Process of Low–Grade Oxidized Zinc and Lead Ores.

Treatment of iron-titanium ores

Blaszkiewicz, G., Borowiec, K. TiO_2—Concentrates from Ilmenite.
Buckenham, M. H., MacArthur, N. A., Watson, J. L. The Iron–Titanium Ores of New Zealand—Their Evaluation, Development and Processing.
Collins, D. N., Bignell, J. D. The Effects of Ilmenite Alteration Products in the Separation of Zircon–Rutile Ores.
Esch, A., Serbent, H. The Use of Ti—Containing Iron Ores for Steelmaking.
Hammoud, N. S. A Process to Produce Low-Titanium High Grade Iron Concentrates from Egyptian Beach Ironsand.
Krukiewicz, R. Titanium-Bearing Compounds in Polish Titanomagnetite Ores—The Reasons for Their Concentration in Magnetic Separation.
Łuszczkiewicz, A., Lekki, J., Laskowski, J. Flotability of Ilmenite.
Meinander, T., High Intensity Magnetic Preconcentration of Otanmäki Ilmenite with the Hips-Separator.
Smorodinnikov, A. V., Voitsehovich, E. B., Eremin, N. Ya., Romanova, O. A., Pershukov, A. A. Beneficiation Features and Utilization of Low-Grade Titanmagnetite Ores.
Wójtowicz, J., Bortel, R., Seweryński, B., Kubacz, N., Gramała, J. Flotation of Ilmenite by Means of Phosphonic Acid Derivatives.

Beneficiation of clay raw materials

Baburek, J., Stepánek, Z. Upgrading of Kaolin Clays With High Iron- and Titanium Content by High Gradient Magnetic Separation.
Cohen, H. E., Dudeney, A. W. L., Shaw, R. A. Preparation of Active Clay Minerals by Grinding of Igneous Rocks Under Hydrothermal Conditions at 350–450°C.
Derwinis, J., Widaj, B., Pytliński, A. Clays in the Turow Lignite Deposit as a Prospective Raw Material for Ceramic Industry.
Goell, G., Schulze, H., Scheibe, W., Töpfer, E., Schmidt, J., Ziegenbalg, S. New Possibilities for the Dry Treatment of Clay Minerals.
Groudev, S. N., Genchev, F. N. Bioleaching of Aluminium from Clays.
Iannicelli, J. New Developments in the High Extraction Magnetic Filtration of Kaolin Clay.
Klyachin, V. V., Komlev, A. M., Gabdukhaev, R. L. Refractory Clay and Kaolin Treatment Using Suspension With High Solid Content.
Lofthouse, C. H. Scobie, C. W. The Beneficiation of Kaoline Using a Commercial High Intensity Magnetic Separator.
Polesiński, Z., Brzęczkowski, J., Kowalska-Smoleń, J. Perspectives of Many-Purpose Utilization of Kaolin Clays From the Maria III Deposit at Nowogrodziec.
Stoch, L., Badyoczek, H., Kowalska-Smoleń, J., Abgarowicz, E., Strauch, T. Physico-Chemical Methods of Purification of Kaolins and Their Mineral Composition and Origin.
Święcki, Z., Kucharska, L., Cyganek, J. Modification of Kaolin Rheological Properties.
Tikhonov, S. A., Babushkina, N. A. Extraction of the Contaminating Impurites from Kaolins.
Yvon, J., Cases, J. M., Garin, P., Lietard, O., Lhote, F. Kaolinic Clays as Natural Rubber Fillers: Influence of Cristallochemical Properties and Superficial States.

List of selected papers presented at the XIIth International Mineral Processing Congress, NACIONAL-Publicacöes e Publicidade, Sao Paulo, Brazil (1977).

Surface chemistry

1. *Roberts, K. and Osterlund, R.* Selective Agglomeration of Tailings.
2. *Malinovski, V. A., Matveyenko, N. B., Orlov, B. D., Denegina, N. N., Mikhin, B. M., and Urarov, U. P.* Development and Application of Froth Separation Flotation Machines.
3. *Shergold, H. L. and Lofthouse, C. H.* The Purification of Kaolins by Two-Liquid Separation Process.
4. *Nakahiro, Y., Wakamatsu, T., and Mukai, S.,* Study on the Removal of Chromium and Cyanide in Waste Water by the Precipitation–Flotation Method.
5. *Stoev, S. and Kintisheva, R.* Vibroacoustical Methods for Improving Selective Flocculation.
6. *Tutchek, P. and Illie, P.* Flotation of Lead–Zinc Ores with Total Water Recycling.
7. *Fallenius, K., Koivistioninen, P., and Korhenen, O.* The Use of Large Flotation Machines in the Concentrators of Outokumpu Oy Finland.
8. *Ohtsuka, T., Nishida, K., Takeda, R., and Hikosaka, T.* Flotation Treatment of Copper-Bearing Zinc Residue.
9. *Dobrescu, L. and Kheil, O.* Flotoamalgamation, a New Process for Treatment of Low-Grade Gold Ores.
10. *Schubert, H., Serrano, A., and Baldauf, H.* Correlations between reagent Structure and Adsorption in Flotation.
11. *Bogdanov, O. S., Podnek, A. K., Rjaboy, V. I., and Janis, N. A.* Reagent Chemisorption on Minerals as a Process of Formation of Surface Compounds with Coordination Bond. (Some Principles of Flotation Reagents Selection.)
12. *Lekki, J., Laskowski, J., Neczaj-Hruzewicz, J., Szcypa, J., and Drzymala, J.* Physical Chemical Models in the Research of Floatability of Minerals.
13. *Pomianowski, A., Janusz, W., and Szczypa, J.* Adsorption on Sphalerite Using Radioactive Copper and Xanthate Species.
14. *Zambrana, Z. G. and Serrano, B. A.* Cassiterite Flotation in the Presence of Alkanocarboxylic Acids.
15. *Fuerstenau, D. W. and Raghavan, S.* Surface Chemistry of Oxide and Silicate Mineral Flotation.
16. *Stewart, P. S. B. and Jones, M. P.* Determining the Amounts and the Compositions of Composite (Middling) Particles.
17. *Tavares, J. R. P. and Ferreira, R. C. H.* Recovery of Nickel from Laterites by Reduction with High Sulphur Mineral Coal and Tailings from Coal Concentration Plants.
18. *Sheridan, M. J. and Burkin, A. R.* Leaching of Silver-Containing Bornite with Acidic Ferric Sulphate Solutions.
19. *Bruynesteyn, A. and Duncan, D. W.* The Practical Aspects of Biological Leaching Studies.
20. *Van Den Steen, A., Bustos, S., Salvo, C., and Fuentes, J.* Selenium Recovery in the Chilean Gran Mineria.
21. *Fuerstenau, M. C., Elmore, M. R., and Ollivier, P.* Studies on the Sulphatation of Garnierites.

Copper Ores

1. *Esna-Ashari, M., Kausel, E., and Paschen, P.* Vibrating Mill as High-Efficiency Reactor for Simultaneous Grinding, Leaching, Precipitation and Flotation of Oxide/Sulfide Copper Ore.
2. *Sangster, K. J., Presgrave, D. K., and Herbert, I. C.* Some Aspects of the Extraction of Copper from its Ores by Hydrometallurgical Techniques.
3. *Yarar, B. and Dogan, Z. M.* The Effects of Long-Chain Polymeric Flocculants on Flotation Tailings of a Low Grade Copper Ore.
4. *Selby, D. W. and Woodcock, J. T.* Some Effects of Wood Used as Mine Timber on Flotation of Chalcopyrite in Lime Circuits.
5. *Yeropkin, Yn. L.; Kondratenko, I. V., Kuzkin, A. S., Kuchaev, V. A., Larionov, L. A., Pudov, V. P., and Tartaky, A. Ye.* Main Features of the Technology of Flotation of Sulphide Copper and Complex Ores of the Mineralized Sandstone Type.
6. *Cannon, K. J.* Solvent Extraction—Electrowinning Technology for Copper.
7. *Kallioinen, J.* Increasing Capacity of Existing Concentrator by Computer Control Based on a Process Behaviour Model.
8. *Bortel, R., Sewerynski, B., and Wojtowicz, J.* The Possibilities of Eliminations of the Effect of Organic and Argilaceous Components in the Flotation of Polish Copper Ores.
9. *Levin, K. F., Mitrofanov, S. I., Semyoshkin, S. S., Samygin, V. D., Mayorov, A. D., Tchunin, A. F., and Muratov, L. I.* Method of Recovery of Oxidized Copper from Ores: Leaching–Cementation–Magnetic Concentration.
10. *McKinney, W. A. and Groves, R. D.* Copper Hydrometallurgy Research Activities of the U. S. Bureau of Mines.
11. *Sutulov, A.* Oxide Copper Processing in South America.
12. *Bhappu, R. B., Chase, C. K., Maass, H. M., Potter, G. M., and Shirley, J. F.* Treatment of Oxide Copper Ores: Practices and Problems.
13. *Laskorin, B. V., Kovaleva, I. I., Khavskiy, N. N., Tokarev, V. D., Chanturiya, V. A., Yakubovich, I. A., Gol'man, A. M., Alekseyeva, R. K., Mararova, G. N., Gorodetskiy, M. I., Shrader, E. A., and Lebedev, V. D.* Working out of Combined Methods of Precious Component Recovery from Poor Oxide Copper Ores.
14. *Fisher, J. F. C. and Deuchar, A. D.* Some Problems in the Metallurgical Processing of Oxide Copper Ores in Zambia.
15. *Yamaguchi, T., Kawai, T., Nakao, M., and Murakami, Y.* Segregation Process at Katanga Mine in Peru.
16. *Pena, P., de La L., and Espinosa de La, L.* Copper Hydrometallurgy in Mexico.

Iron Ore: Flotation and Flocculation

1. *Bunge, F. H., Morrow, J. B., Trainor, L. W., and Dicks, M. L.* Developments and Realizations in the Flotation of Iron Ore in North America (The Hanna Mining Company).
2. *Houot, R., Polgaire, J. L., Pivette, P., and Vant'Hoff, J.* Amine Type Reagents Applied to the Reverse Flotation of Iron Ore of the Itabirite Type.
3. *Whiten, W. J. and White, M. E.* Modelling and Simulation of High Tonnage Crushing Plants.

4. *Oblad, A. E., and Povoa, F. V.* Process Plant Simulation Applied to the Concentration of Brazilian Iron Ore.
5. *Leach, H. J., Stukel, J. E., Nummela, W., Lindroos, E. W., and Kosky, R. A.* Developments and Realizations in the Flotation of Iron Ores in North America (The Cleveland-Cliffs Iron Company).
6. *Kulkarni, R. D., Somasundaran, P., and Ananthapadma Nabhan, K.* Flotation Mechanism based on Ionomolecular Surfactant Complexes.
7. *Carta, M., Del-Fa, C., Gitiani, M., and Alfano, G.* Hyperpure Concentrates Obtained from Hematite Ores by Means of Various Processing Methods.

Special Publication

1. *Ek., C.* Verification of Some Laws of Discontinuous Sedimentation in Flocculated Pulps.
2. *Concha, F. and Bascur, O.* Phenomenologic Model of Sedimentation.
7. *Soto, H. and Carvajal, R.* Recovery of Potassium Chloride from Atacama Brines by Froth Flotation.
8. *Sresty, G. C. and Somasundaran, P.* Beneficiation of Mineral Slimes Using Modified Polymers as Selective Flocculants.
9. *Moustapha, S. I. A.* Scale-Up Correlation of Power and Energy Requirements in Flotation Cells.
10. *Jones, M. H. and Woodcock, J. T.* Optimization of Chemical Environment in Flotation.
11. *Barbery, G., Cecile, J. L., and Plichon, V.* The Use of Chelates as Flotation Collectors.
12. *Waksmundzki, A. and Zanczuk, B.* Energy Changes of the Sulfur–Water–Air System Under the Influence of a Nonpolar Collectors a Comparison to Sulfur Ore Flotation.
13. *Polkin, M. I., Adanov, E. V., Kurkov, A. V., Vetrov, I. S., and Berger, G. S.* The Theoretical Principles of the Use of Organophosphoric Compounds for Complex Processing of Rare Metals and Tin Ores.
14. *Abramov, A. A., Leonov, S. B., Avdohin, V. A., and Kurscakova, G. M.* Electrochemistry and Thermodynamics of Sulfide Minerals with Sulphydryl, Hydrosulphide and Cation Collectors.
16. *Rinelli, G., Marabini, A. M., and Abbruzzese, C.* Selective Separation of Zinc and Lead from Oxide Ores Through Formation of Organo-Metallic Compounds.
17. *Radino, P. and Burkin, A. R.* Extraction of Copper and Nickel from Ammoniacal Solutions (by Carboxylic Acids).
18. *Botcharov, V. A., Ryskin, M. Y., Filimonov, V. N., Mitrofanov, S. I., Pospelov, N. D., Arzhaninov, G. I., Karyukin, B. M., Belyaev, M. A., Gusarov, R. M., and Malikhov, Z. L.* Problems of Concentration of Pyritic Copper–Zinc Ores and their Solution at Development of Industrial Processes.

Phosphates

1. *Nevsky, B. V., Skrinichenko, M. L., Boldyrev, V. A., Gorskov, A. I., and Fadeyeva, T. P.* Development of Processing Methods Involved in the Beneficiation of Low-Grade and Complex Phosphate Ores.

2. *Awasthi, P. K., Luthra, K. L., Kulkarni, D. V., and Jaggi, T. N.* Beneficiation and Utilization of Low Grade Mussoorie Rock Phosphate.

3. *Hammoud, N. S., Khazbak, A. A., and Ali, M. M.* A Process to Up-Grade the Lean Non-Oxidized Complex Phosphorites of Abu Tartur Plateau—Western Desert (Egypt).

4. *Haynman, V. and Pregerson, B.* Beneficiation by Washing of the Israel Phosphates.

References

Abramov, A. A. (1968), Method for the quantitative determination of the sorption forms of collectors on mineral surfaces, in *Fiz-tekh Probl. Razrab. Polez. iskop. Sib. Otd. Akad. Nauk. SSSR*, No. 4, 75–84.

Abramov, A. A. (1969), Role of sulphidizing agent in flotation of oxidized minerals, *Tsvetn. Metall.* **12**(5), 7–13.

Abramov, A. A., Grosman, L. I., and Perlov, P. M. (1969), Efficient schemes for treating oxidized and mixed copper–lead–zinc ores, in *Trudy Nauch. - Tekh. Konf. Inst. Mekhanobr* (1967) **2**, 320–333.

Abramov, A. A., Solozhenkin, P. M., Kulyashev, Yu. G. and Statsura, P. F. (1974) Investigations of the action of reagents and optimization of their concentration in the flotation pulp, in *Tenth International Mineral Processing Congress, London 1973*, M. J. Jones, ed., Institution of Mining and Metallurgy, London, 633–652.

Adam, N. K. (1941), *The Physics and Chemistry of Surfaces*, Oxford University Press, New York.

Adam, N. K. (1957), Use of the term "Young's equation" for contact angles, *Nature* **180**, 809–810.

Adam, N. K., and Livingston, H. K. (1958), Contact angles and work of adhesion, *Nature* **182**, 128.

Adam, N. K., and Pankhurst, K. G. A. (1946), The solubility of some paraffin chain salts, *Trans. Faraday Soc.* **42**, 523–526.

Adam, N. K., and Schutte, H. L. (1938), Anomalies in the surface tension of paraffin chain salts, *Trans. Faraday Soc.* **34**, 758–765.

Adamson, A. W. (1967), *Physical Chemistry of Surfaces*, Interscience, New York.

Addison, C. C. (1943–1945), The properties of freshly formed surfaces, *J. Chem. Soc.*: Part I, The application of the vibrating-jet technique to surface tension measurements on mobile liquids, p. 535 (1943); Part II, The rate of adsorption of iso-amyl alcohol at the air–water surface, p. 252 (1944); Part III, The influence of chain length and structure on the static and the dynamic surface tensions of aqueous alcoholic solutions, p. 98 (1945).

Adorjan, L. A., Annual Review of Mineral Processing, *Min. J.* (*London*), June.

Agar, G. E. (1961), Ph. D. thesis, MIT, Cambridge, Massachusetts.

Agar, G. E. (1967), The use of fatty chemicals as flotation reagents, *J. Am. Oil Chem. Soc.* **44**, 396A.

Agarwal, J. C., Schapiro, N., and Mallio, W. J. (1976), Process petrography and ore deposits, *Min. Congr. J.*, March, 28–35.

Ahrland, S., Chatt, J., and Davies, N. R. (1958), The relative affinities of ligand atoms for acceptor molecules and ions, *Quart. Revs. (Chem. Soc.)* **12**, 265–276.

AIME (1970), *Lead and Zinc* Vol. I, *Mining and concentrating of lead and zinc*, AIME, New York, pp. 709–811.

Akers, R. J., ed. (1976), *Foams, Proceedings of a Symposium by the Society of Chemical Industry at Brunel University*, September 1975, Academic Press, New York.

Al Attar, A. A. A., and Beck, W. H. (1970), Alkaline earth and lanthanum ion electrodes of the third kind based on the hydrogen ion-responsive glass electrode. Thermodynamic solubility products of long-chain normal fatty acids and their alkaline earth and lanthanum salts in water, *J. Electroanal. Chem.* **27**, 59.

Aldrich, R. G., Keller, D. V., and Sawyer, R. G. (1974), Chemical comminution and Mining of Coal, Syracuse University Research Corp., U.S. Patent No. 3, 815, 826.

Alexandrovich, Kh. M. (1975), *Flotation of Water-Soluble Salts*, Nauka, Moscow.

Alferova, L. A., and Titova, G. A. (1969), Study of the reaction rates and mechanisms of the oxidation of hydrogen sulphide, sodium hydrosulphide and sulphides of sodium, iron and copper in aqueous solutions by atmospheric oxygen (in Russian), *Zh. Prikl. Khim.* **42**, 192–196.

Allison, S. A. (1971), Studies of the stability of xanthates and dixanthogen in aqueous solution and the chemistry of iron xanthates and cobalt xanthates, NIM Report No. 1125, Johannesburg, South Africa.

Allison, S. A. (1974), The Role of Sulphur in the Flotation of Sulphide Minerals, NIM Report No. 1597, Johannesburg, South Africa.

Allison, S. A., and Finkelstein, N. P. (1971), Study of the products of reaction between galena and aqueous xanthate solutions, *Trans. IMM* **80**, C235–C239.

Amsden, M. P. (1973), Computer control of flotation at Ecstall, in *Proceedings of the 5th Meeting of Canadian Mineral Processors*, Department of Energy, Mines and Resources, Ottawa, pp. 385–415.

Amsden, M. P., Chapman, C., and Reading, M. G. (1973), Computer control of flotation at the Ecstall concentrator, in *Proceedings of the 10th International Symposium, Application of Computer Methods in the Mineral Industry*, M.D.G. Salamon and F. H. Lancaster, eds., South African IMM, Johannesburg, pp. 331–340.

Amstutz, G. C. (1962), How microscopy can increase recovery in your milling circuit, *Min. World, December 1962*, 19–23.

Andersen, T. N., and Eyring, H. (1970), Principles of electrode kinetics, in *Physical Chemistry*, Vol. IX A, *Electrochemistry,* H. Eyring, ed., Academic Press, New York, Chap. 3, pp. 247–344.

Anfruns, J. F., and Kitchener, J. A. (1977), Rate of capture of small particles in flotation, *Trans. IMM* **86**, C9–C15.

Anon. (1961), 50 Years of flotation, its men, ideas and machines, *Eng. Min. J.* **162**, December, 83–91.

Anon. (1971), Innovative design revamps flotation cells, *Eng. Min. J.*, July, 78–79.

Anon.–Uralmekhanobr (1972), *Practice of ore dressing of non-ferrous, rare and precious metals in the plants of the USSR* (Nedra, 1964, Moscow), transl. C. H. Ahuja, 1972,

published for U.S. Bureau of Mines and National Science Foundation by Amarind Publishing, New Delhi.

Anthony, R. M., Kelsall, D. F., and Trahar, W. J. (1975), The effect of particle size on the activation and flotation of sphalerite, Proceedings of the *Australasian Institute of Mining and Metallurgy*, No. **254**, pp. 47–58.

Apira, P., and Rosenblum, F. (1974), Oxygen demand of flotation pulps, in *Proceedings of the 6th Annual Meeting of Canadian Mineral Processors*, Department of Energy, Mines and Resources, Ottawa, p. 73.

Aplan, F. F. (1966), Flotation, in *Kirk- Othmer Encyclopedia of Chemical Technology*, **9**, 380–398, Interscience Publ., New York.

Aplan, F. F. (1976), Coal flotation, in *Flotation, A. M. Gaudin Memorial Volume*, Vol. II, M. C. Fuerstenau, ed., AIME, New York, pp. 1235–1265.

Aplan, F. F. and De Bruyn, P. L., (1963), Adsorption of hexyl mercaptan on gold, *Trans. AIME* **226**, 235–42.

Aplan, F. F., and Fuerstenau, D. W. (1962), Principles of Nonmetallic Mineral Flotation, in *Froth Flotation—50th Anniversary Volume*, D. W. Fuerstenau, ed., AIME, New York, pp. 170–214.

Arbiter, N., and Harris, C. C. (1962), Flotation kinetics, in *Froth Flotation—50th Anniversary Volume*, D. W. Fuerstenau, ed., AIME, New York, pp. 215–246.

Arbiter, N., Harris, C. C., and Yap, R. (1969), A Hydrodynamic Approach to Flotation Scale-Up, in *Proceedings of the 8th International Mineral Processing Congress*, Mekhanobr Inst., *Leningrad*, 1968, Vol. 1, pp. 588–607.

Aronson, M. P., Zettlemoyer, A. C., Codell, R. B., and Wilkinson, M. C. (1976), The static profiles of several solid/oil/water/vapour configurations. II. Partially submerged drops, *J. Colloid Interface Sci.* **54**, 134–148.

Arsentiev, V. A., and Gorlovskii, S. I., (1973), Application of natural polymers as flotation activators and depressants, *Gorn. Zh.* (Mining Journal) **9**, pp. 156–159.

Arsentiev, V. A., and Gorlovskii, S. I. (1974), Organic modifying agents for flotation of ores containing clay slimes, *Obogashch. Rud* 1(109), 9–13.

Arsentiev, V. A., and Leja, J. (1976), Interactions of alkali halides with insoluble films of fatty amines and acids, in *Colloid and Interface Science*, Vol. V, M. Kerker, ed., Academic Press, New York, pp. 251–270.

Arsentiev, V. A., and Leja, I. (1977), Problems in potash flotation common to ores in Canada and the Soviet Union, *Can. Min. and Met. I.*, March 1–5.

Atademir, M. R., Kitchener, J. A., and Shergold, H. L., (1979), The surface chemistry and flotation of scheelite, *J. Colloid Interface Sci.* **71**(3), 466–476.

Attia, Y. A. (1977), Synthesis of PAMG chelating polymers for the selective flocculation of copper minerals, *Int. J. Miner. Process.* **4**, 191–208.

Attia, Y. A., and Kitchener, J. A. (1975), Development of complexing polymers for the selective flocculation of copper minerals, in *Proceedings of the 11th IMPC*, M. Carta, ed., Università di Cagliari, Cagliari, Italy, pp. 1233–1248.

Austin, M., Bright, B. B., and Simpson, E. A. (1967), The Measurement of the dynamic surface tension of manoxol OT for freshly formed surfaces, *J. Colloid Interface Sci.* **23**, 108–112.

Aveston, J. (1965), Hydrolysis of sodium silicate, ultracentrifugation in chloride solutions, *J. Chem. Soc.* **1965**, 4444–48.

Aveyard, R., and Haydon, D. A. (1973), *An Introduction to the Principles of Surface Chemistry*, Cambridge University Press, Cambridge, England.

Bachmann, R. (1951), Beitrag zur Theorie der Schwimmaufbereitung, *Erzmetall* **4**, 316.

Bachmann, R. (1955), Aufbereitungsprobleme der deutschen Kaliindustrie, in Internationaler Kongress für Erzaufbereitung, Goslar, 1955, *Erzmetall* (*Beihefte*) **8**, 109–114.

Baer, H. J. Dautzenberg, H., and Philipp, B. (1968), Zur Kinetik des Äthylxanthogenatzerfalles in wässriger alkalischer Lösung, *Z. Phys. Chem.* (*Leipzig*) **237**, 145–156.

Bailey, J., and Carson, H. (1975), preprint 75-B-22 at 1975 Annual AIME Meeting.

Baldauf, H., and Schubert, H. (1980), Correlations between structure and adsorption for organic depressants in flotation, in *Fine Particles Processing*, P. Somasundaran, ed., AIME, New York, pp. 767–786.

Ball, B., and Fuerstenau, D. W. (1970), A two-phase distributed-parameter model of the flotation process in *Proceedings of the 9th International Mineral Processing Congress, Prague*, 1970, Ustav Pro Vyzkum Rud, Prague, pp. 199–207.

Ball, B., and Fuerstenau, D. W. (1973), A Review of the measurement of streaming potentials, *Miner. Sci. Eng.* **5**, 267–277.

Ball, B., and Rickard, R. S. (1976), The chemistry of pyrite flotation and depression, in *Flotation, A. M. Gaudin Memorial Volume*, Vol. I, M. C. Fuerstenau, ed., AIME, New York, pp. 458–484.

Bally, R. (1967), Structure Cristalline du Bidiethyldithiocarbamate de Phenylarsine, *Acta Crystallogr.* **23**, 295–306.

Banerjee, B. C., and Palit, S. R. (1950), Dipole moments of metallic soaps. *J. Indian Chem. Soc.* **27**(8), 385–394.

Banerjee, B. C., and Palit, S. R. (1952), Dipole moments of trivalent metallic soaps in benzene and the effect of addition of small quantities of alcohol therein, *J. Indian Chem. Soc.* **29**(3), 175–182.

Bangham, D. H., and Fakhoury, N. (1931), The translational motion of molecules in the adsorbed phase on solids, *J. Chem. Soc.*, 1324–1333.

Bangham, D. H., Fakhoury, N., and Mohamed, A. F. (1932), The swelling of charcoal. Part II—Some factors controlling the expansion caused by water, benzene and pyridine vapours, *Proc. R. Soc.* (*London*) *Ser. A* **138**, 162–183.

Bangham, D. H., Fakhoury, N., and Mohamed, A. F. (1934), The swelling of charcoal. Part III—Experiments with the lower alcohols, *Proc. R. Soc.* (*London*) *Ser. A* **147**, 152–175.

Bansal, V. K., and Biswas, A. K. (1975), Collector-frother interaction at the interfaces of a flotation system, *Trans. IMM* **84**, C131–135.

Baratin, F. (1976), private communication, while at the University of British Columbia on leave from Ecole de Mines, Paris.

Barclay, L. M., and Ottewill, R. H. (1970), Measurement of forces between colloidal particles, *Spec. Discuss. Faraday Soc.* **1**, 138–147.

Barclay, L. M., Harrington, A., and Ottewill, R. H. (1972), The measurement of forces between particles in disperse systems, *Kolloid Z. Z. Polym.* **250**, 655–666.

Barlow, C. A. (1970), The electrical double layer, in *Physical Chemistry*, Vol. IXA, *Electrochemistry*, H. Eyring, ed., Academic Press, New York, pp. 167–246.

Barrett, C. S. (1952), *Structure of Metals*, 2nd ed., McGraw-Hill, New York.

Barsky, G. (1934), Discussion of Wark and Cox paper on principles of flotation, *Trans. AIME* **112**, 236–237.

Bartsch, O. (1924), Beitrag zur Theorie des Schaumschwimmverfahrens, *Kolloidchem. Beih.* **20**, 27–31.

Bascom, W. D., Cottington, R. L., and Singleterry, C. R. (1964), Dynamic surface Phenomena in the spontaneous spreading of oils on solids, in *Contact Angle, Wettability and Adhesion*, R. F. Gould, ed., ACS Advances in Chemistry Series No. 43, American Chemical Society, Washington, D.C.

Bashforth, F., and Adams, J. C. (1883), *An Attempt to Test the Theories of Capillary Action*, Cambridge University Press, Cambridge, England.

Battista, O. A. (1965), Colloidal Macromolecular Phenomena, *Am. Sci.* **53**(2), 151–173; (1969), The 'unhinged' microcrystals, *Ind. Res.*, March, 66–68.

Bayramli, E., and Mason, S. G. (1978), Liquid spreading: edge effect for zero contact angle, *J. Colloid Interface Sci.* **66**(1), 200–202.

Bell, C. F., and Lott, K. A. K. (1963), *Modern Approach to Inorganic Chemistry*, Butterworths, London.

Bellamy, L. J. (1968), *Advances in Infrared Group Frequencies*, Methuen, London.

Belov, V. N., and Sokolov, A. V. (1971), *Extraction and Processing of Potassium Salts*, Khimiya, Leningrad.

Ben-Naim, Ar. (1974), *Water and Aqueous Solutions, Introduction to a Molecular Theory*, Plenum Press, New York.

Berenyi, L. (1920), Prüfung der Polanyischen Theorie der Adsorption, *Z. Phys. Chem.* **94**, 628.

Bernal, J. D., and Fowler, R. H. (1933), A theory of water and ionic solution with particular reference to hydrogen and hydroxyl ions, *J. Chem. Phys.* **1**, 515–548.

Bertocci, U. (1966), The role of crystalline orientation on the behaviour of copper as electrode in chloride solutions, *J. Electrochem. Soc.* **113**(6), 604–608.

Bessonov, C. V. (1954), *Basic Action of Oxygen on Sulphide Minerals and Native Metals in Relation to their Floatability*, Mintsvetmet, Moscow.

Beurskens, P. T., Blaauw, H. J. A., Cras, J. A., and Steggerda, J. J. (1968), Preparation, structure and properties of bis(N,N-di-n-butyl-dithiocarbamato) gold(III) dihalaurate(I), *Inorg. Chem.* **7**, 805–810.

Bickl, H. (1966), thesis, Technische Hochschule, Munich.

Bijsterbosch, B. H., and Lyklema, I. (1969), Structural properties of the silver iodide-aqueous solution interface, in *Hydrophobic Surfaces*, F. M. Fowkes, ed., pp. 164–171, Academic Press, New York.

Bikerman, J. J. (1970), *Physical Surfaces*, Academic Press, New York.

Bikerman, J. J., Perri, J. M., Booth, R. B., and Currie, C. C. (1953), *Foams: Theory and Industrial Applications*, Reinhold, New York.

Bjerrum, N. (1926), Influence of ionic association on the activity of ions at moderate degrees of association, *K. Dan. Videnskab. Selsk. Mat. Fys. Medd.* **7**, (9), 1–48.

Blake, T. D. (1971), Current problems in the theory of froth flotation, in *Symposium on Bubbles and Foams, Nuremberg, West Germany*, September 1971, Verlag des Vereins Deutscher Ingenieure, Düsseldorf.

Blake, A. B., Cotton, F. A., and Wood, J. S. (1964), The crystal, molecular and electronic structures of a binuclear oxomolybdenum(V) santhate complex, *J. Am. Chem. Soc.* **86**, 3024–3031.

Blazy, P. (1970), *La Valorisation des Minerais*, Presses Universitaires de France, Paris.

Blazy, P., Degoul, P., and Houot, R. (1969), The Evaluation by Flotation of a cassiterite ore from a residue of the washing plant; Correlation with physico-chemical parameter, in *Second Technical Conference on Tin, Bangkok*, 1969, W. Fox, ed., International Tin Council, London, pp. 937–960.

Bleier, A., Goddard, E. D., and Kulkarni, R. D. (1976), The structural effects of amine collectors on the flotation of quartz, in *Flotation, A. M. Gaudin Memorial Volume*, Vol. I, M. C. Fuerstenau, ed., AIME, New York, pp. 117–147.

Bleier, A., Goddard, E. D., and Kulkarni, R. D. (1977), Adsorption and critical flotation conditions, *J. Colloid Interface Sci.* **59**(3), 490–504.

Bockris, J. O'M., and Conway, B. E., eds. (1964), *Modern Aspects of Electrochemistry*, Vol. 3, Butterworths, London.

Bockris, J. O'M., and Reddy, A. K. N. (1970), *Modern Electrochemistry*, Vols. I and II, Plenum Press, New York, Chap. 7–10.

Bockris, J. O'M., Devanathan, M. A. V., and Mueller, K. (1963), The structure of charged interfaces, *Proc. Phys. Soc. (London) Ser. A* **274**, 55–79.

Boddy, P. J., and Brattain, W. H. (1963), The distribution of potential at the germanium aqueous electrolyte interface, *J. Electrochem. Soc.* **110**, 570–576.

Bode, H. (1954), Systematic studies on the application of diethyldithiocarbamate in analysis, *Z. Anal. Chem.* **142**, 414–423.

Bogdanov, O. S. (1965), *Issledovanie deystvia flotacjonnych reagentov* (Investigation of the Action of Flotation Reagents), Mekhanobr, Leningrad.

Bogdanov, O. S., Kizevalter, B. V., and Maslova, S. G. (1950), The effects of frothers on the content of air in a flotation pulp, *Bull. Acad. Sci. USSR Tech. Sci. Sec. No. 3*, 412–416.

Bogdanov, O. S., Podnek, A. K., Haynman, V. Ya., and Yanis, I. A. (1959), *Voprosy teorii i tekhnologii flotatsji* (Problems of Flotation Theory and Technology), Mekhanobr Institute, Leningrad.

Bogdanov, O. S., Yeropkin, Y. I., Koltunova, T. E., Khobotova, N. P., and Shtchukina, N. E. (1974), Hydroxamic acids as collectors in the flotation of wolframite, cassiterite and pyrochlore, in *Proceedings of the 10th International Mineral Processing Congress, London 1973*, M. J. Jones, ed., Institution of Mining and Metallurgy, London, p. 553.

Bond, G. C. (1962), *Catalysis by Metals*, Academic Press, London.

Booth, R. B., and Freyberger, W. L. (1962), Froths and frothing agents, in *Froth Flotation—50th Anniversary Volume*, D. W. Fuerstenau, ed., AIME, New York, pp. 258–276.

Born, M. (1920), Volumen and Hydratationswärme der Ionen, *Z. Phys.* **1**, 45–48.

Bortel, R. (1977), Die Flotation der polnischen Kupfererze, *Erzmetall* **30**(9), 396–398.

Bowcott, J. E. L. (1957), Reactions of the xanthate ion at the air/water interface, in *Proceedings of the 2nd International Congress Surface Activity*, Vol. 3, J. H. Schulman, ed., Butterworths, London, pp. 267–271.

Bowden, F. P. (1967), The nature and topography of solid surfaces, in *Fundamentals of gas-surface interactions*, H. Saltsburg, J. N. Smith and M. Rogers, eds. Academic Press, New York.

Bradley, W. F. ed. (1964), *Clays and Clay Minerals*, Proceedings of the Twelfth National Conference on Clays and Clay Minerals, Atlanta, Georgia, 1963, Pergamon Press, New York.

Bragg, L., and Claringbull, G. F. (1965), *Crystal Structures of Minerals*, G. Bell & Sons, London.

Bratby, J. (1980), *Coagulation and Flocculation*, Uplands Press, London.

Brophy, J. H., Rose, R. M., and Wulff, J. (1964), *The Structure and Properties of Materials*, Vol. 2, *Thermodynamics of Structure*, Wiley, New York.

Brown, A. G., Thuman, W. C., and McBain, J. W. (1953), (a) The surface viscosity of detergent solutions as a factor in foam stability, *J. Colloid Sci.* **8**, 491–507; (b) Transfer of air through adsorbed surface films as a factor in foam stability, *J. Colloid Sci.* **8**, 508–519.

Brown, D. J. (1962), Coal flotation, in *Froth Flotation—50th Anniversary Volume*, D. W. Fuerstenau, ed., AIME, New York, pp. 518–538.

Bruce, R. W., and Yacksic, B. (1967), Beneficiation of low-grade tin ores by flotation, *Trans. CIMM.* **70**, 49–53.

Brunauer, S. (1945), *The Adsorption of Gases and Vapours*, Vol. 1, *Physical Adsorption*, Princeton University Press, Princeton, New Jersey.

Brunauer, S. (1961), Solid surfaces and the solid–gas interface, in *Solid Surfaces and the Gas–Solid Interface*, R. F. Gould, ed., Advances in Chemistry Series No. 33, American Chemical Society, Washington, D.C., pp. 5–17.

Brunauer, S., Deming, L. S., Deming, W. E., and Teller, E. (1940), On a theory of the van der Waals adsorption of gases, *J. Am. Chem. Soc.* **62**, 1723–1732.

Brunauer, S., Emmett, P. H., and Teller, E. (1938), Adsorption of gases in multimolecular layers, *J. Am. Chem. Soc.* **60**, 309–319.

Bryner, L. C., Walker, R. B., and Palmer, R. (1967), Some factors influencing the biological oxidation of sulfide minerals, *Trans. AIME* **238**, 56–62.

Buchanan, A. S., and Heymann, E. (1949), The electrokinetic potential of sparingly soluble sulphates. I, *J. Colloid Sci.* **4**, 137–150.

Buckenham, M. H., and Schulman, J. H. (1963), Molecular associations in flotation, *Trans. AIME* **226**, 1–6.

Buff, F. P., and Saltsburg, H. (1957), Curved fluid interfaces. II. The generalized Neumann formula. III. The dependence of the free energy on parameters of external force, *J. Chem. Phys.* **26**, 23–31, 1526–1532.

Bujake, Y. C., and Goddard, E. D. (1965), Surface composition of sodium Lauryl sulphonate and sulphate solutions by foaming and surface tension, *Trans. Faraday Soc.* **61**, 190–195.

Burcik, E. J. (1950), The role of surface tension lowering and its role in foaming, *J. Colloid Sci.* **5**, 421–436.

Burcik, E. I., and Newman, R. C. (1953), A fundamental study of foams and emulsions, Technical Report, Contract No. NONR 656, The Pennsylvania State College, School of Mineral Industries Experiment Station, State College, Pennsylvania.

Burcik, E. J., and Newman, R. C. (1954), The effect of organic additives on the rate of surface tension lowering of sodium dodecyl sulphate solutions, *J. Colloid Sci.* **9**, 498–503.

Burcik, E. J., Sears, J. R., and Tillotson, A. (1954), The surface plasticity of sodium myristate solutions, *J. Colloid Sci.* **9**, 281–284.

Buscall, R., and Ottewill, R. H. (1975), Thin films, in *Colloid Science*, Specialist Periodical Reports, Vol. 2, D. H. Everett, ed., The Chemical Society, London, pp. 191–245.

Busev, A. I., and Ivanyutin, M. I. (1958), Dialkyl and diaryl dithiophosphoric acids as analytical agents, *Zh. Anal. Khim.* **13**, 647–652; (1959), *Chem. Abstr.* **53**, 5954.

Bushell, C. H. G. (1958), Behaviour of xanthates in flotation, *Trans. CIM* **61**, 65–77.

Bushell, C. H. G. (1962), Kinetics of flotation, *Trans. AIME* **223**, 266–278.

Bushell, C. H. G. (1964), On the action of xanthate in flotation, *Trans. CIM* **68**, 327–331.

Bushell, C. H. G., and Malnarich, M. (1956), Reagent control in flotation, *Min. Eng.* (July) [*Trans. AIME* **205**], 734–737.

Bushell, L. A., Lewis, J. M. L., and Lund, B. (1971), A two-stage process for the recovery of flotation concentrates from copper-zinc ores, NIM Report No. 1268, Johannesburg, South Africa.

Butler, J. A. V. (1929), The equilibrium of heterogenous systems including electrolytes. Part. III. The effect of an electric field on the adsorption of organic molecules, and the interpretation of electrocapillary curves, *Proc. R. Soc. London Ser. A* **122**, 399–416.

Butler, J. A. V. (1951), *Electrical phenomena at interfaces*, Methuen, London.

Cadenhead, D. A., and Phillips, M. C. (1968), Molecular interactions in mixed monolayers, in *Molecular Association in Biological and Related Systems*, Advances in Chemistry Series No. 84, R. F. Gould, ed., American Chemical Society, Washington, D.C.

Cambron, A. (1930), The mechanism of the formation of thiuram and xanthogen monosulphides, and observations on thiocarbamyl thiocyanates, *Can. J. Res.* **2**, 341–356.

Cameron, A. W., Kelsall, D. F., Restarick, C. J., and Stewart, P. S. B. (1971), A Detailed Assessment of Concentrator Performance at Broken Hill South Ltd., *Proc. Australas. IMM* **240**, 53–67.

Camp, T. R. (1963), *Water and Its Impurities*, Reinhold, New York.

Canterford, J. H., and Colton, R. (1968, 1969), *Halides of Transition Metals*, Wiley-Interscience, New York.

Capes, C. E., Smith, A. E., and Puddington, I. E. (1974), Economic assessment of the application of oil agglomeration to coal preparation, *CIM Bull.*, July, 115–119.

Carley-Macauly, K. W. (1968), Electrohydraulic crushing, *Chem. and Process Engineering* **49**, September, 87–91, 96.

Carr, J. S. (1953), Radioactive isotopes in mineral dressing research, with particular reference to flotation in *Recent Developments in Mineral Dressing*, IMM, London, pp. 465–501.

Carrai, G., and Gottardi, G. (1960), The crystal structure of xanthates: I. Arsenious xanthate, *Zeitschrift für Kristallographie* **113**, 373–384.

Carta, M., Ghiani, M., and Massacci, P. (1973), Flotation of fluorspar and barite ores with particular reference to the treatment of tailings, *Ind. Miner. Mineralurgie* **2**, 103–11, *S-Etienne, Fr.*, (in French).

Carta, M., Ciccu, R., Delfà, C., Ferrara, G., Ghiani, M., and Massacci, P. (1974), Improvement in electric separation and flotation by modification of energy levels in surface layers, in *Proceedings of the 10th International Mineral Processing Congress* (1973), *London*, M. J. Jones, ed., IMM, London, pp. 349–376.

Carta, M., Alfano, G. B., Del Fà, C., Ghiani, M., Massacci, P., and Satta, F. (1975), Investigation on beneficiation of ultrafine fluorite ore from Latium, in *Eleventh International Mineral Processing Congress, Cagliari*, 1975, pp. 1187–1212, Istituto di Arte Mineraria, Università di Cagliari, Cagliari, Italy.

Cartmell, E. (1971), *Principles of Crystal Chemistry*, Monographs for Teachers, Royal Institute of Chemistry, London.

Cartmell, E., and Fowles, G. W. A. (1961), *Valency and Molecular Structure*, Butterworths, London.

Cases, J. M. (1967), Les phénomènes physico-chimiques à l'interface. Application au procédé de la flottation, theses, L'Université de Nancy.

Cases, J. M. (1968), Adsorption et condensation des Chlorhydrates d'Alkylamine a l'interface Solide-Liquide. Application au procèdé de la flottation des micas, paper

No. S13, in 8th International Mineral Processing Congress, Leningrad. 1968, Mekhanobr Institute, Leningrad.

Cases, J. M. (1979), Adsorption des tensio-actifs à l'interface solide–liquide: thermodynamique et influence de l'hétérogénèite des adsorbants *Bull. Mineral.* **102**, 684–707.

Casimir, H. B. G., and Polder, D. (1948), The Influence of Retardation on the London–van der Waals Forces, *Phys. Rev.* **73**, 360–372.

Cassie, A. B. D., and Baxter, S. (1945), *J. Text. Inst.* **36**, T67.

Castro, S., Goldfarb, J., and Laskowski, J. (1974), Sulphidizing reactions in the flotation of oxidized copper minerals, Part I, *Int. J. Miner. Process.* **1**(2) 141–149; Castro, S., Soto, H., Goldfarb, J., and Laskowski, J. (1974), Part II, *Int. J. Miner. Process.* **1**(2), 151–161.

Castro, S., Gaytan, H., and J. Goldfarb (1976), The stabilizing effect of Na₂S on the collector coating of chrysocolla, *Int. J. Miner. Process.* **3**(1), 71–82.

Chadwick, B. M., and Sharpe, A. G. (1966), Transition metal cyanides and their complexes, in *Advances in Inorganic Chemistry and Radiochemistry*, Vol. 8, Emeleus, H. J., and Sharpe, A. G., eds., Academic Press, New York, pp. 83–176.

Chander, S., and Fuerstenau, D. W. (1975), On the floatability of sulfide minerals with thiol collectors: the chalcocite–diethyldithiophosphate system, in *Proceedings of the XIth International Mineral Processing Congress, Cagliari*, 1975, Istituto di Arte Mineraria, Università di Cagliari, Cagliari, Italy, pp. 583–604.

Chander, S., and Fuerstenau, D. W. (1979), Interfacial properties and equilibria in the apatite–aqueous solution system, *J. Colloid Interface Sci.* **70**(3), 506–516.

Chapman, D. L. (1913), A contribution to the theory of electrocapillarity, *Philos. Mag.* **25**, 475–481.

Charles, G. E., and Mason, S. G. (1960a), The mechanism of partial coalescence of liquid drops at liquid/liquid interfaces, *J. Colloid Sci.* **15**, 105–122.

Charles, G. E., and Mason, S. G. (1960b), The coalescence of liquid drops with flat liquid/liquid interfaces, *J. Colloid Sci.* **15**, 236–267.

Chatenever, A., and King, C. V. (1949), Kinetics of the decomposition of ethyl xanthic acid, *J. Am. Chem. Soc.* **71**, 3787–3591.

Chatt, J., Duncanson, L. A., and Venanzi, L. M. (1956), Electronic structures of dithiocarbamates and xanthates, *Nature* **177**, 1042–1043; Dithiocarbamates, infrared spectra and structure, *Suomen Kemi* **29B**(2), 75–84.

Chessick, J. J., Zettlemoyer, A. C., Healey, F. H., and Young, G. J. (1955), Interaction energies of organic molecules with rutile and graphon surfaces from heats of immersion, *J. Chem.* **33**, 251–258.

Chichibabin, A. E. (1953), Osnovnye Nachala organicheskoi khimii (*Principles of organic chemistry*), Khimizdat, Moscow.

Choi, H. S., Kim, Y. S., and Paik, Y. S. (1967), Flotation characteristics of ilmenite, *CIM Bull.* **60**, 217–220.

Christman, L. J. (1930), Chemistry and the flotation process, American Cyanamid Company Tech. Paper No. 17.

Church, T. M. (1975), *Marine Chemistry in the Coastal Environment*, American Chemical Society Symposium Series No. 18, American Chemical Society, Washington, D.C.

Clark, A. (1970), *The Theory of Adsorption and Catalysis*, Physical Chemistry, a series of monographs, Vol. 18, Academic Press, New York.

Clement, M., Surmatz, H., and Huttenhain, H. (1967), Report on the flotation of barite, *Erzmetall* **20**, 512–522.

Clement, M., Harms, H., and Trondle, H. M. (1970), Über das Flotationsverhalten verschiedener Mineralarten unter besonderer Berücksichtigung der Kornfeinheit, in *Proceedings of the IXth International Mineral Processing Congress, Prague* 1970, Ustav Pro Vyzkum Rud, Prague pp. 179–187.

Clement, M., Brennecke, K., and Bonjer, J. (1973), Floatability of nonsulphidic minerals such as barite, celestine and fluorspar, *Erzmetall* **26**(5), 225–229 (in German).

Clift, R., Grace, J. R., and Weber, M. E. (1978), *Bubbles, Drops and Particles*, Academic Press, New York.

Clint, J. H., Corkill, J. M., Goodman, J. F., and Tate, J. R. (1969), Adsorption of *n*-alkanols at the air/aqueous solution interface, in *Hydrophobic Surfaces*, F. M. Fowkes, ed., Academic Press, New York.

Coelho, E. M. (1972), Flotation of Oxidized Copper minerals: An infrared spectroscopic study, Ph. D. thesis, University of British Columbia, Vancouver.

Coelho, E. M., and Poling, G. W. (1973), Mechanism of adsorption of flotation reagents on oxidized copper minerals, in *Proceedings of the Congress of Brazilian Association of Metals*, Savador, Bahia.

Coghill, W. H. (1916), Molecular forces and flotation, *Min. Sci. Press* **113**, 341–349.

Collins, D. N. (1967), Investigation of collector systems for the flotation of cassiterite, *Trans. IMM* **76**, C77–93.

Collins, D. N., and Read, A. D. (1971), The treatment of slimes, *Miner. Sci. Eng.* **3**, April, 19–31.

Collins, D. N., Kirkup, J. L., Davey, M. H., and Arthur, C. (1968), Flotation of cassiterite: Development of a flotation process, *Trans. IMM* **77**, C1–13.

Collins, G. L., and Jameson, G. J. (1976), Experiments on the flotation of fine particles, *Chem. Eng. Sci.* **31**, 985–991.

Collins, G. L., Motarjemi, M., and Jameson, G. J. (1978), A method for measuring the charge on small gas bubbles, *J. Colloid Interface Sci.* **63**(1), 69–75.

Collins, R. E., and Cooke, C. E. (1959), Fundamental basis for the contact angle and capillary pressure, *Trans. Faraday Soc.* **55**, 1602–1606.

Connor, G. B. (1977), private communication, North Broken Hill Co., Australia.

Connor, P., and Ottewill, R. H. (1971), The adsorption of cationic surface active agents on polystyrene surfaces, *J. Colloid Interface Sci.* **37**(3), 642–651.

Conway, B. E. (1952), *Electrochemical Data*, Elsevier, Amsterdam, pp. 221–232.

Conway, B. E. (1965), *The Theory and Principles of Electrode Processes*, Ronald Press, New York.

Conway, B. E. (1970), Some aspects of the thermodynamic and transport behaviour of electrolytes, in *Physical Chemistry*, Vol. IXA, *Electrochemistry*, H. Eyring, ed., Academic Press, pp. 1–166.

Conway, B. E. (1976), Electrochemical studies in surface science, *Chemistry in Canada*, September 28–32, Chemical Institute of Canada, Ottawa.

Cook, M. A. (1968), Hydrophobicity control of surfaces by hydrolytic adsorption, *J. Colloid Interface Sci.* **28**, 547–556.

Cook, M. A. (1969), Hydrophobicity control of surfaces by hydrolytic adsorption, in *Hydrophobic Surfaces*, F. M. Fowkes, ed., Academic Press, New York.

Cooke, S. R. B. (1950), Flotation, in *Advances in Colloid Science*, vol. III, H. Mark and E. J. W. Verwey, eds., pp. 321–374, Interscience, New York.

Cook, M. A., and Nixon, J. C. (1950), Theory of water repellent films on solids formed by adsorption from aqueous solutions of heteropolar compounds, *J. Phys. Colloid. Chem.* **54**, 445–459.

Cooke, C. E., and Schulman, J. H. (1965), The effect of different hydrocarbons on the formation of microemulsions, in *Surface Chemistry, Proceedings of the 2nd Scandinavian Symposium on Surface Activity, Stockholm, 1964*, P. Ekwall, K. Groth, and V. Runnström-Reio, eds., Academic Press, New York, pp. 231–251.

Cooke, S. R. B., and Talbot, L. (1955), Fluorochemical collectors in flotation, Trans. AIME **202**, 1149–52.

Cooke, S. R. B., Iwasaki, I., and Choi, H. S. (1960), Effect of temperature on soap flotation of iron ore, *Trans. AIME* **217**, 76–83.

Cooper, H. R. (1966), Feedback process control of mineral flotation, Part I, Development of a model for froth flotation, *Trans. AIME* **235**, 439, and discussion of this paper in *Trans. AIME* **238**, 479.

Corkill, J. M., Goodman, J. F., Haisman, D. R., and Harrold, S. P. (1961), The thickness and composition of thin detergent films, *Trans. Faraday Soc.* **57**, 821–828.

Corkill, J. M., Goodman, J. F., Ogden, C. P., and Tate, J. R. (1963), The structure and stability of black foam films, *Proc. R. Soc. London* Ser. A **273**, 84–102.

Corrin, M. L., and Rutkowski, C. P. (1954), Discontinuities in adsorption isotherms, *J. Phys. Chem.* **58**, 1089–1090.

Cotton, F. A., and Wilkinson, G. (1972), *Advanced Inorganic Chemistry*, Interscience, New York.

Cottrell, A. H. (1967), *An Introduction to Metallurgy*, Arnold, London.

Cottrell, T. L. (1958), *The Strengths of Chemical Bonds*, 2nd ed., Butterworths, London.

Coucouvanis, D. (1970), Chemistry of the dithioacid and 1,1-dithiolate complexes, in *Progress in Inorganic Chemistry*, 1970, Vol. 11, S. J. Lippard, ed., Interscience, New York, pp. 273–371.

Coucouvanis, D., and Fackler, J. P., (1967), Square-planar sulphur complexes. VI. Reactions of bases with xanthates, dithiocarbamates and dithiolates of nickel(II), *Inorg. Chem.* **6**, 2047–2053.

Coulson, C. A. (1947), Representation of simple molecules by molecular orbitals, *Q. Rev. Chem. Soc. London* **1**, 144–178.

Courant, R. A., Ray, B. J., and Horne, R. A. (1972), *Zhur. Struk. Khim.* **4**, 481.

Crabtree, E. H., and Vincent, J. D. (1962), Historical Outline of Major Flotation Developments, in *Froth Flotation—50th Anniversary Volume*, D. W. Fuerstenau, ed., AIME, New York, pp. 39–54.

Cross, J., ed. (1977), *Anionic Surfactants—Chemical Analysis*, Marcel Dekker, New York

Cuming, B. D., and Schulman, J. H. (1959), Two-layer adsorption of dodecyl sulphate on barium sulphate, *Aust. J. Chem.* **12**, 413–423.

Cusack, B. L. (1967), Some aspects of the role of oxygen in sulphide mineral flotation, *Proc. Australas. IMM* **224**, 1–7.

Cusack, B. L. (1968), The development of the Davcra flotation cell, in *Broken Hill Mines, 1968, Monograph No. 3*, Australasian IMM, Melbourne, pp. 481–487.

Cusack, B. L. (1971), Innovative design revamps flotation cells, *Eng. Min. J.* **172**(7), 78–80.

Daasch, L. W., and Smith, D. C. (1951), Infrared spectra of phosphorus compounds, *Anal. Chem.* **23**, 853–868.

Daee, M., Lund, L. H., Plummer, P. L. M., Kassner, J. L., and Hale, B. N. (1972), Theory of nucleation of water. I. Properties of some clathrate-like cluster structures, *J. Colloid Interface Sci.* **39**(1), 65–78.

Dalang, F., and Stumm, W. (1976), Comparison of the adsorption or hydrous oxide surfaces of aquo metal ions with that of inert complex cations, in *Colloid & Interface Science*, M. Kerker, ed., Vol. IV, Academic Press, New York, pp. 157–168.

Davenport, J. E., Carroll, F., Kieffer, G. W., and Watkins, S. C. (1969), Beneficiation of Florida hard-rock phosphate. Selective flocculation, *Ind. Eng. Chem. Process. Des. Dev.* **8**(4), 527–533.

Davidsohn, A., and Milwidsky, B. M. (1978), *Synthetic Detergents*, George Goodwin, London, and J. Wiley & Sons, New York.

Davidson, D. W. (1973), Clathrate hydrates, in *Water, A Comprehensive Treatise*, Vol. 2, F. Franks, ed., Plenum Press, New York, pp. 115–234.

Davies, C. W. (1962), *Ion Association*, Butterworths, London.

Davies, J. T. (1957), A study of foam stabilizers using a new "viscous-traction" surface viscometer, in *Proceedings of the 2nd International Congress of Surface Activity*, Vol. 1, J. H. Schulman, ed., Butterworths, London, pp. 220–224.

Davies, J. T., and Rideal, E. K. (1961), *Interfacial Phenomena*, Academic Press, New York, pp. 404–408.

Davis, R. E. (1968), Mechanisms of sulphur reactions, in *Inorganic Sulphur Chemistry*, G. Nickless, ed., Elsevier, Amsterdam, pp. 85–135.

Davis, W. J. N. (1964), The development of a mathematical model of the lead flotation circuit at The Zinc Corporation Ltd., *Proc. Australas. IMM.* **212**, 61–89.

Dean, R. (1944), *The formation of bubbles*, J. Appl. Phys. **15**, 446–451.

Dean, R. S., and Ambrose, P. M. (1944), Development and use of certain flotation reagents, *U.S. Bur. Mines Bull.* No. 449.

de Boer, J. H. (1936), The influence of van der Waals forces and primary bonds on binding energy, strength and orientation with special reference to some artificial resins, *Trans. Faraday Soc.* **32**, 10–38.

de Boer, J. H. (1953), *The Dynamical Character of Adsorption*, Clarendon Press, Oxford.

De Bruyn, P. L., and Agar, G. E. (1962), Surface chemistry of flotation, in *Froth Flotation —50th Anniversary Volume*, D. W. Fuerstenau, ed., AIME, New York, pp. 91–138.

Debye, P. (1929), *Polar Molecules*, Chemical Catalog, New York.

Debye, P. (1949), Light scattering in soap solutions, *J. Phys. Colloid Chem.* **53**, 1–8.

Debye, P. and Anacker, E. W. (1951), Micelle shape from dissymmetry measurements, *J. Phys. Colloid Chem.* **55**, 644–655.

Debye, P. and Hückel, E. (1923a), The theory of electrolytes. I. Lowering of freezing point and related phenomen, *Z. Phys.* **24**, 185–206.

Debye, P. and Hückel, E. (1923b), Theory of electrolytes. II. The limiting law of electrical conductivity, *Z. Phys.* **24**, 305–325.

De Cuyper, J. (1977), Flotation of copper oxide ores, *Erzmetall* **30**, H.(3): 88–94.

Defay, R., and Hommelen, J. (1958), Critical bibliographical study on the methods of measurement of dynamic surface tensions, *Ind. Chim. Belge* **23**, 597–614.

de Feijter, J. A., and Vrij, A. (1975), Contact angles in Newton black soap films drawn from solutions containing sodium dodecyl sulphate and electrolyte, in *Adsorption at Interfaces*, ACS Symposium Series No. 8, K. L. Mittal, ed., American Chemical Society, Washington, D.C., pp. 183–190.

de Freijter, J. A., and Vrij, A. (1978), Contact angles in thin liquid films. II Contact angle measurements in Newton Black Soap Films, *J. Colloid Interface Sci.* **64**(2), 269–277.

de Freijter, J. A., Rijnbout, J. B., and Vrij, A. (1978), Contact angles in thin liquid films. I. Thermodynamic description, *J. Colloid Interface Sci.* **64**(2), 258–268.

Deju, R. A., and Bhappu, R. B. (1966), (1) A chemical interpretation of surface phenomena in silicate minerals, *Trans. AIME* **235**, 329–332; (2) A modified electrophoresis apparatus, *Trans. AIME* **235**, 88–90.

Delahay, P. (1965), *Double Layer and Electrode Kinetics*, Interscience, New York.

Derjaguin, B. V. (1937), The theory of interaction of particles carrying double electrical layers and of the aggregative stability of hydrophobic colloids and disperse systems, *Bull. Acad. Sci. URSS, Classe Sci. Math. Nat. Ser. Chim.*, **5**, 1153–1164.

Derjaguin, B. V. (1955), The concept and determination of the magnitude of the wedge effect and its importance for the statics and kinetics of thin liquid layers, *Colloid J. USSR* **17**, 207–214.

Derjaguin, B. V. (1973), The state of the arts in liquids modification by condensation, in *Recent Advances in Adhesion*, Lieng Huang Lee ed., Gordon and Breach, London and New York.

Derjaguin, B. V., and Churaev, N. V. (1974), Structural component of disjoining pressure, *J. Colloid Interface Sci.* **49**, 249–255.

Derjaguin, B. V., and Dukhin, S. S. (1961), Theory of flotation of small and medium-size particles, *Trans. IMM* **70**, 221–246.

Derjaguin, B. V., and Dukhin, S. S. (1974), Equilibrium (and non-equilibrium) double layer and electrokinetic phenomena, in *Surface and Colloid Science*, Vol. 7, E. Matijevic, ed., J. Wiley & Sons, New York, pp. 49–335.

Derjaguin, B. V., and Kussakov, M. M. (1939), Anomalous properties of thin polymolecular films, Part V, *Acta Physicochim. URSS* **10**, 153–174.

Derjaguin, B. V., and Landau, L. D. (1941), Theory of the Stability of Strongly charged Lyophobic Sols and of the Adhesion of Strongly Charged Particles in Solutions of Electrolytes. *Acta. Physiocochim. URSS* **14**, 633–622.

Derjaguin, B. V., and Samygin, V. D. (1962), On one of the reasons for poor flotation of slime-sized particles, Research Reports of GINCVETMET No. 19, Moscow 1962, pp. 240–254.

Derjaguin, B. V., and Titievskaya, A. S. (1953), The wedge effect of free liquid films and its importance for the stability of foam, *Kolloid. Zh.* **15**, 416–425.

Derjaguin, B. V., and Titievskaya, A. S. (1957), Static and kinetic stability of free films and froths, in *Proceedings of the 2nd International Congress of Surface Activity*, Vol. 1, J. H. Schulman, ed., Butterworths, London, pp. 211–219.

Derjaguin, B. V., and Zorin, Z. M. (1957), Optical study of the adsorption and surface condensation of vapours in the vicinity of saturation on a smooth surface, in *Proceedings of the 2nd International Congress Surface Activity*, Vol. 2, J. H. Schulman, ed., Butterworths, London, pp. 145–152.

Derjaguin, B. V., Titievskaya, A. S., Abricossova, I. I., and Malkina, A. D. (1954), Investigations of the forces of interaction of surfaces in different media and their application to the problem of colloid stability, *Discuss. Faraday Soc.* **18**, 24–41.

Derjaguin, B. V., Dukhin, S. S., and Lisichenko, V. A. (1960), Kinetics of the attachment of mineral particles to bubbles in flotation. I and II., *J. Phys. Chem. (USSR)* **33**, 2280–2287; **34**, 524–529 (in Russian).

Dervichian, D. G. (1958), The existence and significance of molecular associations in monolayers, in *Surface Phenomena in Chemistry and Biology*, J. F. Danielli, K. G. A. Pankhurst, A. C. Riddiford, eds., Pergamon Press, New York, pp. 70–87.

Desnoes, A., and Testut, R. (1954), L'emploi des silicones comme collecteur en flottation, *Rev. Ind. Miner.* **35**(606), 211–215.

Desseigne, G. (1945), Hydrolysis of alkyl sulphates in acid medium, *Ind. Corps Gras* **1**, 136–139.

Dettre, R. H., and Johnson, R. E. (1967), in *Wetting*, Society of Chemical Industry, Monograph No. 25, SCI, London, and Gordon and Breach, New York.

de Vries, A. J. (1957), Foam stability, a fundamental investigation of the factors controlling the stability of foams, *Meded. Rubber Sticht. Delft*, No. 326.

de Vries, A. J. (1958), Foam stability, I. Structure and stability of foams; II. Gas diffusion in foams; III. Spontaneous foam destabilization resulting from gas diffusion; IV. Kinetics and activation energy of film rupture; V. Mechanism of film rupture, *Rec. Trav. Chim.* **77**, 81–91; 209–223; 283–296; 383–399; 441–461.

Devereux, O. F., and de Bruyn, P. L. (1963), *Interaction of Plane Parallel Double Layers*, M.I.T. Press, Cambridge, Massachusetts.

Diamond, R. W. (1967), The development of flotation at Sullivan, *CIM Bull.*, September, 999–1007.

Dibbs, H. P., Sirois, L. L., and Bredin, R. (1974), Some electrical properties of bubbles and their role in the flotation of quartz, *Can. Metall. Q.* **13**(2), 395–404.

Digre, M., and Sandvik, K. L. (1968), Adsorption of amine on quartz through bubble interaction, *Trans. IMM* **77**, C61–64.

Dippenaar, A. (1978), The effect of particles on the stability of flotation froths, NIM Report No. 1988, 30 November 1978, Johannesburg, South Africa.

Dixit, S. G., and Biswas, A. K. (1973), Studies on zircon–sodium oleate flotation system (2)—pH dependence of collector adsorption and critical contact phenomena, *Trans. IMM* **82**, C202–206.

Dobbie, J. W., Evans, R., Gibson, D. V., Smitham, J. B., and Napper, D. H. (1973), Enhanced steric stabilization, *J. Colloid Interface Sci.* **45**, 557–565.

Dobias, B. (1968), Floatability and electrokinetic properties of fluorite and barite, *Erzmetall* **21**, 275–281 (in German).

Dobias, B., and Spurry, J. (1960), The use of some alkylsulfates and sulfonates for the selective flotation of fluorite and barite, in *3rd International Congress of Surface Active Substances, Köln*, Vol. 4, pp. 396–403, Verlag der Universitätsdruckerei, Mainz, West Germany.

Dobias, B. Spurny, J., and Skrivan, P. (1960), Electrokinetic analysis of the flotation of apatite, *Rudy*, **8**(12), 407–414.

Dogan, Z. M. (1973), Concentration of Bati Kef chromium ore of Guleman District, Turkey, *Can. Metall. Quart.* **12**(2), 191–199.

Domenicano, A., Torelli, L., Vaciago, A., and Zambonelli, L. (1968), The crystal and molecular structure of cadmium(II) *N,N*-diethyl-dithiocarbamate, *J. Chem. Soc. A* **1968**, 1351–1361.

Doren, A., Vargas, D., and Goldfarb, J. (1975), Non-ionic surfactants as flotation collectors, *Trans. IMM* **84**, C34–41.

Doss, S. K. (1976), Adsorption of dodecyltrimethylammonium chloride on alumina and its relation to oil–water flotation, *Trans. IMM.* **85**, C195–199.

Drost-Hansen, W. (1967), The structure of water and water–solute interactions, in *Equilibrium Concepts in Natural Water Systems*, R. F. Gould, ed., Advances in Chemistry Series No. 67, American Chemical Society, Washington, D.C., pp. 70–120.

Drost-Hansen, W. (1971a), Role of water structure in cell-wall interactions, *Proc., Fed. Am. Soc. Exp. Biol.* **30**, 1539–1548.

Drost-Hansen, W. (1971b), Structure of water near solid interfaces, in *Chemistry and Physics of Interfaces II*, American Chemical Society, Washington, D.C., pp. 203–241.

Dukhin, S. S. (1960), The kinetics of the attachment of mineral particles to bubbles during flotation. III. Secondary electrical double layer in the vicinity of the moving bubble surface, *J. Phys. Chem. (USSR)* **34**, 501–504.

Dukhin, S. S. (1974), Development of notions as to the mechanism of electrokinetic phenomena and the structure of the colloid micelle, in *Surface and Colloid Science*, Vol. 7, E. Matijevic, ed., Wiley-Interscience, New York, pp. 1–47.

Dukhin, S. S., and Derjaguin, B. V. (1974), Equilibrium (and non-equilibrium) double layer and electrokinetic phenomena, in *Surface and Colloid Science 7*, E. Matijevic, ed., J. Wiley and Sons, New York, pp. 49–335.

Dunkin, H. H. (1953), *Ore dressing methods in Australia*, Australasian Institute of Mining and Metallurgy, Melbourne.

Dupré, A. (1867), Sixieme Memoire sur la Theorie Mecanique de la Chaleur, *Ann. Chim. Phys.* **11**, 194–220.

du Rietz, C. (1953), Chemical problems in flotation of sulphide ores, *Iva* **24**, 257–266.

du Rietz, C. (1957), Xanthate analysis by means of potentiometric titration; some chemical properties of the xanthates, *Svensk. Kem. Tidskr.* **69**, 310–27.

du Rietz, C. (1958), Fatty acids in flotation, in *Progress in Mineral Dressing*, Proceedings Fourth International Mineral Processing Congress, Stockholm, 1957, sponsored and edited by Svenska Gruvforeningen and Jernkontoret, Almqvist and Wiksell, Stockholm, pp. 417–439.

du Rietz, C. (1965), Chemisorption of collectors in surface chemistry, *Proceedings of the 2nd Scandinavian Symposium on Surface Activity*, Stockholm, 1964, Per Ekwall, K. Groth, and V. Runstrom-Reio, eds., Academic Press, New York, pp. 21–27.

du Rietz, C. (1976), Chemisorption of collectors in flotation, in *Proceedings of the 11th International Mineral Processing Congress*, Cagliari, 1975, Istituto di Arte Mineraria, Università di Cagliari, Cagliari, Italy, pp. 375–403.

Dzienisiewicz, J., and Pryor, E. J. (1950), An investigation into the action of air in froth flotation, *Bull. IMM*, No. 521, 1–22.

Dzyaloshinski, E. I., Lifshits, E. M., and Pitaevskii, L. P. (1959), Van der Waals forces in liquid films, *Zh. Eksptl. Teoret. Fiz.* **37**, 229–241.

Dzyaloshinski, E. I., Lifshits, E. M., and Pitaevskii, L. P. (1961), General theory of the Van der Waals forces, *Usp. Fiz. Nauk* **73**, 381–422.

Eadington, P., and Prosser, A. P. (1966), Surface oxidation products of lead sulphide, *Trans. IMM* **75**, C125.

Eadington, P., and Prosser, A. P. (1969), Oxidation of lead sulphide in aqueous suspensions, *Trans. IMM.* **78**, C74.

Eckert, G. (1957), On the use of di-substituted dithiocarbamates for analytical separations, *Z. Anal. Chem.* **155**, 23–35.

Edelman, C. H., and Favejee, J. C. L. (1940), On the crystal structure of montmorillonite and halloysite, *Z. Krist.* **102**, 417–431.

Edwards, G. R., and Ewers, W. E. (1951), The adsorption of sodium cetyl sulphate on cassiterite, *Aust. J. Sci. Res.* **4**(4), 627–643.

Edwards, J. O. (1954), Correlation of relative rates and equilibria with a double basicity scale, *J. Am. Chem. Soc.* **76**, 1540–1547.

Ehrlich, G. (1963a), Adsorption and surface structure, in *Metal Surfaces: Structure, Energetics and Kinetics*, American Society for Metals, Cleveland, pp. 221–258.

Ehrlich, G. (1963b). Adsorption on clean surfaces, *Ann. N.Y. Acad. Sci.* **101**, 722–755.

Eick, J. D., Good, R. J., and Neumann, A. W. (1975), Thermodynamics of contact angles II. Rough solid surfaces *J. Colloid Interface Sci.* **53**(2), 235–248.

Eigeles, M. A. (1950), Kinetics of mineralization of air bubbles in selective flotation and the effect of flotation reagents on it, *Trudy Soveshchaniya Teorii Flotatsion. Obogashcheniya, Moscow 1948, Rol Gazov i Reagentov v Protsessakh Flotatsii*, 63–84.

Eigeles, M. A., and Donchenko, V. A. (1966), Comparative evaluation of some frothing agents in the selective flotation of polymetallic ores, *Tsvet. Metall.* **39**(11), 10–14.

Eigeles, M. A., and Volova, M. L. (1960), Kinetic investigation of effect of contact time, temperature and surface condition on the adhesion of bubbles to mineral surfaces, in *Proceedings of the Fifth International Mineral Processing Congress, London, 1960*, IMM, London, pp. 271–284.

Eigeles, M. A., Kuznetsov, V. P., Volova, M. L., Sokolov, Yu. F., Lyubimov, E. I., and Grebnev, A. N. (1975), Principles of theory and perspectives for the development of beneficiation methods for finely dispersed ores, in *11th International Mineral Processing Congress, Cagliari 1975*, pp. 1213–1232, Istituto di Arte Mineraria, Università di Cagliari, Cagliari, Italy.

Eischens, R. P. (1960), Chemisorption and catalysis, in *The Surface Chemistry of Metals and Semiconductors*, Symposium, Columbus, Ohio, H. G. Gatos, J. W. Faust, Jr., and W. J. Lafleur, eds., Wiley, New York, pp. 421–438.

Eischens, R. P., and Pliskin, W. A. (1958), The infrared spectra of adsorbed molecules, *Adv. Catal.* **10**, 1–56.

Eisenberg, D., and Kauzmann, W. (1969), *The Structure and Properties of Water*, Clarendon Press, Oxford.

Eisenberg, R. (1970), Structural systematics of 1, 1- and 1, 2-dithiolato chelates, in *Progress in Inorganic Chemistry*, Vol. 12, S. J. Lippard, ed., Interscience, New York, pp. 295–369.

Eitel, W. (1964), *Silicate Science*, Vol. 1, *Silicate Structures*, Academic Press, New York.

Eitel, W. (1975), *Silicate Science*, Vol. 6, *Silicate Structures and Dispersoid Systems*, Academic Press, New York.

Ekwall, P. (1962), Properties and structures of systems containing association colloids, in *First Scandinavian Symposium on Surface Activity, Finland 1962*, P. Ekwall and K. Fontell, eds., pp. 58–89.

Ekwall, P., Danielsson, I., and Stenius, P. (1972), Aggregation in surfactant systems, in *Surface Chemistry and Colloids*, M. Kerker, ed., Physical Chemistry Series One, Vol. 7, consultant ed. A. D. Buckingham, Butterworths, London, and University Park Press, Baltimore, pp. 97–145.

Elworthy, P. H., and Mysels, K. J. (1966), The surface tension of sodium dodecylsulfate solutions and the phase separation model of micelle formation, *J. Colloid Interface Sci.* **21**, 331–347.

Elworthy, P. H., Florence, A. T., and Macfarlane, C. B. (1968), *Solubilization by surface active agents*, Chapman and Hall, London.

Epstein, M. B., Ross, J., and Jakob, C. W. (1954a), The observation of foam drainage transitions, *J. Colloid Sci.* **9**, 50–59.

Epstein, M. B., Wilson, A. Jakob, C. W., Conroy, L. E., and Ross, J. (1954b), Film drainage transition temperatures and phase relations in the system sodium lauryl sulphate, lauryl alcohol and water, *J. Phys. Chem.* **58**, 860–864.

Erberich, G. (1960), Zusammenwirken von Sammlern und Schäumern bei der Flotation sulfidischer Erze, *Erzmetall* **14**(2), 73–76.

Erdey-Gruz, Tibor (1974), *Transport Phenomena in Aqueous Solutions*, Adam Hilger, London.

Eucken, A. (1948), Assoziation in Flussigkeiten, *Z. Elektrochem.* **52**, 255–269.

Evans, L. F. (1954), Bubble–mineral attachment in flotation, *Ind. Eng. Chem.* **46**, 2420–2424.

Evans, L. F., and Ewers, W. E. (1953), The process of bubble–mineral attachment, in *Recent Developments in Mineral Dressing*, Proceedings of the First International Mineral Processing Congress, London, 1952, IMM, London, pp. 457–464.

Evans, L. F., Ewers, W. E., and Meadows, F. (1962), The flotation of cassiterite, *Aust. J. Appl. Sci.* **13**(2), 113–146.

Evans, R. C. (1966), *An Introduction to Crystal Chemistry*, Cambridge University Press, Cambridge, England.

Everett, D. H. (1959), *An Introduction to the Study of Chemical Thermodynamics*, Longmans, Green, New York.

Everett, D. H. (1967), Adsorption hysteresis, in *The Solid–Gas Interface*, E. A. Flood, ed., Marcel Dekker, New York.

Everett, D. H. (1972), Appendix II. Definitions, terminology and symbols in colloid and surface chemistry (adopted by IUPAC), reprinted from *Pure Appl. Chem.* **31**(4), 577–638.

Everett, D. H., ed. (1973, 1975), *Colloid Science*, Vol. I (1973), Vol. II (1975), Chemical Society, London.

Everett, D. H., and Ottewill, R. H., eds. (1970), *Proceedings of the International Symposium on Surface Area Determination*, 1969, Butterworths, London.

Evrard, L. and DeCuyper, J. (1975), Flotation of copper–cobalt oxide ores with alkylhydroxamates, in *Proceedings of the XIth International Mineral Processing Congress, Cagliari, 1975*, Istituto di Arte Mineraria, Università di Cagliari, Cagliari, Italy, pp. 655–670.

Evtushenko, N. P. (1965), Infrared spectra and the structure of metal diethylthiocarbamates (Na^+, Pb^{++}, Bi^{3+}, Cd^{++}, Zn^{++}, Cu^{++}), *Ukr. Khim. Zh.* **31**(6), 618–620 (*Chem. Abstr.* **63**, 7770).

Exerova, D., and Sheludko, A. (1964), Black spots and foam stability, *Izv. Inst. Fizikokhim. Bulgar. Akad. Nauk* **4**, 175–183.

Exerova, D., Khristov, Khr., and Penev, I. (1976), Some techniques for the investigation of foam stability, in *Foams*, R. J. Akers, ed., Academic Press, London, New York.

Eyring, E. M., and Wadsworth, M. E. (1956), Differential infrared spectra of adsorbed monolayers-*n*-hexanethiol on zinc minerals, *Trans. AIME* **205**, 531–535.

Falk, M., and Giguère, P. (1957), Infrared spectrum of the H_3O^+ ion in aqueous solutions, *Can. J. Chem.* **35**, 1195.

Farnand, J. R., Smith, H. M., and Puddington, I. E. (1961), Spherical agglomeration of solids in liquid suspension, *Can. J. Chem. Eng.* **39**, 94–97.

Farnsworth, H. E., and Madden, H. H. (1961), Mechanism of chemisorption, place exchange and oxidation on a (100) nickel surface, *J. Appl. Phys.* **32**, 1933–1937.

Faviani, A., Costa, M., and Bordi, S. (1973), Surface potential and nonionic surfactants, *Electroanal. Chem. Interfac. Electrochem.* **47**, 147–154.

Fieser, F., and Fieser, M. (1961), *Advanced Organic Chemistry*, Reinhold, New York.

Finar, I. L. (1959), *Organic Chemistry*, Vol. 1 and 2, Longmans, London.

Finch, J. A., and Lyman, G. J. (1976), The Marangoni effect in oil droplet/solid attachment, *J. Colloid Interface Sci.* **56**(1), 181–183.

Finch, J. A., and Smith, G. W. (1972), Dynamic surface tension of alkaline dodecylamine acetate solutions in oxide flotation, *Trans. IMM* **81**, C213–218.

Finch, J. A., and Smith, G. W. (1973), Dynamic surface tension of alkaline dodecyclamine solutions, *J. Colloid Interface Sci.* **45**, 81–91.

Finch, J. A., and Smith, G. W. (1975), Bubble–solid attachment as a function of bubble surface tension, *Can. Metal. Q.* **14**(1), 47–51.

Finch, J. A., and Smith, G. W. (1979), Contact angle and wetting, *Miner. Sci. Eng.* **11**(1), 36–63.

Finkelstein, N. P. (1967), Kinetic and thermodynamic aspects of the interaction between potassium ethyl xanthate and oxygen in aqueous solution, *Trans. IMM* **76**, C51–C59.

Finkelstein, N. P. (1968), The Bathochromic shift of the 381nm absorption peak in aqueous solutions of potassium ethyl xanthate containing dodecytrimethyl ammonium bromide, NIM Report No. 437, Johannesburg, South Africa.

Finkelstein, N. P. (1969a), Influence of alkyltrimethyl ammonium halides on the stability of potassium ethyl xanthate in aqueous solution, *J. Appl. Chem.* **19**, 73–76.

Finkelstein, N. P. (1969b), Quantitative Aspects of the interaction between galena, xanthate, and oxygen at a pH value of 10, NIM Report No. 527, Johannesburg, South Africa.

Finkelstein, N. P., and Allison, S. (1971), Studies of the stability of xanthates and dixanthogen in aqueous solution and the chemistry of iron xanthates and cobalt xanthates, Report No. 1125, National Institute for Metallurgy, Johannesburg, South Africa, 5 March 1971.

Finkelstein, N. P., and Allison, S. A. (1976), The chemistry of activation, dewatering and depression in the flotation of zinc sulphide, A review in Flotation, A. M. Gaudin Memorial Volume, ed. M. C. Fuerstenau, ed., Vol. I, AIME, New York, pp. 414–457.

Finkelstein, N. P., and Poling, G. W. (1977), The role of dithiolates in the flotation of sulphide minerals, *Miner. Sci. Eng.* **9**, 177–197.

Fisher, B. B., and McMillan, W. G. (1958), Transitions in adsorbed monolayers. I. Experimental, *J. Chem. Phys.* **28**, 549–571.

Fleming, M. G. (1952), Effects of alkalinity in the flotation of lead minerals, *Trans. AIME* **193**, 1231–1236.

Fleming, M. G. (1953), Effects of soluble sulphide in the flotation of secondary lead minerals, in *Recent Developments in Mineral Dressing*, Proceedings of the First International Mineral Processing Congress, London, 1952, IMM, London, pp. 521–528.

Fleming, M. G. (1957), Flotation of vanadium ore from the Abenab West Mine of the South West Africa Co., in *Extraction and Refining of the Rarer Metals*, Proceedings of the Symposium, London, 1956, IMM, London, pp. 212–239.

Flint, L. R. (1973), Factors influencing the design of flotation equipment, *Miner. Sci. Eng.* **5**(3), 232–241.

Flint, L. R., and Howarth, W. J. (1971), The collision efficiency of small particles with spherical air bubbles, *Chem. Eng. Sci.* **26**, 1155–1168.

Flood, E. A., (1967), *The Solid–Gas Interface*, Vols. 1 and 2, Marcel Dekker, New York.

Folman, M., and Yates, D. J. C. (1958), Expansion–contraction effects in rigid adsorbents at low coverages, *Trans. Faraday Soc.* **54**, 429–440.

Ford-Smith, M. H. (1964), *The Chemistry of Complex Cyanides*, Her Majesty's Stationary Office, London.

Forslind, E., and Jacobsson, A. (1975), Clay–water systems, in *Water, a Comprehensive Treatise*, Vol. 5, ed. F. Franks, pp. 173–248, Plenum Press, New York.

Foss, O. (1950), Prevalence of unbranched sulfur chains in polysulfides and polythionic compounds, *Acta Chem. Scand.* **4**, 404–415.

Foss, O. (1961), Stereochemistry of disulfides and polysulfides, in *Organic Sulphur compounds*, Vol. I, N. Kharasch, ed., Pergamon Press, New York, pp. 75–82.

Fowkes, F. M. (1962), Determination of interfacial tensions, contact angles, and dispersion forces in surfaces by assuming additivity of intermolecular interactions in surfaces, *J. Phys. Chem.* **66**, 382.

Fowkes, F. M. (1964), Attractive forces at interfaces, *Ind. Eng. Chem.* **56**(12), 40–52.

Fowkes, F. M., ed. (1969), *Hydrophobic Surfaces*, Academic Press, New York.

Fowler, R. H. (1935), A statistical derivation of Langmuir's adsorption isotherm, *Proc. Camb. Philos. Soc.* **31**, 260–264.

Frank, H. S., and Wen, W. Y. (1957), III. Ion–solvent interaction. structural aspects of ion–solvent interaction in aqueous solutions: A suggested picture of water structure, *Discuss. Faraday Soc.* **24**, 133–140.

Franks, F., ed. (1972–1975, etc.), *Water, a Comprehensive Treatise*, Vols. I–V, Plenum Press, New York.

Franks, F. (1975a), The hydrophobic interaction, in *Water, A Comprehensive Treatise*, Vol. 4, F. Franks, ed., Plenum Press, New York, pp. 1–94.

Franks, F. (1975b), The hydrophobic interactions, in *Water, A Comprehensive Treatise*, Vol. 4, F. Franks, ed., Plenum Press, New York, pp. 1–94.

Franks, F., and Reid, D. S. (1973), Thermodynamic properties, in *Water, A Comprehensive Treatise*, Vol. 2, F. Franks, ed., Plenum Press, New York, pp. 323–380.

Franzini, M. (1963), The crystal structure of nickelous xanthate, Zeitschrift für Kristallographie **118**, 393–403.

Fraser, K. A., and Harding, M. M. (1967), The structure of bis(*N,N*-dimethyldithiocarbamato)pyridine zinc, *Acta. Crystallogr.* **22**, 75–81.

Freudenberg, K. (1966), Analytical and biochemical background of a constitutional scheme of lignin, in *Lignin Structure and Reactions*, Advances in Chemistry Series No. 59, R. F. Gould, ed., pp. 1–21, American Chemical Society, Washington, D.C.

Freund, H., ed. (1966), *Applied Ore Microscopy, Theroy and Technique*, Collier-MacMillan, New York.

Friend, J. P., and Kitchener, J. A. (1973). Some physico-chemical aspects of the separation of finely-divided minerals by selective flocculation, *Chem. Eng. Sci.* **28**, 1071–1080.

Frumkin, A. N. (1926a), Significance of the electrocapillary curve, *Colloid Symp. Ann.* **7**, 89–104 [*Chem. Abst.* **24**, 3150 (1930)].

Frumkin, A. N. (1926b), Über die Beeinflussung der Adsorption von Neutralmolekulen durch ein elektrisches Feld, *Z. Phys.* **35**, 792–802.

Frumkin, A. N. (1933), Ion adsorption on metals and charcoal, *Phys. Z. Sowjetunion* **4**, 239–261.

Frumkin, A. N. (1938). On the phenomena of wetting and the adhesion of bubbles. 1, *Zh. Fiz. Khim.* **12**, 337–345.

Frumkin, A. N. (1955), Adsorptionserscheinungen und Elektrochemische Kinetik, *Z. Elektrochem.* **59**, 807–822.

Frumkin, A. N. (1957), Adsorption of organic compounds on metal–electrolyte solution interfaces: its effect on electrochemical processes, *Nova Acta Leopoldina Neue Folge* **19**(132), 1–19.

Frumkin, A. N., and Damaskin, B. B. (1964), Adsorption of organic compounds at electrodes, in *Modern Aspects of Electrochemistry*, Vol. 3, J. O'M Bockris and B. E. Conway, eds., Butterworths, London, pp. 149–223.

Frumkin, A. N., and Gorodetzkaya, A. (1938), On the phenomena of wetting and the adhesion of bubbles. II. The mechanism of the adhesion of bubbles to a mercury surface, *Acta Physicochim. USSR* **9**, 327–340.

Frumkin, A. N., Gorodetskaya, A., Kabanov, B., and Nekrasov, N. (1932), Electrocapillary phenomena and the wetting of metals by electrolytic solutions. I. *Phys. Z. Sowjetunion* **1**, 255–284.

Fuerstenau, D. W. (1956), Measuring zeta potentials by streaming potential techniques, *Trans. AIME* **205**, 834–836.

Fuerstenau, D. W. (1957), Correlation of contact angles, adsorption density, zeta potentials and flotation rate, *Trans. AIME* **208**, 1365–1367.

Fuerstenau, D. W., ed. (1962), *Froth-Flotation, 50th Anniversary Volume*, AIME, New York.

Fuerstenau, D. W., and Fuerstenau, M. C. (1956), Ionic size in flotation collection of alkali halides, *Trans. AIME* **205**, 302–307.

Fuerstenau, D. W., and Healy, T. W. (1972), Principles of Mineral Flotation, in *Adsorptive Bubble Separation Techniques*, R. Lemlich, ed., Academic Press, New York, pp. 91–121.

Fuerstenau, D. W., and Raghavan, S. (1976), Some aspects of the thermodynamics of flotation, in *Flotation, A. M. Gaudin Memorial Volume*, Vol. I, M. C. Fuerstenau, ed., AIME, New York, pp. 21–65.

Fuerstenau, D. W., and Wayman, C. H. (1958), Effect of chemical reagents on the motion of single air bubbles in water, *Trans. AIME* **211**, 694–699.

Fuerstenau, D. W., and Yamada, B. J. (1962), Neutral molecules in flotation collection, *Trans. AIME* **223**, 50–52.

Fuerstenau, D. W., Gaudin, A. M., and Miaw, H. L. (1958), Iron oxide slime coatings in flotation, *Trans. AIME* **211**, 792–795.

Fuerstenau, D. W., Healy, T. W., and Somasundaran, P. (1964), The role of hydrocarbon chain of alkyl collectors in flotation, *Trans. AIME* **229**, 321–325.

Fuerstenau, M. C. (1975), Role of metal ion hydrolysis in oxide and silicate flotation systems, in *Advances in Interfacial Phenomena of Particulate/Solution/Gas Systems*; *Applications to Flotation Research*, P. Somasundaran and R. B. Grieves, eds., AIChE Symposium Series No. 150, Vol. **71**, pp. 16–23.

Fuerstenau, M. C., ed. (1976), *Flotation, A. M. Gaudin Memorial Volume*, Vols. I and II, AIME, New York.

Fuerstenau, M. C., and Atak, S. (1965), Lead activation in sulfonate flotation of quartz, *Trans. AIME* **232**, 24–28.

Fuerstenau, M. C., and Cummins, Jr., W. F. (1967), The role of basic aqueous complexes in anionic flotation of quartz, *Trans. AIME* **238**, 196–200.

Fuerstenau, M. C., and Miller, J. D. (1967), The role of the hydrocarbon chain in anionic flotation of calcite, *Trans. AIME* **238**, 153–160.

Fuerstenau, M. C., and Palmer, B. R. (1976), Anionic flotation of oxides and silicates, in *Flotation, A. M. Gaudin Memorial Volume*, M. C. Fuerstenau, ed., AIME, New York, pp. 148–196.

Fuerstenau, M. C., and Rice, D. A. (1968), Flotation Characteristics of Pyrolusite, *Trans. AIME* **241**, 453–457.

Fuerstenau, M. C., Martin, C. C., and Bhappu, R. B. (1963), The role of hydrolysis in sulphonate flotation of quartz, *Trans. AIME* **226**, 449–454.

Fuerstenau, M. C., Miller, J. D., Pray, R. E., and Perinne, B. F. (1965), Metal ion activation in xanthate flotation of quartz, *Trans. AIME* **232**, 359–365.

Fuerstenau, M. C., Kuhn, M. C., and Elgillani, D. A. (1968), The role of dixanthogen in xanthate flotation of pyrite, *Trans. AIME* **241**, 148–156.

Fuerstenau, M. C., Elgillani, D. A., Miller, J. D. (1970), Adsorption mechanisms in nonmetallic activation systems. *Trans. AIME* **247**, 11–14.

Fuerstenau, M. C., Huiatt, J. L., and Kuhn, M. C. (1971), Dithiophosphate vs. xanthate flotation of chalcocite and pyrite, *Trans. AIME* **250**, 227–231.

Fuoss, R. M. (1934), Distribution of ions in electrolytic solutions, *Trans. Faraday Soc.* **30**, 967–980.

Gaines, G. L. (1966), *Insoluble Monolayers at Liquid/Gas Interfaces*, Interscience, New York.

Gammage, R. B., and Glasson, D. R. (1976), The effect of grinding on the polymorphs of calcium carbonate, *J. Colloid Interface Sci.* **55**(2), 396–401.

Garbacik, J., Najbar, J., and Pomianowski, A. (1972), Kinetics of reaction of xanthate with hydrogen peroxide, *Roczniki Chemii* **46**, 85–97.

Gardner, J. R., and Woods, R. (1973), The use of a particulate bed electrode for the electrochemical investigation of metal and sulphide flotation. *Aust. J. Chem.* **26**, 1635–1644.

Garner, F. G., and Hammerton, D. (1954), Circulation inside gas bubbles, *Chem. Eng. Sci.* **3**(1), 1–11.

Garrels, R. M., and Christ, C. L. (1965), *Solutions, Minerals and Equilibria*, Harper and Row, New York.

Gasparri, G. F., Nardelli, M., and Villa, A. (1967), The crystal and molecular structure of nickel bis(dithiocarbamate), *Acta Crystallogr.* **23**, 384–391.

Gatos, H. C. ed. (1960), *The surface chemistry of metals and semiconductors*, Proceedings of the symposium on the above topic held in Columbus, Ohio, October 1959, John Wiley & Sons, New York.

Gatos, H. C. (1962), Crystalline structure and surface reactivity, *Science* **137**, 311–322.

Gatos, H. C. ed. (1964), *Solid Surfaces*, Proceedings of the International Conference on The Physics and Chemistry of Solid Surfaces, held at Brown University, Providence, June 1964, North-Holland Publ. Co., Amsterdam.

Gauci, G., and Cusack, B. L. (1971), Metallurgical effects associated with the use of cemented backfill at the Zinc Corporation Ltd. and New Broken Hill Consolidated Ltd., *Australas. Institute of Mining and Metallurgy, Proceedings No. 237*. March, 33–40.

Gaudin, A. M. (1928), Flotation mechanism, a discussion of the functions of flotation reagents, *Trans. AIME* **79**, 50–77.

Gaudin, A. M. (1957), *Flotation*, McGraw-Hill, New York.

Gaudin, A. M., and Fuerstenau, D. W. (1955), Streaming potential studies, quartz flotation with anionic collectors, *Trans. AIME* **202**, 66–72.

Gaudin, A. M., and Schuhmann, R. (1936), The action of potassium *n*-amyl xanthate on chalcocite, *J. Phys. Chem.* **40**, 257–275.

Gaudin, A. M., Groh, J. O., and Henderson, H. B. (1931), Effect of particle size on flotation, *AIME Tech. Publ.* **414**, 3–23.

Gaudin, A. M., Dewey, F., Duncan, W. E., Johnson, R. A., and Tangel, O. F. (1934), Reactions of xanthates with sulphide minerals, *Trans. AIME* **112**, 319–347.

Gaudin, A. M., Schuhmann, R., Jr., and Brown, E. G. (1946), Making tin flotation work—I. Canutillos Ore., *Eng. Min. J.* **147**(10), 54–59.

Gaudin, A. M., Fuerstenau, D. W., and Miaw, H. L. (1960), Slime coatings in galena flotation, *Trans. CIM* **63**, 668–671.

Geisee, O.W. (1942), Annales del Primer Congreso Panamericano de Ingeniera de Minas y Geologia, Vol. 4, p. 1743, Santiago, Chile.

Gerbacia, W., and Rosano, H. L. (1973), Microemulsions: formation and stabilization, *J. Colloid Interface Sci.* **44**, 242–248.

Gerischer, H. (1961), Semiconductor electrode reactions, in *Advances in Electrochemistry and Electrochemical Engineering*, Vol. 1, P. Delahay and C. W. Tobias, eds. Interscience, New York, pp. 139–232.

Gerischer, H. (1970), Semiconductor electrochemistry, in *Physical Chemistry*, Vol. IXA, *Electrochemistry*, H. Eyring, ed., Academic Press, New York, Chap. 5, pp. 463–542.

Gerischer, H., and Mindt, H. (1968), The mechanisms of the decomposition of semiconductors by electrochemical oxidation and reduction, *Electrochim. Acta* **13**, 1329–1341.

Gerischer, H., Kolb, D. M., and Sass, J. K. (1978), The study of solid surfaces by electrochemical methods, *Adv. Phys.* **27**(3), 437–498.

Germer, L. H., and MacRae, A. U. (1962a), Adsorption of hydrogen on a (110) nickel surface, *J. Chem. Phys.* **37**, 1382–1386.

Germer, L. H., and MacRae, A. U. (1962b), Oxygen–nickel structures on the (110) face of clean nickel, *J. Appl. Phys.* **33**, 2923–2932.

Gibbs, W. (1878, 1928), *Collected Works*, Vol. I, *Thermodynamics, VII. Electrochemical Thermodynamics*, Yale University Press, New Haven, Connecticut.

Gierst, L. (1958), Cinetique d'approche et reactions d'electrodes irreversibles, thesis, University of Bruxelles.

Gileadi, E., Kirowa-Eisner, E., and Penciner, J. (1975), *Interfacial Electrochemistry, An Experimental Approach*, Addison-Wesley, Reading, Massachusetts.

Giles, C. H., MacEwan, T. H., Nakhwa, S. N., and Smith, D. (1960), Studies in adsorption isotherms and its use in diagnosis of adsorption mechanisms and in measurement of specific surface areas of solids. *J. Chem. Soc.*, 3973–3993.

Gillespie, R. J., and Nyholm, R. S. (1957), Inorganic stereochemistry, *Q. Rev. Chem. Soc. London* **11**, 339–380.

Gilmour, A., Nelson, S. M., and Pink, R. C. (1953), Dipole moments of metal oleates, *Nature* **171**, 1075.

Girifalco, L. A., and Good, R. J. (1957), A theory for the estimation of surface and interfacial energies. I. Derivation and application to interfacial tension, *J. Phys. Chem.* **61**, 904–909.

Gjostein, N. A. (1967), Surface self-diffusion in FCC and BCC metals: a comparison of theory and experiment, in *Surfaces and Interfaces I*, J. J. Burke, N. L. Reed, and V. Weiss, eds. Syracuse University Press, Syracuse, New York.

Glembotsky, V. A. (1953), The time of attachment of bubbles to solid particles in flotation and its measurement, *Izv. Akad. Nauk USSR Otdel. Tekhn. Nauk*, 1524–1531.

Glembotsky, V. A., and Bechtle, G. A. (1964), *Flotation of iron ores*, Nedra, Moscow.

Glembotsky, V. A., Klassen, V. I., and Plaksin, I. N. (1963), *Flotation*, Engl. transl., Primary Sources, New York.

Glembotsky, V. A., Dimitrieva, G. N., and Sorokin, M. M. (1968), Issledovanie deistvija flotatsionnykh reagentov, Nauka, Moscow; translated as *Flotation Agents and Effects* by the Israel Program for Scientific Translations, Jerusalem, 1970.

Goates, J. R., and Hatch, C. V. (1953), Standard adsorption potentials of water vapour on soil colloids, *Soil Sci.* **75**, 275–278.

Goddard, E. D. (1974), Ionizing monolayers and pH effects, *Advances in Colloid and Interface Science*, **4**(1), 45–78.

Goddard, E. D., ed. (1975), *Monolayers*, Advances in Chemistry Series No. 144, American Chemical Society, Washington, D.C.

Goddard, E. D., and Ackilli, J. A. (1963), Monolayer properties of fatty acids, *J. Colloid Sci.* **18**, 585–595.

Goddard, E. D., and Schulman, J. H. (1953), Molecular interaction in monolayers. Part I. Complex formation, Part II. Steric effects in the nonpolar portion of the molecules, *J. Colloid Sci.* **8**, 309–328; 329–340.

Goddard, E. D., Goldwasser, S., Golikeri, G., and Kung, H. C. (1968), Molecular association in fatty acid–potassium soap systems, in *Molecular Association in Biological and Related Systems*, Advances in Chemistry Series No. 84, R. F. Gould, ed., American Chemical Society, Washington, D.C.

Goldfinger, G., ed. (1970), *Clean Surfaces* (based on a symposium at North Carolina State University at Raleigh), Marcel Dekker, New York.

Goldstick, T. K. (1959), Electrochemical investigations of ethyl xanthate, M.Sc. thesis, Massachusetts Institute of Technology.

Golikov, A. A., and Nagirnyak, F. I. (1961), Catalytic oxidation of xanthates in aqueous solution in the presence of sulphide minerals, *Tsvet. Metally.* **34**(4), 9–11.

Golovanov, G. A., Zhelnin, V. S., Kotilevsky, V. I., and Makarov, A. M., (1968), Processing of apatite ores at the ore-dressing plants of "Apatite" complex, in *Proceedings of the Eighth International Mineral Processing Congress, Leningrad 1968*, pp. 489–502, Mekhanobr Institute, Leningrad.

Good, R. J., and Elbing, E. (1971), Generalisation of theory for estimation of interfacial energies, in *Chemistry and Physics of Interfaces II*, American Chemical Society, Washington, D.C., pp. 71–96.

Good, R. J., and Girifalco, L. A. (1960), A theory for estimation of surface and interfacial energies of solids from contact angle data, *J. Phys. Chem.* **64**, 561–565.

Good, R. J., and Koo, M. N. (1979), The effect of drop size on contact angle, *J. Coll. Interface Sci.* **71**(2), 283–292.

Good, R. J., and Stromberg, R. R., eds. (1979), *Colloid and Surface Science*, Vol. 11, Plenum Press, New York.

Good, W. R. (1973), A comparison of contact angle interpretations, *J. Coll. Interface Sci.* **41**(1), 63–71.

Goodman, R. H., and Trahar, W. J. (1977), Flotation of cassiterite at the Renison Tin Mine, Renison Bell, Australia, presented at the International Tin Symposium, La Paz, Bolivia, November 1977.

Goodrich, F. C. (1957), Molecular interaction in mixed monolayers, in *Proceedings of the 2nd International Congress of Surface Activity*, Vol. I, J. H. Schulman, ed., Butterworths, London, pp. 85–91.

Goodrich, F. C. (1969), The thermodynamics of fluid interfaces, in *Surface and Colloid Science*, Vol. 1, E. Matijevic, ed., Wiley-Interscience, New York, pp. 1–38.

Goold, L. A. (1972), The reaction of sulphide minerals with thiol compounds, NIM Report No. 1439, National Institute for Metallurgy, Johannesburg, South Africa.

Goold, L. A., and Finkelstein, N. P. (1969), An infrared investigation of the potassium alkyl xanthate and alkyltrimethylammonium bromide mixed-collector system, NIM Report 498, Johannesburg, South Africa.

Gorodetskaya, A. (1949), The rate of rise of bubbles in water and aqueous solutions at great Reynolds numbers, *Zh. Fiz. Khim* **23**, 71–77.

Gorodetskaya, A., and Kabanov, B. (1934), Electrocapillary phenomena and wetting of metals by electrolytic solutions, II, *Phys. Z. Sowjetunion* **5**, 418–431.

Gössling, H. H., and McCulloch, H. W. A. (1974), Fluorspar—the rise of a non-metallic mineral, *Miner. Sci. Eng.* **6**(4), 206–222.

Gottardi, G. (1961), The crystal structure of xanthates. II. Antimonious xanthate, *Zeitschrift für Kristallographie* **115**, 451–459.

Gould, R. F., ed. (1964), *Contact Angle, Wettability, and Adhesion*, Advances in Chemistry Series No. 43, American Chemical Society, Washington, D.C.

Gould, R. F. ed. (1966), *Coal Science*, Advances in Chemistry Series No. 55, American Chemical Society, Washington, D.C.

Gouy, G. (1910), Constitution of the Electric Charge at the Surface of an Electrolyte. *J. Phys. (Paris)* **9**(4), 457–467.

Gowans, W. K. (1972), *A Study of Oxygen Effects in Flotation*, Special Report, Cominco Ltd. and Department of Mineral Engineering, University of British Columbia, Vancouver.

Grahame, D. C. (1941), Properties of the electrical double layer at a mercury surface. I. Methods of measurement and interpretation of results, *J. Am. Chem. Soc.* **63**, 1207–1215.

Grahame, D. C. (1947), The electrical double layer and the theory of electrocapillarity, *Chem. Revs.* **41**, 441–501.

Grahame, D. C. (1949), Measurement of the capacity of the electrical double layer at a mercury electrode, *J. Am. Chem. Soc.* **71**, 2975–2978.

Grahame, D. C. (1954), Differential capacity of mercury in aqueous sodium fluoride solutions, I. Effect of concentration at $25\,^\circ$C, *J. Am. Chem. Soc.* **76**, 4819–4823.

Grahame, D. C., and Whitney (1962), *J. Am. Chem. Soc.* **64**, 1548.

Grahame, D. C., Coffin, E. M., Cummings, J. I., and Poth, M. A. (1952), The potential of the electrocapillary maximum of mercury, II, *J. Am. Chem. Soc.* **74**, 1207–1211.

Grainger-Allen, T. J. N. (1970), Bubble generation in froth flotation machines, *Trans. IMM* **79**, C15–22.

Granville, A., Finkelstein, N. P., and Allison, S. A. (1972), Review of reactions in the flotation system galena–xanthate–oxygen, *Trans. IMM* **81**, C1–C30.

Greenler, R. G. (1962), An infrared investigation of xanthate adsorption by lead sulphide, *J. Phys. Chem.* **66**, 879–883.

Gregg, S. J. (1972), The physical adsorption of gases, in *Surface Chemistry and Colloids*, Vol. 7, M. Kerker, ed., Physical Chemistry Series One, consultant ed. A. D. Buckingham, Butterworths, London, and University Park Press, Baltimore, pp. 189–223.

Gregg, S. J., and Sing, K. S. W. (1967), *Adsorption, Surface Area and Porosity*, Academic Press, New York.

Gründer, W. (1955), Tin stone ore dressing, *Z. Erzbergb. Metallhüttenw.* **8**, 152–157.

Gründer, W. (1960), Discussion on developments in the treatment of Malayan tin ores, in *Proceedings of the Fifth International Mineral Processing Congress, London*, 1960, IMM, London, p. 652.

Guggenheim, E. A. (1950), *Thermodynamics*, North-Holland, Amsterdam.

Gundry, P. M., and Tompkins, F. C. (1956), Chemisorption of gases in nickel films, Part I.—Kinetic studies, *Trans. Faraday Soc.* **52**, 1609–1617.

Gurvich, L. G. (1915), Physico-chemical attractive force, *Zh. Russ. Fiz-Khim. Obshchestva Chem.* **47**, 805–827.

Gurvich, L. G. (1923), Physico-chemical attractive forces. III. The dispersion of solids by shaking with liquids, *Kolloid-Zh.* **33**, 321–324.

Gutierrez, C. (1973), The mechanism of flotation of galena by xanthates, *Miner. Sci. Eng.* **5**, 108–118.

Gutman, V. ed. (1967), *Halogen Chemistry*, Vols. 1, 2, and 3, Academic Press, New York.

Gwathmey, A. T., and Lawless, K. R. (1960), The influence of crystal orientation on the oxidation of metals, in *The Surface Chemistry of Metals and Semiconductors*, Symposium, Columbus, Ohio, 1959, H. C. Gatos, ed., J. Wiley & Sons, New York, pp. 483–521.

Haberkorn, R. (1967), thesis, Technische Hochschule, Munich.

Haberman, W. L., and Morton, R. K. (1953), An experimental investigation of the drag and shape of air bubbles rising in various liquids, U.S. Navy Department, David W. Taylor Model Basin, Report. No. 802, September.

Hagihara, H. (1952a), Surface oxidation of galena in relation to its flotation as revealed by electron diffraction, *J. Phys. Chem.* **56**, 610–615.

Hagihara, H. (1952b), Mono- and multilayer adsorption of aqueous xanthate on galena surfaces *J. Phys. Chem.* **56**, 616–621.

Hagihara, H., and Yamashita, S. (1966), The crystal structure of lead ethylxanthate, *Acta Crystallogr.* **21**, 350–358.

Hagihara, H., Watanabe, Y., and Yamashita, S. (1968), The crystal structure of lead *n*-butylxanthate. I. Disordered structure, *Acta Crystallogr.* **B24**, 960–966.

Hagni, R. D. (1978), Ore microscopy applied to beneficiation, *Mining Eng.* October, 1437–1447.

Haig-Smillie, L. D. (1974), Sea water flotation in *Proceedings 6th Annual Meeting of Canadian Mineral Processors*, 1974, EMR, Ottawa, pp. 263–281.

Hahn, H. H., and Stumm, W. (1968), Coagulation by Al(III); The role of adsorption of hydrolyzed aluminum in the kinetics of coagulation, in *Adsorption from Aqueous Solution*, Advances in Chemistry Series No. 79, 91–111, American Chemical Soc., Washington, D.C.

Hair, M. L. (1967), *Infrared Spectroscopy in Surface Chemistry*, Marcel Dekker, New York.

Hall, E. S. (1965), The zeta potential of aluminum hydroxide in relation to water treatment coagulation, *J. Appl. Chem.* **15**, 197–205.

Hall, P. G., and Rose, M. A., (1978), Dielectric properties of water adsorbed by kaolinite clays, *J. Chem. Soc. Faraday Trans. 1* **74**(5), 1221–1233.

Halliwell, H. F., and Nyburg, S. C. (1963), Enthalpy of hydration of the proton, *Trans. Faraday Soc.* **59**, 1126–1140.

Hamaker, H. C. (1937), The London–van der Waals attraction between spherical particles, *Physica (Utrecht)* **4**(10), 1058–1072.

Hamilton, I. C., and Woods, R. (1979), The effect of alkyl chain on the aqueous solubility and redox properties of symmetrical dixanthogens, *Austr. J. Chem.* **32**, 2171–2179.

Hamilton, W. C., and Ibers, J. A. (1968), *Hydrogen Bonding in Solids*, W. A. Benjamin, New York.

Hanna, H. S., and Somasundaran, P. (1976), Flotation of salt-type minerals, in *Flotation*, *A. M. Gaudin Memorial Volume*, Vol. I, M. C. Fuerstenau, ed., AIME, New York, pp. 197–272.

Hannay, N. B., and Smyth, C. P. (1946), The dipole moment of hydrogen fluoride and the ionic character of bonds, *J. Am. Chem. Soc.* **68**, 171–173.

Hansen, R. S., and Mann, J. A. (1964), Propagation characteristics of capillary ripples. I. The theory of velocity dispersion and amplitude attenuation. *J. Appl. Phys.* **35**, 152–161.

Hansen, R. J., and Toong, T. Y., (1971), Dynamic contact angle and its relationship to forces of hydrodynamic origin, *J. Colloid Interface Sci.* **37**(1), 196–207.

Hansen, R. S., and Wallace, T. C. (1959), The kinetics of adsorption of organic acids at the water–air interface, *J. Phys. Chem.* **63**, 1085–1091.

Hansen, R. S., Kelsh, D. J., and Grantham, D. H. (1963), The interference of adsorption from differential double layer capacitance measurements. II. Depencence of surface charge density on organic nonelectrolyte surface excess, *J. Phys. Chem.* **67**, 2316–2326.

Hansen, R. S., Lucassen, J., Bendure, R. L., and Bierwagen, G. P. (1968), Propagation characteristics of interfacial ripples, *J. Colloid Interface Sci.* **26**, 198–208.

Harbaum, K. L., and Houghton, G. (1960), Effects of sonic vibrations on the rate of absorption of gases from bubble beds, *Chem. Eng. Sci.* **13**, 90–92.

Harkins, W. D. (1952), *The Physical Chemistry of Surface Films*, Reinhold, New York.

Harkins, W. D., and Boyd, E. (1941), The states of monolayers, *J. Phys. Chem.* **45**, 20–43.

Harkins, W. D., and Mittelmann, R. (1949), X-Ray investigations of the structure of colloidal electrolytes. IV. A new type of micelle formed by film penetration, *J. Colloid Sci.* **4**, 367–381.

Harris, C. C. (1976), Flotation machines, in *Flotation, A. M. Gaudin Memorial Volume*, Vol. 2, M. C. Fuerstenau, ed., AIME, New York, pp. 753–815.

Harris, C. C., Jowett, A., and Morrow, N. R. (1964), Effect of contact angle on the capillary properties of porous masses, *Trans IMM* **73**, 335–351.

Harris, C. C., and Lepetic, V. (1966), Flotation cell design, *Mining Eng.*, September, 67–72.

Harris, P. J., and Finkelstein, N. P. (1975), Interactions between sulphide minerals and xanthates. The formation of nonothiocarbonate at galena and pyrite surfaces, *Int. J. Miner. Process.* **2**, 77–100.

Hartley, G. S. (1936), *Aqueous Solutions of Paraffin-Chain Salts*, Hermann et Cie, Paris.

Hassialis, M. D., and Myers, C. G. (1951), Collector mobility and bubble contact, *Min. Eng.* **3**, 961–968.

Hasted, J. B. (1972), Liquid water: Dielectric properties, in *Water, A Comprehensive Treatise*, Vol. I, F. Franks, ed., Plenum Press, New York, pp. 255–309.

Hawsley, J. R. (1972), *Use, Characteristics and Toxicity of Mine Mill Reagents*, Ministry of Environment, Ontario, Canada.

Hayakawa, T. (1957), The low temperature adsorption of non-polar gases on octahedral potassium chloride, *Bull. Chem. Soc. Japan* **30**, 343–349.

Haydon, D. A. (1964), The electrical double layer and electrokinetic phenomena, in

Recent Progress in Surface Science, Vol. 1, J. F. Danielli, K. G. A. Pankhurst, and A. C. Riddiford, eds., Academic Press, New York, pp. 94–158.

Haynman, V. J. (1975), Fundamental model of flotation kinetics, in *Proceedings of the XI International Mineral Processing Congress, Cagliari*, 1975, Istituto di Arte Mineraria, Università di Cagliari, Cagliari, Italy, pp. 537–560.

Healy, T. W., and Moignard, M. S. (1976), A review of electrokinetic studies of metal sulphides, in *Flotation, A. M. Gaudin Memorial Volume*, Vol. I, M. C. Fuerstenau, ed., AIME, New York, pp. 275–297.

Healy, T. W., James, R. O., and Cooper, R. (1968), The adsorption of aqueous Co(II) at the silica–water interface, in *Adsorption from Aqueous Solution*, R. F. Gould, ed., Advances in Chemistry Series No. 79, American Chemical Society, Washington, D.C., pp. 67–73.

Healy, T. W., Wiese, G. R., Yates, D. E., and Kavanagh, B. V. (1973), Heterocoagulation in mixed oxide colloidal dispersions, *J. Colloid Interface Sci.* **42**, 647–649.

Heinrich, E. Wm. (1965), *Microscopic identification of minerals*, McGraw-Hill, New York.

Heitler, W., and London, F. (1927), Interaction of neutral atoms and homopolar binding according to the quantum mechanics, *Z. Phys.* **44**, 455–472.

Helmholtz, H. L. F. (1879), *Wiss. Abh. Phys. Tech. Reichsanst.* **1**, 925.

Henning, A. (1958), A critical survey of volume and surface measurement in microscopy, Zeiss Werkzeitschrift No. 30, 1958.

Hepel, T., and Pomianowski, A. (1973), Thermodynamics of Cu–xanthate–H_2O flotation system, in *Physico-Chemical Problems of Mineral Processing*, No. 7 (in Polish), p. 43, Technical University, Wroclaw.

Hepel, T., and Pomianowski, A. (1977), Diagrams of electrochemical equilibria of the system copper–potassium ethylxanthate–water at 25°C, *Int. J. Miner. Process.* **4**(4), 345–361.

Hergt, H. F. A., Rogers, J., and Sutherland, K. L. (1946), Principles of flotation—Flotation of cassiterite and associated minerals, *Trans. AIME* **169**, 448–465.

Herring, C. (1951), Some theorems on the free energies of crystal surfaces, *Phys. Rev.* **82**, 87–93.

Herzberg, G. (1944), *Atomic Spectra and Atomic Structure*, Dover, New York.

Heyes, G. W., and Trahar, W. J. (1977), The natural floatability of chalcopyrite, *Int. J. Miner. Process.* **4**, 317–344.

Hiemenz, P. C. (1977), *Principles of Colloid and Surface Chemistry*, Marcel Dekker, New York.

Hiemstra, S. A. (1969), Photomicrography, *Miner. Sci. Eng.* **1**(2), 34–50.

Hildebrand, Joel, H. (1916), Principles underlying flotation, *Min. Sci. Press* **113**, 168–170.

Hill, T. L. (1960), *An Introduction to Statistical Thermodynamics*, Addison-Wesley, Reading, Massachusetts, pp. 140–143.

Hill, T. L. (1963), *Thermodynamics of Small Systems*, Part 1, W. A. Benjamin, New York.

Hill, T. L. (1968), *Thermodynamics for Chemists and Biologists*, Addison-Wesley, Reading, Massachusetts, Chap. 7.

Hines, P. R. (1962), Before flotation, in *Froth Flotation—50th Anniversary Volume*, D. W. Fuerstenau, ed., AIME, New York, pp. 5–10.

Hines, P. R., and Vincent, J. D. (1962), The early days of froth flotation, in *Froth Flotation—50th Anniversary Volume*, D. W. Fuerstenau, ed., AIME, New York, pp. 11–38.

Hirschfelder, J. O., Curtiss, C. F., and Bird, R. B. (1954), *Molecular Theory of Gases and Liquids*, Wiley, New York, p. 967.

Hirth, J. P., and Lothe, J. (1967), *Theory of Dislocations*, McGraw-Hill, New York.

Ho, F. C., and Conway, B. E. (1978), "Electrochemical behavior of the surface of lead sulfide crystals—as revealed by potentiodynamic, reflectance, and rotating-electrode studies", *J. Coll. Inerface Stci.* **65**, 19–35.

Hoerr, C. W., and Ralston, A. W. (1943), Studies on high molecular weight aliphatic amines and their salts. XI. Transference numbers of some primary amine hydrochlorides in aqueous solution and their significance in the interpretation of the micelle theory, *J. Am. Chem. Soc.* **65**, 976–983.

Hoerr, C. W., McCorkle, M. R., and Ralston, A. W. (1943), Studies on high molecular weight aliphatic amines and their salts. X. Ionization constants of primary and symmetrical secondary amines in aqueous solution, *J. Am. Chem. Soc.* **65**, 328–329.

Hoffman, E. J., Boyd, G. E., and Ralston, A. W., (1942), Studies on high molecular weight aliphatic amines and their salts. V. Soluble and insoluble films of the amine hydrochlorides, *J. Am. Chem. Soc.* **64**, 498.

Hofmann, U., Endell, K., and Wilm, D. (1933), Kristallstructur, und Quellung von Montmorillonit, *Z. Krist.* **86**, 340–348.

Hofmann-Perez, M., and Gerischer, H. (1961), Charge distribution on the surface of germanium electrodes in aqueous solutions of electrolytes, *Z. Elektrochem.* **65**, 771–775.

Holmes, H. F., Fuller, E. L., Gammage, R. B., and Secoy, C. H. (1968), The effect of irreversibly adsorbed water on the character of thorium oxide surfaces, *J. Colloid Interface Sci.* **28**, 421–429.

Homylev, P. I. (1936), The behaviour of aqueous solutions of potassium ethyl and iso-amyl xanthates, *Mekhanobr Inst.* **2**, ONTI; results quoted in Klassen's and Mok-rousov's *Theory of Flotation*, Butterworths, London, 1963, pp. 238–239.

Hopstock, D. M. (1969), Theory of the flotation of sulfide minerals, Progress Report, Mines Experimental Station University Minnesota, No. 18, 14–20.

Hopstock, D. M. (1970), Chemical properties of xanthates and dithiocarbamates, Progress Report, University Minnesota Mineral Resources Center, No. 21, 92–112.

Horne, R. A. ed. (1972), *Water and Aqueous Solutions. Structure, Thermodynamics and Transport Processes*, Wiley-Interscience, New York.

Horne, R. A. (1978), *The Chemistry of Our Environment*, Wiley-Interscience, New York.

Horne, R. W., Ottewill, R. H., and Watanabe, A. (1960), Precipitation of metallic salts of tetradecyl sulphate, in *Proceedings of the 3rd International Congress of Surface Activity, Cologne*, 1960, Vol. 1 (Vorträge in Originalfassung des III Internationalen Kongresses für Grenzflachenaktive Stoffe), Verlag der Universitatsdruckerei, Mainz GMBH, 203–208.

Hornsby, D. (1975), A high speed (20 min.) film on bubble coalescence and particle–bubble attachment, presented at the Conference of Metallurgists, CIM, Edmonton, Alberta.

Hosking, K. F. G. (1964), Rapid identification of mineral grains in composite samples, *Min. Mag.*, January, 30–37.

Hovenkamp, S. G. (1963), Sodium dithiocarbonate as a by-product in xanthating reactions. A contribution to the chemistry of viscose. *J. Polym. Sci. Part C*, No. 2, 341–355.

Hückel, E. (1934), Aromatic and Unsaturated Molecules: Contributions to the Problem of Their Constitution and Properties. *Int. Conf. Phys., London Phys. Soc. Vol. 2*: 9-35.

Huddleston, R. W., and Smith, A. L. (1976), Electric charge at the air/solution interface, in *Foams*, R. J. Akers, ed., Academic Press, New York, pp. 163–177.

Hughes, M. N. (1974), *The Inorganic Chemistry of Biological Processes*, Wiley, New York.

Huh, C., and Mason, S. G. (1977), Effects of surface roughness on wetting (theoretical), *J. Colloid Interface Sci.* **60**(1), 11–38.

Huiatt, J. L., (1969), A study of the oxidation of the flotation collectors potassium ethyl xanthate and ammonium diethyl dithiophosphate, M.Sc. thesis, Department of Metallurgy, University of Utah.

Hukki, R. T. (1973), Hot flotation improves selectivity and raises mineral recoveries, *World Min.*, March, 74–76.

Hume-Rothery, W. (1936), *The Structure of Metals and Alloys*, Institute of Metals, London.

Hume-Rothery, W., and Raynor, G. V. (1962), *The Structure of metals and alloys*, 4th ed., Institute of Metals, London.

Hummel, D. (1964), *Identification and Analysis of Surface Active Agents by Infrared and Chemical Methods*, transl. I. M. Wulkow, Interscience Wiley-Interscience, New York (79 pp. of spectra).

Hund, F. (1931), Zur Frage der chemischen Bindung, *Z. Phys.* **73**, 1–30.

Hunt, J. P. (1963), Metal ions in aqueous solution, W. A. Benjamin, London.

Hyatt, D. E. (1976), Chemical basis of techniques for the decomposition and removal of cyanides, *Trans. AIME* **260**, 204–208.

Hyslop, W. F. (1975), The use of multiple beam laser interferometry in the study of diffusion of gases in liquids, Ph.D thesis, University of Victoria, Victoria, British Columbia, Canada.

Ikeda, T., and Hagihara, H. (1966), The crystal structure of zinc ethyl–xanthate, *Acta Crystallogr.* **21**, 919–927.

Iler, R. K. (1955), *The Colloid Chemistry of Silica and Silicates*, Cornell University Press, Ithaca, New York.

Iler, R. K. (1973), Colloidal Silica, in *Surface and Colloid Science*, Vol. 6, E. Matijevic, ed., J. Wiley & Sons, New York, pp. 1–100.

Iler, R. K. (1979), *The Chemistry of Silica*, Wiley, New York.

Imaizumi, T., and Inoue, T. (1965), Kinetic consideration of froth flotation, in *Mineral Processing*, A. Roberts, ed., Proceedings of the Sixth International Mineral Processing Congress, Cannes, France, 1963, pp. 581–589, Pergamon Press, New York.

Ingerson, E. ed. (1964), Clays and clay minerals, Pergamon Press, New York.

Ingri, N. (1959), Equilibrium studies of polyanions, IV. Silicate ions in NaCl medium, *Acta Chem. Scand.* **13**, 758.

Israelachvili, J. N., and Ninham, B. W. (1977), Intermolecular forces—the long and short of it, in *Colloid and Interface Science*, M. Kerker, A. C. Zettlemoyer and R. L. Lovell, eds., Vol. I, pp. 15–26, Academic Press, New York.

Israelachvili, J. N., and Tabor, D. (1973), Van der Waals forces. Theory and experiment, in *Progress in Surface and Membrance Science*, Vol. 7, ed. J. F. Danielli, M. D. Rosenberg and D. A. Cadenhead, eds., Academic Press, New York, pp. 1–55.

Ives, K. J. ed. (1978), *The Scientific Basis of Flocculation*, Proceedings of the NATO Advanced Study Institute, held in Cambridge, England, July 1977, Sijthoff and Noordhoff, Alpher aan den Rijn, The Netherlands.

Iwasaki, I., and Cooke, S. R. B. (1957), Absorption spectra of some sulphydryl compounds, *Trans. AIME* **208**, 1267.

Iwasaki, I., and Cooke, S. R. B. (1958), The decomposition of xanthate in acid solution, *J. Am. Chem. Soc.* **80**, 285–288.

Iwasaki, I., and Cooke, S. R. B. (1959), Dissociation constant of xanthic acid as determined by spectrophotometric method, *J. Phys. Chem.* **63**, 1321–1322.

Iwasaki, I., and Cooke, S. R. B. (1964), Decomposition mechanism of xanthate in acid solution as determined by a spectrophotometric method, *J. Phys. Chem.* **68**, 2031–2033.

Iwasaki, I., and de Bruyn, P. L. (1958), The electrical double layer on silver sulphide at pH 4.7, *J. Phys. Chem.* **62**, 594.

Iwasaki, I., Cooke, S. R. B., and Choi, H. S. (1960a), Flotation characteristics of hematite, goethite, and activated quartz with 18-carbon aliphatic acids and related compounds, Trans. AIME **217**, 237–244.

Iwasaki, I., Cooke, S. R. B., and Colombo, A. F. (1960b), Flotation characteristics of goethite, *U.S. Bur. Mines, Rep. Invest.* **5593**.

Iwasaki, I., Cooke, S. R. B., Harraway, D. H., and Choi, H. S. (1962a), Iron wash ore slimes—some mineralogical and flotation characteristics, *Trans. AIME* **223**, 97–108.

Iwasaki, I., Cooke, S. R. B., and Kim, Y. S. (1962b), Some surface properties and flotation characteristics of magnetite, *Trans. AIME* **223**, 113–120.

Janssen, M. J. (1960), Physical properties of organic thiones, Part 1—Electronic absorption spectra of nitrogen-containing thione groups, *Rec. Trav. Chim.* **79**, 454; Part II—Electronic absorption spectra of compounds containing the thiocarbonyl group, *Rec. Trav. Chim.* **79**, 464.

Jaycock, M. J., and Ottewill, R. H. (1963), Adsorption of ionic surface active agents by charged solids, *Bull. IMM* No. 677; *Trans IMM* **72**, 497–506.

Jaycock, M. J., Ottewill, R. H., and Rastogi, M. C. (1960), The adsorption of cationic surface active agents, in *Proceedings of the Third International Congress of Surface Activity, Cologne* 1960, Vol. 2 (Vorträge in Originalfassung des III Internationalen Kongresses für Grenzflachenaktive Stoffe), Verlag der Universitätsdruckerei, Mainz GMBH, pp. 283–287.

Jaycock, M. J., Ottewill, R. H., and Tar, I. (1964), Adsorption of sodium dodecyl sulphate on activated and non-activated stannic oxide, *Bull. IMM*, No. 686, 255–266.

Jeffrey, G. A. and McMullen, R. K. (1967), The clathrate hydrates, in *Progress in Inorganic Chemistry*, Vol. 8, F. A. Cotton, ed., pp. 43–108.

Jenkins, L. H., and Bertocci, U. (1965), On the equilibrium properties of single crystalline copper electrodes, *J. Electrochemical Soc.* **112**(5), 517–520.

Jenkins, L. H., and Durham, R. B. (1970), Galvanostatic overpotential transients and electrocrystallization processes on copper single crystals in solutions of cupric sulphate, *J. Electrochemical Soc.* **117**(12), 1506–1512.

Jenkins, L. H., and Stiegler, J. O. (1962), Electrochemical dissolution of single crystalline copper, *J. Electrochemical Soc.* **109**(6), 467–475.

Johnson, A. I., Besik, F., and Hamielec, A. E. (1969), Mass transfer from a single rising bubble, *Can. J. Chem. Eng.* **47**, 559–564.

Johnson, R. E. (1959), Conflicts between Gibbsian thermodynamics and recent treatments of interfacial energies in solid–liquid–vapour systems, *J. Phys. Chem.* **63**, 1655–1661.

Johnson, R. E., and Dettre, R. H. (1964), Contact angle hysteresis. III. Study of an idealized heterogenous surface, *J. Phys. Chem.* **68**, 1744–1750.

Johnson, R. E., and Dettre, R. H. (1969), Wettability and contact angles, in *Surface and Colloid Science*, Vol. 2, E. Matijevic, ed., Wiley-Interscience, New York, pp. 85–153.

Johnson, R. E., Dettre, R. H., and Brandreth, D. A. (1977), Dynamic contact angles and contact angle hysteresis, *J. Colloid Interface Sci.* **62**, 205–212.

Johnston, D. L. (1969), *The flotation of apatite and dolomite in orthophosphate solution*, thesis, University of British Columbia, Vancouver.

Johnston, D. L., and Leja, J. (1978), Flotation behaviour of calcium phosphates and carbonates in orthophosphate solution, *Trans. IMM* **87**, pp. C237–242.

Johonott, E. S. Jr. (1906), The black spot in thin liquid films, *Philos. Mag.* **11**(6), 746–753.

Jones, G., and Wood, L. A. (1948), The measurement of potentials at the interface between vitreous silica and solutions of potassium chloride by the streaming potential method, *J. Chem. Phys.* **13**, 106–121.

Jones, M. H., and Woodcock, J. T. (1969), Spectrophotometric determination of Z-200 (Isopropyl Ethylthionocarbamate) in flotation liquors, *Proc. Australas. IMM* **231**, 11–18.

Jones, M. H., and Woodcock, J. T. (1973a), *Ultraviolet spectrometry of flotation reagents with special reference to the determination of xanthate in flotation liquors*, the Institution of Mining and Metallurgy, London.

Jones, M. H., and Woodcock, J. T. (1973b), UV spectrophotometric determination of 2-mercaptobenzothiazole (MBT) in flotation liquors, *Can. Metall. Q.* **12**, 497–505.

Jones, M. H., and Woodcock, J. T. (1978), Perxanthates—a new factor in the theory and practice of flotation, *Int. J. Miner. Process.* **5**, 285–296.

Jones, M. P., and Fleming, M. G. (1965), *Identification of mineral grains*, Elsevier, Amsterdam.

Jorgensen, C. K. (1962), *Absorption Spectra and Chemical Bonding in Complexes*, Pergamon, London.

Jowett, A. (1980), Formation and disruption of particle–bubble aggregates in flotation in *Fine Particles Processing*, Vol. 1, P. Somasundaran, ed., AIME, New York.

Joy, A. S., and Robinson, A. J. (1964), Flotation, in *Recent Progress in Surface Science*, Vol. 2, J. F. Danielli, K. G. A. Pankhurst, and A. C. Riddiford, eds., Academic Press, New York, pp. 169–260.

Joy, A. S., Watson, D., and Botten, R. (1965), An improved streaming potential apparatus, *Res. Tech. Instrum.* **1**(1), 6–9.

Jungermann, E. (1970), *Cationic Surfactants*, Marcel Dekker, New York.

Kabanov, B., and Frumkin, A. (1933), The size of electrolytically generated gas bubbles, *Z. Phys. Chem. Abt. A* **165**, 433–452.

Kabanov, B., and Ivanishenko, N. (1937), Electro-capillary phenomena and the wetting of metals. III. Influence of surface-active substances on wetting properties of multimolecular layers, *Acta Physicochim. USSR*, **6**, 701–718.

Kaertkemeyer, L. (1957), Determination du Pouvoir Moussant et Phenomenes de Synergie, in *Proceedings of the 2nd International Congress Surface Activity*, Vol. I, J. H. Schulman, ed., Butterworths, London, pp. 231–241.

Kakovsky, I. A. (1957), Physicochemical properties of some flotation reagents and their salts with ions of heavy non-ferrous metals, in *Proceedings of the 2nd International Congress of Surface Activity*, Vol. IV, J. H. Schulman, ed., Butterworths, London, pp. 225–237.

Kakovsky, I. A., and Komkov, V. D. (1970), Flotation properties of dithiophosphates, *Izv. Vyssh. Ucheb Zaved., Gorn. Zh.* **13**(11), 181–186.

Kamienski, B., (1931), So-called flotation, *Przem. Chem.* **15**, 201–202.

Kampe, P., and Tredmers, L. (1952), Beobachtungen in Wasser Aufsteigenden Gasblasen, *Phys. B.* **12**, Mosbach, Baden.

Kane, P. F., and Larrabee, G. R., eds. (1974), *Characterization of solid surfaces*, Plenum Press, New York.

Karrer, P. (1950), *Organic Chemistry*, Elsevier, New York.

Kavanau, J. L. (1964), *Water and Solute–Water Interactions*, Holden-Day, San Francisco.

Kavanau, J. L. (1965), *Structure and Function in Biological Membranes*, Vol. 1, Holden-Day, San Francisco.

Kelsall, D. F. (1961), Application of probability in the assessment of flotation systems, *Trans. IMM* **70**, 191–204.

Kelvin, Lord (Sir William Thomson) (1871), On the equilibrium of vapour at a curved surface of liquid, *Philos. Mag.* **42**(4), 448–452.

Kerker, M. ed. (1976), *Colloid and Interface Science*, Vols. I–V, Proceedings of the International Conference on Colloids and Surfaces, San Juan, Puerto Rico, June 1976, Academic Press, New York.

Kerner, H. T. (1976), Foam control agents, Noyes Data Corp., Park Ridge, New Jersey.

King, H. G. C., and White, T. (1957), Tannins and polyphenols of schinopsis (Quebracho) species: their genesis and interrelation, *J. Soc. Leather Trades Chem.* **41**, 368–383.

King, R. P. (1972), Flotation research work of the NIM Research Group and the Dept. of Chem. Eng., Univ. of Natal, *J. S. Afr. Inst. Mining Metall. April*, 135–145.

King, R. P. (1973), Model for the design and control of flotation plants, in Proceedings of the 10th International Symposium, *Application of Computer Methods in the Mineral Industry*, M. D. G. Salamon and F. H. Lancaster, eds., South African Institute of Mining and Metallurgy, Johannesburg, pp. 341–350.

King, R. P. (1976), The use of simulation in the design and modification of flotation plants, in *Flotation, A. M. Gaudin Memorial Volume*, Vol. 2, M. C. Fuerstenau, ed., AIME, New York, pp. 937–962.

King, R. P., Hatton, T. A., and Hulbert, D. G. (1974), Bubble loading during flotation, *Trans. IMM* **83**, C112–115.

Kipling, J. J. (1965), *Adsorption from Solutions of Non-Electrolytes*, Academic Press, New York.

Kirchberg, H., and Wottgen, E. (1964), Untersuchungen zur Flotation von Zinnstein mit Phosphon-, Arson- und Stibonsäuren, in *Proceedings of the Fourth International Congress on Surface Active Substances, Brussels 1964*, Vol. 3, C. Paquot, ed., Gordon and Breach, London, pp. 693–704.

Kirkwood, J. G. (1932), Polarisierbarkeiten, Suzeptibilitaten und van der Waalssche Kräfte der Atome mit mehreren Elektronen, *Phys. Z.* **33**, 57–60.

Kiselev, A. V., and Poshkus, D. P. (1958), Calculation of the energy of adsorption of hydrocarbons on magnesium oxide, *Zh. Fiz. Khim.* **32**, 2824–2834.

Kitchener, J. A. (1960), Gravity, the Angle of Contact and Young's Equation: a critical review, in *Proceedings of the Third International Congress of Surface Activity, Cologne 1960*, Vol. 2 (Vorträge in Originalfassung des III Internationalen Kongresses für Grenzflächenaktive Stoffe), publ. Verlag der Universitätsdruckerei, Mainz GMBH, pp. 426–432.

Kitchener, J. A. (1963), Surface forces in thin liquid films, *Endeavour* **22**, 118–122.

Kitchener, J. A. (1964), Foams and free liquid films, in *Recent Progress in Surface Science*,

Vol. 1, J. F. Danielli, K. G. A. Pankhurst and A. C. Riddiford, eds., Academic Press, New York, pp. 51–93.

Kitchener, J. A. (1972), Principles of action of polymeric flocculants, *Brit. Polym. J.* **4**, 217–229.

Kitchener, J. A. (1978), Flocculation in mineral processing, in *The Scientific Basis of Flocculation*, K. J. Ives, ed., Sijthoff and Noordhoff, Alphen aan den Rijn, The Netherlands.

Kitchener, J. A., and Cooper, C. F. (1959), Current concepts in the theory of foaming, *Q. Rev. (London)* **13**, 71–97.

Kitchener, J. A., and Haydon (1959), in Discussion on electric double layer in Colloid Science, *Nature* **183**, 78.

Klassen, V. I. (1948a), Theoretical reasons for intensification of the flotation process—activation of floated minerals by air precipitating from solution *Zh. Fiz. Khim.* **22**(8), 991–998.

Klassen, V. I. (1948b), Effect of gas adsorption on the floatability of minerals, *Gornyi Zhur.* **122**(9), 32–34.

Klassen, V. I. (1952), Attachment of mineral particles to air bubbles, *Proceedings 2nd Conf. Mekhanobr Inst. (Papers)*, *Metallurgizdat*, Leningrad.

Klassen, V. I. (1960), Theoretical basis of flotation by gas precipitation, *Proceedings of the Fifth International Mineral Processing Congress, London 1960*, IMM, London, pp. 309–322.

Klassen, V. I. (1963), *Flotation of coals*, Gosgortekhizdat, Moscow.

Klassen, V. I., and Krokhin, S. J. (1963), Contribution to the mechanism of action of flotation reagents (concentration of collectors at the contact perimeter and activation of hematite flotation by sodium silicate), in *Mineral Processing*, A. Roberts, ed., Proceedings of the Sixth International Mineral Processing Congress, Cannes 1963, Pergamon Press, London, pp. 397–406.

Klassen, V. I., and Mokrousov, V. A. (1963), *An Introduction to the Theory of Flotation*, Engl. transl. by J. Leja and G. W. Poling, Butterworths, London.

Klassen, V. I., and Plaksin, I. N. (1954), The mechanism of action of some reagents and of aeration during the flotation of bituminous coals, *Izvest. Akad. Nauk USSR Otdel Tekh. Nauk* No. 3, 62–71.

Klein, E., Bosarge, J. K., and Norman, I. (1960), Spectrophotometric determination of fast xanthate decomposition kinetics, *J. Phys. Chem.* **64**, 1666–70.

Klier, K., and Zettlemoyer, A. C. (1976), Water at interfaces: molecular structure and dynamics, in *Colloid and Interface Science*, Vol. I, M. Kerker, ed., Academic Press, New York, pp. 231–244.

Klimpel, R., and Manfroy, W. (1977), Development of chemical grinding aids and their effect on selection-for-breakage and breakage distribution parameters in the wet-grinding of ores, in *Proceedings of the XIIth International Mineral Processing Congress, Sao Paulo 1977*, Nacional Publicacöes e Publicidade, Sao Paulo.

Klotz, I. M. (1950), *Chemical Thermodynamics*, Prentice-Hall, Englewood Cliffs, New Jersey.

Kloubek, J. (1970), unpublished results obtained at the University of British Columbia.

Klymowsky, I. B., and Salman, T. (1970), The role of oxygen in xanthate flotation of galena, pyrite and chalcopyrite, *CIM Bull.* **63**, 683–688.

Koch, H. P. (1949), Absorption spectra and structure of organic sulphur compounds. III. Vulcanisation accelerators and related compounds, *J. Chem. Soc.*, 401–408.

Konev, A. S., and Debrivnaja, L. S. (1958), Separation of bulk sulphide concentrates by flotation, in *Progress in Mineral Dressing, Proceedings of the Fourth International Mineral Processing Congress, Stockholm 1957*, Almquist and Wiksell, Stockholm, p. 541.

Koryta, J. (1953), The influence of dyes of the eosin group on reversible oxidation-reduction at the dropping mercury electrode, *Czech. chem. communiqués* **18**, 206–213.

Kosolapoff, G. M. (1958), *Organophosphorus Compounds*, Wiley, New York.

Kossel, W. (1927), The theory of crystal growth, *Nachr. Ges. Wiss. Goettingen Math. Phys. Kl.*, 135–143.

Kowal, A. and Pomianowski, A. (1973), Cyclic voltammetry of ethyl xanthate on natural copper sulphide electrode, *J. Electroanal. Chem. Interfac. Electrochem.* 46, 411–420.

Krescheck, G. C. (1975), Surfactants, in *Water, A Comprehensive Treatise*, Vol. 4, F. Franks, ed., Plenum Press, New York, pp. 95–167.

Kruyt, H. R. ed. (1952), *Colloid Science*, Elsevier, Amsterdam.

Kubaschewski, O., and Hopkins, B. E. (1962), *Oxidation of Metals and Alloys*, Butterworths, London, p. 278.

Kuhn, M. C., Arbiter, N., and Kling, H. (1974), Anaconda's Arbiter process for copper, *CIM Bull.* **67**, 62–73.

Kuhn, W. E., ed. (1963), *Ultrafine Particles*, John Wiley & Sons, New York.

Kulkarni, R. D., and Somasundaran, P. (1975), Kinetics of oleate adsorption at the liquid/air interface and its role in hematite flotation, *AIChE Symp. Ser.* No. 150, **71**, pp. 124–133.

Kulkarni, R. D., and Somasundaran, P. (1977), Effects of pretreatment on the electrokinetic properties of quartz, *Int. J. Miner. Process.* **4**, 89–98.

Kulkarni, R. D., Goddard, E. D., and Kanner, B. (1977), Mechanism of antifoaming action, *J. Colloid Interface Sci.* **59**(3), 468–476.

Kung, H. C., and Goddard, E. D. (1963), Studies of molecular association in pairs of long-chain compounds by differential thermal analysis. I. Lauryl and myristyl alcohols and sulfates, *J. Phys. Chem.* **67**, 1965–1969.

Lagerström, G. (1959), Equilibrium studies of polyanions. The silicate ions in $NaClO_4$ medium, *Acta Chem. Scand.* **13**, 722.

Lambert, B., and Peel, D. H. P. (1934), Studies on gas–solid equilibria. Part V. Pressure–concentration equilibria between silica gel and (1) oxygen, (2) nitrogen (3) mixtures of oxygen and nitrogen, determined isothermally at 0°C, *Proc. R. Soc. London Ser. A* **144**, 205–225.

Lamont, W. E., Spruiell, E. C., Jr., Brooks, D. R., and Feld, I. L., (1972), Laboratory flotation studies of Tennessee phosphates in the presence of slimes, *U.S. Nat. Tech. Inform. Serv., PB Rep.*, No. 206891.

Landolt, H. H., and Börnstein, R. (1923), *Physikalischchemische Tabellen*, p. 763, Springer, Berlin.

Lange, E., and Crane, P. W. (1929), Die elektrische Ladung des Silberjodids in gestättigen Silberjodidlösungen, *Z. Phys. Chem.* **141**, 225–248.

Langmuir, D. (1965), Stability of Carbonates in the System $MgO–CO_2–H_2O$, *J. Geol.* **73**(5), September, 730–754.

Langmuir, I. (1918), The adsorption of gases on plane surfaces of glass, mica and platinum, *J. Am. Chem. Soc.* **40**, 1361–1403.

Langmuir, I. (1920), The mechanism of the surface phenomena of flotation, *Trans. Faraday Soc.* **15**, 1–13.

Langmuir, I. (1938), (a) Repulsive forces between charged surfaces in water and the cause of the Jones–Ray effect, *Science* **88**, 430–432; (b) Surface electrification due to the recession of aqueous solutions from hydrophobic surfaces, *J. Am. Chem. Soc.* **60**, 1190–1194.

Lapidot, M., and Mellgren, O. (1968), Conditioning and flotation of ilmenite ore, *Trans. IMM* **77**, C149–165.

Laskowski, J. (1966), The flotation of naturally hydrophobic minerals in solutions with a raised concentration of inorganic salts, Habilitation thesis, Slask Politech. Gliwice, *Zesz. Nauk. Politech. Slask. Gorni.*, No. 149.

Laskowski, J. (1974), Particle–bubble attachment in flotation, *Miner. Sci. Eng.* **6**(4), 223–235.

Laskowski, J., ed. (1979), *Preprints of papers, XIII International Mineral Processing Congress, Warsaw, 1979*, Polish Scientific Publishers, Wroclaw.

Laskowski, J., ed. (1981), *Mineral Processing*, Proceedings of the XIII International Congress, Elsevier Scientific Publ. Co., Amsterdam, and Polish Scientific Publishers, Warszawa.

Laskowski, J., and Kitchener, J. A. (1969), The hydrophilic–hydrophobic transition on silica, *J. Colloid Interface Sci.* **29**, 670–679.

Last, G. A. (1951), The mechanism of collector–depressant equilibria at mineral surfaces, Ph.D. thesis, University of Utah.

Last, G. A., and Cook, M. A. (1952), *Collector-depressant equilibria in flotation, J. Phys. Chem.* **56**, 637–642.

Latimer, W. M. (1952), *Oxidation Potentials*, Prentice-Hall, Englewood Cliffs, New Jersey.

Le Fèvre, R. J. W. (1953), *Dipole moments*, Methuen, London.

Leja, J. (1956), On the floatability of malachite, RMSL, Kitwe, Zambia, pamphlet 622, 765, 72, July 1956.

Leja, J. (1968), On the mechanisms of surfactant adsorption, in Proceedings of the VIII International Mineral Processing Congress, Leningrad 1968, Mekhanobr Institute, Leningrad.

Leja, J. (1973), Some electrochemical and chemical studies related to froth flotation with xanthates, *Miner. Sci. Eng.* **5**, 278–286.

Leja, J., and Bowman, C. W. (1968), Application of thermodynamics to the (separation of) Athabasca tar sands, *Can. J. Chem. Eng.* **46**, 479–481.

Leja, J., and Nixon. J. C. (1957), Ethylene oxide and propylene oxide compounds as flotation reagents, in *Proceedings of the Second International Congress of Surface Activity*, Vol. 3, J. H. Schulman ed., Butterworths, London, pp. 297–307.

Leja, J., and Poling, G. W. (1960), On the interpretation of contact angle, in *Proceedings of the Fifth International Mineral Processing Congress, London 1960*, IMM, London, pp. 325–332.

Leja, J., and Schulman, J. H. (1954), Flotation theory: molecular interactions between frothers and collectors at solid/liquid/air interfaces, *Trans. AIME* **199**, 221–228.

Leja, J., Little, L. H., and Poling, G. W. (1963), Xanthate adsorption studies using infrared spectroscopy, *Trans. IMM* **72**, 407–423.

Lekki, J., and Laskowski, J. (1971), On the dynamic effect of frother collector joint action in flotation, *Trans. IMM* **80**, C174–180.

Lekki, J., and Laskowski, J. (1975), A new concept of frothing in flotation systems and general classification of flotation frothers, in *Proceedings of the 11th IMP Congress*,

Cagliari, M. Carta, ed., Istituto di Arte Mineraria, Università di Cagliari, Cagliari, Italy, pp. 427–448.

Lemlich, R. A., ed. (1972), *Adsorptive Bubble Separation Techniques*, Academic Press, New York.

Lennard-Jones, J. E. (1929), The electronic structure of some diatomic molecules, *Trans. Faraday Soc.* **25**, 668–686.

Lennard-Jones, J. E. (1932), Processes of adsorption and diffusion on solid surfaces, *Trans. Faraday Soc.* **28**, 333–359.

Leonard, J. W., and Mitchell, D. R., eds. (1968), *Coal Preparation*, AIME, New York, Chaps. 1–19 (part 3 of Chap. 10 is devoted to froth flotation, pp. 10-66 to 10-90).

Levich, V. G. (1949), Motion of bubbles at high Reynolds' numbers, *J. Exp. Theor. Phys. Moscow* **19**(1), 79–83.

Levich, V. G. (1962a), Movement of drops and bubbles in liquid media, *Int. Chem. Eng.* **2**, 78–89.

Levich, V. G. (1962b), *Physico-chemical Hydrodynamics*, Prentice-Hall Englewood Cliffs, New Jersey.

Levich, V. G. (1970), Kinetics of reactions with charge transfer, in *Physical Chemistry*, Vol. IXB, *Electrochemistry*, H. Eyring, ed., Academic Press, New York, Chap. 12, pp. 985–1074.

Levine, S., and Bell, G. M. (1962), The discretness of charge effect in electrical double layer theory, *J. Colloid Sci.* **17**, 838–843.

Levine, S., and Bell, G. M. (1963), The discrete-ion and colloid stability theory. Two parallel plates in a symmetrical electrolyte, *J. Phys. Chem.* **67**, 1408–1419.

Levine, S., Bell, G. M., and Calvert, D. (1962), The discretness-of-charge effect in electrical double layer theory, *Can. J. Chem.* **40**, 518–538.

Lewis, G. N., and Randall, M.; revised by Pitzer, K. S., and Brewer, L. (1961), *Thermodynamics*, McGraw-Hill, New York.

Liebermann, L. (1957), Air bubbles in water, *J. Appl. Phys.* **28**, 205–211.

Lifshits, E. M. (1955), The theory of molecular attractive forces between solids, *Sov. Phys. (J.E.T.P.)* **29**, 94–110.

Lin, I. J., and Metzner, A. (1974), Co-adsorption of paraffinic gases in the system quartz–dodecyl ammoniumchloride and its effect on froth flotation, *International J. Mineral Process.* **1**, 319–334.

Lin, I. J., and Somasundaran, P. (1972), Alterations in properties of samples during their preparation by grinding, *Powder Technol.* **6**, 171–179.

Lin, I. J., Nadiv, S., and Grodzian, D. J. M. (1975), Changes in the state of solids and mechano–chemico reactions in prolonged comminution processes, *Miner. Sci. Eng.* **7**(4), 314–336.

Linfield, W. M. ed. (1976), *Anionic Surfactants*, Marcel Dekker, New York.

Linsen, B. G. (1970), *Physical and Chemical Aspects of Adsorbents and Catalysts*, Academic Press, New York.

Lintern, P. A., and Adam, N. K. (1935), The influence of adsorbed films on the potential difference between solids and aqueous solutions, with special reference to the effect of xanthates on galena, *Trans. Faraday Soc.* **31**, 564–574.

Linton, M., and Sutherland, K. L. (1957), Dynamic surface forces, drop circulation and liquid/liquid mass transfer, in *Proceedings of the Second International Congress Surface Activity*, Vol. 1, J. H. Schulman ed., Butterworths, London, pp. 494–502.

Lister, M. W. (1955), Cyanic acid and cyanates, *Can. J. Chem.* **33**, 426.

Little, L. H. (1966), *Infrared Spectra of Adsorbed Species*, Academic Press, New York.

Livingston, H. K. (1947), The relationship between the Braunauer–Emmett–Teller adsorption isotherm and the new isotherm of Jura and Harkins, *J. Chem. Phys.* **15**, 617–624.

Livingstone, S. E. (1965), Metal complexes of ligands containing sulphur, selenium or tellurium as donor atoms, *Q. Rev.* **19**, 386–425.

London, F. (1930), (1) Theory and systematics of molecular forces, *Z. Phys.* **63**, 245–279, (2) Properties and applications of molecular forces, *Z. Phys. Chem.* **11**, 222–251.

Lorenz, W., Möckel, F., and Müller, W. (1960), Zur Adsorptionsisotherme organischer Moleküle und Molekülionen an Quecksilberelektroden, I, *Z. Phys. Chem. (N.F.)*, **25**, 145.

Lorenz, W., and Müller, W. (1960), Zur Adsorptionsisotherme organischer Moleküle und Molekülionen an Quecksilberelektroden, II, *Z. Phys. Chem. (N.F.)* **25**, 161.

Loveday, B. K. (1966), Analysis of froth flotation kinetics, *Trans. IMM.* **75**, C219–225.

Loveday, B. K., and Marchant, G. R. (1973), Simulation of multicomponent flotation plants, in Proceedings of the 10th International Symposium, *Application of Computer Methods in the Mineral Industry*, M. D. G. Salamon and F. H. Lancaster, eds., S.A.fr. IMM, Johannesburg, pp. 325–330.

Lovell, V. M. (1976), Froth characteristics in phosphate flotation, in *Flotation, A. M. Gaudin Memorial Volume*, M. C. Fuerstenau, ed., AIME, New York, pp. 597–621.

Lowry, H. H., and Olmstead, P. S. (1927), The adsorption of gases by solids with special reference to the adsorption of carbon dioxide by charcoal *J. Phys. Chem.* **31**, 1601.

Lui, A. W., and Hoey, G. R. (1975), Mechanisms of corrosive wear of steel balls in grinding hematite ore, *Can. Metall. Q.* **14**(3), 281–285.

Lukkarinen, T., and Heikkila L. (1974), Beneficiation of chromite ore, Kemi, Finland, in *Proceedings of the Xth International Mineral Processing Congress, London* 1973, M. J. Jones, ed., IMM, London, pp. 869–884.

Lund, H. F. (1971), *Industrial Pollution Control Handbook*, McGraw-Hill, New York.

Lynch, A. J. (1977), *Mineral Crushing and Grinding Circuits, Their Simulation Optimisation, Design and Control*. Elsevier, Amsterdam.

Lyon, C. C., and Fewings, J. H. (1971), Control of the flotation rate of carbonaceous material in Mount Isa chalcopyrite ore, *Proc. Australas. IMM* **237** 41–46.

Mackenzie, J. M. W. (1966), Zeta potential of quartz in the presence of ferric iron, *Trans. AIME* **235**, 82–88.

Mackenzie, J. M. W. (1970), Interactions between oil drops and mineral surfaces, *Trans. AIME* **247**, 202–208.

Mackenzie, J. M. W. (1971), Zeta potential studies in mineral processing: Measurement techniques and applications, *Miner. Sci. Eng.* **3**(3), 25–43.

Mackenzie, J. M. W., and O'Brien, R. T. (1969), Zeta potential of quartz in the presence of nickel(II) and cobalt(II), *Trans. AIME* **244**, 168–173.

Madelung, E. (1912), Das Elektrische Feld in Systemen von Regelmässig Angeordneten Punktladungen, *Phys. Z.* **19**, 524–533.

Mahne, E. J., (1971), Foam separation processes, *Chem. Can.* March 1971, 32–33.

Majima, H. (1961a), Fundamental studies on the collection of sulphide minerals with xanthic acids. I. On the dissociation and decomposition of xanthic acids, *Sci. Rept. Res. Inst. Tohoku Univ. Ser. A* **13**, 183–197.

Majima, H. (1961b), Fundamental studies on the collection of sulphide minerals with xanthic acids. (II). Formation of heavy metal complexes with xanthic acids, *Sci. Rep. RITU* A-13, 433–47.

Majima, H. (1969), How oxidation affects selective flotation of complex sulphide ores, *Can. Metall. Q.* **8**, 269–273.

Majima, H., and Peters, E. (1966), Oxidation rates of sulfide minerals by aqueous oxidation at elevated temperatures, *Trans. Met. Soc. AIME* **236**, 1409–1413.

Majima, H., and Peters, E. (1968), Electrochemistry of sulphide dissolution in hydrometallurgical systems, in *Proceedings of the VIII International Mineral Processing Congress, Leningrad 1968*, Mekhanobr Institute, Leningrad, pp. E7–13.

Majima, H., and Takeda, M. (1968), Electrochemical studies of the xanthate–dixanthogen system on pyrite, *Trans. AIME* **241**(12), 431–436.

Malatesta, L. (1940), The relation between the solubilities of sulphides and xanthates, or thiocarbamates, of heavy metals, *Chim. Ind. (Milan)* **23**, 319–321.

Malatesta, L., and Bonati, F. (1969), *Isocyanide Complexes of Metals*, Wiley-Interscience, New York.

Malinowskii, V. A., Matveenko, N. V., Knaus, O. M., Uvarov, Y. P., Teterina, N. N. and Boiko, N. N. (1974), Technology of froth separation and its industrial application, *Proceedings of the 10th International Mineral Processing Congress, London 1973*, M. J. Jones, ed., IMM, London, 717.

Manegold, E. (1953), *Schaum*, Strassenbau, Heidelberg.

Mann, F. G. (1970), The heterocyclic derivatives of phosphorous, arsenic, antimony and bismuth, 2nd ed., Wiley-Interscience, New York.

Mann, J. A., and Du, G. (1971), Dynamic surface tension: prediction of surface viscosity relaxation effects on capillary ripple properties, *J. Colloid Interface Sci.* **37**(1), 2–13.

Manser, R. M. (1975), *Handbook of Silicate Flotation*, Warren Spring Laboratory, Stevenage, England.

Marangoni, C. G. M. (1871), *Ann. Phys. (Poggendorff)* **43**, 337.

Marcus, Y., and Kertes, A. S. (1969), *Ion Exchange and Solvent Extraction of Metal Complexes*, Wiley-Interscience, London.

Markley, K. S. (1960, 1961), *Fatty Acids: Their Chemistry, Properties, Production and Uses*, 2nd ed., Parts 1 & 2, Interscience, New York.

Maroudas, N. G. (1967), Electrohydraulic crushing, *Brit. Chem. Engineering* **12**(4), 558–562.

Martin, D. F. (1968, 1970), *Marine Chemistry*, Vols. I and II, Marcel Dekker, New York.

Martinez, E., Haagensen, R. B., and Kudryk, V. (1975), Applications of new techniques in developing a barite flotation process, *Trans. AIME* **258**, 27–30.

Matalon, R. J., and Schulman, J. H. (1947), Penetration of sodium cetyl sulphate into cetyl alcohol: effect of salt on the time–penetration curves, *Trans. Faraday Soc.* **43**, 479–485.

Matijevic, E. (1967), Charge reversal of lyophobic colloids, in *Principles and Applications of Water Chemistry*, S. D. Faust, and J. V. Hunter, eds., Wiley, New York, pp. 328–369.

Matijević, E., ed. (1969–1979), Surface and Colloid Science, Vols. 1–10, Wiley-Interscience, New York.

Matijević, E. (1973a), Colloid stability and complex chemistry, Kendall Award Address, ACS, *J. Colloid Interface Sci.* **43**, 217–245.

Matijević, E. (1973b), Colloids—the world of neglected dimensions, *Chem. Technol.* **3**(11), 656–662.

Matijević, E. (1976), Preparation and characterization of monodispersed metal hydrous oxide sols, *Progress Colloid Polymer Sci.* **61**, 24–35.

Matijević, E. (1978), Preparation and properties of monodispersed colloidal metal hydrous oxides, *Pure Appl. Chem.* **50**, 1193–1210.

Matijević, E. (1979), Preparation and characterization of model colloidal corrosion products, *Corrosion* **35**, 264–273.

Matijević, E., and Pethica, B. A. (1958), The properties of ionized Monolayers, Part 1, Sodium dodecyl sulphate at the air/water interface, *Trans. Faraday Soc.* **54**, 1382–1389.

Matijević, E., Kratohvil, J. P., and Kerker, M. (1961), A simple demonstration of some precipitation and solubility effects, *J. Chem. Educ.* **38**, 397–399.

Matijević, E., Leja, J., and Nemeth, R. (1966), Precipitation phenomena of heavy metal soaps in aqueous solutions. I. Calcium oleate. *J. Colloid Interface Sci.* **22**, 419–429.

Maxwell, J. R. (1972), Large flotation cells in Opemiska concentrator, *Trans. SME AIME* **251**(1), 95–98.

Mazzi, F., and Tadini, C. (1963), The crystal structure of potassium ethyl xanthate (with an appendix on rubidium xanthate), *Zeitschrift für Kristallographie* **118**, 378–392.

McBain, J. W. (1939), Soaps and similar long-chain derivatives as simple half-strong electrolytes in dilute solution, *J. Phys. Chem.* **43**, 671–679.

McBain, M. E. L., and Hutchinson, E. (1955), *Solubilization and Related Phenomena*, Academic Press, New York.

McClellan, A. L. (1963), *Tables of Experimental Dipole Moments*, Freeman, San Francisco.

McClellan, A. L. (1974), *Tables of Experimental Dipole Moments*, Rahara Enterprises, El Cerito, California.

McClellan, A. L., and Harnsberger, H. F. (1967), Cross-sectional areas of molecules adsorbed on solid surfaces, *J. Colloid Interface Sci.* **23**, 577–599.

McCutcheon, J. W., Inc., and Morristown, J. (1963), *Detergents and Emulsifiers*, McCutcheon Division, M.C. Publishing Company, Ridgewood, New Jersey.

McCutcheon, J. W., Inc. (1975), *McCutcheon's Detergents and Emulsifiers*, North American Edition, McCutcheon Division, M.C. Publishing Company, Ridgewood, New Jersey.

McKee, D. J., Fewings, J. H., Manlapig, E. V., and Lynch, A. J. (1976), Computer control of chalcopyrite flotation at Mount Isa Mines Ltd., in *Flotation, A. M. Gaudin Memorial Volume*, Vol. 2, M. C. Fuerstenau, ed., AIME, New York, pp. 994–1025.

McIvor, R. A., Grant, G. A., and Hubley, C. E. (1956), Infrared studies of sulphur-containing organic derivatives of phosphorus pyroacids, *Can. J. Chem.* **34**, 1611–1640.

McLachlan, A. D. (1963), Retarded dispersion forces between molecules, *Proc. R. Soc. London Ser. A* **271**, 387–401.

McLachlan, A. D. (1965), Effect of the medium on dispersion forces in liquids, *Discuss. Faraday Soc.* **40**, 239–245.

McLeod, N. (1934), Solubility of lead xanthates in acetone—water mixtures, thesis, Montana School of Mines.

McNutt, J. E., and Andes, G. M. (1959), Relationship of the contact angle to interfacial energies, *J. Chem. Phys.* **30**, 1300–1303.

Meadus, F. W., Mykytiuk, A., and Puddington, I. E. (1966), The upgrading of tin ore by continuous agglomeration, *Trans. CIM* **69**, 303–305.

Melik-Gaikazyan, V. I. (1952), Formation of multilayers at the mercury–solution interface and its effect on the differential capacity of the double layer, *Zh. Fiz. Khim.* **26**, 1184–1190.

Melik-Gaykaryan, V. I., Yemelyanova, E. P., Teptin, V. F., Glazunova, Z. I., Voron-chikin, G. A., and Bichevina, N. G. (1975), On the joint action of xanthate and alcohols in the flotation of galena, *Obogashch. Rud* **5**(121), 26–30.

Mellgren, O. (1966), Heat of adsorption and surface reactions of potassium ethyl xanthate on galena, *Trans. AIME* **235**, 46–53.

Mellgren, O., and Lapidot, M. (1968), Determination of oleic acid, tall oil and fuel oil adsorbed on ilmenite flotation products, *Trans. IMM* **77**, C140–C148.

Mellgren, O., and Lwakatare, S. L. (1968), Desorption of xanthate ions from galena with sodium sulphide, *Trans. IMM* **77**, C101–C104.

Mellgren, O., and Rao, S. R. (1968), Heat of adsorption and surface reactions of potassium diethyldithiocarbamate on galena, *Trans. IMM* **77**, C65–C71.

Mellgren, O., and Rau, M. G. S. (1963), Adsorption of ethyl xanthate on galena, *Bull. IMM* **676**, 425–442.

Mellgren, O., and Shergold, H. L. (1966), Method for recovering ultrafine mineral particles by extraction with an organic phase, *Trans. IMM* **75**, C267–C268.

Mellgren, O., Gochin, R. J., Shergold, H. L., and Kitchener, J. A. (1974), Thermochemical measurements in flotation research, in *Proceedings of the Tenth International Mineral Processing Congress, London 1973*, M. J. Jones, ed., IMM, London, pp. 451–472.

Mellor, J. W. (1928), *Comprehensive Treatise on Inorganic and Theoretical Chemistry*, Vol. 8, Longmans, Green, London, p. 194.

Meltzer, Y. L. (1979), *Water-Soluble Polymers*, Noyes Data Corp., Park Ridge, New Jersey.

Memming, R., and Neumann, G. (1969), Electrochemical reduction and hydrogen evolution on Germanium electrodes, *J. Electroanal. Chem. Interfac. Electrochem.* **21**, 295–305.

Miagkova, T. M. (1955), *The effect of frothers on the dispersion of air and effectiveness of flotation*, Leningrad Mining Institute, Leningrad.

Mikhail, R. Sh., Youssef, A. M. and El-Nabarawy, T. (1979), Surface properties and catalytic activity of pure and silica-coated aluminas, *J. Colloid Interface Sci.* **70**(3), 467–474.

Miles, G. D. (1945), Minima in surface–tension and interfacial–tension curves, *J. Phys. Chem.* **49**, 71–76.

Miles, G. D., and Shedlovsky, L. (1944), Minima in surface tension–concentration curves of solutions of sodium alcohol sulphates, *J. Phys. Chem.* **48**, 57–62.

Miles, G. D., Ross, J., and Shedlovsky, L. (1950), Film drainage: a study of the flow properties of films of solutions of detergents and the effect of added materials, *J. Am. Oil Chem. Soc.* **27**, 268–273.

Miller, K. J. (1975), Coal–Pyrite flotation, *Trans. AIME* **258**, 30–33.

Mirnik, M. (1970), Fixed charge double layer potential equations—a derivation. *Croat. Chem. Acta* **42**, 49–56.

Mitrofanov, S. I. (1967), *Differential Flotation*, Nedra, Moscow.

Mitrofanov, S. I., and Kushnikova, V. G. (1958), Adsorption of diethyldithiophosphate and butyl xanthate by sulphides, in *Progress in Mineral Dressing*, Proceedings of the Fourth International Mineral Processing Congress, Stockholm 1957, sponsored and edited by Svenska Gruvforeningen and Jernkontoret, Almqvist and Wiksell, Stockholm, pp. 461–473.

Mitrofanov, S. I., and Kushnikova, V. G. (1959), The adsorption of diethyl dithiophosphate and butyl xanthate on the surface of sulphides in acid medium, *Sb. Nauchn. Tr. Gos. Nauchn. Issled. Inst. Tsvetno. Met.* **16**, 9–24.

Mitrofanov, S. I., and Rozin, E. E., (1955), Testing of alkyl sulphate for flotation of cassiterite, *Sb. Nauchn. Tr. Gos. Nauchn. Issled. Inst. Tsvetn. Met.* **10**, 120–123. [*Chem. Abstr.* **52**, 197d (1958)].

Mitrofanov, S. I., Kushnikova, V. G., Strigin, I. A., and Rozhavskii, G. S. (1955), Sulfidization reactions of oxidized minerals, *Sb. Nauchn. Tr. Gos. Nauchno Issle. Inst. Tsvetn. Met.* **10**, 7–29.

Mittal, K. L. ed. (1979), *Surface Contamination*, Proceedings of the Symposium on Surface Contamination, Washington D.C., September 1978, Plenum Press, New York.

Modi, H. J., and Fuerstenau, D. W. (1960), Flotation of corundum—an electrochemical interpretation, *Trans. AIME* **217**, 381–387.

Moffat, E. G., Pearsall, G. W., and Walff, J. (1964), *The Structure and Properties of Materials*, Vol. 1, *Structure*, Wiley, New York.

Möller, H. G. (1908), Electrolytic phenomena at the surfaces of electrodes, *Z. Phys. Chem.* **65**, 226–254.

Moncrieff, A. G., Noakes, F. D. L., Viljoen, D. A., Davey, J. M., and Boulter, G. N. (1973), Development and operation of cassiterite flotation at mines of the Consolidated Gold Fields Group, in *Proceedings of the Tenth International Mineral Processing Congress, London* 1973, M. J. Jones, ed., IMM, London, p. 565.

Morgan, J. J. (1967), Applications and limitations of chemical thermodynamics in water systems, in *Equilibrium Concepts in Natural Water Systems*, R. F. Gould, ed., Advances in Chemistry Series ACS No. 67, American Chemical Society, Washington, D.C., pp. 1–29.

Morrison, R. T. and Boyd, R. N. (1966), *Organic Chemistry*, 2nd. ed., Allyn and Bacon, Inc. Boston.

Morrison, S. R. (1971), Surface phenomena associated with the semiconductor/electrolyte interface, *Progr. Surf. Sci.* **1**(2), 105–154.

Mott, N. F., and Watts-Tobin, R. J. (1961), The interface between a metal and an electrolyte, *Electrochim. Acta* **4**, 79–107.

Mott, N. F., and Watts-Tobin, R. J. (1962), *Philos. Mag.* **7**(8), 483.

Mukai, S., and Wakamatsu, T. (1962), Electrochemical study on flotation, *Mem. Fac. Eng. Kyoto Univ.* **24**, Part IV, 389–410.

Mukai, S., Wakamatsu, T., and Takahashi, K. (1972), (a) Mutual interaction between xanthate collectors and alcohol frothers. Its effect on flotation phenomena, *Suiyokai-Shi* **17**, 278–282; (b) Mutual interaction between collectors and frothers in flotation, *Mem. Fac. Eng. Kyoto Univ.* **34**, Part III, 279–288.

Mukerjee, P. (1967), The nature of the association equilibria and hydrophobic bonding in aqueous solutions of association colloids, *Adv. Colloid Interface Sci.* **1**, 241–275.

Mukerjee, P., and Mysels, K. J. (1971), Critical micelle concentrations of aqueous surfactants systems, *Not. Stand. Ref. Data Ser. Nat. Bur. Stand.*, 36.

Mular, A. L. (1972), Empirical modelling and optimization of mineral processes, *Miner. Sci. Eng.* **4**, 30–42.

Mular, A. L. (1976), Optimization in flotation plants, in *Flotation, A. M. Gaudin Memorial Volume*, Vol. 2, M. C. Fuerstenau, ed., AIME, New York, pp. 895–926.

Mular, A. L., and Bhappu, R. B. (1978), *Mineral Processing Plant Design and Practice*, AIME, New York.

Mular, A. L., and Bull, W. R. (1969, 1970, 1971), *Mineral Processes, Their Analysis, Optimization and Control*, The Canadian Institute of Mining and Metallurgy, Montreal.

Mular, A. L., and Bull, W. R. (1973), *Mineral Processes, Their Analysis, Optimization and Control*, Department of Mineral Engineering, University of British Columbia, Vancouver.

Mular, A. L., and Puddington, I. E. (1968), A technically feasible agglomeration–separation process, *CIM Bull.* **61**, 726–730.

Mular, A. L., and Roberts, R. B. (1966), A simplified method to determine isoelectric points of oxides, *Trans. CIM* **69**, 438–439.

Müller, A. (1936), The van der Waals potential and the lattice energy of a *n*-CH- chain molecule in a paraffin crystal, *Proc. R. Soc. London Ser. A* **154**, 624.

Müller, H., Friberg, S., and Hellsten, M. (1970), Investigations on the interaction between metal ions and alkyl phosphate monolayers at the air–water interface by means of surface pressure determination and infrared spectroscopy, *J. Colloid Interface Sci.* **32**, 132–141.

Mulliken, R. S. (1932), Electronic structures of polyatomic molecules and valence, II. General considerations, *Phys. Rev.* **41**, 49–71.

Mullins, W. M. (1963), Solid surface morphologies governed by capillarity, in *Metal Surfaces: Structure, Energetics and Kinetics*, American Society for Metals, Cleveland, pp. 17–66.

Murphy, C. N., and Winter, G. (1973), Monothiocarbonates and Their Oxidation to Disulphides, CSIRO, Division of Mineral Chemistry, Port Melbourne, Australia.

Murr, L. E. (1975), *Interfacial Phenomena in Metals and Alloys*, Addison-Wesley, Reading, Massachusetts.

Murray, D. J., Healy, T. W., and Fuerstenau, D. W. (1968), The adsorption of aqueous metal on colloidal hydrous manganese oxide, in *Adsorption from Aqueous Solution*, W. J. Weber and E. Matijevic, eds., Advances in Chemistry Series No. 79, pp. 74–81. American Chemical Society, Washington, D.C.

Murray, R. C., and Hartley, G. S. (1935), Equilibrium between micelles and simple ions with particular reference to the solubility of long-chain salts, *Trans. Faraday Soc.* **31**, 183–189.

Mysels, K. J., and Florence, A. T. (1970), Techniques and criteria in the purification of aqueous surfaces, in *Clean Surfaces: Their Preparation and Characterization for Interfacial Studies*, G. Goldfinger, ed., Marcel Dekker, New York, pp. 227–268.

Mysels, K. J., and Florence, A. T. (1973), The effect of impurities on dynamic surface tension—basis for a valid surface purity criterion, *J. Colloid Interface Sci.* **43**(3), 577–582.

Mysels, K. J., Shinoda, K., and Frankel, S. (1959), *Soap Films—Studies of Their Thinning*, Pergamon Press, New York.

Mysels, K. J., Cox, M. C., and Skewis, J. D. (1961), The measurement of film elasticity, *J. Phys. Chem.* **65**, 1107–1111.

Mysels, K. J., Huisman, H. F., and Razouk, R. I. (1966), Measurement of contact angle between thin film and bulk of same liquid, *J. Phys. Chem.* **70**, 1339–1340.

Nagy, L. G., and Schay, G. (1960), Determination of the surface area of adsorbents from the adsorption isotherms of binary liquid mixtures, *Magy. Kem. Foly.* **66**, 31–37.

Nakamoto, K., Fujita, J., Condrate, R. A., and Morimoto, Y. (1963), Infrared spectra of metal chelate compounds. IX. A normal coordinate analysis of dithiocarbamato complexes, *J. Chem. Phys.* **39**, 423–427.

Nakatsuka, K., Matsouka, I., and Shimoizaka, J. (1970), On the flotation of ilmenite from magnetite sand, *Proceedings of the Ninth International Mineral Processing Congress, Prague* 1970, Ustav Pro Vyzkum Rud, Prague, pp. 251–258.

Nancollas, G. H. (1966), *Interactions in Electrolyte Solutions*, Elsevier, Amsterdam.

Nanjo, M., and Yamasaki, T. (1966), A polarographic study of the reaction between a xanthate and cadmium ion, *Chem. Ind.*, 1530.

Nanjo, M., and Yamasaki, T. (1969), Spectrophotometric studies of ethyl xanthate complexes in aqueous solutions, *Bull. Chem. Soc. Japan* **42**, 968–972.

Napper, D. H., and Hunter, R. J. (1972), Hydrosols, in *Surface Chemistry and Colloids*, Vol. 7, MTP International Review of Science, M. Kerker, ed., Butterworths, London.

Natarajan, K. A., and Iwasaki, I. (1973), Practical implications of E_h measurements in sulfide flotation circuits, *Trans. AIME* **254**, 323–328.

Natarajan, K. A., and Iwasaki, I. (1974), Significance of mixed potentials in E_h measurements with Pt electrodes, *Trans. AIME* **256**, 82–86.

Natarajan, K. A., and Iwasaki, I. (1975), Adsorption mechanism of sulphides at a Pt/solution interface, in *Advances in Interfacial Phenomena of particulate/solution/gas systems; applications to flotation research*, P. Somasundaran and R. B. Grieves, eds., American Institute of Chemical Engineering Symposium No. 150, Vol. 71, pp. 148–156.

Nemeth, R., and Matijević, E. (1971), Precipitation and electron microscopy of calcium and barium oleate sols, *Kolloid Z. Z. Polym.* **245**, 497–507.

Nemethy, G., and Scheraga, H. A. (1962), Structure of water and hydrophobic bonding in proteins. I. A model for the thermodynamic properties of liquid water. II. Model for the thermodynamic properties of aqueous solutions of hydrocarbons, *J. Chem. Phys.* **36**, 3382–3400; 3401–3417.

Neumann, A. W. (1974), Contact angles and their temperature dependence: thermodynamic status, measurement, interpretation and application, *Adv. Coll. Interface Sci.*, **4**(2, 3), 105–191.

Neumann, A. W. (1978), Contact angles, in *Wetting, Spreading and Adhesion*, J. F. Padday, ed., Academic Press, New York.

Neumann, A. W., and Good, R. J. (1972), Thermodynamics of contact angles. I. Heterogeneous solid surfaces, *J. Colloid Interface Sci.* **38**, 341–358.

Neunhoeffer, O. (1943), Synthesis of a flotation agent for cassiterite—verification of chemical principles of flotation, *Met. Erz.* **40**, 174–176.

Neunhoeffer, O. (1944), Lyophobic association, *Kolloid Z.* **107**, 104–107.

Ney, P. (1973), *Zeta Potentiale und Flotierbarkeit von Mineralen*, Springer-Verlag, Vienna.

Nickless, G. ed. (1968), *Inorganic Sulphur Chemistry*, Elsevier, Amsterdam.

Nixon, J. C. (1957), Discussion, in *Proceedings of the 2nd International Congress on Surface Activity*, Vol. 3, J. H. Schulman, ed., p. 369, Butterworths, London.

Nixon, J. C., and Moir, D. N. (1957), The assessment of flotation results, *Trans. IMM* **66**, 453–456.

Noll, W. (1968), *Chemistry and Technology of Silicones*, Academic Press, New York.

Noller, C. R. (1965), *Chemistry of organic compounds*, Saunders, Philadelphia.

Nutting, G. C., and Long, F. A. (1941), The change with time of the surface tension of sodium laurate solutions, *J. Am. Chem. Soc.* **63**, 84.

O'Brien, R. N., and Hyslop, W. F. (1975), A Fabry–Perot interferometer for monitoring gas–liquid exchange, in *Chemistry and Physics of Aqueous Gas Solutions*, W. A. Adams, ed., The Electrochemical Society, Princeton, New Jersey, pp. 326–336.

O'Donovan, P., Laflamme, J. H. G., and Pinard, R. G. (1972), Polishing procedures for ore minerals, mill products and synthetic materials in the mineralogy group, Mineral Sciences Division, Mines Branch, Canada, Information Circular 286.

Oktawiec, M. T., and Olender, K. (1969), Diacetone alcohol, a new frother for copper ores (in Polish), *Rudy Met. Niezelaz.* **R**:14(4), 192–195.

Oldroyd, J. G. (1955), The effect of interfacial stabilizing films on the elastic and viscous properties of emulsions, *Proc. R. Soc. London Ser. A* **232**, 567–577.

Olin, A. (1961), *Sven. Kem. Tidskr.* **73** [as quoted in Stumm (1967)].

Oliver, J. F., and Mason, S. G. (1977), Microspreading studies on rough surfaces by scanning electron microscopy, *J. Colloid Interface Sci.* **60**, 480–487.

Oliver, J. F., Huh, C., and Mason, S. G. (1977), Resistance to spreading of liquids by sharp edges, *J. Colloid Interface Sci.* **59**(3), 568–581.

Osborne, D. G. (1978), Recovery of slimes by a combination of selective flocculation and flotation, *Trans. IMM* **87**, C189–193.

Osipow, L. I. (1962), *Surface Chemistry, Theory and Industrial Applications*, Reinhold, New York.

Ottewill, R. H. (1973), Particulate dispersions, in *Colloid Science*, Vol. I, D. H. Everett, ed., The Chemical Society, London.

Ottewill, R. H., and Rastogi, M. C. (1960), The stability of hydrophobic sols in the presence of (cationic) surface active agents, *Trans. Faraday Soc.* **56**, 866–879.

Otto, E. W. (1966), Hydrodynamics of liquid surfaces, Part 1, *Cryogenic Eng. News*, March, 24–28.

Overbeek, J. Th. G. (1950), Quantitative interpretation of the electrophoretic velocity of colloids, in *Advances in Colloid Science*, Vol. 3, H. Mark and E. J. W. Verwey, eds., Interscience, New York and London, pp. 97–135.

Overbeek, J. Th. G. (1952), Electrokinetic phenomena, in *Colloid Science*, Vol. I, H. R. Kruyt, ed., Elsevier, Amsterdam.

Padday, J. F. (1967), The effect of surfactants on van der Waals' forces at a solid-liquid interface, in SCI Monograph No. 25, *Wetting*, p. 234.

Padday, J. F. (1970), Cohesive properties of thin films of liquids adhering to a solid surface, *Spec. Discuss. Faraday Soc.* **1**, pp. 64–74. Thin liquid films and boundary layers.

Padday, J. F., ed. (1978), *Wetting, Spreading and Adhesion*, Academic Press, London.

Palmer, B. R., Gutierrez, B. G., and Fuerstenau, M. C. (1975a), Mechanisms involved in the flotation of oxides and silicates with anionic collectors, Part 1; Part 2 by Palmer, B. R., Fuerstenau, M. C., and Aplan, F. F., *Trans. AIME* **258**, 257–263.

Palmer, B. R., Fuerstenau, M. C., and Aplan, F. F. (1975b), Mechanisms involved in the flotation of oxides and silicates with anionic collectors, Part II, *Trans. AIME* **258**, 261–263.

Parks, G. A. (1965), The isoelectric points of solid oxides, solid hydroxides, and aqueous hydroxo complex systems, *Chem. Rev.* **65**, 177–198.

Parks, G. A. (1967), Aqueous surface chemistry of oxides and complex oxide minerals. Isoelectronic point and zero point of charge, in *Equilibrium Concepts in Natural Water Systems*, R. F. Gould, ed., Advances in Chemistry Series No. 67, American Chemical Society, Washington, D.C., pp. 121–160.

Parreira, H. C. (1965), Automatic recording apparatus for measurements of streaming potentials, *J. Colloid Sci.* **20**, 1–6.

Parsegian, V. A. (1975), Long range van der Waals forces, in *Physical Chemistry: Enriching Topics from Colloid and Surface Science*, H. van Olphen and K. J. Mysels, eds., Theorex, La Jolla, California.

Parsons, R. (1954), Equilibrium properties of electrified interphases, in *Modern Aspects of Electrochemistry*, Vol. 1, J.O'M. Bockris and B. E. Conway, eds., Butterworths, London, pp. 103–179.

Pashley, D. W. (1970), Recent developments in the study of epitaxy, in *Recent Progress in Surface Science*, Vol. 3, J. F. Danielli, A. C. Riddiford, and M. D. Rosenberg, eds., Academic Press, New York, pp. 23–69.

Pashley, R. M., and Kitchener, J. A. (1979), Surface forces in adsorbed multilayers of water on quartz, *J. Colloid Interface Sci.* **71**(3), 491–500.

Patel, I. A., Patel, K. U., Obrecht, M. F., and DeWitt, C. C. (1956), Flotation studies with alkyl chelate type collectors, paper presented at the AIME National Meeting, February 1956.

Pauling, L. (1948), *The Nature of the Chemical Bond*, Cornell University Press, Ithaca, New York.

Pauling, L. (1957), The structure of water, in *Hydrogen Bonding*, D. Hadzi and H. W. Thompson, eds., Pergamon Press, New York, pp. 1–6.

Pawlikowska-Czubak, J., and Leja, J. (1972), Impedance of the dropping mercury electrode in potassium ethyl xanthate solutions, Parts I & II, *Roczm. Chem.* **46**, 1567–1576; 1789–1800.

Payne, R. (1973), Double layer at the mercury–solution interface, in *Progress in Surface and Membrane Science*, Vol. 6, J. F. Danielli, M. D. Rosenberg, and D. A. Cadenhead, eds., pp. 51–123.

Pearson, R. G. (1963), Hard and soft acids and bases, *J. Am. Chem. Soc.* **85**, 3533–3539.

Peck, A. S., and Wadsworth, M. E. (1964), Infrared study of the depression effect of fluoride, sulfate, and chloride on chemisorption of oleate on fluorite and barite, in *Proceedings of the Seventh International Mineral Processing Congress*, N. Arbiter, ed., Gordon and Breach, New York, pp. 259–265.

Peck, A. S., Raby, L. H., and Wadsworth, M. E. (1966), An infrared study of the flotation of hematite with oleic acid and sodium oleate, *Trans. AIME* **235**, 301–306.

Perrin, J. (1904), Mécanisme de l'éctrisation de contact et solutions colloidales, *J. Chim. Phys.* **2**, 601–651.

Perrin, J. (1918), The stratification of liquid films, *Ann. Phys.* (*N.Y.*) **10**, 160–184.

Peskov, N. P. (1940), *Course in Colloid Chemistry*, Goskhimizdat, Moscow.

Petersen, H. D., Fuerstenau, M. C., Rickard, R. S., and Miller, J. D. (1965), Chrysocolla flotation by the formation of insoluble surface chelates, *Trans. AIME* **232**, 388–392.

Petersen, P. E. (1967), The development of selective flotation, *CIM Bull.*, April, 415–417.

Pethica, B. A. (1960), Micelle formation, in *Proceedtngs of the Third International Congress of Surface Activity, Cologne*, 1960 (Vorträge in Originalfassung des III Internationalen Kongresses für Grenzflachenaktive Stoffe), Verlag der Universitätsdruckerei, Mainz GMBH, A/II/3, pp. 212–226.

Pethica, B. A. (1977), The contact angle equilibrium, *J. Colloid Interface Sci.* **62**, 567–569.

Peyronell, G., and Pignedoli, A. (1967), The crystal and molecular structure of nickel(II) bis(*N*, *N*-di-*n*-propyldithiocarbamate), *Acta Crystallogr.* **23**, 398–410.

Philipp, B., and Fichte, C. (1960), Kinetic studies on the decomposition of xanthate, *Faserforsch. Textiltechn.* **11**, 118–24, 172–79.

Philippoff, W. (1950), Micelles and x-rays, *J. Colloid Sci.* **5**, 169–191.

Philippoff, W. (1952), Some dynamic phenomena in flotation, *Trans. AIME* **193**, 386–390.

Phillips, C. S. G., and Williams, R. J. P. (1965), *Inorganic Chemistry*, Vol. 1, Oxford University Press, New York.

Phillips, C. S. G., and Williams, R. J. P. (1966), *Inorganic Chemistry*, Vol. II. *Metals*, Oxford University Press, New York.

Pickett, D. E., ed. (1977), *Mining Practice in Canada*, CIM Special Volume 16, CIM, Montreal.

Pierotti, R. A. (1963), The solubility of gases in liquids, *J. Phys. Chem.* **67**, 1840–1845.

Pierotti, R. A. (1965), Aqueous solutions of nonpolar gases, *J. Phys. Chem.* **69**, 281–288.

Pierotti, R. A. (1967), On the scaled-particle theory of dilute aqueous solutions, *J. Phys. Chem.* **71**, 2366–2367.

Pilpel, N. (1963), Properties of organic solutions of heavy metal soaps, *Chem. Rev.* **63**, 221–234.

Pimentel, G. C., and McClellan, A. L. (1960), *The Hydrogen Bond*, Freeman, San Francisco.

Plaksin, I. N., and Anfimova, E. A. (1954), A study of the interaction of xanthates with surfaces of sulphide minerals, *Proc. Min. Inst. Acad. Sci. USSR.* **1**, 225–234.

Plaksin, I. N., and Bessonov, S. V. (1957), Role of gases in flotation reactions, in *Proceedings of the Second International Congress of Surface Activity*, Vol. 3, Butterworths, London, pp. 361–367.

Plaksin, I. N., and Khaginskaya, G. N. (1956), The collector action of some foaming agents during sphalerite flotation, *Izvest. Akad. Nauk USSR Otdel. Tekh. Nauk*, No. 9, 121–123.

Plaksin, I. N., and Khaginskaya, G. N. (1957), Pyrrhotite flotation, *Izvest. Akad. Nauk USSR Otdel. Tekh. Nauk*, No. 2, 91–97.

Plaksin, I. N., and Shafeyev, R. S. (1958), On the effect of the electric potential on the distribution of xanthates over sulphide surfaces, *Dokl. Ak. Nauk USSR.* **118**(3), 546–548.

Plaksin, I. N., Bessonov, C. V., and Tiurnikova, V. I. (1957), Autoradiographic method applied to studies on distribution of flotation reagents on sulphide minerals, *Bull. Acad. Sci. SSSR, Tech. Sc. Sec.*, No. 3.

Plante, E. C., and Sutherland, K. L., (1949), Effects of oxidation of sulfide minerals on their flotation properties, *Trans. AIME* **183**, 160–188.

Plateau, J. (1873), *Statique Experimentale et Theoretique des Liquides Soumis Aus Seulles Forces Moleculaires*, Gauthier-Villars, Paris.

Polanyi, M. (1932), Theories of the adsorption of gases. A general survey and some additional remarks, *Trans. Faraday Soc.* **28**, 316–33.

Poling, G. W. (1961), Infrared spectroscopy of xanthate compounds in the solid, solution and the adsorbed states, M.Sc. thesis, University of Alberta, Edmonton, Canada.

Poling, G. W. (1963), Infrared studies of adsorbed xanthates, Ph.D. thesis, University of Alberta, Edmonton, Canada.

Poling, G. W. (1976), Reactions between thiol reagents and sulphide minerals, in *Flotation, A. M. Gaudin Memorial Volume*, Vol. 1, M. C. Fuerstenau, ed., AIME, New York, pp. 334–363.

Poling, G. W., and Leja, J. (1963), Infrared study of xanthate adsorption on vacuum deposited films of lead sulphide and metallic copper under conditions of controlled oxidation, *J. Phys. Chem.* **67**, 2121–2126.

Polkin, S. I., and Berger, G. S. (1968), On forms of fixation and flotation action of long chain collectors, VIII International Mineral Processing Congress, Leningrad, 1968, Paper S-7, Mekhanobr Institute, Leningrad.

Pol'kin, S. I., Laptev, S. F., Matsuev, L. P., Adamov, E. V., Krasnukhina, A. V., and Purvinskii, O. F. (1974), Theory and practice in the flotation of cassiterite fines, in *Proceedings of the Tenth International Mineral Processing Congress*, M. J. Jones, ed., IMM, London, p. 593–614.

Pomianowski, A. (1957), Electric phenomena accompanying the process of flotation, in *Proceedings of the Second International Congress Surface Activity*, Vol. 3, J. H. Schulman, ed., Butterworths, London, pp. 332–342.

Pomianowski, A. (1967), Electrical and surface characteristics in mercury–xanthate–air systems, Dozent thesis, Institute of Physical Chemistry PAN, Krakow.

Pomianowski, A. (1967), Differential capacity of electric double layer on mercury polarized in potassium ethylxanthate solutions, *Roczn. Chem.* **41**, 775–790.

Pomianowski, A., and Leja, J. (1963), Spectrophotometric study of xanthate and dixanthogen solutions, *Can. J. Chem.* **41**, 2219–2230.

Pomianowski, A., and Leja, J. (1964), Equimolar solutions of xanthate and alkyl trimethylammonium bromide adsorption on copper, nickel and sphalerite powders, *Trans. AIME* **229**, 307–312.

Pomianowski, A., and Liszka, R. (1962), Physical chemistry of model flotation of mercury, *Zesz. Nauk. Univ. Jagiellonsk Pr. Chem.* **7**, (52), 13–30.

Pomianowski, A., and Pawlikowska-Czubak, J. (1967), Electrical and surface characteristics of the mercury/solution/air system containing xanthates and dodecyltrimethylammonium bromide, Pt. I., *Przemysl Chem.* **46**(8), 481–485.

Pomianowski, A., Najbar, J., and Kruk, J. (1968), Electrical and surface characteristics of the mercury/solution/air system containing xanthates and dodecyltrimethylammonium bromide, Pt. II., *Zesz. Nauk. Univ. Jagiell. (Krakow)* **13**, 119–126.

Popiel, W. J. (1978), *Introduction to Colloid Science*, Exposition Press, Hicksville, New York.

Pople, J. A. (1951), *Proc. R. Soc. London Ser. A* **205**, 163.

Popov, S. I. *et al.* (1971), Use of alkyl aerofloats as collectors, *Tsvet. Metal.* **44**(4), 80–81.

Pourbaix, M. J. N. (1949), *Thermodynamics of Dilute Aqueous Solutions*, E. Arnold, London.

Predali, J. J. (1968), Flotation of carbonates with salts of fatty acids: role of pH and the alkyl chain, *Trans. IMM* **77**, C140–147.

Predali, J. J., (1971), Adsorption des sels d'acides gras a l'interface solide-liquide et flottation des carbonates, PhD. thesis, Université de Nancy, France.

Predali, J. J., and Cases, J. M. (1974), Thermodynamics of the adsorption of collectors, in *Proceedings of the 10th IMPC, London*, 1973, M. J. Jones, ed., IMM, London, pp. 473–492.

Preston, W. C. (1948), Some correlating principles of detergent action, *J. Phys. Colloid Chem.* **52**, 84–97.

Price, C. C., and Oae, S. (1962), *Sulphur Bonding*, Ronald Press, New York.

Prigogine, I., and Defay, R. (1954), *Chemical Thermodynamics*, Longmans, Green, New York.

Prince, L. M., ed. (1977), *Microemulsions, Theory and Practice*, Academic Press, New York.

Prins, W. (1955), Studies on some long-chain sodium alkyl-sulphates, Dr.Sc. thesis, Rijks University, Leiden.

Prins, W. (1967), Equilibrium swelling due to sorption, in *The Solid–Gas Interface*, E. A. Flood, ed., Marcel Dekker, New York, pp. 667–690.

Prutton, M. (1975), *Surface Physics*, Clarendon Press, Oxford.

Pryor, E. J. (1961), Flotation's early years, in Fiftieth Anniversary of Froth Flotation in the U.S.A., *Colo. Sch. Mines Q. Golden, Colo.* **56**(3), 217–239.

Pryor, E. J. (1965), *Mineral Processing*, 3rd ed., Elsevier, Amsterdam.

Pryor, E. J., and Wrobel, S. A. (1951), Studies in cassiterite flotation, *Trans. IMM.* **60**, 201–237.

Purcell, G., and Sun, S. C. (1963), Significance of double bonds in fatty acid flotation— An electrokinetic study, *Trans. AIME* **226**, 6–17.

Raghavan, S., and Fuerstenau, D. W. (1977), Characterization and pore structure analysis of a copper ore containing chrysocolla, *Int. J. Miner. Process.* **4**, 381–394.

Rahman, A., and Stillinger, F. H. (1971), Molecular dynamics study of liquid water, *J. Chem. Phys.* **55**, 3336–3359.

Raison, M. (1957), The Krafft point of binary mixtures of sodium alkyl sulphates, in *Proceedings of the 2nd International Congress of Surface Activity*, Vol. I, J. H. Schulman, ed., Butterworths, London, p. 374.

Ralston, A. W. (1948), *Fatty Acids and Their Derivatives*, Wiley, New York.

Ralston, A. W., and Hoerr, C. W. (1942a), Studies on high molecular weight aliphatic amines and their salts. VI. Electrical conductivities of aqueous solutions of the hydrochlorides of octyl-, decyl-, tetradecyl- and hexadecyl-amines, *J. Am. Chem. Soc.* **64**, 772–776.

Ralston, A. W., and Hoerr, C. W., (1942b), The solubilities of the normal saturated fatty acids, *J. Org. Chem.* **7**, 546–555.

Ralston, A. W., and Hoerr, C. W. (1945), Solubilities of binary mixtures of the saturated fatty acids, *J. Org. Chem.* **10**, 170–174.

Ralston, A. W., and Hoerr, C. W. (1946), The electrical behaviour of hexyl- and dodecylammonium chlorides in various dilutions of aqueous ethanol, *J. Am. Chem. Soc.* **68**, 2460–2464.

Ralston, A. W., Hoffman, E. J., Hoerr, C. W., and Selby, W. M. (1941), Studies on high molecular weight aliphatic amines and their salts. I. Behaviour of the hydrochlorides of dodecylamine and octadecylamine in water, *J. Am. Chem. Soc.* **63**, 1598–1601.

Ralston, A. W., Hoerr, C. W., Pool, W. O., and Harwood, H. J. (1944), Solubilities of high molecular weight normal aliphatic primary amines, *J. Org. Chem.* **9**, 102–112.

Ralston, A. W., Eggenberger, D. N., and Broome, F. K. (1949), The effect of inorganic electrolytes upon the conductivity of aqueous solutions of dodecylammonium chloride, *J. Am. Chem. Soc.* **71**, 2145–2149.

Ralston, O. C., and Allen, G. L. (1916), The flotation of oxidized ores, *Min. Sci. Press* **113**, 171–174.

Ramdohr, P. (1969), *The Ore Minerals and Their Intergrowths* (transl. by Chr. Amstutz), Pergamon Press, New York.

Randall, H. M., Fowler, R. G., Fuson, N., and Dangl, J. R. (1949), *Infrared Determination of Organic Structures*, Van Nostrand, New York.

Rao, S. R. (1962), Kinetics of decomposition of xanthates in the presence of lead salt, *J. Sci. Ind. Res. (India)* **21D**, 125–128.

Rao, S. R. (1969), The collector mechanism in flotation, *Sep. Sci.* **4**, 357–411.

Rao, S. R. (1971), *Xanthates and Related Compounds*, Marcel Dekker, New York.

Rao, S. R., and Patel, C. C. (1960), Kinetics of decomposition of xanthates in presence of ferric salts, *J. Mines Met. Fuels* **8**(7), 15–17.

Rao, S. R., and Patel, C. C. (1961), Kinetics of decomposition of xanthates in the presence of copper salt, *J. Sci. Ind. Res. (India)* **20D**, 299–303.

Rao, S. R., Moon, K. S., and Leja, J. (1976), Effect of grinding media on the surface reactions and flotation of heavy metal sulphides, in *Flotation, A. M. Gaudin Memorial Volume*, M. C. Fuerstenau, ed., AIME, New York, pp. 509–527.

Read, A. D., and Hollick, C. T. (1976), Selective flocculation techniques for recovery of fine particles, *Miner. Sci. Eng.* **8**(3), 202–213.

Read, A. D., and Kitchener, J. A. (1967), The thickness of wetting films, in *Wetting*, Society of Chemical Industry Monograph No. 25; also, Gordon & Breach, New York, pp. 300–313.

Read, A. D., and Kitchener, J. A. (1969), Wetting films on silica. *J. Colloid Interface Sci.* **30**, 391–398.

Read, A. D., and Manser, R. M. (1972), Surface polarizability and flotation: study of the effect of cation type on the oleate flotation of three orthosilicates, *Trans. IMM* **81**, C69–C78.

Read, A. D., and Manser, R. M. (1975), The action of fluoride as a modifying agent in silicate flotation, Warren Spring Laboratory, Stevenage, England.

Read, A. D., and Whitehead, A., Treatment of mineral combinations by selective flocculation, in *10th International Mineral Processing Congress, London* 1973, IMM, London, pp. 949–957.

Reay, D., and Ratcliff, G. A. (1973), *Removal of fine particles from water by dispersed air flotation: effects of bubble size and particle size on collection efficiency*, Can. J. Chem. Engineering, vol. 51, pp. 178–185.

Reed, R. M., and Tartar, H. V. (1936), A study of salts of higher alkyl sulfonic acids, *J. Am. Chem. Soc.* **58**, 322–332.

Reid, E. E. (1962), Thiocarbonic acids and derivatives, in *Organic Chemistry of Bivalent Sulphur*, Vol. IV, Chemical Publ. Co., New York, Chap. 2, pp. 131–195.

Renou, J., Francois-Rossetti, J., and Imelik, B. (1960), Étude des Solides Poreux, III. Isotherms D'Adsorption Irregulieres, *Bull. Soc. Chim. Fr.*, 446–450.

Reuter, B., and Stein, R. (1957), The oxidation of lead sulphide at low temperatures. I. Chemical and x-ray investigations, *Z. Elektrochem.* **61**, 440–449.

Revnivtsev, V. I., Khopunov, E. A., Shelegin, V. I., Shatailov, U. Z., Tomilo, V. M., and Goldobin, U. S. (1977), Selective liberation of minerals, in *Proceedings of the XIIth International Mineral Processing Congress, Sao Paulo* 1977, Nacional Publicačoes e Publicidade, Sao Paulo.

Rey, M. (1954), Flotation of oxidized ores of lead, copper and zinc, *Trans. IMM.* **63**, 541–547.

Rey, M. (1958), Differential flotation of lead–zinc ores, in *Progress in Mineral Dressing*, in Proceedings of the Fourth International Mineral Processing Congress, Stockholm 1957, sponsored and edited by Svenska Gruvforeningen and Jernkontoret, Almqvist and Wiksell, Stockholm, pp. 525–536.

Rey, M. (1979, 1980), Memoirs of milling and process metallurgy, *Trans. IMM*, Pt. I, **88**, C245–C250; Pt. II, **89**, C1–C6; Pt. III, **89**, C65–C70.

Rey, M., and Formanek, V. (1960), Some factors affecting selectivity in the differential flotation of lead–zinc ores, particularly in the presence of oxidised lead minerals, *Proceedings (5th) International Mineral Processing Congress* 1960, IMM, London, pp. 343–53.

Rey, M., Sitia, G., Raffinot, P., and Formanek, V. (1954), Flotation of oxidized zinc ores, *Trans. AIME* **199**, 416–420.

Rey, M., De Merre, P., Mancuso, R., and Formanek, V. (1961), Recent research and developments in flotation of oxidized ores of copper, lead and zinc, *Q. Colo. Sch. Mines* **56**(3), 163–175.

Rhodin, T. N. (1950), Low temperature oxidation of copper, *J. Am. Chem. Soc.* **72**, 5102–5106; also, in *Advances in Catalysis*, Vol. 5, W. G. Frankenbrug, V. I. Komarewsky, and E. K. Rideal, eds., Academic Press, New York, 1953, p. 39.

Rich, A., and Davidson, N., eds. (1968), *Structural Chemistry and Molecular Biology*, Freeman, San Francisco.

Richardson, P. E., and Maust, E. E. Jr. (1976), Surface stoichiometry of galena in aqueous electrolytes and its effect on xanthate interactions, in *Flotation, A. M. Gaudin Memorial Volume*, Vol. I, M. C. Fuerstenau, ed., AIME, New York, pp. 364–392.

Richmond, P. (1975), The theory and calculation of van der Waals forces, in *Colloid Science*, Vol. II, D. H. Everett, ed., The Chemical Society, London.

Rietveld, H. M., and Maslen, E. N. (1965), The crystal structure of cadmium *n*-butyl xanthate, *Acta Crystallogr.* **18**, 429–436.

Rinelli, G., and Marabini, A. M. (1973), Flotation of zinc and lead oxide–sulphide ores with chelating agents, in *Proceedings of the Tenth International Mineral Processing Congress, London* 1973, M. J. Jones, ed., IMM, London, p. 493.

Rinelli, G., Marabini, A. M., and Alesse, V. (1976), Flotation of cassiterite with salicylaldehyde as a collector, in *Flotation, A. M. Gaudin Memorial Volume*, Vol. I, M. C. Fuerstenau, ed., AIME, New York, pp. 549–560.

Ripan, R., Eger, I., and Mirel, C. (1963), Infrared spectra of some salts of the *O,O*-diethyl ester of dithiophosphoric acid (in French), *Rev. Chim. Acad. Rep. Pop. Roum.* **8**(2), 163–168.

Roberts, J. D., Stewart, R., and Caserio, M. C. (1971), *Organic Chemistry: Methane to Macromolecules*, W. A. Benjamin, New York.

Rochester, C. H. (1976), Infrared spectroscopic studies of powder surfaces and surface–adsorbate interactions including the solid/liquid interface, a review, *Powder Technol.* **13**, 157–176.

Rockett, J. (1962), The infrared spectra of metal dialkylphosphorodithioates, *Appl. Spectrosc.* **16**, 39–40.

Roentgen, W. C. (1891), *Wied. Ann. Physik.* **45**, 91.

Rogers, D. W., and Poling, G. W. (1978), Compositions and performance characteristics of some commercial polyacrylamide flocculants, *CIM Bull.*, May, 152–158.

Rogers, J. (1962), Principles of sulphide mineral flotation, in *Froth Flotation—50th Anniversary Volume*, D. W. Fuerstenau, ed., AIME, New York, pp. 139–169.

Rogers, J., and Schulman, J. H. (1957), A mechanism of the selective flotation of soluble salts in saturated solutions, in *Proceedings of the 2nd International Congress of Surface Activity, London*, 1957, J. H. Schulman, ed., Vol. III, pp. 243–251, Butterworths, London.

Roman, R. J., Fuerstenau, M. C., and Seidel, D. C. (1968), Mechanisms of soluble salt flotation, Pt. I, *Trans. AIME* **241**, 56–64.

Rosano, H. L., Breindel, K., Schulman, J. H., and Eydt, A. J. (1966), Mechanism of ionic exchange with carrier molecules through non-aqueous liquid membranes, *J. Colloid Interface Sci.* **22**, 58–67.

Rosen, M. J. (1978), *Surfactants and Interfacial Phenomena*, Wiley, New York.

Ross, S. (1967), The heterogeneity of solid substrates, in *Surfaces and Interfaces 1*, J. J. Burke, N. L. Reed, and V. Weiss, eds., Syracuse University Press, Syracuse, New York, pp. 169–196.

Ross, S., and Boyd, G. E. (1947), New observations on two-dimensional condensation phenomena, MDCC Document 662, Office of Technical Services, Department of Commerce, Washington, D.C.

Ross, S., and Butler, J. N. (1956), The inhibition of foaming. VII. Effects of antifoaming agents on surface-plastic solutions, *J. Phys. Chem.* **60**, 1255–1258.

Ross, S., and Cutillas, M. J. (1955), The transmission of light by stable foams, *J. Phys. Chem.*, **59**, 863–866.

Ross, S., and Haak, R. M. (1958), Inhibition of foaming. IX. Changes in the rate of attaining surface tension equilibrium in solutions of surface-active agents on addition of foam inhibitors and foam stabilizers, *J. Phys. Chem.*, **62**, 1260–1264.

Ross, S., and Olivier, J. P. (1964), *On Physical Adsorption*, Interscience, New York.

Ross, S., and Winkler, W. (1954), On physical adsorption. V. Two-dimensional condensation of ethane on surfaces of solids at 90°K, *J. Am. Chem. Soc.* **76**, 2637–40.

Rubio, J., and Goldfarb, J. (1975), Separation of chrysocolla from quartz by selective flocculation by polyacrylamide-type flocculants, *Trans. IMM* **84**, C123–C127.

Rubio, J., and Kitchener, J. A. (1977), New basis for selective flocculation of mineral slimes, *Trans. IMM*, **86**, pp. C97–C100.

Rubio, J., and Kitchener, J. A. (1976), The mechanism of adorption of poly(ethylene oxide) flocculant on silica, *J. Colloid Interface Sci.* **57**, 132–142.

Rybczynski, W. (1911), Ueber die Fortschreitende Bewegung einer Fluessigen Kugel in einem Zachen Medium, *Bull. Acad. Sci. Cracovie Ser. A*, 40–46.

Sagert, N. H., Quinn, M. J., Cribbs, S. C., and Rosinger E. L. J. (1976), Bubble coalescence in aqueous solutions of *n*-alcohols, in *Foams*, R. J. Akers, ed., Academic Press, New York, pp. 147–162.

Salamon, M. D. G., and Lancaster, F. H. (1973), in Proceedings of the 10th International Symposium, *Application of Computer Methods in the Mineral Industry*, South African IMM, Johannesburg.

Salamy, S. G., and Nixon, J. C. (1953), The application of electrochemical methods to flotation research, in *Recent Developments in Mineral Dressing, Proceedings of the First International Mineral Processing Congress*, London 1952, IMM, London, pp. 503–518.

Salamy, S. G., and Nixon, J. C. (1954), Reaction between a mercury surface and some flotation reagents: an electrochemical study, *Aust. J. Chem.* **7**, 146–156.

Samoilov, O. Ya (1946), *Zh. Fiz. Khim.* **20**, 1411.

Samoilov, O. Ya (1957), *Structure of Aqueous Electrolyte Solutions and the Hydration of Ions* (trans. by D. J. G. Ives), Consultants Bureau, New York (1965).

Samorjai, G. A. (1972), *Principles of Surface Chemistry*, Prentice-Hall, Englewood Cliffs, New Jersey.

Sandvik, K. L. (1977), Die Oxydation in basischen Sulfid-Trüben und die Wirkung der entstehenden Thiosulfate auf die Flotationseigenschaften einiger Sulfide, *Erzmetall* **30**(9), 391–395.

Schay, G. (1969), Adsorption from solutions of nonelectrolytes, in *Surface and Colloid Science*, Vol. 2, E. Matijevic, ed., Wiley-Interscience, New York, pp. 155–212.

Schay, G. (1975), Thermodynamics of adsorption from solution, in *Physical Chemistry: Enriching Topics from Colloid and Surface Science*, H. V. Olphen and K. J. Mysels, eds. Theorex, La Jolla, California, pp. 229–249.

Schick, M. J., ed. (1967), *Nonionic Surfactants*, Marcel Dekker, New York.

Schick, M. J., and Fowkes, F. M. (1957), Foam stabilizing additives for synthetic detergents; interaction of additives and detergents in mixed micelles, *J. Phys. Chem.* **61**, 1062–1068.

Schindler, P. W., Fürst, B., Dick, R., and Wolf, P. U. (1976), Ligand properties of surface silanol groups, I. Surface complex formation with Fe^+, Cu^+, Cd^+ a-d Pb^+, *J. Colloid Interface Sci.* **55**(2), 469–475.

Schouten, C. (1962), *Determination Tables for Ore Microscopy*, Elsevier, New York.

Schubert, H. (1965), Beitrag zur Theorie der Flotation von KCl und NaCl, *Bergakademie* **17**(8), 485–491, Mitteilung aus dem Institut für Aufbereitung der Bergakademie Freiberg, Sa.

Schubert, H. (1966), Zur Theorie der Alkalisalzflotation, *Arbeitungs Tech.* No. 6, 305–313.

Schubert, H. (1967), *Aufbereitung Fester Mineralischer Rohstoffe*, Vol. II, VEB Deutscher Verlag, Leipzig, pp. 311–312.

Schubert, H. (1972), *Die Rolle der Assoziation der unpolaren Gruppen bei der Sammleradsorption*, VEB Deutscher Verlag fur Grundstoffindustrie, Leipzig.

Schubert, H., and Schneider, W. (1968), Role of non-polar groups association in collector adsorption, in Proceedings VIII International Mineral Processing Congress, Leningrad, Vol. 2, Paper S-9, Mekhanobr Institute, Leningrad, pp. 315–324.

Schuhmann, R., Jr., and Prakash, B. (1950), Effects of activators and alizarin dyes on soap flotation of cassiterite and fluorite, *Trans. AIME* **187**, 601–608.

Schulman, J. H., ed. (1957), *Proceedings of the 2nd International Congress of Surface Activity*, London, 1957, vols. I–IV, Butterworths, London.

Schulman, J. H., and Friend, J. A. (1949), Penetration and complex formation in monomolecular layers (in German), *Kolloid Z.* **115**, 67–75.

Schulman, J. H., and Hughes, A. H. (1935), Monolayers of proteolytic enzymes and proteins: (a) Enzyme reactions and penetration of monolayers; (b) Mixed unimolecular films, *Biochem. J.* **29**, 1236–1242; 1243–1252.

Schulman, J. H., and Leja, J. (1954), Molecular interactions at solid–liquid interfaces, *Koll. Ztsch.* **136**, 107–119.

Schulman, J. H., and Leja, J. (1958), Static and dynamic attachment of air bubbles to solid surfaces, in *Surface Phenomena in Chemistry and Biology*, J. F. Danielli, K. G. A. Pankhurst, and A. C. Riddiford, eds. Pergamon Press, New York, pp. 236–245.

Schulman, J. H., and Rideal, E. K. (1937), Molecular interaction in monolayers, Part 1, Complexes between large molecules, Part II, The action of haemolytic and agglutinating agents on lipoprotein monolayers, *Proc. R. Soc. London* **122**B, 29–57.

Schulman, J. H., and Smith, T. D. (1953), Selective flotation of metals and minerals, in *Recent Developments in Mineral Dressing*, Proceedings of the First International Mineral Processing Congress, London 1952, IMM, London, pp. 393–414.

Schulman, J. H., and Stenhagen, E. (1938), Molecular interaction in monolayers, Part III, Complex formation in lipoid monolayers, *Proc. R. Soc. London* **126**B, 356–369.

Schulman, J. H., Waterhouse, R. B., and Spink, J. A. (1956), Adhesion of amphipathic molecules to solid surfaces, *Kolloid-Z.* **146**, 77–95.

Schulze, H. J. (1977), New theoretical and experimental investigations on stability of bubble/particle aggregates in flotation: a theory on the upper particle size of floatability, *Int. J. Miner. Process.* **4**, 241–255.

Schwartz, A. M. (1972), The physical chemistry of detergency, in *Surface and Colloid Science*, Vol. V, pp. 195–244, E. Matijević, ed., J. Wiley, New York.

Schwartz, A. M. (1975), The dynamics of contact angle phenomena, *Advances in Colloid and Interface Science* **4**(4), 349–374.

Schwartz, A. M. (1979), Research techniques in detergency, in *Surface and Colloid Science*, Vol. 11, pp. 305–334, Experimental Methods, R. J. Good, and R. R. Stromberg, eds., Plenum Press, New York.

Schwartz, A. M., and Perry, J. W. (1949), *Surface Active Agents*, Interscience, New York.

Schwartz, A. M., and Tejada, S. B. (1972), Studies of dynamic contact angles on solids, *J. Colloid Interface Sci.* **38**(2), 359–375.

Schwartz, A. M., Perry, J. W., and Berch, J. (1966), *Surface Active Agents and Detergents*, Vol. II, Interscience, New York.

Schwarzenbach, G. (1961), The general, selective and specific formation of complexes by metallic cations, in *Advances in Inorganic Chemistry and Radiochemistry*, Vol. 3, H. J. Emeléus and A. G. Sharpe, eds., Academic Press, New York, pp. 257–285.

Scobie, A. G., and Wyslouzil, D. M. (1968), Design, construction and operation of the Lake Dufault Treatment Plant, metallurgical testing, *Trans. CIM* **71**, 81–87.

Scott, A. F., ed. (1964), *Survey of Progress in Chemistry*, Vol. 2, Academic Press, New York, p. 219.

Scott, J. W., and Poling, G. W. (1973), Chrysocolla flotation, *Can. Metall. Q.* **12**, 1–8.

Scott, T. R., and Dyson, N. F. (1968), The catalyzed oxidation of zinc sulphide under acid pressure leaching conditions, *Trans. AIME* **242**, 1815–1830.

Scowen, R. V. (1966), unpublished work at the University of Alberta, Edmonton.

Scowen, R. V., and Leja, J. (1967), Spectrophotometric studies on surfactants. I. Interactions between cationic and anionic surfactants. II. Infrared study of adsorption from solutions of single and mixed surfactants, on copper substrates, *Can. J. Chem.* **45**, 2821–2835.

Sebba, F. (1962), *Ion Flotation*, Elsevier, Amsterdam.

See, J. B. (1976), Fluorspar and fluorine compounds in high temperature smelting and refining of metals, *Miner. Sci. Eng.* **8**(4), 217–241.

Seidel, D. C., Roman, R. J., and Fuerstenau, M. C. (1968), Mechanisms of soluble salt flotation, Pt. II, *Trans AIME* **241**, 64–70.

Seidell, A. (1959), *Solubilities; Inorganic and Metal-Organic Compounds*, 4th ed., revised by W. F. Linke, Van Nostrand, New York.

Seitz, F. (1940), *The Modern Theory of Solids*, McGraw-Hill, New York.

Sěr, F., MacDonald, I. D., Whyte, R. M., and Hillary, J. E. (1970), Sulphydric flotation of previously sulphidized oxide copper minerals of Nchanga Copper Mines, *Rudy (Prague)* **18**, 167–174.

Shafeev, R. Sh. (1966), Relationship between the semiconductor characteristics of minerals and the action of flotation reagents, in *Flotation Properties of Semiconductor Minerals*, Nauka, Moscow.

Shah, D. O., Bansal, V. K., Chan, K. S., and Hsieh, W. C. (1977), The structure, formation and phase-inversion of microemulsions, in *Improved Oil Recovery by Sur-*

factant and Polymer Flooding, D. O. Shah and R. S. Schechter, eds., Academic Press, New York, pp. 293–337.

Shankaranarayana, M. L., and Patel, C. C. (1961), Infrared spectra and the structures of xanthates and dixantogens, *Can. J. Chem.* **39**, 1633–1637.

Shankaranarayana, M. L., and Patel, C. C. (1965), The electronic spectra of some derivatives of xanthic, dithiocarbamatic and trithiocarbamic acids, *Acta. Chem. Scand.* **19**, 1113–1119.

Sharpe, A. G., and Chadwick, B. M. (1966), Transition metalcyanides and their complexes, *Adv. Inorg. Chem. Radiochem.* **8**, 83–176.

Shaw, D. J. (1966), *Introduction to Colloid and Surface Chemistry*, Butterworths, London.

Shaw, D. J. (1969), *Electrophoresis*, Academic Press, New York.

Sheikh, N. (1972), *The Chemical Stability of Heavy Metal Xanthates*, Ph.D. thesis, University of British Columbia, Vancouver.

Sheikh, N., and Leja, J. (1973), Stability of lead ethyl xanthate in aqueous systems, *Trans. AIME* **254**, 260-264.

Sheikh, N., and Leja, J. (1974), Precipitation and stability of copper ethyl xanthate in hot acid and alkaline solutions, *J. Colloid Interface Sci.* **47**, 300–308.

Sheikh, N., and Leja, J. (1977), Mossbauer spectroscopy of Fe xanthates, *Sep. Sci.* **12**(5), 529–540.

Sheka, Z. A., and Kriss, I. I. (1959), Metal xanthates, *Rab. Khim. Rastvorov Kompleksn. Soedin., Akad. Nauk Ukr. SSR*, **2**, 135–162.

Sheludko, A. (1957), Über das Ausfliessen der Lösung aus Schaumfilmen, *Kolloid Z.* **155**, 39–44.

Sheludko, A. (1958), Spontaneous thinning of thin, bilateral liquid films, *Dokl. Akad. Nauk* **123**, 1074–1076.

Sheludko, A. (1962), Sur Certaines Particularités des Lames Mousseuses, *Proc. Kon. Ned. Akad. Weten. Ser. B* **65**, 76–108.

Sheludko, A. (1963), Zur Theorie der Flotation, *Kolloid Z.* **191**, 52-58.

Sheludko, A. (1966a), *Colloid Chemistry*, Elsevier, Amsterdam.

Sheludko, A. (1966b), Elasticity of adsorption films, *Abh. Deut. Akad. Wiss. Berlin Kl. Chem. Geol. Biol. No.*, 6, 531–541.

Sheludko, A. (1967), Thin liquid films, *Advances in Colloid Interface Sci.* **1**, J. T. G. Overbeek, W. Prins, and A. C. Zettlemoyer, eds., Elsevier, Amsterdam pp. 391–464.

Sheludko, A., and Exerova, D. (1959), Electrostatic pressure in foam films of aqueous electrolyte solutions, *Kolloid Z.* **165**, 148–151.

Sheludko, A., and Platikanov, D. (1961), Investigation of thin liquid layers on mercury, *Kolloidn. Zh.* **175**, 150–158.

Sheludko, A., and Polikarova, R. (1956), Influence of thickness and concentration of soap on the tear of soap membranes, *Ann. Univ. Sofia Fac. Sci. Phys. Math.* **49**(2), 15–24.

Sheludko, A., Dessimirov, G., and Nikolov, K. (1956), Flow of solution through foam membranes, *Ann. Univ. Sofia Fac. Sci. Phys. Math.* **49**(2), 127–141.

Sheludko, A., Radoev, B., and Kolarov, T. (1968), (a) Surface tension of a thin layer and the contact angle between the layer and bulk liquid, *God. Sofii. Univ. Khim. Fak. 1966–1967* **61**, 137–154; (b) Tension of liquid films and contact angles between film and bulk liquid, *Trans. Faraday Soc.* **64**, 2213–2220.

Sheludko, A., Toshev, B. V., and Bojadiev, D. (1976), Attachment of particles to a liquid surface, *Faraday Transactions I* **72**, 2815–2828.

Sheludko, A., Exerova, D., and Platikanov, D. (1970a), Thin liquid films, *Izv. Otd. Khim. Nauki Bulg. Akad. Nauk.* **2**, 499–509.

Sheludko, A., Toshev, B. V., and Bojadiev, D. (1976), Attachment of particles to a liquid surface, *Faraday Transactions I* **72**, pp. 2815–2828.

Sheludko, A., Tschaljowska, Sl., and Fabrikant, A. (1970b), Contact between a gas bubble and a solid surface and froth flotation, in *Thin liquid films and boundary layers, Spec. Discuss. Faraday Soc.* **1**, 112–117.

Shergold, H. L. (1976), Two-liquid flotation for the treatment of mineral slimes, *Ind. Miner. (S.-Etienne, Fr.) Mineralurgie*, November 1976, 192–205.

Shergold, H. L., and Mellgren, O. (1969), Concentration of minerals at the oil–water interface: hematite–isoctane–water system in the presence of sodium dodecyl sulphate, *Trans. IMM* **78**, C121–C132.

Shergold, H. L., and Mellgren, O. (1970), Concentration of minerals at the oil/water interface, *Trans. AIME* **247**, 149–159.

Shergold, H. L., and Mellgren, O. (1971), Concentration of hematite at the isoctane–water interface with dodecylamine as a collector, *Trans. IMM* **80**, C60–C68.

Shergold, H. L., and Parkins, E. J. (1977), Effect of oxygen on conditioning and flotation of an ilmenite ore, *Trans. IMM* **86**, C41–C43.

Shergold, H. L., Mellgren, O., and Kitchener, J. A. (1966), Demountable electrophoretic cell for mineral particles, *Trans. IMM* **75**, C331–C332.

Shergold, H. L., Prosser, L. P., and Mellgren, O. (1968), New region of floatability in the hematite–dodecylamine system, *Trans. IMM* **77**, C166.

Shimoiizaka, J., Usui, S., Matsuoka, I., and Sasaki, H. (1976), Depression of galena flotation by sulphite or chromate ion, in *Flotation, A. M. Gaudin Memorial Volume*, Vol. I, M. C. Fuerstenau, ed., pp. 393–413.

Shinoda, K., ed., (1967), *Solvent properties of surfactant solutions*, Marcel Dekker, New York.

Shinoda, K., and Friberg, S. (1975), Microemulsions: colloidal aspects, *Adv. Colloid Interface Sci.* **4**, 281–300.

Shinoda, K., Nakagawa, T., Tamamushi, B., and Isemura, T. (1963), *Colloidal Surfactants*, Academic Press, New York.

Shopov, D., Ivanov, S., Kateva, I., and Karshalykov, K. (1970), Infrared spectra of dithiophosphates, *Izv. Otd. Khim. Nauki Bulg. Akad. Nauk.* **3**(1), 33–40 [*Chem. Abstr.* 74, 7989 (1971)].

Shorsher, I. N. (1946), Flotation of cassiterite, *Tsvet. Metall.* **19**(6), 13–19 [*Chem. Abstr.* **41**, 3410b (1947)].

Short, M. N. (1940), Microscopic determination of the ore minerals, *U.S. Geol. Surv. Bull.* **914**, 1–314.

Shuttleworth, R. (1950), The surface tension of solids, *Proc. Phys. Soc. London* **63A**, 444–457.

Sillén, L. G. (1959), Quantitative studies of hydrolytic equilibriums, *Q. Rev.* **13**, 146–168.

Sillén, L. G., and Martell, A. E. (1964), *Stability Constants of Metal–Ion Complexes*, Special Publication No. 17, The Chemical Society, London.

Sillén, L. G., and Martell, A. E. (1971), Stability constants of metal-ion complexes, Supplement No. 1, *Special Publication No. 25*, The Chemical Society, London.

Simard, G. L., Chupak, J., and Salley, D. J. (1950), Radiotracer studies on the interaction of dithiophosphate with galena, *Trans. AIME* **187**, 359–364.

Singer, P. C., and Stumm, W. (1969), Oxygenation of ferrous iron, in Federal Water

Quality Administration, Water Pollution Control Research Series 14010-06/09, Washington D.C. 20242, June 1969 [reproduced in Hill, R. D., and Wilmoth, R. C., Limestone treatment of acid mine drainage, *Trans. AIME* **250**, 162–166, (June 1971)].

Singewald, A. (1961), Zum gegenwärtigen Stand der Erkenntnisse in der Salzflotation, *Chem. Ing. Tech.* **33**(8), I; 376–393; II, 558–572; III, 676–688.

Singewald, A. (1961), An investigation of the mechanisms of selective salt flotation and discussion of underlying theories in the Proceedings of the Symposium on the fiftieth Anniversary of Froth Flotation in the U.S.A., *Q. Colo. Sch. Mines* **56**(3), 65–88.

Sirianni, A. F., Capes, C. E., and Puddington, I. E. (1969), Recent experience with the spherical agglomeration process, *Can. J. Chem. Eng.* **47**, 166–170.

Siu, R. G. H. (1951), *Microbial decomposition of cellulose*, Reinhold, New York.

Slater, R. W., Clark, J. P., and Kitchener, J. A. (1969), Chemical factors in the flocculation of mineral slurries with polymeric flocculants, *Proc. Brit. Ceram. Soc.* **13**, 1.

Smani, M. S., Blazy, P., and Cases, J. M. (1975), Beneficiation of sedimentary Moroccan phosphate ores, parts I-IV, *Trans. AIME* **258**, 168–182.

Smith, H. H. (1916), The theory of flotation, *Min. Sci. Press* **113**, 16–19.

Smith, H. M., and Puddington, I. E. (1960), Spherical agglomeration of barium sulphate, *Can. J. Chem.* **38**, 1911–1916.

Smith, H. W. (1976), Computer control in flotation plants, in *Flotation Plants, A. M. Gaudin Memorial Volume*, Vol. 2, M. C. Fuerstenau, ed., AIME, New York, pp. 963–993.

Smith, R. W. (1971), Relations among equilibrium and nonequilibrium aqueous species of aluminum hydroxy complexes, in *Nonequilibrium Systems in Natural Water Chemistry*, Advances in Chemistry Series No. 106, American Chemical Society, Washington, D.C.

Smith, R. W. (1973), Effect of amine structure in cationic flotation of quartz, *Trans. AIME* **254**, 353–357.

Smith, R.W., and Akhtar, S. (1976), Cationic flotation of oxides and silicates, in *Flotation, A. M. Gaudin Memorial Volume*, Vol. I, M. C. Fuerstenau, ed., AIME, New York, pp. 87–116.

Smith, R. W., and Lai, R. W. M. (1966), On the relationship between contact angle and flotation behaviour, *Trans. AIME* **235**, 413–418.

Smith, R. W., Haddenham, R., and Schroeder, C. (1973), Amphoteric surfactants as flotation collectors, *Trans. AIME* **254**, 231–235.

Smith, T., and Lindberg, G. (1978), Effects of acoustic energy on contact angle measurements, *J. Colloid Interface Sci.* **66**(2), 363–366.

Smolders, C. A. (1961), Contact angles, Wetting and de-wetting of mercury, Ph. D. thesis, B. Centen-s Uitgeversmaatschappij, Hilversum, Holland.

Smolders, C. A., and Duyvis, E. M. (1961), Contact angles, Wetting and de-wetting of mercury. Part 1. A Critical examination of surface tension measurement by the sessile drop method, *Rec. Trav. Chim. Pays-Bas* **80**, 635.

Smoluchowski, M. (1903), *Krak. Anz.* 182.

Sobieraj, S., and Laskowski, J. (1973), Flotation of chromite, (1) Early research and recent trends, (2) Flotation of chromite and surface properties of spinel minerals, *Trans. IMM* **82**, C207.

Solnyshkin, V. I., ed. (1968), *Flotation Agents and Effects*, engl. transl. C. Nisenbaum (1970), Israel Program for Scientific Translations, Jerusalem.

Solozhenkin, N. M., and Zinchenko, Z. A. (1970), in Discussion of Paper by Prasad and Rao, *Proceedings Ninth International Mineral Processing Congress, Prague, 1970*, 3rd part, Ustav Pro Vyzkum Rud, Prague, p. 109.

Somasundaran, P. (1972), Foam separation methods, in *Separation and Purification Methods, A Supplement to The Journal of Separation Science*, Vol. I(1), 117–198.

Somasundaran, P., ed. (1980), *Fine Particles Processing*, Proceedings of the International Symposium in Las Vegas, Nevada, February 1980, Vols. I and II, AIME, New York.

Somasundaran, P., and Agar, G. E. (1967), The zero point of charge of calcite, *J. Colloid Interface Sci.* **24**, 433–440.

Somasundaran, P., and Fuerstenau, D. W. (1966), Mechanisms of alkyl sulfonate adsorption at the alumina–water interface, *J. Phys. Chem.* **70**, 90–96.

Somasundaran, P., and Fuerstenau, D. W. (1968), On incipient flotation conditions, *Trans. AIME* **241**, 102–104.

Somasundaran, P., and Grieves, R. B., eds. (1975), Advances in interfacial phenomena on particulate/solution/gas systems, Application to flotation research, *AIChE Symp. Ser.*, No. 150.

Somasundaran, P., and Hanna, H. S. (1977), Physico-chemical aspects of adsorption at solid/liquid interfaces, in *Improved Oil Recovery by Surfactant and Polymer Flooding*, D. O. Shah and R. S. Schechter, eds., Academic Press, New York, pp. 205–274.

Somasundaran, P., and Kulkarni, R. D. (1973), Effect of chain length of perfluoro-surfactants as collectors, *Trans. IMM* **82**, C164–C167.

Somasundaran, P., and Moudgil, B. M. (1974), The effect of dissolved hydrocarbon gases in surfactant solutions on froth flotation of minerals, *J. Coll. Interface Sci.* **47**(2), 290–299.

Somasundaran, P., Healy, T. W., and Fuerstenau, D. W. (1964), Surfactant adsorption at the solid/liquid interface—dependence of mechanism on chain length, *J. Phys. Chem.* **68**, 3562–3566.

Somasundaran, P., Ramachandran, S., and Kulkarni, R. D. (1977), Measurement of streaming potentials, *J. Electrochem Soc. India* **26**(2), 7–13.

Sparrow, G., Pomianowski, A., and Leja, J. (1977), Soluble copper xanthate complexes, *Sep. Sci.* **12**(1), 87–102.

Spedden, H. R., and Hannan, W. S. (1948), Attachment of mineral particles to air bubbles in flotation, *Min. Technol.* **12**(2), AIME Tech. Publ. 2354.

Spira, P., and Rosenblum, F. (1974), The oxygen demand of flotation pulps, in *Proceedings of the 6th Annual Meeting of Canadian Mineral Processors*, Department Energy, Mines and Resources, Ottawa, Canada, p. 73.

Sproule, K., Harcourt, G. A., and Renzoni, L. S. (1961), Treatment of nickel–copper matte, in *Extractive Metallurgy of Copper, Nickel and Cobalt*, P. Queneau, ed., Interscience, New York pp. 33–54.

Stamboliadis, E., and Salman, T. (1976), Solubility product of metal dithiophosphates, *Trans. AIME* **260**, 250–253.

Stepanov, B. A., Kakovsky, I. A., and Serebryakova, N. V. (1959), Oxidation–reduction potentials of xanthates, *Nauch. Dokl. Vyssh. Shk. Khim. Khim. Technol.*, No. 2, 277–279.

Sterling, C. V., and Scriven, L. E. (1959), Interfacial turbulence: hydrodynamic instability and the Marangoni effect, *Am. Inst. Chem. Eng. J.* **5**, 514–523.

Stern, O. (1924), The theory of the electrolytic double layer, *Z. Elektrochem.* **30**, 508–516.

Stillinger, F. H., and Rahman, A. (1972), Molecular dynamics study of temperature effects on water structure and kinetics, *J. Chem. Phys.* **57**, 1281–1292.

Stillinger, F. H., and Rahman, A. (1974), Molecular dynamics study of liquid water under high compression, *J. Chem. Phys.* **61**, 4973.

Stirton, A. J. (1962), α-Sulfo fatty acids and derivatives. Synthesis, properties and use, *J. Am. Oil Chem. Soc.* **39**, 490–496.

Stoev, S. M., and Pirinkov, S. (1966), The influence of sonic vibrations on the secondary froth concentration, *Ugol Ukr.* **10**(9), 45–47.

Stoev, S. M., and Watson, D. (1967), Influence of ultrasonic vibrations on secondary concentration of minerals in a froth, *Trans. IMM* **76**, C284–C286.

Stokes, G. C. (1851), On the effect of the internal friction of fluids on the motion of pendulums, *Proc. Cambridge Philos. Soc.* I, 1843–63, 104–106; *Trans. Cambridge Philos. Soc.* IX, Part II, 8–106.

Stumm, W. (1967), Metal ions in aqueous solutions, in *Principles and Applications of Water Chemistry*, S. D. Faust and J. V. Hunter, eds., Wiley, New York, pp. 520–560.

Stumm, W., and Morgan, J. J. (1970), *Aquatic Chemistry*, Wiley-Interscience, New York, Chap. 5, pp. 180–196; 227–228.

Stumpf, A., and Berube, Y. (1973), Aqueous oxidation of molybdenite in chalcopyrite concentrates, *Trans. AIME* **254**, 305.

Sun, S. C. (1943), The mechanism of slime-coating, *Trans. AIME* **153**, 479–492.

Sun, S. C. (1952), Frothing characteristics of pine oils in flotation, *Trans. AIME* **193**, 65–71, 1082.

Sun, S. C. (1954a), Effects of oxidation of coals on their flotation properties, *Trans. AIME* **199**, 396–401.

Sun, S. C. (1954b), Hypothesis for different floatabilities of coals, carbons and hydrocarbon minerals, *Trans. AIME* **199**, 67–75.

Sun, S. C., and Zimmerman, R. E. (1950), The mechanism of coarse coal and mineral froth flotations, *Trans. AIME* **187**, 616–622.

Sun, S. C., Snow, R. E., and Purcell, V. I. (1957), Flotation characteristics of a Florida leached zone phosphate ore with fatty acids, *Trans. AIME* **208**, 70–75.

Sutherland, K. L. (1948), Physical chemistry of flotation. XI. Kinetics of the flotation process, *J. Phys. Chem.* **52**, 394–424.

Sutherland, K. L. (1950), Ph.D. thesis, London University, also Rideal, E. K., and Sutherland, K. L., *Trans. Faraday Soc.* **48**, 1109 (1952).

Sutherland, K. L. (1951), The change of surface and interfacial tensions of solutions with time, *Rev. Pure Appl. Chem.* **1**(1): 35–50.

Sutherland, K. L. (1954), The oscillating jet method for the measurement of surface tension, *Aust. J. Chem.* **7**, 319–328.

Sutherland, K. L., and Wark, I. W. (1955), *Principles of Flotation*, Australasian IMM, Melbourne.

Sutulov, A. (1974), *Copper Porphyries*, University of Utah Printing Services, Salt Lake City, Utah.

Suwanasing, P., and Salman, T. (1970), Particle size in flotation studies, *Can. Mining J.* **91**(12), 55–62.

Sven-Nillson, I. (1934), Effect of contact time between mineral and air bubbles on flotation, *Kolloid Z.* **69**, 230–232.

Swisher, R. D. (1970), *Surfactant Biodegradation*, Marcel Dekker, New York.

Syrkin, Y. K., and Dyatkina, M. E. (1964), *Structure of Molecules and the Chemical Bond*, Dover, New York.

Szczypa, J., Chibowski, St., and Kuspit, K. (1979), Mechanism of adsorption of sodium laurate by calcium carbonate, *Trans. IMM* **88**, C11–13.

Szeglowski, Z. (1960), Electric potential of local galvanic elements on the galena surface and their influence upon the adsorption of potassium xanthate, in *Proceedings of the Third International Congress of Surface Activity, Cologne 1960* Vol. 2 (Vorträge in Originalfassung des III Internationalen Kongresses für Grenzflachenaktive Stoffe), publ. Verlag der Universitätsdruckerei, Mainz GMBH, **2**: 110-111.

Szymonska, J., and Czarnecki, J. (1978), Interaction of dodecyltrimethylammonium and ethylxanthate ions at the micelle surface, *J. Coll. Interface Sci.* **64**(2), 228–236.

Szyszkowski, B. V. (1908), Experimentelle Studien Über Kapillare Eigenschaften der wässerigen Lösungen von Fettsäuren, *Z. Phys. Chem.* **64**, 385–414.

Taggart, A. F. (1945, 1947), *Handbook of Mineral Dressing*, J. Wiley & Sons, New York.

Taggart, A. F., and Arbiter, N. (1943), Collector coatings in soap flotation, *Trans. AIME* **153**, 500–507.

Taggart, A. F. and Beach, F. E. (1916), An explanation of the flotation process, *Bull. AIME*, Aug., 1373–1389, also *Trans. AIME* **55**, 547–562.

Taggart, A. F., and Gaudin, A. M. (1923), Surface tension and adsorption phenomena in flotation, *Trans. AIME* **68**, 479–535.

Taggart, A. F., and Hassialis, M. D. (1946), Solubility product and bubble attachment in flotation, *Trans. AIME* **169**, 259–265.

Taggart, A. F., Taylor, T. C., and Knoll, A. F. (1930), Chemical reactions in flotation, *Trans. AIME* **87**, 217–260.

Taggart, A. F., del Giudice, G. R. M., and Ziehl, O. A. (1934), The case for the chemical theory of flotation, *Trans. AIME* **112**, 348–381.

Takahashi, K., and Iwasaki, I. (1969), Inductive effect of polar groups on methyl stretching vibrations of alkyl groups and its implication in flotation chemistry, *Trans. AIME* **244**, 66–71.

Talmud, D. L., and Lubman, N. M. (1930), New micro-method for measuring contact angles, *Zh. Fiz. Khim.* **1**(3).

Tamamushi, B., and Tamaki, K. (1957), Adsorption of long-chain electrolytes at the solid/liquid interface, in *Proceedings 2nd International Conference of Surface Activity*, Vol. III, J. H. Schulman, ed., Butterworths, London, pp. 449–456.

Tanila, J., Paakkinen, U., and Huhtelin, T. (1973), Outokumpu Oy's Pyhasalmi plant. Computer control pays off at Finnish copper flotation plant, *Eng. Min. J.* **174**(4), 112–117.

Tartar, H. V., and Cadle, R. D. (1939), *J. Phys. Chem.* **4**, 1173–1179.

Tartar, H. V., and Wright, K. A. (1939), Studies of sulfonates. III Solubilities, micelle formation and hydrates of the sodium salts of the higher alkyl sulfates, *J. Am. Chem. Soc.* **61**, 539–544.

Taube, H. (1970), *Electron Transfer Reactions of Complex Ions in Solution*, Academic Press, New York.

Teichmann, H. (1964), *Semiconductors*, transl. L. F. Secretan, Butterworths, London.

Temkin, M. (1938), Transition state in surface reactions, *Acta Physicochim. USSR* **8**, 141–170.

Thomas, J. G. N., and Schulman, J. H. (1954), Metal ion–monolayer interactions, effect of uranyl nitrate on myristic acid monolayers, *Trans. Faraday Soc.* **50**, 1128–1147.

Thomas, L. C. (1957), in Discussion of the Paper by D.E.C. Corbridge, Infrared analysis of phosphorus compounds, in *J. Appl. Chem.* **6**, 456–465; *Chem. Ind.*, 198.

Thomas, W. D. E., and Potter, L. (1975), Solution/air interfaces. I. An oscillating jet relative method for determining dynamic surface tensions, *J. Colloid Interface Sci.* **50**(3), 397–412.

Thorn, G. D., and Ludwig, R. A. (1962), *The Dithiocarbamates and Related Compounds*, Elsevier, Amsterdam.

Thunaes, A., and Abell, L. (1935), Private Reports, Consolidated Mining and Smelting Company of Canada Ltd., cited from Gaudin (1957), *Flotation*, McGraw-Hill, New York, p. 229.

Tipman, R. N. (1970), The reactions of potassium ethyl xanthate in aqueous solution, Ph.D. thesis, University of British Columbia.

Tipman, R. N., and Leja, J. (1975), Reactivity of xanthate and dixanthogen in aqueous solutions of different pH, *Colloid and Polymer Sci.* **253**, 4–10.

Tipman, R. N., Agar, G. E., and Paré, L. (1976), Flotation chemistry of the INCO matte separation process, in *Flotation, A. M. Gaudin Memorial Volume*, Vol. I, M. C. Fuerstenau, ed., AIME, New York, pp. 528–548.

Titoff, A. (1910), Adsorption von Gasen durch Kohle, *Z. Phys. Chem.* **74**, 641.

Tolun, R., and Kitchener, J. A. (1964), Electrochemical study of the galena–xanthate–oxygen flotation system, *Trans. IMM* **73**, 313–22.

Toperi, D., and Tolun, R. (1969), Electrochemical study and thermodynamic equilibria of the galena–xanthate flotation system, *Trans. IMM* **78**, C191–C197.

Töpfer, E. (1960), Cassiterite flotation tests using oleic acid, *p*-toluenearsonic acid and phosphates, *Freiberger Forschungsh. A* **163**, 63–75.

Töpfer, E. (1967), Studies on the flotation of cassiterite, *Trans. IMM* **76**, C192–195.

Törnell, B. (1966), Xanthate decomposition in acid media, Part I. Experimental technique and a study of the decomposition of ethyl xanthate, *Sven. Papperstidning* **69**(19), 658–663.

Torza, S., Cox, R. G., and Mason, S. G., (1971), Electrohydrodynamic deformation and burst of liquid drops, *Philos. Trans. R. Soc. London Ser. A* **269**, 295–319.

Toshima, S. (1971), The anode electrolyte interface, *Progr. Surf. Memb. Sci.* **4**, 231–297.

Trahar, W. J. (1965), The present status of cassiterite flotation, in *Proceedings VIII Commonwealth Mining and Metallurgical Congress, Australia and New Zealand*, Vol. 6, pp. 1125–1132, Australas. IMM, Melbourne.

Trahar, W. J. (1970), Cassiterite flotation, *Trans. IMM* **79**, C64–65.

Trahar, W. J. (1972), private communication, CSIRO, Melbourne.

Trahar, W. J. (1976), The selective flotation of galena from sphalerite with special reference to the effects of particle size, *Int. J. Miner. Process.* **3**, 151–166.

Trahar, W. J., and Warren, L. J. (1976), The floatability of very fine particles—a review, *Int. J. Miner. Process.* **3**, 103–131.

Traube, J. (1891), Über die Capillaritätsconstanten Organischer Stoffe in Wasserigen Lösungen, *Ann. Chem. Justus Liebigs* **265**, 27–55.

Tverdovskiy, I. P., and Frumkin, A. N. (1947), Relation between the wetting of mercury and the nature of the solvent, *Zh. Fiz. Khim.* **21**, 819–824.

Tyagai, V. A. (1965), Current–voltage characteristics of quasi-equilibrium electrochemical systems, *Elektrokhimiya* **1**(6), 685–688.

Usoni, L., Rinelli, G., Marabini, A. M., and Ghigi, G. (1968), Selective properties of flocculants and possibilities of their use in flotation of fine minerals, in *8th Internatio-*

nal Mineral Processing Congress, Leningrad, 1968, Vol. 1, Paper D-13, Institute Mekhanobr, Leningrad, pp. 514–533.

Usui, S., and Iwasaki, I. (1970a), Adsorption studies of dodecylamine at the mercury/solutions interface through differential capacity and electrocapillary measurements and their implication in flotation, *Trans. AIME* **247**, 213–219.

Usui, S., and Iwasaki, I. (1970b), Effect of pH on the adsorption of dodecylamine at the mercury/solution interface, *Trans. AIME* **247**, 220–225.

Usui, S., and Iwasaki, I. (1971), Inductive effect of polar groups on C–H and C–F Stretching vibrations of alkyl and perfluoroalkyl groups and its relation to adsorption, *J. Colloid Interface Sci.* **357**(4), 553–559.

Usui, S., and Sasaki, H. (1978), Zeta potential measurements of bubbles in aqueous surfactant solutions, *J. Coll. Interface Sci.* **65**, 36–45.

Uytenbogaardt, W. (1968), *Tables for Microscopic Identification of Ore Minerals*, Hafner Publishing Co., New York.

Vail, J. G. (1952), *Soluble Silicates, Their Properties and Uses*, Vol. 1, Reinhold, New York.

Van Lamsweerde-Gallez, D., Bisch, P. M., and Sanfeld, A. (1979), Hydrodynamic stability of monolayers at fluid-fluid interfaces, *J. Colloid Interface Sci.* **71**(3), 513–521.

Van Olphen, H. (1963), *An Introduction to Clay Colloid Chemistry*, Interscience, New York.

Van Olphen, H. (1965), Thermodynamics of interlayer adsorption of water in clays. I. Sodium vermiculite, *J. Colloid Sci.* **20**, 822–837.

Van Olphen, H., and Mysels, K. J. (1975), *Physical Chemistry, Enriching Topics from Colloid and Surface Science*, Theorex, La Jolla, California.

van Wazer, J. R. (1958), *Phosphorus and its Compounds*, Vol. 1, Interscience, New York.

van Wazer, J. R., and Callis, C. F. (1958), Complexing of metals by phosphate, *Chem. Rev.* **58**, 10–11.

Varbanov, R., Nikolov, D., and Nishkov, I. (1979), New static flotation technique increases mineral recovery and quality, *Min. Eng.* October 1979, 1455–1457.

Vazquez, L. A., Ramachandran, S., and Grauerholz, N. L. (1976), Selective flotation of scheelite, in *Flotation, A. M. Gaudin Memorial Volume*, Vol. I, M. C. Fuerstenau, ed., AIME, New York, pp. 580–596.

Venable, R. L., and Nauman, R. V. (1964), Micellar weights of and solubilitation of benzene by a series of tetradecylammonium bromides. The effect of the size of the charged head, *J. Phys. Chem.* **68**, 3498–3503.

Verwey, E. J. W. (1941), The charge distribution in the water molecule and the calculation of the intermolecular forces, *Rec. Trav. Chim.* **60**, 887.

Verwey, E. J. W. (1942), The interaction of ion and solvent in aqueous solutions of electrolytes, *Rec. Trav. Chim.* **61**, 127.

Verwey, E. J. W., and Overbeek, J. Th.G. (1948), *Theory of the Stability of Lyophobic Colloids*, Elsevier, Amsterdam.

Vetter, K. J. (1961), *Elektrochemische Kinetik*, Springer, Berlin.

Vetter, K. J., and Thiemke, G. (1960), Überspannung und Kinetik der Tl^{3+}/Tl^{+} Redoxelektrode, *Z. Elektrochem.* **64**, 805–812.

Vifian, A., and Iwasaki, I. (1968), Mineralogical beneficiation studies of the copper-nickel bearing Duluth Gabbro, *Trans. AIME* **241**, 421–431.

Vladimirova, M. G., and Kakovsky, I. A. (1950), Physiocochemical constants characteristic of the formation and composition of the lowest cuprous cyanide complex, *Zh. Priklad. Khim.* **23**, 580–598.

Volmer, M. (1932), The migration of adsorbed molecules on surfaces of solids, *Trans. Faraday Soc.* **28**, 359–363.

Vostrčil, J., and Juračka, F. (1976) *Commercial organic flocculants*, Noyes Data Corp., Park Ridge, New Jersey.

Vrij, A. (1966), Possible mechanism for the spontaneous rupture of thin, free liquid films, *Dicuss. Faraday Soc.* **42**, 23–33.

Wada, M. (1960), The wetting of solid in solutions of surface active substances as a function of solute concentration, in *Proceedings of the Fifth International Mineral Processing Congress, 1960*, IMM, London.

Wada, M., and Majima, H. (1963), Flotation of molybdenite. III. Adsorption of dithiocarbonate and dithiophosphate by molybdenite, *Tokoku Daigaku Senko Seiren Kenkyusho Iho* **19**(1), 21–26.

Wadsworth, M. E. (1951), Acid and base adsorption on solids from aqueous solutions of strong electrolytes, Ph.D. thesis, University of Utah.

Wakamatsu, T., and Fuerstenau, D. W. (1968), The effect of hydrocarbon chain length on the adsorption of sulfonates at the solid/water interface, in *Adsorption from Aqueous Solution*, Advances in Chemistry Series, No. 79, American Chemical Society, Washington, D.C., pp. 161–172.

Waksmundzki, A., and Maruszak, E. (1964), Investigations on some parameters influencing the adhesion forces between quartz particles and air bubbles in flotation systems (in Polish), *Rocz. Chem. Ann. Soc. Chim. Polonorum* **38**, 835–842.

Waksmundzki, A., and Maruszak, E. (1966), Variation of the contact angle during detachment of the air bubble from a solid surface, (in Polish), *Rocz. Chem. Ann. Soc. Chim. Polonorum* **40**, 649–656.

Waksmundzki, A., and Szczypa, J. (1961), Potentials of fluorite powder electrodes in solutions of sodium oleate, *Przem. Chem.* **40**(6), 330–332.

Waksmundzki, A., and Szczypa, J. (1962), On the possibility of application of powder electrodes for the determination of floatability of minerals, *Folia Societatis Scientiarum Lublinensis* **2**, 155–160 (publ. M. Curie-Sklodowska University, Lublin, Poland).

Waksmundzki, A., and Szymanski, E. (1965), The destruction of the structure of the sediment bed as a method for investigation of the adhesion forces between particles of mineral suspensions (in Polish) *Rocz. Chem.* **39**, 731–736.

Waksmundzki, A., Chojnacka, G., and Szymanski, E. (1965), On the parameters influencing adhesion forces between mineral particles in aqueous suspensions (in Polish), *Rocz. Chem. Ann. Soc. Chim. Polonorum* **39**, 895–900.

Waksmundzki, A., Neczaj-Hruzewicz, J., and Planik, M. (1971), Slime coatings formation and their properties in the sulphur–calcite system, *Trans. IMM* **80**, C200–203.

Waksmundzki, A., Neczaj-Hruzewicz, J., and Planik, M. (1972), Mechanism of carryover of gangue slimes during flotation of sulphur ore, *Trans. IMM* **81**, C249–251.

Wang, S. C. (1927), The mutual influence of the two atoms of hydrogen, *Phys. Z.* **28**, 663–666.

Wark, I. W. (1938), *Principles of Flotation*, Australasian Institute of Miningand Metallurgy, Melbourne.

Wark, I. W., and Cox, A. B. (1934), Principles of flotation. I. An experimental study of the effect of xanthates on the contact angles at mineral surfaces, *Trans. AIME* **112**, 189–244.

Warren, L. J., and Kitchener, J. A. (1972), Role of fluoride in the flotation of feldspar: adsorption on quartz, corundum and potassium feldspar, *Trans. IMM* **81**, C137–147.

Watson, D., and Manser, R. M. (1968), Some factors affecting the limiting conditions in cationic flotation of silicates, *Trans. IMM* **77**, C57–60.

Watt, G. W., and McCormick, B. J. (1965), The synthesis and characterization of methyl and ethyl xanthato-complexes of Pt(II), Pd(II), Ni(II), Cr(III) and Co(III), *J. Inorg. Nucl. Chem.* **27**, 898–900.

Weber, W. J., and Matijevic, E., eds. (1968), *Adsorption from Aqueous Solution*, Advances in Chemistry Series, No. 79, American Chemical Society, Washington, D.C.

Wells, A. F. (1962), *Structural Inorganic Chemistry*, Clarendon Press, Oxford.

Wells, P. F. (1973), Unpublished report of investigation, Department of Mineral Engineering, University of British Columbia, Vancouver.

Wells, P. F., Nagy, E., and van Cleave, A. B. (1972), The mechanism of the adsorption of alkyl xanthate species on galena, *Can. J. Chem. Eng.* **50**, 81–85.

Wenzel, R. N. (1949), Surface roughness and contact angle, *J. Phys. and Colloid Chem.* **53**, 1466–1467.

Whelan, P. F. (1953), Froth-flotation reagents for coal: Distillation fractions of commercial oils and some simple phenolic compounds, *J. Appl. Chem. (London)* **3**, 289–301.

Whelan, P. F., and Brown, D. J. (1956), Particle–bubble attachment in froth flotation, *Bull. IMM* **591**, 181–192.

Whistler, R. L., and Paschall, E. F., eds. (1965), *Starch: Chemistry and Technology*, Vol. 1, Academic Press, New York.

Whiteside, J. A. C., ed. (1974), *Instrumentation in the Mining and Metallurgical Industries*, ISA Publ., Pittsburgh.

Wilkomirsky, I. A., Watkinson, A. P., and Brimacombe, J. K. (1977), Kinetics of oxidation of molybdenite, *Trans. IMM* **86**, C16–22.

Will, G. (1969), Crystal structure analysis by neutron diffraction I, *Angew. Chem. Int. Ed. Engl.* **8**, 356–369.

Williams, H. E. (1948), *Cyanogen Compounds*, E. Arnold, London.

Winter, G., and Woods, R. (1973), The relation of collector redox potential to flotation efficiency: monothiocarbonates, *Sep. Sci.* **8**, 261–267.

Wolstenholme, G. A., and Schulman, J. H. (1950), Metal–monolayer interactions in aqueous systems, Parts I, II, *Trans. Faraday Soc.* **46**, 475–480.

Wolstenholme, G. A., and Schulman, J. H. (1951), Metal–monolayer interactions in aqueous systems, Part III, Steric effects with branched chain fatty acid monolayers, *Trans. Faraday Soc.* **47**, 788–794.

Woodburn, E. T. (1970), Mathematical modelling of flotation processes, *Miner. Sci. Eng.* **2**(2), 3–17.

Woodburn, E. T., King, R. P., and Colborn, R. P. (1971), The effect of particle size distribution on the performance of a phosphate flotation process, *Metall. Trans. N.Y.* **2**(11), 3163–74.

Woodcock, J. T., and Jones, M. H. (1969), Oxygen concentrations, redox potentials, xanthate residuals and other parameters in flotation plant pulps, in *Proceedings of the 9th Commonwealth Mining & Metallurgical Congress 1969*, IMM, London.

Woodcock, J. T., and Jones, M. H. (1970), Chemical environment in Australian lead–

zinc flotation plant pulps: I, pH, redox potentials, and oxygen concentrations, *Proc. Australas. Inst. Min. Met.* **235**, 45–60, II. Collector residuals, metals in solution, and other parameters, *Proc. Australas. Inst. Min. Met.* **235**, 61–76.

Woodruff, D. P. (1973), *The solid–liquid interface*, Cambridge University Press, Cambridge, England.

Woods, R. (1971), The oxidation of ethyl xanthate on platinum, gold, copper, and galena electrodes. Relation to the mechanism of mineral flotation, *J. Phys. Chem.* **75**, 354–362.

Woods, R. (1972a), Electrochemistry of sulphide flotation, *Proc. Australas. IMM* **241**, 53–61.

Woods, R. (1972b), The anodic oxidation of ethyl xanthate on metal and galena electrodes, *Aust. J. Chem.* **25**, 2329–2335.

Woods, R. (1976), Electrochemistry of sulphide flotation, in *Flotation, A. M. Gaudin Memorial Volume*, Vol. 1, M. C. Fuerstenau, ed., pp. 298–333.

Wottgen, E. (1967), Theory of the cumulative effect when floating sulphide minerals, *Tenside* **4**, 248–252.

Wottgen, E. (1969), Adsorption of phosphonic acids on cassiterite, *Trans. IMM* **78**, C91–97.

Wright, A. J., and Prasser, A. P. (1964), Study of the reactions and flotation of chrysocolla with alkali xanthates & sulphides, *Trans. IMM* **74**, 259–279.

Wrobel, S. A. (1969), Amphoteric flotation collectors, *Min. Miner. Eng.*, April, 35–40.

Wrobel, S. A. (1970), Amphoteric collectors in the concentration of some minerals by froth flotation, *Min. Miner. Eng.*, January, 42–45.

Wrobel, S. A. (1971), Activation of cassiterite for concentration by froth flotation, *Chem. Process. Eng.* **52**, 63–65.

Wroblowa, H., Kovac, Z. and Bockris, J. O'M. (1965), Isotherms and related data in the electro-adsorption of certain ions on mercury, *Trans. Faraday Soc.* **61**, 1523–1548.

Wronski, M. (1959), Kinetics of decomposition of xanthogenates in sodium hydroxide solutions, *Rocz. Chem.* **33**, 1071–1080; *Z. Phys. Chem. (Leipzig)* **211**, 113–117.

Wulff, G. (1901), *Zeitschrift für Kristallographie* **34**, 449.

Wyslouzil, D. M. (1970), private communication, Lakefield Research, Ontario.

Yamasaki, T., and Nanjo, M. (1969), Studies of xanthate complexes by solubility method, *Sci. Rep. RITU A* **21**(1), 45–62.

Yarar, B., and Kitchener, J. A. (1970), Selective flocculation of minerals: 1. Basic principles; 2. Experimental investigation of quartz, calcite, and galena, *Trans. IMM* **79**, C23–C33.

Yarar, B., Haydon, D. A., and Kitchener, J. A. (1969), Electrochemistry of the galena-diethyldithiocarbamate-oxygen flotation system, *Trans. IMM* **78**, 181–184.

Yates, D. J. C. (1956), The influence of the polar nature of the adsorbate on adsorption expansion, *J. Phys. Chem.* **60**, 543–549.

Yeager, E., and Kuta, J. (1970), Techniques for the Study of Electrode Processes, in *Physical Chemistry*, Vol. IXA, *Electrochemistry*, H. Eyring, ed., Academic Press, New York, pp. 345–461.

Yigit, E., Johnston, H. A., and Maroudas, N. G. (1969), Selective breakage in electrohydraulic comminution, *Trans. Instn. Chem. Engrg.* **47**, T332–T334.

Yoon, R. H., Salman, T., and Donnay, G. (1979), Predicting points of zero charge of oxides and hydroxides, *J. Colloid Interface Sci.* **70**(3), 483–493.

Young, D. M. (1951), Calculation of the adsorption behaviour of argon on octahedral potassium chloride, *Trans. Faraday Soc.* **47**, 1228–1233.

Young, D. M., and Crowell, A. D. (1962), *Physical Adsorption of Gases*, Butterworths, London.

Young, F. W., Cathcart, J. V., and Gwathmey, A. T. (1956), The rates of oxidation of several faces of a single crystal of copper as determined with elliptically polarized light, *Acta Metall.* **4**, 145–152.

Yousef, A. A., Arafa, M. A., and Boulos, T. R. (1971), Influence of manganese dioxide slimes on quartz flotation, *Trans. IMM* **80**, C223–227.

Yücesoy, A., and Yarar, B. (1974), Zeta potential measurements in the galena–xanthate–oxygen flotation system, *Trans. IMM* **83**, C96–100.

Yusa, M. (1977), Mechanisms of pelleting flocculation, *Int. J. Miner. Process.* **4**, 293–305.

Zahradnik, R. (1958), The reaction of amino acids with carbon disulphide. VII. Preparation and physico-chemical properties of salts of dithiocarbamido carboxylic acids, *Collect. Czech. Chem. Commun.* **23**, 1443; *Chem. Listy* **50**, 1892–1898.

Zemansky, M. W. (1957), *Heat and Thermodynamics*, McGraw-Hill, New York.

Zettlemoyer, A. C. (1969), Hydrophobic surfaces, in *Hydrophobic Surfaces*, F. M. Fowkes, ed., Academic Press, New York.

Zettlemoyer, A. C., and Hsing, H. H. (1976), Water on Organosilane-Treated Silica Surfaces, in *Colloid & Interface Science*, M. Kerker, ed., Vol. 1, pp. 279–290, Academic Press, New York.

Zettlemoyer, A. C., Chessick, J. J., and Hollabaugh, C. M. (1958), Estimations of surface polarity of solids from heat of wetting measurements, *J. Phys. Chem.* **62**, 489–490.

Zhelnin, A. A. (1973), *Theoretical Principles and Practice of Flotation of Potassium Salts*, Khimiya, Leningrad.

Zimmerman, R. E. (1948), Flotation of bituminous coal, *Trans. AIME* **177**, 338–356.

Zimmerman, R. E. (1968), Froth Flotation, in *Coal Preparation*, 3rd ed., J. W. Leonard and D. R. Mitchell, eds., AIME, New York, pp. 10–66 to 10–90.

Zisman, W. A. (1964), Relation of equilibrium contact angle to liquid and solid constitution, in *Contact Angle, Wettability, and Adhesion*, R. F. Gould, ed., Advances in Chemistry Series No. 43, American Chemical Society, Washington, D.C., pp. 1–51.

Zundel, G. (1970), *Hydration and Intermolecular Interaction*, Academic Press, New York.

Index